Groundwater Lowering in Construction

T0239566

Applied Geotechnics Series

William Powrie (ed.)

Geotechnical Modelling
David Muir Wood

Particulate Discrete Element Modelling
Catherine O'Sullivan

Practical Engineering Geology
Steve Hencher

Practical Rock Mechanics
Steve Hencher

Drystone Retaining Walls
Design, Construction and Assessment
Paul McCombie et al.

Fundamentals of Shield Tunnelling
Zixin Zhang and Scott Kieffer

Soil Liquefaction, 2nd ed
Mike Jefferies and Ken Been

Introduction to Tunnel Construction, 2nd ed
David Chapman et al.

Groundwater Lowering in Construction, 3rd ed
Pat M. Cashman and Martin Preene

For more information about this series, please visit: https://www.routledge.com/Applied-Geotechnics/book-series/APPGEOT

Groundwater Lowering in Construction

A Practical Guide to Dewatering

3rd Edition

Pat M. Cashman and Martin Preene

CRC Press
Taylor & Francis Group
Boca Raton London New York

CRC Press is an imprint of the
Taylor & Francis Group, an **informa** business

3rd edition published 2021
by CRC Press
2 Park Square, Milton Park, Abingdon, Oxon, OX14 4RN

and by CRC Press
6000 Broken Sound Parkway NW, Suite 300, Boca Raton, FL 33487-2742

© 2021 Martin Preene and the late Pat M. Cashman

First edition published by CRC Press 2001
Second edition published by CRC Press 2013

CRC Press is an imprint of Informa UK Limited

British Library Cataloguing-in-Publication Data
A catalogue record for this book is available from the British Library

---

Library of Congress Cataloging-in-Publication Data

Names: Cashman, P. M. (Pat M.), author. | Preene, M. (Martin), author.
Title: Groundwater lowering in construction : a practical guide to
dewatering / P. M. Cashman, M. Preene.
Description: 3rd edition. | Boca Raton : CRC Press, 2021. | Includes
bibliographical references and index.
Identifiers: LCCN 2020007241 | ISBN 9780367504748 (hardback) | ISBN
9781003050025 (ebook)
Subjects: LCSH: Earthwork. | Drainage. | Building sites. | Groundwater
flow. | Water table. | Zone of aeration.
Classification: LCC TA715 .C35 2020 | DDC 624.1/5136--dc23
LC record available at https://lccn.loc.gov/2020007241

ISBN: 978-0-367-50474-8 (hbk)
ISBN: 978-0-367-50475-5 (pbk)
ISBN: 978-1-003-05002-5 (ebk)

Typeset in Sabon
by Deanta Global Publishing Services, Chennai, India

This book is dedicated to the Memory of Joyce Preene (1928–2020)

Contents

SECTION 3
CONSTRUCTION

14 Sump Pumping   429

15 Wellpoint Systems   443

28 The Future   789

TOBY ROBERTS

Preface to the Third Edition

This third edition has been extensively revised from previous editions, with considerable quantities of new material. To aid the reader, the main text is essentially divided into three sections – Principles, Design and Construction.

The Principles section covers the fundamentals of groundwater flow as it relates to civil engineering excavations. The digital revolution has not changed the basic principles, and even now that sophisticated numerical modelling tools are widely available, going back to Darcy's law and the concept of permeability can be illuminating, even with complex problems.

The Design section includes a general broadening of coverage from previous editions, including much more detail on groundwater control for tunnelling projects (including shafts and cross passages) and on permeability assessment methods.

The chapters on Construction include the complete life cycle of a groundwater control scheme, including monitoring, maintenance and decommissioning. This section includes 11 case histories from the Cashman and Preene casebook.

Finally, I would like to reiterate that, true to Pat Cashman's intent when he started writing nearly 30 years ago, this third edition is intended to be a *practical* book. While much theory and analysis will be found herein, I hope the advice and experience distilled here is of some help when trying to control groundwater out there in the field.

Acknowledgements to the Third Edition

I would like to thank Tony Moore and the editorial team at Taylor & Francis for convincing me to update this book to a third edition and for supporting me through the highs and lows of the process.

I am grateful to the following colleagues, who provided useful and informed comments on parts of the text of the third edition: Nick Armstrong, Steve Buss, Jono Cook, Bob Essler, David Hartwell, Ian Heath, Adrian Koe, Laurence Miles, William Powrie, Andrew Ridley, Dave Terry and Jim Usherwood. I would also like to thank my friend of more than 30 years' standing, Toby Roberts, for kindly updating Chapter 28 on the future and providing some stimulating discussions during the preparation of this edition.

A large number of new illustrations have been prepared for this edition. I am grateful to Chris Warren and Lucy Preene for their hard work in preparing the new figures.

I would like to thank my family, and especially my wife, Pam, for her continuing support and increasingly surprising tolerance over the 20 years since the I started work on the first edition of this book.

I would like to thank all those who provided images and other materials for this book.

Illustrations and photographs provided by others, or from published works, are acknowledged in the captions.

I am grateful to Debra Francis and the staff of the Institution of Civil Engineers Library for their assistance in obtaining some of the references used in this book.

The epigraphs at the start of chapters and section dividers are reproduced courtesy of the following permissions:

Section divider (Principles): David Greenwood quotation by permission of ICE Publishing: from Greenwood, D A. (1994). Engineering solutions to groundwater problems in urban areas. *Groundwater Problems in Urban Areas* (Wilkinson, W B, ed). Thomas Telford, London, pp369–387.

Chapter 1: Pat Cashman quotation by permission of Ground Engineering: from Cashman, P M. (1973). Groundwater control. *Ground Engineering*, 6, September, pp26–29.

Chapter 3: Extract of King James Bible reproduced by permission of Cambridge University Press through PLSclear.

Chapter 5: Bryan Skipp quotation by permission of ICE Publishing: from Skipp, B O. (1994). Keynote paper: setting the scene. *Groundwater Problems in Urban Areas* (Wilkinson, W B, ed). Thomas Telford, London, pp3–16.

Chapter 6: Epigraph quoting Sir Harold Harding used with the permission of Tilia Publishing UK/Amanda Davey: from Harding, H J B and Davey, A. (2015). *It's Warmer Down Below: the Autobiography of Sir Harold Harding, 1900–1986*. Tilia Publishing UK, Sussex.

Section divider (Design): Vic Milligan quotation by permission of Golder Associates (2019).

Chapter 10: Epigraph quoting Sir Harold Harding used with the permission of Tilia Publishing UK/Amanda Davey: from Harding, H J B and Davey, A. (2015). *It's Warmer Down Below: the Autobiography of Sir Harold Harding, 1900–1986*. Tilia Publishing UK, Sussex.

Chapter 11: Rudolph Glossop quotation by permission of ICE Publishing: from Glossop, R. (1968). The rise of geotechnology and its influence on engineering practice. *Géotechnique*, 18, 2, pp107–150.

Section divider (construction): Karl Terzaghi quotation by permission of ICE Publishing: from Terzaghi, K. (1939). Soil mechanics - a new chapter in engineering science, the 45th James Forrest Lecture, 1939. *Journal of the Institution of Civil Engineers*, 12, 7, June, pp106–142.

Chapter 14: J Patrick Powers quotation by permission of John Wiley and Sons: from Powers, J P, Corwin A B, Schmall, P C and Kaeck, W E. (2007). *Construction Dewatering and Groundwater Control: New Methods and Applications*, 3rd Edition. Wiley, New York.

Chapter 15: William Powrie quotation by permission of William Powrie: from Powrie, W. (1990). *Legal Aspects of Construction Site Dewatering for Temporary Works*. Unpublished MSc thesis, University of London (King's College).

Chapter 26: Sir Ove Arup quotation reproduced by permission of Arup Group Limited.

Chapter 27: Sir Harold Harding quotation by permission of Golder Associates, from: Harding, H J B. (1981). *Tunnelling History and My Own Involvement*. ©1981 Golder Associates, Toronto.

Martin Preene
Wakefield, 2019

Preface to the Second Edition

In the decade since the first edition of this book was published I have perceived that groundwater lowering and dewatering activities are, thankfully, being increasingly integrated into the wider ground engineering schemes on major projects, rather than being a standalone activity. There has also been a very significant increase in the number of projects where potential environmental impacts are of concern, and require assessment during the development of the project.

This second edition reflects those changes. Although, as the title suggests, the primary focus of the book is still on 'groundwater lowering' or 'dewatering', new material is included to provide information on more wide ranging 'groundwater control' solutions which may not rely solely on dewatering by pumping.

Specifically, new chapters have been added on cut-off methods used for groundwater exclusion (Chapter 12) and on the issues associated with permanent or long term groundwater control systems (Chapter 14). Another notable addition is the new section in Chapter 11 which looks at groundwater control technologies used on contaminated sites – an increasingly common application.

Probably the biggest change is the need to understand, predict and mitigate the potential environmental impacts that can be caused by groundwater control works. This is covered in the new Chapter 15.

The remainder of the text has been updated to reflect relevant changes in technology and practice.

Acknowledgements to the Second Edition

Preparing this edition has been a rewarding, if at times challenging, experience. I would like to thank the editorial team at Taylor & Francis for their encouragement to complete this work.

I am grateful for the constructive and intelligent comments on parts of the text that were provided by Bob Essler, Dave Hall, Bridget Plimmer, Dan Taylor and David Whitaker. I would also like to thank my good friend Toby Roberts for kindly updating the afterword, and providing support and comment during the preparation of this edition.

Illustrations and photographs provided by others, or from published works, are acknowledged in the captions, and I would like to thank all those who provided such material and permissions.

Many new figures and illustrations have been included in this edition, and James Welch and Tomasz Jasinski have done a great job in preparing many of the new figures. I am also grateful for the help of Dyfed Evans with some of the photographic images and Lesley Power for her help with preparation of the manuscript.

Finally, as always I would like to thank my family and especially my wife, Pam, for her continuous support and amazing tolerance during the gestation and birth of this book.

Martin Preene
Wakefield, 2011

Acknowledgements to the First Edition

It is difficult to make fulsome and comprehensive acknowledgments to the many who have encouraged me and thereby helped me to persist with the task of writing this book. My wife has been a steadfast 'encourager' throughout its very lengthy gestation period. It was Daniel Smith who first persuaded me to compile this work. I have lost track of his present whereabouts but I hope that he will approve of the finished work. I am grateful for his initial stimulation.

I owe many thanks to my near neighbour, Andy Belton. Without his patient sorting out of my computer usage problems, this text would never have been fit to send to a publisher. Andy will be the first to admit that this is not a subject that he has knowledge of, but having read some of the draft text and many of my faxes; on occasions he asked some very pertinent questions. These made me think back to first principles and so have – I am sure – resulted in improved content of this work.

I thank the many organisations that have so kindly allowed use of their material, photographs and diagrams. Where appropriate, due acknowledgment is made in the text beneath each photo or diagram. I hope that this will be acceptable to all who have helped me and given their permissions for their material to be reproduced; a complete list of each and every one would be formidable indeed: I do ask to be forgiven for not so doing. Following this philosophy it would be invidious to single out and name a few of the major assistors for fear of upsetting others.

Pat M. Cashman
Henfield, West Sussex, 1996

At the time of Pat Cashman's death in 1996 the manuscript for this book was well advanced. I am grateful to William Powrie and the editorial team at E & F N Spon for encouraging me to complete his work. I would also like to acknowledge the support and contribution of David Hartwell, who contributed closely with Pat on an earlier draft of the text, and who has provided information and assistance to me during the preparation of the manuscript.

Valuable comments on parts of the text were provided by Lesley Benton, Rick Brassington, Steve Macklin, Duncan Nicholson, Jim Usherwood and Gordon Williams. I am also grateful to Toby Roberts for kindly writing the afterword.

Illustrations and photographs provided by others, or from published works, are acknowledged in the captions, and I would like to thank all those who provided such material and permissions. The libraries of the Institution of Civil Engineers and Ove Arup & Partners have also been of great assistance in the gathering together many of the references quoted herein.

Finally, I would like to thank my wife, Pam, for her invaluable and unstinting support throughout this project.

Martin Preene
Wakefield, 2001

Pat M. Cashman

Pat M. Cashman, the leading British exponent of groundwater control of his generation, died on 25 June 1996. For more than 40 years, during the growth of soil mechanics into the practice of geotechnical engineering, Pat was responsible, through the organizations he ran and later as a consultant, for maintaining a practical and straightforward approach to the art of groundwater control. This book, the manuscript of which was well advanced at the time of his death, sets out that approach.

Following war service with the Royal Engineers, Pat graduated from the University of Birmingham and joined Soil Mechanics Limited, soon transferring to the Groundwater Lowering Department, so beginning his lifelong interest in this field. He became head of the department in 1961 and later became responsible for the joint venture with Soletanche that introduced French techniques into the UK. In 1969, he became contracts director for Soletanche (UK) Limited.

In 1972. Pat joined Groundwater Services Limited (later Sykes Construction Services Limited) as managing director. Over a 10 year period, he designed and managed a huge number and range of groundwater lowering projects. Commercial and financial success was achieved alongside technical innovation and practical advancements. In the 1980s, he joined Stang Wimpey Dewatering Limited as managing director. Again, he achieved commercial success as well as introducing American ideas into British practice. During this period, Pat made a major contribution to the production of the Construction Industry Research and Information Association (CIRIA) Report 113, *Control of Groundwater for Temporary Works* – the first comprehensive dewatering guide produced in the United Kingdom in the modern era.

In 1986, he 'retired' and commenced an active role as a consultant, often working closely with Ground Water Control Limited, a contracting company formed by men who had worked for Pat during the Sykes years. His practical approach to problems meant that he was always in demand, particularly by contractors when they were in trouble. Although a practical man with a healthy suspicion of arcane theory and in particular computer modelling of problems, he took to the computer to document his own experience. This book is a record of a singular approach to a challenging business.

Martin Preene

Authors

Martin Preene is a civil engineer with a bachelor's degree from the University of Bristol and a PhD in geotechnical engineering from the University of London, and is qualified as a Chartered Engineer and Chartered Geologist. For more than 30 years, he has specialized in groundwater engineering, working for contractors and consultants. He has experience of design and construction on dewatering and groundwater control projects worldwide, including on construction projects such as basements, metro systems, shafts, tunnels and dry docks, and on mining projects, including open pit and underground mines. He is based in the UK and works internationally as an independent consultant through his professional practice Preene Groundwater Consulting.

The late **Pat M. Cashman** was the leading British exponent of groundwater control for his generation, championing a practical and straightforward approach for more than forty years.

Chapter 1

Groundwater Lowering
A Personal View and Introduction
by Pat M. Cashman

> Over a thousand years ago King Canute learned by experience that control of water
> cannot be achieved by words alone.
>
> Pat M. Cashman

Many engineering projects, especially major ones, entail excavations into water-bearing soils. For all such excavations, appropriate system(s) for the management and control of the groundwater and surface water run-off should be planned before the start of each project. In practice, this can only be done with knowledge of the ground and groundwater conditions likely to be encountered by reference to site investigation data. The control of groundwater (and also surface water run-off) is invariably categorized as 'temporary works' and so is often regarded by the client and their engineer or architect as the sole responsibility of the contractor and of little or no concern to them. In many instances, this philosophy has been demonstrated to be short-sighted and ultimately costly to the client.

Sometimes, as work proceeds, the actual soil and groundwater conditions encountered may differ from what was expected. Should this happen, all concerned should be willing to consider whether to modify operations and construction methods as the work progresses and as more detailed information is revealed. Based upon this philosophy, I advocate, particularly for large projects, frequent 'engineering-oriented' reappraisal meetings between client or owner, or both, and contractor (as distinct from 'cost-oriented' meetings). This will afford the best assurance that the project will be completed safely, economically and within a realistic programme time and cost.

On a few occasions, I have been privileged to be involved in the resolution of some difficult excavation and construction projects when the engineer succeeded in persuading the client to share the below-ground risks with the contractor. During the progress of the contract, there were frequent engineering-oriented meetings with the contractor to discuss and mutually agree how to proceed. I believe that the engineers concerned with these complex projects realized that it would not be in the best interests of their client to adhere rigidly to the traditional view that the contractor must take all of the risks. They were enlightened and had a wealth of practical experience, and so had a realistic awareness that the soil and groundwater conditions likely to be encountered were complex. Also, they realized that the measures for effecting stable soil conditions during construction might not be straightforward. The few occasions when I have experienced this joint risk-sharing approach have, without exception in my view, resulted in sound engineering solutions to problems that needed to be addressed: they were resolved sensibly, and the projects were completed within realistic cost to the client. Furthermore, claims for additional payments for dealing with unforeseen conditions were not pressed by the contractor.

I found these experiences most interesting and enlightening, and I learned much by having direct access to different points of view of the overall project as distinct from my own view as a specialist contractor. I find it encouraging that in recent years, the target form of contract – the client and the contractor sharing the risks of unforeseen conditions – is being implemented more frequently. Thereby, the contractor is confident of a modest but reasonable profit, and the client is not eventually confronted with a multitude of claims for additional payments, some of which may be spurious, but all requiring costly and time-consuming analysis and investigation.

There are three groups of methods available for temporary works control of groundwater:

a) Lowering of groundwater levels in the area of construction by means of water abstraction, in other words groundwater lowering or dewatering
b) Exclusion of groundwater inflow to the area of construction by some form of very low-permeability cut-off wall or barrier (e.g. sheet-piling, diaphragm walls, artificial ground freezing)
c) Application of a fluid pressure in confined chambers such as tunnels, shafts and caissons to counterbalance groundwater pressures (e.g. compressed air, earth pressure balance tunnel boring machines)

Rudolf Glossop (1950) stated:

> The term drainage embraces all methods whereby water is removed from soil. It has two functions in engineering practice: permanent drainage is used to stabilise slopes and shallow excavations; whilst temporary drainage is necessary while excavating in water-bearing ground.

This book principally addresses the subject of temporary drainage, though many of the principles are common to both temporary and permanent requirements.

The book is intended for use by the practical engineer (either contractor or consultant or client); but it is intended particularly for the guidance of the specialist 'dewatering practitioner' or advisor. In addition, it is commended to the final-year graduate or master's student reading civil engineering or engineering geology as well as to the civil engineering-oriented hydrogeologist. It is deliberately addressed to the practitioner involved in the many day-to-day small- to medium-scale dewatering projects for which a simplistic empirical approach is usually adequate. It is anticipated that the typical reader of this work will be one quite comfortable with this philosophy but one who is aware of the existence of – though perhaps wishing to avoid – the purist hydrogeologist philosophy and the seemingly unavoidable high-level mathematics that come with it.

We, the writers and the readers, are pragmatic temporary works engineers – or, in the case of some readers, aspiring to be so – seeking the successful and economical completion of construction projects. For the small- and medium-size projects (which are our 'bread and butter'), there seems to be little practical justification for the use of sophisticated and time-consuming techniques, when simpler methods can give serviceable results. The analytical methods described in this book are based on much field experience by many practitioners from diverse countries and have thereby been proven to be practicable and adequate for most temporary works assessment requirements. I consider that J P Powers stated a great dewatering truism: 'The successful practitioner in dewatering will be the person who understands the theory and respects it, but who refuses to let theory overrule judgement' (Powers *et al.*, 2007).

Extensive use is made of the Dupuit–Forchheimer analytical approaches. I am conscious that purists will question this simplistic approach. My riposte – based on some 30 or more years of dealing with groundwater lowering problems – is that in my experience and that of many others, this empirical philosophy has resulted in acceptably adequate pumping installations; always provided, of course, that due allowance is made for the often limited reliability of available ground and groundwater information. Acquired practical field experience is required to assess the quality of the site investigation data. Whenever possible, reference should be made to other excavations in adjacent areas or in similar soil conditions to verify one's proposals. Towards the end of the book, in Chapter 27, there are some brief descriptions of a number of relevant case histories that the authors have dealt with in the past.

I readily acknowledge that for a groundwater lowering system design pertinent to large-scale and/or long-term projects – for example construction of a dry dock, a nuclear power station or an open cast mining project – more sophisticated methods of analysis will be appropriate. These can provide reassurance that the pragmatic solution is about right, but do we ever know the 'permeability value' to a similar degree of accuracy?

The underlying philosophy of this publication is to address the pragmatic approach. It follows that three questions arise:

- How does water get into the ground, and how does it behave whilst getting there and subsequently behave whilst there?
- What is the inter-relationship between the soil particles and the groundwater in the voids between them?
- How can groundwater and surface water run-off be controlled and so prevented from causing problems during excavation and construction?

A thorough site investigation should go a long way towards providing the answers to these questions. Unfortunately, experience indicates that many engineers responsible for specifying the requirements for project site investigations consider only the *designer's* requirements and do not address the other important considerations, namely – *how can this be built?* Often, the site investigation is not tailored to obtaining data pertinent to temporary works design requirements or to problems that may occur during construction.

The contractor should not expect always to encounter conditions exactly as revealed by the site investigation. Soils, due to the very nature of their deposition and formation, are variable and rarely, if ever, isotropic and homogeneous, as is assumed in many of the analytical methods. The contractor should carry out the works using their professional skills and abilities and should be prepared to adjust if changed circumstances are revealed as the work proceeds.

Throughout the planning, excavation and construction phases of each project, safety considerations must be of paramount importance. Regrettably, the construction industry historically has a poor safety performance record.

Let us consider another professional discipline! Hopefully, no surgeon would contemplate commencing an operation on a patient without carrying out a thorough physical examination and having the information from x-ray; ECG; urine, blood and other test results; and any pertinent scans available beforehand. The surgeon will realize that these may not indicate everything and that during the operation complications may occur, but the possibilities of such 'surprise' occurrences will have been minimized by having reliable site investigation data concerning the patient.

Likewise, if the client's engineer/designer provides comprehensive ground and groundwater information at tender stage, the 'surprise' occurrences during construction should be minimized. An experienced contractor, with the co-operation of an experienced client/engineer, should be able to agree how to adjust working techniques to deal adequately with the changed circumstances as, or if, they are revealed, and this at realistic final cost.

1.1 STRUCTURE OF THE REST OF THE BOOK

At the commencement of each chapter, the introduction acts as a summary of the subject matter to be covered therein. I hope that this approach will enable the reader to decide speedily which chapters to read forthwith when seeking guidance on how to deal with their individual requirements, and which chapters may be deferred till later.

The first part of this book deals with fundamental principles of groundwater flow in the context of excavations for construction purposes:

- Chapter 2 contains a brief historical review of the principal theories concerning seepage towards wells and of the technologies used to apply them. Many readers will probably consider this as superfluous and omit it from their initial reading. As their interest in this subject becomes further stimulated, I hope that they will turn back to it. I derived great pleasure when researching this aspect some years ago.
- Chapter 3 contains a very brief summary of the hydrological cycle (i.e. how does water get into the ground?), and the concepts of aquifers and aquitards are introduced. The theoretical concepts of Darcy's law are presented, which are much used in a practical sense later. Groundwater chemistry is also briefly discussed.
- Chapter 4 discusses groundwater flow and permeability (hydraulic conductivity) as it applies to soil and rock. Chapter 5 discusses groundwater models in all their forms – conceptual, analytical and numerical – especially in relation to flow to wells.
- Chapter 6 discusses the objectives of groundwater control, including preventing groundwater-induced instability. An understanding of these issues can be vital when selecting a groundwater control method. Chapters 7 and 8 discuss in more detail the mechanisms of groundwater instability problems for excavations in soil and rock, respectively.

The second part of this book addresses design issues:

- Chapter 9 presents, in summary form, the principal features of the methods available for control of groundwater by exclusion and by pumping. Understanding the capabilities and limitations of the various techniques is an essential part of the design process.
- Chapter 10 discusses, in some detail, groundwater control problems and solutions for the particular case of tunnelling projects, including tunnels, shafts and cross passages.
- Chapter 11 addresses site investigation requirements, but only those specific to groundwater control, and does not detail the intricacies of the available methods of site investigation. This chapter leads directly into Chapter 12, which covers the determination of permeability in the field and laboratory. Guidance is also given on the relative reliability of permeability estimated by the various methods.
- Chapter 13 describes various empirical and simple design methods for assessing the discharge flow rates required for groundwater lowering installations and for determining the number of wells, etc. Several simple design examples are included in an appendix.

The third and final part of this book deals with the practical application of groundwater control methods in the field:

- Chapters 14 to 17 address the various groundwater lowering methods: sump pumping; wellpointing; deep wells; and less commonly used methods.

- Chapter 18 describes the various techniques that can be used to form groundwater cut-off walls or barriers in order to exclude groundwater from excavations.
- Chapter 19 describes the types of pumps suitable for the various systems.
- Chapter 20 describes the particular issues and challenges associated with permanent (or at least very long-term) dewatering and groundwater control systems.
- Chapter 21 deals with some environmental impacts of groundwater lowering, including ground settlements. Various approaches to assessment and mitigation of environmental impacts are also discussed.
- Chapters 22 and 23 present appropriate methods of monitoring and maintenance, respectively, to ensure that groundwater lowering systems operate effectively when they are first installed and after extended periods of operation.
- Chapter 24 describes methods that can be used to decommission the elements of groundwater control systems when they are no longer needed.
- Chapter 25 covers safety, contractual and environmental regulation issues (using British convention but based on principles of good practice that are applicable in other countries).
- Chapter 26 discusses the challenge of trying to develop 'optimal' groundwater control solutions.
- Chapter 27 presents 11 case histories that highlight various aspects of practical groundwater control problems.
- Chapter 28 is a short afterword (contributed by the eminent dewatering engineer Toby Roberts) looking at the future of groundwater control.

The book ends with a comprehensive list of references and a glossary of terms pertinent to the subject matter and also a summary of the symbols and notations used. Six appendices are included, providing detailed background information on various subjects.

1.2 SOME FINAL POINTS

Finally, I reiterate my earlier statement: this book is intended for the guidance of practical engineers and those – many, I hope! – desirous of joining us. It is worth considering the words of the eminent geotechnical engineers Hugh Golder and John Seychuk of Golder Associates, who in their paper (Golder and Seychuk, 1967) described, hopefully in an ironic fashion, the attributes needed to succeed in this field. They said:

> The limitations of the different systems which are available are practical rather than theoretical and the design of a system is no task for an optimist. A sound engineer with a melancholy outlook, whose life has been a series of unhappy trials, is the best man to plan a water-lowering system.

I trust that there are several new aspirants who will realize after reading this book that groundwater lowering is a challenging scientific field and to some extent an applied art form as well, wherein the requirement for practical experience is paramount. No two sites have the same requirements – this is one of its fascinations!

SECTION I

PRINCIPLES

Sound engineering concepts cannot be derived without detailed knowledge of the immediate ground fabric and properties, especially those controlling the flow of water.

David Greenwood

Chapter 2

The History of Groundwater Theory and Practice

Study the past if you would define the future.

Confucius, ancient Chinese philosopher

2.1 INTRODUCTION

Man has been aware of groundwater since prehistory, long before Biblical times. Over the centuries, the mysteries of groundwater have been solved, and man has developed an increasing capability to manipulate it to his will. This chapter describes some of the key stages in the development of the understanding and control of groundwater. The history of some of the technology now used for groundwater lowering is also discussed, especially in relation to early applications in the United Kingdom. Detailed knowledge of the history of groundwater control might not be considered essential for a practical engineer working today. Nevertheless, study of the past can be illuminating, not least by showing that even when theories are incomplete and technology untried, the application of scientific principles and engineering judgement can still allow groundwater to be controlled.

2.2 FROM THE EARLIEST TIMES TO THE SIXTEENTH CENTURY

The digging of wells for the exploitation of water and primitive implementation of water management date back to Babylonian times and even earlier.

The source of the water flowing from springs and in streams was a puzzling problem and the subject of much controversy and speculation. It was generally held that the water discharged from springs could not be derived from rain. Ingenious hypotheses were formulated to account for the occurrence of springs. Some early writers suggested large inexhaustible reservoirs, while others recognized that there must be some form of replenishment of the supplying reservoirs. The Greek hypotheses, with many incredible embellishments, were generally accepted until near the end of the seventeenth century. The theory of rainfall infiltration was propounded by only a very few writers.

Though the Romans and other early cultures indulged in quite sophisticated water management projects, building imposing aqueducts to channel water from spring and other sources to centres of population, they had no understanding of the sources of replenishment of groundwater. The aqueducts of the Romans were remarkable and showed great appreciation of the value of water, but their methods of measuring or estimating water quantities were crude. Generally, they appear to have lacked any semblance of knowledge of either surface or groundwater hydrology.

According to Tolman (1937),

> Centuries have been required to free scientists from superstitions and wild theories handed down from the dawn of history regarding the unseen sub-surface water, … even in this century there is still much popular superstition concerning underground water.

An elemental principle, that gravity controls the motion of water underground as well as at the surface, is not appreciated by all. A popular belief exists that 'rivers' of underground water pass through solid rock devoid of interconnected interstices and flow under intervening mountain ranges.

Marcus Vitruvius, a Roman architect who lived about the time of Christ, produced a book describing the methods of finding water. He wrote of rain and snow falling on mountains and then percolating through the rock strata and reappearing as springs at the foot of the mountains. He gave a list of plants and of other conditions indicative of groundwater, such as colour and dampness of the soil and mists rising from the ground early in the morning. Vitruvius and two contemporaries of his, Cassiodorus and Pliny, were the first to make serious efforts to list practical methods for locating water, and this when geology was yet unknown!

2.3 FROM THE RENAISSANCE PERIOD TO THE NINETEENTH CENTURY

At the beginning of the sixteenth century, Leonardo da Vinci directed his attention to the occurrence and behaviour of water. He correctly described the infiltration of rain and the occurrence of springs and concluded that water goes from rivers to the sea and from the sea to the rivers and thus is constantly circulating and returning. About the same time, Palissy, a French Huguenot, presented a clear and reasoned argument that all water from springs is derived from rain.

The latter part of the seventeenth century was a watershed in the beginnings of an understanding of the replenishment of groundwater. Gradually, there arose the concept of a 'hydrological cycle'. This presumed that water was returned from the oceans by way of underground channels below the mountains. The removal of salt was thought to be either by distillation or by percolation, and there were some highly ingenious theories of how water was raised up to the springs.

2.4 PROGRESS FROM A QUALITATIVE TO A QUANTITATIVE SCIENCE

It was in the seventeenth century that the quantitative science of hydrology was founded by Palissy, Pérrault and Mariotté in France, Ramazzini in Italy and the astronomer Halley in Britain.

Palissy, a sixteenth-century potter and palaeontologist, stated that rain and melt snow were the only source of spring and river waters, and that rain water percolates into the earth, following 'a downward course until they reach some solid and impervious rock surface over which they flow until they make some opening to the surface of the earth'.

Pérrault made rainfall run-off measurements and demonstrated the fallacy of the long-held view that the rainfall was not sufficient to account for the discharge from springs. He also measured and investigated evaporation and capillarity. Mariotté verified Pérrault's results and showed that the flow from springs increased in rainy weather and decreased in times of drought. Halley made observations of the rate of evaporation from the Mediterranean ocean and showed that this was adequate to supply the quantity returned to that sea by its rivers. His measurements of evaporation were conducted with considerable care, but his estimates of stream flow were very crude.

Towards the end of the eighteenth century, La Metherie extended the researches of Mariotté and brought them to the attention of meteorologists. He also investigated permeability and explained that some rain flows off directly (surface water run-off), some infiltrates into the top soil layers only and evaporates or feeds plants, whilst some rain penetrates underground, whence it can issue as springs (i.e. infiltration or groundwater recharge). This is the first recorded mention of 'permeability' and so is the first link between hydrology and seepage to wells.

2.4.1 Seepage towards Wells

The robust Newcomen engine greatly influenced mining practice during the eighteenth century, but it was far too cumbersome for construction works. Generally speaking, until the early nineteenth century, civil engineers, by the use of timber caissons (Figure 2.3) and other devices, avoided pumping whenever possible. However, where there was no alternative, pumping was usually done by hand – a very onerous task – using a rag and chain pump (Figure 2.1), known also as 'le chaplet' (the rosary).

The use of pumps for mining in the sixteenth century is described in two seminal references by Agricola and Ramelli. *De Re Metallica* (Latin for *On the Nature of Metals*) by Georgius Agricola (translated by Hoover and Hoover, 1950) was a review of the state of art in mining and remained a key text on mining for nearly two centuries. Agricola describes several designs of piston pumps, which are either man or animal powered or powered by waterwheels. The suction lift limits of these pumps meant that multiple stages of pumps were required for the deepest mines. *Le diverse et artificiose machine del Capitano Agostino Ramelli* (1588) translates as *The Various and Ingenious Machines of Captain Agostino Ramelli*. Ramelli, an Italian military engineer, spent the latter half of his life and career in France as *Ingénieur du Roi* (Engineer to the King) in the service of Henri III and Henri IV. His book describes several machines and methods to pump water from excavations (Figure 2.2) and even includes the use of timber cofferdams to exclude groundwater (Figure 2.3).

Some idea of the magnitude of pumping problems is given by de Cessart in his book *Oeuvres Hydrauliques*. Speaking of the foundations for the abutment of a bridge at Saumur in 1757, he says that 45 chain pumps were in use, operated by 350 soldiers and 145 peasants. Work on this type of pump was, of course, most exhausting, and the men could only work in short spells. Pryce, in his *Mineralogea Cornubiensis* in 1778, said that work on pumps of this sort led to a great many premature deaths among Cornish miners.

For permanent installations such as graving docks, large horse-driven chaplet pumps were used. Perronnet, the famous French bridge builder, made use of elaborate pumping installations in the cofferdams for the piers of his larger bridges; for example, under-shot water wheels were used to operate both chaplet pumps and Archimedean screws.

According to Crèsy, writing in 1847, the first engineer to use steam pumps on bridge foundations was Rennie, who employed them on Waterloo Bridge in 1811. In the same year, Telford, on the construction of a lock on the Caledonian Canal at Clachnacarry, at first used a chain pump worked by six horses but replaced it by a 9 horsepower steam-engined pump. From then on, steam pumps were used during the construction of all the principal locks on that canal. In 1825, Marc Brunel used a 14 horsepower steam engine when sinking the

Figure 2.1 Rag and chain pump, manually operated, 1556. (From Bromehead, C N, Mining and quarrying in the seventeenth century. In *A History of Technology* (Singer, C, Holmyard, E J, Hall, A R and Williams, T L, eds), 1956, by permission of Oxford University Press.) The balls, which are stuffed with horsehair, are spaced at intervals along the chain and act as one-way pistons when the wheel revolves.

shafts for the Thames Tunnel. By this date, steam pumping seems to have been the common practice for dealing with groundwater, so below-ground excavations for construction were less problematical.

2.4.2 Land Drainage in the Eighteenth and Nineteenth Centuries

Notable contributions to British practice from the later eighteenth century through to the mid-nineteenth century were the work of British land drainers such as Joseph Elkington, John Wedge and James Anderson (Stephens and Stephens, 2006). Their approach involved using an understanding of shallow groundwater conditions to drain waterlogged ground. The methods they developed are comparable to sumps and relief wells (to bleed ground-water pressure from confined layers), which would be recognized by practitioners today. Interestingly, because of the successful and widespread application of his knowledge to horizontal and vertical drainage in the latter part of the eighteenth century, Stephens and Ankeny (2004) make the case that Joseph Elkington was perhaps the first professional consulting hydrogeologist.

Figure 2.2 Device for two men to drain water from a foundation: mechanics: hand crank; force pump. (Plate 103 from Ramelli, A., *Le Diverse et Artificiose Machine del Capitano Agostino Ramelli*, Paris, France, 1588. With permission from Science History Institute.)

2.4.3 Kilsby Tunnel, London to Birmingham Railway

For civil engineering excavations, there appear to have been no important advances in pumping from excavations until the construction in the 1830s by the renowned civil engineer Robert Stephenson of the Kilsby Tunnel south of Rugby on the London to Birmingham Railway (Preene, 2004). He pumped from two lines of wells sited parallel to and on either side of the line of the tunnel drive (Figure 2.4).

It is clear from Stephenson's own *Second Report to the Directors of the London, Westminster and Metropolitan Water Company*, 1841, that he was the first to observe and explain the seepage or flow of water through sand to pumped wells. The wells were sited just outside the periphery of the construction so as to lower the groundwater level in the area of the work by pumping from these water abstraction points. This is most certainly the first temporary works installation of a deep well groundwater lowering system in Britain, if not

Figure 2.3 Cofferdam used to exclude water from an excavation. Workmen use buckets to drain water from the section walled off by the cofferdam in order to secure foundations for construction (Plate III from Ramelli, A., *Le Diverse et Artificiose Machine del Capitano Agostino Ramelli*, Paris, France, 1588. With permission from Science History Institute.)

in the world. The following extract (from Boyd-Dawkins, 1898, courtesy of the Institution of Civil Engineers Library) quotes from the report and shows that Stephenson had understood the mechanism of groundwater flow towards a pumping installation:

> The Kilsby Tunnel, near Rugby, completed in the year 1838, presented extreme difficulties because it had to be carried through the water-logged sands of the Inferior Oolites, so highly charged with water as to be a veritable quicksand. The difficulty was overcome in the following manner. Shafts were sunk and steam-driven pumps erected in the line of the tunnel. As the pumping progressed the most careful measurements were taken of the level at which the water stood in the various shafts and boreholes; and I was soon much surprised to find how slightly the depression of the water-level in the

Figure 2.4 Pumps for draining the Kilsby Tunnel. (From Bourne, J C, *Drawings of the London and Birmingham Railway*, 1839, with permission from the Institution of Civil Engineers.) A pumping well is shown in the foreground, with the steam pumphouse in the distance.

one shaft, influenced that of the other, notwithstanding a free communication existed between them through the medium of the sand, which was very coarse and open. It then occurred to me that the resistance which the water encountered in its passage through the sands to the pumps would be accurately measured by the angle or inclination which the surface of the water assumed towards the pumps, and that it would be unnecessary to draw the whole of the water off from the quicksands, but to persevere in pumping only in the precise level of the tunnel, allowing the surface of the water flowing through the sand to assume that inclination which was due to its resistance.

The simple result of all the pumping was to establish and maintain a channel of comparatively dry sand in the immediate line of the intended tunnel, leaving the water heaped up on each side by the resistance which the sand offered to its descent to that line on which the pumps and shafts were situated.

As Boyd-Dawkins then comments,

The result of observations, carried on for two years, led to the conclusion that no extent of pumping would completely drain the sands. Borings, put down within 200 yards [185 m] of the line of the tunnel on either side, showed further, that the water-level had scarcely been reduced after 12 months continuous pumping and, for the latter six months, pumping was at the rate of 1,800 gallons per minute [490 m³/h]. In other words, the cone of depression did not extend much beyond 200 yards [185 m] away from the line of pumps.

In this account, ... it is difficult to decide which is the more admirable, the scientific method by which Stephenson arrived at the conclusion that the cone of depression was small in range, or the practical application of the results in making a dry [the authors would have used the word 'workable' rather than 'dry'] pathway for the railway between the waters heaped up [in the soil] ... on either side.

It is astonishing that neither Robert Stephenson nor any of his contemporaries realized the significance of this newly discovered principle: that by sinking water abstraction points, and more importantly placing them clear of the excavation so that the flow of water in the ground would be away from the excavation rather than towards it, stable ground conditions were created. For many decades, this most important principle was ignored.

2.4.4 Early Theory – Darcy and Dupuit

In the 1850s and early 1860s, Henri Darcy made an extensive study of the problems of obtaining an adequate supply of potable water for the town of Dijon (Freeze, 1994). He is famous for his Darcy's law (Darcy, 1856) postulating how to determine the permeability of a column of sand of selected grading when the rate of water flow through it is known (Figure 2.5). In fact, Darcy's experiments formed only a small part of his treatise. He compiled a very comprehensive report (two thick volumes) in which he analysed the available sources of water from both rivers and wells – some of them artesian – and how economically to harness all these for optimum usage.

Darcy investigated the then current volume of supply of water per day per inhabitant for about 10 municipalities in Britain – Glasgow, Nottingham and Chelsea, among others – as well as Marseille, Paris and many other French towns. He concluded that the average water provision in Britain was 80–85 litres/inhabitant/day and more than 60 litres/inhabitant/day for Paris. Darcy designed the water supply system for Dijon on the basis of 150 litres/inhabitant/day – no doubt his Victorian contemporaries this side of 'la Manche' would have applauded this philosophy.

In the mid-1860s, Dupuit (1863), using Darcy's law to express soil permeability, propounded his equations for determining flow to a single well positioned in the middle of an island. Dupuit made certain simplifying assumptions, and having stated them (i.e. truly horizontal flow), he then discounted their implications! For this, Dupuit has been much castigated by some later purists, but most accept that the Dupuit concept, later slightly modified by Forchheimer, is acceptable and adequate in many practical situations.

Exchange of information was not as simple in Darcy's and Dupuit's time as it is now. Much of the fundamental work of these two French engineers was duplicated by independent developments shortly afterwards in Germany and Austria and a little later in the United States.

In about 1883, Reynolds demonstrated that for linear flow – i.e. flow in orderly layers, commonly known as laminar flow – there is a proportionality existing between the hydraulic gradient and the velocity of flow. This is in keeping with Darcy's law, but as velocity increased, the pattern of flow became irregular (i.e. turbulent), and the hydraulic gradient approached the square of the velocity. Reynolds endorsed the conclusion that Darcy's law gives an acceptable representation of the flow within porous media – i.e. the flow through the pore spaces of soils will remain laminar save for very rare and exceptional circumstances. However, this may not always be true of flow through fractured rocks (e.g. karstic limestone formations).

2.5 LATER THEORETICAL DEVELOPMENTS

In his Rankine Lecture, Glossop (1968) suggested that 'classical soil mechanics' was founded on the work of Terzaghi, which dates from his first book published in 1925, and that it is strongly influenced by geological thinking. In 1913, Terzaghi published a paper dealing with the hydrology of the Karst region of Yugoslavia after studying the geology for

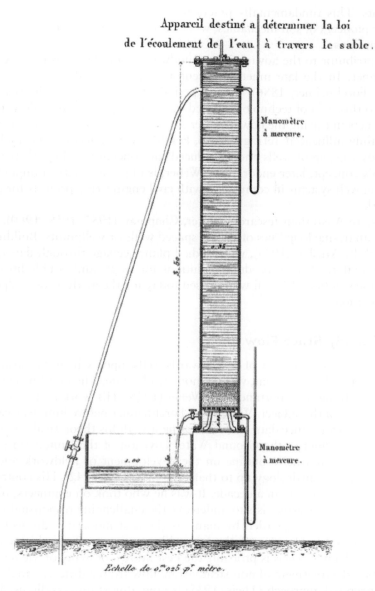

Figure 2.5 Darcy's original sand column apparatus. (From Darcy, H., *Les Fontaines Publiques de la Ville de Dijon*, Dalmont, Paris, 1856.)

a hydro-electric scheme in Croatia. He soon realized that geology would be of far more use to engineers if the physical constants of rocks and soils were available for design (Terzaghi, 1960). This was the first positive marrying of civil engineering and geology.

2.5.1 Verifications and Modification of Darcy

In 1870, a civil engineer in Dresden, Adolph Theim, reviewed Darcy's experiments and went on to derive the same equations as Dupuit for gravity and artesian wells. Theim was the first to collect extensive field data in support of his and Dupuit's equations. Thus, he was the first researcher to apply practical field experience to test the validity of the analytical pronouncements. This fundamentally practical philosophy was to be typical of Terzaghi and other later pioneers of soil mechanics, where reliable field measurements were essential to verify assumptions.

The next contributor to the advance of groundwater flow theory was an Austrian engineer, Forchheimer. In the late nineteenth century, he published his first paper on flow towards wells (Forchheimer, 1886). Over a period of half a century, he published many contributions to this field of technology. Based on an analogy with heat flow, he developed the use of flow nets in tracing the flow of water through sands. His results were published in 1917 and certainly influenced Terzaghi. Also, Forchheimer undertook the analysis of gravity flow towards a group of wells and introduced the concept of a hypothetical equivalent single well. This concept, later endorsed by Weber, is of great practical importance to this day in analysing well systems in connection with civil engineering projects for construction temporary works.

In the 1950s, an Australian research worker, Chapman (1957, 1959, 1960), investigated the problem of analysing long lines of closely spaced wells or wellpoints. Building on earlier theoretical work by Muskat (1935), who studied plane seepage through dams, Chapman's modelling and analytical work produced solutions for single and double lines of wells or wellpoints. These solutions are still widely used today for the analysis of wellpoint systems for trench excavations.

2.5.2 Non-Steady-State Flow

An important aspect of the theory of flow towards wells, upon which the various investigators in Europe were relatively silent, was the non-steady state. The first European investigator to produce anything of importance was Weber (1928). His work is still one of the most complete treatments of the subject. Weber, like Forchheimer before him, was very thorough in gathering field performance data and correlating it with his theoretical analyses.

In 1942, Meinzer, chief of the Ground Water Division of the United States Geological Survey, edited a comprehensive outline on the development of fieldwork and theoretical analysis in groundwater hydrology up to that time (Meinzer, 1942). His contributions had already extended over more than a decade. It was he who took on engineers, physicists and mathematicians as well as geologists to undertake the challenging responsibilities delegated to his division. Meinzer is regarded by many as the first modern hydrogeologist, and in this specialist field is considered its 'father', in a like manner as Terzaghi is considered the 'father' of soil mechanics. Theis is probably the best known of many of Meinzer's protégés. Theis approached the treatment of non-steady-state flow from a different angle than that of Weber. His conceptual approach (Theis, 1935) is now almost universally used as the basis for non-steady-state flow analysis.

During the early 1930s, M Muskat, chief physicist for the Gulf Research and Development Company, was the leader of a team who compiled a comprehensive and scholarly volume treating all phases of flow of fluids through homogeneous porous media (Muskat, 1937). The work was concerned primarily with the problems involved in the flow of oil and oil–gas mixtures through rocks and sands. Muskat's approach was consistently that of a theoretical physicist rather than that of a field engineer. His appraisals of analytical methods of test procedures tended to be more those of the scientific purist than of one concerned with the practical pragmatic needs in the field. However, the accomplishments of Muskat and his colleagues have been of immeasurable value to an understanding of seepage flow.

Muskat made much use of electric analogue models and of sand tank models. He investigated the effects of stratification and anisotropy and developed the transformed section. He made extensive studies of multiple confined wells. He seems to have given much serious thought to the limited validity of the Dupuit equations and pronounced: 'Their accuracy, when valid, is a lucky accident.' No one before or since has been so intolerant of Dupuit. Boulton (1951) and many other investigators do not subscribe to Muskat's dismissiveness of the practical usefulness of the Dupuit–Forchheimer approach.

2.6 GROUNDWATER MODELLING

In a pedantic context, groundwater modelling involves the use of models or analogues to investigate or simulate the nature of groundwater flow (see Chapter 5). In modern parlance, groundwater modelling invariably refers to numerical models run on computers. In fact, these are not true models but iterative mathematical solutions to a model or mesh defined by the operator; a solution is considered to be acceptable when the errors reach a user-defined level. These models have been developed by many organizations from original esoteric research tools into the current generation of easier-to-use models with excellent presentation of results, which when used appropriately, can clearly demonstrate what is happening to the groundwater flow. The use of numerical modelling is described in Sections 5.5.3 and 13.12.

The origins of groundwater models and analogues are to be found not in groundwater theory, but in other scientific fields such as electricity and heat flow. In the mid-nineteenth century, Kirchhoff studied the flow of electrical current in a thin disk of copper. Just who recognized that the mathematical expressions or laws governing the flow of electricity were analogous to those governing the flow of thermal energy and groundwater is not clear.

This dawning of the electrical analogy to groundwater flow was given a major impetus in North America, where the growing interest in the theoretical aspects of oil reservoir development led to major advances. As noted earlier, the significant contributions of Muskat during the 1930s included electrical analogues and sand tank models. Wyckoff, who had worked with Muskat, published in 1935 the first conductive paper model study of groundwater flow through a dam. The use of conductive paper was a direct development from Kirchhoff's original work, nearly a century earlier, and became the practical basis for much of the two-dimensional isotropic analogue modelling for the next 50 years before numerical modelling finally superseded it. The conductive paper technique was given a further boost with the development of Teledeltos paper in 1948. This aluminium-coated carbon-based paper enabled rapid two-dimensional studies to be undertaken.

Karplus (1958) presents an extensive discussion on the use and limitations of conductive paper models, as well as introducing resistance networks where the uniform conducting layer is replaced by a grid of resistors. The advantage of this system was that it could be developed into three dimensions, and so the problems of making measurements within a solid were overcome. Herbert and Rushton (1966) developed resistance networks to introduce switching techniques to determine a free water surface, and transistors to simulate storage, and also evolved time-variant solutions. Case histories of the practical application of resistance networks have been published for pressure relief wells under Mangla Dam in Pakistan (Starr et al., 1969) and for a deep well system at Sizewell B Power Station in England (Knight et al., 1996).

Other analogues include electrolytic tanks, where a thin layer of conducting fluid is used instead of the conducting paper, or a stretched rubber membrane, which, when distorted at right angles to the surface, forms a shape analogous to the cone of depression formed by a well. Neither of these techniques or others appears to have been used extensively for the solution of practical groundwater problems.

A different class of models is the sand tank type (see Cedergren, 1989). These physical models consist of two parallel plates of clear material such as glass, closely spaced and filled with sand to represent the aquifer. Physical impediments to flow, such as dams or wells, can be inserted and the model saturated. If a tracer or dye is injected on the upstream side, the flow path(s) can be observed. The viscous flow or Hele-Shaw technique – so named after the man who first used this technique in England – is a variation of the sand tank. These techniques have been used regularly to demonstrate the form of groundwater flow, particularly beneath dams (Wesley and Preene, 2019), but have limited application to groundwater lowering in construction.

2.7 EARLY DEWATERING TECHNOLOGY IN BRITAIN

Much of the technology that forms the basis of modern dewatering methods was developed in the United States or Germany and was introduced to Britain in the early part of the twentieth century. Up to that time, any groundwater lowering required was achieved by the crude (but often effective) method of pumping from timbered shafts or from open-jointed subdrains laid ahead and below trench or tunnel works.

During the nineteenth century, early generations of groundwater exclusion methods had been developed in the mining industry in the form of close-fitting 'tubbing' used to line shafts and keep water out. These tubbings were initially made from timber and later from cast iron. In the very early years of the twentieth century, British mines made the first use of artificial ground freezing (whereby chilled brines were used to freeze groundwater) and the 'cementation' method, whereby cement-based grouts were used to prevent groundwater flow through pores and fissures in soils and rocks (Younger, 2004). These techniques were soon added to the repertoire of civil engineers.

Interlocking cast iron sheet-piles (the forerunner of modern steel sheet-piles) were used as a groundwater exclusion method for the construction of docks in London from the early nineteenth century onwards (Borthwick, 1836). Compressed air methods for shaft sinking and tunnelling (Glossop, 1976) and injection methods of grouting (Glossop, 1960, 1961) were also developed in civil engineering in the late nineteenth century and the early twentieth century.

The wellpoint method is probably the oldest of the modern dewatering pumping techniques. Originally, wellpoints were a simple form of driven tube, first widely deployed by the British general Sir Robert Napier on his march to Magdala in 1868 during his Abyssinian campaign. Each Abyssinian tube well, as they were known, was driven to depth using a sledgehammer. Rose (2012) provides a more detailed description of Abyssinian wells – by that time known as Norton tube wells, after an early patent – as used by the British military during the First World War:

> 'Driven' or 'drive' wells (from about 1860 onwards patented as 'Norton tube wells') comprised narrow diameter (typically 40–60 mm) iron tubes, with a steel point welded to the lower end, the lower part of the tube perforated with small holes above the point. These were driven < 7 m into unconsolidated sediments by blows applied at the top of the tube by an iron weight.

If water was found, the tube was then equipped with a conventional village-type hand pump (Figure 2.6).

According to Powers *et al.* (2007), wellpoints were used for dewatering in North America from 1901, but the modern form of the method probably derives from equipment developed by Thomas Moore in New Jersey in 1925 (Figure 2.7). His equipment was an advance

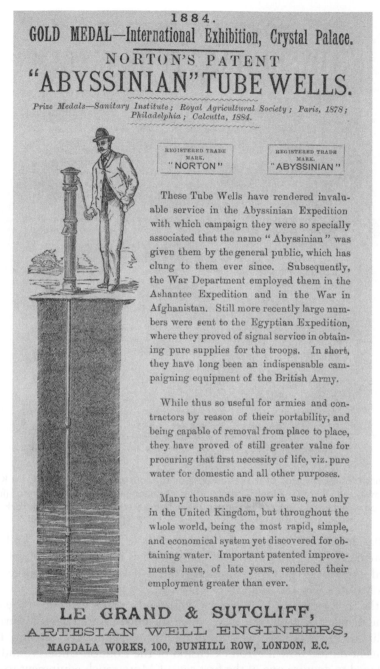

Figure 2.6 Abyssinian tube well. (From the collection of the British Library, with permission. © The British Library Board, Evanion 7442.) The image shows an 1884 advertisement for Abyssinian tube wells, so named from their use by British troops campaigning in Abyssinia (modern Ethiopia) during the nineteenth century, but also known as Norton tube wells from an early patent.

Figure 2.7 Thomas Moore, inventor of the jetted wellpoint. (Courtesy of Moretrench – a Keller company, Rockaway, NJ.) Thomas Moore (rear) and his engineer-in-charge Mr. Schwartz (front) and their crew on a partially completed sewer project in Salina Street, Syracuse, New York.

in that installation was by jetting to form a clean hole, which was backfilled with filter sand. His system, known as 'Moretrench' equipment, is identical in principle to that used today – indeed, the Moretrench company is still in the dewatering business nearly a century after Thomas Moore's original innovation.

In Britain, the civil engineer H J B (later Sir Harold) Harding was one of the leading practitioners in the new art of geotechnology, which included groundwater lowering (Harding and Davey, 2015). In the 1930s, working for John Mowlem & Company, Harding was contractor's agent on the Bow–Leyton extension of the London Underground central line. Here, he managed to acquire one of the first sets of Moretrench equipment to enter Britain and used it on sewer diversion work (Figure 2.8). Harding and his colleagues became expert in the method and contributed to the development of British alternatives to the American equipment.

One unexpected result of Harding's expertise was that during the Second World War, he and his crews often assisted British Army Royal Engineer units with dewatering of excavations in the search for unexploded German bombs. Harding and Davey (2015) report one case where an unexploded bomb ended its swerving path below ground under the only tank for testing ship models, at the National Physical Laboratory, a facility critical to the British

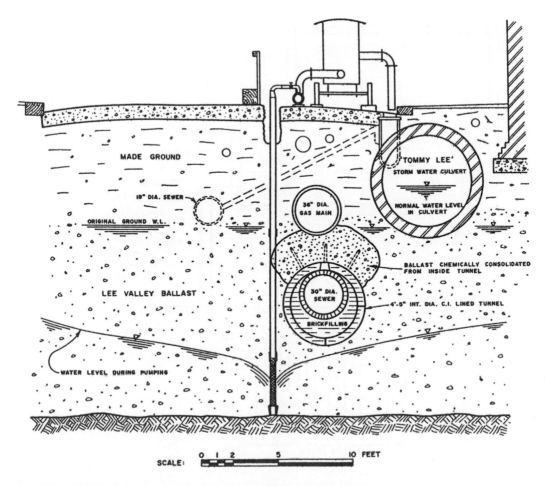

Figure 2.8 Sewer diversion under gas main using wellpoints and chemical injections. (From Harding, H J B, *Tunnelling History and My Own Involvement*, Golder Associates, Toronto, 1981. With permission from Golder Associates, Toronto, ON.) This shows an early British wellpoint application on the Bow–Leyton extension of the London Underground central line in the 1930s. The dewatering allowed tunnel works beneath an existing gas main.

war effort. The bomb had passed through a 6 m thickness of gravels containing 3 m depth of groundwater before coming to rest in the London Clay Formation 10 m below ground level. Harding assisted the Royal Engineers Bomb Disposal Section in installing a set of the, then new and unfamiliar, Moretrench wellpoint equipment to dewater for a shaft sunk through the gravels and into the clay. Once the shaft was sufficiently deep below the top of the clay to reduce the risk of inundation from the gravels, a timber-lined heading (tunnel) was driven along the trace until the bomb was reached so that it could be defused and made safe (Figure 2.9).

A very early application of groundwater lowering in Britain was a joint venture in 1934 by Edmund Nuttall Sons & Company and John Mowlem & Company at Bentalls Department Store in Kingston-on-Thames, London (Harding, 1935). The basement of the existing building was being deepened in permeable gravels of the River Terrace Deposits (known locally as 'Thames Ballast'). A series of suction wells (see Section 16.10) was successfully used to reduce groundwater levels in the gravels (Figure 2.10).

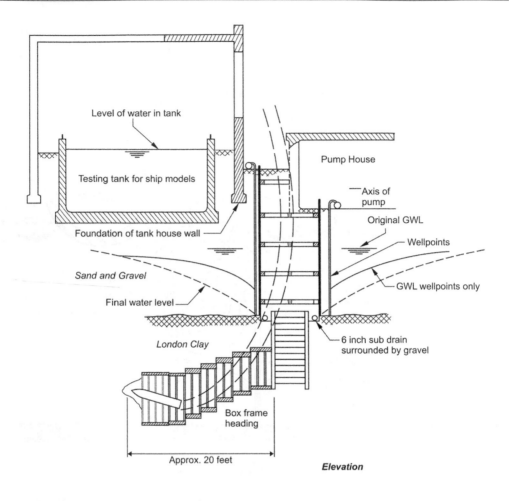

Level of water in tank

Testing tank for ship models

Pump House

Axis of pump

Original GWL

Wellpoints

Foundation of tank house wall

Sand and Gravel

GWL wellpoints only

Final water level

London Clay

6 inch sub drain surrounded by gravel

Box frame heading

Approx. 20 feet

Elevation

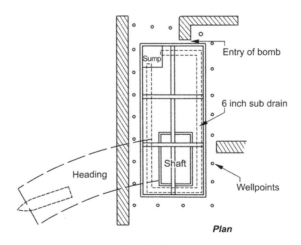

Entry of bomb

Sump

6 inch sub drain

Heading

Shaft

Wellpoints

Plan

Figure 2.9 Groundwater lowering by wellpoints for bomb recovery at National Physical Laboratory, Teddington, United Kingdom. (Re-drawn from Harding, H J B and Davey, A., *It's Warmer Down Below: The Autobiography of Sir Harold Harding, 1900–1986*, Tilia Publishing UK, Sussex, 2015, with permission from Tilia Publishing UK/Amanda Davey.)

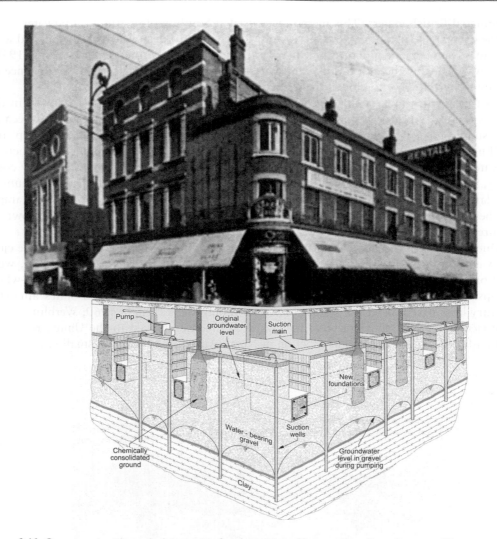

Figure 2.10 Cross section through dewatering for deepening of basement at Bentalls store, Kingston upon Thames, London. (Based on Harding, H J B., *Concrete & Constructional Engineering*, 30, 3–12, 1935.) This dewatering system, installed in 1934, was the first large-scale application of suction wells in the UK.

In the early 1930s, Harding was also instrumental in the introduction of the modern deep well method to Britain. Mowlem, with Edmund Nuttall Sons & Company, was awarded the contract to construct the King George V graving dock at Southampton to accommodate the liner *Queen Mary*, which was being built on Clydeside. The dock was to be 100 feet (31 m) deep. The docks engineer at Southampton was, according to Harding (1981), 'a wise and experienced man, he carried out his site investigation to unusual depths'. This revealed beds of Bracklesham sands containing water under artesian head, which would reach to above ground level.

At the time, large deep wells using the recently developed submersible pump had been used for groundwater lowering in Germany from 1896 onwards, initially for the construction of the Berlin U-Bahn underground railway, but this was not a method recognized in Britain. In 1932, Mowlem had obtained licencing agreements with Siemens Bau-Union to use their patents for, among other methods, groundwater lowering by deep wells, with Harding as the nominated British expert. This method was used at Southampton, and groundwater levels in the deep artesian aquifer were successfully lowered by 10 deep wells, each equipped with a Siemens submersible pump, run from a central control (McHaffie, 1938; Harding, 1938). This was probably the first rational application of the deep well method in Britain since its use by Robert Stephenson for the construction of the Kilsby Tunnel a century earlier.

Further applications of the deep well method with submersible pumps followed in the 1940s (Figure 2.11), now unencumbered by licence agreements, and the method became an established technique for the control of groundwater (Harding, 1946; Glossop and Collinridge, 1948, Williams and Norbury, 2008; Williams, 2010). The pioneering practical dewatering work on the Mowlem contracts was continued by Mowlem's subsidiary company, Soil Mechanics Limited, whose groundwater lowering department carried out numerous large-scale projects on power stations and docks in the 1950s and 1960s. An example of the wellpoint equipment used in the United Kingdom during that period is shown in Figure 2.12.

One of the most recent techniques to be introduced into the United Kingdom is the ejector well dewatering method. Jet pumps, which are the basis of the ejector method, were first proposed in the 1850s by Thomson (1852) for the removal of water from water-wheel sumps. Dewatering systems using ejectors were developed in the United States almost a century later based on jet pumps used in water supply wells (Prugh, 1960; Werblin, 1960). The ejector method does not seem to have been much employed in the United Kingdom until a few decades after its introduction in North America, although a small-scale ejector

Figure 2.11 Early application of deep wells in Britain. A submersible pump is being prepared for installation into a well on site in the 1940s.

Figure 2.12 Example of wellpoint system in the United Kingdom, early 1970s. (Courtesy of J K White.) The connection from the header pipe to each wellpoint is by short lengths of metal pipe, linked by elbow and union fittings (flexible plastic pipes were not widely used at this time).

system was used in England in 1962 to dewater the Elm Park Colliery drift (Greenwood, 1989). During the 1980s, British engineers and contractors used ejector wells on projects in Asia, such as the HSBC headquarters in Hong Kong (Humpheson *et al.*, 1986) and at the Benutan Dam in Brunei (Cole *et al.*, 1994). However, it was not until the late 1980s (Powrie and Roberts, 1990) that a large-scale ejector system was used in the United Kingdom, for the casting basin and cut-and-cover sections of the River Conwy Crossing project in North Wales (Figure 2.13).

2.8 PRACTICAL PUBLICATIONS

As described in Chapter 1, the control of groundwater is a practical problem, where theory is only part of the picture – how the theory is put into practice is vital. Historically, the best practical guidance came from in-house dewatering manuals produced by companies such as Geho Pumpen in Holland or the Moretrench American Corporation in North America. One of the first more practical dewatering texts to be widely published was Mansur and Kaufman (1962), which formed a chapter of the book *Foundation Engineering*, edited by G A Leonards. This is a detailed statement-of-the-art of the time, with a strong bias towards the analytical but with some reference to practical considerations. Although it may seem a little dated, this book is essential reading for all who aspire to be specialist dewatering practitioners and wish to understand some of the more accessible theory. Most of its content lives on in the form of the 'NAVFAC Guide' (Unified Facilities Criteria 2004), a US government construction guidance document.

The role of practical engineers in the development of groundwater control technology cannot be overstated. Because many of the techniques were developed from a practical rather than a theoretical basis, much of the experience became concentrated in specialist

Figure 2.13 The A55 Conwy Crossing under construction (west Cut-and-Cover section shown). This dewatering system, operated between 1987 and 1989, was the first large-scale application of ejector wells in the UK.

contracting companies instead of consulting engineers or academic bodies. In the United Kingdom, companies such as Soil Mechanics Limited carried out numerous large-scale groundwater control projects in Britain in the 1950s and 1960s, and many of their staff ultimately moved on and spread their experience around several successful groundwater control contractors in the 1970s, 1980s and beyond.

The improvements in groundwater control practice were supported by some publications that selflessly shared some of the hard-won experience acquired by contractors. First, in 1981, J P Powers, then of the Moretrench American Corporation, produced his book *Construction Dewatering: A Guide to Theory and Practice*, which is understandably oriented towards American practice. Now in its third edition as Powers *et al.* (2007), this remains a thorough and readable book. In the United Kingdom, in 1986 the Construction Industry Research and Information Association (CIRIA) produced Report 113 *Control of Groundwater for Temporary Works* (Somerville, 1986). It was aimed at the non-specialist engineering designer and site staff and was largely based on the experience of Pat M. Cashman from his work with Soil Mechanics Limited, Sykes and other organizations. In the late 1990s, CIRIA produced another, more detailed report. This was Report C515 *Groundwater Control Design and Practice* (Preene *et al.*, 2000), later updated as Report C750 (Preene *et al.*, 2016), again based on the experience of a specialist contractor, this time WJ Groundwater Limited. A more wide-ranging book by Woodward (2005) also provides some useful information on groundwater control techniques in the context of the range of available ground treatment technologies.

Useful information and case histories can sometimes be found in groundwater conference proceedings. In the 1980s and 1990s, relevant conferences were held on *Groundwater in Engineering Geology* (Cripps *et al.*, 1986), *Groundwater Effects in Geotechnical Engineering* (Hanrahan *et al.*, 1987) and *Groundwater Problems in Urban Areas* (Wilkinson, 1994).

This book is intended to complement and augment these texts and will concentrate, in the main, on the practical requirements for groundwater control for temporary works.

Chapter 3

Principles of Groundwater Flow

> All the rivers run into the sea; yet the sea is not full; unto the place from whence the rivers come, thither they return again.
>
> Ecclesiastes 1:7

3.1 INTRODUCTION

To allow even a basic approach to the control of groundwater, the practitioner should be familiar with some of the principles governing groundwater flow. Similarly, the specialist terms and language used must be understood. This chapter describes briefly the background to the flow of water through the ground. Chapter 4 will discuss permeability and hydraulic conductivity in more detail.

This chapter outlines the hydrological cycle and introduces the concepts of aquifers and permeability (a measure of how easily water can flow through a porous mass). Darcy's law – which is used to describe most groundwater flow regimes – is described.

The importance of aquifer types, geological structure, at both large and small scale, and the effect of aquifer boundaries on groundwater flow are discussed. Finally, this chapter discusses groundwater temperatures, basic groundwater chemistry and the effect of climate and weather.

3.2 HYDROLOGY AND HYDROGEOLOGY

The study of water, and its occurrence in all its natural forms, is called 'hydrology'. This branch of science deals with water, its properties and all of its behavioural phases. It embraces geology, soil mechanics, meteorology and climatology, as well as hydraulics and the chemistry and bacteriology of water.

Terminology may vary from country to country, but in general, those professionals who specialize in the hydrology of groundwater are known as 'hydrogeologists'. The study and practice of hydrogeology are of increasing importance both in developed and developing countries where groundwater is used as a resource, to supply water for the population and industries; for example, around one-third of the United Kingdom's drinking water is obtained from groundwater. The reader interested in hydrogeology is recommended to an introductory text by Price (1996) and more theoretical treatments by Fetter (2000) or Younger (2007), among the copious literature on the subject.

For the dewatering practitioner, detailed knowledge of some of the more arcane areas of hydrogeology is not necessary. However, if a rational approach is to be adopted, it is vital that the basic tenets of groundwater flow are understood in principle. Further study, beyond basic principles, will require some understanding of higher mathematics and analysis – this can be very useful but will not be for everyone. The references given at the end of this book should allow interested and motivated readers to pursue the subject as far as they wish.

It is worth mentioning that there is sometimes a little professional friction between dewatering practitioners (generally from an engineering background) and hydrogeologists (often with a background in earth sciences). For their part, the dewatering engineers are often pragmatic. They have an unstable excavation and need to control groundwater accordingly – they see it very much as a local problem. Hydrogeologists, on the other hand, are trained to view groundwater as part of the wider environment. They see that a groundwater lowering system should not be considered in isolation – but sometimes, they may not fully appreciate the practical limits of available dewatering methods. As with many professional rivalries, the wise observer can learn from both camps and gain a more rounded view of the problem at hand. This is the approach that the authors wish to encourage. Some of the misunderstandings between dewatering practitioners and hydrogeologists have resulted from slightly different terminology used by each group. This book unashamedly prefers the engineering terms used in dewatering practice but introduces and explains alternative forms where appropriate.

3.2.1 The Hydrological Cycle

The hydrological cycle is now so widely accepted that it is difficult to conceive of any other concept. The hydrological cycle is based on the premise that the volume of water on earth is large but finite. Most of this water is in continuous circulation. Water vapour is taken up into the atmosphere from surface water masses (principally the oceans) to form clouds; later, cold temperatures aloft cause the water to fall as precipitation (dew, fog, rain, hail or snow) on the earth's surface, whence it is eventually returned to the oceans (Figure 3.1), from where the cycle is repeated.

Soil is made up of mineral particles in contact with each other. Rock is formed from mineral grains or crystals cemented together. The soil and rock masses contain voids, either

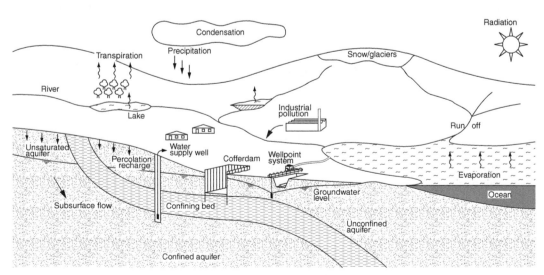

Figure 3.1 The hydrological cycle. (From Preene, M, Roberts, T O L and Powrie, W, *Groundwater Control – Design and Practice*, 2nd edition, 2016, Construction Industry Research and Information Association, CIRIA Report C750, London. Reproduced with permission from CIRIA: www.ciria.org)

widely distributed in the form of pores or locally concentrated as fractures. In simple terms, the water contained in the voids of the soil and rock is known as groundwater. The water is held in the 'void space' or 'pore space' within the material. A useful term is porosity (defined as the fraction of void space in soil or rock, and typically given the symbol n).

Generally, a 'water table' will exist at some depth below the ground surface; below the water table, the soil and rock pores and fractures are full of water (and are said to be saturated). Above the water table, in a free-draining soil such as a sand or gravel, the pores and fractures are unsaturated – that is, they contain both water and air. However, in very fine-grained soils such as silts or sands, soil capillary forces may retain water in the soil pores above the 'water table'. In such cases, a capillary saturated zone is said to exist above the water table (see Section 3.3.5).

The hydrological cycle illustrates clearly that precipitation replenishes the groundwater in the voids of the soils and rocks and that gravity flow causes movement of groundwater through the pores or fractures of soil and rock masses, towards rivers and lakes and eventually to the oceans.

It is unlikely that all of the precipitation that falls on landmasses will percolate downwards sufficiently far to replenish groundwater:

a) Some precipitation may fall onto trees or plants and evaporate before ever reaching the ground.
b) Some precipitation will contribute to surface run-off. This proportion will depend mainly upon the composition of the ground surface. For instance, an urban paved surface area will result in almost total surface run-off with little percolation into the ground.
c) Of that precipitation which does enter the ground, not all of it will percolate sufficiently far to reach the water table. Some will be taken up by plant roots in the unsaturated zone just below the ground surface and will be lost as evapotranspiration from the plants.

One unavoidable conclusion from the hydrological cycle is that most groundwater is continually in motion, flowing from one area to another. Under natural conditions (i.e. without interference by man in the form of pumping from wells), groundwater flow is generally relatively slow, with typical velocities in the range of a few metres per day in high-permeability soils and rocks down to a few millimetres per year in very low-permeability deposits. Larger groundwater velocities generally only exist in the vicinity of pumped wells or sumps. The exception to this is in karst aquifers, where much higher velocities (in excess of 1 km per day) can occur when aquifers are pumped (groundwater velocities are discussed in Section 3.3.4).

3.2.2 Geology and Soil Mechanics

The successful design and application of groundwater lowering methods do not just depend on the nature of the groundwater environment (such as where the site is within the hydrological cycle) but is also critically influenced by the geology or structure of the soils and rocks through which water flows. This is especially true when trying to assess the effect of groundwater conditions on the stability of engineering excavations (as will be discussed in Chapters 6 to 8). The eminent soil mechanics engineer Karl Terzaghi wrote in 1945: 'It is more than mere coincidence that most failures have been due to the unanticipated action of water, because the behaviour of water depends, more than on anything else, on minor geological details that are unknown' (Peck, 1969). This is still true today, but perhaps the last sentence should be changed to 'minor geological details that are often overlooked during site investigation' – good practice for site investigation will be discussed in Chapter 11.

The reader wishing to apply groundwater lowering methods will need to be familiar with some of the aspects of hydrogeology, soil mechanics (including its applied form, geotechnical engineering) and engineering geology; texts by Powrie (2013) and Blyth and De Frietas (1984), respectively, are recommended as introductions to the latter two subjects. However, as has been stated earlier, study of theory is only part of the learning process. Many soils and rocks may not match the idealized homogeneous isotropic conditions assumed for some analytical methods. Judgement will be needed to determine the credibility of results of analyses based on such models.

An important geological distinction is made between uncemented 'drift' deposits (termed 'soil' by engineers) and cemented 'rock'. Drift deposits are present near the ground surface and consist of sands, gravels, peat, clays and silts, which may have resulted from weathering or from glacial or alluvial processes. Groundwater flow through drift deposits is predominantly intergranular – that is, through the pores in the mass of the soil (Section 4.3). Groundwater problems in these materials are discussed in Chapter 7.

Rock may be exposed at the ground surface or may be covered by layers of drift and can consist of any of the many rock types that exist, such as sandstone, mudstone, limestone, basalt, gneiss and so on. Groundwater flow through rock is often not through pores (which may be too small to allow significant passage of water) but along fractures, bedding planes and joints within the rock mass (Section 4.4). This means that groundwater flow through rock may be concentrated locally where the deposition and subsequent solution or tectonic action has created fracture networks. Groundwater problems for excavations in rock are discussed in Chapter 8.

3.3 PERMEABILITY, HYDRAULIC CONDUCTIVITY AND GROUNDWATER FLOW

Understanding the concepts and principles of groundwater flow is a fundamental pre-requisite for success in the design and management of groundwater control systems. This does not require detailed knowledge of mathematical equations (although that is useful for more complex design work). It does, however, require an understanding of the factors that cause groundwater to flow.

3.3.1 Drivers for Groundwater Flow

The hydrological cycle introduced earlier implies that groundwater is continuously in motion, albeit slowly. This can be apparent even to a non-expert observing groundwater seepage from springs. This leads to some common misconceptions about groundwater, including the often-heard one that groundwater flows in 'underground rivers'. Even a basic knowledge of geology will indicate that this is rarely the case, and groundwater typically flows through the void spaces in the pores of soil or fractures, bedding planes and joints of rock.

But why does groundwater move, and what controls its speed and direction? This is a key question for dewatering practitioners, because if we can understand the principles of groundwater flow, then perhaps we can manipulate the groundwater regime around our project and create an effective groundwater lowering system.

The starting point for understanding groundwater flow is permeability. This is a critical parameter for the assessment of how water flows through soil and rocks (it is so important that it will be discussed in much more detail in Chapter 4, but here the basic concepts are introduced).

The precise meaning of the term 'permeability' is sometimes a cause of confusion between engineers and hydrogeologists. Civil and geotechnical engineers are interested almost exclusively in the flow of water through soils and rocks and use the term 'coefficient of permeability', given the symbol k. For convenience, k is generally referred to simply as 'permeability', and throughout this book this terminology will be used. Therefore, for groundwater lowering purposes, permeability, k, will be defined as 'a measure of the ease or otherwise with which groundwater can flow through the pores of a given soil or rock mass'.

For idealized, homogeneous soils, permeability depends primarily on the properties of the soil, including the size and arrangement of the soil particles, and the resulting pore spaces formed when the particles are in contact. In rock, the permeability depends on the number, nature and interconnectivity of the various fractures and discontinuities. Permeability is discussed in much greater detail in Chapter 4.

A slight complication is that the permeability of porous mass is dependent not only on the nature of the porous media but also on the properties of the permeating fluid. In other words, the permeability of a soil to water is different from the permeability of the soil to another fluid, such as air or oil. Hydrogeology references highlight this by calling the engineer's permeability 'hydraulic conductivity' to show that it is specific to water. Often, if the term 'permeability' appears in hydrogeology references, it actually means the permeability of the porous media independent of the permeating fluid, sometimes also known as intrinsic permeability (see Section 4.2.1).

The permeability values used throughout this book will be reported in m/s (metres per second), which is the convention in engineering and soil mechanics, and is useful when carrying out dewatering calculations. This is in contrast to hydrogeological references, which generally report permeability (or hydraulic conductivity) in m/d (metres per day). Conversion factors between various units are given at the end of this book.

3.3.2 Darcy's Law

The modern understanding of flow of groundwater through permeable ground originates with the research of the French hydraulics engineer, Henri Darcy (1856) (see Section 2.4.4). He investigated the purification of water by filtration through sand and developed an equation of flow through a granular medium based on the earlier work of Poiseuille concerning flow in capillary tubes. His studies resulted in Darcy's law, which is still used as the basis for the solutions of the vast majority of groundwater problems, from the simple to the very complex.

Precise analysis of the flow of groundwater through soil is probably one of the more complex mathematical problems faced in routine geotechnical analysis. Yet Darcy's conclusions can be expressed algebraically as the deceptively simple Equation 3.1, universally referred to as Darcy's law:

$$Q = -k\left(\frac{\Delta h}{l}\right)A \tag{3.1}$$

where (see Figure 3.2 which schematically shows Darcy's experiment):
Q = the volumetric flow of water per unit time (the 'flow rate')
A = the cross-sectional area through which the water flows
l = the length of the flow path between the upstream and downstream ends
Δh = the difference in total hydraulic head between the upstream and downstream ends
k = the permeability of the porous medium through which the water flows

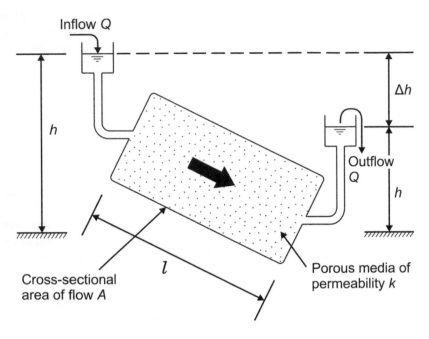

Figure 3.2 Darcy's experiment.

$\Delta h/l$ is the change in hydraulic head divided by the length of the flow path; it is termed the hydraulic gradient and given the symbol i. The negative term is necessary in Equation 3.1 because flow occurs down the hydraulic gradient – that is, from high head to low head.

The total hydraulic head at a given point is the sum of the 'pressure head' and the 'elevation head' at that point, as shown in Figure 3.3. The elevation head is the height of the measuring point above an arbitrary datum, and the pressure head is the pore water pressure u, expressed as metres head of water.

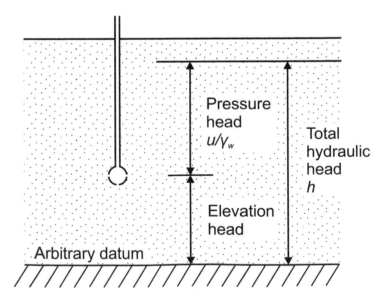

Figure 3.3 Definition of hydraulic head.

Total hydraulic head is important because it controls groundwater flow. Water will flow from high total hydraulic head to low total hydraulic head. This important concept is often misunderstood; it means that water does not necessarily flow from high pressure to low pressure or from high elevations to low elevations—it will only flow in response to differences in total hydraulic head, not pressure or elevation considered in isolation, as illustrated in Figure 3.4.

Although groundwater heads are the driver for groundwater flow, the pressure element of the head is also relevant. In soil mechanics and rock mechanics, this is normally expressed as a pressure, termed the pore water pressure u. This is an important factor in effective stress (see Section 6.5), which controls the stability of soil and rock. Pore water pressures are often presented as pore water pressure distributions with depth.

Figure 3.5 shows some simple pore water pressure distributions, where:

- If there is no vertical groundwater flow, pore water pressures increase by approximately 10 kPa for every metre of depth below groundwater level (Figure 3.5a). This is known as a 'hydrostatic' condition. In this case, the total hydraulic head is constant with depth, and monitoring wells installed at different depths show the same level.
- If there is upward vertical groundwater flow, pore water pressures increase by more than 10 kPa for every metre of depth below groundwater level; that is, groundwater pressures at a given level will be greater than hydrostatic (Figure 3.5b). In this case, the total hydraulic head is greater at depth, so deeper monitoring wells will show higher groundwater levels than shallower wells.
- If there is downward vertical groundwater flow, pore water pressures increase by less than 10 kPa for every metre of depth below groundwater level; that is, groundwater pressures at a given level will be less than hydrostatic (Figure 3.5c). In this case, the total hydraulic head is lower at depth, so deeper monitoring wells will show lower groundwater levels than shallower wells.

Examples of more complex pore water pressure distributions are shown in Figures 3.17 and 3.21.

A term that is used frequently in any discussion of groundwater lowering is 'drawdown'. Drawdown is the amount of lowering (in response to pumping) of total hydraulic head and is a key, measurable performance target for any dewatering operation. Drawdown is equivalent to:

(i)   The amount of lowering of the 'water table' in an unconfined aquifer.
(ii) The amount of lowering of the 'piezometric level' in a confined aquifer.
(iii) The reduction (expressed as metres head of water) of pore water pressure observed in a monitoring well or piezometer (see Section 11.7.6). In this case, the drawdown of total hydraulic head can be estimated directly from the change in pressure head, since the level of the piezometer tip does not change, so the change in elevation head is zero. If the reduction in pore water pressure is Δu, the drawdown is equal to $\Delta u / \gamma_w$, where γ_w is the unit weight of water.

Darcy's law is often written in terms of the hydraulic gradient i, which is the change in hydraulic head divided by the length of the flow path ($i = \Delta h / l$). Equation 3.1 then becomes

$$Q = -kiA \qquad\qquad (3.2)$$

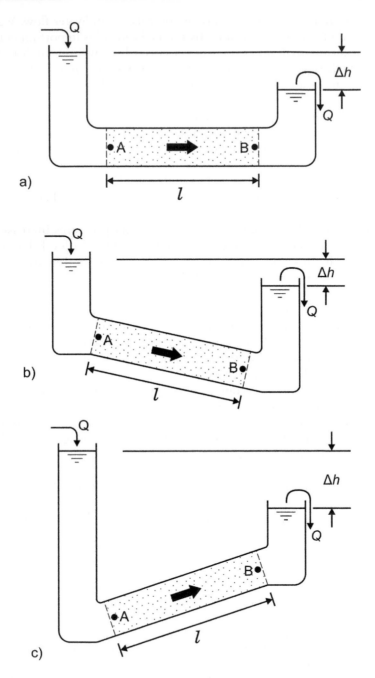

Figure 3.4 Groundwater flow driven by differences in total hydraulic head. (a) There is no change in elevation between A and B – groundwater is flowing from high-pressure to low-pressure areas. (b) Due to the change in elevation, the pressure at A is lower than at B, and groundwater flows from low-pressure to high-pressure areas. (c) The excess head Δh drives groundwater flow from A to B and means that water can flow 'uphill' along upward physical gradients.

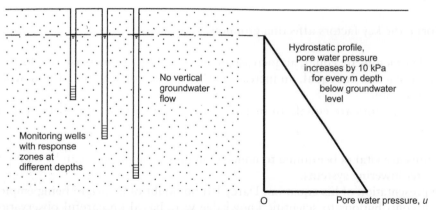

a) Hydrostatic conditions (no vertical flow of groundwater)

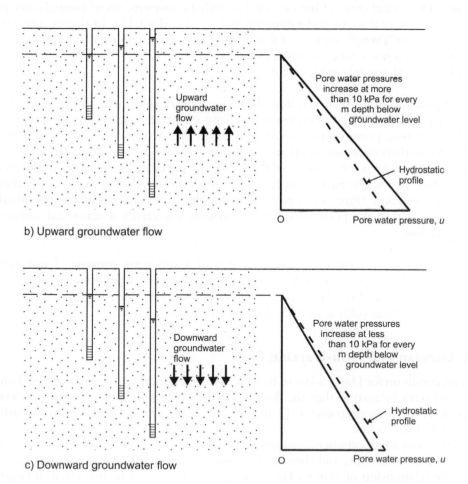

b) Upward groundwater flow

c) Downward groundwater flow

Figure 3.5 Pore water pressure distribution with depth. (a) Hydrostatic conditions (no vertical flow of groundwater). (b) Upward groundwater flow. (c) Downward groundwater flow.

In this form, the key factors affecting groundwater flow are obvious:

  (i) If other factors are equal, an increase in permeability will increase the flow rate.
 (ii) If other factors are equal, an increase in the cross-sectional area of flow will increase the flow rate.
(iii) If other factors are equal, an increase in hydraulic gradient will increase the flow rate.

These points are vital in beginning to understand how groundwater can be manipulated by groundwater lowering systems.

In the presentation of his equation, Darcy left no doubt of its origin being empirical. His important contributions to scientific knowledge were based on careful observation in the field and in the laboratory and on the conclusions that he drew from these. Permeability is in fact only a theoretical concept but one vital to realistic assessments of groundwater pumping requirements, and so an understanding of it is most desirable. In theory, permeability is the notional (or 'Darcy') velocity of flow of pore water through unit cross-sectional area, considered to be a homogeneous porous medium.

Ineson (1956) eloquently explains this:

> As opposed to uni-directional flow through capillary tubes, the path of an individual fluid particle passing through the multiplicity of intercommunicating capillary channels comprising porous media is complex due to the tortuosity of these channels, rapid changes in their cross-sectional area, range in flow conditions from streamline to turbulent and rotational movement of the individual particle. The analysis of the mechanics of flow of a single particle under these conditions would appear to be incapable of unique resolution. Darcy's law is concerned with the summation of the individual flow processes within the porous media and presents essentially a statistical summary of viscous flow.

In fact, the majority of the cross-sectional area of a soil mass consists of soil particles, through which pore water cannot flow. The actual pore water flow velocity is greater than the 'Darcy velocity' and is related to it by the soil porosity n (porosity is the ratio of voids, or pore space, to total volume). This is discussed in Section 3.3.4.

3.3.3 Darcian and Non-Darcian Flow

The main condition for Darcy's law to be valid is that groundwater flow should be 'laminar', a technical term indicating that the flow is smooth (Figure 3.6). This is termed 'Darcian flow', where, for a given geometry, Q and Δh have a linearly proportional relationship, as expressed by Darcy's law.

At higher Reynolds numbers (essentially at higher hydraulic gradients), flow becomes turbulent or non-Darcian, and there is no longer a linear relationship between Q and Δh (hence, the relationship of Darcy's law no longer applies). This is illustrated in Figure 3.7, which shows that at higher hydraulic gradients, where non-Darcian flow occurs, the flow rate increases under-proportionally with hydraulic gradient.

It is generally accepted that, for the range of hydraulic gradients encountered in most geotechnical problems, Darcian flow will dominate. However, in rock where more open fractures are present, and if hydraulic gradients are high, non-Darcian (turbulent) flow will occur, and Q will increase under-proportionally with Δh as energy is lost to turbulence.

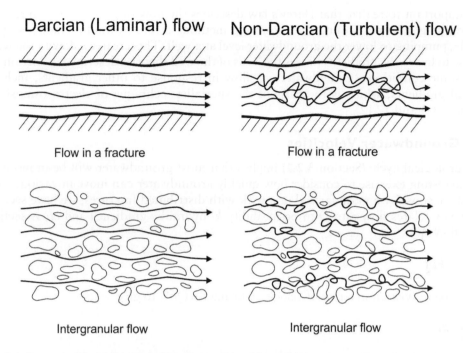

Darcian (Laminar) flow   Non-Darcian (Turbulent) flow

Flow in a fracture   Flow in a fracture

Intergranular flow   Intergranular flow

Figure 3.6 Darcian and non-Darcian groundwater flow. During Darcian (laminar) conditions, the flow paths are essentially parallel, and little energy (head) is lost. During non-Darcian (turbulent) conditions, the flow paths are more complex and irregular, and more energy (head) is lost.

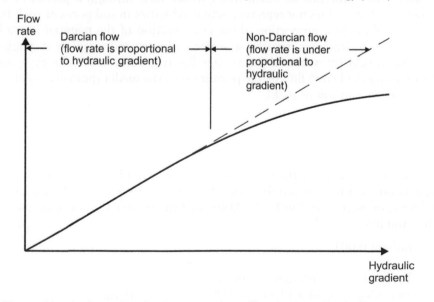

Flow rate

Darcian flow
(flow rate is proportional
to hydraulic gradient)

Non-Darcian flow
(flow rate is under
proportional to
hydraulic
gradient)

Hydraulic gradient

Figure 3.7 Flow and head relationships for Darcian and non-Darcian groundwater flow.

Examples where non-Darcian flow may occur include packer permeability tests in rock with very open fractures (Preene, 2019), or immediately around high-flow-rate wells pumping from coarse gravel aquifers. In most other geotechnical problems, hydraulic gradients are low enough to ensure that Darcian flow conditions prevail, which is the underlying assumption in most methods of analysis and modelling used for the design of groundwater control systems.

It is important to realize that Darcy's law describes the flow of groundwater in response to hydraulic gradients. Groundwater lowering methods create hydraulic gradients by, for example, pumping to lower the groundwater level at a well. This creates a flow to the well in response to hydraulic gradients (this is the basis of the well theories described in Section 5.8). But in some conditions, groundwater can flow in response to other gradients, including electrical gradients, which is the basis of the specialist technique of electro-osmosis (see Sections 4.2.3 and 17.12).

3.3.4 Groundwater Velocities

The hydrological cycle (Section 3.2.1) implies that most groundwater will be in motion. It is an interesting exercise to consider how quickly groundwater can move in various conditions. If we consider a granular soil or a rock with distributed fractures, we have seen that Darcy's law relates flow rate Q to permeability k, hydraulic gradient i and cross-sectional area of flow A:

$$Q = -kiA \tag{3.3}$$

This can be re-written in terms of velocity v (using $v = Q/A$) to give

$$v = -ki \tag{3.4}$$

where v is the Darcy velocity (sometimes known as the Darcy flux), defined as the velocity equivalent to the flow per unit cross-sectional area of a porous medium. It is an apparent velocity, equivalent to the rate at which water would flow through a porous medium if it were an open conduit; it does not represent actual velocities in soil pores or rock fractures.

In a soil or rock, only a proportion of the cross section of the material is available for flow (the soil pores or fractures in rock), so water will move more rapidly through the pore spaces. The mean velocity is termed the average linear groundwater velocity v_x and is calculated by dividing the Darcy flux by the porosity n of the media (porosity is defined as the fraction of void space in soil or rock).

$$v_x = -\frac{ki}{n} \tag{3.5}$$

This equation can be used to estimate natural rates of groundwater flow. In the absence of pumping, it is unusual for a relatively permeable stratum to have a hydraulic gradient (for horizontal flow) of more than 1 in 1000 (0.001) to 1 in 100 (0.01). Assuming a soil porosity of 0.25, this implies:

Hydraulic gradient 0.001

> Permeability $k = 1 \times 10^{-5}$ m/s (equivalent to a fine sand),
> groundwater velocity $v_x = 4 \times 10^{-8}$ m/s $= 0.003$ m/d $= 1.2$ m/year
> Permeability $k = 5 \times 10^{-3}$ m/s (equivalent to a coarse gravel),
> groundwater velocity $v_x = 2 \times 10^{-5}$ m/s $= 1.7$ m/d $= 630$ m/year

Hydraulic gradient 0.01

> Permeability $k = 1 \times 10^{-5}$ m/s (equivalent to a fine sand),
> groundwater velocity $v_x = 4 \times 10^{-7}$ m/s $= 0.03$ m/d $= 12$ m/year

Permeability $k = 5 \times 10^{-3}$ m/s (equivalent to a coarse gravel),
groundwater velocity $v_x = 2 \times 10^{-4}$ m/s = 17 m/d = 6.3 km/year

These example calculations show that under natural conditions, groundwater flow is typically very slow – where higher hydraulic gradients exist, such as near saturated slopes, these tend to be in lower-permeability material, and the groundwater velocity (the product the of hydraulic gradient and permeability) will still be very low.

Pumping from a well or sump can generate hydraulic gradients much greater than occur naturally. A limiting hydraulic gradient i_{max} near a pumped well was proposed by Sichardt (see Section 13.9.1):

$$i_{max} = \frac{1}{15\sqrt{k}} \tag{3.6}$$

For a coarse gravel where $k = 5 \times 10^{-3}$ m/s, $i_{max} = 0.94$. Using Equation 3.5, this implies a groundwater velocity close to the well of approximately $v_x = 0.2$ m/s = 68 m/h = 1.6 km/d. While these velocities are much higher than occur naturally, they are still relatively low; as a comparison, even close to the well, horizontal groundwater velocity is much less than walking pace.

The key exception to the general presumption that natural groundwater velocities are very low is in karst aquifer conditions. In these conditions, groundwater is concentrated in enlarged flow paths that have developed within soluble rocks such as Limestone. MacDonald *et al.* (1998) report tracer tests in a karstic zone of the Chalk aquifer in Hertfordshire, United Kingdom that gave natural groundwater velocities in specific flow zones of up to 6 km/d. This is orders of magnitude higher than in a granular soil, but still lower than walking pace, as an example. Karst aquifer systems are discussed in Sections 4.6.2 and 8.3.5.

3.3.5 Effect of Groundwater on Soil and Rock Properties

It is a fundamental tenet of theoretical soil mechanics and rock mechanics that the presence of groundwater has a significant effect on the properties of soils and rocks (this is primarily due to the influence of effective stress on soil behaviour, as discussed in Section 6.5). Of course, anyone who has seen the difference in site conditions between an initially waterlogged site and a site after groundwater levels have been lowered will need no convincing of this effect!

It is also worth considering how water moves out of soil or rock when groundwater levels are lowered by pumping. If we visualize a zone of soil or rock in our excavation that is below groundwater level, the voids in the soil/rock are at positive pore water pressures ($u > 0$) and are saturated (i.e. full of water). If by pumping we lower groundwater levels below this zone, the pore water pressures will fall to zero, and water will try to drain out of the voids, to be replaced by air – if this occurs, the soil/rock will become unsaturated.

The line of zero pore water pressure is known as the phreatic surface, and it is sometimes assumed that this forms the boundary between the unsaturated zone (above the phreatic surface) and the saturated zone (below the phreatic surface). In an unconfined aquifer (see Section 3.4.1), the phreatic surface is synonymous with the water table.

In a coarse soil – such as a sandy gravel – which might be described as 'free-draining', the phreatic surface may approximate to the boundary between saturated and unsaturated material. This soil pores are relatively large, and water will drain from the pores (and air will enter the pores) when the pore water pressure reaches zero (this is described as having an 'air entry value' of zero). However, even in a coarse-grained, free-draining soil, not all of the water will drain out when the water table is lowered. Some water will be retained in the smaller soil pores by capillary forces.

Finer-grained soils, such as silts, clays, and silty and clayey sands, behave very differently. The void spaces in these soils are so small that surface tension effects, known as capillary forces, resist the drainage of water under gravity, and soil pores can remain fully saturated with water even at significant negative pore water pressures. The role of capillary effects in soil pores of different sizes can be illustrated by the simple experiment shown in Figure 3.8a, where tubes of different diameter are placed in water – the finer the tube diameter, the higher the water will rise in the tube. This is analogous to the height to which water is retained in soil pores under the action of capillary forces (Figure 3.8b).

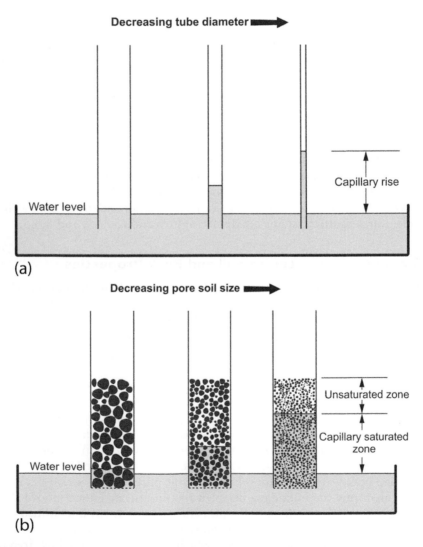

Figure 3.8 Water retained in soil pores by capillary forces. (a) Capillary rise in open-ended glass tubes placed in water. Water will rise higher in the smaller-diameter tubes. Surface tension forces retain water in the tubes, acting against gravity. (b) Capillary forces retaining water in soil pores. In fine-grained soils, the capillary saturated zone extends higher above the water table. Surface tension forces retain water in the pores, acting against gravity.

This phenomenon means that fine-grained soil will be saturated with capillary water for a height above the phreatic surface (this is the capillary saturated zone; Figure 3.9). The height of the capillary saturated zone above the phreatic surface is dependent on the effective size of the pores – the smaller the pores, the greater the height. At the upper surface of the capillary saturated zone, the surface tension forces between the water molecules and the soil particles are only just sufficient to prevent air being drawn into the soil pores. Above the top of the capillary saturated zone, air can enter the soil, which becomes unsaturated.

A soil can desaturate only when the pore water pressure u falls below the air entry pressure u_a (u_a is lower than atmospheric and so has a negative pore water pressure [suction] when expressed as 'gauge pressure' relative to atmospheric). Terzaghi et al. (1996) provide an empirical formula for air entry pressure u_a related to the particle size of a soil:

$$-\frac{0.05\gamma_w}{eD_{10}} < u_a < -\frac{0.01\gamma_w}{eD_{10}} \tag{3.7}$$

Where e is the void ratio (the ratio of solids to voids within a soil), D_{10} is the 10% particle size in millimetres, and γ_w is the unit weight of water in kilonewtons per cubic metre. The height of the capillary saturated zone is approximately equal to u_a/γ_w. Figure 3.10 shows the range of u_a/γ_w evaluated from Equation 3.7 for e = 0.4–0.6 with permeability estimated from D_{10} using Hazen's rule ($k = 0.01D_{10}{}^2$; k is in metres per second and D_{10} is in millimetres: Hazen, 1892).

While the correlation between D_{10} and permeability is only very approximate for fine-grained soils, Figure 3.10 illustrates that the height of the capillary saturated zone may be several metres in low-permeability soils such as clays and silts.

It is well established that the permeability of a soil (or a fractured rock) reduces dramatically once the soil becomes unsaturated (Fredlund et al., 1994); see Figure 3.11. Therefore, if the soil or rock is unsaturated above the phreatic surface, that surface becomes the effective boundary of the zone of water flow, since the permeability above the phreatic surface is very low. This is a typical assumption when modelling groundwater lowering systems in free-draining soil or rock, but it is a less valid assumption for seepage problems for slopes and embankments in clays and silts, where significant seepage may occur above the phreatic

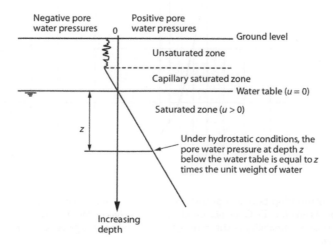

Figure 3.9 Groundwater conditions above the water table.

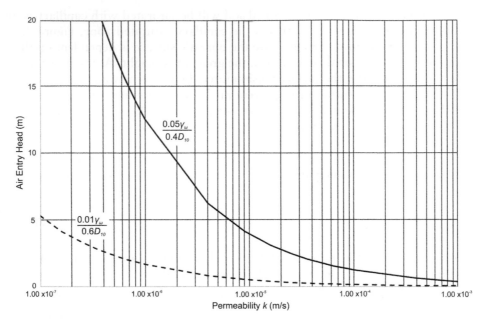

Figure 3.10 Example of variation of air entry head with permeability *k*. (Based on Powrie and Preene, *Géotechnique*, 44, 1, 1994a.) Air entry head $= u_a \big/ \gamma_w$

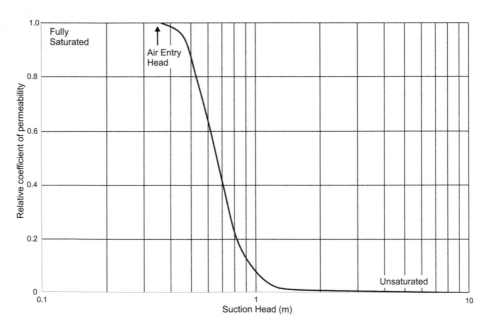

Figure 3.11 Example relationship between permeability of a soil and pore water pressure for a silty sand. (Based on Fredlund, D G *et al.*, *Canadian Geotechnical Journal*, 31, 533–546, 1994.) Relative coefficient of permeability is the ratio of permeability at a given suction to the fully saturated permeability.

surface, where the soil remains saturated and is of similar permeability above and below the phreatic surface (Wesley, 2014).

3.4 AQUIFERS, AQUITARDS AND AQUICLUDES

'Aquifer' is a useful term that appears whenever groundwater is discussed, but it must be used carefully when discussing groundwater lowering systems. As used by hydrogeologists, an aquifer might be defined as 'a stratum of soil or rock that can yield groundwater in economic or productive quantities'. Almost all wells used for water supply purposes are drilled into, and pump from, aquifers. Examples of aquifers in the United Kingdom include the Chalk or Sherwood Sandstone. By this definition, strata that yield water at flow rates too small to be used for supply are not aquifers and might be considered 'non-aquifers' (in the United Kingdom, the regulatory bodies sometimes use the term 'unproductive strata'). Examples of non-aquifers might include alluvial silts, glacial lake deposits or unfractured mudstones.

From a groundwater lowering point of view, the hydrogeologists' definition of an aquifer in terms of 'productive quantities' is unhelpful. The groundwater in many strata that yield just a little water (and so are non-aquifers) can cause severe problems for excavation stability (see Chapter 6) – in practice, many large-scale groundwater control schemes have been deployed in strata that hydrogeologists might class as 'non-aquifers'. For the purposes of this book, the definitions of CIRIA Report C750 (Preene *et al.*, 2016) will be adopted, where the reference to productive quantities is omitted. This definition is given here, together with those for aquiclude and aquitard. The relationship between these strata types is discussed in the following sections.

> *Aquifer.* Soil or rock forming a stratum, group of strata or part or stratum that is water-bearing (i.e. saturated and permeable).
> *Aquiclude.* Soil or rock forming a stratum, group of strata or part or stratum of very low permeability, which acts as a barrier to groundwater flow.
> *Aquitard.* Soil or rock forming a stratum, group of strata or part or stratum of intermediate to low permeability, which yields only very small groundwater flows.

3.4.1 Unconfined Aquifers

An unconfined aquifer is probably the simplest for most people to visualize and understand. The soil or rock of the aquifer contains voids, pores or fractures. These voids are saturated (i.e. full of groundwater) up to a certain level, known colloquially as the 'water table' (Figure 3.12a), which is open to the atmosphere. A monitoring well (see Section 11.7.6) installed into the saturated part of the aquifer will show a water level equivalent to the water table. The analogy that can easily be drawn is that of digging a hole in the sand on a beach. Water does not enter the hole until the water table is reached, at which point water will enter the hole and stay at that level unless water is pumped out.

The water table can be defined as the level in the aquifer at which the pore water pressure is zero (i.e. equal to atmospheric); the line of zero pore water pressure is also known as the phreatic surface. Below the water table, the soil voids are at positive pore water pressures and are saturated. Above the water table, the pressure in the voids will be negative (i.e. less than atmospheric), and, at some height above the water table, that height depending on the nature of the soil or rock, they may be unsaturated and contain both water and air (see Section 3.3.5).

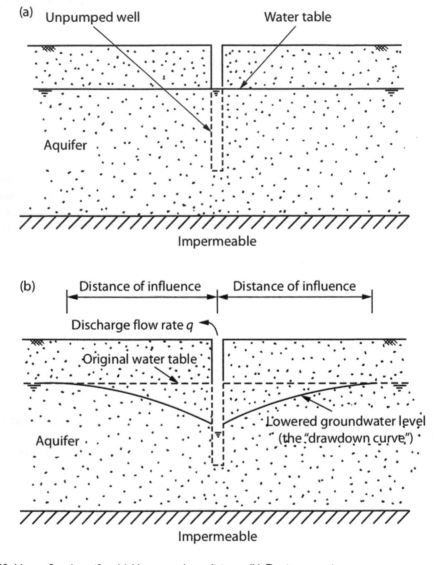

Figure 3.12 Unconfined aquifer. (a) Unpumped conditions. (b) During pumping.

If groundwater is pumped (or abstracted) from an unconfined aquifer, it is intuitively apparent that the water table will be lowered locally around the well, and a 'drawdown curve' will be created (Figure 3.12b) – which in three dimensions forms an inverted cone. In simple terms, the drawdown curve is the new, curved shape of the water table. Pore water will drain out of the soil above the new lowered water table and in free-draining soils or rock, will be replaced by air – this soil/rock will become unsaturated. The amount of water contained in a soil/rock will depend on the porosity of the soil, but it is important to note that not all of the water in an unconfined aquifer will drain out when the water table is lowered. Some water will be retained in the smaller soil pores by capillary forces. The proportion of water that can drain from an unconfined aquifer is described by the specific yield S_y, which is generally lower than the porosity.

When precipitation such as rain or snow falls on the ground surface above a confined aquifer, a proportion of this water will eventually infiltrate downwards through the unsaturated zone and reach the saturated zone of the aquifer. The percentage of surface recharge that reaches the aquifer will depend on how much of the water evaporates, is transpired by vegetation, is absorbed by any moisture deficit in very shallow soils, or runs off over the top of low-permeability surface layers. However, the important principle is that water levels in unconfined aquifers can vary in response to infiltration from the ground above them.

3.4.2 Confined Aquifers

The distinction between unconfined and confined aquifers is important, because they behave in quite different ways when pumped. In contrast to an unconfined aquifer, where the top of the aquifer is open to the atmosphere and an unsaturated zone may exist above the water table, a confined aquifer is overlain by a very low-permeability layer known as an 'aquitard' or 'aquiclude', which forms a confining bed. A confined aquifer is saturated throughout, because the water pressure everywhere in the aquifer is above atmospheric. A monitoring well drilled into the aquifer would initially be dry when drilled through the confining bed. When the borehole penetrates the aquifer, water will enter the borehole and rise to a level above the top of the aquifer. Because the pore water pressures are everywhere above atmospheric, a confined aquifer does not have a water table. Instead, its pressure distribution is described in terms of the 'piezometric level', which represents the height to which water levels will rise in monitoring wells installed into the aquifer (Figure 3.13a).

If a confined aquifer is pumped, the piezometric level will be lowered to form a drawdown curve, which represents the new, lower, pressure distribution in response to pumping (Figure 3.13b). Provided the piezometric level is not drawn down below the top of the aquifer (i.e. the base of the confining bed), the aquifer will remain saturated. Water will not drain out of the soil pores to be replaced by air in the manner of an unconfined aquifer. Instead, a confined aquifer yields water by compression of the aquifer structure (reducing pore space) and expansion of the pore water in response to the pressure reduction. The proportion of water that can be released from a confined aquifer is described by the storage coefficient S.

If the water pressure in the confined aquifer is sufficiently high, a well drilled through the confining bed into the aquifer will be able to overflow naturally at ground level and yield water without pumping (Figure 3.14). This is known as the flowing artesian condition – artesian is named after the Artois region of France, where such conditions were first recorded. Flowing artesian conditions are possible from wells drilled in low-lying areas into a confined aquifer that is recharged from surrounding high ground. Flowing artesian aquifers are a special case of confined aquifers – confined aquifers that do not exhibit flowing conditions are sometimes known as artesian aquifers or occasionally (and incorrectly) as sub-artesian aquifers.

3.4.3 Aquicludes

An aquiclude is a very low-permeability layer that will effectively act as a significant barrier to groundwater flow, for example as a confining bed above an aquifer. It need not be completely 'impermeable' to act as an aquiclude but should be of sufficiently low permeability that during the life of the pumping system, only negligible amounts of groundwater will flow through it. The most common forms of aquitard are layers of relatively unfractured clay or rock with permeabilities of 10^{-9} m/s or lower.

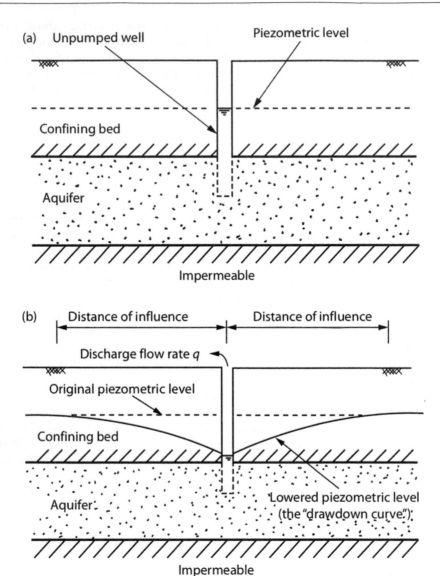

Figure 3.13 Confined aquifer. (a) Unpumped conditions. (b) During pumping.

In addition to having a very low permeability, a stratum should meet two other criteria before it can be considered to act as an aquiclude:

(i) It must be continuous across the area affected by pumping; otherwise, water may be able to bypass the aquiclude.
(ii) It must be of significant thickness. A thin layer of extremely low-permeability material may be less effective as an aquiclude compared with a much thicker layer of greater, but still low, permeability. The thicker a layer of clay or rock, the more likely it is to act as an aquiclude.

It is important to remember that no geological material is truly 'impermeable', so even an aquiclude can transmit groundwater, albeit very slowly. Therefore, there is no clear

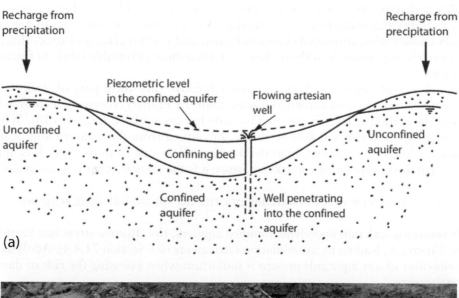

(a)

(b)

Figure 3.14 Flowing artesian conditions. (a) Schematic cross section through confined aquifer with flowing artesian conditions. (b) Flow from a flowing artesian well. (Courtesy of Dales Water Services Limited, Ripon, UK.)

definition between considering a stratum an aquiclude or an aquitard; the characterization will depend on the hydrogeological setting.

In contrast to unconfined aquifers, little of the precipitation falling on the ground surface above a confined aquifer will infiltrate into the aquifer. Confined aquifers receive their infiltration from sections of the aquifer that are unconfined and are termed 'recharge zones'. When dewatering pumping is carried out at a location in a confined aquifer, it is possible that the recharge zones may be located many kilometres from the location of the dewatering wells.

3.4.4 Aquitards and Leaky Aquifers

An aquitard is a stratum of intermediate (but still low) permeability with properties between that of an aquifer and an aquiclude. In other words, it is of sufficiently low permeability that it is unlikely that anyone would consider installing a well to yield water, but it is not of such low permeability that it can be considered effectively impermeable. Soil types that may form aquitards include silts, laminated clays/sands/silts, and certain clays and rocks that, while being relatively impermeable in themselves, contain a more permeable fabric of fractures or laminations.

Aquitards are of interest to hydrogeologists because they form part of 'leaky aquifer' systems. Such a system (also known as a semi-confined aquifer) consists of a confined aquifer where the confining layer is not an aquiclude but an aquitard (Figure 3.15). When the aquifer is pumped, water will flow vertically downward from the aquitard and 'leak' into the aquifer, ultimately contributing to the discharge flow rate from the well. It is apparent that the term 'leaky aquifer' is a misnomer, since it is the aquitard that is actually doing the leaking.

Aquitards are relevant to the dewatering practitioner for the following reasons:

a) If aquitards leak into underlying pumped aquifers, the effective stress (see Section 6.5) will increase, leading to consolidation settlements (see Section 21.4.4). Analysis of the behaviour of any aquitards present is important when assessing the risk of damaging settlements.

b) While aquitards may not yield enough water to form a supply, construction excavations into aquitards are likely to encounter small but problematic seepages and instability problems. Many applications of pore water pressure control systems using some form of vacuum wells (see Section 9.5.4) are carried out in soils that would be classified as aquitards.

3.4.5 Aquifer Parameters

The concept of permeability k has been introduced earlier (see Section 3.3) and is an important parameter used to describe aquifer properties. The thickness D of an aquifer is also important, since thicker aquifers of a given permeability will yield more water than thinner ones. These two terms can be combined into the hydrogeological term 'transmissivity', T.

$$T = kD \tag{3.8}$$

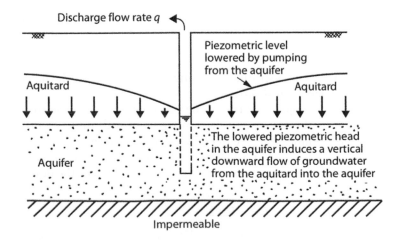

Figure 3.15 Leaky aquifer system.

For SI units. k and D will be in metres per second and metres, respectively, so T will have units of metres squared per second. Results of pumping tests (see Section 12.8.5) are sometimes reported in terms of transmissivity; Equation 3.8 allows these to be converted to an average permeability over the aquifer thickness. In unconfined aquifers where the aquifer thickness reduces as a result of pumping, transmissivity will reduce in a similar manner.

The amount of water released from an aquifer as a result of pumping is described by the storage coefficient. This is defined as the volume of water released from storage, per unit area of aquifer, per unit reduction in head; it is a dimensionless ratio. Because of the different way that water is yielded by confined and unconfined aquifers, storage coefficient is dealt with differently for each.

For an unconfined aquifer, the storage coefficient is termed the specific yield, S_y. This indicates how much water will drain out of the soil, to be replaced by air, under the action of gravity. Coarse-grained aquifers, such as sands and gravels, yield water easily from their pores when the water table is lowered. Finer-grained soils, such as silty sands, have smaller pores, where capillary forces may retain much of the pore water even when the water table is lowered (see Section 3.3.5); S_y may be much lower than for gravels. Typical values of S_y are given in Table 3.1. Surface tension forces may also mean that the water may not drain out of the pores instantaneously when the water table is lowered; it may drain out slowly with time. This phenomenon is known as 'delayed yield' and can affect drawdown responses in unconfined aquifers.

In a confined aquifer, since the aquifer remains saturated, there is no specific yield, and the storage coefficient S is used to describe the aquifer behaviour. Since water is only released by compression of the aquifer and expansion of the pore water, typical values of S will be small, perhaps of the order of 0.0005–0.001. More compressible confined aquifers will yield more water under a given drawdown, and tend to have a greater storage coefficient, than stiffer aquifers. If the piezometric level in the aquifer is lowered sufficiently that it falls below the top of the aquifer, unconfined conditions will develop, and a value of S_y will apply to the unconfined part of the aquifer.

Most of the design methods presented later in this book (see Chapter 13) are based on simple steady-state methods commonly used for temporary works construction applications. The storage coefficient does not appear in those calculations, because by the time steady state has occurred, all the water will have been released from storage. For relatively small construction excavations, this is a reasonable assumption, because steady state is generally achieved within a few days or weeks, so the volumes from storage release are only a

Table 3.1 Typical Values of Specific Yield

Aquifer	Specific yield (S_y)
Gravel	0.15–0.30
Sand and gravel	0.15–0.25
Sand	0.10–0.30
Chalk	0.01–0.04
Sandstone	0.05–0.15
Limestone	0.005–0.05

Based on data from Sterrett, R, *Groundwater and Wells*, 3rd edition, Johnson Division, St Paul, Minnesota, 2008 and Oakes, D B, Theory of groundwater flow. *Groundwater, Occurrence, Development and Protection* (Brandon, T W, ed.), Institution of Water Engineers and Scientists, Water Practice Manual No. 5, London, 1986.

concern during the initial drawdown period. Storage volumes may be more of a concern for large quarrying or opencast mining projects, when steady state may take a much longer time to develop (see Section 9.7).

3.5 AQUIFERS AND GEOLOGICAL STRUCTURE

The aquifer types described in the foregoing sections are a theoretical ideal. At many sites, more than one aquifer may be present (perhaps separated by aquicludes or aquitards), or the aquifers may be of finite extent and be influenced by their boundaries. The particular construction problems resulting from the presence of aquifers at a site will be strongly influenced by the stratification and geological structure of the soil or rock.

This section will illustrate the importance of an appreciation of geological structure to the execution of groundwater lowering works. Two case histories will be presented: the London basin shows how large-scale geological structure can allow multiple aquifer systems to exist; and a problem of base heave in a small trench excavation illustrates the importance of smaller-scale geological details.

3.5.1 Multiple Aquifers beneath London

The city of London is founded on river gravels and alluvial deposits associated with the River Thames, which are underlain by the very low-permeability London Clay Formation. These gravels, known as River Terrace Deposits, form a shallow (generally less than 10 m thick) water-bearing layer, termed the Upper Aquifer. Construction of utility pipelines, basements and other shallow structures often requires groundwater lowering to be employed; well-pointing and deep wells have proved to be effective expedients in these conditions. However, the geology beneath London allows two other, deeper, aquifers to exist below the city, largely isolated from the Upper Aquifer.

Beneath the London Clay Formation lie a series of sands and clays principally comprising the Lambeth Group stratum (formerly known as the Woolwich and Reading Beds) and the Thanet Sand Formation. These are underlain by the Chalk, a fractured white or grey limestone, which rests on the very low-permeability Gault Clay. The overall geological structure is a syncline forming what is often called the 'London Basin'.

It has long been known that the Chalk, Thanet Sand Formation and parts of the Lambeth Group together form an aquifer – the Lower Aquifer (Woods *et al.*, 2004). The upper 60–100 m of the Chalk is probably the dominant part of the aquifer, where significant fracture networks in the rock readily yield water to wells. The sands above the Chalk are of moderate permeability and generally do not yield as much water as the Chalk. The overlying London Clay Formation acts as an aquiclude or confining bed of very low vertical permeability, effectively separating the Lower Aquifer from the Upper Aquifer.

The Chalk has a wide exposure on the North Downs to the south of London and on the Chilterns to the north and occurs as a continuous layer beneath the Thames Valley (Figure 3.16). Rain falling on central London may ultimately reach the gravel of the Upper Aquifer, but the London Clay Formation prevents it from percolating down to the Chalk. The Chalk of the Lower Aquifer obtains its recharge from rain falling on the North Downs and the Chilterns many miles from the city. Ultimately, this water forms part of the reservoir of water in the Chalk aquifer. If the recharge exceeds the discharge from the aquifer (either from wells or natural discharge to springs and the River Thames), the water pressure in the aquifer will rise slowly. If discharges exceed recharge, the water pressure will fall.

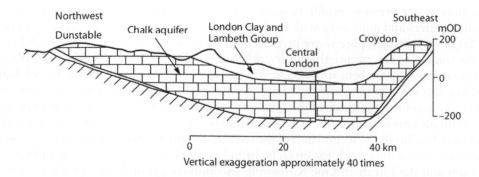

Figure 3.16 Chalk aquifer beneath London. (After Sumbler, M G, *British Regional Geology: London and the Thames Valley*, 4th edition, HMSO, London, 1996.) The Chalk aquifer extends beneath the London basin and receives recharge from the unconfined areas to the north and south. The London Clay deposits act as a confining layer beneath central London. Prior to the twentieth century, flowing artesian conditions existed in many parts of the city.

Before London developed as a city, the natural rates of recharge and discharge meant that the Lower Aquifer had sufficient water pressure for it to act as a confined aquifer (see Section 3.4.2). In the lower-lying areas of the city, there was originally sufficient pressure in the aquifer to allow a well drilled through the London Clay Formation into the Chalk to overflow naturally as a flowing artesian well. In fact, in central London there are still a few public houses called the *Artesian Well*, indicating that in earlier days the locals were probably supplied with water from a flowing well.

This availability of groundwater led to a large number of wells being drilled into the Lower Aquifer (where the water quality was more 'wholesome' than in the gravels of the Upper Aquifer). Rates of groundwater pumping increased during the eighteenth, nineteenth and early twentieth centuries. This resulted in a significant decline in the piezometric level of the Lower Aquifer. Artesian wells ceased to flow, pumps had to be installed to allow water to continue to be obtained, and over the years, the pumps had to be installed lower and lower to avoid running dry. By the 1960s, the water level in wells in some areas of London was 90 m below the ground surface – a huge drop relative to the original flowing artesian conditions. In some locations, the water pressure was reduced below the base of the London Clay Formation, so the formerly confined aquifer became unconfined. The deeper water levels increased pumping costs and made well supplies less cost-effective compared with mains water. This, together with a general re-location of large water-using industries away from central London, has resulted in a significant reduction in groundwater abstraction. As a result, the piezometric level in the Lower Aquifer has recovered since then (at more than 1 m/year at some locations in the 1980s) – see Figure 25.1. By the 1990s, the piezometric level in many areas was within 55 m of ground level.

This continuing rise of water pressures is a major concern, because much of the deep infrastructure beneath London (deep basements, railway and utility tunnels) was built during the first half of the twentieth century, when water pressures in the Lower Aquifer were at an all-time low. Several studies (see for example Simpson *et al.*, 1989) have addressed the risk of flooding or overstressing of existing deep structures if water levels continue to rise. The management of water levels beneath London (and, indeed, beneath other major cities around the world) is an important challenge to be faced by groundwater specialists during the first half of the twenty-first century. The use of permanent dewatering systems as part of the strategy to deal with rising groundwater levels is discussed in Chapter 20.

The pore water pressure profile beneath London is complex because of the presence of multiple aquifers and aquitards, with the low vertical permeability of the aquitards causing quite large variations in pore water pressure over short vertical distances (Figure 3.17).

In the middle of the twentieth century, when groundwater control works were planned, the groundwater system beneath London was considered to essentially be the Upper Aquifer (the River Terrace Deposits) and the Lower Aquifer (the Chalk and associated sands), separated by the aquitard/aquiclude of the London Clay Formation. Little water was expected in the zone (many tens of metres thick) between the base of the Upper Aquifer and the top of the Lower Aquifer. But in the second half of the twentieth century and at the start of the twenty-first century, tunnelling work for projects, including the London Underground Jubilee Line Extension and the Elizabeth Line (Crossrail), encountered groundwater problems in sandy and silty beds in the Lambeth Group between the Upper and Lower Aquifers. This is now considered to be a third aquifer beneath some parts of London – the Intermediate Aquifer. The presence of these permeable zones can cause problems during shaft sinking and the

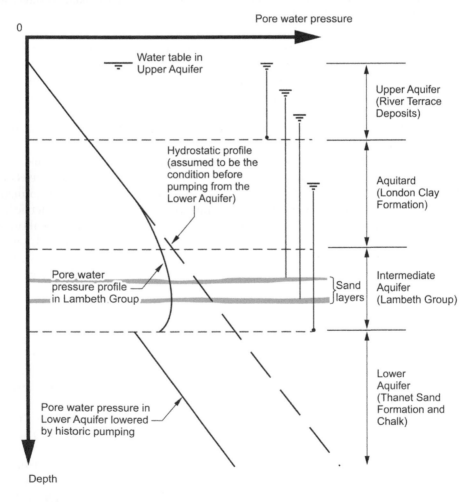

Figure 3.17 Schematic of pore water pressure distribution in multiple aquifer system below London. Historic pumping from the Lower Aquifer has lowered the piezometric level (and hence reduced pore water pressures) below original hydrostatic conditions. However, the low vertical permeability of clay beds within the Intermediate Aquifer means that pore water pressures at that level remain high relative to the Lower Aquifer.

construction of tunnel cross passages, requiring the use of wellpoint or ejector dewatering from within the tunnel (Preene and Roberts, 2002; Roberts, Smith, *et al.*, 2015). Further details on groundwater control for cross passages can be found in Section 10.6.

In a construction context, an appreciation of the aquifer system is vital to ensure that deep structures are provided with suitable temporary works dewatering. Figure 3.18 shows a typical arrangement that might be used for a deep shaft structure in central London (in an area where the Intermediate Aquifer is absent). Important points to note are:

(i) The structure penetrates two aquifers, separated by an aquiclude. Groundwater will need to be dealt with separately in each aquifer.

(ii) In London, it is common to deal with the Upper Aquifer by constructing a cut-off wall (see Chapter 18), penetrating to the London Clay Formation, to exclude the shallow groundwater. This is possible because the London Clay Formation is at relatively shallow depth. If clay were present only at greater depth, any cut-off would need to be deeper, and it might be more economic to dewater the Upper Aquifer.

(iii) Wells are used to pump from the Lower Aquifer to reduce the piezometric level to a suitable distance below the excavation. Because the wells must be relatively deep (perhaps up to 100 m), and therefore costly, it is important to design the wells and pumps to have the maximum yield possible, so that the number of wells can be minimized.

(iv) Although the Lower Aquifer consists of both the Chalk and the various sand layers between the top of the Chalk and the base of the London Clay Formation, wells are often designed to be screened in the Chalk only and are sealed from the sands using

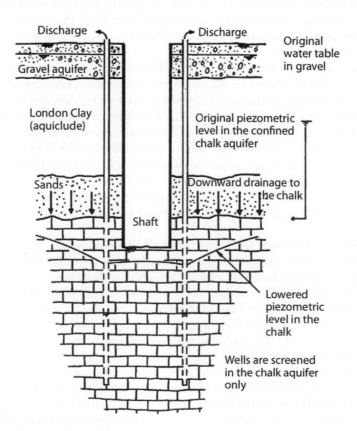

Figure 3.18 Groundwater control in multiple aquifers.

casing. This is because it can be difficult to construct effective well filters in the sands (which are fine-grained and variable), yet wells screened in the Chalk are simpler to construct and can be very efficient, especially if developed by acidization (see Section 16.7). This is the approach successfully adopted for some structures on the London Underground Jubilee Line Extension project described by Linney and Withers (1998). Excavations in the Thanet Sand Formation were dewatered without pumping directly from the sand but by pumping purely from the underlying Chalk. This might seem a rather contradictory approach but is an example of the 'underdrainage' method. This is a way of using the geological structure to advantage by pumping from a more permeable layer beneath the layer that needs to be dewatered; the upper poorly draining layer will drain down into the more permeable layer (see Section 13.6.2).

3.5.2 Water Pressures Trapped beneath a Trench Excavation

Dr W H Ward reported some construction difficulties encountered by a contractor excavating a pipeline trench near Southampton (Ward, 1957). The trench excavation was made through an unconfined aquifer of sandy gravel overlying the clays of the Bracklesham Beds. The contractor dealt with the water in the sandy gravel by using steel sheet-piling to form a cut-off on either side of the trench and exclude the groundwater. The clay in the base of the excavation was not yielding water, so dewatering measures were not adopted.

The trench was approximately 6.1 m deep to allow placement of a 760 mm diameter pipe, which was laid on a 150 mm thick concrete slab in the base of the excavation. The construction difficulties encountered consisted of uplift of the bottom of the trench (called 'base heave'; see Section 7.5.1), often occurring overnight while the trench was open. At one location, the trench formation rose by almost 150 mm before the concrete slab was cast and a further 50 mm after casting.

When Dr Ward and his colleagues at the Building Research Station were consulted, they suggested that the problem might be due to a high groundwater pressure in a water-bearing stratum below the base of the trench. This was proved to be the case when a small borehole was drilled in the base of the trench. This borehole overflowed into the trench, with the flowing water bringing fine sand with it. The water pressure in the borehole was later determined to be at least 1.3 m above trench formation level, but it is likely that the original piezometric level was even higher, since the flowing discharge from the borehole may have reduced pressures somewhat. Once the problem had been identified, the contractor was able to complete the works satisfactorily by installing a system of gravel-filled relief wells (see Section 17.8) in the base of the trench to bleed off the excess groundwater pressures (Figure 3.19).

This case history illustrates the importance of identifying the small-scale geological structure around an excavation. It appears that in this case the trench was excavated through an upper aquifer (the sandy gravel), which was dealt with using a sheet-pile cut-off, and the base of the trench was dug into low-permeability clay, which is effectively an aquiclude. The problems occurred because there was a separate confined aquifer beneath the aquiclude, which contained sufficient water pressure to lift the trench formation. Once identified and understood, the problem was solved easily using relief wells. However, because the problem was not identified in the site investigation before work started, time and money were wasted in changing the temporary works as well as making good the damaged pipelines. A classic failing of site investigations for groundwater lowering projects is that the boreholes are not taken deep enough to identify any confined aquifers that may exist beneath the proposed excavations. Guidelines on suitable depths for boreholes are given in Section 11.9.

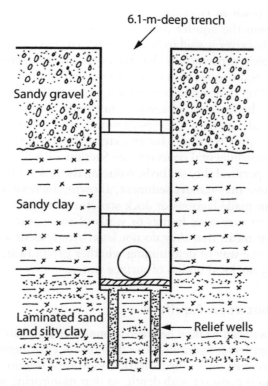

Figure 3.19 Use of simple relief wells to maintain base stability. (After Ward, W H, *Géotechnique*, 7, 134–139, 1957.)

3.6 AQUIFER BOUNDARIES

Civil and geotechnical engineers can readily appreciate the importance of permeability in the design of groundwater lowering systems – permeability is a traditional numerical engineering value. However, engineers are sometimes less proficient in recognizing the less quantifiable effects that aquifer boundary conditions have on groundwater flow to excavations. The following sections will outline some important aquifer boundary conditions.

3.6.1 Interaction between Aquifers and Surface Water

It is obvious from the hydrological cycle (see Section 3.2) that groundwater is inextricably linked with surface waters such as rivers, streams and lakes. The significance of the link between groundwater and surface water at a given site depends on the geological and hydrological setting.

Where surface water flows across or sits on top of an aquifer, water will flow from one to the other – the direction of flow will depend on the relative hydraulic head. The magnitude of the flow will be controlled by Darcy's law and will be affected by the permeability and thickness of any bed sediment, and the head difference between the aquifer and the surface water. Bodies of surface water that are quiet or slow flowing may have low-permeability bed sediments, which may dramatically reduce flow between surface and groundwater. Similarly, if the surface water is sitting on an aquitard or aquiclude, it may be effectively isolated from groundwater in underlying aquifers.

Where watercourses (such as rivers and streams) are connected to aquifers, they commonly receive water from the aquifer (the water entering the river from groundwater is termed 'baseflow'). Such a river is said to be a 'gaining' river (Figure 3.20a). This is perhaps contrary to many people's expectations that rivers should feed groundwater, rather than vice versa. 'Losing' rivers do sometimes exist, especially in unconfined aquifers with deep water tables (Figure 3.20b). If groundwater lowering is carried out near a 'gaining' river, the hydraulic gradients may be reversed, changing a normally 'gaining' river to a 'losing' one. Even if the hydraulic gradients do not reverse, groundwater lowering may reduce baseflow to a gaining river. If pumping continues for an extended period, this may reduce river flow and perhaps result in environmental concerns (see Section 21.4.8).

Lakes often have low-permeability silt beds, reducing the link with groundwater, although wave action near the shore may remove sediment, allowing increased flow into or out of the groundwater body. Man-made lagoons or dock structures may have silt beds (especially if they are relatively old) or may have linings or walls of some sort – however, just because linings exist, it does not mean that they do not leak! Analysis of a pumping test (Section 12.8.5) can be an effective way of determining whether groundwater lowering will be significantly influenced by any local bodies of surface water.

3.6.2 Interaction between Aquifers

In the same way that there can be flow between groundwater and surface water, groundwater can flow between aquifers if hydraulic head differences exist between them.

Many aquifer systems exist under 'hydrostatic conditions' (Figure 3.5a) – this is the case when the hydraulic head is constant with depth, so that monitoring wells installed at different depths show the same level (Figure 3.21a). In this case, there would be no vertical flow of

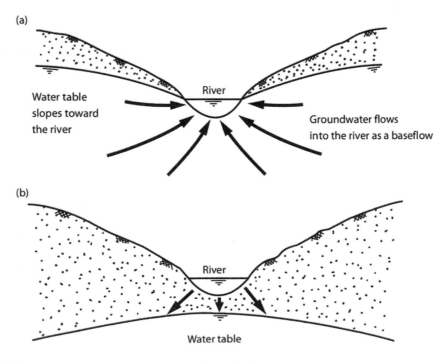

Figure 3.20 Interaction between rivers and aquifers. (a) Cross section through a gaining river. (b) Cross section through a losing river.

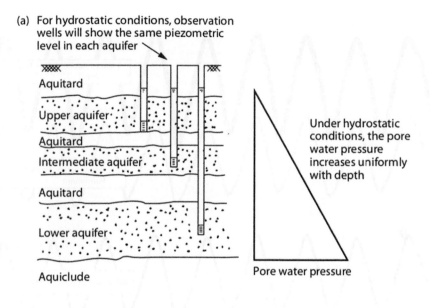

(a) For hydrostatic conditions, observation wells will show the same piezometric level in each aquifer

Aquitard

Upper aquifer

Aquitard

Intermediate aquifer

Aquitard

Lower aquifer

Aquiclude

Under hydrostatic conditions, the pore water pressure increases uniformly with depth

Pore water pressure

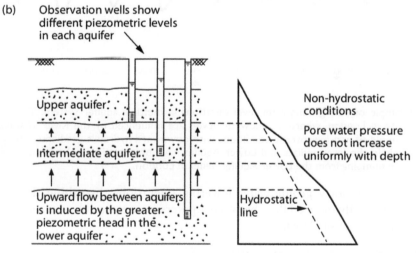

(b) Observation wells show different piezometric levels in each aquifer

Upper aquifer

Intermediate aquifer.

Upward flow between aquifers is induced by the greater. piezometric head in the lower aquifer

Hydrostatic line

Non-hydrostatic conditions

Pore water pressure does not increase uniformly with depth

Figure 3.21 Interaction between aquifers. (a) Hydrostatic conditions. (b) Flow between aquifers.

groundwater between shallow and deep aquifers separated by an aquitard. But sometimes, non-hydrostatic conditions exist. In Figure 3.21b, the hydraulic head in the deep aquifer is greater than the shallow aquifer, so water will flow upwards from the deep to the shallow aquifer. The flow may be very slow if the aquitard is of low permeability; if the stratum between the aquifers is an aquiclude, the flow will be so small that it is often ignored in the analysis of short-term groundwater lowering installations.

Figure 3.22 shows groundwater level data from a site with multiple aquifers. In this case, the groundwater levels are lower in the deeper aquifer than in the two shallower aquifers. This implies non-hydrostatic conditions with water flowing downwards from the shallow to the deeper aquifers.

Inter-aquifer flow may be a long-term phenomenon, perhaps sustained by different recharge sources for each aquifer. On the other hand, it may be a short-term condition

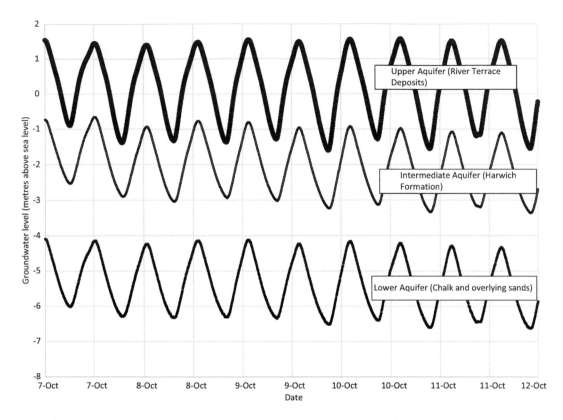

Figure 3.22 Example of tidal groundwater level variations in multiple aquifer system. The data are from a site in London, where groundwater levels are influenced by the River Thames. This is an example of non-hydrostatic conditions with downward hydraulic gradients. The groundwater levels in the Intermediate and Lower Aquifer have been artificially lowered by historic long-term pumping from the Lower Aquifer beneath London.

resulting from temporary groundwater lowering operations disturbing the groundwater regime (this artificially induced flow between aquifers is one possible environmental impact of dewatering; see Section 21.6).

3.6.3 Tidal Groundwater Conditions

During work on sites near estuaries or the sea, groundwater levels can exhibit tidal response, where water levels are influenced by tides – this is illustrated in Figure 3.22, which shows groundwater levels at a site with multiple aquifers.

Typically, tidally influenced groundwater levels exhibit a diurnal variation in groundwater levels, which are influenced by, but not the same as, the water level variation in the estuary or sea:

- The amplitude of the groundwater level variation will be lower than the water level variation in the estuary or sea.
- The timing of each peak/trough in groundwater levels will lag behind the corresponding tidal peak/trough in the surface waters.
- The level of the groundwater peaks/troughs will vary on an approximately monthly cycle, influenced by the spring and neap tides in the estuary/sea, which are controlled by the lunar cycle.

It is sometimes mistakenly assumed that significant tidal groundwater level variations imply a significant hydraulic connection between the estuary/sea and the aquifer. While tidal groundwater conditions can be observed when there is a hydraulic connection, they can also occur in confined aquifers with no significant hydraulic connection to the estuary/sea. In the latter case, it is the loading effect of the weight of the tidal water on top of the confining aquitard that causes the water level variations in the deeper confined aquifer. Tidal groundwater phenomena are discussed by White and Roberts (1993).

3.6.4 Recharge Boundaries

Zones or features where water can flow into an aquifer are termed 'recharge boundaries', some commonly occurring examples of which are shown in Figure 3.23. If recharge boundaries exist within the distance of influence, they can have a significant effect on the behaviour of dewatering schemes. They may cause the cone of depression to become asymmetric (since the extent of the cone will be curtailed where it meets the recharge source). The flow rate that must be pumped by a dewatering system will often be increased by the presence of a recharge boundary. It is essential that any potential recharge boundaries be considered during the investigation and design of dewatering works.

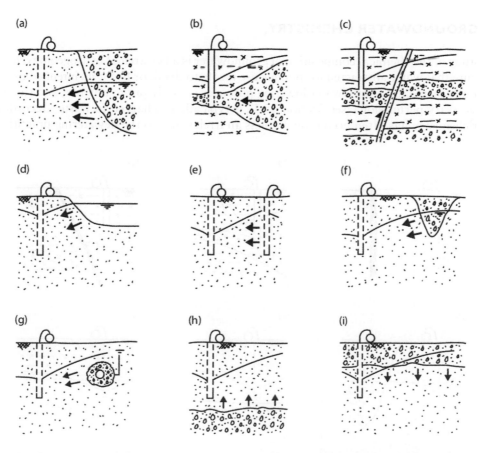

Figure 3.23 Potential aquifer recharge boundaries. (a) High-permeability zone. (b) Increase in aquifer thickness. (c) Fault. (d) Surface water in hydraulic connection with the aquifer. (e) Recharge wells. (f) Gravel lens or channel. (g) Flooded land drain or sewer bedding. (h) Underlying high-permeability stratum. (i) Overlying high- permeability stratum

3.6.5 Barrier Boundaries

Real aquifers cannot be of infinite extent and are often bounded by features that form barriers to groundwater flow. Figure 3.24 shows some commonly occurring barrier boundaries. The presence of barrier boundaries will tend to reduce the pumped flow rate necessary to achieve the required drawdown.

3.6.6 Discharge Boundaries

Water can sometimes be discharged naturally from aquifers. Water will flow from an unconfined aquifer if the water table intersects the ground surface. Diffuse discharges are called seepages, or if the flow is very localized (perhaps at a fault or fracture), the discharge is termed a spring. Flowing artesian aquifers can also discharge if faults or fractures allow water a path to the surface. Water flowing between aquifers may also constitute a discharge boundary condition.

Man's influence, in the form of pumping from wells (either for supply or for groundwater lowering for construction, quarrying, mining, etc.) can also create discharge boundaries. Figure 3.25 shows some examples of discharge boundaries.

3.7 GROUNDWATER CHEMISTRY

The study of the chemical composition of groundwater is a key area of hydrogeology, helping to ensure that water abstracted for potable use is safe to drink or fit for its intended use. The water abstracted for groundwater lowering purposes is rarely put to any use, and subject to the necessary permissions (see Section 25.4), is typically discharged to waste. Accordingly, water chemistry has often been thought of as an irrelevance to the dewatering practitioner.

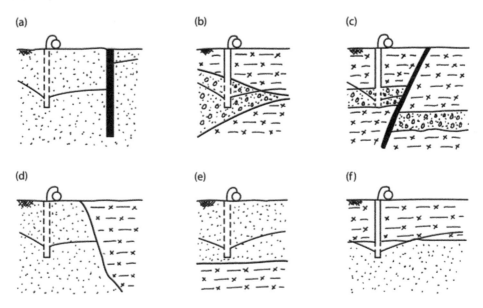

Figure 3.24 Potential aquifer barrier boundaries. (a) Partial cutoff wall. (b) Reduction in aquifer thickness. (c) Fault. (d) Low-permeability zone. (e) Underlying low- permeability stratum. (f) Overlying low-permeability stratum.

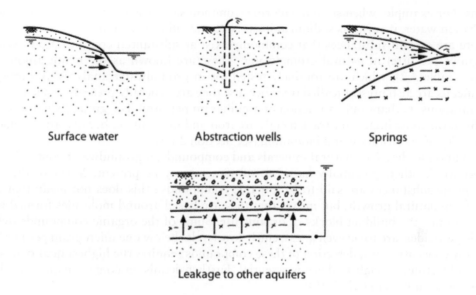

Surface water   Abstraction wells   Springs

Leakage to other aquifers

Figure 3.25 Potential aquifer discharge boundaries.

However, a basic knowledge of groundwater chemistry can be useful for the following reasons:

(i) Discharge consents and permissions. In many locations, permission to dispose of the discharge water to a sewer or surface watercourse will only be granted if it can be demonstrated that the water is of adequate quality.

(ii) Corrosion and encrustation. Water chemistry will influence well clogging, encrustation, corrosion and biofouling (see Section 23.5), which may affect systems in long-term operation. Clogging and encrustation can be a particular problem for artificial recharge systems (see Section 17.13).

(iii) The effect of groundwater lowering on water quality. Abstracting water from an aquifer may affect existing groundwater quality. Monitoring may be necessary to determine whether these effects are occurring and to control any mitigation measures.

If a project is likely to require detailed study of water chemistry, specialist advice should be obtained. This section is intended only to highlight some of the important practical issues of water chemistry relevant to groundwater lowering. Readers interested in the field are recommended to Lloyd and Heathcote (1985) as an introductory text.

3.7.1 Chemical Composition of Groundwater

Water is a powerful solvent, and a wide range of substances will dissolve in it to some degree. Almost all water in the hydrological cycle will contain some dissolved minerals or other substances. Even rainwater contains dissolved carbon dioxide and some sodium chloride lifted into the atmosphere from the oceans.

Substances dissolved in water exist as electrically charged atoms or molecules known as ions. Positively charged ions are known as anions, and negatively charged ions are known as cations. For example, when sodium chloride (common salt, NaCl, which is highly soluble) dissolves in water, it exists as sodium anions (Na^+) and chloride cations (Cl^-).

There are certain substances that commonly exist at substantial concentrations (several milligrams per litre) in natural groundwater. These are known as the major anions and cations (Table 3.2), which are routinely tested for in groundwater analyses. A variety of trace metals are also present (called trace because they are generally present in much smaller concentrations, perhaps only a fraction of a milligram per litre). For groundwater lowering purposes, the most important trace metals are iron and manganese, because they influence the severity of encrustation and biofouling (see Section 23.5).

In addition to dissolved natural minerals and compounds, if groundwater contamination has occurred, other, potentially harmful, substances may be present. Many of the most problematic substances are said to be 'organic compounds'; this does not mean that they result from natural growth, but means that they are based around molecules formed from carbon atoms, the building blocks of organic life. Some of the organic compounds (which include pesticides) are toxic even at extremely low levels (below one microgram per litre). At these low concentrations, detection of these compounds requires the highest quality of sampling and testing; specialist advice is essential. Further details on contamination problems can be found in Fetter *et al.* (2018).

It is apparent that almost no groundwater is 'pure', if by pure we mean containing nothing but H_2O! Nevertheless, water that contains relatively little dissolved material is said to be 'fresh'. Most groundwater abstracted for drinking use is classified as fresh and requires little treatment other than basic sterilization to kill any harmful bacteria present. Even water suitable for industrial use (such as steam raising in a boiler) tends to be relatively fresh, because the lower mineral content reduces the build-up of scale deposits within the pipework. In many developed countries, permissible limits are set for the chemical composition of water that can be used for human consumption; guidelines are also available for water to be used for cultivation of crops or livestock (see Lloyd and Heathcote, 1985, chapter 10; Brassington, 1995, chapter 6).

3.7.2 Field Monitoring of Groundwater Chemistry

In principle, the most obvious way to investigate groundwater chemistry is to obtain a sample of groundwater, seal it into a bottle and send it to a suitably equipped laboratory for testing. It might be appropriate to test for major anions and cations, selected trace metals and any other substances of interest at the site in question.

For groundwater lowering systems, obtaining a sample is relatively straightforward. It may be possible to fill a sample bottle directly from the pumped flow at the discharge tank

Table 3.2 Major Anions and Cations in Groundwater

Major anions	Major cations
Sodium (Na^+)	Bicarbonate (HCO_3^-)
Potassium (K^+)	Carbonate (CO_3^-)
Calcium (Ca^{2+})	Sulphate (SO_4^{2-})
Magnesium (Mg^{2+})	Chloride (Cl^-)
	Nitrate (NO_3^-)
	Nitrite (NO_2^-)
	Ammonium (NH_3)
	Phosphate (PO_4^{2-})

or via a sample tap at the wellhead. However, when groundwater samples are taken from the discharge flow, the following factors should be considered:

(i) Try to minimize the exposure of the sample to the atmosphere. Ideally, obtain it directly from the discharge of the pump or dewatering system. Totally fill the bottle, and try to avoid leaving any air inside when it is sealed. If the pump discharge is 'cascading' before the sampling point, the water will become aerated, and oxidation may occur. The discharge arrangements should be altered so that the sample can be obtained before aeration occurs.

(ii) Be aware that samples may degrade between sampling and testing. The samples should be tested as soon as possible after they are taken and, ideally, should be refrigerated in the meantime. The bottles used for sampling should be clean with a good seal. However, the sample may degrade while in the bottle (for example by trace metals oxidizing and precipitating out of solution). Specialists may be able to advise on the addition of suitable preservatives to prevent this from occurring. The choice of sample bottle (glass or plastic) should also be discussed with the laboratory, since some test results can be influenced by the material of the sample bottle.

(iii) Use an accredited, experienced laboratory.

Sometimes, groundwater samples may be required from a site when there is no dewatering pumping taking place – perhaps for pre- or post-construction background monitoring. In that case, a sampling pump will have to be used to obtain a sample from a monitoring well. The water standing in the well has been exposed to the atmosphere and is unlikely to represent the true aquifer water chemistry. Therefore, it is vital to 'purge' the well before taking a sample. A common approach to purging involves pumping the monitoring well at a steady rate until at least three 'well volumes' of water have been removed (a 'well volume' is the volume of water originally contained inside the well liner). Specialist sampling pumps should be used in preference to airlifting, since the latter method may aerate the sample, increasing the risk of oxidation of trace metals and other substances. Further details on groundwater sampling are given in Section 11.10.3.

If samples are taken to an off-site laboratory, it may be several days before the results are ready. Even if the tests are rushed through as priority work, some of the actual procedures may take a week or more. Comprehensive testing of samples is not cheap, either; a reasonably complete suite of testing may cost several hundred pounds per sample at 2019 prices. A good way of reducing costs, and obtaining rapid results, is to take and test water samples on a periodic basis (perhaps monthly) but to carry out daily (or, using a datalogger, continuous) monitoring of the 'wellhead chemistry'.

Wellhead chemistry is a hydrogeological term used to describe certain parameters, which are best measured as soon as the water is pumped from the well – that is, at the wellhead, or within the pipework system before the water is discharged. It is best to measure these parameters here, as they are likely to change during sampling and storage, which makes laboratory-determined values less representative. Typical wellhead chemistry parameters include:

a) Specific conductivity, EC
b) Water temperature
c) pH
d) Redox potential, E_H
e) Dissolved oxygen, DO

These parameters are sometimes included in routine monitoring of groundwater control systems (see Section 22.8).

Perhaps the most commonly measured wellhead parameter for groundwater lowering systems is specific conductivity, EC. Specific conductivity is a measure of the ability of the water to conduct electricity and is a function of the concentration and charge of the dissolved ions; it is reported in units of microsiemens per centimetre corrected to a reference temperature of 25 °C. EC is useful in that it can be related to the amount of total dissolved solids (TDS) of the water (Lloyd and Heathcote, 1985).

$$TDS = k_e EC \qquad\qquad (3.9)$$

where

TDS = total dissolved solids (mg/l)

EC = specific conductivity (μS/cm at 25 °C)

k_e = a calibration factor with values between 0.55 and 0.80 depending on the ionic composition of the water.

TDS and EC have been related to each other for various water classifications in Table 3.3. Fresh water will have a low TDS (most water supply boreholes for potable use produce water with a TDS of no more than a few hundred milligrams per litre). The higher the TDS, the less fresh the water. The term 'saline' is used for convention's sake and does not necessarily imply that a high TDS is the result of saline intrusion (see Section 21.4.11). A high TDS may be an indicator of highly mineralized waters that have been resident in the ground for very long periods, slowly leaching minerals from the soils and rocks.

In the Middle East, due to the low recharge and high rates of evaporation, groundwater can be highly mineralized with very high salinity. Roberts *et al.* (2009) describe a dewatering system in Dubai, United Arab Emirates, where the sodium chloride concentration of groundwater was approximately four times that of typical seawater. This is not unusual for dewatering systems in very arid climates and presents challenges for corrosion of pumps and equipment (see Section 23.5.4).

Daily monitoring of EC using a conductivity probe has proved useful on sites where there was concern that water of poorer quality (e.g. from saline intrusion or from deeper parts of the aquifer) might be drawn towards the pumping wells. Readings of conductivity were taken every day and reviewed for any sudden or gradual changes in EC, which would have indicated that poor-quality water was reaching the wells. In effect, the EC monitoring was used as an early warning or trigger to determine when more detailed water testing was required.

3.8 GROUNDWATER TEMPERATURES

The temperature of groundwater is largely influenced by groundwater velocities. Although the hydrological cycle (Section 3.2) implies that groundwater is constantly in motion, in

Table 3.3 Classification of Groundwater Based on Total Dissolved Solids (**TDS**)

Classification of groundwater	Total dissolved solids, TDS (mg/l)	Specific conductivity, EC (μS/cm)
Fresh	<1,000	<1,300–1,700[a]
Brackish	1,000 to 10,000	1,300–1,700 to 13,000–17,000[a]
Saline	10,000 to 100,000	13,000–17,000 to 130,000–170,000[a]

[a] Approximate correlation; precise value depends on ionic composition of water.

most natural hydrogeological situations, groundwater velocities are very low (Section 3.3.4). This means that groundwater typically has sufficient residence time for temperatures to come into equilibrium with the host soil or rock.

Price (1996) states that in temperate climates, the temperature of groundwater below a few metres' depth tends not to vary with the seasons and remains remarkably constant at around the mean annual air temperature. In the United Kingdom, this is around 10–14 °C. This is obvious on site if you have to take a water sample from the pumped discharge of a dewatering system. On a frosty winter morning, the pumped groundwater will seem pleasantly warm on your hands, but take a similar sample on a warm summer day, and the discharge water will seem icy cold. At greater depths in bedrock, the temperature stays constant with time but increases with depth. This is known as a geothermal temperature gradient, mainly the result of heat generated by the decay of radioactive materials in the rock. As an example, in the Chalk (Lower Aquifer) beneath London, the geothermal gradient apparent from groundwater temperature measurements is around 1.8 °C per 100 m depth (Pike et al., 2013).

The relatively constant year-round temperature of the ground and groundwater at a site can be harnessed via ground energy systems in order to provide heating and cooling for buildings or industrial processes. Using either open loop or closed loop ground collectors (Figure 3.26), heat energy can be exchanged between a building and the ground. This heat energy is then used in place of traditional heating and cooling systems (such as gas-fired boilers or electrically powered cooling systems), thereby reducing carbon dioxide emissions from buildings. Further details of ground energy systems (also known as geothermal systems) are given in Banks (2012) and Preene and Powrie (2009).

Occasionally, much hotter groundwater is encountered in dewatering systems for deep mining projects; this is especially likely to occur in tectonically active regions. Williamson and Vogwill (2001) report that dewatering for the Lihir gold mine in New Guinea had to manage water at temperatures of 90 to 100 °C, and Popielak et al. (2013) describe groundwater at 70 to 160 °C in the Cerro Blanco underground mine in Guatemala (groundwater can exist at temperatures of greater than 100°C under pressure at depth in an aquifer but may flash to steam if exposed to atmospheric pressure at the surface).

3.9 EFFECT OF CLIMATE AND WEATHER

The hydrological cycle (Section 3.2) shows that groundwater is part of the same process as rain, snow and other forms of precipitation. In some hydrogeological settings, short-term and long-term variations in precipitation can result in significant variations of groundwater level, which may affect the design and implementation of groundwater lowering systems.

The effect of rainfall and other precipitation on groundwater levels is a matter of the weather and climate. Weather is the short-term variation of atmospheric conditions, such as temperature, humidity, precipitation, cloudiness, visibility and wind.

Climate is the pattern of weather events at a location averaged over a period of time, perhaps over several decades, and can encompass extreme events (e.g. droughts, severe storms) within a more general trend of weather patterns. The difference can be illustrated by considering the planning and interpretation of groundwater level monitoring. If you are thinking about whether last week's heavy rain will have caused groundwater levels to rise above typical levels, then you are dealing with the weather. If you are trying to assess whether the groundwater levels you measure this week are representative of long-term levels, and whether they can be used to design a long-term groundwater control system, you are probably thinking about climate.

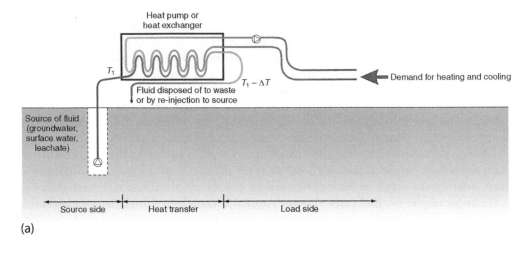

(a)

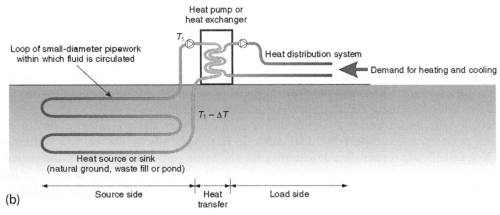

(b)

Figure 3.26 Ground energy system. (a) Open loop system. (b) Closed loop system. (From Preene, M and Powrie, W, *Proceedings of the Institution of Civil Engineers, Energy*, 162, 77–84, 2009. With permission.) (a) Groundwater is abstracted from the source (typically one or more boreholes), passed through a heat pump or heat exchanger, and disposed of either to waste (sewer or watercourse) or by re-injection to the source (typically by one or more aquifer re-injection boreholes). (b) A thermal transfer fluid is circulated through a closed circuit of pipework embedded in the ground, thereby allowing the building heat pump system to reject or extract heat from the ground. The ground loop can be configured into shallow trenches or an array of vertical boreholes, or can be incorporated into the building piles and other foundations.

In relation to weather, it may be prudent to make major or minor contingency plans for the effect of precipitation. For instance, in some parts of the world with a tropical climate it is known that there is a risk that monsoons, typhoons and hurricanes will occur at certain seasons of the year; the consequential high-volume surface run-off may be very significant. Less obviously, sites in cold climates that experience high snowfalls can experience very high rates of surface water run-off during spring snow melt. Hence, planning for such areas would dictate that appropriate measures to deal with sudden influxes of surface water, prevent slope erosion and deal with short-term rises in groundwater levels must be incorporated into the excavation plan at the design stage. Particularly when one is working in shallow unconfined aquifers, intense rainfall events can have a major effect on lowered water levels. Figure 3.27 shows groundwater level and rainfall data taken during the rainy season at

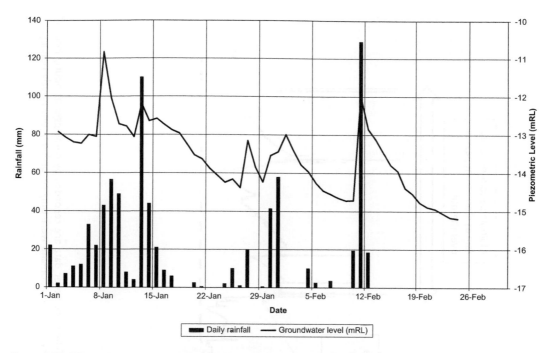

Figure 3.27 Effect of extreme rainfall events on groundwater levels. Data are from monitoring of a wellpoint system operating on a site in South-East Asia during the rainy season.

a site in South-East Asia. A wellpoint system was used to lower water levels in a very silty sand and had achieved a drawdown of several metres. Whenever there was heavy rain, the water levels rose rapidly and only fell back to their original, lowered levels slowly under the influence of pumping.

Exposed excavated platforms and slopes can be protected from monsoon-type erosion by laying an appropriate geotextile membrane (a type of low-permeability surface cover, as described in Section 18.13). A judgement should be made at the design stage on whether to cater for the once in 10-year occurrence, the 50-year occurrence or more onerous conditions.

Even in moderate climates, heavy rain can cause short-term rises in groundwater levels. Figure 3.28 shows groundwater level data (recorded hourly by datalogger) from standpipe piezometers on a groundwater lowering project in the United Kingdom. The project involves groundwater lowering by deep wells in a relatively low permeability rock (permeability 10^{-7} to 10^{-6} m/s), and the drawdown effect of pumping from wells can clearly be seen. However, short-term 'spikes' (sudden rises) in groundwater level can be seen, superimposed on the drawdown trend. These spikes follow shortly after heavy rainfall events. Individual rainfall events cause a short rise that quickly dissipates. However, multiple rainfall events (such as when a major stormfront passes overhead) cause longer-persisting rises in groundwater level. In this case, the probable cause is surface water falling on the exposed rock on horizontal platforms (both in the base of the excavation and on the surrounding ground), which infiltrates quickly into exposed fractures in the rock. The rock has a very low specific yield (see Section 3.4.1), and so even modest rates of water infiltration at the surface can result in significant rises in groundwater level. This problem can be mitigated by covering any exposed rock with low-permeability material (such as blinding concrete) to encourage run-off instead of infiltration (Section 18.13) and by having effective systems of surface water control in and around the excavation (Section 9.2).

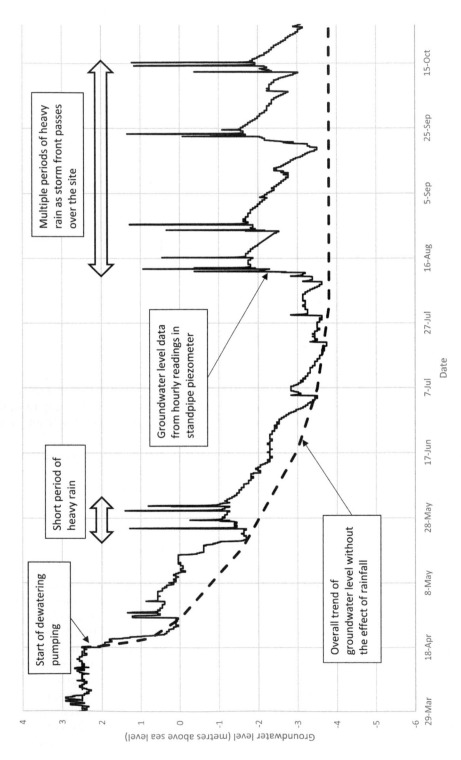

Figure 3.28 Example of groundwater levels during dewatering influenced by heavy rainfall. The piezometer data are recorded by datalogger at hourly intervals. Some of the highest 'peaks' of groundwater level after heavy rainfall were of very short duration and were only recorded in one hourly reading.

In terms of climate, it is well established that groundwater levels can be subject to long-term variations in response to climate effects (Jackson *et al.*, 2015). The most obvious example is low groundwater levels (below long-term trends) during a period of drought (a prolonged period of abnormally low rainfall). If a groundwater lowering system is being planned for long-term operation using recent groundwater level data, or if it is being designed based on groundwater level data from the past (previous years, or even from a different season), the designer should check whether they can be confident that the groundwater level data are representative of the range of possible conditions during construction.

The authors are aware of several cases where groundwater levels gathered a year or more in advance of construction indicated that groundwater level was below the deepest proposed excavation, and hence groundwater lowering was not required. But subsequently, when construction started, groundwater was encountered above dig level, and delays and extra costs resulted from deploying groundwater control measures that had not been originally planned. In most of these cases, post-construction review of the groundwater level data suggested that the groundwater levels used in design were associated with either a dry summer or a longer, dryer period lasting more than 1 year. Therefore, the data used were unrepresentatively low, and long-term groundwater levels were above excavation dig levels, as was discovered during construction.

An interesting example is given by Lazarus (2013), which describes a deep basement structure in England founded, in part, in the Chalk (a fractured limestone), constructed in the 1990s. Based on groundwater levels from the 1980s/1990s, the proposed basement was assessed to be several metres above groundwater level, and the concrete structure was provided with only nominal structural waterproofing. However, the autumn of 2000 – September to November – was the wettest on record in England since records began in 1766, and the basement suffered from water ingress problems. Investigations indicated that the water ingress was caused by groundwater levels in the Chalk, which had risen approximately 11 m during the immediately preceding months as a consequence of the heavy and prolonged rainfall. Review of historic maps and other sources revealed that the area had a history of watercourses, including springs and ponds lost during urban expansion. During and after the period of heavy rainfall, some of the historic watercourses were reported as re-establishing, if only for a limited time.

The lesson is that where below-ground structures are planned, the site investigation desk study (Section 11.6) should try to collate groundwater level data over as long a period as possible to aid understanding of the likely range of water level variations.

Chapter 4

Permeability of Soils and Rocks

To see a world in a grain of sand

William Blake, *Auguries of Innocence*

4.1 INTRODUCTION

Permeability (also known as hydraulic conductivity) is a parameter used widely in geotechnical analysis. It is of particular relevance to groundwater control projects, where it has a dominant effect on groundwater pumping rates and seepage quantities into excavations below groundwater level. Unfortunately, permeability is complex, both in concept and when applied in geotechnical design. Natural materials can have a very wide range of permeability values, and the factors controlling permeability are different in, say, a sedimentary sequence of glacial deposits from those in a weathered and fractured rock. Permeability is also difficult to assess by in situ or laboratory methods, with many test procedures having significant limitations. This chapter will discuss the fundamentals of permeability of soils and rocks, including different scales of permeability (further information on groundwater problems in soils and rocks is given in Chapters 7 and 8, respectively). Available estimation methods to determine permeability will be introduced. Permeability testing methods are covered in more detail in Chapter 12.

4.2 WHAT IS PERMEABILITY?

Permeability was introduced in Chapter 3 as part of the formulation of Darcy's law. In essence, permeability is a measure of the ease or otherwise with which a fluid passes through a porous medium. A complication is that the ease of flow is dependent not only on the nature of the porous media but also on the properties of the permeating fluid. In other words, the permeability of a soil or rock to water is different from the permeability to another fluid, such as air or oil (the permeability parameter, independent of the fluid, is known as the intrinsic permeability of a material). Hydrogeology references highlight this by using the term 'hydraulic conductivity' to show that the permeability parameter used is specific to water. As a reminder of terminology, this book follows geotechnical engineering practice and uses the term 'permeability' to mean hydraulic conductivity, coefficient of permeability or Darcy's permeability, which applies specifically in relation to the flow of water through porous media.

4.2.1 Intrinsic Permeability

Darcy's law (Section 3.3) relates the flow rate Q through a cross-sectional area A due to a hydraulic gradient i. The formulation familiar to most geotechnical engineers uses the hydraulic conductivity version of permeability and is specific to the flow of water.

$$Q = -kiA \qquad (4.1)$$

Darcy's law can also be formulated in terms of the intrinsic permeability k_i and the properties of the permeating fluid (density ρ and dynamic viscosity μ):

$$Q = -\frac{k_i \rho g}{\mu} iA \qquad (4.2)$$

and intrinsic permeability k_i can be related to hydraulic conductivity permeability k:

$$k = \frac{k_i \rho g}{\mu} \qquad (4.3)$$

Intrinsic permeability is independent of the permeating fluid and has the SI units of metres squared, although it is commonly described in the non-SI darcy units (where 1 darcy = 1 × 10^{-12} m^2).

Even when dealing solely with water as the permeating fluid, Equation 4.3 shows that the properties of water can have an influence. The viscosity of water varies with temperature, so hydraulic conductivity will also vary with temperature. However, in most cases, the range of groundwater temperature variations in geotechnical problems is small enough that any resulting changes in hydraulic conductivity can reasonably be neglected.

Occasionally, on deep mining projects, testing methods from the oil and gas industry are used for hydrogeological testing, and the results of permeability tests are reported in darcy units or metres squared. As an order of magnitude conversion, 1 darcy = 9.6 × 10^{-6} m/s for fresh water at 20°C. At other temperatures or for water with different fluid properties (e.g. highly saline water), hydraulic conductivity and intrinsic permeability can be related by Equation 4.3 using the properties of the fluid at the relevant temperature.

4.2.2 Hydraulic Conductivity

Civil and geotechnical engineers are also interested almost exclusively in the flow of water through soils and rocks and use the term 'permeability' to mean hydraulic conductivity (or coefficient of permeability or Darcy's permeability), and this terminology will be used in this book.

Like many geotechnical parameters, permeability is simple in concept but has some very complex aspects in practice, especially when trying to obtain realistic measurements or estimates of properties. Mathematically, permeability is a coefficient in Darcy's law (Section 3.3), which relates water flow to hydraulic gradient under laminar flow conditions. This is easy to understand for flow through an isotropic block of porous media, as you might see in a textbook, where hydraulic conductivity is the same at all points (uniform and homogeneous) and in all directions (isotropic).

Of course, the flow of water through soils or rocks is anything but homogeneous and is rarely isotropic. Water flows through geological materials via interconnected voids.

- In soils, the structure is made up of mineral particles in contact to form the soil skeleton, with a network of interconnected pores in the space between, and water will flow through the pores (Figure 4.1a). The flow of water through the pores of a soil mass is sometimes called 'primary permeability' or intergranular flow.
- In contrast, in most fractured rocks, the principal way that groundwater flows is not through the spaces between the mineral particles forming the rock (the rock mass itself tends to have a very low hydraulic conductivity). Instead, the water must pass along fissures, fractures or discontinuities within the rock mass (Figure 4.1b). This type of flow is sometimes called 'secondary permeability' or fracture flow.

On a micro-scale, the voids (soil pores and rock fractures) through which water flows are a vastly complex hydrodynamic environment. If it were possible to visualize what they really looked like, the scene would probably seem like an alien world out of a science fiction movie. The pragmatic solution for practising engineers and hydrogeologists is to

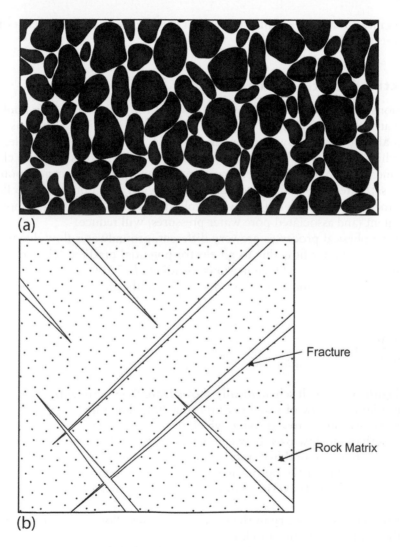

Figure 4.1 Groundwater flow in soils and rocks: (a) soil, (b) rocks.

'zoom out' and not to try and discern micro-scale properties but to look for 'average' or 'representative' parameters or depictions of soil or rock properties. This is essentially the approach embodied in Darcy's law and results in the permeability values routinely used in dewatering calculations.

Water must take an often tortuous path along irregular pores, spaces or fractures. The use of Darcy's law and the concept of permeability is justified by a macroscopic approach and treating blocks of soil as being relatively homogeneous porous media, but it is important to realize that groundwater flow in soils can be very complex at small scale.

- In soil, groundwater flow can be affected by soil structure or fabric, such as layering, laminations or weathering.
- In rock, groundwater flow can be affected by fracture direction, frequency and opening width, which may be affected or controlled by rock structure and stresses. Where the rocks are water soluble (such as Chalk and limestones), fractures may have become enlarged by millennia of natural groundwater flow.

These effects mean that permeability values can vary depending on the scale of the test or assessment (Section 4.6).

4.2.3 Electro-Osmotic Permeability

The discussions in the previous sections (and indeed the majority of this book) are based on groundwater flow occurring in response to hydraulic gradients – this is the basis of operation of almost all groundwater lowering systems. The exception is the electro-osmosis technique, which can be used in very low-permeability soils such as silts and clays (Section 17.12). The method is based on the observation that applying an electrical potential (i.e. a voltage) to a saturated fine-grained porous medium will cause pore water to flow towards the cathode (negative electrode). If the water collecting at the cathode is pumped away, the moisture content (and associated pore water pressures) will reduce.

Although the physical processes are very different, groundwater flow due to a hydraulic gradient and groundwater flow due to an electro-potential gradient (electro-osmotic flow) have very similar mathematical formulations (Asadi *et al.*, 2013).

We have seen that groundwater flow due to hydraulic gradients can be described by Darcy's law:

$$Q = -k \left(\frac{\Delta h}{l} \right) A \tag{4.4}$$

where (see Figure 3.2, which schematically shows Darcy's experiment)
Q = the volumetric flow of water per unit time (the 'flow rate')
A = the cross-sectional area through which the water flows
l = the length of the flow path between the upstream and downstream ends
Δh = the difference in total hydraulic head between the upstream and downstream ends
k = the permeability of the porous medium through which the water flows (in this section, we will use the term 'Darcy's permeability' to avoid confusion)

The negative term is necessary in the equations because flow occurs down the hydraulic gradient – that is, from high head to low head.

Casagrande (1952) presented an analogous formulation for electro-osmotic flow:

$$Q = -k_e \left(\frac{\Delta E}{l} \right) A \qquad (4.5)$$

where
ΔE = the difference in electrical potential between the upstream and downstream ends
k_e = the electro-osmotic permeability of the porous medium through which the water flows

In a similar way to Darcy's law, the negative term is necessary in the equations because flow occurs down the potential gradient – that is, from high electrical potential to low electrical potential (Figure 4.2).

Electro-osmotic flow occurs because of the electrochemical properties of soil particles and pore water (the electrolyte). A colloidal soil particle in water will have a negatively charged surface. The particle is surrounded by a 'double layer' of positively charged ions (cations, so named because they will be attracted to a cathode – a negatively charged electrode). The inner surface layer of cations is relatively firmly fixed to the soil particles, but further away the attractive forces reduce, and the outer layer of cations is more mobile. When an electrical potential is applied to the saturated soil, the positively charged cations will migrate to the cathode, dragging with them the mobile pore water that is further from the surface of the soil particles.

This phenomenon is of practical use because of the nature of electro-osmotic permeability compared with Darcy's permeability. As discussed in Section 4.3, Darcy's permeability is strongly influenced by particle size and can vary by more than 10 orders of magnitude. Conversely, electro-osmotic permeability k_e is much less sensitive to particle size and varies only by one order of magnitude for natural soils. For initial assessments, Farmer (1975) suggests an average value of $k_e = 5 \times 10^{-3}$ mm²/Vs (millimetres squared per volt second).

The relevance to geotechnical practitioners of the arcane phenomenon of electro-osmotic flow relates to drainage and pore water pressure reduction in very fine-grained soils such as clays and silts. The Darcy permeability of these soils is so low that even if large head differences (drawdowns) are applied, the water flow rates that can be generated are infinitesimally small. But because electro-osmotic permeability does not vary significantly with particle size, better drainage can be achieved by applying relatively modest potential gradients (of the order of 50 V/m). Farmer (1975) indicates that if Darcy's permeability is lower than approximately 1×10^{-7} m/s, there may be advantages in using electro-osmotic techniques in place of conventional methods. The application of electro-osmosis techniques is discussed in Section 17.12.

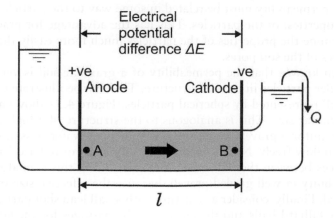

Figure 4.2 Electro-osmotic flow

4.3 PERMEABILITY IN SOILS

Soil is a very complex medium. Conceptually, it comprises a skeleton of mineral particles in contact with each other, leaving a more or less interconnected system of pore spaces between them (Figure 4.1a). When fluid flows through a soil (and if we assume the soil is saturated, then that fluid is water), the flow occurs through the pore space (in the vast majority of strata, the soil grains themselves can be considered impermeable). The concept of soils as being a 'porous medium' is fundamental to Darcy's law and many methods of analysis used in the fields of geotechnical engineering and hydrogeology.

When we consider a mass of soil (or rock) as a porous medium, a key parameter is the porosity n, which is the ratio of voids to the total volume of soil:

$$n = \frac{\text{Volume of voids}}{\text{Total volume}} \tag{4.6}$$

In soil mechanics, a slightly different term – void ratio – is used to define the quantity of voids in a soil:

$$e = \frac{\text{Volume of voids}}{\text{Volume of solids}} \tag{4.7}$$

Porosity can be related to void ratio by

$$n = \frac{e}{(1+e)} \tag{4.8}$$

Intuitively, it is easy to accept that the ability of a soil to transmit water (i.e. permeability) is controlled, at least in large part, by the nature of the soil pores (the viscosity of water, which will vary with temperature, is also a factor, but for typical temperature ranges, this effect is small compared with the soil type). Features of the soil pores that may have an influence on the flow of water include the size distribution of the pore space; the tortuosity of the pore space; and the shape and roughness of soil particles forming the edges of the pore space.

In routine geotechnical analyses, it is hard to measure the size and properties of the soil pores, but the nature and properties of the pore space will be strongly influenced by the size, shape, roughness and other properties of the soil particles themselves. It is therefore a logical step to think that permeability must be related in some way to the particle size distribution (and the other properties) of the particles. This has the advantage for practising engineers that we can determine the properties of the particles much more easily than we can determine the properties of the soil pores.

It has long been known that the permeability of a granular soil is very strongly influenced by the smaller particles in the soil structure. This can be illustrated by considering a very idealised 'soil' represented by spherical particles. Figure 4.3a shows an assemblage of billiard balls of similar size. This is analogous to the structure of a high-permeability soil (such as a coarse uniform gravel) where the voids (or pore water passages) are large, and the pore water can flow freely. Next, consider an assemblage of billiard balls with marbles placed in the spaces between the billiard balls (Figure 4.3b); this is analogous to a soil of moderate permeability (a well graded gravel), because the effective size of the pore water passages is reduced. Finally, consider a structure with small lead shot particles placed in the voids between the billiard balls and the marbles – the passages for the flow of pore water

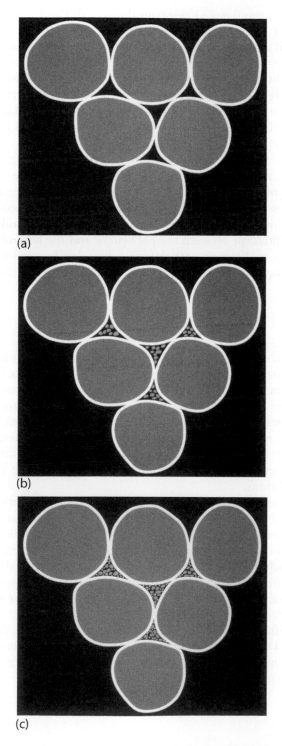

(a)

(b)

(c)

Figure 4.3 Soil structure and permeability. (a) High permeability. (b) Medium permeability. (c) Low permeability.

are further reduced; this simulates a low-permeability soil (Figure 4.3c). This analogy shows that the 'finer' portion of a sample dominates permeability. The coarser particles are just the skeleton of the soil and may have little bearing on the permeability of a sample or of a soil mass in situ.

The proportion of different particle sizes in a sample of soil (as might be recovered from a borehole) can be determined by carrying out a particle size distribution (PSD) test; see Head (2006). The results of the test are normally presented in a similar format to Figure 4.4, which shows typical PSD curves for a range of soils. The methodology of particle size analysis is described in Appendix 1.

PSD curves are sometimes known as grading curves or sieve analyses, after the sieving methods used to categorize the coarser particle sizes. The PSD curve represents the cumulative percentage of soil mass retained on each sieve size and allows the estimation of particle size parameters such as D_{10}, D_{15}, D_{50}, D_{60} (and so on), which are used in later design methods. D_n is the particle sieve aperture through which n per cent of a soil sample will pass and is determined by the intercept between the PSD curve and the horizontal line of n per cent passing. Combinations of different particle size values can be used to categorize different soil types. One commonly applied relationship is the uniformity coefficient U, defined as

$$U = \frac{D_{60}}{D_{10}} \tag{4.9}$$

When visualizing soils, it is tempting to think of the structure purely in terms of the orderly arrangements of idealized spheres discussed earlier. However, recent studies at Imperial College London (Taylor *et al.*, 2016) have provided an insight into more realistic soil structures. Figure 4.5 shows images from micro X-ray computed tomography (micro-CT) scans of samples of a relatively uniform sand that has been artificially compacted in the laboratory. What is striking is the small-scale variability apparent in these samples, which are formed from relatively uniform sands (U of 4) deposited under controlled conditions. We can only imagine the even greater variability in soil structure that exists in natural soils deposited in real conditions.

Over the last hundred years or so, many researchers and practitioners have developed empirical correlations between permeability and certain particle size characteristics. Some of these methods are still in use today and are described in detail in Section 12.6.1, along with the drawbacks and limitations of such correlations. It is interesting to note that some of the correlations, including Hazen's method, are not for soils at all; in fact, they are for granular filter media for water treatment systems. Presumably, at some point an enterprising person applied this to a geotechnical problem in sandy soil, liked the results, and the rest is history.

The relationship best known to geotechnical engineers is Hazen's formula (developed by Allen Hazen; Hazen, 1892), which dates from the late nineteenth century. Hazen's method is a simple example of how permeability can be related to particle size and states that permeability k may be estimated as

$$k = C(D_{10})^2 \tag{4.10}$$

where
　　C is a calibration coefficient
　　D_{10} is the 10 per cent particle size taken from the particle size distribution curves (Figure 4.4)

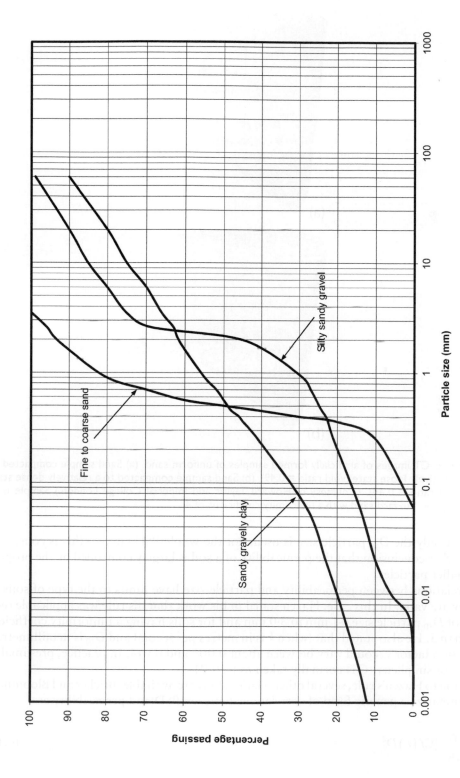

Figure 4.4 Particle size distribution.

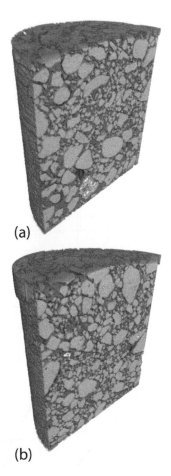

(a)

(b)

Figure 4.5 Micro-CT images of artificially formed samples of uniform sand. (a) Sand sample compacted to a relatively loose state (void ratio = 0.49). (b) Sand sample compacted to a relatively dense state (void ratio = 0.44). (Courtesy of Howard Taylor and Imperial College London.) Sample is of Leighton Buzzard sand with a uniformity coefficient of 4.

Note that only the D_{10} particle size is used, not the whole PSD curve – this is consistent with the earlier conclusion that the permeability of a soil is largely controlled by the proportion of smaller particles.

All correlations between permeability and particle size have limits to the type of soils in which they are valid. In this case, Hazen stated in his work that his rule was applicable over the range of D_{10} particle size 0·1 mm to 3·0 mm and for soils having a uniformity coefficient U lower than 5. He also stated that (when k is in metres per second and D_{10} is in millimetres) his calibration factor C could vary between about 0.007 and 0.014. In practice, presumably for reasons of simplicity, C is normally taken to be 0.01.

In addition to Hazen's rule, several other relationships are available; Bricker and Bloomfield (2014) summarize a generic form of correlations between PSD and permeability as

$$k = \left(\frac{\gamma_w}{\mu_w} \right) Cf(n) D_e^2 \tag{4.11}$$

where

    γ_w is the unit weight of water
    μ_w is the dynamic viscosity of water
    C is a sorting coefficient
    n is porosity
    $f(n)$ is a porosity function
    D_e is effective size (typically D_{10} or D_{15})

Some formulae, such as Hazen's method, conflate multiple aspects of the correlation to produce a simpler equation, requiring only limited input data. Other formulae, such as Kozeny–Carman (Carrier, 2003), apply the separate aspects of Equation 4.11 and therefore require estimates of porosity and other factors.

It is clear that different PSD correlations will produce different estimates of permeability for a given sample. Furthermore, the process of sampling and PSD testing may render the samples less representative of the in situ conditions, introducing further error into the permeability estimates. This is discussed in more detail in Section 12.6.1.

Equation 4.11 tells us that in a granular soil, the permeability varies as the square of the effective void or grain size. Therefore, the 5–6 orders of magnitude range of grain sizes between boulders (>200 mm) and clay (<2 μm) leads to a 10–12 orders of magnitude variation in permeability. No other engineering material property, not even the variation in strength or stiffness between a soft clay and a hard steel, exhibits such a range.

4.4 PERMEABILITY IN ROCK

The flow of groundwater in rock is fundamentally different from flow in soils. The granular nature of soils means that there is an expectation that groundwater will flow in a relatively diffuse manner through the soil pores – on average, the vast majority of the cross-sectional area will be involved in the flow (Figure 4.1a). In contrast, groundwater flow in rock is often concentrated in fractures, and large parts of the area of the rock transmit very little water (Figure 4.1b). The term 'fracture' is used here to describe all discontinuities that can allow groundwater flow, including joints, bedding planes, faults and so on (Figure 4.6). In karst rocks (see Section 4.6.2), groundwater is further concentrated in enlarged flow paths (much larger than conventional fractures) that have developed within soluble rocks such as limestone.

Some rock formations (such as igneous rocks) are dominated by fracture flow, where there is almost no flow through the blocks of rock between the fractures. The unfractured rock may contain saturated pore spaces, but the pore water is essentially immobile, at least at the time scales relevant to groundwater lowering for construction projects. The porosity of fractured hard rocks (the ratio of fracture space to total volume) can be much lower than for soils. Porosities of 0.1 per cent to 3 per cent are not unusual.

Other rock formations are known as 'dual porosity', where fracture flow is important, but the pore water in the rock mass itself is potentially mobile. This can occur in weathered hard rocks that have locally been broken down into sandy and clayey materials, or rocks that have a potential for intergranular flow, such as fractured sandstones. In these settings, much of the groundwater flow occurs primarily by fracture flow, but when groundwater levels are lowered, pore water can be released from storage in the rock mass.

In many cases where groundwater control is required, the fracture network within the rock will be sufficiently distributed and connected that, on a large scale, the rock can be

Figure 4.6 Groundwater flow via fractures in rock. The photograph shows an exposed face of fractured limestone with two principal groups of fractures approximately perpendicular to each other. The field notebook is used to provide scale.

considered to act as an equivalent porous medium. Groundwater flow can therefore be modelled by the same methods as used in soils, based on values of permeability determined from pumping tests and other methods. Groundwater control for excavations in rock is covered in Chapter 8.

While it is common for analysis methods for groundwater control in rock to ignore the effect of individual fractures and model the rock mass as a whole, occasionally a high-yielding fracture might be encountered in an excavation or test borehole. It is illustrative to consider the potential permeability of a fracture.

If a fracture is idealized as being bounded by a pair of smooth parallel plates, the Navier–Stokes equation leads to the conclusion that under laminar flow conditions, the flow rate through the idealized fracture is proportional to b^3, where b is the fracture aperture (Klimczak *et al.*, 2010). This formulation is commonly known as the 'cubic law', which gives the transmissivity T_f of a fracture of aperture b as

$$T_f = \frac{\rho_w g b^3}{12\mu_w} \tag{4.12}$$

where
   g is the acceleration due to gravity
   ρ_w is the density of water
   μ_w is the dynamic viscosity of water

The permeability k_f of the fracture is given by

$$k_f = \frac{\rho_w g b^2}{12\mu_w} \tag{4.13}$$

The cubic law accurately describes flow between smooth-walled plates. However, real fractures will be rough walled and non-parallel, with the aperture narrowing at choke points, reducing total flow. In this case, the equivalent aperture through which fluids can flow, called the 'hydraulic aperture', is smaller than the actual fracture aperture.

The cubic law only addresses the water-transmitting capacity of a fracture; it does not address the 'supply' element – where is the water coming from? Price (1994) indicates that for water supply boreholes in non-karstic aquifers, a single fracture is unlikely to contribute an equivalent transmissivity of more than around 1000 m²/d (1×10^{-2} m²/s). From Equation 4.12 (assuming fresh water at 10°C), this is equivalent to a fracture opening of only 2.6 mm, much smaller than the fracture openings observed in CCTV surveys of high-yielding boreholes in rock. Price concludes that a high-yielding fracture intercepted by a borehole is effectively acting as a conduit collecting water from the wider fracture network and rock matrix, providing a low-head-loss pathway for water to reach the pumped well. The practical outcome of this is that flow estimates from cubic law should be viewed as upper bound estimates, with the flow rate most likely being limited by the water availability in the surrounding aquifer (see Section 8.3.2). As noted in Section 3.3.3, if high-capacity fractures are influenced by pumping from groundwater control systems, turbulent flow conditions may occur, further reducing the maximum flow capacity of a fracture.

4.5 PROBLEMS WITH ASSESSING PERMEABILITY

In practice, meaningful values of permeability are more difficult to visualize than in Darcy's experiment. Darcy's concept of permeability assumes that the soil permeability is homogeneous (i.e. it is the same everywhere) and isotropic (i.e. it has the same properties in all directions). Even a basic study of geology will show that many soils and rocks exhibit properties far removed from these assumptions. Many strata are more or less heterogeneous in character, so even in an apparently consistent stratum, the permeability may vary greatly between one part and another. This may result from inhomogeneities such as fractures, erosion features, or sand and clay lenses. Also, the nature of deposition of soils can introduce anisotropy, where the soil permeability is not the same in all directions. Soils laid down in water may have a bedded, layered or laminated structure to form a 'fabric' of material of different permeability – as a result, the permeability in a horizontal direction is usually significantly greater than that in the vertical direction (Figure 4.7).

The influence of soil fabric on permeability is discussed in detail by Rowe (1972). Similarly, fracture systems in rock often have one or more preferential directions, with the resulting permeability being much greater in the direction of the fractures than perpendicular to it.

Despite these complications, any rational attempt at groundwater lowering will require permeability to be investigated and assessed. Some of the design methods discussed in Chapter 13 assume, for simplicity, isotropic and homogeneous permeability conditions – on a theoretical basis, this is clearly unrealistic. Yet, these methods are established and field-proven methods and if applied appropriately, can give useful results. This highlights the principle that in ground engineering it is often necessary to tolerate some simplification to make a problem more amenable to analysis. The key is to apply critical judgement to the parameters used in analysis and not to blindly accept the results of the analysis until they are corroborated or validated by other data or experience.

4.6 WORKING WITH PERMEABILITY IN THE REAL WORLD

It is clear that estimating realistic values of permeability can be problematic. As an engineering parameter, permeability is unusual because of the tremendous range of possible

Figure 4.7 Example of bedding in a gravel deposit creating anisotropic permeability conditions. The photograph shows a temporary cut face in a sand and gravel deposit. While the deposit is relatively permeable and water-bearing, subtle bedding layers within the gravel mean that the vertical permeability k_v is lower than horizontal permeability k_h.

values for natural soils and rocks; there can be a factor of perhaps 10^{10} between the most permeable gravels and almost impermeable intact clays. It is not unknown for the results of permeability tests to be in error by a factor of 10, 100, 1000 or even more. This often stems purely from the limitations of the test and interpretation methods themselves and can occur even if the tests are carried out in an exemplary fashion by experienced personnel.

Table 4.1 provides some general guidance on the permeability of typical soil types. This table is intended to be used for comparison purposes with permeability test results, to look for inconsistencies between soil descriptions (from borehole logs) and reported test results. Nevertheless, this table should be used with caution, especially in mixed soil types where the soil fabric (which is sometimes not adequately indicated in the soil description) may play a dominant role.

Table 4.1 Typical Values of Permeability of Soil

Soil type	Typical classification of permeability	Permeability (m/s)
Clean gravels	high	$>1 \times 10^{-3}$
Clean sand and sand/gravel mixtures	high to moderate	1×10^{-3} to 5×10^{-4}
Fine and medium sands	moderate to low	5×10^{-4} to 1×10^{-4}
Silty sands	low	1×10^{-4} to 1×10^{-6}
Sandy silts, very silty fine sands and laminated or mixed strata of silt/sand/clay	low to very low	1×10^{-5} to 1×10^{-8}
Fissured or laminated clays	very low	1×10^{-7} to 1×10^{-9}
Intact clays	practically impermeable	$<1 \times 10^{-9}$

It is difficult to provide general guidance on the permeability of different rock types. This is because the permeability will depend on both the type of rock and the fractures (width, frequency, orientation, etc). As an example, Freeze and Cherry (1979) indicate that unfractured metamorphic and igneous rocks could have very low permeabilities, down to 10^{-10} to 10^{-14} m/s. At the other end of the scale, carbonate rocks (such as limestones) that exhibit karst conditions (see Section 4.6.2) could have permeabilities as high as 10^{-2} m/s. Groundwater problems in rock are discussed in Chapter 8 and permeability testing methods in Chapter 12.

4.6.1 Permeability at Different Scales

A key issue that affects permeability measurement is that in heterogenous materials, measured values of permeability will vary with the scale of the test – in effect, with the size of the volume of soil or rock within which water flow is induced. Considering Darcy's law via Figure 3.2, it is apparent that permeability cannot be meaningfully associated with a single 'point'. Rather, it is an average value associated with a particular volume through which flow is occurring.

Rowe (1972) addressed this from a geotechnical testing perspective (for laboratory testing of consolidation properties). His study identified that larger-diameter samples gave higher values of permeability because there was a greater likelihood of permeable fabric being captured in such samples. Other studies such as Chapuis (2013) addressed scale effects in hydrogeological testing. Typically, well pumping tests give higher permeability values than borehole tests, because a well pumping test will influence a much greater volume of a stratum and is more likely to connect with preferential flow paths. Examples of preferential flow paths include coarser beds in soils (where intergranular flow dominates) or permeable fracture networks in rock.

Figure 4.8 is a schematic illustration of scale in permeability testing. Test zones show the volume of material significantly hydraulically influenced – either ex situ (in the laboratory) or in situ.

At very small (laboratory test) and small (borehole test) scale, the test location may or may not intercept significant numbers of preferential flow paths. Tests at these scales tend to have a large scatter between maximum and minimum values and may be biased towards lower values, as they do not fully hydraulically interact with the wider network of permeable fabric or fractures.

At the large and very large scale, the zone affected is more likely to be a representative average of the host soil or rock, including any permeable pathways. Typically, these scales of tests will produce higher values of permeability than small- and very small-scale tests.

Dewatering and groundwater control systems tend to interact with very significant volumes of the ground, and the most useful permeability estimates are large and very large scale. Unfortunately, such data are often not available. Whatever the source of the permeability dataset, it is important that the designer or analyst understands the potential limitations of the different test methods and views the test results with a critical eye. Assessment of permeability values for use in design is discussed in Section 12.9.

4.6.2 Karstic Flow in Rocks

In the discussions in the foregoing sections on intergranular flow in soils and fracture flow in rock, there is a presumption that if we 'zoom out' far enough, the distribution of groundwater flow is relatively homogeneous across the mass of soil or rock. This is an implicit

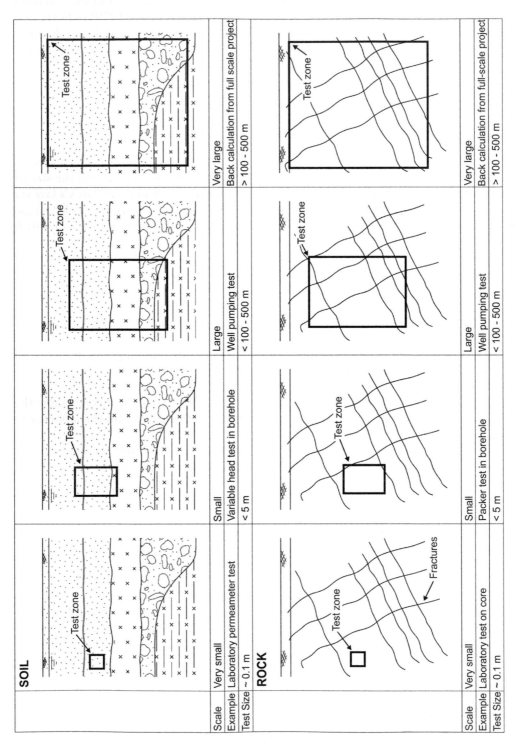

Figure 4.8 Scale effects on measurement of permeability in soil and rock. (From Preene, M and Powrie, W, *Proceedings of the XVII ECSMGE, Geotechnical Engineering: Foundation of the Future*, Reykjavik, September 2019. With permission.)

assumption in many methods of design and analysis. However, special mention should be made of rocks where 'karstic flow' occurs, when groundwater flow can be concentrated in a limited number of very permeable preferential channels.

Karst is a geomorphological term for terrains associated with carbonate rocks, including limestones and dolomites, that are significantly water soluble. In such areas, the infiltration of rainfall and surface waters can selectively dissolve parts of the rock. Typically, water will initially flow via existing fractures, joints and bedding planes, along which dissolution will occur. This will enlarge the flow path, allowing more water to pass and allowing the dissolution and enlargement to continue. In hydrogeological terms, this means that groundwater is concentrated in, and flows rapidly through, a network of fractures and 'solution features' including conduits (significantly enlarged fractures) and caves (conduits that are large enough to be explored physically by humans). An example of karst flow conduits exposed in a limestone excavation is shown in Figure 4.9. In some cases, the solution features can be completely or partly filled by sediment or redeposited minerals, significantly reducing their capacity to transmit water. If pumping is carried out for groundwater control purposes, there is a risk that the water flow induced in the solution features will erode and wash away the infill, increasing the flow that can pass through the channels.

When groundwater control works in potentially karstic rocks are planned and investigated, a key challenge is that even extensive borehole drilling programmes may fail to intercept any significant karst features, possibly implying that the bedrock is massive and of low permeability. Desk studies (Section 11.6) carried out by suitably experienced geologists and geomorphologists can be very useful if karst conditions are a concern.

Further information on the geology of karst conditions can be found in MacDonald *et al.* (1998) and Waltham (2016). Some problems that may be encountered when excavating in karst geology are discussed in Section 8.3.5.

Figure 4.9 Karstic features within the Simsima Limestone in Doha, Qatar. (Courtesy of T O L Roberts.)

4.7 METHODS OF DETERMINING PERMEABILITY

A range of methods are available to assess permeability. These can be grouped as:

- Non-quantitative assessment methods
- Quantitative methods:
    - Empirical assessment
    - Ex situ methods
    - In situ methods

These methods are reviewed briefly in the following sub-sections.

4.7.1 Non-Quantitative Assessment Methods

The design of groundwater lowering systems typically needs a design 'value' (or range of values) for permeability. It is tempting, therefore, for designers to focus on quantitative methods, because these give 'values' of permeability. However, as discussed earlier, many test methods have limitations and may give different scales of permeability values. These results are often not adequately validated against non-quantitative methods.

A simple validation that is often overlooked is to assess the permeability values against published ranges based on classification of the soil or rock (such as Table 4.1 for soils). Such values must necessarily be generic, but experience shows that there is a strong correlation between permeability and the nature (including stress state and weathering) of the soil or rock in question. Such correlations can be used to exclude unrepresentative test results from the permeability dataset.

Other non-quantitative methods that can provide information on permeability include:

- Water strike (inflow) records from boreholes
- Water levels recorded in boreholes at the start and end of drilling shifts
- Drilling fluid (flush) loss records, voids or tool drops observed during borehole drilling
- Geophysical fluid logging carried out in boreholes

Typically, these methods can be used to identify the presence of localized high-permeability zones at specific depths in boreholes that may not be apparent from small-scale tests or from large-scale tests that report average permeability (for example from a well pumping test). Non-quantitative methods are discussed in more detail in Section 12.5.

4.7.2 Quantitative Assessment Methods

Quantitative values of permeability can be obtained by a wide variety of techniques; each method has limitations and may estimate permeability values at different scales. Possible methods are summarized in Table 4.2, which also directs the reader to the relevant section of the book that discusses each method in more detail.

Table 4.2 Common Methods for Quantitative Assessment of Permeability

Method	Typical scale (Figure 4.8)	Characteristics and limitations	For further details see
Empirical methods			
Particle size distribution (PSD) correlations	Very small	Elements of soil grading (PSD) curves are used in empirical correlations. Each correlation method (e.g. Hazen; Kozeny–Carman) has a limited range of PSDs to which it should be applied; use outside those ranges can result in gross errors. Samples may be unrepresentative because a) sampling and the PSD testing process result in homogenization of samples and loss of structure and fabric; and b) fine particles may be lost from the sample (typically flushed out into the borehole fluid).	Section 12.6.1
Ex situ methods			
Permeameter testing	Very small	Water flow is induced through a specimen of soil or rock. Sample size is limited to a few hundred millimetres. Disturbance (disruption of structure and fabric) and changes in stress state due to sampling mean that permeability values may not be representative.	Section 12.7.1
Oedometer testing	Very small	Similar to permeameter testing.	
In situ methods			
Variable head tests in boreholes	Small	Common method (also known as slug test) for testing discrete zones during borehole drilling, especially in soils. Falling head tests cause water to flow out of the borehole, and rising head tests induce inflow to the borehole. In high-permeability strata, water levels change rapidly during the test, and the rate of change can be difficult to record manually (the problem can be overcome by the use of dataloggers). In low-permeability soils, a test may need to last several hours as excess heads slowly dissipate. The presence of a disturbed zone (caused by drilling) around the borehole can result in unrepresentative responses. Different methods of analysis can give different values of permeability for the same test response.	Section 12.8.1
Variable head tests in standpipe piezometers	Small	Similar to variable head tests in boreholes, but carried out in standpipe piezometers in boreholes. Can allow tests to be carried out with better-defined response zones compared with borehole tests during drilling.	Section 12.8.2
Packer tests in boreholes	Small	Widely used for testing of boreholes in fractured rock, typically by injecting water to induce flow out of a borehole. Produces an average permeability for the test section; it can be difficult to determine whether the permeability is associated with many distributed fractures or a smaller number of discrete fractures.	Section 12.8.3
Geophysical flowmeter logging	Small	A specialist technique that can be used in boreholes in fractured rock. A profile of vertical flow velocity within the borehole can be used to determine vertical variations in permeability.	Section 12.8.7

(Continued)

Table 4.2 (Continued) Common Methods for Quantitative Assessment of Permeability

Method	Typical scale (Figure 4.8)	Characteristics and limitations	For further details see
Well pumping tests	Large	Water is pumped from a well while pumped flow rate and drawdown in neighbouring monitoring wells are monitored. Provided that the pumped flow rate is sufficient to create a large drawdown in the strata around the well, and if pumping continues for an extended period (typically several days), a large volume of the ground can be influenced. The large-scale permeability values from this method are a good match for the needs of dewatering designers. May also provide information on hydraulic boundary conditions.	Section 12.8.5
Groundwater control trials	Large/very large	Similar to well pumping tests, but typically pumping from multiple wells. Potentially affects even greater volumes of the ground.	Section 12.8.6
Back calculation from full-scale projects	Very large	Pumping rate and groundwater level data are used to assess permeability from full-scale dewatering. Can only be applied during or after significant phases of a project are in progress, provided that adequate monitoring is in place. Useful to refine designs for later phases.	Section 12.6.3

Based on Preene, M and Powrie, W, *Proceedings of the XVII ECSMGE, Geotechnical Engineering: Foundation of the Future*, Reykjavik, September 2019. With permission.

Chapter 5

Groundwater Models

The challenge is to use the theory and modelling wisely, observe assiduously, rate the data sensibly, assemble and characterize experience and communicate liberally.

Bryan Skipp

5.1 INTRODUCTION

This chapter deals with groundwater modelling in its various forms. When thinking about groundwater modelling, it is tempting to focus on numerical modelling and imagine an 'expert' modeller sitting at a computer working on complex numerical modelling problems. But this is much too narrow a view. Groundwater modelling includes conceptual models, empirical models, analogue models and analytical models as well as numerical models. In reality, every time a dewatering designer or practitioner tries to use the information at hand to understand groundwater conditions and to then develop designs or predict system performance, they are, perhaps without realising it, using groundwater models. The remainder of this chapter will describe the different types of groundwater models and discuss how they can be used when developing groundwater lowering systems.

5.2 GROUNDWATER MODELLING

Groundwater modelling is often misunderstood, so it is useful to begin with some definitions. One possible definition of a model is:

A model is a representation of a system or process, with the aim to aid understanding of how it works.

Specifically for groundwater models, a definition might be:

A groundwater model is a simplified representation of a groundwater system.

Notable points about these definitions are that mention is made of simplification (so a model need not attempt to be a perfect representation of reality) and of the model having the aim of aiding understanding. Most importantly, the type of model is not prescribed, including whether it is quantitative (i.e. intended to produce numbers for, say, dewatering flow rates) or whether it is qualitative (i.e. not intended to produce numerical values).

The value of a groundwater model is that it can link the present to the future (predictive models) or to the past (reactive models).

i. Predictive models look forward, using the information we have now, to help us make predictions, such as estimates of dewatering pumping rates or required numbers of wells. These models are clearly of value in dewatering design, which is essentially a predictive process.
ii. Reactive models look back from the current information to help us understand why things have happened. This can be useful when applying the observational method (Section 13.3.2) to dewatering design or when trying to understand why a groundwater lowering system has problems (Section 23.6).

There is a range of model types that can be useful, and these are described later in this chapter. A key distinction is between conceptual models and analytical models:

1. A conceptual model (Section 5.4) is a largely qualitative description of key elements of a groundwater system, without the use of mathematics.
2. An analytical model (Section 5.5) is a quantitative representation of key elements of a groundwater system. Analytical models include empirical models, mathematical models using closed-form solutions and numerical models.

Other, less commonly used model types are outlined in Section 5.6, including graphical models, physical models and analogue models.

5.3 COMPLEXITY IN MODELLING

A fundamental point about groundwater modelling is that no model should try to perfectly simulate reality – that would be impossible. Instead, the aim should be to develop a model that helps us understand the engineering problem being addressed. As the statistician George Box once said, 'All models are wrong but some are useful.' Our objective should be to produce a useful model.

A useful model will be relevant – in other words, it will help answer the pertinent question, and it will also, inevitably, involve some simplification or approximation of conditions to make the problem more tractable.

When one is developing a groundwater model of any type, the need for simplifications should be accepted. However, it is useful to remember the guidance of CIRIA Report C750 (Preene *et al.*, 2016), which said:

> In geotechnical engineering it is usually necessary to make simplifications to arrive at a conceptual model of a real situation, which is amenable to analysis. The secret of success is not to ignore any factor, which could destroy the applicability of the conceptual model adopted.

5.3.1 Relevance of Models

For any groundwater model to be of use, it must be relevant to the question being asked or the engineering problem that has given rise to the need for the model. In relation to groundwater control projects, common objectives of groundwater models include:

- Assessment of pumped flow rates for dewatering design
- Assessment of the required depths and thicknesses of groundwater cut-off walls to act as effective groundwater barriers
- Assessment of the size of the zone of influence (the area where significant drawdowns occur) around a dewatering system
- Assessment of potential external impacts (e.g. ground settlement, migration of contamination) caused by groundwater lowering

A given modelling problem may have multiple objectives. These may be addressed by one modelling approach or by multiple approaches.

A key point is not to seek precision unnecessarily. As noted earlier, we should not be trying to simulate reality. For our purposes. a groundwater model should be an aid to decision making as part of an engineering problem. The required model precision should be matched to the question being addressed (see Section 13.5).

5.3.2 Typical Simplifications Used in Modelling

As discussed in Chapter 3, groundwater flow in soils and rocks is complex, often involving flow in three dimensions with anisotropic permeability and varying with time. The mathematical formulations to even approximately describe these conditions can be complex, and it is not unusual to simplify conditions to make the problem more tractable.

The most commonly used generic mathematical formulation for groundwater flow is in the form of a second-order differential equation. Hydrogeology textbooks (such as Rushton, 2003) show how the governing equation of groundwater flow in three dimensions can be derived by combining Darcy's law (Section 3.3.2) with the principle of continuity (which states that the flow of mass into and out of an elemental volume must be equal, once any change of mass stored in the volume has been accounted for). The equation for three-dimensional groundwater flow is

$$\frac{\partial}{\partial h}\left(k_x\frac{\partial h}{\partial x}\right)+\frac{\partial}{\partial h}\left(k_y\frac{\partial h}{\partial y}\right)+\frac{\partial}{\partial h}\left(k_z\frac{\partial h}{\partial z}\right)=S_s\frac{\partial h}{\partial t} \tag{5.1}$$

where
> x, y, z are the cartesian coordinates
> t is time
> h is total head
> k_x, k_y and k_z are the permeability values in the x, y and z directions, respectively
> S_s is the specific storage (the volume of water released from a unit volume of aquifer for a unit fall in head)

This equation is the basis of the mathematical solutions in numerical models that address three-dimensional transient (i.e. time-variant) problems. However, many of the solutions used in practice and discussed in the book are based on simplifications of the governing equation. Equation 5.1 can be simplified for isotropic conditions where $k_x = k_y = k_z = k$:

$$k\left(\frac{\partial h^2}{\partial x^2}+\frac{\partial h^2}{\partial y^2}+\frac{\partial h^2}{\partial z^2}\right)=S_s\frac{\partial h}{\partial t} \tag{5.2}$$

A further simplification is to assume steady-state flow rate (where the change in head with time is zero, and therefore the change in water storage is zero), and the equation becomes

$$k_x \frac{\partial h^2}{\partial x^2} + k_y \frac{\partial h^2}{\partial y^2} + k_z \frac{\partial h^2}{\partial z^2} = 0 \qquad (5.3)$$

which for isotropic conditions ($k_x = k_y = k_z = k$) becomes a form of the Laplace equation:

$$\frac{\partial h^2}{\partial x^2} + \frac{\partial h^2}{\partial y^2} + \frac{\partial h^2}{\partial z^2} = 0 \qquad (5.4)$$

A very common simplification in geotechnical engineering is to reduce three-dimensional problems to two dimensions, which gives (for steady-state conditions)

$$\frac{\partial h^2}{\partial x^2} + \frac{\partial h^2}{\partial y^2} = 0 \qquad (5.5)$$

For the case of radial flow to a well (see Section 5.8), in isotropic conditions the flow rate to the well will be the same in all radial directions, so the three-dimensional flow equation for steady-state flow becomes

$$\frac{\partial h^2}{\partial r^2} + \frac{1}{r} \frac{\partial h}{\partial r} + \frac{\partial h^2}{\partial z^2} = 0 \qquad (5.6)$$

where
   r is the radial coordinate (centred on a well)
   z is the vertical cartesian coordinate

   The reason for presenting the fundamental equations here is not to imply that a detailed knowledge of mathematics is required to understand groundwater models. Rather, the intention is to highlight that some of the commonly used analytical models are based on significant simplifications. An understanding of these simplifications can be useful when selecting a suitable groundwater modelling approach.

5.4 CONCEPTUAL MODELS

A conceptual model is described by Brassington and Younger (2010) as 'a theory-based description that represents the phenomena being studied that is founded on a set of variables with logical and quantitative relationships'. In plain English, a conceptual model is a simplified non-mathematical representation of a groundwater system to help understand and communicate groundwater conditions.
   Until the conceptual model is developed, it is not possible to make rational decisions about which of the analytical modelling methods is suitable for the problem. In order to develop a conceptual model, the designer needs to be familiar with the concepts of groundwater flow, aquifers, boundary conditions, permeability and so on (see Chapters 3 and 4).
   If a conceptual model is a poor match for actual conditions, then much of the subsequent analytical modelling work may be of dubious value for dewatering design. Even if analytical work is diligently and carefully carried out, if it is based on an inappropriate conceptual model, the design may be going in totally the wrong direction – and this is likely to lead to

poorly performing dewatering systems. Developing the conceptual model at an early stage also forces the designer to review the available data. If the designer has insufficient reliable data to formulate a model, this could be a sign that further site investigation is needed.

A very wide range of factors are relevant to hydrogeological conceptual models, as described by Brassington and Younger (2010). But, as described earlier, models must be relevant, and for the purposes of dewatering design, key factors to be considered in conceptual models include:

(i) Aquifer type(s) and properties. Aquifers (the water-bearing strata in which groundwater levels are to be lowered) may be classified as confined or unconfined. These types behave in quite different ways; the aquifer type(s) must be identified. It is essential that the likely permeability range of each aquifer be identified, including any potential anisotropy affecting horizontal and vertical permeability. If transient analyses are to be carried out, the storage coefficient must also be estimated.

(ii) Aquifer depth and thickness. These dimensions must be estimated and a judgement taken as to whether they are effectively constant across the area affected by the dewatering, or whether the aquifer thickness varies significantly.

(iii) Presence of aquitards and aquicludes. Very low-permeability silt or clay layers may act as barriers to vertical groundwater flow. The presence of such strata may necessitate well screens above and below the aquiclude or aquitard.

(iv) Distance of influence and aquifer boundaries. Is all the pumped water likely to be derived purely from storage in the aquifer? If not, are there any nearby sources of recharge water? The presence of any barrier boundaries (such as steeply dipping faults in bedrock aquifers) can also influence groundwater flow.

(v) Initial groundwater level and pore water pressure profile. The initial groundwater level determines the amount of drawdown required. Complications may arise if the groundwater level slopes across the site or if groundwater levels vary with tidal or seasonal influences.

(vi) Presence of compressible strata. If present in significant thickness, this indicates that potentially damaging consolidation settlements may occur.

There are also some factors that are not directly related to ground conditions or geology. Where relevant, these need to be included in the conceptual model.

(vii) Geometry of the proposed works. Different depths and size of the excavation will interact with groundwater conditions in different ways.

(viii) Groundwater lowering technique. Different dewatering methods may interact with the groundwater regime in quite different ways, including pumping and exclusion methods (see Section 5.9).

(ix) Period for which groundwater lowering is required. Longer periods of pumped dewatering will potentially influence wider regions of the aquifer.

(x) Depth of proposed wells. In very deep aquifers, dewatering wells may not penetrate to the base of the aquifer. Instead, shallower, partially penetrating wells may be more suitable, and this may affect the conceptual model.

(xi) Environmental constraints. The location or nature of the site and its surroundings may give rise to limits on discharge rates or drawdown being imposed by the client or environmental regulator. There may be zones of contaminated ground or contaminated groundwater that may be affected or mobilized by dewatering pumping.

A final point is that uncertainty should be considered in conceptual models. The designer should be clear that no conceptual model can ever be 'certain'. For example, site investigations can only sample and test a tiny percentage of the volume of soil or rock that will be affected by a groundwater control system; they cannot reveal all the potential complexities and variability of the strata. Similarly, groundwater levels are often only monitored for a finite period and may not identify true maximum and minimum groundwater levels. If these factors could have a significant effect on potential groundwater control designs – such as selection of the depth of cut-off walls or the required dewatering pumping rate – such uncertainties must be explicitly identified in the conceptual model. This can help designers address uncertainty by carrying out further site investigation or by adopting a conservative approach in design (e.g. by using worst credible parameters). Conceptual models are outlined in each of the design examples given in Appendix 5.

5.5 ANALYTICAL MODELS

Analytical models should build on conceptual models with the aim of providing quantitative assessment of some elements of groundwater flow. Most (but not all) involve some use of mathematics. The principal types of analytical model are:

- Empirical models (based on experience and rules of thumb)
- Mathematical models using closed-form solutions (which can be solved directly without iteration)
- Numerical models, where the volume to be studied is typically divided up into different elements and the mathematics of the groundwater flow problem solved iteratively

Each model type is described in the following sub-sections.

5.5.1 Empirical Models

Empirical models are not derived from theory or analysis of the physical laws of a problem; rather, they are based on observation and previous experience. They are sometimes known as 'rules of thumb'. Atkinson (2008) states:

> The origins of the term 'rule of thumb' are obscure. Apparently Roman bricklayers used the tip of the thumb from the knuckle as a unit of measure. Brewers used their thumb to test the temperature of fermenting ale … Nowadays, rule of thumb implies a rough estimate based on experience rather than formal calculation.

Given the wide use of mathematical models, it is tempting to think that empirical models have little value in modern practice. However, there is a long record of empirical methods being used to aid the understanding of groundwater problems – for example, Darcy's law, the cornerstone of most practical methods of groundwater analysis, is empirical in origin (Section 3.3.2). When used appropriately, they can still have valid uses.

One of the challenges is that empirical models are based on a finite underlying dataset and/or series of assumptions. An empirical relationship based on a finite range of observations can only be used with confidence within the limits of the observations upon which it is based; although, of course, some models, such as Darcy's law, can have a validity beyond the original range of observations.

A key risk is that these models may be applied (knowingly or unknowingly) in cases where the model is not valid. Such applications can result in grossly unrepresentative modelling outcomes. To avoid this, it is important that the user has a good understanding of the basis of the empirical model and its limitations. Atkinson (2008) identifies three types of empirical models, based on earlier work by Wroth (1984):

- Class 1 empirical models: They have a sound theoretical basis and as a result are generally applicable everywhere. They can be derived from theory alone without the need for empirical observations and typically can be used in any combination of consistent units. Darcy's law is in this class.
- Class 2 empirical models: They have a sound theoretical basis and are generally applicable everywhere, but they require empirical correlations. Typically, the models or equations require certain combinations of units to be applied in order for valid results to be obtained. Hazen's formula (to estimate permeability from the particle size distribution of a soil; see Section 4.3) is in this class because it requires specific units to be used (D_{10} in millimetres to estimate permeability in metres per second).
- Class 3 empirical models: These models are based purely on observation, they do not involve an appreciation of the physical processes, and there is no underlying theory. Examples are rules of thumb based on 'local knowledge', where a contractor might know from previous jobs that 'running sand' conditions (see Section 7.4) encountered during trenchworks in a certain part of a town can be stabilized by a certain arrangement of wellpoint dewatering.

This classification is useful, because it highlights that Class 1 models have a theoretical basis and can be used with confidence in a wide range of conditions. Class 3 models are entirely empirical and have no apparent theoretical basis and should be used only in the context within which they were developed. Class 2 models fall between these two extremes.

5.5.2 Mathematical Models Using Closed-Form Solutions

Closed-form solutions are a type of mathematical model that can be written as a discrete equation and solved directly without significant iteration. They are distinct from numerical models, which commonly involve large numbers of iterations. Examples include many of the commonly used well flow rate equations, such as the Dupuit–Forchheimer equation (Equation 13.5). Closed-form solutions can be solved by hand calculation or simple spreadsheet applications.

The closed-form solutions provide a quick, yet effective, way of understanding groundwater behaviour via quantitative methods. The various solutions available are based on simplifications of groundwater flow systems to allow the problem to be analysed (this is discussed in Section 5.8.2 for flow to wells). Provided the underlying simplifications, assumptions and limitations are understood, this type of model can be used to predict groundwater behaviour under a wide range of conditions.

Different groundwater flow regimes will require different mathematical solutions, and a large number of equations are available for different conditions. The key challenge is selecting an appropriate solution – essentially, the method chosen should be appropriate to the conceptual model (this is discussed in Chapter 13 on design of groundwater lowering systems).

5.5.3 Numerical Models

Numerical models typically require the groundwater regime to be simplified and divided into a series of different 'elements' to which material properties and boundary conditions are allocated (Figure 5.1). Computer software is then used to solve multiple iterations of

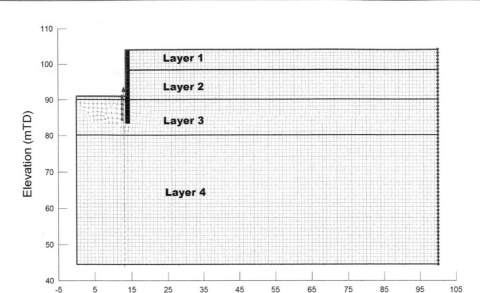

Figure 5.1 Example of finite element mesh for a two-dimensional numerical groundwater model. (Courtesy of WJ Groundwater Limited, Kings Langley, UK.) The model is for seepage to the excavation shown at the left hand edge, with a low-permeability cut-off wall penetrating below the excavation dig level.

the governing groundwater flow equations (between neighbouring elements and the model external boundaries) until the degree of change between successive iterations is sufficiently small not to influence the results. Suitable constituted numerical models can estimate dewatering flow rates, external drawdowns, groundwater quality changes and many other factors.

Numerical modelling of groundwater flow problems has been carried out since the 1960s, but it is only since the 1980s and 1990s that advances in computer hardware and software have made this approach viable for a wide range of groundwater lowering problems. Computers are now so ubiquitous in geotechnical engineering that not only is there a computer on every desk but they are also in many site engineers' field bags, and powerful modelling software is available at a cost similar to the price of this book! Numerical modelling as part of the design of dewatering and groundwater control systems is here to stay. But it is important to identify when it can be used most effectively and to be aware of the potential pitfalls.

This section will concentrate on the use of numerical groundwater models applied to dewatering problems. It will describe the general approach that should be adopted and will not go into the detail of the modelling process, which is described in texts such as Anderson *et al.* (2015). While numerical modelling packages are becoming easier to apply, it is vital that anyone contemplating their use understands the theoretical basis and limitations of the program in question. In complex or unusual situations, help should be obtained from an experienced groundwater modeller.

Essentially, a numerical model breaks down the overall problem, its geometry and its boundary conditions into a number of discrete smaller mathematical problems that can be solved individually. These are solved iteratively, whereby the solution to each smaller problem is adjusted until there is acceptable agreement at the boundaries between the smaller problems. It is sometimes stated that numerical models produce 'approximate' solutions. This is true in the mathematical sense, because there will be a small difference between an analytical solution and the numerical results, but in an engineering sense, the numerical output is a pretty accurate reflection of the input data and the groundwater model that has been formulated.

The key point is that the numerical package is only following instructions given to it by the user. If there are errors in the input data, or more importantly, if the conceptual model (on which the numerical model is based) is unrealistic, gross errors may result. It is an old cliché, but the phrase 'garbage in, garbage out' – meaning that the results can only be as good as the input and instructions – is very true for groundwater modelling. The conceptual model (see Section 5.4) is the critical starting point for any modelling exercise. If the conceptual model is not a good match for actual conditions, then the output will be of questionable value.

It is important to select a numerical modelling package appropriate to the problem in hand. Some packages were originally developed for use in water resources modelling of large areas of aquifers for regional-scale studies. These packages can be useful for large dewatering works in highly permeable aquifers with large distances of influence but may be less applicable for smaller-scale seepage problems. Another group of packages were developed for geotechnical problems such as seepage beneath cofferdams or through earth embankments; these may be appropriate for small-scale problems.

The modelling package used must be capable of solving the relevant type of problem. Most packages can solve steady-state problems, but not all are designed for transient time-dependent seepage. Most groundwater flow problems are three-dimensional to some degree, but in some cases it may be possible to simplify conditions to two-dimensional flow without significant error; some packages can model three-dimensional flow, but many are limited to two dimensions only.

5.6 OTHER TYPES OF MODELS

Occasionally, other methods are used to model groundwater flow regimes for engineering purposes. Examples are graphical models; physical models; and analogue models.

5.6.1 Graphical Models

Flow net analyses are a graphical representation of a solution of a given two-dimensional groundwater flow problem and its associated boundary conditions. A flow net (Figure 5.2) is a network of 'flow lines' and 'equipotentials' that, when developed correctly, provide a graphical solution to a steady-state groundwater flow regime in two dimensions.

- Flow lines represent the paths along which water can flow within a two-dimensional cross section.
- Equipotentials are lines of equal total head.

Consideration of Darcy's law (Section 3.3.2) shows that groundwater flows from higher total head to lower total head. Because each equipotential represents constant total head, there can be no flow along an equipotential. Therefore (under isotropic conditions), each flow line and equipotential must intersect at right angles. Practical guidance on constructing flow nets is given in Cedergen (1989), including their application under anisotropic permeability conditions. Hand sketching of flow nets can be used to obtain solutions to certain flow problems, considered either in plan or in cross section, for isotropic or anisotropic conditions.

A typical problem where flow nets are used is illustrated in Figure 5.2 for seepage into an excavation where the presence of partial cut-off walls alters the groundwater flow paths (see

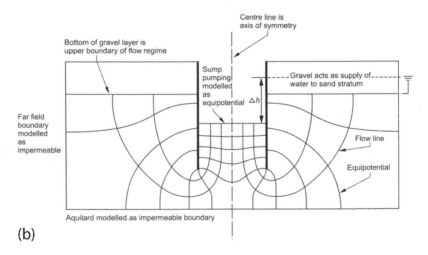

Figure 5.2 Flow net into an excavation, where groundwater flow is affected by low-permeability cut-off walls forming a cofferdam. (a) Conceptual model on which the graphical model is based. (b) Graphical model in the form of a flow net.

Williams and Waite, 1993). Experience has shown that for such a problem, if a hand-drawn flow net is constructed by someone competent in the technique, the calculated flow rate and head distribution from the flow net will agree very closely with the results of numerical modelling. This illustrates that there is no fundamental difference between hand-drawn flow nets and two-dimensional numerical modelling – each approach develops a solution to the same mathematical problem, but using different methods.

5.6.2 Physical Models

Very rarely, physical models are used to analyse groundwater flows. Such models are simplified scaled physical representations of a groundwater flow problem – such as seepage through a dam or into an excavation cofferdam. Typical examples include 'sand tank' models, where granular materials are placed within a tank and groundwater is added or removed to create the desired groundwater flow patterns. Physical models require careful design to take account of scale effects to achieve a representative model, and are rarely used outside teaching and research.

5.6.3 Analogue Models

Analogue models are an application of the fact that several physical processes (which can easily be physically realized and measured) are controlled by mathematical relationships that are analogous to groundwater flow (see Table 5.1). This means that physical models of these analogous processes can be used to predict and interpret certain groundwater flow cases. In the past, analogue models were used more commonly to analyse complex problems, but in recent times, advances in numerical modelling methods have made these techniques largely obsolete.

Examples include:

- Electrical resistance or resistance-capacitance analogues. Such models, as described in Rushton and Redshaw (1979), are based on the analogy between electrical flow and groundwater flow. In the past, before the widespread availability of numerical models, these methods were used more commonly to analyse complex problems. A rare recent electrical analogue application from the 1980s is described by Knight *et al.* (1996).
- Viscous flow analogues. Commonly known as Hele-Shaw models. This analogue approach uses the slow flow of a viscous fluid in the narrow space between two parallel plates to simulate groundwater problems. This approach is rarely used outside teaching and research.

5.7 THE ROLE OF GEOLOGICAL STRUCTURE IN MODELS

Chapter 3 described how the boundaries and structure of aquifers and aquitards can complicate the flow of groundwater. The presence of recharge boundaries or multiple layered aquifers will have a significant effect on the groundwater lowering requirements for an excavation.

Although geological structures and boundaries can make dewatering more difficult, conversely, a designer can try to use these features to advantage. For example, if the conceptual model highlights the presence of a more permeable layer within the aquifer sequence, by far the most effective approach is to design dewatering wells to abstract water directly from the

Table 5.1 Conduction Phenomena Analogous to Groundwater Flow

Variable	Groundwater flow	Flow of electricity	Heat flow
Driving potential	total hydraulic head [m]	voltage [Volt]	temperature [K]
Flow	groundwater flow rate [m³/s]	electric current [Ampere]	heat [Watt]
Physical property controlling flow	hydraulic conductivity [m/s]	electrical conductivity [S]	thermal conductivity [W/mK]
Storage parameter	specific storage S_s [1/m]	capacitance [Farad]	specific heat capacity [J/m³K]
Relationship controlling flow	Darcy's law	Ohm's law	Fourier's law

Relevant SI units shown in square brackets: J = Joule; K = Kelvin; S = Siemens (formerly mhos).

permeable layer. This will maximize well yields and will induce the adjacent, less permeable layers to drain to the permeable layer. Where the permeable layer is at depth, this approach of pumping preferentially from the most permeable layer is known as underdrainage (see Section 13.6.2 and Figure 13.9).

Sometimes, an understanding of the geological structure can have a very beneficial effect when dealing with groundwater problems. An example is given in the following.

Figure 5.3 shows a 12 m deep excavation being constructed in fractured bedrock of inter-bedded limestone and mudstone, where the bedding of the strata was not horizontal but 'dipped' gently to the east – this means that the rock bedding sloped very slightly (2 to 3° below horizontal) eastward. The planned groundwater control included a system of pumped ejector wells (Section 17.2). However, a sensitive archaeological site was present a few hundred metres to the east (within the predicted zone of influence), and there was concern that drawdown from the dewatering pumping could cause adverse impacts (Section 21.4).

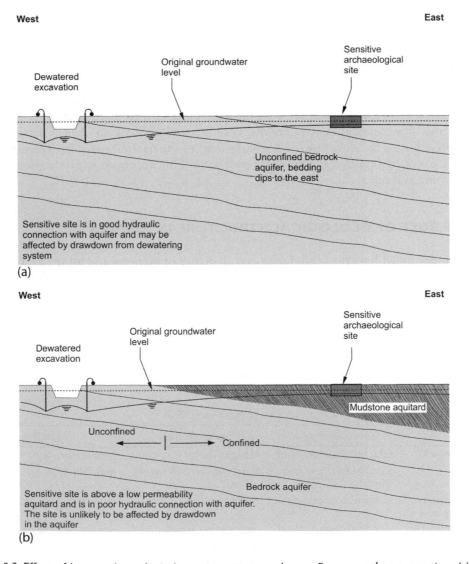

Figure 5.3 Effect of large-scale geological structure on groundwater flow around an excavation. (a) Local ground conditions only considered. (b) Regional geology considered.

At the excavation location, the bedrock formed an unconfined aquifer, with groundwater level close to ground level. If similar groundwater conditions were present at the archaeological site, then that site would have been impacted by drawdown of groundwater levels (Figure 5.3a). This could have required mitigation measures, such as artificial recharge, to minimize groundwater lowering beneath the sensitive site. However, a geological desk study (Section 11.6) revealed that as the bedrock aquifer dipped to the east, an overlying aquitard of very low-permeability mudstone layer was observed, gradually increasing in thickness. At the location of the archaeological site, the bedrock aquifer was confined below the mudstone aquitard and was effectively hydraulically isolated from the shallow archaeological site (Figure 5.3b). This indicated that there was a low risk of drawdown impacts affecting the archaeological site; and that expensive mitigation measures were not needed. It is interesting to note that this conclusion would not have been reached based on the investigation boreholes on the excavation site alone, since the confining mudstone layer was absent at that location. This is a powerful illustration of the value that can be gained from an understanding of the wider geological setting of a site.

5.8 FLOW TO WELLS AND SLOTS

Groundwater models can be an interesting field of study in their own right. But our interest in groundwater is more than purely academic; we need to be able to control groundwater conditions to allow construction to proceed. The principal approach is to install a series of wells and to pump or abstract water from these wells – this is the essence of groundwater lowering or dewatering. Hence, the commonly used models that describe flow towards wells must be understood.

Some definitions will be useful. For the purposes of this book, a 'well' is any drilled or jetted device that is designed and constructed to allow water to be pumped or abstracted. These definitions are, again, slightly different from those used by some hydrogeologists. In hydrogeology, a 'well' is often taken to mean a large-diameter (greater than, say, 1.8 m) well or shaft, such as may be dug by hand in developing countries. A smaller-diameter well, constructed by a drilling rig, is often termed a 'borehole', or in developing countries a 'tube well'.

A well will have a well screen (a perforated or slotted section that allows water to enter from certain strata) and a well casing (which, conversely, prevents water entering from certain strata); see Figure 5.4.

In dewatering terminology, wellpoints (Chapter 15), deep wells (Chapter 16) and ejectors (Section 17.2) are all types of wells, categorized by their method of pumping. A 'sump' (Chapter 14) is not considered a well, since it is effectively a more or less crude pit to allow the collection of water.

A 'slot' is a linear feature that is used to pump groundwater – a trench drain being an obvious example. An important point is that the geometry of groundwater flow to a slot is notably different from flow to wells, and this affects the methods of analysis that should be used. Flow to slots is of interest to dewatering designers, because it is sometimes useful to simplify a line of closely spaced wells into a slot.

5.8.1 Planar Flow to Slots

The geometry of a long linear slot means that when it is pumped, groundwater flow lines to the slot are parallel to each other and perpendicular to the slot (Figure 5.5). These parallel flow lines are an example of 'planar' or 'plane' flow.

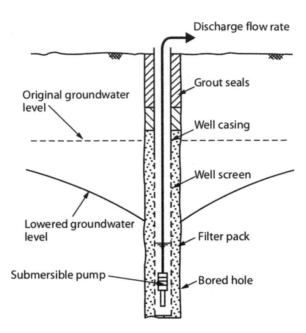

Figure 5.4 Principal components of a groundwater lowering well.

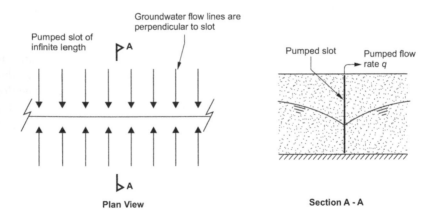

Figure 5.5 Planar flow to a long linear slot.

A drawdown curve is generated around the slot as a result of pumping; in three dimensions around the well, the drawdown curve describes a 'trough'-like depression in the groundwater level, with the slot in the centre of the trough (Figure 5.5). The limit of the drawdown defines the 'zone of influence'. If the slot is of finite length, as well as planar flow conditions in the main part of the slot, there may be an element of radial flow towards the two ends of the slot (Figure 5.12c).

5.8.2 Radial Flow to Wells

One of the defining features of flow through an aquifer towards a well is that flow converges radially as it passes through an ever smaller cross-sectional area of aquifer, resulting in a corresponding increase in flow velocity as the well is approached (Figure 5.6).

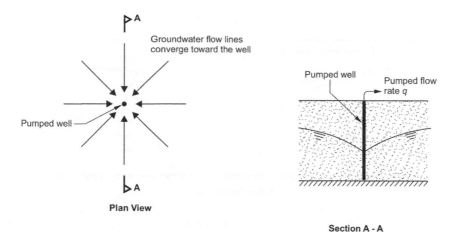

Plan View

Section A - A

Figure 5.6 Radial flow to a well.

A drawdown curve is generated around the well as a result of pumping; in three dimensions, the drawdown curve describes an inverted conic shape known as the 'cone of depression' with the well in the lowest point at the centre of the inverted cone (Figure 5.6). The limit of the cone of depression defines the 'zone of influence'.

The effect of converging flow is compounded in unconfined aquifers because the drawdown at the well effectively reduces the aquifer thickness at the well (Figure 5.7), further reducing the area available for groundwater flow. This is one of the reasons why flow rate equations for unconfined aquifers are more complex than for confined aquifers.

Convergence of flow, and the associated high groundwater velocities, leads to the phenomenon of 'seepage face' and 'well loss', where the water level inside a pumped well may be significantly lower than in the aquifer immediately outside.

The regime around a well in an unconfined aquifer is sometimes simplified by the use of the 'Dupuit approximation', whereby groundwater flow is assumed to be horizontal (i.e. any vertical groundwater flow is neglected). This is illustrated in Figure 5.8, which shows that the true flow regime has curved equipotentials close to the well, where there is some downward groundwater flow. The simplified Dupuit regime has vertical equipotentials.

The Dupuit approximation is useful in that it allows the flow to a well to be analysed using relatively straightforward analytical solutions such as the Dupuit–Forchheimer equation

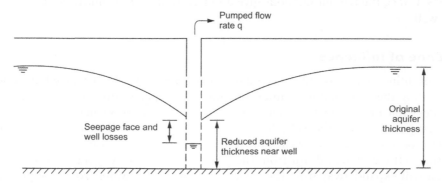

Figure 5.7 Reduction of effective aquifer thickness close to a pumped well in an unconfined aquifer. Seepage face and well losses mean that the water level in the pumped well may be lower than in the aquifer immediately outside the well.

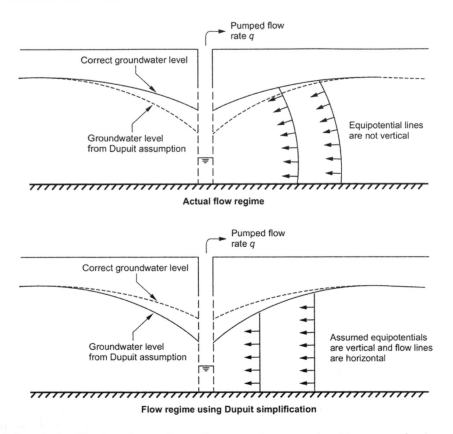

Figure 5.8 Dupuit simplification of groundwater flow towards a pumped well in an unconfined aquifer.

(Equation 13.5). While the Dupuit flow regime is not a perfect representation of the actual flow regime, studies such as Hantush (1962) have shown that the Dupuit–Forchheimer equation can be used to provide accurate predictions of the flow rate to wells. However, larger errors occur when Dupuit methods are used to predict the groundwater level forming the drawdown curve around a well. Close to the well, at radial distances r of less than $1.5H$ (where H is the initial saturated thickness of the aquifer), Dupuit methods will overestimate drawdown (i.e. predict a lower groundwater level than will actually occur). At greater distances ($r > 1.5H$), the Dupuit methods give good estimates of groundwater levels around a pumped well.

5.8.3 Zone of Influence

The zone of influence is a useful conceptual simplification to help visualize how a well or a slot is affecting the surrounding aquifer. Imagine a well penetrating an aquifer that has an initial water table or piezometric level at the same elevation everywhere (Figure 5.9). When water is first pumped from the well, the water level in the well will be lowered, and flow will occur from the aquifer into the well. This water will be released from storage in the aquifer around the well (i.e. it will drain from the pores in the soil). As time passes, the cone of depression will expand away from the well, releasing further water from storage. The zone of influence will continue to increase with time, but at a diminishing rate, until either an aquifer boundary is reached or the infiltration into the aquifer (from surface recharge and/or leakage from overlying or underlying aquitards) within the zone of influence is sufficient to

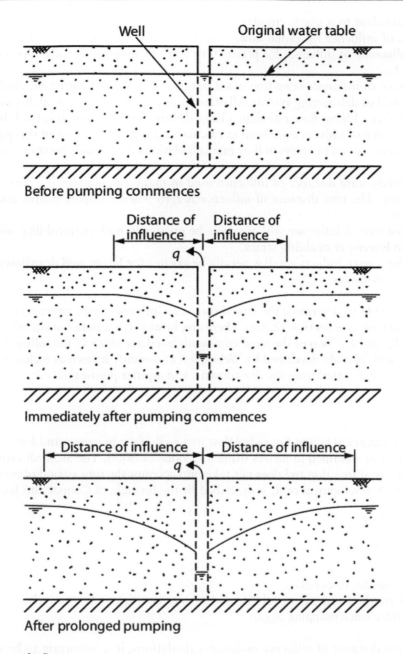

Figure 5.9 Zone of influence.

supply the yield from the well. An idealized zone of influence is perfectly circular. The size of the zone is described by the 'distance of influence' or 'radius of influence', which is the distance from the centre of the well to the outside edge of the zone where drawdown occurs.

Even if no source of recharge is encountered by the expanding zone of influence, after some time of pumping it will be expanding so slowly that it is effectively constant and is said to have reached a quasi-steady state. This is the approach assumed in several of the design methods described in Chapter 13. This simplifies all the sources of water (whether from water stored in the aquifer, surface infiltration, leakage from aquitards or external recharge) as being equivalent to a single circular source of water at distance R_o from the well. R_o is the distance of influence for radial flow, sometimes known as the 'radius of influence'. The equivalent distance of influence for planar flow (e.g. for flow to line of wellpoints) is given the symbol L_o.

The distance of influence is an important parameter when estimating the discharge flow rate from a well or dewatering system. All other things being equal, a small distance of influence will predict a higher flow rate than a large distance of influence. It is vital that realistic values of distance of influence are used in calculations – poor selection of this parameter is one of the prime causes of errors in flow rate calculations. Important points to consider are:

(i) The steady-state distance of influence used in many calculations is a theoretical concept only. The true distance of influence is zero when pumping begins and increases with time.

(ii) The distance of influence will generally be greater in high-permeability soil and rock than in lower-permeability strata.

(iii) The distance of influence will generally be greater for larger well drawdowns than for small drawdowns.

It may be possible to determine the true distance of influence from appropriately analysed well pumping tests (see Section 12.8.5). There are, however, a number of equations available to estimate R_o and L_o. Two of the most commonly used are given in the following.

The empirical formula developed by Weber, but commonly known as Sichardt's formula, allows R_o to be estimated from the drawdown s and aquifer permeability k:

$$R_0 = Cs\sqrt{k} \qquad (5.7)$$

where C is an empirical factor. For radial flow to a well, if s is in metres and k is in metres per second, R_o can be obtained in metres using a C value of 3000. The Sichardt estimate of R_o is a quasi-steady-state value and does not take into account the time period of pumping. The time-dependent development of R_o is described by the formula of Cooper and Jacob (1946):

$$R_o = \sqrt{\frac{2.25kDt}{S}} \qquad (5.8)$$

where
  D is the aquifer thickness
  k and S are the aquifer permeability and storage coefficient, respectively,
  t is the time since pumping began

When using distance of influence in design calculations, it is important to be vigilant for unrealistic values of R_o and L_o, especially very small or very large values. In the authors' experience, values of less than around 30 m or more than 5000 m are rare and should be viewed with caution. It may be appropriate to carry out sensitivity analyses using a range of distance of influence values to see the effect on calculated flow rates.

Another problem can occur if the system is designed using the long-term R_o or L_o. This will predict a much lower flow rate than may be generated during the initial period of pumping, when R_o is small as the cone of depression is expanding. It may be appropriate to design for a smaller, short-term R_o, or to design using the long-term R_o, but provide spare pumping capacity, over and above that predicted, to deal with the higher flows during the initial drawdown period.

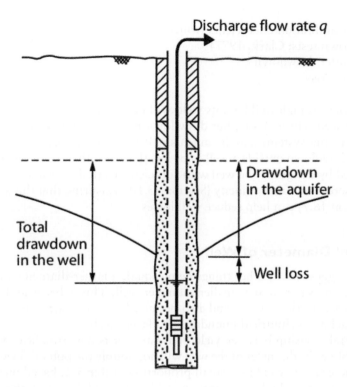

Figure 5.10 Well losses.

5.8.4 Well Losses

In general, the water level inside the screen or casing of a pumped well will be lower than in the aquifer immediately outside the well (Figure 5.7). This difference in level is known as the 'well loss' and results from the energy lost as the flow of groundwater converges towards the well. The drawdown observed in a pumped well has two components (Figure 5.10):

(i) The drawdown in the aquifer. This is the drawdown resulting from laminar (smooth) flow in the aquifer and is sometimes known as the aquifer loss. This component is normally assumed to be proportional to flow rate q.

(ii) The well loss. This is the drawdown resulting from resistance to turbulent flow in the aquifer immediately outside the well and through the well screen and filter pack. This component is normally assumed to be proportional to q^n and so increases rapidly as the flow velocity increases.

Well loss means that the water levels observed in a pumped well are unlikely to be representative of the aquifer in the vicinity of the well. Drawdowns estimated from pumped wells will tend to overestimate drawdowns. It is preferable to have dedicated monitoring wells, or at least to observe drawdown in some of the wells that are not being pumped.

Jacob (1946) suggested that in many cases $n = 2$. This is now widely accepted and allows the drawdown s_w in a well to be expressed as

$$s_w = Bq + Cq^2 \tag{5.9}$$

where

B and C are calibration coefficients (which can be determined from the results of step drawdown tests; Clark, 1977)

Bq is the aquifer drawdown

Cq^2 is the well loss

If a well experiences high well loss, its yield will be limited. At high flow rates, the drawdown inside the well will be large, but drawdown in the aquifer (which is the aim of any groundwater lowering system) may be much smaller. In poorly performing wells, it is not unusual for the drawdown outside a well to be less than half that inside the well. Well losses can be minimized by designing the well with sufficient 'wetted screen length' (Section 13.9) and with a low screen entrance velocity (Section 13.10). Ensuring that the well is adequately developed (Section 16.7) can help reduce well losses.

5.8.5 Effect of Diameter of Well

Common sense suggests that, other things being equal, a larger-diameter well will have the ability to yield more water than a smaller-diameter well. This is because the well will have more contact area with the aquifer, and also because less convergence of the flowing water is required to reach the cylindrical boundary of a larger well.

However, the relationship between yield and diameter is not straightforward. It is worth stating that doubling the diameter of the well will not double the potential yield from the well and may only increase the yield by a small proportion. Table 5.2, based on work by Ineson (1959), shows the estimated relation between yield and bored diameter (the drilled diameter through the aquifer, not the diameter of the well screen). This indicates, for example, that in a homogeneous aquifer, increasing the well diameter by 50 per cent from approximately 200 to approximately 300 mm will increase the yield by only 11 per cent. The increase in yield is more marked in fractured aquifers, because a larger-diameter well has an increased chance of intercepting water-bearing fractures. The true relationship may be more complex than this, especially for wells with high losses. In those cases, the increase in diameter, which will reduce groundwater flow velocity, may help reduce well losses, and attainable well yield may increase rather more than shown in Table 5.2.

In practice, the economics of drilling and lining wells means that temporary works groundwater lowering wells are rarely constructed at diameters of greater than 300–450 mm. Well diameters are often chosen primarily to ensure that the proposed size of electro-submersible pump can be installed in the well (see Section 16.3).

5.8.6 Equivalent Wells and Lines of Wells

The foregoing discussions have concentrated on a single pumped well in isolation. In reality, most groundwater lowering is carried out using several closely spaced pumped wells (be they wellpoints, deep wells or ejectors) *acting in concert*. With this approach, the cone of

Table 5.2 Effect of Well Diameter on Yield

Bored diameter of well through aquifer (mm)	203	305	406	457	610
Homogeneous aquifer (intergranular flow)	1.00	1.11	1.21	1.23	1.32
Fractured aquifer	1.00	1.29	1.52	1.61	1.84

Based on data from Ineson, J, *Proceedings of the Institution of Civil Engineers*, 13, 299–316, 1959.

depression of each well overlaps (or interferes), creating additional drawdown. This effect is known as the superposition of drawdown (see Section 13.8.2) and can create groundwater lowering over wide areas, which is ideal to allow excavations to be made below original groundwater levels (Figure 5.11).

Wells are normally installed either as rings around an excavation (Figure 9.5a) or as lines alongside one or more sides of the dig (Figure 9.5b). It is theoretically possible to consider the influence of each well individually and determine the complex interaction between the wells, but this can be tedious, especially when large numbers of wells are involved. A more practical, and empirically very effective, approach is to consider a dewatering system of many wells to act on a gross scale as a large equivalent well or slot.

Circular or rectangular rings of wells can be thought of as large equivalent wells (Figure 5.12a) to which flow is, on a gross scale, radial. Similarly, lines of closely spaced wells (such as wellpoints alongside a trench) can be thought of as equivalent slots (Figure 5.12b) to which flow is predominantly planar to the sides. In some geometries, both planar and radial flow can occur (Figure 5.12c).

Once this conceptual leap is made, the well and slot formula given in Chapter 13 can be used to model the overall behaviour of dewatering systems without having to consider each well individually. Formulae for estimating the size of an equivalent well from the dimensions of a dewatering system are given in Section 13.6.1.

5.9 MODELLING GROUNDWATER CONTROL TECHNOLOGY

The obvious element of groundwater models is the 'natural' element – the ground and groundwater conditions. But for models used as part of dewatering design, there will also be an 'artificial' element – groundwater control technology such as wells, sumps and groundwater barriers or cut-off walls. A common limitation of groundwater models is that dewatering technology is not modelled appropriately. Some examples of modelling errors include:

- Modelling a greater drawdown in the dewatering wells than can be achieved by the selected technology. An example is where dewatering pumping is planned to use wellpoint methods, where the practical limit of drawdown that can be achieved is approximately 4 to 6 m below the level of the pump. If this limitation is not included in the model, then the outcomes will not be representative.

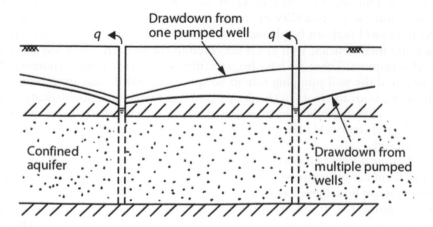

Figure 5.11 Superposition of drawdown.

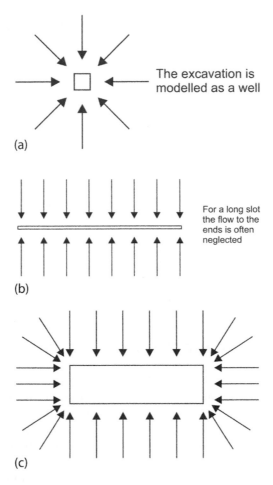

The excavation is modelled as a well

(a)

For a long slot the flow to the ends is often neglected

(b)

(c)

Figure 5.12 Planar and radial flow. (a) Radial flow to an equivalent well. (b) Planar flow to a long slot. (c) Combined planar and radial flow to a rectangular excavation.

- Modelling unrealistic dewatering well yields. As discussed in Section 13.9, the yield (maximum pumping rate) of individual wells will be controlled by the geometry of the well and the permeability of the surrounding soil and rock. A given well has a maximum yield that can be physically achieved. Furthermore, in practice, due to well losses and inefficiencies, well yields observed in the field are often less than the theoretical maximum. There are have been examples where the input to numerical models has required the well pumping rate to be specified, and the use of unrepresentatively high values leads to errors in the design outputs. Some more sophisticated numerical models are able to predict the pumping rates for individual wells, but it is important that the predicted yields are validated against estimates from other sources (such as Section 13.9).
- Modelling unrealistic properties for groundwater cut-off walls. Each of the methods used to form cut-off walls or zones of treated ground has lower limits on the permeability that can be achieved as well as limits on maximum depths and thicknesses that can be constructed. If cut-off walls are modelled, it is important that the assumed geometry and properties of the wall can actually be constructed by the proposed techniques.

It is important that the key performance limitations of various dewatering and cut-off technologies are understood when developing models. Information on pumped groundwater control methods can be found in Chapters 14 to 17, and for groundwater exclusion methods in Chapter 18.

5.10 SELECTING AND APPLYING AN APPROPRIATE MODEL

This chapter has highlighted the various types of groundwater model available, including the commonly used models for groundwater flow to wells and slots. It is clear that there are several different approaches that can be applied to the modelling of groundwater systems. The challenge for the dewatering designer is to select an appropriate approach and to use representative concepts, parameters and boundary conditions. Furthermore, modelling is not 'design'; it is one of the tools we use in dewatering design. Dewatering design is discussed in detail in Chapter 13.

Chapter 6

Objectives of Groundwater Control

Below ground 'water is the root of all evil'

Sir Harold Harding

6.1 INTRODUCTION

To avoid troublesome conditions during excavation and construction, measures must be taken to control groundwater flows and pore water pressures in water-bearing soils and rocks. Surface water run-off must also be effectively managed. An understanding of how an excavation may be affected will assist in assessing which groundwater control measures are necessary to ensure stability.

This chapter discusses the objectives of groundwater control and introduces the two principal approaches – pumping and exclusion – that can be used. The potential for inadequately controlled groundwater to cause unstable ground conditions is discussed. The interaction between pore water pressures, effective stress and stability is introduced, and the problems associated with surface water in excavations are also discussed. Particular groundwater problems associated with excavations in soil and in rock are addressed in more detail in Chapters 7 and 8, respectively.

6.2 GROUNDWATER CONTROL – THE OBJECTIVES

The fundamental requirement of groundwater control measures is that they should ensure stable and workable conditions throughout, so that excavation and construction can take place economically and under safe conditions at all times.

Groundwater may be controlled by exclusion methods; one or more types of pumping systems commonly termed 'dewatering'; or a combination of pumping plus exclusion techniques. This book is primarily concerned with groundwater control by pumping or dewatering, although techniques used to form cut-off barriers for groundwater exclusion schemes are discussed in Chapter 18.

A correctly designed, installed and operated dewatering system ensures that construction work can be executed safely and economically by:

(i) Local lowering of groundwater levels and interception of any seepages due to perched water tables or residual water that might otherwise emerge on the exposed slopes or in the base of the excavation(s).

(ii) Improving the stability of the excavation slopes and preventing material being removed from them by erosion due to seepage. Effective control of groundwater may allow slopes to be steepened and the area of excavation reduced.

(iii) Preventing hydraulic failure of the base of an excavation. Base failure can take several forms, including heave, blow-outs or the development of 'running sand' or 'quick' conditions in the floor of the excavation. All forms of base failure will have detrimental effects on the bearing capacity at formation level.

(iv) Draining the soil or rock to improve excavation, haulage, trafficking and other characteristics of the material involved. This may allow use of the excavated spoil for some purposes that would not be possible if it was dug in a wet 'undewatered' condition.

(v) Reducing the lateral loads on temporary support systems such as steel sheet-piling, concrete diaphragm walls, etc.

The practical upshot of objective (i) is that by lowering groundwater levels and controlling any local seepages, the excavation will not be flooded by groundwater, and might be said to be 'dewatered'. This is perhaps the most obvious aim of groundwater control – prevention of flooding – but objectives (ii) to (v) can be equally important in improving the stability of excavations. This is discussed further in the following sections and in Chapter 7 (for soils) and Chapter 8 (for rock).

The reader may note that the word 'dry' does not appear anywhere in the objectives of groundwater control outlined earlier. The authors are not in favour of any specifications or contracts stating that the aim of dewatering is to provide 'dry' conditions. After all, a heavy rain shower may cause 'wet' conditions irrespective of how well groundwater is controlled! A much more useful target is for the dewatering measures to provide 'workable conditions'.

6.3 GROUNDWATER CONTROL AS PART OF TEMPORARY WORKS

The focus of this book is for groundwater control to provide suitable working conditions in excavations and tunnels during the construction phase of an engineering project. As such, the groundwater control measures will usually form part of the 'temporary works'.

On a construction project, the 'permanent works' are those things that are being constructed to remain in place and be used for a long time. Focusing specifically on the below-ground elements, on a building project, the permanent works would be the basement sub-structure; on a tunnel project, the permanent works would be the final, lined tunnels and shafts, and so on. Almost all construction projects require some form of 'temporary works' whose role is to aid or facilitate the safe and economic construction of the permanent works.

Groundwater control is one type of temporary works. There are many other types, including access scaffolds, excavation propping and shoring, falsework and formwork for concrete construction, etc. Often the temporary works are removed or discontinued when they are no longer needed, or in some cases they can be incorporated into the permanent works. Further details on a wide range of temporary works can be found in Pallett and Filip (2018).

Methods based on groundwater control by pumping are truly temporary (since groundwater levels will recover when pumping is stopped) and rarely form part of the permanent works. Occasionally, groundwater control systems are intended for long-term or 'permanent' use; this is discussed in Chapter 20. Groundwater exclusion methods lend themselves more readily to incorporation into permanent works – steel sheet-piles, concrete diaphragm walls and bored pile walls sometimes serve a dual temporary works and permanent works role.

The fundamental point about temporary works is that it is essential that the same efforts of design, investigation and construction supervision are applied to temporary works as for permanent works. As temporary works are needed for only a limited time and have no visible role in the final constructed works, there is a risk that some in the project team may assume they are less important. This is a false and dangerous assumption. Failure or ineffectiveness of temporary works can place the workforce and the public at risk and can cause delays and cost overruns for a project. This applies as much to groundwater control systems as it does to more visible elements of temporary works such as excavation propping systems or scaffold towers.

6.4 GROUNDWATER MANAGEMENT IN EXCAVATIONS

Groundwater management for an excavation should address all the objectives listed in Section 6.2. However, in many cases, water management focuses on objective (i), essentially to 'dewater' the excavation and remove any water that enters, without giving due consideration to the destabilizing effect of groundwater seepage into the excavation.

This direct water management and removal approach is essentially the sump pumping technique described in Chapter 14. This can be effective if used where the strata around the excavations are not easily destabilized or otherwise affected by groundwater inflow. Examples where this approach can work include hard fractured rocks or coarse gravelly soils with little sand content. However, there are many types of soil and soft rock where trying to directly manage water inflow can cause serious problems. Figure 6.1 shows a 360° excavator sunk into the base of an excavation in a fine sandy soil. Note that the excavation was not 'flooded' by groundwater. A simple pumping system was able to keep water levels down at excavation formation level. The problem was that the groundwater seepage destabilized the base of the excavation and meant that the bearing capacity of the soil became too low to support the weight of the excavator. This situation could perhaps have been avoided if a system of pumped wells had been used to lower groundwater levels in advance of the excavation. This would have reduced pore water pressures and controlled levels of effective stress. As will be described in the following section, controlling effective stress is key to ensuring the stability of excavations below groundwater level.

Figure 6.1 A case of groundwater management that could have been done better. (Courtesy of Ferrer SL, Museros, Spain.)

6.5 GROUNDWATER, EFFECTIVE STRESS AND INSTABILITY

The concept of effective stress is fundamental to understanding the interrelation between groundwater and the stability of excavations in soil and rock. The principle of effective stress was proposed by Karl Terzaghi in the 1920s and is described in detail in soil mechanics texts such as Powrie (2013).

Consider first the case of soils, where the flow of groundwater is essentially intergranular (see Section 4.3). Soil has a skeletal structure of solid material (mineral grains) with an interconnecting system of pores. The pores may be saturated (wholly filled with water), as is the case in confined aquifers or below the water table in unconfined aquifers. Above the water table, the pores may be unsaturated (filled with a mixture of water and air).

As a saturated soil mass is loaded, the total stress is carried by both the soil skeleton and the pore water. The pore pressure acts with equal intensity in all directions. The stress carried by the soil skeleton alone (by interparticle friction) is thus the difference between the total applied stress σ and the water pressure set up in the pores u. This is termed the effective stress σ' and is expressed by Terzaghi's equation

$$\sigma' = \sigma - u \tag{6.1}$$

Because water has no significant shear strength, the soil skeleton may deform while the pore water is displaced (a process known as pore water pressure dissipation). This action continues until the resistance of the soil structure is in equilibrium with the external forces. The rate of dissipation of the pore water will be dependent on the permeability of the soil mass and the physical drainage conditions.

The ability of a soil to resist shear stress τ is primarily due to interparticle friction and so is dependent on the effective stress. The shear stress at failure τ_f can be expressed by the Mohr–Coulomb failure criterion:

$$\tau_f = \sigma' \tan \varphi' \tag{6.2}$$

where ϕ' is the angle of shearing resistance of the soil.

These two simple equations show that reducing pore water pressures (as a result of drawing down the groundwater level) will increase the effective stress within the soil in the area affected. This produces a corresponding increase in the shear strength of the soil, which will improve the stability of the soils around and beneath an excavation.

A useful idealized example is to imagine a seepage apparatus similar to that used to illustrate Darcy's law. Figure 6.2 shows a test cell where upward groundwater flow is induced through a sample of sand by application of an excess head to the base of the sample. A load (weight) is placed on the upper surface of the sand. When the applied excess head Δh is small (Figure 6.2a), water will flow upwards and emerge around the load, but the sand will be stable and will bear the load. However, if the excess head Δh is increased significantly, the sand will be much less stable, and the top may 'boil', water and sand may be washed away and the load will sink (just like the excavator in Figure 6.1). This is easy to explain in effective stress terms – the strong upward seepage means there are relatively high pore water pressures. By reference to Equation 6.1, we can see that immediately below the surface of the sand the total stress will be low, and when the pore water pressures are high, effective stress becomes very low, even zero. Equation 6.2 then shows that if effective stresses are low, the soil shear strength will be very low – the soil will lose its strength, and the load will sink. In extreme cases, the soil can lose almost

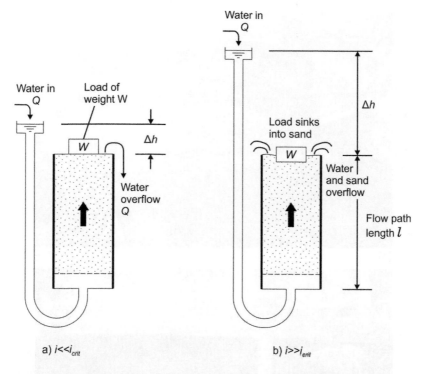

a) $i << i_{crit}$ b) $i >> i_{crit}$

Figure 6.2 The effect of pore water pressure on the stability of soils. Upward water flow is induced in the test cell. In case (a), the upward hydraulic gradient i is lower than the critical hydraulic gradient i_{crit}, and the sand in the test cell is stable and can bear the weight of the load W. In case (b), the upward hydraulic gradient i is higher than the critical hydraulic gradient i_{crit}, and the sand in the test cell is unstable and becomes fluid. Water and sand will 'boil' over the top of the cell and cannot bear the weight of the load W, which sinks into the sand. (Critical hydraulic gradients are discussed in Section 7.5.2.)

all its strength and become fluid – this is the so-called 'quick' condition, also known as quicksand or 'running sand'. This phenomenon is discussed in terms of critical hydraulic gradients in Section 7.5.2.

A practical example of the effect of reduction of pore water pressures on the stability of soil is shown in Figure 6.3. Groundwater problems for excavations in soils are discussed in more detail in Chapter 7.

Consider now excavations in rock, where groundwater flow is predominantly along fractures (see Section 8.3). Unlike soil, where groundwater is within the interconnected pore space of the soil skeleton, in rock, the mobile groundwater (i.e. that which can easily flow) is typically contained in discrete fractures. When pore water pressures are changed (for example by a groundwater pumping system), the principal change will be seen in saturated fractures. High pore water pressures in a fracture reduce the frictional forces along the joint. This can be visualized as the water pressures acting to force apart the two sides of a fracture. The reduced friction can increase the risk of displacement along a fracture when the rock is stressed by excavation or tunnelling. If the geometry of a fracture (or a network of fractures) allows movement of one or more blocks of rock, then a 'failure' can occur, where rock movements lead to the collapse of a slope or tunnel (see Section 8.4).

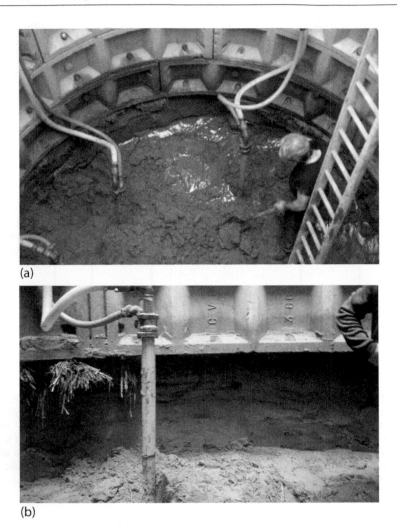

(a)

(b)

Figure 6.3 The effect of pore water pressure reduction on the stability of soils. (a) Unstable conditions before pumping: A small diameter (4.6 m) shaft was being sunk by underpinning methods in a stratum of silty fine sand. When the excavation reached groundwater level, the base of the excavation became wet, soft and unworkable – the wet area was too soft to stand on. It was not possible to progress the shaft any deeper, because the cut faces slumped in when underpinning was attempted. A series of ejector wells were installed in the base of the shaft as shown here (at the time of the photograph, the ejector wells had not started pumping). (b) Stable conditions during pumping: Pumping from the ejector wells successfully lowered the groundwater level at the shaft location, thereby reducing pore water pressures and increasing effective stresses. This transformed the behaviour of the sand, allowing stable vertical faces to the cut for underpinning, as shown here.

In a manner analogous to soils, reducing groundwater pressures in rock will improve stability. This is achieved by reducing water pressures in the fractures, which increases shear resistance along the fractures. The mathematics of effective stress in rock is discussed in Gray (2017). Groundwater problems for excavations in rock are discussed in more detail in Chapter 8.

The increase in effective stresses that results from groundwater lowering is also an important factor in the potential for ground settlements around dewatering systems. This is discussed in Section 21.4.4.

6.6 GROUNDWATER BEHAVIOUR IN HIGH- AND LOW-PERMEABILITY SOILS AND ROCKS

The design and analysis of groundwater control systems require an understanding of the factors driving groundwater flow, as outlined in Chapters 3 to 5. We have also seen, when effective stress theory was introduced in Section 6.5, that groundwater pressures influence the available shear strength of soil – this is the basis of soil mechanics analysis. Karl Terzaghi, widely considered the father of modern soil mechanics, said: 'In engineering practice difficulties with soils are almost exclusively due not to the soils themselves but to the water contained in their voids. On a planet without any water there would be no need for soil mechanics' Terzaghi (1939).

A detailed knowledge of all aspects of soil mechanics is not necessary for the successful implementation of groundwater control measures and is not the purpose of this book. The interested reader wanting to learn more is directed to accessible textbooks on soil mechanics, such as Powrie (2013), and similar texts on rock mechanics, such as Hencher (2015). However, there is a particular aspect of soil behaviour that is worth discussing here – 'drained' versus 'undrained' behaviour – because of its profound effect on the temporary stability of excavations in low- and very low-permeability strata.

6.6.1 Drained and Undrained Conditions

It is useful to start with some definitions. Duncan *et al.* (2014) highlight that the soil mechanics terms 'drained' and 'undrained' should not be taken literally. In non-technical usage, drained would mean dry or emptied and undrained would mean not dry or not emptied. The soil mechanics usage of these terms actually relates to the transient pore water response to load changes. Duncan *et al.* (2014) give the following useful definitions:

- Drained is the condition under which water is able to flow into or out of a mass of soil as rapidly as the soil is loaded or unloaded. Under drained conditions, changes in load do not cause changes in pore water pressure within the soil.
- Undrained is the condition under which there is no flow of water into or out of a mass of soil in response to load changes. Under undrained conditions, changes in load cause changes in the pore water pressure, because the water is unable to move in or out of the soil as rapidly as the soil is loaded or unloaded. The changes in pore water pressure, relative to original groundwater conditions, are termed 'excess' pore water pressures.

When dealing with groundwater control problems, our main interest is in excavations dug below groundwater level, such as that shown in Figure 6.4. Consider point I (at depth Z below original ground level), located immediately below the excavation. Before the excavation is dug (Figure 6.4a), the pore water pressure u will be

$$u = \gamma_w \left(Z - H_w \right) \tag{6.3}$$

where
  H_w is the depth to groundwater level
  γ_w is the unit weight of water

The vertical effective stress σ'_v is the vertical total stress σ_v minus the pore water pressure u:

$$\sigma'_v = \sigma_v - u \tag{6.4}$$

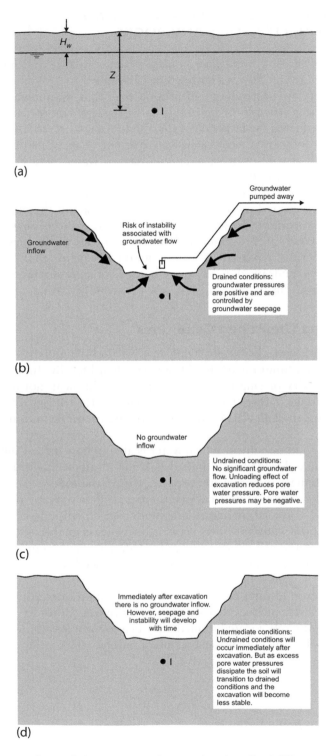

Figure 6.4 Different groundwater behaviour around excavations in soils of different permeability. (a) Initial conditions. (b) Excavation in high-permeability strata. (c) Excavation in very low-permeability strata. (d) Excavation in strata of intermediate permeability.

which for hydrostatic groundwater conditions can be expressed as

$$\sigma'_v = \gamma_s Z - \gamma_w (Z - H_w) \tag{6.5}$$

where γ_s is the unit weight of soil.

Consider now the pore water pressure change that will occur in response to the unloading of the soil due to excavation. The speed of unloading is related to the rate of excavation. For small excavations (or where larger excavations are dug in several cells or sections), rapid deepening is possible, and excavation rates of approximately 2 to 10 m vertically per day are possible. Assuming the unit weight of the soil is 20 kN/m³, this gives an unloading rate of 40 to 200 kPa/day. If excavations of large plan area are dug in a single operation across the whole area, the logistical constraints of digging and removing spoil will often slow rates of excavation to around 0.5 to 2 m/day – this is an unloading rate of approximately 5 to 40 kPa/day. As we will see, the unloading effect will cause a different pore water response in a highly permeable soil compared with a soil of much lower permeability.

6.6.2 Drained Pore Water Pressure Responses during Excavation

Drained responses during excavation typically apply to soils of relatively high permeability that might be termed 'water-bearing'. To examine this case in soil mechanics terms, consider Figure 6.4b, which shows an excavation in a sandy gravel, where permeability will be in the order of 10^{-4} m/s. In this case, groundwater in the soil can flow rapidly compared with the rate of excavation and unloading – therefore, the soil will act in a drained manner, and water will almost immediately flow into the excavation. The unloading effect will not generate any excess pore water pressure, and the groundwater regime will be controlled by the seepage into the excavation.

In Figure 6.4b, the simple expedient of sump pumping is used to remove water from the excavation, and water will seep upwards into the base of the excavation. This means that at point I, there will be a pore water pressure greater than zero (i.e. a water pressure greater than atmospheric) associated with the seepage. However, point I will also have experienced considerable unloading due to excavation of the depth d of soil. This will reduce the vertical total stress σ'_v by γ_s times d. Effective stress theory (Section 6.5) shows that by decreasing the total stress in a zone of relatively high pore water pressures, effective stresses will reduce to low levels (which could reach zero). Low effective stresses reduce the shear strength that the soil can mobilize, and can lead to the excavation base and side slopes becoming unstable. Such instability is sometimes known as 'quick' conditions or 'running sand'. In water-bearing soils, this type of instability can be avoided by deploying pre-drainage groundwater control methods to lower groundwater levels (Section 9.5.1). This will reduce pore water pressures, preventing effective stress falling to very low levels and hence avoiding instability.

6.6.3 Undrained Pore Water Pressure Responses during Excavation

Drained conditions in highly permeable water-bearing soils are easy to understand. Essentially, groundwater will flow into an excavation at the same time as the excavation is dug. In contrast, undrained conditions that can occur when excavating in very low-permeability soils are more complex and often misunderstood. The unloading effect will generate excess pore water pressure, and at least in the short term, this will control groundwater conditions around the excavation.

Figure 6.4c shows an excavation made in a deposit of silty clay that does not contain any permeable fabric (e.g. laminations of silt or sand), where the permeability will be in the order of 10^{-9} m/s. In this case, groundwater flow will be very slow (due to the very low permeability of the soil), and water flow is unable to keep pace with the rate of unloading during excavation.

In perfectly undrained conditions, the rate of excavation is sufficiently rapid, relative to the soil permeability, that no groundwater flow occurs during excavation. In these conditions, effective stress as expressed in Equation 6.4 will not change, and for every kilopascal that the total stress is reduced by unloading, the pore water pressure will reduce below the pre-excavation values. The pore water pressure changes caused by unloading are termed excess pore water pressures, and in this case will be negative. At point I, the unloading caused by excavation of the depth d of soil will reduce the vertical total stress σ'_v by $\gamma_s d$, and (in the absence of groundwater flow) the pore water pressure u will reduce by a corresponding amount. If the unloading exceeds the pre-excavation pore water pressure at point I, then the pore water pressure will become negative (i.e. less than atmospheric pressure). Negative pore water pressures are known as 'soil suctions'.

Because effective stresses are not reduced, the shear strength that the soil can mobilize is unchanged in the short term. This can allow excavations to be made in which the sides of the excavation may be temporarily stable at relatively steep angles, and where the base of the excavation can resist some upward groundwater pressures from deeper strata. No water seepage will occur into the excavation in the short term, because the pore water pressures in the soil bounding the excavation are negative (in the fully undrained condition) due to the effect of unloading. However, from a safety and risk perspective, it is vital to understand that the undrained condition is temporary, and as time passes, the excavation will become less stable, as explained in the following.

The reduced pore water pressures will create hydraulic gradients towards the excavation, which will draw groundwater in from the surrounding soil. This will cause the negative excess pore water pressures to dissipate, and water pressures around the excavation will slowly increase until they come into equilibrium with the surrounding ground. Increased pore water pressures will reduce the effective stresses; instability of the excavation base and/or side slopes will follow if measures are not taken to artificially keep the pore water pressures low or to stabilize the excavation by other means.

In very low-permeability soils such as the silty clay in Figure 6.4c, excess pore water pressure dissipation may be very slow, occurring over many years or decades, but it is inevitable. As time passes, the soil will move towards a drained condition with reduced soil shear strength. It is now well established by soil mechanics analysis that cases of the failure of clay slopes for road and rail cuttings, occurring many decades after they were first formed, are typically due to the slopes slowly transitioning from undrained conditions to the less stable drained condition (Skempton, 1964).

6.6.4 Intermediate Conditions between Drained and Undrained Response

The physical principles of soil behaviour are the same for both drained and undrained responses – the difference is time. In the drained case, groundwater can move so quickly (relative to the rate of excavation) that there is no need to consider transient conditions, and equilibrium pore water pressures apply almost instantaneously. Conversely, in the perfectly undrained case, it is assumed that there is insufficient time for any drainage to occur, and the focus is on the excess pore water pressures generated by the unloading effects of excavation.

Drained and undrained conditions are two extremes. There are many cases where, during the timescales of a construction project, the soil around an excavation will be neither truly drained nor undrained, but its state will be in the continuum between the two extremes. Figure 6.4d shows an excavation made in a soil of relatively low permeability intermediate between the two preceding cases. This condition can occur in laminated silt and clay deposits that may contain sand partings, where the horizontal permeability is in the order of 10^{-7} or 10^{-8} m/s.

Immediately after excavation, the soil will exhibit an undrained response, with negative excess pore water pressures caused by unloading, and the excavation side slopes and base may be temporarily stable. However, in this range of permeability, the excess pore water pressures can potentially dissipate during the period while the excavation is open and the permanent works are being constructed. As the excess pore water pressures dissipate, the soil will lose its ability to resist shear. If an excavation has been designed assuming undrained conditions, and the permeability of the soil is high enough to allow some drainage, significant instability can occur days, weeks or even months after the excavation is first formed. Instability can include softening and slumping of side slopes, excessive pressures on retaining walls and shoring systems, and softening and heave of the base of the excavation.

6.6.5 Relevance of Drained and Undrained Conditions to Groundwater Control Problems

The vast majority of cases where groundwater control is required are in relatively permeable strata – such as sands, gravels and highly fractured rock. These are materials that an engineer would describe as 'water-bearing' and a hydrogeologist would classify as an aquifer. In such materials, it can be assumed that excess pore water pressures will not be generated by the unloading from excavations, and drained conditions will always apply.

For the specific case of excavations for civil engineering and mining projects, drained conditions are likely to apply for permeabilities greater than around 10^{-6} m/s. Note that for more rapid loading or unloading (for example the ground motion induced by an earthquake), the lower permeability limit given earlier, where purely drained conditions can be assumed, will be higher, and undrained conditions and resultant excess pore water pressures can occur in relatively permeable sands – this is one of the causal factors in earthquake-induced liquefaction (Madabhushi, 2007).

Undrained conditions are more relevant when we consider the lower limits of permeability at which dewatering is required. Consider again the excavation shown in Figure 6.4c, dug into a clay of permeability of around 10^{-9} m/s. Any experienced excavator operator will know that a trial pit dug into such clay will not need dewatering, and they would report the pit as 'dry', by which they mean that no water flowed in and no pumping was needed. It is possible that the operator may assume that the pit is dry because it is above groundwater level, but if we consider undrained behaviour, we can see that this need not be the case. In the United Kingdom, where the copious annual rainfall means that groundwater level is often close to ground level, pits dug into clay will be apparently 'dry' despite being below groundwater level. What is happening is that the unloading effect of excavation generates negative pore water pressures in the clay around the pit, so even though the clay is fully saturated with water, there is no hydraulic gradient into the pit, and there can be no water inflow until the excess pore water pressures dissipate significantly.

Undrained behaviour explains why groundwater control by pumping is not required for temporary excavations in very low-permeability soils and rocks, even many metres below groundwater level. Some engineers might describe such soils (erroneously unless the soil grains are bonded) as 'cohesive', and a hydrogeologist would classify these types of soil and rock as aquicludes. The key aspect is that the permeability should be sufficiently low that the excess pore water pressures (caused by the unloading effect of excavation) do not dissipate significantly during the construction period; this requirement is typically satisfied when the permeability is lower than 10^{-8} to 10^{-10} m/s.

Some of the most challenging groundwater control problems occur in soils of relatively low permeability that are not normally considered water-bearing, but where undrained conditions cannot be relied upon during the period for which the excavation is open. Immediately after excavation, there will be some negative excess pore water pressures, but these will dissipate, and problematic seepage and instability will follow. These conditions typically occur in soils where the permeability is in the range of 10^{-6} to 10^{-8} m/s. This permeability range is tentative. In fine-grained soils, structure and fabric have a great influence on the permeability, especially in the horizontal direction. If the soil structure consists of thin alternating layers or laminations of coarser and finer soils, groundwater flow will be more rapid, and the onset of drained conditions (and the increased risk of instability) will occur sooner after excavation. Such instability is often avoided by the use of pore water pressure control techniques such as ejector wells, vacuum wellpoints and deep wells with vacuum (Section 9.5.4).

6.7 POSSIBLE APPROACHES TO GROUNDWATER CONTROL

There are two principal approaches to groundwater control:

- Groundwater control by exclusion (Section 9.4), where low-permeability cut-off walls are used to exclude groundwater from the excavation (Figure 6.5). The cut-off wall may be formed from artificial elements introduced into the ground (e.g. a pile wall) or may involve a zone of ground that has been treated or modified to reduce its permeability (e.g. where grouting is used). The only groundwater pumping requirement will be to drain the water trapped within the soil or rock in the area enclosed by the cut-off, and to deal with leakages through the wall and through the impermeable stratum (pumping for surface water control may also be required; see Sections 6.8 and 9.2).

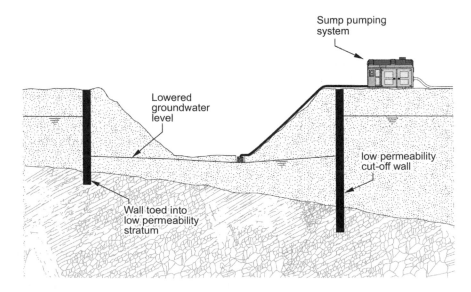

Figure 6.5 Groundwater control by exclusion. Cut-off walls toe into stratum of very low permeability.

- Groundwater control by pumping (Section 9.5), where groundwater is abstracted from an array of wells or sumps to temporarily lower groundwater levels (Figure 6.6). Groundwater control by pumping is also known as groundwater lowering, construction dewatering or simply dewatering.

Pumping and exclusion methods may be used in combination (Section 9.6).

6.8 SURFACE WATER PROBLEMS

This book is primarily concerned with the control of groundwater, but since the aim is to provide workable conditions for construction, the importance of surface water control must not be forgotten. Surface water may arise in many ways: as direct precipitation into the excavation; as precipitation run-off from surrounding areas; as leakage through cut-off walls used to exclude groundwater; as waste water from construction operations such as concreting; or even from washing down of plant and equipment. Since the excavation is obviously going to form a low point in the site, surface water, if given free rein, is likely to collect in the bottom of the excavation.

Whatever the source, if surface water is allowed to pond in the excavation, it will impede efficient excavation and construction. Figure 6.7 shows an example of an excavation before and after surface water was suitably dealt with. Neglecting the management of surface water will undoubtedly cost the project time and money. Methods for control of surface water are described in Section 9.2 and Section 14.3.

6.9 CHANGING GROUNDWATER CONDITIONS

It has been discussed in Chapter 3 that groundwater conditions are part of the hydrological cycle and therefore are dynamic and may change with time. It is important when considering the potential need for groundwater control that the possibility of 'extreme' conditions should be considered.

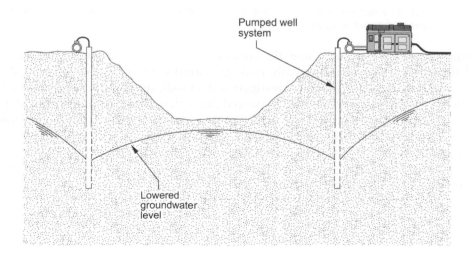

Figure 6.6 Groundwater control by pumping.

(a)

(b)

Figure 6.7 Control of surface water in excavations. (a) Without adequate surface water control. (b) With effective surface water control.

An important case is to assess how the groundwater control requirements would change should the site experience an unusually prolonged and/or intense period of rainfall. This includes the case where the ground investigation data indicate that groundwater levels are below the proposed excavation level. This would imply that the excavation will be 'dry' and groundwater control is not needed. It can be useful to review the local hydrogeology (including historic groundwater level data if available) to assess whether, following extended periods of wet weather, the excavation may be below groundwater level (which could be either a general water table or perched water). This issue is discussed in more detail in Section 3.9, and in relation to desk studies and monitoring in Chapters 11 and 22, respectively.

Chapter 7

Groundwater Problems for Excavations in Soils

> I leaped headlong into the Sea, and thereby have become more acquainted with the Soundings, the quicksands, and the rocks, than if I had stayed upon the green shore, and piped a silly pipe, and took tea and comfortable advice.
>
> John Keats, *Letter to James Augustus Hessey*, 1818

7.1 INTRODUCTION

Excavation and tunnelling below groundwater level present particular problems due to groundwater flows and pore water pressures in water-bearing soils. This chapter discusses the various large-scale and small-scale stability problems that can occur when working in soil, including instability of excavation side slopes and bases. The particular groundwater problems associated with tunnelling in soils are also discussed. The various methods of groundwater control available to stabilize excavations will be outlined in Chapter 9 for excavations and Chapter 10 for tunnels.

7.2 GROUNDWATER MANAGEMENT PROBLEMS IN SOILS

In the context of the design and implementation of groundwater control, the term 'soil' means uncemented deposits of mineral particles such as gravel, sand, silt and clay, where the flow of groundwater will be predominantly intergranular. This is distinct from rock, where groundwater flow is dominated by fracture flow (soils and rock are discussed in more detail in Sections 4.3 and 4.4, respectively).

It is important to realize that for geotechnical purposes, the distinction between soils and rocks is based on the behaviour of a stratum, not its origin or its age alone. In groundwater control design, materials would be classified as a geotechnical soil if groundwater flow is intergranular and if the material can become readily disturbed or eroded under the action of groundwater flow in an excavation. So for example, in the United Kingdom, the Thanet Sand Formation (Menkiti *et al.*, 2015) typically behaves as an uncemented sand and is analysed as a soil despite being part of the bedrock or 'solid' geology below London (the Thanet Sand Formation is more than 23 million years old). The underlying Chalk Group is analysed as rock because of the more cemented nature of the dominant limestones and marls; groundwater flow is mainly via fractures (Preene and Roberts, 2017).

Younger 'drift' or superficial deposits (often associated with the Quaternary period of geological time) are almost invariably analysed as soil. Found in many parts of the world, these Quaternary deposits are dominated by materials deposited under glacial and periglacial environments (Griffiths and Martin, 2017). Such glacial deposits have been the cause of groundwater problems on many engineering projects (see Case Histories 27.3 and 27.7).

The concept of groundwater management in excavations was introduced in Section 6.4 and is essentially the approach of 'dewatering' an excavation by removing the groundwater that enters without considering the destabilizing effect of groundwater seepage into the excavation. The problem with this simple approach when applied in soils is that even small flow rates of seepage (which can easily be pumped away by a small sump pump) can lead to instability of the sides and base of an excavation (Figure 7.1).

In a geotechnical context, soils are typically categorized in relation to the dominant particle sizes; formal schemes of visual description (Norbury, 2015) have been developed to allow consistent reporting of soil type by different observers. As was discussed in Section 4.3, the particle size distribution of a soil has a strong influence on permeability.

Different countries use slightly different sub-divisions of particle size for soil classification, but a common division is between coarse-grained and fine-grained soils:

- Coarse-grained soils are dominated by particles of sand size (2 to 0.06 mm) and gravel size (60 to 2 mm) and larger. Examples of this type of soil are sandy gravels and fine sands.
- Fine-grained soils mainly consist of particles of silt size (0.06 to 0.002 mm) and clay size (less than 0.002 mm). Examples of this soil type include sandy silts and silty clays.

Most groundwater control problems in soils occur within coarse-grained materials, where groundwater inflows will be significant, but which can be drained easily by pumping methods – such soils are colloquially termed 'water-bearing' and might have permeabilities from 10^{-2} m/s (for gravels and cobbles) down to around 10^{-6} m/s (for silty sands). In terms of soil mechanics terminology (Section 6.6), these soil types can be expected to act in a drained manner.

Figure 7.1 Slope instability resulting from seepage into an excavation, where the slope is formed of fine sand. Despite the rate of seepage being modest, and being pumped away by sump chambers at the toe of the slope, seepage through the slope causes instability and erosion of fine particles.

Fine-grained soils are more difficult to categorize in terms of groundwater control requirements:

- Soils with very high clay contents (such as silty clays) are often of very low permeability and only allow groundwater flow at very slow rates. Such soils typically exhibit behaviour that is considered 'undrained' in soil mechanics terms (Section 6.6) and are sometimes described as 'cohesive'. Permeabilities of such soils might be less than 10^{-8} to 10^{-10} m/s. Typically, groundwater control measures of the type described in this book are not required in such low-permeability soils.
- Soils with a greater proportion of silt and sand-sized particles (such as sandy silt) or where there are permeable sand laminations within a generally silty or clayey soil mass will typically have low permeability. These materials (where the permeability might be between 10^{-6} and 10^{-8} m/s) can present some of the most challenging groundwater control problems. Groundwater flow is restricted by the low permeability (which means that inflows to excavations and dewatering well yields will be very low), but pore water pressures can lead to instability of the sides and base of excavations. When an excavation is first made, it will be relatively stable (as the soil acts in an undrained manner), but it will become less stable with time as soil suctions dissipate and drained conditions develop (Section 6.6). Such instability is often avoided by the use of pore water pressure control techniques such as ejector wells, vacuum wellpoints and deep wells with vacuum (Section 9.5.4).

Because of the complex relationship between soil description and the potential for groundwater flow, information from permeability testing (Chapter 12) can be very valuable for the design of groundwater lowering systems.

7.3 LARGE-SCALE AND SMALL-SCALE INSTABILITY CAUSED BY GROUNDWATER

Inadequately controlled groundwater can cause stability problems on a large and a small scale via a variety of mechanisms; the problems likely to be prevalent at a given site and for a given excavation geometry will be largely controlled by ground conditions. These mechanisms can be explained in terms of effective stress and pore water pressures (introduced in Section 6.5).

Large-scale instability is discussed in Sections 7.4 and 7.5 in terms of two classic cases of excavation instability for excavations in soil: first, groundwater-induced slope instability in an excavation with battered side slopes dug into a sandy soil forming an unconfined aquifer; second, base instability of an excavation caused by the presence of uncontrolled groundwater. The mechanisms of potential instability will be quite different in each case.

Small-scale or localized groundwater problems are discussed in Section 7.6. Examples of this type of instability include slope instability and problems with perched water tables.

7.4 SLOPE INSTABILITY

Consider an excavation with sloping sides dug in a bed of silty fine sand below the standing groundwater level. If the inflow water is pumped from a sump within the excavation, the sides will slump in when a depth of about 0.5 to 1.0 m below the original standing water level is reached. As digging proceeds, the situation will get progressively worse, and the edges of the excavation will recede. The bottom will soon fill with a sand slurry in an almost liquid condition, which will be constantly renewed by material slumping from the side slopes. The collapse of the side slopes results from the presence of positive pore water and seepage pressures, which are developed in the ground by the flow of water to the pumping sump of the open excavation (Figure 7.2).

The mechanisms causing this unstable condition can be explained in terms of effective stress.

Above the water table, assuming that the soil is dry (i.e. has zero pore water pressure), then it can stand in stable slopes of up to ϕ' to the horizontal. In reality, the pore water

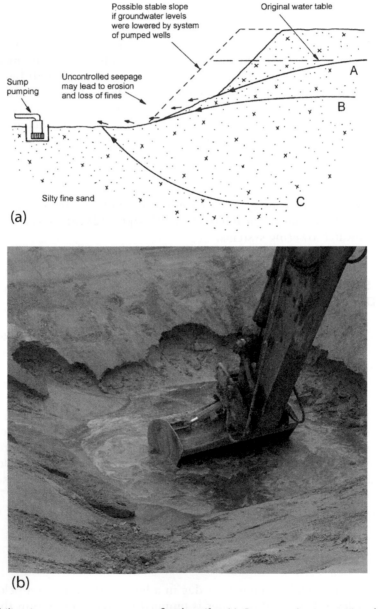

(a)

(b)

Figure 7.2 Instability due to seepage in an unconfined aquifer. (a) Conceptual cross section through an excavation being excavated below groundwater level in a fine sand. (b) Example of a small excavation in fine sand being excavated without groundwater control.

pressure above the water table may not be zero; in fine-grained soils such as silty sands, negative pore water pressures may exist temporarily due to capillary effects (Section 3.3.5). As a result, stable slopes even steeper than ϕ' may temporarily be possible. However, in time, the negative pore pressure will dissipate as the capillary water exposed on and near the face dries out. After some finite period of time (which may be no more than a few hours), the over-steepened slope will eventually crumble to that of the long-term stable slope angle.

Where seepage emerges from the slope, there will be positive pore water pressures. Positive pore water pressures will reduce effective stress, the shear strength of the soil will reduce in turn, and the soil will not stand at slopes as steep as in dry soils. This is the mechanism leading to the 'slumping' effect seen when excavating below the groundwater level.

Considering Figure 7.2a in detail, flow line 'A' represents the flow line formed by the water table or phreatic surface. Seepage into the excavation will cause a slight lowering of the water table, so flow line 'A' curves downwards and emerges almost parallel to the surface of the excavation. Immediately below its point of emergence, the positive pore water pressures generated by the seepage mean that the soil can no longer support a slope of ϕ'. Below the emergence of the seepage line, the soil slope will slump to form shallower angles. At the point of emergence of the almost horizontal flow line 'B', the sand will stand at $\frac{1}{2}\phi'$ or less. Where there is upward seepage into the excavation base (flow line 'C'), the effective stress may approach zero, and the soil cannot sustain any slope at all. In these circumstances, the soil may 'boil' or 'fluidize' and lose its ability to support anything placed on it – this is the so-called 'quicksand' case (see Section 6.5 and Section 7.5.2).

'Running sand' is another term used to describe conditions when a granular soil becomes so weak that it cannot support any slope or cut face, and becomes an almost liquid slurry. The term is often used as if it were a property of the sand itself. Actually, it is the flow of groundwater through the soil and the resultant low values of effective stress that cause this condition. Effective groundwater lowering can change 'running sand' into a stable and workable material.

In addition to the loss of strength, seepage of groundwater through slopes may cause erosion and undermining of the excavation slopes. This is a particular problem in fine-grained sandy soils and is discussed in Section 7.6.

Any solution to groundwater-induced slope stability problems will need to reduce pore water pressures in the vicinity of the excavation. The most commonly used expedient is to adopt the 'pre-drainage' approach of lowering groundwater levels in advance of excavation (Section 9.5.1). This can be achieved by installing a system of pumped dewatering wells (Figure 7.3) around the excavation to lower groundwater levels to below the base of the excavation, and to ensure that seepage does not emerge from the side slopes. A typical target is to lower the groundwater level a short distance (say 0. to 1.0 m) below the deepest excavation formation level.

If the excavation is to penetrate only a short distance (say less than 1.5 m) below the original groundwater level, a sump pumping system (Chapter 14) used in combination with slope drainage (see Sections 7.6 and 17.10) might also be effective. Extreme care must be taken when using sump pumping in this way to avoid destabilizing seepages into the excavation; the risk of this increases for excavations further below the original groundwater level.

7.5 BASE INSTABILITY

If the sides of the excavation are supported by physical cut-off walls such as steel sheet-piles or concrete diaphragm walls (see Section 9.4), then the risk of instability of side slopes is not normally a concern. However, the risk of base instability remains. Base instability caused by

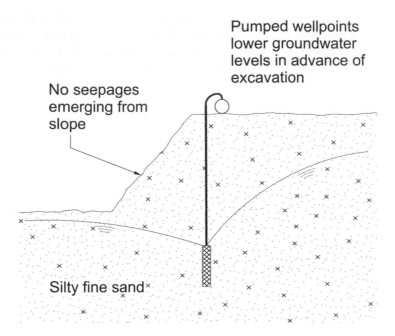

No seepages
emerging from
slope

Pumped wellpoints
lower groundwater
levels in advance of
excavation

Silty fine sand

Figure 7.3 Shallow wells used to improve stability in an excavation below groundwater level in a fine sand. A line of closely spaced wells behind the slope intercepts horizontal seepage and reduces pore water pressures in the slope.

groundwater effects can be a problem even if the base of the excavation is founded in very low-permeability soils such as clay.

There are several potential mechanisms of base instability, and the wide range of terminology used by practitioners can be confusing. Hartwell and Nesbit (1987) present a useful summary of terms, and the following list is based on their work:

- Blow: The rupture or fracture of a stratum due to underlying groundwater pressure
- Boil: Upward flow of groundwater causing loss of strength and rapid erosion
- Erosion: Carrying away of soil particles by water movement
- Heave: Upward elastic or plastic movement of the base of an excavation due to underlying groundwater pressures, and/or stress relief
- Bed separation: The intact movement of a low-permeability stratum, which allows a reservoir of fluid to develop below the stratum

These terms, while evocative and descriptive of what is observed in the field, are imprecise and do not necessarily allow the instability mechanism to be clearly identified.

In this book, we will follow the terminology of Eurocode 7 (BS EN 1997-1:2004) and adopt the generic term 'hydraulic failure' to refer to the disturbance or disruption of the base of an excavation caused by groundwater flows and pressures. Eurocode 7 identifies four key mechanisms of base instability (these are termed Hydraulic Failure in Eurocode 7):

   i. Buoyancy uplift (Figure 7.4a)
  ii. Fluidization due to upward seepage gradients (Figure 7.4b)
 iii. Internal erosion
  iv. Piping (Figure 7.4c)

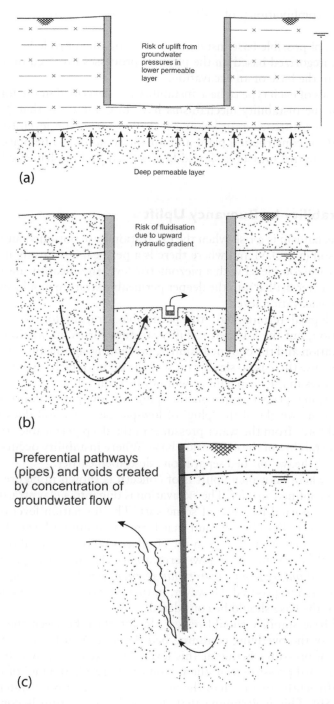

(a)

(b)

Preferential pathways
(pipes) and voids created
by concentration of
groundwater flow

(c)

Figure 7.4 Base instability of excavations. (a) Buoyancy or uplift failure. Upward groundwater pressure from the deep permeable layer exceeds the deadweight and shear resistance of the plug of low-permeability soil in the base of the excavation. Heave or uplift may result. (b) Fluidization due to upward seepage gradients. Upward groundwater flow into the base of an excavation results in very low effective stresses. 'Boils' or quicksand conditions may result. (c) Piping. Groundwater flow results in the movement of particles within the soil, creating voids or 'pipes' within which flow of groundwater is concentrated.

It is possible for combinations of different categories of hydraulic failure to occur simultaneously.

These four types of possible base instability caused by uncontrolled or poorly controlled groundwater are categorized based on the physical processes involved in disrupting or disturbing the strata in the base of an excavation.

The characteristics of each type of base instability are described in the following sections. One thing that all the instability mechanisms have in common is that by the soils being disturbed, loosened or softened at or below excavation formation level, the bearing capacity will be reduced for construction plant and for the structure to be placed in the excavation. Groundwater control measures should be designed and implemented to prevent base instability.

7.5.1 Base Instability by Buoyancy Uplift

This instability mechanism is a risk when the excavation base bottoms out in a layer of very low-permeability soil (e.g. clay) and where there is a permeable stratum at a shallow depth below excavation formation level with a piezometric level significantly above dig level. This means that a piezometer drilled into the deeper permeable stratum would show a water level significantly above excavation level.

Using hydrogeological terminology, the permeable stratum forms a confined aquifer below the excavation, but the low permeability stratum forms an 'aquitard' or 'aquiclude' between the excavation and the confined aquifer, preventing significant vertical seepage into the excavation. Because there is no significant seepage up into the base of the excavation, the groundwater pressures in the deep aquifer will not be reduced.

Upward groundwater pressures act on the 'plug' of low-permeability soil beneath the excavation floor. If the weight of the 'plug' of low-permeability soil does not significantly exceed the upward force from the water pressures in the deep permeable stratum, buoyancy forces may cause instability of the excavation base. Where instability occurs, this is typically observed as a 'blow', 'heave' or 'bed separation', described earlier.

The classic case where buoyancy uplift is of concern is shown in Figure 7.5. It shows an excavation above a confined aquifer. The excavation is dug into a clay stratum with the sides of the excavation supported by walls of some sort. The formation level of the excavation is in a very low-permeability clay stratum that forms a confining layer above a permeable confined aquifer. The piezometric level in the aquifer is considerably above the base of the clay layer. Because the excavation is in low-permeability clay, direct seepage into the excavation will be small, and *purely from a water management perspective*, simple sump pumping would allow excavation to proceed. However, in this type of ground conditions, the risk of hydraulic failure of the base should be considered.

Assessments of base stability should review the interface between the confined aquifer and the underside of the overlying confining bed (level A–A in Figure 7.5a). Consider the two separate sets of pressures acting at this level. The excavation will be stable and safe provided the downward pressure from the weight of the residual 'plug' of unexcavated clay (above A–A) is sufficiently greater than the upward pressure of pore water confined in the aquifer (Figure 7.5b). This is assuming that the confining stratum is consistently of very low permeability, and is competent and unpunctured (e.g. it is not penetrated by any poorly sealed investigation boreholes, which could form water pathways).

If excavation is sunk further into the clay without any reduction of pore pressures in the sand aquifer, there will come a time when the clay plug is 'buoyant'. In other words, the upward water pressure in the aquifer will exceed the downward forces keeping the clay plug

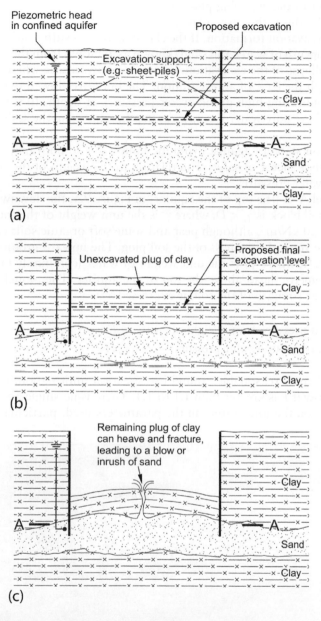

Figure 7.5 Instability of excavation by buoyancy uplift due to confined aquifer below dig level. (a) Before excavation. Excavation is to be made through low-permeability clay, but a more permeable stratum forms a confined aquifer at depth. The piezometric head in the confined aquifer is significantly above final excavation level. (b) Initial, shallow stages of excavation (resisting forces > uplift groundwater pressures) – stable. For a relatively shallow excavation, the weight of the unexcavated plug of clay is greater than the upward water pressure from the confined aquifer. (c) Later, deeper stages of excavation (resisting forces < uplift groundwater pressures) – unstable. For a deeper excavation, the upward water pressure from the confined aquifer exceeds the weight of the clay plug, leading to base heave and, ultimately, a 'blow'.

in place (the weight of the clay plug plus any contribution from the strength of the clay to resist deformation) – this is the case in Figure 7.5c. There will then be an upward movement or 'heave' of the excavation formation. If the clay plug heaves sufficiently, the clay may rupture, allowing an uprush of water and sand (known as a 'blow') from the aquifer, which may even lead to the collapse of the excavation. Figure 7.6 shows a case where a deeper bore was constructed through the base of an excavation. The bore overflowed (without pumping) at excavation level, indicating high groundwater pressures at depth. In this case, the flowing well is acting to relieve pore water pressures at depth, improving the stability of the excavation base.

A simple check on stability against hydraulic failure due to buoyancy uplift can be done by assessing a 'global' factor of safety F in relation to the weight of the soil plug compared with the upward water pressures.

Considering level A–A (the top of the aquifer) in Figure 7.5, the downward pressure from the weight of the soil block is $\gamma_s \times D$, where γ_s is the unit weight of the soil (γ_s is typically in the range of 18 to 20 kN/m^3, although peat and some soft organic soils can be notably less dense) and D is the vertical thickness of the soil plug. The upward groundwater pressure at the same level is $\gamma_w H$, where γ_w is the unit weight of water (commonly taken as 10 kN/m^3) and H is the piezometric head above the top of the aquifer. The global factor of safety F (neglecting the shear strength of the soil) is expressed as

$$F = \frac{\text{Downward pressure from weight of soil}}{\text{Upward groundwater pressure}} = \frac{\gamma_s D}{\gamma_w H} \tag{7.1}$$

Typically, where there is reasonable ground investigation information, the risk of instability is likely to be low if F is greater than 1.1 to 1.4. There is no single acceptable value of F, because it depends on the uncertainty in the parameters used, particularly the level of the

Figure 7.6 Groundwater overflowing from a bore within an excavation. A low-permeability aquitard exists below the excavation. The bore penetrates down to a deeper aquifer, below the aquitard, which has not been depressurized by the shallow dewatering system within the excavation. (Courtesy of Ferrer SL, Museros, Spain.)

top of the confined aquifer and the piezometric head in the aquifer. In many cases it is useful to think of F as a factor of uncertainty rather than a factor of safety; its role in this type of analysis is often to provide conservatism in case the 'true' parameters differ unfavourably from those used in the calculations. In practice, where excavations have failed due to buoyancy uplift (and there are many examples, including Ward, 1957, Hartwell and Nesbit, 1987 and Case History 27.6), it is often the case that either the confined aquifer was not recognized or the stability was not analysed, rather than the failure occurring due to errors in the parameters used in analyses.

Modern design methods for hydraulic failure often take a 'partial factor' approach to stability calculations (where separate factors are applied to different parameters in calculations) rather than using 'unfactored' parameters to obtain a global factor of safety (as was done in Equation 7.1). This can be illustrated by considering the approach used in Eurocode 7 (BS EN 1997-1:2004).

Stability against hydraulic failure under the partial factor approach of EC7 is based on demonstrating that at any given level of permeable stratum, the factored vertical stabilizing forces $V_{stabilizing}$ (soil weight plus any contribution from soil strength or shear resistance) are greater than the factored destabilizing forces $V_{destabilizing}$ (upward groundwater pressures).

Stability is assessed via an analysis such as Equation 7.2, where the factored forces (i.e. the assessed forces multiplied by a partial factor) are compared. For stability, the factored vertical destabilizing forces must be less than or equal to the factored vertical stabilizing forces due to soil weight, plus any factored additional resistance to uplift, R_{uplift}, from soil strength. Therefore, stability requires

$$V_{destabilizing} \leq V_{stabilizing} + R_{uplift} \tag{7.2}$$

The forces are usually expressed and compared as pressures. The factored destabilizing forces are the uplift groundwater pressures multiplied by the partial factor $F_{destabilizing}$, which is greater than 1.0 (the UK National Annex to EC7 [BS EN 1997-1:2004 + A1:2013] specifies a value of 1.1). Using the geometry of Figure 7.7,

$$V_{destabilizing} = F_{destabilizing} \times \gamma_w \times H \tag{7.3}$$

The factored stabilizing forces are the weight of the soil plug multiplied by the partial factor $F_{stabilizing}$, which is less than 1.0 (the UK National Annex to EC7 [BS EN 1997-1:2004 + A1:2013] specifies a value of 0.9):

$$V_{stabilizing} = F_{stabilizing} \times \gamma_s \times D \tag{7.4}$$

The factored 'additional resistance' to uplift, R_{uplift}, is normally considered to derive from the undrained shear strength of the soil in the plug, and any shear resistance at the side of the excavation between the soil plug and any retaining walls. When calculating the factored resistance, the relevant shear strengths and soil parameters are multiplied by partial factors of less than 1.0.

In theory, the additional resistance from soil strength and shear strength can mean that the base of excavations may be stable when the global factor of safety based on water pressures and soil weight alone (Equation 7.1) is less than 1.0. Finite element modelling by Hong et al. (2015) indicates that for excavations of width b that are narrow (relative to their thickness D of the soil plug below excavation level), the additional resistance can be significant.

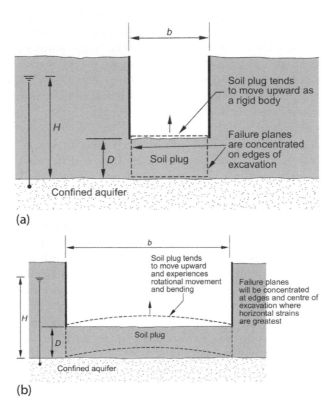

Figure 7.7 Different modes of base failure due to buoyancy uplift in excavations of various widths. (a) Narrow excavation (*b/D* < 4) assuming a continuous aquitard beneath the excavation. (b) Wide excavation (*b/D* > 4) assuming a continuous aquitard beneath the excavation. (Based on modelling from Hong, Y., *et al.*, *Canadian Geotechnical Journal*, 52, 599–608, 2015.)

The modelling of Hong *et al.* (2015) identified a different failure mechanism for narrow excavations compared with wider excavations (Figure 7.7):

- Narrow excavations, where *b/D* is less than approximately 4, tend to fail by the soil plug moving upwards in a rigid body manner, with failure by shear along vertical planes at the edge of the excavation (Figure 7.7a). The undrained shear strength of the soil means that the plug may not 'fail' until the excess water pressure is significantly greater than the weight of the soil plug. For very narrow excavations, the modelling indicated that failure may not occur until the global factor of safety *F* is as low as around 0.7 (when calculated by Equation 7.1).
- In wider excavations, where *b/D* is greater than approximately 4, the soil plug acts as a beam and is subject to rotational movement and bending, which creates horizontal strain. The horizontal strain is concentrated in the centre and edges of the excavation (Figure 7.7b). The undrained shear strength of the soil makes no significant contribution to stability, and the plug 'fails' when the excess water pressure is equal to the weight of the soil plug (i.e. the global factor of safety *F* of 1.0 when calculated by Equation 7.1).

Although in narrow excavations the effect of undrained shear strength of the soil plug means that excavations may be theoretically stable if the global factor of safety (based on excess water pressure and soil weight alone) is less than 1.0, the authors consider that this often will not be an acceptable strategy for design, because there must be some displacement of the soil plug to mobilize the shear strength. The modelling of Hong *et al.* (2015) indicates that the displacement to mobilize the shear strength can be significant and would result in unacceptable pre-construction heave of the base of the excavation. The authors consider that in most cases, basal stability against buoyancy should be calculated based on the balance of the excess water pressure and the weight of the soil plug only, using appropriate factors of safety (on either a global or a partial factor basis). Assessment of basal stability is discussed in Section 13.7 in the chapter on design methods.

It is important to remember that these discussions are based on the assumption that the soil plug is uniform and homogeneous and of very low permeability and is not penetrated by permeable vertical features:

- The assumption of very low permeability is critical for base stability. Such low-permeability materials will act in an undrained manner (see Section 6.6), which means that negative excess pore water pressures (soil suctions) will develop in the plug in response to the unloading caused by excavation. Problems can occur if there are some zones of slightly higher permeability (e.g. silt or sand inclusions) within the soil plug; these can cause the excess pore water pressures to dissipate more rapidly. This can allow the soil in the plug to move to a drained condition, which will result in reduced shear strength, increased seepage and potential base instability.
- There have been cases where excavation bases have failed due to localized seepage pathways (e.g. sand beds within Glacial Till, or poorly sealed investigation boreholes), even though stability calculations indicated that there was an adequate clay plug below formation level. Where localized vertical seepage pathways exist, seepage into the base of the excavation often starts as minor seepage but can become progressively worse as softening and erosion occur. Seepage rates can increase, leading to a more significant and widespread failure of the base of the excavation.

Instability by buoyancy uplift can be avoided by:

a) Adequately reducing the pore water pressure in the confined aquifer by pumping from suitable dewatering wells inside the excavation footprint. Alternatively, the wells may be sited outside the excavation but will be less hydraulically efficient. The wells should lower the aquifer water pressure so that the downward forces exceed the upward pressures by a suitable factor of safety.

b) Reducing the groundwater pressure in the confined aquifer by installing relief wells (Section 17.8) within the excavation footprint. The relief wells provide a preferential pathway for upward groundwater flow. As the excavation is made below piezometric level, the relief wells will overflow, and the water can be pumped away by sump pumps. This will hold piezometric level at the excavation dig level at the relief well locations. The use of relief wells is appropriate to target permeable zones of limited thickness and/or limited recharge, where total flow from the relief wells will be modest. Provided the relief wells are installed on a grid pattern at a suitably close spacing, this is typically sufficient to achieve stability.

c) Increasing the depth of the physical cut-off wall sufficiently to penetrate below the base of the confined aquifer. Provided the cut-off walls are watertight, this will prevent further recharge, thus leaving only the water pressure contained in the aquifer within the cut-off walls to be dealt with. This can be affected by installing relief wells prior to excavation, from which limited flow is anticipated. The economics of this alternative will depend primarily on the depth needed to secure a seal and the effectiveness of that seal.

  d) Increasing the downward pressure on the base of the excavation by keeping it partly topped up with water during the deeper stages of work. Excavation is made underwater, and a tremie concrete plug is used to seal the base on completion (this is the basis of the 'wet caisson' method used for shaft construction; see Section 10.5.4).

The preceding discussion relates to buoyancy uplift resulting in relatively large-scale 'blow' or 'heave' conditions where a significant confined aquifer exists below excavation. Buoyancy uplift can also sometimes occur in nominally unconfined aquifers that contain discrete (but perhaps very thin) very low-permeability clay layers. Figure 7.8 shows a case where dewatering wells are used to lower groundwater levels around an excavation but do not penetrate the clay layer below the excavation. The dewatering system has lowered groundwater pressures above the clay layer, but the original high groundwater pressures remain beneath. These high water pressures can cause uplift. This mechanism is sometimes known as 'bed separation', because as the clay layer moves upwards, a lens of water will develop beneath. This form of uplift can be avoided by using deeper dewatering wells to control pore water pressures at depth.

7.5.2 Base Instability by Fluidization Due to Upward Seepage Gradients

Where an excavation is made through a thick water-bearing stratum (such as sand or gravel), groundwater control requirements can be reduced if the sides of the excavation are supported by physical cut-off walls. The cut-off walls remove the need to consider the instability of side slopes, and particularly for relatively small excavations such as shafts or cofferdams, it may be possible to lower groundwater levels simply by sump pumping from within the excavation. Methods used to form cut-off walls are described in Chapter 18.

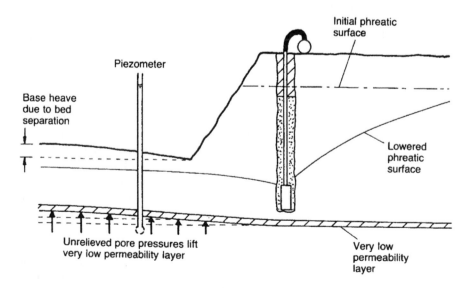

Figure 7.8 Pore water pressures not controlled beneath thin very low-permeability layer – unstable condition. (From Preene, M and Powrie, W, *Proceedings of the Institution of Civil Engineers, Geotechnical Engineering*, 107, 17–26, 1994. With permission.)

However, when sump pumping is performed inside an excavation contained within a cut-off wall, there is a risk that upward hydraulic gradients due to pumping will significantly reduce effective stresses in the soil. In extreme cases, effective stresses will approach zero, and the soil will fluidize and lose all strength – in these circumstances, the soil is said to be 'quick' (see Section 6.5). This can lead to catastrophic structural failure of the excavation and associated cut-off walls (which will lose much of the passive support from the soil below excavation level). The aftermath of a failure by this mechanism is shown in Figure 21.1.

The risk of fluidization of the base due to upward seepage is related to the upward hydraulic gradient generated by pumping (Figure 7.9). Fluidization will theoretically occur when the upward hydraulic gradient exceeds a critical value i_{crit}.

$$i_{crit} = \frac{(\gamma_s - \gamma_w)}{\gamma_w} \tag{7.5}$$

where

γ_s is the unit weight of soil
γ_w is the unit weight of water

In general, the density of soil is about twice the density of water (peat being a notable exception, which can be significantly less dense), so generally, $i_{crit} \approx 1$. This is the hydraulic gradient at which fluidization will occur, so it is important that this value is not approached. In design, it is normal to limit the predicted hydraulic gradients to be significantly lower than i_{crit} in order to provide a suitable factor of safety F. This implies that an acceptable hydraulic gradient for design purposes would be i_{crit}/F.

As noted during the earlier discussion of buoyancy uplift failure, it is useful to think of F as a factor of uncertainty rather than a factor of safety. Van Zyl and Harr (1981) discuss factors of safety on seepage gradients that are reported in literature from studies of seepage effects between the 1930s and the 1960s, where the recommended factors of safety on

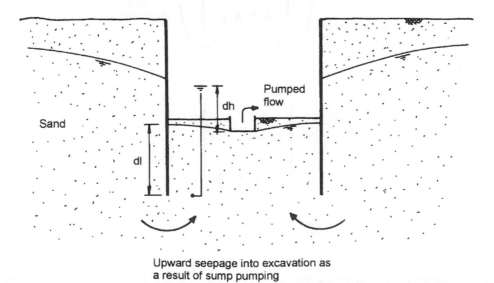

Upward seepage into excavation as
a result of sump pumping

Figure 7.9 Fluidization of excavation due to sump pumping creating upward seepage gradients in an unconfined aquifer. Mean upward hydraulic gradient below excavation floor = dh/dl.

hydraulic gradients vary between 1.5 and more than 10, with F commonly in the range from 4 to 6. Van Zyl and Harr (1981) note that lower values of F (lower than 4 to 6) may be only be justified where site conditions (permeability and seepage boundary conditions) are known with a high degree of confidence, hence reducing uncertainty.

There are different examples of conditions that can lead to fluidizations of the base of an excavation due to upward seepage gradients:

a) Unconfined aquifer comprising relatively uniform permeable stratum. In this case, the driving head for the seepage gradients is controlled by the groundwater level in the permeable stratum (Figure 7.10a).
b) Layered aquifer system with shallow high-permeability stratum. The driving head for the seepage gradients is controlled by the groundwater level in the shallow, very permeable stratum, which effectively acts as a close source of recharge (Figure 7.10b).

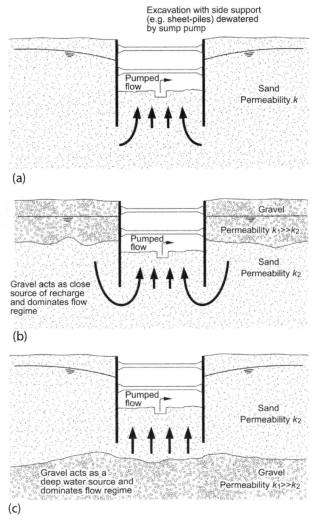

Figure 7.10 Examples of soil conditions that can lead to fluidization of excavation due to upward seepage. (a) Unconfined aquifer comprising relatively uniform permeable stratum. (b) Layered aquifer system with shallow high-permeability stratum. (c) Layered aquifer system with deeper high-permeability stratum.

c) Layered aquifer system with deeper high-permeability stratum. In this case, the driving head is controlled by the groundwater level in the deeper permeable stratum, which acts as a source of recharge at depth (Figure 7.10c).

The latter two cases present the most significant risk of basal instability because of the presence of close sources of recharge to act as drivers for destabilizing hydraulic gradients. However, instability can still occur in case (a) in uniform soil conditions if sump pumping is carried out inside an excavation where the cut-off walls do not penetrate sufficiently deeply to reduce hydraulic gradients to acceptable levels.

Conditions shown in Figure 7.10 are idealized and imply relatively uniform seepage gradients across the base of an excavation. In reality, soil conditions are not uniform, and seepage flows and gradients will vary correspondingly. If upward seepage gradients are high, the onset of instability may be localized, and the base of the excavation can experience 'boiling' or 'quick' conditions, where upward flow of groundwater causes loss of strength and rapid erosion – examples are shown in Figure 7.11. Typically, slight variations in ground conditions will allow seepage to break out into the floor of the excavation in one or more discrete locations, which are known as individual 'boils' (Figure 7.11a). In some cases, this can result in very localized heave and ejection of material to form a conical feature known as a 'sand volcano' like the one visible in Figure 7.11b. Boils and sand volcanos indicate that material is being eroded, which implies that vertical permeability is increased locally. This can lead to a feedback loop where water flow becomes concentrated in that area, causing more erosion, leading to higher flow rates and so on. This can lead to erosion and piping problems, described in Sections 7.5.3 and 7.5.4.

The mean hydraulic gradient for upward seepage is defined as the head difference dh divided by the flow path length dl (see Figure 7.9). Equations to estimate upward hydraulic gradients in the base of excavations enclosed by vertical cut-off walls are given in Section 13.8.4. The hydraulic gradient can be controlled in two ways:

a) By increasing dl. This can be achieved by ensuring that the cut-off walls penetrate a sufficient depth below formation level. This is an important part of the design of cut-off walls in unconfined aquifers (see Kavvadas *et al.* (1992) and Williams and Waite, 1993).

b) By reducing dh. This requires reducing the groundwater head difference between the inside and the outside of the excavation enclosed within the cut-off walls. The most obvious way of doing this is to use a system of dewatering wells outside the excavation to lower the external groundwater level and so reduce the head difference to an acceptable level while still sump pumping from within the excavation. This approach might be taken a step further; the external wells could be used to lower the groundwater level to below formation level, avoiding the need for sump pumping altogether. An alternative approach to reducing dh is to keep the excavation partly or fully topped up with water and carry out excavation underwater (for example using grabs) and then place a tremie concrete plug in the base. This method is sometimes used for the construction of shafts and cofferdams and is often known as the 'wet caisson' method (see Section 10.5.4).

Alternatively, a system of relief wells (see Section 17.8) could be installed within the excavation to provide preferential, engineered flow paths for groundwater to enter the excavation. This will dramatically reduce the hydraulic gradients within the excavation and, provided the relief wells are spaced closely enough, will prevent fluidization of soil in the base of the excavation.

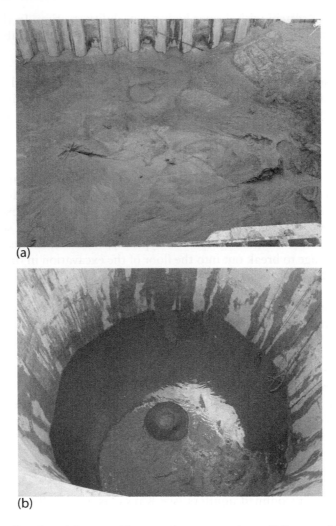

Figure 7.11 Examples of basal instability caused by upward seepage gradients. (a) Base instability involving multiple boils (localized upward flow of groundwater) in the base of a sheet-piled excavation. (b) Base instability involving a single major boil, resulting in a 'sand volcano'; other secondary boils may be present (a small boil is visible at the toe of the wall at the top of the picture). (Courtesy of K Vaughan.)

Figure 7.11 shows that instability due to upward hydraulic gradients often starts with local problems, which can get bigger and bigger. The whole base of the excavation rarely turns to 'quick sand' or 'running sand' immediately, but if uncontrolled seepage and erosion were allowed to continue, then instability might affect large areas of excavation (as shown in Figure 6.1). If remedial measures to reduce upward hydraulic gradients cannot be implemented rapidly, then it may be appropriate to flood the excavation (by diverting the dewatering discharge back into the excavation) up to the original groundwater level. This will stop the problematic seepage and erosion, and allow more time to plan and implement a solution.

7.5.3 Base Instability by Internal Erosion

Internal erosion involves the movement of soil particles under the influence of flowing groundwater. While Eurocode 7 (BS EN 1997-1:2004) identifies internal erosion as a separate mode of instability, in practice it is often associated with instability caused by fluidization due to upward seepage gradients (Figure 7.4b) and can be associated with piping (Section 7.5.4).

Internal erosion may occur where hydraulic gradients are high (for example where ground-water flow is concentrated where it enters the excavation or flows into a sump). Soils that are prone to erosion are uncemented coarse-grained soils that are uniformly graded (where most of the soil particles are with a narrow size range) or are internally unstable (Skempton and Brogan, 1994; Fannin and Moffat, 2006), where the finer particles (fine sand/silt/clay) can move within the skeleton of larger particles.

The removal of fine particles (termed 'loss of fines') can result in the loosening of the soils and the creation of voids, which may collapse. Loss of fines is a particular risk where sump pumping is used (see Section 14.8), because it is often the case that adequate filters are not in place around the sump.

The problems caused by the removal of fines during dewatering of excavations vary depending on the nature of the soil, particularly the particle size distribution (PSD):

1. In soils of uniform grading (i.e. where the majority of the PSD is in a narrow range – such as a fine sand), once erosion starts, localized slumping and collapse will occur almost immediately. This is because there is no coarser fraction to form a temporary 'filter' that might allow the soil to sustain some loss of fine particles without collapse. These types of conditions are illustrated in Figure 7.2 and Figure 7.13.

2. In soils of non-uniform grading (i.e. where PSD contains a wide range of particles sizes – such as sandy gravel), it may be possible for the coarser particles in the soil skeleton to form a stable 'washed out' soil structure as the fine particles are eroded away. This can allow loss of fines to occur without collapse of the soil; however, the permeability of the washed out zone can increase significantly, leading to dramatic increases in dewatering pumping rates. The term 'suffusion' is sometimes used to describe the loss of fine particles due to seepage that does not result in the collapse of the resulting eroded soil structure.

3. Alternatively, it is also possible for soils of non-uniform grading (i.e. where PSD contains a wide range of particle sizes – such as sandy gravel) to be subject to internal erosion, where transport of the finer particles by groundwater seepage flow causes the remaining soil structure to collapse. 'Suffosion' is sometimes used to describe the loss of fine particles due to seepage that results in the collapse of the resulting eroded soil structure.

Internal erosion due to seepage has been studied in detail in relation to the stability of embankment dams and water retaining structures. Useful background on the phenomenon is given in Bulletin 164 of the International Commission on Large Dams (ICOLD, 2016). All of the preceding cases will generate 'dirty' or sediment laden water that must be disposed of. Water disposal and water treatment are discussed in Sections 14.9 and 21.9.

The measures used to control hydraulic gradients and reduce the risk of base instability from fluidization due to upward seepage gradients (see Section 7.5.2) are generally also suitable for use to reduce the risk of internal erosion in the base of an excavation. Additionally, the risk of erosion of fine soil particles from the zones where groundwater is pumped within the excavation can be reduced by using sumps with adequate filters around them, or by replacing sump pumping with wellpoints or deep wells equipped with suitable filter media.

7.5.4 Base Instability by Piping

Although identified in Eurocode 7 (BS EN 1997-1:2004) as a separate mode of instability, piping is most commonly the result of instability caused by fluidization due to upward seepage gradients (Figure 7.4b) in combination with internal erosion (Section 7.5.3). In these circumstances, high hydraulic gradients and potentially mobile granular soils can result in the washing out of fine particles from the soil to create preferential flow paths. Water will naturally tend to take the easiest path, so water flow is further concentrated in these zones, causing further erosion of soils. This can continue until most of the water is entering the excavation through a small number of discrete open voids – termed 'pipes' – created by the washing out of fine particles along the preferential flow paths (Figure 7.4c). Where piping channels have occurred, it is likely that inflows to the excavation will be significantly increased compared with non-piping conditions. Piping channels are often highly unstable and may collapse unexpectedly, threatening the stability of the excavation and support structures.

The measures used to control hydraulic gradients and reduce the risk of base instability from fluidization due to upward seepage gradients (see Section 7.5.2) are generally also suitable for use to reduce the risk of piping instability in the base of an excavation.

7.6 LOCALIZED GROUNDWATER PROBLEMS

In addition to groundwater effects that can cause large-scale failure of an excavation, groundwater can also cause more localized problems in excavations. Such localized problems can vary greatly in scope, from minor inconveniences caused by a perched water table to severe problems resulting from the collapse of side slopes due to spring sapping – localized groundwater problems should not be ignored. A number of possible mechanisms are described in the following sections.

7.6.1 Drainage of Slopes and Formation

The slopes of open cut excavations should be designed so that instability due to seepage and associated continuous removal of fines does not occur. Where the slope is too steep or the hydraulic head too large, seepage can emerge on the slope and cause localized washout and erosion (Figure 7.12) or more problematic slope failure (Figure 7.13).

Often, the solution is to have a flatter slope, sometimes together with a toe drain backfilled with filter media to prevent emerging water continuously removing fines and so causing

Figure 7.12 Localized seepage erosion from a slope. Seepage from a concentrated permeable lens in a relatively low-permeability glacial deposit has washed fine particles from the soil, partly burying dewatering pipework.

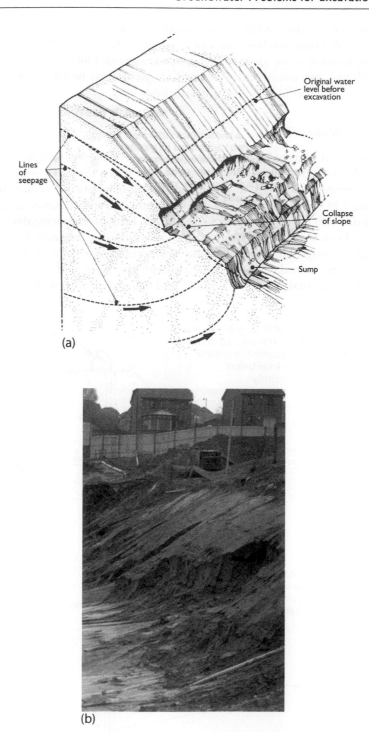

(a)

(b)

Figure 7.13 Effect of seepage on excavation. (a) Collapse of excavation caused by seepage from steep slopes. (b) Unstable excavation slope resulting from seepage. Seepage has eaten back into the slope, and an outwash fan of saturated washed out material has formed at the base of the slope. (Panel (a) reproduced from Somerville, S H, *Control of Groundwater for Temporary Works*. Construction Industry Research and Information Association, CIRIA Report 113, London, 1986. With permission from CIRIA: www.ciria.org)

damage leading to instability. Alternatively, more formal engineered slope and toe drains can be used. Slope drainage methods are discussed in Section 17.10.

Observance of good practice applicable to slope grades and filter design criteria will help achieve stable slopes, but during excavation, it is essential to be on the lookout for local trouble spots, generally due to variations in soil conditions.

7.6.2 Perched Water Table

Often, within permeable granular strata, isolated lenses of significantly lower permeability occur, locally inhibiting downward drainage of water. When the general groundwater level is lowered by pumping from wells, a 'perched water table' is likely to result from this phenomenon (Figure 7.14). This means that a residual body of water remains just above the lens of low-permeability soil even though the groundwater level in the permeable strata beneath has been drawn down lower. This troublesome condition can be dealt with effectively by forming vertical drainage holes through the low-permeability lens. This can sometimes be done by the simple expedient of digging through the layer with an excavator bucket. Alternatively, relatively small-diameter vertical drains may be appropriate (see Section 17.9).

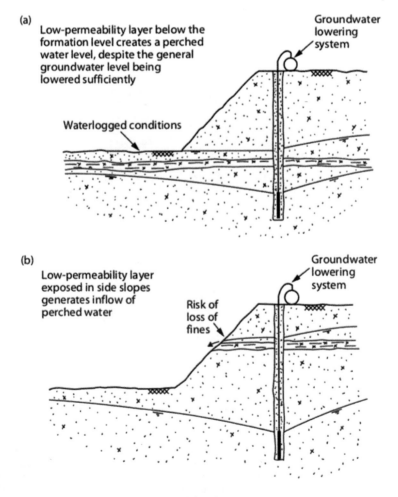

Figure 7.14 Perched water table. (a) Low-permeability layer below formation level. (b) Low-permeability layer above formation level and exposed in side slope.

Where seepage from perched water tables persists during excavation, this is known as residual seepage or 'overbleed' seepage. If the rates of overbleed seepage are low, the water will merely constitute a nuisance rather than lead to a serious risk of flooding. Nevertheless, even at small rates of seepage, if fines are being transported by the water, this should be counteracted immediately. Possible measures include the placement of sand bags, a granular drainage blanket or geotextile mesh to act as filters. A specific problem of overbleed during shaft sinking is discussed in Section 7.7.3.

An extensive perched water table in an unconfined aquifer of gravelly sand is shown in Figure 7.15. Readings in several monitoring wells indicated that the pumping from a system of deep wells had drawn the groundwater level down to the target level, which was several metres below the perched water level visible in the figure. The contractor was advised to continue excavation by dragline at one location. Within a few hours of continuous digging, the water started to swirl, and there was a loud gurgling sound as the perched water drained away.

7.6.3 High-Permeability Zones

Sometimes a lens of significantly more permeable water-bearing soil, not previously detected, may be revealed during excavation. This will act as a source of copious recharge requiring additional pumping capacity to be installed to abstract groundwater from the lens. Such permeable gravel lenses are not uncommon in fluvio-glacial or alluvial soils, which were laid down in water. The permeable zones may result from paths of former streams or flow channels within the geological deposits. Like modern-day watercourses, these gravel lenses may follow an irregular, difficult-to-predict path. Long, thin gravel lenses are known as 'shoestrings'; these may easily pass undetected during site investigation. Case History 27.5 is an example of problems caused by such high-permeability features.

The same recharge effect as from a natural gravel lens can result from man-made features such as the permeable gravel bedding or sub-drains associated with sewers. There have been cases where dewatering adjacent to existing sewers has proved difficult until the flow path

Figure 7.15 Perched water table. The groundwater level in monitoring wells was several metres below the standing water visible in the photograph. When the clay layers below this level were punctured by excavating, the perched water drained away. (Courtesy of ESG Soil Mechanics.)

formed by the permeable material around the sewer was blocked or the flow intercepted directly by sump pumping from the gravel bedding.

The principle to be remembered is that whenever a local source of more copious recharge is encountered, additional wells or sumps should ideally draw groundwater direct from the recharge source (i.e. the permeable zone). It is preferable to intercept the flow upstream of the exposed excavation, but sometimes this may not be practicable. In such circumstances, the additional wells or sumps must be installed as close as possible to the area of groundwater emergence into the works.

7.6.4 Spring Sapping and Internal Erosion

This mechanism, which can occur in sands and silty fine sands, is not uncommon but is not well understood. The sides of an excavation in silty fine sand may show gully structures from which sand slurry can be seen to flow into the bottom of the excavation. This is due to internal erosion. The water flowing along the upper surface of the lowered phreatic line carries sand slurry with it and thus tunnels backwards beneath the overlying sand, which has some strength due to capillary tension. Before this tunnelling effect has gone far, the top collapses, and the collapsed material is also carried away. This tunnelling phenomenon has often been observed in fine sands and especially in silty fine sands. It tends to occur immediately above a thin and less permeable layer of finer material that has caused a perched water table, though this condition may only be temporary. This phenomenon is quite well known in geology under the name of spring sapping or seepage erosion.

In the 1960s, an interesting experiment into this phenomenon was carried out at Imperial College London. A slope of fine sand was formed in a glass-sided tank, and a small source of water was allowed to flow in through the base of the tank. The base of the tank was at a flat angle. A very small flow of water was introduced beneath the sand, and erosion rapidly started at the point of exit. One would intuitively expect a cone of eroded material to build up and the process of erosion to come to an end, but in fact, the flow of sand was continuous – as it was in a case described by Ward (1948). It seems that this flow can only happen if excess pore pressures exist in eroded material, but their cause is still unexplained, and it is hard to see why they are not quickly dissipated.

The site studied by Ward (1948) was near Newhaven in Sussex, at a point where Woolwich and Reading Beds (now known more correctly as the Lambeth Group) overlie the Thanet Sand Formation and Chalk Group. A thin bed of fine sand in the Woolwich and Reading Beds carried a small flow of groundwater towards the sea, and although the hydraulic gradient was only about 1 in 10, internal erosion on a large scale had continued for centuries, and debris flows remained active and carried the sand and the collapsed mass of the overlying clay down to the beach some 55 m below. This state of affairs was eventually controlled by a simple system of filters and drains.

The curious thing about this process is that it takes place when the flow of water is very small and the gradient very low, yet because it is continuous, in time very large quantities of material are removed.

Figure 7.16 shows a close-up view of a small example of spring sapping. A thin lens of predominantly shells was present within an extensive stratum of fine sand. The levels recorded in nearby monitoring wells indicated that the water had been drawn down to some 4 m below this lens, but because the permeability of the lens of shelly material was significantly greater, there was still water flowing, continuously transporting fines and so causing local spring sapping. This was evidenced by the two cavities in the centre of the photograph and the outwash fans of almost liquid sand in the foreground.

Figure 7.16 Spring sapping. Local spring sapping has caused the two cavities in the centre of the photograph with outwash fans of almost liquid sand in the foreground.

Variable soil conditions may be the cause of local trouble spots being revealed as excavation proceeds. Drift deposits are often heterogeneous, and so the possibility of exposing local trouble spots in the sides of excavations is ever present. Often, these trouble spots are due to a layer or lens of slightly lower-permeability material within the stratum being dewatered. Due to the difference in permeability, the pore pressures in the layer or lens are higher than in the surrounding stratum, and despite the general effectiveness of the installed groundwater lowering system, there is some residual flow from the lens, which causes transportation of fines. If this condition is allowed to persist, it will result in instability due to seepage erosion.

7.7 GROUNDWATER PROBLEMS WHEN TUNNELLING IN SOILS

A tunnel is a challenging working environment at the best of times. Any groundwater ingress is only going to add to the difficulties. The tunnel face is in a confined space, deep below ground, and access is often difficult if there is a need to deploy additional pumps or ground treatment equipment. It is essential that groundwater ingress is prevented or managed in a safe and appropriate manner. Furthermore, soil instability, which in a surface excavation might be a minor inconvenience, could be a major catastrophe in the confined space of a tunnel – ensuring the stability of exposed soils is vital.

The primary elements constructed during tunnelling are the tunnels themselves – linear sub-horizontal features, often, but not always, cylindrical. On transportation projects, there are often two parallel tunnels (known as twin tunnels). In addition to the tunnels themselves, there are other underground works, including shafts (to access the tunnel from the surface), cross passages to link twin tunnels, adits and connector tunnels, and so on.

Just like surface excavations, tunnels and other underground works typically require groundwater control to achieve two principal objectives:

1. Groundwater management, to provide a workable environment in the tunnel and prevent the works being 'flooded out'
2. Prevention of groundwater-induced instability of soil exposed in the faces or other working areas

In relation to objective 1 – groundwater management – even in relatively permeable soils, the rates of groundwater ingress may not be so large and can be managed by simply pumping away from the face along the tunnel to the pit bottom of a shaft and thence to the surface.

However, objective 2 – preventing groundwater-induced instability – is also important. Modern tunnelling contracts are generally planned on the basis of rapid and efficient tunnel progress, which requires stable and controllable ground behaviour at the face, and tunnel spoil (colloquially called 'muck') that is easy to handle. The presence of uncontrolled groundwater can seriously slow down tunnel progress. Water seeping through the face can cause localized instability and collapse (especially in silts and sands), and the presence of groundwater in the tunnel will make the 'muck' wet and difficult to handle. Effective control of groundwater can have huge benefits to the efficiency of tunnelling operations and pay back the cost of groundwater control measures several times over.

7.7.1 Groundwater Instability Effects in Tunnel Faces

When working in soils, many tunnelling methods construct a relatively watertight support system (the tunnel primary lining) immediately behind the working face; hence, typically, at any given time, only a small area of tunnel face is exposed to water-bearing ground. Groundwater control methods for tunnelling works are described in Chapter 10.

Most tunnel excavation in soils is carried out within a 'shield', a strong cylindrical sheath that is pushed forwards into the ground, while soil is excavated from within it. The shield provides circumferential support to the soil ahead of the permanent lining and ensures that the only soil exposed is in the face itself.

A shield is a form of open face tunnel boring machine (TBM). Open face TBMs allow groundwater to enter the tunnel, from where it can be pumped away. The tunnel will act as a drain (Figure 7.17) and will create some reduction of pore water pressures (and drawdown of groundwater levels) above the vicinity of the working face. As the tunnel progresses, the working face moves forward, and the centre of the zone of drawdown will move forward with it. When the watertight lining is in place, the groundwater levels should return to their original levels – although long-term seepage into transportation tunnels is known to lower groundwater levels locally in several cities.

Many cases of tunnelling by open face TBMs are in soils of low to moderate permeability – silts, silty sands and sands – where the quantities of water inflow are modest and can be easily managed. The key challenge is instability of tunnel faces that are subject to groundwater inflow. Doran *et al.* (1995) presented an elegant analysis of this stability problem, based on the work of Professor R E Gibson, summarized later. Figure 7.18 shows an open face tunnel being advanced using a shield and permanent lining. The tunnel is at atmospheric pressure (the so-called 'free air' condition), so the tunnel internal pressure σ_t is zero, the depth to tunnel axis is Z and the depth to groundwater level is H_w

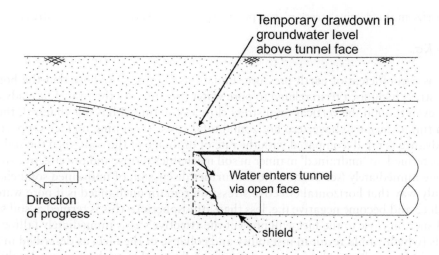

Figure 7.17 Interaction between tunnel construction and groundwater for an open face tunnel with shield under free air conditions (tunnel at atmospheric pressure).

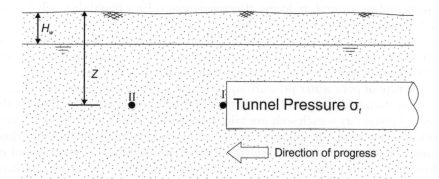

Figure 7.18 Geotechnical stability of tunnel faces in soils of low to moderate permeability. (After Doran, SR, et al., *Proceedings of the 11th European Conference on Soil Mechanics and Foundation Engineering, Copenhagen, Denmark, 1995.*)

Before tunnelling, the pore water pressure u at point II (remote from point I at the tunnel face) at the level of the tunnel axis is

$$u = \gamma_w \left(Z - H_w \right)$$

(7.6)

and the vertical effective stress σ'_v is the vertical total stress minus the pore water pressure:

$$\sigma'_v = \sigma_v - u = \gamma_s Z - \gamma_w \left(Z - H_w \right)$$

(7.7)

where
    γ_w is the unit weight of water
    γ_s is the unit weight of soil

The horizontal effective stress σ'_h is a proportion of the vertical effective stress:

$$\sigma'_h = K\sigma'_v \qquad (7.8)$$

where K is a coefficient of earth pressure. Equations 7.7 and 7.8 show that the horizontal effective stress includes a term related to pore water pressure. As the tunnel face advances to point II, even though the vertical total stress close to the tunnel has not changed, the soil at the open tunnel face will have zero horizontal total stress (assuming no face support). If the tunnel advance is rapid (compared with the rate of drainage of the soil), then the soil will act in what is termed an 'undrained' manner in soil mechanics terms, and the effective stress will not change immediately (see Section 6.6 for further discussion of undrained soil behaviour).

The only way that horizontal effective stress can remain unchanged is if pore water pressures reduce and become negative (i.e. less than atmospheric) – this is termed a 'soil suction'. This soil suction close to the face will prevent the effective stress immediately falling to very low levels (which would result in instability and collapse of the face, as described in Section 6.5). It is vital to understand that in any unsupported tunnel face, the improved stability due to suctions resulting from horizontal unloading is temporary, and the face will become less stable as time passes. The reduced pore water pressures around the face will create hydraulic gradients towards the face, which will draw groundwater in from the surrounding soil. This will increase pore water pressures at the face as the soil near the face comes into equilibrium with the surrounding ground, reducing effective stresses. Face instability will follow if support or stabilization measures are not taken.

The rate at which this takes place is controlled primarily by the excess head of groundwater (the difference in head between the tunnel depth and the piezometric level in the wider groundwater regime) and by the permeability of the soil. In a very low-permeability soil (such as a clay), the rate of pore water pressure equilibration may be very slow, allowing the cycle of tunnel excavation and placement of the tunnel lining to be safely carried out without significant risk of instability – such soils are said to have long 'stand-up times'. Conversely, in a more permeable fine sand, the soil suctions caused by excavation of the face may dissipate in a few minutes or seconds, resulting in the rapid onset of instability – a short 'stand-up time'. Soils of low permeability (such as silts or clays with permeable fabric) are an intermediate case, where soil exposed in a tunnel face may not become unstable immediately but may not be stable for long enough to allow the tunnel construction cycle to be completed safely.

When working with open face TBMs in soils with short stand-up times, mitigation measures include pumping methods to reduce groundwater level to below tunnel invert (ensuring the pore water pressures around the face are zero or lower) or ground treatment (such as grouting or artificial ground freezing) to reduce permeability and increase soil strength. Groundwater control measures for tunnels are discussed in Chapter 10.

In soils where the stand-up time of a soil is short, it is possible to apply a counter pressure to the face to exclude groundwater and stabilize the face – this is equivalent to applying an internal tunnel pressure σ_t in Figure 7.18. This can be achieved by closed face TBMs (Figure 7.19), where the counter pressure can be provided by an earth paste (earth pressure balance [EPB] TBMs) or a bentonite or polymer slurry (slurry TBMs). A variation on this approach is to use an open face TBM but pressurize the tunnel with compressed air to counteract groundwater pressures. These tunnelling methods are discussed in Section 10.4.

The work of Professor R E Gibson (Doran *et al.*, 1995) indicates that at a tunnel face (point I in Figure 7.18) in a permeable soil, the horizontal effective stress σ'_h can be expressed in terms of the horizontal total stress σ_h and the pore water pressure u:

$$\sigma'_h = \sigma_h - u = \sigma_h - \gamma_w\left(Z - H_w\right) \qquad (7.9)$$

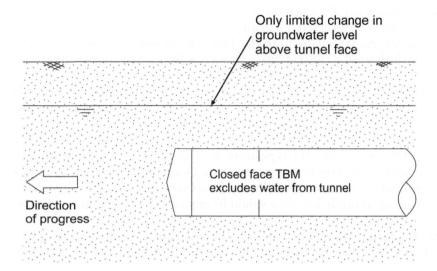

Figure 7.19 Interaction between tunnel construction and groundwater for a closed face tunnel boring machine (TBM). The TBM applies a counter pressure at the face to avoid or minimize
groundwater inflow.

In a tunnelling method that applies a counter pressure σ_t, the horizontal total stress at
the face is σ_t. Gross instability (in the form of 'running sand' or 'quick' conditions; see
Section 6.5) will occur if effective stress reaches zero. Therefore, Equation 7.9 indicates that
stable face conditions require that the counter pressure is greater than the pore water pressure at the face.

$$\sigma_t > \gamma_w \left(Z - H_w \right) \tag{7.10}$$

This implies the following:

i. The applied tunnel counter pressure (from compressed air or the support fluid) must
   exceed the pore water pressure in the soil.
ii. If the pore water pressure at the distant point II is unaffected by tunnelling, and has the
    same hydrostatic pore water pressure, then there must be an outward hydraulic gradient
    from the face. This implies a net outflow of support fluid (air or slurry) from the face.

The preceding discussion is simplified because it compares conditions at points I and II that
are at the same depth and have the same initial pore water pressure. In reality, initial pore
water pressures will vary vertically across the face. Assuming a hydrostatic profile, the pore
water pressure will vary by 10 kN/m² for every vertical metre of the tunnel face. This is a
challenge for setting pressures for face support fluids and compressed air working. This is
discussed in more detail in Section 10.4 on tunnelling methods.

7.7.2 Buoyancy Uplift in Tunnels

The previous section relates primarily to instability during construction of tunnel faces due
to unbalanced groundwater pressures. However, where tunnel faces are in low-permeability
materials such as clay, if permeable zones exist close below invert and contain high groundwater pressures, then there can be a risk of 'hydraulic failure' of the invert and lower sections

of the tunnel excavation. This risk is relevant during the period after excavation and before the primary lining is completed, and can affect tunnel and cross passage construction by the sequential excavation method (SEM). This type of instability is directly analogous to the failure of the base of a conventional excavation due to buoyancy uplift (Section 7.5.1).

When forming a tunnel in 'free air' (i.e. at atmospheric pressure) close above a confined aquifer, the invert of the tunnel (before placement of the primary lining) can only be stable if there is sufficient thickness of clay below invert to resist the upward groundwater pressures (Figure 7.20a). If the thickness of clay is insufficient to resist the upward groundwater pressures, then the invert may 'heave' or 'blow', and there is a risk of significant inflow of water and soil (Figure 7.20b). The typical circular geometry of tunnels means that there can be some 'arching' of the clay below invert, and undrained shear strength can provide some resistance to uplift. Therefore, it may be the case that the required thickness of clay below tunnel invert is significantly less than would be required to ensure stability below a conventional flat-bottomed excavation (Figure 7.5).

If hydraulic failure occurs in a tunnel, the consequences can be severe. An example from the United Kingdom in the 1980s is described by Clarke and Mackenzie (1994) and

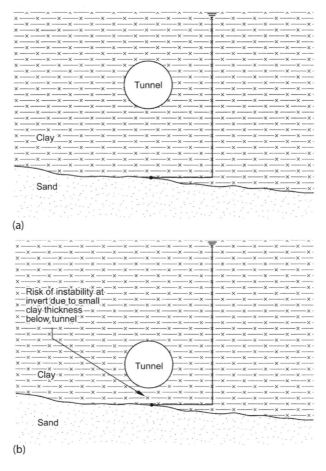

(a)

(b)

Figure 7.20 Instability of tunnel excavation due to buoyancy uplift. (a) Stable conditions – sufficient thickness of clay is present to resist uplift groundwater pressures on tunnel invert and prevent direct seepage into the tunnel from confined aquifer. (b) Unstable conditions – thickness of clay below tunnel invert is insufficient to resist uplift groundwater pressures from confined aquifer.

Newman (2009), where the Tooting Bec Common section of the London Water Ring Main (of 2.54 m i.d.) was being tunnelled through clay soils within the Lambeth Group stratum. The tunnel was constructed under compressed air working, but the air pressure was significantly lower than the groundwater pressure in the Thanet Sand Formation, which formed a confined aquifer only a few metres below tunnel invert. In this case, Newman (2009) reports that a probing exercise (to determine the depth of clay cover between invert and the confined aquifer)

> facilitated a pathway between the Thanet Sand aquifer and the tunnel that, as a result, became inundated by what was described as a sudden inrush of water from the rear end of the tunnelling shield via the annulus between the tailskin and extrados of the last completed ring. The tunnel was evacuated and allowed to fill up with water.

Subsequent recovery works included the construction of two emergency shafts and the use of artificial ground freezing to allow tunnelling to commence from the shaft in the vicinity of the failure. Newman (2009) reports: 'During the recovery operation, it was discovered that over 100 m³ of sand material had been forced into the tunnel through the completed lining, which had been badly damaged in the process'. The tunnel was ultimately completed but with very significant delays and additional costs. This example highlights the importance of understanding the risk of buoyancy uplift of the inverts of tunnels and other underground works such as cross passages.

If the risk of buoyancy uplift is a concern, geotechnical analyses (allowing for the undrained shear strength of the soil and the cylindrical tunnel geometry) should be carried out – an example is described in Soler *et al.* (2016). The objective should be to determine whether the thickness of clay below tunnel invert is adequate to resist upward groundwater pressures from the confined aquifer.

Consideration should also be given to how confident the tunnel designers are that the following two conditions do not occur: a) natural variations in the clay thickness below the tunnel where the clay thins significantly; or b) vertical pathways (such as boreholes or wells) that would allow water to flow directly into the tunnel from the pressurized zone. Either of those conditions could be problematic for tunnel stability.

Possible mitigation measures to reduce the risk of this type of instability during tunnel construction include:

1. Use of pumped wells alongside the tunnel alignment to depressurize the confined aquifer.
2. Use of in-tunnel pumped wells or relief wells, drilled downwards to allow water from the confined aquifer to flow into the tunnel in a controlled manner, from where the water is pumped away. The use of in-tunnel wells is appropriate to target permeable zones of limited thickness and/or limited recharge, where total flow rate will be modest.
3. Use of tunnelling methods that apply a counter pressure (such as compressed air or closed face TBMs) to balance the uplift pressures.
4. Use of ground treatment methods (such as grouting or artificial ground freezing) to improve the strength of the soil in a zone below tunnel invert, thereby providing additional resistance to uplift pressures.

7.7.3 Groundwater Instability Effects on Shaft Construction

Tunnel projects commonly involve the construction of shafts (vertical excavations, often, but not always, circular) between which the tunnel is driven. Other shafts may be needed for access or ventilation as part of the completed scheme. Shafts are at risk of the same types of instability (of the sides and the base) as conventional excavations, and groundwater control methods using pumping and exclusion are routinely deployed. However, some methods have been developed that are specific to shaft construction, including compressed air working and the 'wet caisson' method. Methods of shaft construction and associated groundwater control are discussed in Section 10.5.

One of the primary challenges with shaft sinking is that working space in the base of the shaft is limited, and some excavation and lining methods, such as the underpinning technique, require vertical faces to be cut in the soil (see Figure 6.3). Achieving workable and stable conditions is crucial to safe and efficient excavation and construction. While sump pumping is commonly used during shaft sinking (where soil conditions allow), the sump and associated equipment can be an impediment to construction, and it is often preferable to use external pumped wells to lower groundwater levels in advance of excavation. Perimeter exclusion methods (see Chapter 18) such as steel sheet-piles, diaphragm walls and secant pile walls are also often used to reduce the requirement for groundwater pumping.

Shafts are often relatively deep and may be excavated through soils of different permeability. Stability problems sometimes occur if shaft excavation progresses downwards from a relatively permeable stratum (such as a sandy gravel) into a stratum of much lower permeability (such as a clay). Figure 7.21 shows a shaft being sunk in these conditions, with a system of external deep wells used to lower groundwater levels by pumping from the permeable

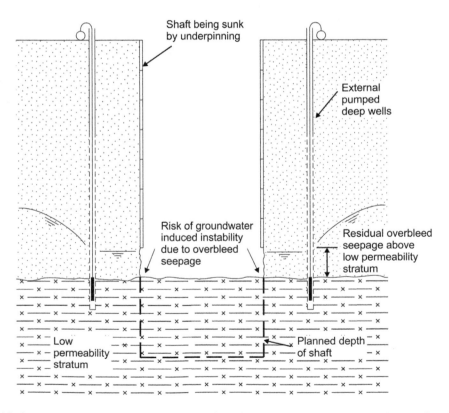

Figure 7.21 Geotechnical instability during shaft sinking due to overbleed seepage at the interface between a permeable and a less permeable stratum. Pumping from wells cannot completely 'dewater' the gravel stratum, and residual overbleed seepage will occur just above the low-permeability stratum. This can cause localized instability in this zone when soil is excavated as part of the underpinning process.

gravel stratum. However, this will not completely 'dewater' the gravels, and some residual 'overbleed' seepage will occur at the gravel/clay interface and can cause local instability. Because of the vertical excavated faces and the limited space, it can be difficult to deploy the usual stabilization measures for localized seepage (Section 7.6). If stability in this zone becomes a problem, groundwater exclusion methods such as steel sheet-pile walls or permeation grouting of the gravel may be necessary. Occasionally, a ring of in-shaft wells (Section 10.5.3.3) may be used to better control groundwater at the interface.

7.7.4 Groundwater Instability Effects on Other Underground Works

In addition to tunnels and shafts, tunnelling projects often include other 'secondary' excavations such as cross passages, connector tunnels, adits and station caverns. These excavations often have irregular geometries and are constructed by mechanized or hand mining without the protection of shields or TBMs.

Depending on the geometry of a given excavation, the types of instability that can occur may be similar to tunnels or shafts (or a combination of both). Particular applications of groundwater control techniques have been developed for these geometries, as described in Sections 10.5 and 10.6.

Chapter 8

Groundwater Problems for Excavations in Rock

The facility with which the water will pass from one part of the sandstone to the other, depends principally on the size of the fissures, their character and direction

Robert Stephenson, *Report on the Supply of Water to the Town of Liverpool*, 1850

8.1 INTRODUCTION

Excavation and tunnelling in rock below groundwater level present different challenges to excavations in soils. Groundwater flows and water pressures still cause problems, but the mechanisms are rather different. This chapter discusses the various large-scale and small-scale stability problems that can occur in rock, including instability of excavation side slopes and bases. The particular groundwater problems associated with tunnelling in rock are also discussed. The available methods of groundwater control will be outlined in Chapter 9 for excavations and Chapter 10 for tunnels.

8.2 GROUNDWATER MANAGEMENT PROBLEMS IN ROCK

In the context of the design and implementation of groundwater control, the term 'rock' means geological deposits formed from mineral grains or crystals cemented together – this is distinct from the uncemented soil deposits discussed in Chapter 7. Typically, in rock, the flow of groundwater will be predominantly through fractures (which can be formed by joints, solution features and other discontinuities), although intergranular flow can occur in some rock types and in weathered rock (differences between groundwater flow in soils and rock are discussed in more detail in Sections 4.3 and 4.4). In a geotechnical context, the distinction between rock and soils is based on the behaviour of a stratum, not its origin or age alone. For the purposes of groundwater control design, materials would be classified as rock if there was a significant element of groundwater flow through the stratum via fracture flow (which is sometimes termed secondary permeability), rather than being entirely by intergranular flow (primary permeability). During excavation, weaker rocks may be dug by heavy duty conventional excavation plant. Stronger rocks may be difficult to excavate by conventional plant alone and may require the use of hydraulic breakers or explosives.

The engineering behaviour of rocks is controlled by the nature of the rock (its composition and mineralogy), the effect of weathering/decomposition of the rock and small-scale and large-scale fractures. From a groundwater control perspective, it is tempting to think that the purely 'geological' nature of a given rock provides little useful information to a designer. However, this is not true, because identifying and understanding the origins and stress history of the rock can aid prediction of its behaviour; useful background on geology can be found in Blyth and de Freitas (1984). There are three principal geological classifications of rock types:

- Igneous rocks. These include the oldest rocks on earth and are formed through the cooling and solidification of magma or lava. Examples of this rock type include granite and basalt. These rocks can contain fractures formed by cooling joints, which are apparently randomly oriented. Igneous rocks can also occur as intrusions within other rock types.
- Sedimentary rocks. These are formed by the deposition and cementation of mineral or organic particles, and often have distinctive layering or bedding from their deposition. The particles forming the rock may be mineral fragments eroded from older rocks or can be the organic remains of tiny organisms. Examples of this rock type include mudstone, sandstone and limestone. Bedding layers in sedimentary rocks can be laterally persistent and have an effect on groundwater flow (see Section 8.3.1). Layers of different permeability can occur in these types of rock (see Section 8.7).
- Metamorphic rocks. These are formed by the alteration of pre-existing rocks by the action of geological processes creating heat and pressure. Examples of this rock type include marble and dolomite. Faults and shear zones are a common occurrence in metamorphic rocks.

Across the whole range of excavations below groundwater level for engineering (construction, tunnelling and mining) projects, groundwater problems can occur in all three rock types. However, due to their presence over around three-quarters of the Earth's surface, sedimentary deposits are the most common type of rock where groundwater causes problems for excavations.

It is important to realize that some weathered rocks may lose their structure and can act, for geotechnical purposes, like soils. Preene and Roberts (2017) describe groundwater control problems in the Chalk, a fractured limestone that forms a major aquifer in parts of the United Kingdom. They note that while the less weathered grades of Chalk are dominated by fracture flow and can be treated as rock, more weathered material may be 'structureless' with little fracture flow and will often act in the same geotechnical manner as soils. Weathering of rocks is discussed in Section 8.3.4.

The concept of groundwater management in excavations was introduced in Section 6.4 and is essentially the approach of 'dewatering' an excavation by removing the groundwater that enters, without considering the destabilizing effect of groundwater seepage into the excavation. Chapter 7 discussed how when applied in soils, this simple approach can lead to instability of the sides and base of an excavation. In contrast, groundwater-induced instability is much less of a problem for excavations in rocks, and simple water management can often be the primary objective of groundwater control.

In relatively stable and unweathered rock, an excavation below groundwater level will encounter localized inflows where fractures are intercepted. Provided the rock around the fractures is competent, these inflows may not cause stability problems. In such cases, the aim is to prevent the inflows from affecting construction operations, and sump pumping is often used to manage water within the excavations (Figure 8.1).

Figure 8.1 Seepage into an excavation in rock. The excavation is in fractured mudstone. Groundwater inflow is predominantly via fractures. The excavation is relatively stable, and sump pumping can be used effectively. Some 'ravelling' of the excavation can be observed where the block size is small.

In addition to general water management to keep rock excavations workably dry, there can be other groundwater control requirements:

i. Reduction of water pressures in slopes and tunnel faces. This improves effective stresses and increases stability, and can allow excavation sides slopes to be formed at steeper angles (Section 8.5).
ii. Pressure relief of the strata below the base of the excavation to reduce the risk of hydraulic failure of the base (Section 8.6).
iii. Diversion of seepages (for example using drainage holes) to collect groundwater inflow in convenient areas of the excavation, avoiding problematic seepages that might interfere with construction activities (Section 8.7).

8.3 GROUNDWATER EFFECTS ON EXCAVATIONS IN ROCK

The destabilizing effect of groundwater on excavations in soils was described in Chapter 7 and can be understood in terms of the principles of effective stress. Groundwater can destabilize rock excavations in a directly comparable way, again controlled by effective stress (Section 6.5). Therefore, reducing water pressures can improve the stability of excavations in rock.

The principal difference is that in soil, groundwater flow is typically distributed across the majority of the mass of a soil stratum, with the permeability controlled, in large part, by the size of the soil pores (which in turn is related to the particle size distribution). In contrast, the undisturbed matrix of rock is typically of very low permeability and will only allow groundwater to pass at very low rates. However, most rock strata contain fractures, which, depending on the fracture opening, spacing and orientation, can allow groundwater to flow more freely (permeability in rock is discussed in more detail in Section 4.4). Apart from in weathered rock, groundwater control systems in rock invariably target groundwater flowing in the fracture network.

8.3.1 Groundwater and Small-Scale Fractures

It has been discussed previously that the permeability of rock is controlled largely by fractures on various scales (Figure 8.2). On a small scale, the fracture networks that might be observed in core recovered from investigation boreholes have an important effect on groundwater flow; this is discussed later. On a larger scale, geological structures such as faults and shear zones can also control groundwater flow; geological structures are discussed in Section 8.3.2.

In this book, 'fracture' is the general term used to describe all types of discontinuities caused by mechanical stresses in bedrock (rather than those caused by dissolution or erosion). The most common types of fracture are classified as 'joints', which are fractures formed by tension or compression but with no significant shear movement in the plane of the joint. The geometry of a planar fracture is defined by the strike, dip and dip direction (Figure 8.2c). The dip of a fracture in relation to the proposed excavation has a key effect on groundwater inflows to an excavation or tunnel (Figure 8.2). Discussion of the fundamental properties of fractures can be found in texts on rock mechanics and rock engineering, such as Hencher (2015) and Palmström and Stille (2015).

In most sedimentary rocks, bedding planes play a key role. These are the result of changes in deposition when the sediment forming the rock was deposited and are typically laterally persistent. Often, the bedding planes can cause significant anisotropy in the permeability of the rock (Figure 8.3). When first formed, bedding planes will be essentially horizontal, but if strata are folded or faulted, then bedding planes may acquire a 'dip' inclination to the horizontal or may become locally disrupted.

When considering groundwater problems, we are interested in the hydraulically significant fracture networks – in other words, those that can influence groundwater pressures and inflows that an excavation might experience. Groundwater inflow will be influenced by the basic geometry of a fracture where it intercepts the excavation (Figure 8.2) as well as by the opening width (at right angles to the plane of the fracture), the roughness of the fracture surfaces and any infill material (sediment or weathered rock) within the fracture. Furthermore, for a set of fractures to be hydraulically significant, they must be laterally persistent and connected to a source of groundwater. There have been cases during construction of deep tunnels in fractured rock where, once small volumes of water have drained from the rock immediately around the tunnel, the fractures intercepting the tunnel were 'dry' despite the tunnel being more than 100 m below groundwater level. The explanation for this is twofold. First, at depth, the rock stresses will be high, and this tends to keep fractures 'tight' or closed, reducing the permeability of individual fractures. Second, the rock around the tunnel may be separated hydraulically from surface recharge of water by many metres of lower-permeability rock acting as an aquitard. When water local to the tunnel is drained, there may be no long-term supply of groundwater to maintain seepage into the tunnel.

Given the importance of fracturing to the engineering behaviour of rock, various 'rock mass classification systems' have been developed to describe, in standard terms (for a given

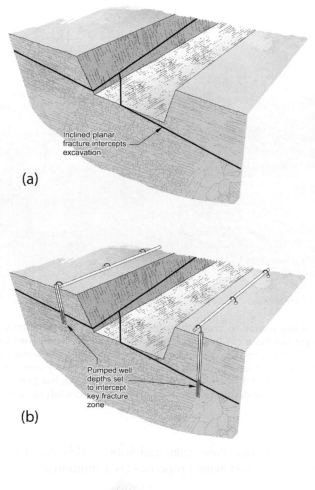

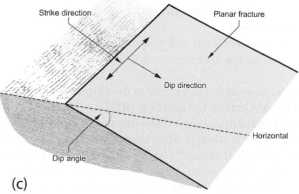

Figure 8.2 Effect of fracture orientation on groundwater inflow to excavations and tunnels. (a) Intersection between excavation and a dipping planar fracture. The fracture potentially controls the inflow of groundwater into the excavation. (b) Pumped well groundwater control system with wells targeted to key fracture zone. (c) Geometry of planar fractures.

Figure 8.3 Effect of bedding planes to create anisotropic permeability conditions in sedimentary rock. A vertical face has been cut in rock of sedimentary origin that comprises alternating beds of mudstone and limestone. The permeability is significantly greater along the bedding planes than at right angles to the plane. The bedding planes will have originally been approximately horizontal when the rock was laid down. Over geological time, the rock has been folded, and the bedding planes (and direction of dominant permeability) now dip from left to right in the photo at around 10 degrees.

system), the rock characteristics. Palmström and Stille (2015) state that these classification systems typically address the following properties (as a minimum):

- Rock type and material, including strength.
- Weathering and alteration.
- Fracture occurrence, orientation and properties.
- Weakness zones, including faults.

Several different classification systems have been developed for various engineering applications. Some (such as Rock Quality Designation [RQD] – Deere *et al.*, 1966) are essentially factual descriptions of fracturing observed in core recovered from boreholes. Other methods, such as the Rock Mass Rating (RMR) (Bieniawski, 1973), Q Classification (Barton *et al.*, 1974), and the Geological Strength Index (GSI) (Hoek, 1994), produce rock parameters that can be used directly in some engineering design methods. Some of the classifications, including the RMR and Q systems, include an element of assessing the potential for groundwater flow in the rock.

Palmström and Stille (2015) highlight that during ground investigations for rock engineering projects, there is often a strong focus on detailed description of rock cores recovered from boreholes. Such fracture information from core is valuable and necessary, but it is essential that this information is viewed in the context of a conceptual model (Section 5.4 and 13.4) that includes the wider geological structures and information from permeability tests (Section 12.8). Figure 8.4 shows a fractured limestone bed exposed in an excavation. It is clear that in

Figure 8.4 Discrete fractures in a limestone bed exposed in an excavation. The matrix of the rock is of much lower permeability than the fractures. In this case, groundwater flow will be dominated by the relatively widely spaced fractures. If a cored borehole missed the fractures, description of the core may give the misleading impression that this layer is of very low permeability. Note the rock in the photographs is exposed within an excavation, and the visible fracture openings may be wider than when the rock is in situ at depth.

this case, a cored borehole could pass through the massive (unfractured) part of the rock and miss the significant fractures. Taken in isolation, such a core would imply that this layer was of very low permeability, whereas in reality, water can pass freely via the widely spaced open fractures. This highlights why natural outcrops or larger exposures of rock (e.g. in quarries) are mapped during large-scale ground investigations on major projects. Such mapping can give a better perspective on the overall nature of fracture networks.

While groundwater flow within fractures is important in almost all rock types, if the rock matrix is potentially porous and permeable, then the flow system may be 'dual porosity', where groundwater flow occurs both in fractures (secondary permeability) and in the rock mass (primary permeability) itself. Dual porosity conditions can occur in weathered hard rocks that have locally been broken down into sandy and clayey materials, or rocks that have a potential for intergranular flow in the rock matrix, such as fractured sandstones (Figure 8.5). In these settings, much of the groundwater flow occurs by fracture flow, but when groundwater levels are lowered, pore water can be released from storage in the rock mass.

8.3.2 Effect of Isolated Highly Permeable Fractures

As described in Section 4.4, the potential water flow rate through a fracture is proportional to the third power of the fracture opening – the so-called 'cubic law'. A doubling of the fracture opening theoretically results in an eightfold increase of water flow. There have been many cases of excavation and tunnelling in rock where a small number of open fractures have dominated groundwater flow and resulted in very high inflows. Hartwell (2015) describes a shaft sinking project in Croydon in the United Kingdom, where a pumped flow rate of 225 l/s was ultimately required when a shaft was formed through Chalk (a type of limestone). Investigations concluded that this flow was probably largely due to the presence of a single fracture of approximately 100 mm opening that intercepted the base of the shaft.

Figure 8.5 Vertical cut face in sandstone. A vertical face of sandstone is cut below the toe of the sheet-pile wall. The groundwater flow in the sandstone is dual porosity and occurs in a combination of closely spaced fractures (secondary permeability) and intergranular flow (primary permeability) through the rock matrix.

When investigating and planning works below groundwater level in fractured rock, it is appropriate to try to identify any 'hydraulically significant' fractures with large apertures that may contribute significant groundwater flows. Unfortunately, it can be difficult to identify such fractures based on core descriptions alone. The best approach is to use methods that induce groundwater flow, such as borehole geophysics under pumped conditions (Section 12.8.7) or packer permeability tests (Section 12.8.3).

While individual or multiple open and highly permeable fractures are often associated with large groundwater flow rates to excavations and dewatering wells, it is important to recognize that such open fractures do not 'create more water'. In fact, the role of such fractures is to provide preferential pathways to draw water to the zone of drawdown from a wider area than otherwise. This is illustrated by Figure 8.6, which shows a shaft being sunk through fractured rock. The sides of the shaft are lined with a layer of sprayed concrete lining (SCL). To aid the SCL works, water pressures and inflows through the sides of the shaft are reduced by a system of pumped deep wells external to the shaft.

a) Figure 8.6a shows a case where the fracture network around the shaft is dominated by a large number of relatively 'tight' small aperture fractures, each of which can only transmit a small flow rate to a well. So, the average yield of a dewatering well will be relatively small, and a large number of closely spaced, low-yielding wells will be required to lower groundwater levels around the shaft.

b) Figure 8.6b shows a fracture network that contains a small number of fractures of wider aperture that can transmit much greater flow rates. Well A, which intercepts a wider fracture, will have a much larger yield than Well B. The network of smaller fractures is connected into the wider fracture and feeds water into it when water pressures are lowered by pumping. It provides a hydraulically efficient means for Well A to collect groundwater from a much larger volume of rock than Well B. In these conditions, the yield of dewatering wells will vary depending on how each well connects into the fracture network; wells that intercept the more hydraulically significant fractures may

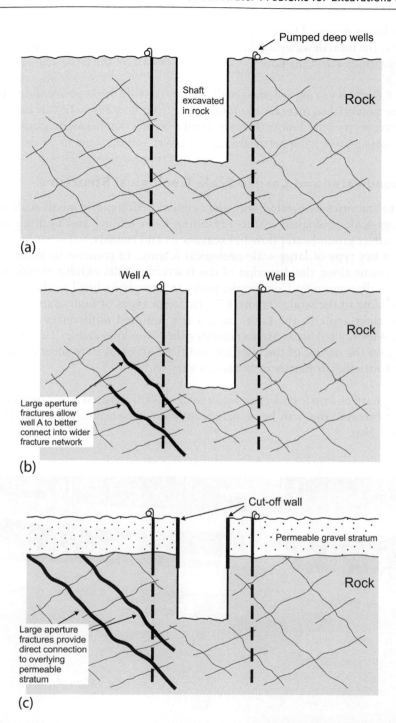

Figure 8.6 Effect of hydraulically significant fractures. (a) Dewatering wells around the shaft connect into a large number of small-aperture fractures. A large number of low-yielding wells is required. (b) Some dewatering wells (Well A) around the shaft connect into a few large-aperture fractures. These wells will have large yields. The total dewatering flow rate will be similar to case (a), but fewer wells may be required. (c) Large-aperture fractures provide a direct connection to the overlying permeable stratum, which acts as a source of recharge. In this case, the large-aperture fractures may increase the total dewatering flow rate.

require larger pumps. However, when comparing Figure 8.6a and b, it is important to note that the total dewatering flow rate will be essentially the same in both cases. The wider fractures affect the flow to wells but do not affect the large-scale permeability of the rock.

c) Figure 8.6c shows a case where a small number of wide, high-permeability fractures do have the potential to increase the total dewatering flow rates. In this case, the permeable fractures provide a more direct hydraulic connection between a source of recharge (in this case an overlying permeable gravel layer) and the zone to be dewatered.

8.3.3 Groundwater and Larger-Scale Geological Structure

In addition to fractures associated with the original bedding and small-scale structure of the rock, large-scale geological features resulting from folding and faulting of rocks can significantly affect groundwater flow to excavations and tunnels.

Faults are a key type of large-scale geological feature. In contrast to joints (where there is no displacement along the direction of the fracture), faults exhibit significant displacement along the discontinuity. Commonly, faults are steeply inclined to the vertical and can cut across bedding of the strata (Figure 8.7). Different types of faults can vary in thickness (measured at right angle to the fault) from a few hundred millimetres to several tens of metres. Very wide faulted areas are sometimes called 'weakness zones' or 'shear zones'.

Depending on the nature of the bedrock and the geological movements that gave rise to faulting, the hydraulic properties of faults can vary widely:

i. Faults in hard or brittle rocks can result in intensely fractured zones (sometimes known as 'fault breccia') that can be of significantly greater permeability than the host rock (Figure 8.18a).

Figure 8.7 Steeply dipping features. A vertical face of sandstone is cut below the toe of the sheet-pile wall. Several steeply dipping large fractures (possibly faults) are visible.

ii. Many faults contain 'clay gouge' – a very low-permeability material created from the host rock by the stress and disturbance of the faulting. This can cause a fault to act as a low-permeability feature, which can form a barrier to horizontal groundwater flow (Figure 8.8a).
iii. Vertical displacement of alternating higher- and lower-permeability beds can offset the beds, impeding groundwater flow along the beds (Figure 8.8b).

Faults with the hydraulic properties of cases (ii) and (iii) can act as significant barriers to groundwater flow. This is sometimes described as 'compartmentalization' of groundwater flow, where pumping from dewatering wells in one compartment may have very little drawdown effects in other compartments, which are separated from the wells by faults. At the boundaries between compartments, there can be large changes in piezometric level over short horizontal distances (Figure 8.8a).

8.3.4 Weathered Rocks

Many engineering projects deal with rocks that are characterized as being weathered. Weathering describes the range of processes whereby the properties of rock are altered due to exposure to the wider environment. Weathering processes can be categorized as:

- Physical weathering – primarily disintegration processes caused by water, wind and temperature effects. This often includes an increased frequency of fractures; with more intense weathering, rock can become so degraded it may act as 'soil' for geotechnical purposes.
- Chemical weathering – decomposition processes caused by the breakdown and transformation of minerals into new compounds by chemical agents in the air and water. Chemical weathering caused by water is often the most severe near fractures.
- Biological weathering – disruption of the rock due to the actions of plants and animals.

As a general rule, weathering degrades rock and reduces strength. However, the spatial distribution and nature of weathering can vary greatly. Weathering profiles are controlled by many factors, including the original rock type, geological structures, groundwater conditions, topography and climate. Further background on weathered rock can be found in Price (1995) and Hencher and McNicholl (1995).

From the perspective of groundwater control, weathering is relevant in two principal ways:

- Reduction in strength due to weathering. Typically, this makes weathered materials more prone to groundwater-induced instability compared with more competent, less weathered rock.
- Changes in permeability. Weathering can cause significant changes in permeability compared with the unweathered host rock. There are no simple guidelines on how permeability will be affected by weathering; in some rocks, weathered materials can be more permeable than the original material, while in other conditions, weathered rock may be of reduced permeability. This is illustrated by the example of Chalk (a type of limestone), discussed in Preene and Roberts (2017). They report that the softer chalk in the south of the United Kingdom typically weathers to a very low-permeability 'structureless chalk' (permeability 10^{-7} to 10^{-9} m/s), but similarly weathered 'structureless chalk' derived from the harder chalks in the north of the country can be several orders of magnitude more permeable (permeability 10^{-5} to 10^{-3} m/s).

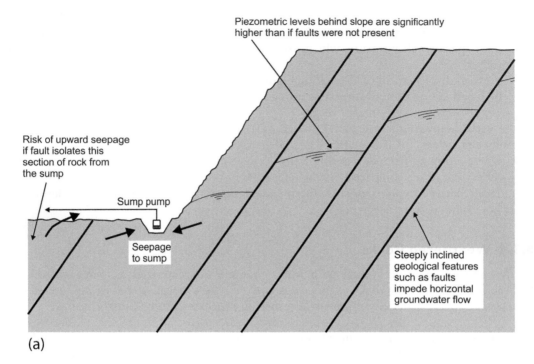

Piezometric levels behind slope are significantly higher than if faults were not present

Risk of upward seepage if fault isolates this section of rock from the sump

Sump pump

Seepage to sump

Steeply inclined geological features such as faults impede horizontal groundwater flow

(a)

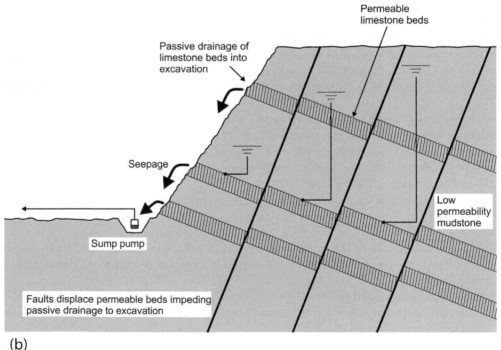

Permeable limestone beds

Passive drainage of limestone beds into excavation

Seepage

Sump pump

Low permeability mudstone

Faults displace permeable beds impeding passive drainage to excavation

(b)

Figure 8.8 Hydraulic compartmentalization of groundwater in slopes. (a) Barriers to horizontal groundwater flow caused by steeply inclined faults. (b) Disruption of alignment of permeable beds caused by steeply inclined faults.

8.3.5 Soluble Rock and Karst Groundwater Conditions

Karst groundwater conditions were discussed in Section 4.6.2. These conditions can occur in soluble rocks (typically carbonate rocks) or rocks that contain beds or zones of soluble evaporites (such as gypsum or halite). The flow of groundwater can cause dissolution of material. This can enlarge fractures to form 'solution features' including conduits (significantly enlarged fractures) and caves (conduits that are large enough to be explored physically by humans). In these conditions, groundwater flow can become very concentrated in a limited number of enlarged flow channels (Figure 4.9), and very large groundwater velocities (up to several kilometres per day) can occur under the influence of pumping (MacDonald *et al.*, 1998).

If karst conditions may be present on a groundwater control project, it should be recognized that even extensive borehole drilling programmes may fail to intercept any significant karst solution features and could mistakenly imply that the bedrock was massive and of low permeability. Desk studies (Section 11.6) carried out by suitably experienced geologists and geomorphologists can be very useful, and the use of geophysical techniques should be considered to try to identify any potential solution features. Further information on the geology of karst conditions can be found in MacDonald *et al.* (1998) and Waltham (2016).

Groundwater control projects in karst aquifer conditions are at risk of very high pumping rates, particularly if significant networks of permeable solution features are in the zone affected by the project. There have been cases where excavations were initially dewatered with large, but manageable, pumped flow rates. In these cases, problems occurred when the flow rate required to hold down groundwater levels increased over a period of weeks. The rate of increase is often slow at first but gradually accelerates until all pumping efforts are overwhelmed and the excavation floods. These effects are often attributed to the groundwater flow caused by pumping being concentrated in solution features that are partly filled by sediment or redeposited minerals. The high groundwater flow velocities can erode and wash away the infill, increasing the flow that can pass through the channels and dramatically increasing the flow rates of groundwater reaching the excavation.

8.4 GROUNDWATER-INDUCED INSTABILITY IN ROCK

For excavations in rock, instability effects can be caused by groundwater pressures and groundwater flow. Groundwater control solutions typically try to manipulate these two phenomena to improve stability.

Methods based on pumping have the objective of reducing water pressures in saturated fractures (and in the rock mass itself in weathered rock or in dual porosity conditions). This will increase effective stress, improving stability. Groundwater control methods based on exclusion reduce groundwater flow towards the excavation so that when the excavation is dug, passive drainage will be more effective in reducing water pressures.

There are several specific problems that may result from groundwater inflows into rock excavations:

(i) In very large rock excavations, there may be a risk of 'block failure', where part of the rock mass slides or moves into the excavation, often moving along fractures that slope into the excavation. This is discussed in Section 8.5.

(ii) If the rock possesses bedding that is roughly horizontal, or dipping at only a shallow angle, groundwater pressures below the base of the excavation may remain trapped beneath lower-permeability layers, leading to the 'buoyancy uplift'-type base stability problem. This is discussed in Section 8.6.

(iii) Localized groundwater problems are discussed in Section 8.7, and include:
  a. The concentration of flow where water flows from one or more fractures into the excavation may create high flow velocities. This can give an associated risk of erosion in and around the fracture and of dissolution of material in soluble rocks.
  b. Weathered beds or zones (which may exist within otherwise competent rock) that behave like soil and may slump inward as groundwater flows through them.
  c. Groundwater emerging into the excavation at locations dictated by the rock structure, bedding, fractures, etc. These locations will often be inconvenient for construction operations and can cause problems.

Groundwater problems relevant to tunnelling in rock are highlighted in Section 8.8.

8.5 SLOPE INSTABILITY

The stability of rock slopes is affected by groundwater. The potential failure mechanisms will depend on the nature of the rock:

  a) If the rock is very intensely fractured (with very closely spaced fractures), or has lost its structure and strength due to weathering, a slope will fail in a structureless manner. The rock can act as an equivalent porous medium, and failure surfaces will not be controlled by fractures and rock structure (Figure 8.9a).
  b) Where the rock has significant fractures and structure, these features tend to be weaker than the overall rock mass. If the geometry provides a mechanism that allows blocks of rock to move as part of the failure mechanism, slope failures will tend to be controlled by fracturing (Figure 8.9b).
  c) It is also possible to have an intermediate condition where the failure mechanism is partly controlled by fractures and partly by the rock mass. One such failure mechanism is known as 'ravelling', which is the process by which slopes fail by gradual movement of large numbers of relatively small blocks, with no defined failure surface.

Other small-scale groundwater-induced instability problems that may affect slopes are discussed in Section 8.7.

Water pressures acting in the potential failure zones (in the rock mass or on fractures) can have a critical effect on the stability of slopes. The application of measures to reduce water pressures in rock slopes can be very beneficial in improving the stability of existing slopes or allowing proposed slopes to be constructed at steeper angles than otherwise. In the mining industry, where open pits may be several hundred metres deep, there can be huge cost savings associated with steepening the pit slopes by even a few degrees. Many open pit mines have extensive 'pit wall depressurization' programmes whereby groundwater drainage measures are used to ensure the stability of open pit slopes in rock. Further details are given in Beale and Read (2013).

Commonly used approaches to reduce water pressures within rock slopes are:

  a) Passive drainage into the excavation by natural drainage pathways (Figure 8.10a). This approach allows the slope to drain into the excavation via the natural fractures and other drainage pathways; the water is collected in the excavation and pumped away by sump pumps. This can be effective if the geometry of drainage pathways is favourable to allow water to flow to the excavation and provided there is no close source of recharge that could maintain high water pressures in the slope.

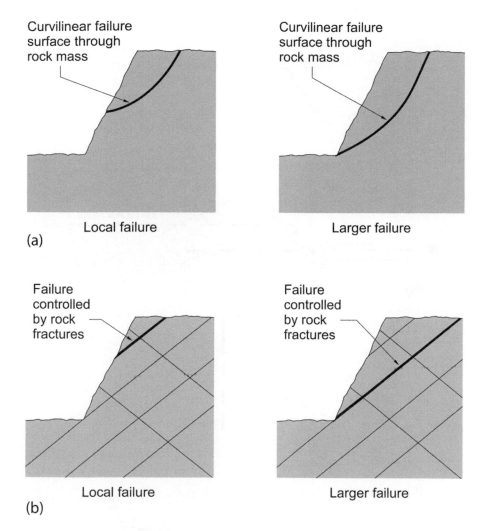

Figure 8.9 Potential failure mechanisms in rock slopes. (a) Rock is very closely fractured (or has been weakened and destructured by weathering), and failure surfaces are not controlled by fractures. (b) Failure controlled by fractures and rock structure.

b) Passive drainage into the excavation assisted by artificial drainage pathways. (Figure 8.10b). If the fracture network (or barriers caused by large-scale geological features such as faults) does not allow water to drain easily into the excavation, artificial drainage pathways can be created to aid drainage. Sub-horizontal drains (Section 17.5) are often used.

c) Active drainage by vertical wells (Figure 8.10c). Vertical pumped wells can be used to lower groundwater levels in a similar approach to that used in soils. The success of this approach relies on the wells connecting into the same fracture network that controls water pressures in the slope.

d) Low-permeability cut-off walls used to aid drainage by reducing lateral recharge (Figure 8.10d). Where near-surface deposits (weathered rock or drift deposits) are highly permeable and feed water to slopes, cut-off walls can be used to reduce lateral groundwater flow. This can allow drainage measures at the slopes to be more effective.

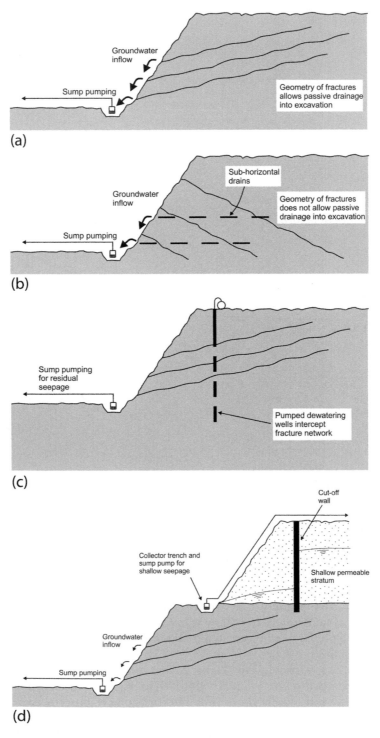

Figure 8.10 Groundwater control strategies to improve the stability of rock slopes. (a) Passive drainage into the excavation by natural drainage pathways. (b) Passive drainage into the excavation assisted by artificial drainage pathways. (c) Active drainage by vertical wells. (d) Low-permeability cut-off walls used to aid drainage by reducing lateral recharge.

e) Where depressurization of large volumes of rock is required, drainage tunnels or adits (Section 17.7) can be driven behind or below a slope to drain groundwater. Due to the high cost and long timescales required for tunnel construction, this approach is applied only rarely.

Where the bedding of the rock is approximately horizontal, it is often the case that anisotropic conditions exist, where permeability is much lower in the vertical direction (at right angles to bedding planes) compared with the horizontal direction (along bedding planes). This means that even with active depressurization by pumped wells and passive drainage from the face, the slope cannot be made completely 'dry' (Figure 8.11). Some residual seepage will remain at the base of each permeable bed with a small residual water pressure.

8.6 BASE INSTABILITY

Basal instability of excavations in soil was discussed in Section 7.5, which identified several mechanisms by which groundwater inflow or groundwater pressures can cause instability. In general, unweathered rocks are stronger and more cemented than soils, and basal stability is often not a major concern in rock excavations. However, there are examples where basal instability has been a problem in rock excavations, and designers should include this aspect in their assessments.

Weathered and very weakly cemented rocks effectively act in the same geotechnical manner as soils and can be at risk of the mechanisms of buoyancy uplift, fluidization due to upward seepage gradients, internal erosion and piping, as outlined in Section 7.5. In

Figure 8.11 Rock excavation in sub-horizontally bedded limestone. The darker zone in the rock slopes is due to residual seepage from a permeable bed that is underlain by a lower permeability layer. (Courtesy of M Bickley.)

contrast, in relatively strong rock, the two most relevant failure mechanisms affecting the base of an excavation are:

i. Buoyancy uplift (Section 8.6.1)
ii. Enlargement/erosion of water pathways (Section 8.6.2)

Different types of failure can occur in mixed rock conditions, such as when more weathered and less weathered rock are both present below an excavation. The failure mechanism may be controlled by the weaker or more weathered rock.

8.6.1 Base Instability by Buoyancy Uplift

This type of base instability for excavations in soils was described in detail in Section 7.5.1. This mechanism is a risk where excavation formation level bottoms out in a very low-permeability stratum but there is a more permeable 'aquifer' zone at depth containing high groundwater pressures. If the weight of the 'plug' of ground (plus the resisting forces from the shear strength of the rock) is not sufficient to resist the upward groundwater pressures, then the plug of ground can move upward, leading to instability and failure.

Even though rocks are typically significantly stronger than soils, and so the contribution to resisting forces due to shear strength may be greater than in soils, this type of base instability can still occur in rock excavations. Haydon and Hobbs (1977) describe the uplift of base of a deep excavation for a construction of a new power station in the United Kingdom. The base of the excavation was underlain by alternating beds of low-permeability mudstone and more permeable fractured limestone. Investigations indicated that heave was caused by high groundwater pressures in the limestone beds. An attempt made during excavation to prevent hydraulic uplift by drilling relief wells was only partly successful due to the difficulty of intersecting the water-bearing fractures in the limestone.

Examples of where buoyancy uplift may be of concern for excavations in rock are shown in Figure 8.12. The formation level of the excavation is in a very low-permeability rock stratum (such as a mudstone with no significant fracture network). This low-permeability stratum forms a confining layer above a more permeable zone (a confined 'aquifer') that could be a separate 'bed' of rock (Figure 8.12a) or may be a more fractured zone within the same rock type (Figure 8.12b). In these examples, the piezometric level in the aquifer is considerably above excavation formation level, but the excavation is in very low-permeability rock, and direct seepage into the excavation will be small. *Purely from a water management perspective*, simple sump pumping would allow excavation to proceed. However, it would not address the risk of hydraulic failure of the base.

Base stability assessments should review the shortest vertical distance between the confined aquifer and the excavation (level A–A in Figure 8.12) and assess the two separate sets of pressures acting at this level. If we take the conservative approach of ignoring the contribution from the shear strength of the rock, the base of the excavation will be stable provided the downward pressure from the weight of the residual 'plug' of unexcavated rock is sufficiently greater than the upward pressure of groundwater confined in the aquifer. This is assuming that the confining stratum is consistently of very low permeability and is competent and unpunctured (e.g. there are no significant vertical fractures through this zone, and it is not penetrated by any poorly sealed investigation boreholes, which could form water pathways).

If excavation is dug deeper into the very-low permeability rock without any reduction of groundwater pressures in the aquifer zone, there will come a time when the remaining plug

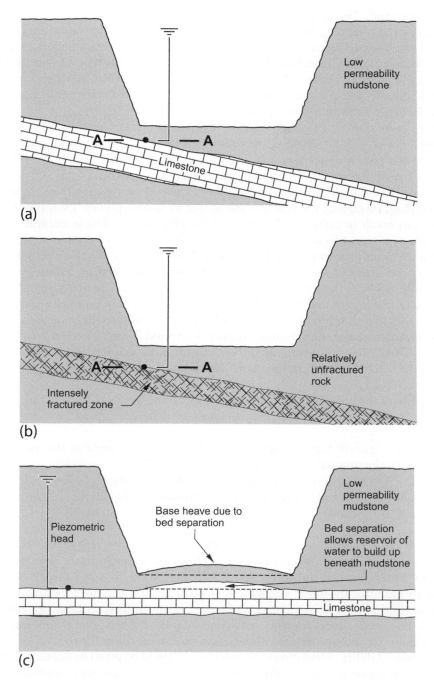

Figure 8.12 Example of base instability due to buoyancy uplift for an excavation in rock. (a) Confined aquifer formed by a more permeable bed of rock is present below excavation. (b) A more permeable fractured zone of rock forms a confined aquifer below the excavation. (c) High groundwater pressures below excavation can cause 'bed separation', where the rock below the excavation moves upwards and allows a reservoir of water to build up beneath.

of rock is 'buoyant'. The upward water pressure in the aquifer will exceed the downward forces keeping the rock plug in place (the weight of the plug plus the contribution from the strength of the rock to resist deformation). There will then be an upward movement or 'heave' of the excavation formation. In rock excavations, the upward movement will often result in 'bed separation', where the rock strength means that the plug of rock can deform upward and initially remain intact (without rupture), allowing a reservoir of water to build up below the low-permeability rock (Figure 8.12c).

However, even though the rock strength can, in some circumstances, allow limited upward movement without rupture, the rock plug will become highly stressed. For wide excavations, the plug will act as a beam (Figure 8.12c) and will be subject to horizontal strains and tensile stresses. Furthermore, the reduction in vertical total stress caused by excavation will cause stress relief and reduce shear strength along any pre-existing fractures in the rock plug, increasing the risk of these acting as failure planes. Even limited upward movement can result in failure and rupture of the plug of rock below excavation, leading to an inrush of water. Moore and Longworth (1979) report an example of the hydraulic failure of a 29 m deep quarry in the United Kingdom. The base of the pit was in very low-permeability mudstone of the Oxford Clay Formation, which was underlain by permeable limestone aquifers with piezometric levels close to ground level. The first stage of failure was approximately 150 mm of uplift of the base of the excavation due to water accumulating by 'bed separation' in a thin limestone layer just beneath the mudstones in the base of the dig. When the base of the excavation finally ruptured (around 3 days after problems were first noted), it released 7000 m^3 of water into the excavation. Subsequent investigations revealed that water had entered the excavation via an area in the base of some 30 m by 10 m that had heaved and broken up.

The stability of the base of an excavation in rock can be assessed by the same methods as for soils, presented in Section 7.5.1, and can take two different approaches:

i. Assessing a 'global' factor of safety F in relation to the weight of the rock plug compared with the upward water pressures (Equation 7.1)
ii. Adopting a 'partial factor' approach to stability calculations (where separate factors are applied to different parameters in calculations; see Equations 7.2 through 7.4), as proposed in Eurocode 7 (BS EN 1997-1:2004)

For approach (ii), it is possible to include the effect of the shear strength of rock as a resisting force to act against groundwater pressures. However, as noted in the discussion of base stability for excavations in soil in Section 7.5.1, there must be some displacement of the rock plug to mobilize the shear strength. It is possible that the displacement to mobilize the shear strength would be significant and would result in unacceptable pre-construction heave of the base of the excavation or could result in failure along pre-existing fractures that have opened up due to stress release. If designers wish to account for the role of rock shear strength in base stability calculations, consideration should be given to numerical modelling of the rock behaviour to analyse the stability of the excavation base, including any pre-failure deformations.

Instability by buoyancy uplift for excavations in rock can be avoided by:

a) Adequately reducing the groundwater pressure by pumping from suitable dewatering wells intercepting the aquifer zone. The wells should lower the aquifer water pressure so that the downward forces exceed the upward pressures by a suitable factor of safety.
b) Reducing the groundwater pressure in the confined aquifer by installing relief wells (Section 17.8) within the excavation footprint. The relief wells provide a preferential

pathway for upward groundwater flow. As the excavation is made below piezometric level, the relief wells will overflow, and the water can be pumped away by sump pumps. This will hold piezometric level at the excavation dig level at the relief well locations. The use of relief wells is appropriate to target permeable zones of limited thickness and/or limited recharge, where total flow from the relief wells will be modest. Provided the relief wells are installed on a grid pattern at a suitably close spacing, this is typically sufficient to achieve stability.

c) Install a very low-permeability physical cut-off wall to penetrate below the base of the confined aquifer. Provided the cut-off walls are watertight, this will prevent further recharge, thus leaving only the water pressure contained in the aquifer within the cut-off walls to be dealt with. This can be affected by installing relief wells prior to excavation, from which limited flow is anticipated. The economics of this alternative will depend primarily on the depth needed to secure a seal and the effectiveness of that seal.

d) Increasing the downward pressure on the base of the excavation by keeping it partly topped up with water during the deeper stages of work. Excavation is made underwater, and a tremie concrete plug is used to seal the base on completion (this is the basis of the 'wet caisson' method used for shaft construction; see Section 10.5.4).

8.6.2 Enlargement/Erosion of Water Pathways

In some types of soil, sump pumping from the base of excavations can lead to instability due to fluidization caused by upward seepage gradients (Section 7.5.2), internal erosion (Section 7.5.3) and piping (Section 7.5.4). Excavations in competent rock tend not to be prone to these failure mechanisms, although these conditions can affect excavations in weathered rock.

Problems caused by upward seepage can still occur in rock excavations but tend to be associated with the enlargement or erosion of pre-existing water pathways. Potential pathways that can initiate this type of failure mechanism include:

a) Natural fractures in the rock, especially sub-vertical fracture networks and larger fractures that are infilled with sediment.
b) Weathered zones within the rock, where the material may be weaker and more erodible.
c) Artificial flow pathways that pre-date the excavation, such as unsealed or poorly sealed investigation boreholes.
d) Artificial flow pathways associated with construction activities, including the boring of piles or wells.

One of the situations in which this problem occurs is when there is significant anisotropy of permeability, so the vertical permeability is much less than in the horizontal direction. This means that significant upward seepage gradients can occur, and the lower permeability zones below the dig can significantly reduce total inflow to the base of the excavation.

Problems occur when the upward hydraulic gradients and the rate of flow are sufficient to open up any pre-existing pathways (listed earlier) by either washing out any existing infill material (clearing out the pathway) or, in weaker rocks, eroding the edges of the pathway (enlarging the pathway). These processes can dramatically increase the water flow rate along the pathway and thence into the excavation. The concentration of flow into a small number of vertical pathways further increases the potential for erosion and enlargement of pathways, which will in turn increase flow rate. If a nearby source of recharge is present, a feedback loop can develop, and an accelerating process of erosion and increased sump pumping rates can occur.

There have been examples of excavations in weak rocks located near to large bodies of surface water (Figure 8.13), where when the excavation is first dug to final formation level, the rates of seepage in the base are relatively modest and can easily be dealt with by sump pumping. Then, over the course of a few weeks, the rates of seepage gradually increase, requiring more sump pumps to be deployed. As time passes, the seepage flows increase at an accelerating rate, and significant volumes of sediment are washed into the base of the excavation. Eventually, the sump pumps are overwhelmed and the excavation flooded. The mechanism is the concentration of flow in eroding pathways beneath the base of the excavation, allowing a very direct hydraulic connection to develop to the body of surface water (Figure 8.13c).

Base failure by enlargement/erosion of water pathways for excavations in rock can be avoided by:

a) Adequately reducing the groundwater pressure below the base of the excavation (below the lower-permeability zone) by pumping from dewatering wells of adequate depth. The wells should lower the groundwater pressures so that there is no upward seepage along the potentially problematic pathways.

b) Installing relief wells (Section 17.8) within the excavation footprint. The relief wells provide a preferential pathway for upward groundwater flow, reducing upward seepage along the potentially problematic pathways. As the excavation is made below piezometric level, the relief wells will overflow, and the water can be pumped away by sump pumps.

c) Install a very low-permeability physical cut-off wall to penetrate sufficiently deeply to cut off the permeable layers. Provided the cut-off walls are deep enough and are watertight, this will prevent further lateral recharge, thus leaving only the water pressure contained within the area enclosed by the cut-off walls to be dealt with. This can be affected by installing relief wells prior to excavation, from which limited flow is anticipated. The economics of this alternative will depend primarily on the depth needed to secure a seal and the effectiveness of that seal.

8.7 LOCALIZED GROUNDWATER PROBLEMS

In addition to groundwater effects that can cause large-scale failures, groundwater can also cause more localized problems in excavations in rock. A number of possible problems are described in the following sections.

8.7.1 Localized Erosion Due to High Groundwater Velocities

Inflows into excavations in rock will typically be via fractures. The concentration of flow where water exits from a fracture into the excavation can result in high flow velocities. If the rock is relatively soft or erodible, there may be an associated risk of erosion at the fracture exit. There have been cases in relatively soft mudstones and sandstones where initial inflows were modest but increased significantly with time as the fractures were enlarged by erosion. One solution would be to install a system of external dewatering wells to reduce the groundwater head driving flow into the excavation. This will reduce the potential for erosion. Alternatively, sub-horizontal drains (Section 17.5) could be drilled into the slope to provide a preferential pathway for water to flow into the excavation.

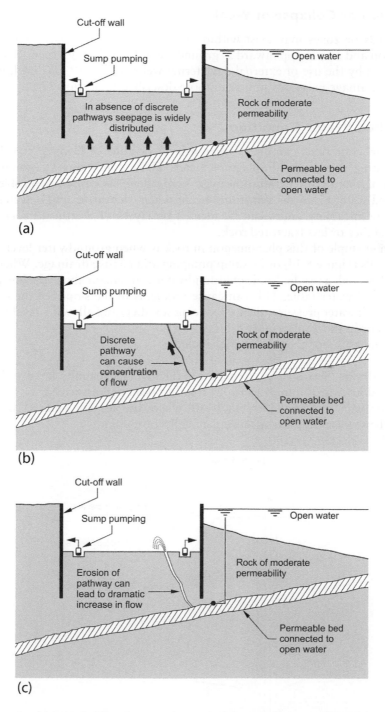

Figure 8.13 Example of base instability due to enlargement/erosion of water pathways in an excavation in rock. (a) In the absence of discrete pathways upwards, vertical seepage from a deep permeable bed is distributed over a wide area. (b) If a discrete vertical pathway exists, this can lead to a concentration of flow in a small area. (c) Concentration of flow can lead to erosion and enlargement of the pathway. A direct hydraulic connection can develop with the nearby open water, leading to a dramatic increase in dewatering flow rate.

8.7.2 Erosion or Collapse of Weak or Weathered Zones

Weathered beds or zones may exist within otherwise competent rock. These zones may behave like soil and may slump inward as groundwater flows through them. These problems may be avoided by the use of external dewatering wells or, if very localized, by the use of passive slope drainage measures similar to those described in Section 17.10.

8.7.3 Perched Water Conditions

'Perched water' conditions can occur when a layer or zone of lower permeability occurs locally, inhibiting downward drainage of water. These conditions are common in soils, where layers of clay or silt within otherwise sandy deposits can cause problems (Section 7.6.2). Perched water problems sometimes occur when excavating and dewatering in rock, where lower-permeability zones can be formed by layers of weathered rock, or beds of lower-permeability or less fractured rock.

A common example of this phenomenon in rock is when groundwater levels are lowered by pumped wells (Figure 8.14) or by sump pumping and passive drainage. When groundwater levels are lowered, a residual zone of groundwater can remain above the low permeability zone – a perched water table. This can cause problems in the slopes of the excavation by maintaining high water pressures (hence reducing stability) and allowing residual 'overbleed' seepage to occur in the slope (Figure 8.11).

Mitigation measure for these conditions include:

a) Vertical drains (Section 17.9) to puncture the low-permeability layer and allow water to drain downwards into deeper rock that has been depressurized.
b) Slope drainage measures (Section 17.10) where the seepage emerges in the slope to direct the seepage water to a convenient collection point.

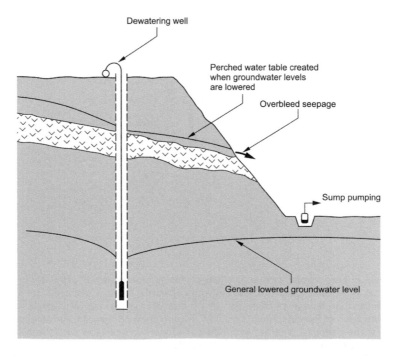

Figure 8.14 Perched water conditions for an excavation in rock drained by pumped wells.

c) Sub-horizontal drains (Section 17.5) to attempt to intercept and drain the perched water into the excavation.

8.7.4 Localized Concentration of Water Inflow

Groundwater will emerge into the excavation at locations dictated by the rock structure, joints, bedding and other fractures. These locations will often be inconvenient for construction operations – trying to place high-quality structural concrete on top of a gushing fracture is unlikely to be efficient! Poor working conditions can also result if groundwater is dripping or running into the base of a deep excavation due to seepages higher up in the excavation faces.

Suitable methods to divert groundwater away from working areas in rock excavations include:

- Sub-horizontal drainage (Section 17.5) holes drilled into vertical or steeply inclined rock faces (Figure 8.15a). These act to relieve groundwater pressures behind any linings (such as sprayed concrete) applied to the face and provide a preferential outlet for any residual seepage water. If any drains have a continuous flow, they can be fitted with a small-diameter flexible hose to discharge the water directly to a sump in the base of the excavation.
- Improvised gutters or channels fitted along the edge of the excavation with the objective of intercepting small quantities of water dripping or running down the rock face (Figure 8.15b). This provides drier working conditions for construction activities deeper in the excavation.
- Relief wells (Section 17.8) drilled through the base of the excavation to try to intercept major fractures and to provide a preferential pathway to allow groundwater to enter the excavation at a more convenient location (Figure 8.15c). The water from the relief wells is directed to a sump pump.

8.7.5 Water Trapped by Rock Structure

In some cases, the structure of the rock (dip of bedding, folding and faulting) can cause some areas of rock to drain poorly, even if the overall excavation has been successfully dewatered and depressurized. This is termed hydraulic compartmentalization; see Figure 8.8.

Knowledge of the rock structure, together with monitoring of groundwater levels and water pressures, can help identify zones where water may be trapped. Typically, conditions can be improved by installing sub-horizontal drains (Section 17.5) to provide preferential flow pathways linking to the excavation or to depressurized zones of rock (Figure 8.16).

8.8 GROUNDWATER PROBLEMS WHEN TUNNELLING IN ROCK

Groundwater inflow to tunnels in rock typically causes fewer problems than for tunnels constructed in soil (Section 7.7). Often, groundwater inflow through tunnel faces in unweathered rock will not cause significant instability. In many cases, the primary concern is managing inflows and pumping water away from the face.

The primary elements in tunnelling projects are the tunnels themselves – linear sub-horizontal features, often but not always cylindrical. On transportation projects, there are often two parallel tunnels (known as twin tunnels). Tunnelling projects often include other

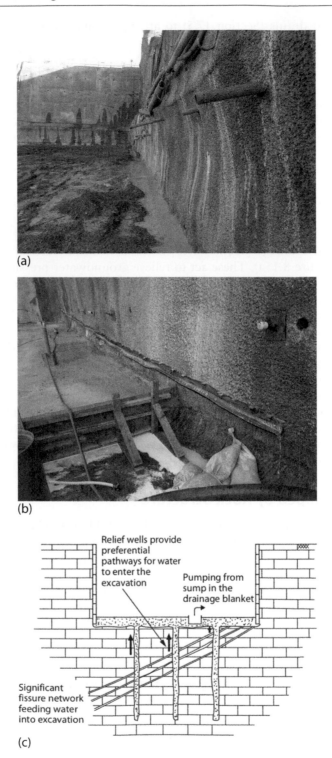

(a)

(b)

(c)

Figure 8.15 Measures to deal with localized concentration of groundwater inflow in rock excavations. (a) Sub-horizontal drainage holes drilled into vertical rock face covered in sprayed concrete lining (SCL). (b) Gutters fitted along the edge of the excavation with the objective of intercepting small quantities of water dripping or running down the face. (c) Relief wells used in rock excavations to intercept flow in discrete fractures.

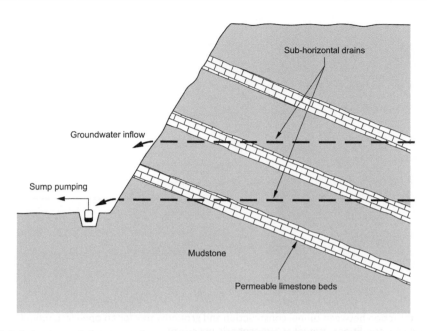

Figure 8.16 Sub-horizontal drains used to provide preferential flow pathways. The geometry of the limestone beds is unfavourable for natural drainage into the excavation. The sub-horizontal drains provide a better pathway for the permeable limestone beds to drain.

underground works, including shafts (to access the tunnel from the surface), cross passages to link twin tunnels, adits, connector tunnels and caverns.

Tunnels and other underground works typically require groundwater control to achieve two principal objectives:

1. Groundwater management to provide a workable environment in the tunnel and prevent the works being 'flooded out'
2. Prevention of groundwater-induced instability of rock exposed in the faces or other working areas

When tunnelling in rock, objective 1 – groundwater management – is often the focus of groundwater control requirements. However, in weaker or weathered rock, objective 2 – preventing groundwater-induced instability – can also be important.

8.8.1 Groundwater Management during Tunnel Construction

Underground excavations in rock are typically more stable than comparable works dug in soil, and this influences the commonly used tunnelling methods. In harder and more competent rocks, tunnelling may not use a 'shield' (a strong cylindrical sheath to protect and support the ground being excavated). In this case, a much larger area of unsupported ground is temporarily exposed around the tunnel face compared with shield tunnelling in soil.

During construction, groundwater can enter the tunnel through the exposed rock, and the tunnel will act as a drain (Figure 8.17). This will cause reduction of water pressures (and drawdown of groundwater levels) above the vicinity of the working face. As the tunnel progresses, the working face moves forward, and the centre of the zone of drawdown will move forward with it. If the tunnel is to be constructed with a watertight lining, then groundwater

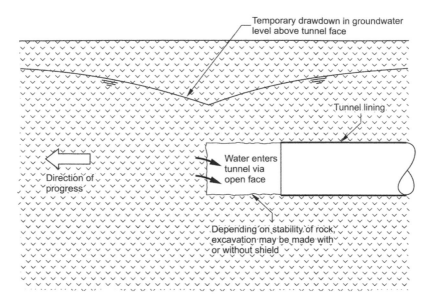

Figure 8.17 Interaction between tunnel construction and groundwater for an open face tunnel in rock under free air conditions (tunnel at atmospheric pressure).

levels should return to their original levels. However, in order to reduce the thickness of the tunnel lining, rock tunnels are sometimes designed to have permeable linings. This allows some groundwater to enter the tunnel in the long term, which reduces external pressure on the lining. It does, however, mean that the tunnel may act as a drain in the long term.

Rock tunnelling without a shield can be progressed by 'drill and blast' methods using explosives (typically in very strong rocks) or by mechanized excavation using 'road headers' or 'continuous miners' (boom mounted rotary cutters). Often, the face is not advanced as a single front across the whole tunnel face, but is advanced by a sequential excavation method (SEM). This involves excavating and supporting different areas of the face in a sequential manner, for example pilot headings and benches, with localized support such as rock bolts and sprayed concrete linings deployed as required.

In weaker rocks, tunnels may be formed using a shield to protect the excavation zone ahead of the permanent lining. Shields can be used as part of an open face tunnel boring machine (TBM) where the face is dug by a road header or backactor excavator within the shield. Full face TBMs (where a rotary cutting head is used to attack the entire face simultaneously) can also be used in weak rock; these are still classified as 'open face' because they do not attempt to use a counter pressure to exclude groundwater (see Section 7.7.1). Closed face TBMs of the earth pressure balance (EPB) type or slurry type are sometimes used in weak rocks, but unless the rock is so weak that it is effectively acting as 'soil', open face TBMs typically offer better rates of progress in rock.

When working with open face TBMs in weak rocks, or on tunnel drives with variable rock conditions that could include weak and unstable zones, possible mitigation measures include:

- Drilling of an array of long horizontal drain holes ahead of the tunnel to partially pre-drain the ground ahead of the advancing face and bring the water from the drain hole into the tunnel face in a controlled manner
- Ground treatment (such as grouting or artificial ground freezing) ahead of the advancing face to reduce permeability and increase rock strength

- Groundwater lowering to reduce groundwater level below tunnel invert (to decrease water inflows at the face and improve face stability)

Groundwater control methods for tunnelling works are described in Chapter 10.

8.8.2 Sudden Inflows

One of the challenges of tunnelling in relatively stable rock is that to allow efficient and economic progress, the working methods are often optimized for 'typical' ground conditions along the tunnel drive. Severe problems can occur if the tunnel encounters a more unstable water-bearing zone, where the working methods (water management and face support) are inadequate. If the tunnel is driven without knowledge of the ground conditions ahead of the face, there is a risk of sudden water inflows and instability, which could have severe consequences. Examples of such ground conditions are given in Figure 8.18 and Section 10.8.

On long tunnel projects, ground investigation boreholes may be spaced hundreds of metres apart, and it is possible that problematic zones such as those shown in Figure 8.18 may be 'missed' and not identified in advance. A commonly used strategy during tunnelling in rock is to drill an array of probe holes ahead of the face (Figure 8.19). Probe holes can provide information on the geology and rock parameters and if water is encountered, can allow potential water inflows to be assessed.

Probe drilling sequences and geometries must be developed based on the expected ground conditions and tunnel geometry. Typical lengths of probe drilling are 10 to 50 m per probe hole depending on the equipment available. If flowing water is encountered, data can be gathered by:

i. Fitting the probe hole with a well head (sometimes called a 'collar' or 'standpipe') or an inflatable packer where it emerges from the tunnel face. This can allow water flow from the hole to be stopped temporarily. A 'shut off' valve and pressure gauge or sensor can be installed to determine the groundwater pressure at tunnel level.
ii. Carrying out packer permeability testing (Section 12.8.3) in probe holes to estimate permeability in sections of probe hole.
iii. Measuring water flow rate from probe holes. For low flow rates, this can be done by capturing the flow rate in a container of known volume and estimating the flow rate from the time to fill the container. For higher flow rates, if the probe hole is sealed by a collar or packer, it may be possible to use a flowmeter (Figure 8.20).
iv. For probe holes sealed into the face, a 'pressure release' test can be carried out by using a valve to stop the inflow and 'shut in' the water pressure and record how quickly the pressure builds up. When the pressure has stabilized at its maximum value, the valve is opened rapidly, and the flow rate and water pressure are recorded in a manner analogous to a pumping test (Section 12.8.5). The data from such tests can provide some approximate information on permeability.

8.8.3 Groundwater Problems during Shaft Sinking in Rock

Tunnel projects also commonly involve the construction of shafts (vertical excavations, often, but not always, circular) between which the tunnel is driven. Other shafts may be needed for access or ventilation as part of the completed scheme. Shafts in rock have similar water management problems as conventional excavations, and groundwater control methods using pumping and exclusion (typically grouting) are routinely deployed. Methods of shaft construction and associated groundwater control are discussed in Section 10.5.

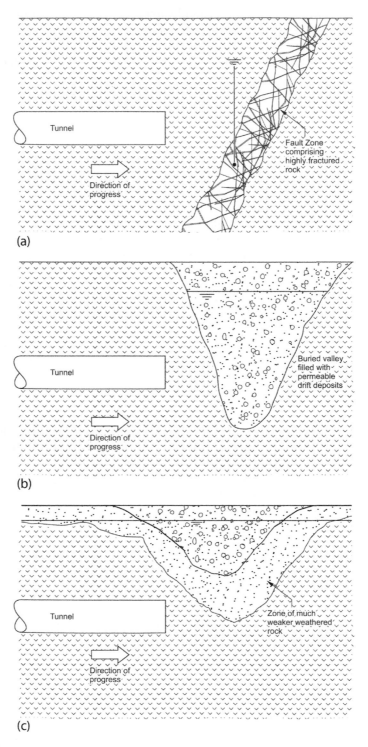

Figure 8.18 Tunnel drives in rock encountering significantly different ground conditions on sections of the route. (a) Tunnel drive passes through fault zone. (b) Tunnel drive passes through buried valley filled with permeable drift deposits. (c) Tunnel drive passes through more weathered zone with reduced cover to drift deposits above tunnel.

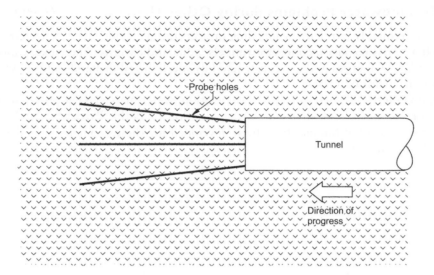

Figure 8.19 Probing ahead of the tunnel working face. The number, length and geometry of probes holes will vary depending on tunnel geometry and expected ground conditions.

Figure 8.20 Flowmeter used to assess flow rate from a probe hole in a tunnel face in rock. (Courtesy of F Chartier.)

Because excavations in rock are typically more stable than excavations in soils, sump pumping is commonly used during shaft sinking in rock. However, the sump and associated equipment can be an impediment to construction, and it is sometimes preferable to use external pumped wells to lower groundwater levels in advance of excavation. Perimeter exclusion methods (see Chapter 18), such as grout curtains and (in weak rock) diaphragm walls and secant pile walls, are also often used to reduce the requirement for groundwater pumping.

8.8.4 Groundwater Problems during Other Underground Works

In addition to tunnels and shafts, tunnelling projects often include other 'secondary' excavations such as cross passages, connector tunnels, adits and caverns. These excavations often have irregular geometries and are constructed by mechanized or hand mining without the protection of shields or TBMs.

Depending on the geometry of a given excavation, water management problems are similar to those for tunnels or shafts (or a combination of both). Particular applications of groundwater control techniques have been developed for these geometries, as described in Sections 10.5 and 10.6.

SECTION 2

DESIGN

The expectation of most people about engineering is that it is an exact science, yet those in practice are all too aware that, much of the time, the civil engineer works 'on the boundary between the calculable and incalculable'.

Vic Milligan

Chapter 9

Methods for Control of Surface Water and Groundwater

Nothing in the world is more flexible and yielding than water. Yet when it attacks the firm and strong, none can withstand it, because they have no way to change it.

Lao Tzu, sixth-century Chinese Philosopher

9.1 INTRODUCTION

As has been discussed in Chapters 6 through 8, the presence of groundwater can lead to troublesome conditions when construction operations are to take place below the original groundwater level in soil and rock. Surface water run-off must also be effectively controlled to prevent interference with excavation and construction works. This chapter briefly outlines some available methods for control of surface water and groundwater.

Techniques for the control of groundwater can be divided into two principal types:

a) Those that exclude water from the excavation (known as exclusion techniques)
b) Those that deal with groundwater by pumping (known as dewatering techniques)

This chapter includes brief discussions and tabular summaries of the various techniques of both groups, including their advantages and disadvantages. Methods specifically applicable to tunnelling projects are outlined in Chapter 10.

This book primarily addresses the dewatering techniques. The practicalities of the more commonly used pumping or dewatering methods are described in detail in later chapters. The techniques used for groundwater exclusion are discussed in Chapter 18.

This chapter, and indeed the rest of the book, focuses primarily on groundwater control methods for construction projects. However, the techniques described here can be used in other fields, including the mining industry. The principles of groundwater control by exclusion or pumping can be used on open pit and underground mines, and this type of application is discussed briefly in Section 9.7.

9.2 CONTROL OF SURFACE WATER

In order to optimize efficiency (and profitability), it is essential that conditions within the area of excavation and construction be workable; the mechanical plant should be able to operate efficiently without getting bogged down. To achieve this, in addition to the lowering of groundwater levels, surface water must be controlled. Thus, both surface water run-off and groundwater should be disposed of expeditiously.

On many construction sites, the measures enacted to control surface water run-off are inadequate, and result in a considerable and unnecessary waste of time and money.

There are three fundamental tenets to efficient control of surface water on construction sites:

  i. Source control
  ii. Water collection
  iii. Water treatment

9.2.1 Collection and Control of Surface Water Run-Off

The basic rule of source control is to deal with surface water before it can become a problem. In other words, collect and control the surface water run-off as soon as, or better still even before, it enters the area of work.

If water is allowed to run over exposed soil or weathered rock, it will pick up fine particles, which will either be deposited elsewhere on site, potentially blocking drains and ditches, or be carried in the water to the discharge point, where the water will require treatment to avoid pollution of the receiving watercourse.

The drainage system should incorporate appropriate measures to intercept and collect run-off from land areas surrounding and adjacent to an excavation to prevent surface water run-off encroaching into the construction area. Adoption of this philosophy requires the installation of adequate interceptor drains sited uphill or upstream of the excavation at original ground level.

Rain water or discharges from other construction activities (such as concreting or washing down of plant) should be prevented from entering an excavation, particularly on a sloping site, by the simple expedient of digging collector ditches or drains on the high ground. The drains should lead the water away to discharge points (which could be pumped by sump pumps lower down the slope). Further details on surface water drainage and sump pumping methods are given in Section 14.3.

A key element in successful control of surface water is often education and training of foremen, excavator operators and other site operatives. If they can be convinced that a little effort in minimizing the generation of surface water run-off, and diverting water away from exposed areas of bare soil, will benefit them by reducing the need to clean out drainage ditches and avoiding pollution at the discharge point, then site works will often go much more smoothly.

9.2.2 Treatment of Surface Water Run-Off

Water collected in surface water drainage systems on construction sites will generally have some suspended solids (clay, silt and sand-sized particles) in it as a result of its transit across site. On most sites, it is not acceptable to discharge this water from the site without some form of water treatment to reduce suspended solids to acceptably low levels. If significant concentrations of solids are discharged into a watercourse, there is significant potential for environmental and ecological harm (see Section 21.9). Settlement lagoons or more sophisticated treatment systems may be required.

9.3 METHODS OF GROUNDWATER CONTROL

The techniques available for control of groundwater fall into two principal groups:

a) Those that exclude water from the excavation (known as exclusion techniques) – Figure 9.1a
b) Those that deal with groundwater by pumping (known as dewatering techniques) – Figure 9.1b

This book is primarily concerned with the second group, the dewatering methods. However, the essential features of groundwater control by exclusion are outlined in the following section, and potential techniques are discussed in Chapter 18. Situations where exclusion and pumping methods might be used in combination are discussed in Section 9.6.

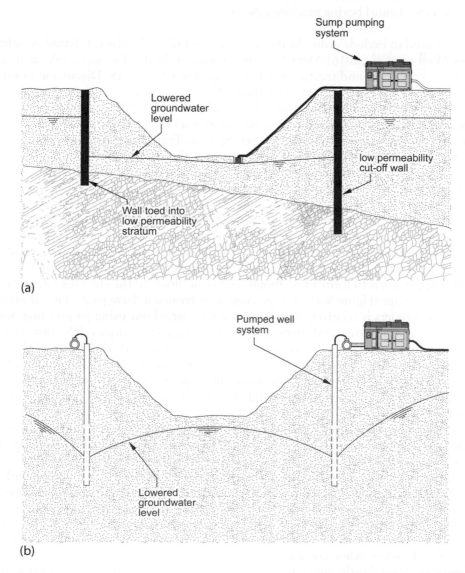

(a)

(b)

Figure 9.1 Categories of groundwater control methods. (a) Groundwater control by exclusion. Low-permeability cut-off walls are used to keep water out of the excavation. Sump pumping is used to deal with residual water within the excavation. (b) Groundwater control by pumping. Pumping from an array of wells is used to lower groundwater levels.

9.4 EXCLUSION METHODS

The aim of groundwater control by exclusion is to prevent groundwater from entering the working area. The methods used can be grouped into three broad categories:

1. Methods where a very low-permeability discrete wall or barrier is physically inserted or constructed in the ground (e.g. sheet-piling, diaphragm walls) – Figure 9.2a
2. Methods that use some form of ground treatment to reduce the permeability of the in situ ground (e.g. grouting methods, artificial ground freezing) – Figure 9.2b
3. Methods that use a fluid pressure in confined chambers such as tunnels or shafts to counterbalance groundwater pressures, such as compressed air (Section 10.4.1) and closed face tunnel boring machines (Section 10.4.2)

Techniques used to exclude groundwater are listed in Table 9.1, which is based on information from Preene *et al.* (2016). Details of the various methods of group 1 (physical cut-off walls) and group 2 (ground treatment) can be found in Chapter 18. Discussion of group 3 (tunnelling methods) can be found in Chapter 10.

One of the most common applications of the exclusion method involves forming a notionally impermeable physical cut-off wall or barrier around the perimeter of the excavation to prevent groundwater from entering the working area. Typically, the cut-off is vertical and penetrates down to a very low-permeability stratum that forms a basal seal for the excavation (Figure 9.2). On tunnelling projects, other geometries of groundwater barriers are sometimes required, including inclined and sub-horizontal cut-offs (see Chapter 10).

The costs and practicalities of constructing a physical cut-off wall are highly dependent on the depth and nature of any underlying permeable stratum. If a suitable very low-permeability stratum does not exist or is at great depth, then upward seepage may occur beneath the bottom of the cut-off wall, leading to a risk of base instability (see Section 7.5). In such circumstances, dewatering methods may be used in combination with exclusion methods (Figure 9.3).

Another approach is to form a horizontal barrier or 'floor' to the cut-off structure to prevent vertical seepage (Figure 9.4) – this is sometimes termed a 'base plug'. The construction of horizontal barriers is relatively rare but has been carried out using jet grouting, mix-in-place, permeation grouting and artificial ground freezing techniques. The objective is to provide a continuous horizontal barrier to seal up against the vertical elements of the cut-off barrier. When the area enclosed by the cut-off walls is dewatered and excavated, it will be buoyant (see Section 7.5.1) and will experience unbalanced upward groundwater forces. The configuration of the basal seal may be influenced by the need to resist buoyancy forces (see Section 18.3.1).

If a complete physical cut-off is achieved, some groundwater will be trapped inside the working area. This will need to be removed to allow work to proceed, either by sump pumping during excavation or by pumping from wells or wellpoints prior to excavation.

One of the attractive characteristics of the exclusion technique is that it allows work to be carried out below groundwater level with minimal effects on groundwater levels outside the site. This means that many potential environmental impacts of dewatering (see Chapter 21) are avoided. In particular, in urban areas, exclusion methods are often used in preference to dewatering methods to reduce the risk of settlement damage caused by lowering of groundwater levels. However, when considering using exclusion techniques to avoid groundwater lowering in areas outside the site, it is essential to remember that almost all cut-off walls will leak to some extent. Leakage may particularly occur through any joints (between columns, panels, piles, etc.) resulting from the method of installation.

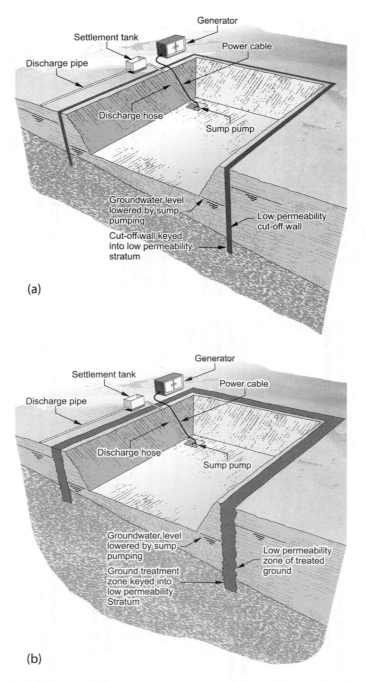

Figure 9.2 Groundwater control by exclusion where there is a shallow very low-permeability stratum.
(a) Groundwater exclusion using physical cut-off walls. Cut-off walls penetrate into very low-permeability stratum. (b) Groundwater exclusion using ground treatment. Ground treatment barriers penetrate into very low-permeability stratum.

Table 9.1 Principal Methods for Groundwater Control by Exclusion

Method	Brief description	Typical applications	Notes
Displacement barriers			
Steel sheet-pile walls	A wall is formed by steel sheet-pile sections driven or pushed into the ground. Each section interlocks with neighbouring piles via 'clutches' to form a continuous wall.	Open excavations and shafts in most soils, but obstructions such as boulders or timber baulks may impede installation. Can support the sides of the excavation with suitable propping.	Can be used to form a permanent cut-off, or used as a temporary cut-off with piles removed at the end of construction. Rapid installation. Seal may not be perfect, especially if obstructions present (which can cause neighbouring piles to 'declutch' and separate). Vibration and noise of driving may be unacceptable on some sites, but 'silent' methods are available where piles are pressed into the ground by hydraulic jacks, with previously installed piles used to provide the necessary reaction force.
Combi-pile walls	A wall is formed by interlocking steel sheet-piles and tubular piles.	Open excavations and shafts in most soils, but obstructions such as boulders or timber baulks may impede installation. Can support the sides of the excavation with suitable propping.	The tubular piles provide the principal bending resistance of the wall, while the sheet-piles provide the groundwater cut-off between the tubular piles. Other aspects are as for steel sheet-pile walls.
Vibrated beam wall	A wall is formed by vibrating an I beam vertically into the ground and then filling the resultant void with grout as the beam is extracted. Overlapping penetrations form a continuous wall.	Open excavations in silts and sands.	Once the I beam has penetrated to full depth, cement-based grout is injected through nozzles at the toe of the I beam as it is extracted to form a thin, low-permeability membrane. Rapid installation. Forms a permanent barrier.
Excavated barriers			
Slurry trench wall using cement-bentonite or soil-bentonite	A trench is excavated under bentonite slurry by backactor, grab or hydromill. The trench is backfilled with low-permeability material.	Open excavations in silts, sands and gravels up to a permeability of about 5×10^{-3} m/s.	Types of trench backfill include cement-bentonite, soil-bentonite or soil-cement-bentonite mixture. The method is typically used to form a low-permeability curtain wall around the excavation. Forms a permanent barrier.

(Continued)

Table 9.1 (Continued) Principal Methods for Groundwater Control by Exclusion

Method	Brief description	Typical applications	Notes
Concrete diaphragm walls	A trench is constructed under bentonite slurry as a series of panels by grab or hydromill. A reinforcing cage is placed in the trench and the bentonite slurry displaced by concrete.	Side walls of excavations and shafts in most soils and weak rocks, but presence of boulders may cause problems. Can support the sides of the excavation with suitable propping.	A continuous reinforced concrete wall is formed by the interconnecting panels, providing a permanent barrier. Often used to form the below-ground structural walls of permanent retaining structures, including shafts. Diaphragm walls are sometimes used without reinforcement where the wall does not serve a structural purpose.
Concrete secant pile walls	Concrete bored piles are constructed along the line of the proposed cut-off wall. In the secant pile arrangement, adjacent piles cut into each other and provide a continuous groundwater cut-off. In the alternative contiguous pile arrangement (these do not act as groundwater exclusion barriers), adjacent piles closely abut each other, and a small gap remains between piles.	As concrete diaphragm walls, but penetration through boulders may be costly and difficult. Can support the sides of the excavation with suitable propping.	Secant pile walls can provide an effective groundwater cut-off, provided there is good control of pile verticality and alignment, so that there are no gaps between piles. Secant pile walls are often used to form the below-ground structural walls of permanent structures, including shafts. Contiguous pile walls do not form an effective groundwater cut-off. Sealing between contiguous piles can be difficult, and additional grouting or sealing of gaps between piles may be necessary.
Injected barriers			
Permeation and rock grouting using cement-based grouts	Liquid grout is injected via drill holes to fill the pore spaces (in soil) and fractures (in rock), reducing the flow of water through the treated zone of soil or rock. Grout is a suspension of cement particles (and other admixtures) in water.	Tunnels, shafts and relatively small excavations in gravels and coarse sands (permeation grouting) and fractured rocks (rock grouting).	The grout comprises a suspension of cement particles (and other admixtures) in water. The cement can be ordinary Portland cement (OPC) or microfine or ultrafine types. Equipment is simple and can be used in confined spaces and be deployed in inclined or horizontal applications. A comparatively thick zone needs to be treated to ensure that a continuous barrier is formed. Multiple stages of treatment may be needed. Forms a permanent barrier.

(Continued)

Table 9.1 (Continued) Principal Methods for Groundwater Control by Exclusion

Method	Brief description	Typical applications	Notes
Permeation and rock grouting using chemical and solution grouts	Liquid grout is injected via drill holes to fill the pore spaces (in soil) and fractures (in rock), reducing the flow of water through the treated zone of soil or rock. Grout is either a solution or a colloidal suspension of very fine particles.	Tunnels, shafts and relatively small excavations in medium sands (chemical grouts), fine sands and silts (resin grouts), and fractured rocks.	The grouts are either solutions or colloidal suspensions of very fine particles; this allows the grout to penetrate smaller pores or fractures. Grout materials can be expensive. Silty soils are difficult to grout, and treatment may be incomplete, particularly if more permeable laminations or lenses are present. Forms a permanent barrier, although some grout types are less durable in the long term compared with cement grouts. Other aspects are as for cement-based grouting.
Jet grouting	High-pressure jets of water/cement-based grout are used to create a series of overlapping cylinders of soil/grout mixture.	Tunnels, shafts and relatively small excavations in most soils and very weak rocks.	Can be deployed in inclined or horizontal applications. Can be messy and create large volumes of waste slurry. Risk of ground heave if not carried out with care. Forms a permanent barrier.
Mix-in-place walls	In situ mixing of soil and injected grout is used to form overlapping columns or panels.	Side walls of excavations and shafts in most soils and very weak rocks.	Columns formed using auger-based equipment, panels formed using cutter soil mixing (CSM) equipment. Produces little spoil. Less flexible than jet grouting. Forms a permanent barrier.
Tunnel exclusion methods			
Compressed air tunnelling	The working area is sealed, and a supply of compressed air is used to increase air pressure (up to 3.5 Bar) on exposed soil/rock faces.	Confined chambers such as tunnels, sealed shafts and caissons in most soils and weak rocks.	The compressed air applies a counter pressure to the exposed ground and temporarily achieves an approximate balance between the tunnel pressure and pore water pressure in the ground, reducing the hydraulic gradient and limiting groundwater inflow. There are potential health hazards to workers in compressed air, and specialist medical supervision is required. Air losses may be significant in strata of high permeability. High set-up and running costs.

(Continued)

Table 9.1 (Continued) Principal Methods for Groundwater Control by Exclusion

Method	Brief description	Typical applications	Notes
Closed face tunnel boring machines (TBM) using earth pressure balance (EPB) or slurry methods	A specialist TBM excavates for the tunnel and temporarily supports the soil/rock and excludes groundwater by maintaining a balancing fluid pressure against the face.	Tunnels in most soils and weak rocks.	A fluid within the head of the TBM applies a counter pressure to the face, stabilizing the soil/rock and limiting groundwater inflow. For EPB TBMs, the fluid is a paste formed of a mixture of soil cuttings, groundwater and conditioning agents (such as polymer or bentonite muds). For slurry TBMs, the fluid is a bentonite or polymer-based slurry circulated in a closed circuit and used to carry cuttings from the face. TBMs need to be carefully selected to deal with prevailing ground conditions; set-up and running costs may be high.
Other types			
Low-permeability surface barriers	Low-permeability materials (including geosynthetic membranes, blinding concrete, sprayed concrete) are used to temporarily seal exposed soil/rock surfaces or faces.	Open excavations and shafts in most soils and weak rocks.	The objective is to reduce infiltration of surface water and/or softening and degradation of exposed material.
Artificial ground freezing using brine or liquid nitrogen	Very low-temperature refrigerant is circulated through pipework in a series of closely spaced boreholes to form a temporary wall of frozen ground (a freezewall).	Tunnels and shafts and relatively small excavations. May not work if groundwater flow velocities are excessive (>2 m/day for brine or >20 m/day for liquid nitrogen).	Circulation of the refrigerant will freeze the ground and cause ice cylinders to develop around each of the closely spaced boreholes. After a period of time, the ice cylinders coalesce to form a temporary wall of frozen ground (a freezewall), which can support the side of the excavation as well as excluding groundwater. Liquid nitrogen freezing is expensive but quick; brine freezing is cheaper but slower. Liquid nitrogen is to be preferred if groundwater velocities are relatively high. Can be deployed in inclined or horizontal applications. High set-up and running costs.

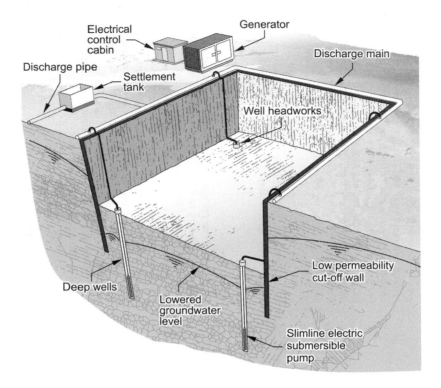

Figure 9.3 Cut-off walls used in combination with dewatering methods where there is no shallow very low-permeability stratum.

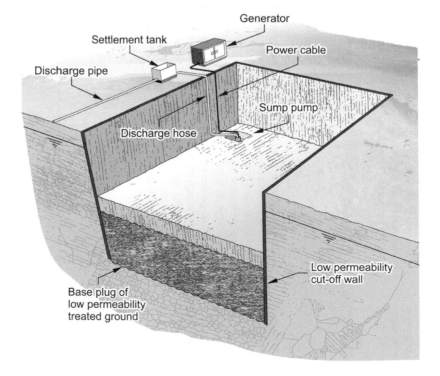

Figure 9.4 Groundwater control by exclusion using physical cut-off walls with horizontal barrier (base plug) to seal base of excavation.

Leakage of groundwater through cut-offs into the excavation or working area can cause a number of problems:

(i) During construction, the seepages may interfere with site operations, necessitating the use of sump pump or surface water control methods to keep the working area free of water.

(ii) The leakage into the excavation may be significant enough to locally lower groundwater levels outside the site, creating the risk of settlement or other side effects.

(iii) If the cut-offs form part of a permanent structure (such as the walls of a deep basement), even very small seepages will be unsightly in the long term and may cause problems with any architectural finishes applied to the walls.

In many cases, the significant seepages that give rise to problems (i) and (ii) can be dealt with by grouting or other treatment. On the other hand, it can be very difficult to prevent or to seal the small seepages of (iii). David Greenwood (1994) has said:

Water penetration is very difficult to oppose. It is comparatively easy to reduce torrents to trickles, but to eliminate trickles is difficult. If it is essential to have a completely dry or leakproof structure, costs rise steeply.

This is an important point to consider if cut-off walls are to be incorporated in the permanent works.

9.5 DEWATERING METHODS

Dewatering methods control groundwater by pumping, effecting a local lowering of groundwater levels (Figure 9.5). The aim of this approach is to lower groundwater levels to a short distance (say 0.5 m) below the deepest excavation formation level.

The two guiding principles for securing the stability of excavations, and especially of slopes, by dewatering or groundwater lowering are:

(i) Do not hold back the groundwater. This may cause a build-up of pore water pressures that will eventually cause catastrophic movement of soil and groundwater.

(ii) Ensure that 'fines' (clay, silt and sand-sized particles) are not continuously transported, since this will result in erosion and consequent instability. Apply a suitable filter blanket to avoid any build-up of pore water pressures and prevent transportation of fines. As a general guide, if the permeability is lower than about 1×10^{-7} m/s, migration of fines ceases. (This is a helpful guide when water testing after grout treatment, say prior to shaft sinking or the like, to assess whether or not further grouting is desirable.)

9.5.1 Open Pumping and Pre-Drainage Methods

An important distinction within groundwater pumping methods is whether they can be considered as 'open pumping' or 'pre-drainage' methods:

i. Open pumping, most commonly carried out by sump pumping, involves allowing groundwater to enter an excavation, from where it is then removed by pumping (Figure 9.6a). This is less than ideal, since it draws water towards the excavation and may promote instability. Nevertheless, this can be acceptable provided that fines are

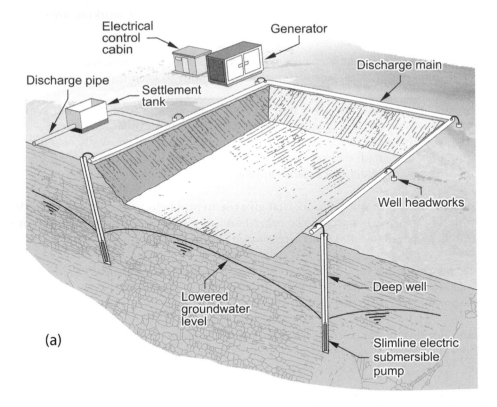

(a)

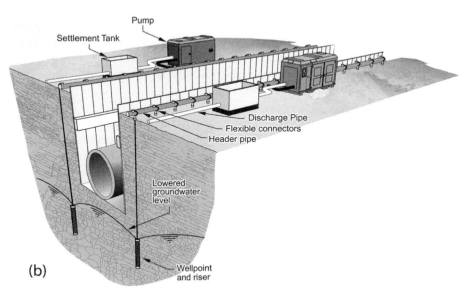

(b)

Figure 9.5 Groundwater control by pumping. (a) Pumped groundwater control system arranged as a ring of dewatering wells around an excavation. The method shown is pumping from deep wells with submersible pumps. This is a pre-drainage method. (b) Pumped groundwater control system arranged as lines of dewatering wells alongside a long trench excavation. The method shown is pumping from wellpoints. This is a pre-drainage method.

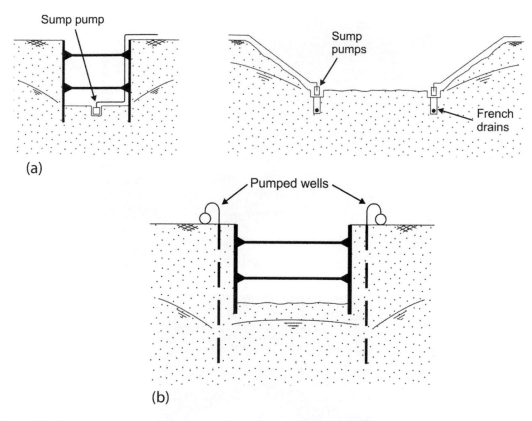

Figure 9.6 Groundwater control by open pumping or pre-drainage. (a) Open pumping methods. (b) Pre-drainage methods.

not removed from the slopes and base of the excavation. While simple in practice, open pumping has the further disadvantage that groundwater levels cannot be lowered in advance of excavation.

ii. Pre-drainage methods (Figure 9.6b) adopt the contrasting approach of using pumped wells to lower groundwater levels in advance of excavation works. Wells are typically installed outside the excavation to form either a ring of wells around the excavation (Figure 9.5a) or lines of wells parallel to long trench excavations (Figure 9.5b). This arrangement of wells draws water towards the wells, not the excavation, avoiding troublesome seepages and resulting instability; the target groundwater level is typically set 0.5 to 1.0 m below the deepest dig level. This group of methods (which includes the methods based on pumped wells, such as wellpoints, deep wells and ejector wells) has the advantage that groundwater can be managed so that water does not enter the excavation, reducing the risk of groundwater-induced instability.

Figure 9.7 shows an interesting comparison of two shaft excavations sunk near to each other in very similar ground conditions of weathered rock comprising weakly cemented calcareous sandstone – one shaft dewatered by open pumping and one shaft by pre-drainage methods. The sides of each shaft were supported by concrete bored piles installed to form

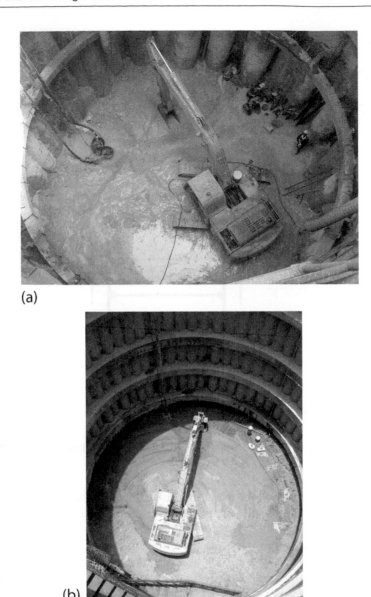

Figure 9.7 Comparison of excavation of shaft within contiguous pile walls dewatered by open pumping or pre-drainage methods. (a) Groundwater control is by sump pumping from within the shaft. (b) Groundwater control is by an array of external dewatering wells (one internal well is evident at the top left). Panel (a): This is an open pumping method, and water is allowed to enter the excavation by flowing through the gaps between the contiguous piles. This causes instability of the weathered rock exposed between the piles and washes in fine sand and silt-size particles out of the ground. The working conditions in the base of the dig are wet and inefficient. Panel (b): This is a pre-drainage method where a system of external wells is used to lower groundwater levels in advance of excavation. The base of the excavation is dry, and no fine particles are being washed out of the ground between the perimeter piles. (Courtesy of T O L Roberts.)

a contiguous wall (Section 18.4). In this arrangement, the bored pile wall does not form an effective groundwater cut-off, and water can pass easily between the piles.

One shaft was sunk by sump pumping (an open pumping method). In this case, water entered the shaft between the piles, washing fine particles out of the weak rock and allowing water and sand slurry to pool in the working area. The bottom of the excavation became an unstable and inefficient working environment (Figure 9.7a). The other shaft was sunk after groundwater levels had been lowered by pumping from an array of deep wells located outside the shaft (i.e. after pre-drainage). No significant groundwater flows entered the excavation as the shaft was sunk, and there was no groundwater-induced loss of fine particles from between the piles. Figure 9.7b shows that drier and much more efficient working conditions were achieved in the excavation.

9.5.2 Passive Drainage Methods

The open pumping and pre-drainage methods discussed in the previous section are classified as 'active' dewatering methods in that they involve direct pumping of groundwater. In some cases, the alternative approach of 'passive' dewatering methods can be used.

Passive drainage methods are arranged so that groundwater flow from an excavation, tunnel or slope is facilitated by gravity flow, i.e. without the requirement for direct pumping from the well or drain (Figure 9.8). The flow of water into the excavation can be driven by cutting an excavation into sloping topography or by groundwater pressures from a depth where the base of an excavation is below groundwater level. Indirect pumping of the water, after it has emerged from the ground or structure by gravity flow, is often necessary to prevent the water causing problems elsewhere on site. Relief wells (Section 17.8) and sub-horizontal drains (Section 17.5) are commonly used methods of passive drainage.

9.5.3 Comparison of Different Pumped Groundwater Control Methods

There are several techniques or methods available for controlling groundwater flow for a construction project. The selection of a technique or techniques appropriate to a particular

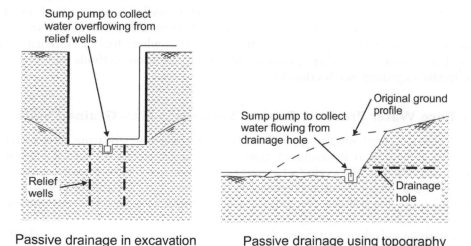

Passive drainage in excavation   Passive drainage using topography

Figure 9.8 Passive drainage of excavations.

project at a particular site or country will depend on many factors. However, the lithology and permeability(ies) of the soil or rock will always be of paramount importance. Other factors to be considered are:

- Extent of the area of construction requiring dewatering
- Depth of deepest formation level below existing ground level and the amount of lowering required
- Proximity of existing structures, the nature of their foundations and the soil strata beneath them

The various dewatering techniques are tabulated in Table 9.2, which is expanded from information in Preene *et al.* (2016).

There are projects where a single method is insufficient and a combination of methods is appropriate. Where the excavation is to penetrate a succession of strata of widely varying lithology, this problem is more likely to arise.

It will be seen, by studying the column of Table 9.2 headed 'Typical applications', that only a few methods are suitable for use in all types of soils and rocks. The ranges of soils that are suitable for treatment by the various dewatering methods are shown in Figure 9.9 in the traditional form of particle size distribution curves. These curves are taken from CIRIA Report 113 (Somerville, 1986) and are based on the earlier work of Glossop and Skempton (1945) and others. Similarly, the ranges of soils suitable for treatment by the various exclusion methods are shown in Figure 9.10. These are tentative economic and physical limits. The emphasis is on the word *tentative*.

An interesting and useful variation of Figure 9.9 was presented by Roberts and Preene (1994a). Figure 9.11 is taken from CIRIA Report C750 (Preene *et al.*, 2016) and is based on the Roberts and Preene (1994a) original *Range of application of construction dewatering systems* paper but has been modified in the light of the joint experiences of the authors (Cashman, 1994) and others. The practical achievements at the lower end of the permeability range are constrained by the physical limitations of applied vacuum. At the upper end of the range of permeability values, the cost of pumping the ensuing large volumes of water is the constraining factor. The greater the permeability, the greater the rate of pumping necessary to achieve the required lowering: the energy costs increase at an alarming rate.

The upper economic limit for the use of a deep well or wellpoint system is of the order of 5×10^{-3} m/s. The pumping costs of dealing with soils of greater permeability are generally uneconomic – except in the almost unheard-of situation where fuel or electrical power is provided at no cost. In such high-permeability strata, exclusion methods may offer a more cost-effective expedient (see Section 9.6).

9.5.4 Pore Water Pressure Control Systems in Fine-Grained Soils

When dewatering methods are applied to coarse-grained water-bearing soils of relatively high permeability, local lowering of groundwater levels occurs by gravity drainage in response to pumping. The period of pumping necessary to lower the water level is quite short, as the pore water is rapidly replaced by air. In fine-grained soils of low and very low permeability, gravity drainage of the pore water is resisted by capillary tension. Such soils drain poorly and slowly by gravity drainage.

Because these soils do not drain easily, any excavation made below the groundwater level will encounter only minor seepages and is unlikely to flood rapidly. Yet, even the small seepages encountered (perhaps less than 1 l/s even for a large excavation) can have a dramatic destabilizing effect. Side slopes may collapse or slump inwards, and the base may

Table 9.2 Principal Methods for Groundwater Control by Pumping

Method	Brief description	Typical applications	Notes
Active drainage methods (open pumping methods)			
Drainage pipes or ditches (e.g. French drains) used with pumped sumps	Pipes, ditches and trenches are pumped to divert or remove surface water or shallow groundwater from the working area, including slopes and batters.	Control of surface water run-off and shallow groundwater (including perched water and residual seepages into excavation) in rock and coarse-grained soils.	Pipes, ditches and trenches may obstruct construction traffic and will not control groundwater at depth. Unlikely to be effective in reducing pore water pressures in fine-grained soils. May generate silt or sediment-laden discharge water, causing environmental problems if pumped water is not adequately treated prior to discharge.
Sump pumping	Water is collected in pits or low points (sumps) within the excavation, from where it is pumped away.	Shallow excavations in clean coarse-grained soils or stable fractured rock for control of groundwater and surface water.	May not give sufficient drawdown to prevent seepage from emerging on the cut face of a slope, possibly leading to loss of fines and instability. May generate silt or sediment-laden discharge water, causing environmental problems if pumped water is not adequately treated prior to discharge.
Active drainage methods (pre-drainage methods)			
Wellpoints	Lines or rings of closely spaced small-diameter wells are installed around an excavation and pumped by a suction system.	Generally shallow, open excavations in sandy gravels down to fine sands and possibly silty sands. Deeper excavations (requiring >5–6 m drawdown) will require multiple stages of wellpoints to be installed.	Quick and easy to install in sands. Suitable for progressive trench excavations. Maximum drawdown is ~5–6 m for a single stage in sandy gravels and fine sands but may only be ~4 m in silty sands.
Horizontal wellpoints laid by trenching machine	Horizontal drainage pipe is laid by specialist trenching machines and pumped by suction pumps.	Generally shallow trench or pipeline excavations or large open excavations in sands and possibly silty sands.	Suitable for long runs of trench excavations outside urban areas, where very rapid installation is possible.
Deep wells with electric submersible pumps	Slimline electrically driven borehole submersible pumps are used to pump from bored wells.	Deep excavations and shafts in sandy gravels to fine sands and water-bearing fractured rocks.	No limit on drawdown in appropriate hydrogeological conditions. Installation costs of wells are significant, but fewer wells may be required compared with most other methods. Close control can be exercised over well screen and filter.

(Continued)

Table 9.2 (Continued) Principal Methods for Groundwater Control by Pumping

Method	Brief description	Typical applications	Notes
Deep wells with electric submersible pumps and vacuum	Slimline electrically driven borehole submersible pumps are used to pump from bored wells with a separate vacuum system used to apply vacuum to the wells.	Deep excavations and shafts in silty fine sands, where drainage from the soil into the well may be slow, or rock of low permeability in which pore water pressure control is required.	Number of wells may be dictated by the requirement to achieve an adequate drawdown between wells, rather than the flow rate, and an ejector well system may be more economical.
Shallow bored wells with suction pumps	Bored wells are pumped by surface suction pumps.	Shallow excavations in sandy gravels to silty fine sands and water-bearing fractured rocks.	Particularly suitable for coarse materials of high permeability, where flow rates are likely to be high. Useful where correct filtering is important, as closer control can be exercised over the well filter than with wellpoints. Drawdowns limited to ~4–8 m depending on soil conditions.
Ejector wells	Low-capacity wells are pumped by a nozzle and venturi system.	Excavations in silty fine sands, silts or laminated or fissured clays, or rock of low permeability in which pore water pressure control is required.	Drawdowns generally limited to 20–50 m depending on equipment. Low energy efficiency, but this is not a problem if flow rates are low. In sealed wells, a vacuum is applied to the soil, promoting drainage. Also known as eductor wells.

Passive drainage methods

Method	Brief description	Typical applications	Notes
Drainage pipes or ditches (e.g. French drains)	Pipes, ditches and trenches are arranged to flow by gravity to divert or remove surface water or shallow groundwater from the working area, including slopes and batters.	Control of surface water run-off and shallow groundwater (including perched water and residual seepages into excavation) in rock and coarse-grained soils.	Pipes, ditches and trenches may obstruct construction traffic and will not control groundwater at depth. Unlikely to be effective in reducing pore water pressures in fine-grained soils.
Relief wells	Boreholes or wells are used to create a vertical upward flowpath for water to enter the excavation from deeper strata.	Relief of pore water pressure in confined aquifers (including fractured rocks) or sand lenses below the base of the excavation to ensure basal stability.	Water overflowing from relief wells must then be directed to sumps and be pumped away. Used for both temporary and permanent applications.

(Continued)

Table 9.2 (Continued) Principal Methods for Groundwater Control by Pumping

Method	Brief description	Typical applications	Notes
Vertical drains	Artificial vertical flow pathways (either wells, sand-filled boreholes or pre-fabricated vertical drains [PVDs]) are used to create a vertical downward flowpath for water to drain from soil or rock into deeper strata where groundwater levels are lower.	Drainage of perched or shallow groundwater downwards into strata with lower groundwater levels to reduce seepages into excavation slopes.	May be applied in combination with active drainage methods used to lower groundwater levels in deeper strata.
Sub-horizontal drains	Sub-horizontal drill holes are used to form artificial flow pathways to allow water to drain into the excavation from slopes or behind excavation faces.	Draining of perched or trapped groundwater in low-permeability soils and rocks.	Water overflowing from drains must then be directed to sumps and pumped away. Used for both temporary and permanent applications.
Slope drainage	Passive drains (of similar design to French drains) are arranged to collect seepages from slopes, to flow by gravity to toe drains, where water may be pumped away by sumps.	Slope drainage and landslide stabilization in soils/rocks of low permeability.	Used for both temporary and permanent applications.
Drainage tunnels and adits	Tunnels, adits or headings are driven below groundwater level and then allowed to continue to drain by gravity (if topography allows) to lower groundwater levels over a wide area.	Long-term groundwater drainage for engineering and mining projects.	Temporary groundwater control may be required during driving of the tunnels, adits or headings. Typically used for large-scale and long-term applications. Drainage holes may be drilled outwards from the main adits. Used for both temporary and permanent applications.
Siphon drains	Self-priming siphon systems are installed in large-diameter wells.	Long-term slope drainage and landslide stabilization in soils and weak rocks of low permeability.	Can allow passive drainage of slopes without the need for pumping. Used for both temporary and permanent applications.
Specialist pumping methods			
Collector wells	Sub-horizontal wells (laterals) are drilled radially outwards from central pumping shaft.	Tunnels or deep excavations in relatively permeable soils (such as sands and gravels and sands) and rock, where surface access does not allow the installation of large numbers of wells.	Each collector well is expensive to install, but relatively few wells may produce large flow rates and be able to dewater large areas.

(Continued)

Table 9.2 (Continued) Principal Methods for Groundwater Control by Pumping

Method	Brief description	Typical applications	Notes
Horizontal directionally drilled (HDD) wells	Specialist HDD rigs are used to drill steered sub-horizontal wells, within which screens and filters are installed.	Groundwater control for linear infrastructure projects or for lowering groundwater levels in areas where surface access for vertical wells is restricted.	Wells can be pumped or be allowed to flow passively if topography allows.
Electro-osmosis	A system of anodes and cathodes is used to apply electrical potential gradients to promote groundwater flow in materials of very low permeability.	Soils of very low permeability, e.g. clays, silts and some peats.	Very specialized technique. Only generally used for pore water pressure control or ground improvement when considered as an alternative to artificial ground freezing. Installation and running costs are comparatively high.
Artificial recharge	Pumped groundwater is re-injected back into the ground, under careful control, via trenches or wells.	Soils of moderate to high permeability and fractured rocks, where lowering of groundwater is to be controlled so that environmental impacts can be mitigated. Can also be used as a means to dispose of dewatering water where other discharge routes are not available.	Typically complex to operate and maintain. Recharge wells and trenches often suffer from clogging due to water chemistry and bacterial growth effects and may require periodic backflushing and cleaning.

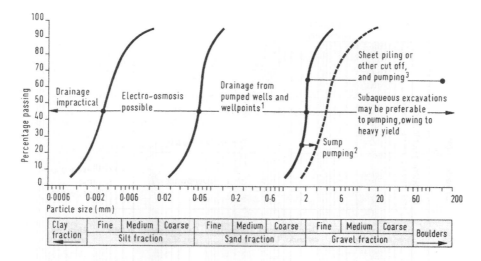

Notes:
1. Wellpoints in fine sands require good vacuum.
2. Zone may be extended in finer soils by using large sumps with gravel filters.
3. To reduce the high water pressure on sheet piling it may be preferable to control pumping as excavation proceeds and to install the support system as the water level is lowered.

Figure 9.9 Tentative ranges for groundwater lowering methods. (From Somerville, S H, *Control of Groundwater for Temporary Works*, Construction Industry Research and Information Association, CIRIA Report 113, London 1986. With permission from CIRIA: www.ciria.org)

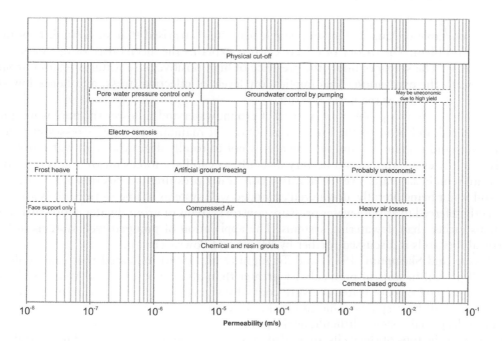

Figure 9.10 Tentative economic ranges for exclusion methods in soils. (Amended after Doran, S R, *et al.*, Storebælt Railway tunnel – Denmark: Design of cross passage ground treatment. *Proceedings of the 11th European Conference on Soil Mechanics and Foundation Engineering*, Copenhagen, Denmark, 1995 and Preene, M, *et al.*, *Groundwater Control – Design and Practice*, 2nd edition, Construction Industry Research and Information Association, CIRIA Report C750, London, 2016.)

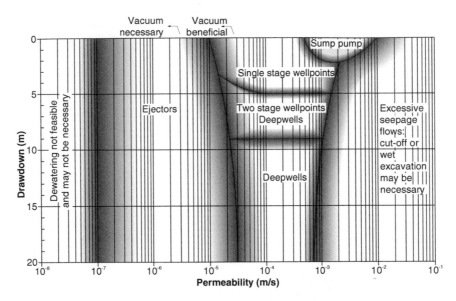

Figure 9.11 Range of application of pumped well groundwater control techniques. (Adapted from Roberts, T O L and Preene, M, Range of application of construction dewatering systems, in *Groundwater Problems in Urban Areas* (Wilkinson, W B, ed), Thomas Telford, London, 1994 and modified after Cashman, P M, Discussion, in *Groundwater Problems in Urban Areas* (Wilkinson, W B, ed), Thomas Telford, London, 1994, from Preene, M, *et al.*, *Groundwater Control – Design and Practice*, 2nd edition, Construction Industry Research and Information Association, CIRIA Report C750, London, 2016. With permission from CIRIA: www.ciria.org)

become unstable or 'quick'. On site, people are often surprised that such small flow rates can be a problem. The theory of effective stress explains the mechanism of instability (see Section 6.5). The seepages imply the presence of high positive pore water pressures around and beneath the excavation. This implies low levels of effective stress, and hence low soil strength – instability is the natural result.

The solution to this problem is to abstract groundwater and so lower the pore water pressures around and beneath the excavation. This will maintain effective stresses at acceptable levels and prevent instability. The aim is not to totally drain the pore water from the soils – in any event, this would be very difficult, as capillary forces mean that fine-grained soils can remain saturated even at negative pore water pressures (see Section 3.3.5). Because the soil is not being literally 'dewatered', pumped well systems in fine-grained soils are more correctly referred to as pore water pressure control systems or groundwater depressurization systems rather than dewatering systems. The application of pumped well systems in low-permeability soils is discussed further by Preene and Powrie (1994).

Glossop and Skempton (1945) and Terzaghi *et al.* (1996) indicated that gravity drainage will prevail in soils of permeability greater than about 5×10^{-5} m/s. Where wells are installed as part of pore water pressure control systems in soils of lower permeability, the well yields will be very low. This can make the continued operation of conventional well-point or deep well systems difficult, as the pumps are prone to overheating at low flow rates. However, if the tops of the wells are sealed, a partial vacuum can be applied to assist drainage. The increase in yield of a well due to the application of vacuum is likely to be of the order of 10 per cent, sometimes up to 15 per cent.

A sealed vacuum wellpoint system (see Section 15.6) operated by vacuum tank pump (see Section 19.5.3) may be effective in soils of permeability down to about 1×10^{-6} m/s. If a vacuum is applied to sealed deep wells (see Section 16.9), the lower limit of effectiveness of deep wells may be extended to about 1×10^{-5} m/s. Ejectors (see Section 17.2) installed in sealed wells will automatically generate a vacuum in the well when yields are low. Ejector well systems can be effective in soils of permeability as low as 1×10^{-7} m/s.

Again, the permeability ranges quoted earlier are tentative. In fine-grained soils, structure and fabric have a great influence on the performance of vacuum well systems. If the soil structure consists of thin alternating layers or laminations of coarser and finer soils, the drainage will be more rapid (see Case History 27.3). This is because the layers of coarse material will more rapidly drain the adjacent layers of finer-grained soil. There have also been cases where pore water pressure control systems have been effective in clays of very low permeability – the success of the method was attributed to the presence of a permeable fissure network in the clay (see for example Roberts *et al.*, 2007).

9.5.5 Some Deep Well and Ejector Projects Deeper Than 20 m

The 20 m depth limit on Figure 9.11 is not restrictive but is merely a convenience for the presentation of the diagram. However, there are insufficient reported case histories to have confidence in extending Figure 9.11 significantly below 20 m depth, but four known case histories are reported in the following to substantiate the view that with care, it is economically feasible to use wells and ejectors to depths of the order of 40 m or more.

At Dungeness A nuclear power station, sited on the south coast of England, 60 deep wells were pumped for more than 2 years. The soil conditions at the site were:

a) Gravel with cobbles, +5.5 metres above Ordnance Datum (mAOD) to –1.5 mAOD, average permeability 3×10^{-3} m/s
b) Original groundwater level from +2.4 mAOD to 0.9 mAOD
c) Gravelly sand, –1.5 mAOD to –9.1 mAOD, average permeability 8×10^{-4} m/s
d) Sand, –9.1 mAOD to –33.5 mAOD, permeability $1·5 \times 10^{-4}$ m/s, decreasing with depth to 5×10^{-5} m/s

The construction excavations were encircled by a continuous girdle of interlocking sheet-piles driven to ⁻9.1 mAOD level so as to exclude recharge from the overlying high-permeability soils. The lowering for the excavations for the Turbine Hall and the Reactors was achieved by pumping 29 wells 20 m deep. The lowering for the Cooling Water Pumphouse and the Syphon Recovery Chamber was achieved by pumping 19 and 12 wells, respectively, each 41 m deep.

Prior to the construction of the East Twin Dry dock in Northern Ireland, site investigation borings had revealed a confined aquifer of Triassic sandstone beneath alluvial and glacial deposits (both mainly clays). The depth to the upper surface of Triassic sandstone varied between 34 m below ground level and 21 m (which was 1 m below the formation level of the entrance structure). There was a high piezometric level confined in the sandstone – to about 3 m above original ground level. Twenty deep wells were installed to depths between 33.5 m and 43 m below ground level using reverse circulation rotary drilling methods. The wells were operated for more than 2 years.

At the Mufulira mine No. 3 dump in northern Zambia, about 200 ejector wells (using twin pipe ejectors) were installed to reduce the moisture content and thereby stabilize slimes lagoon deposits, which were of porridge-like consistency and had broken through the roof of the main adit, causing disastrous loss of life. The Ministry of Mines, Zambia required that the slimes deposits over the cave-in be stabilized before the mine could be permitted to reopen. After 9 months of pumping of the ejector installation, the phreatic surface was drawn down about 10 m. Eventually, the phreatic surface was drawn down some 30 m, resulting in an acceptable increase in the shear strength of the slimes deposits over the cave-in. The pumping lift was in the range from 20 m to over 40 m.

At the Benutan Dam site in Brunei, ejector wells (using single pipe ejectors) were installed to stabilize alluvium comprising heterogeneous loose silty fine sands, very soft clays, thin layers of peat and buried timbers (see Cole *et al.*, 1994). Brunei is in an earthquake zone of moderate seismic activity, so it was judged that there was a risk of liquefaction of these loose alluvial foundation soils beneath the proposed dam unless they were stabilized or replaced with other, more stable materials. As at Mufulira, an ejector system was installed and operated to reduce the pore water pressures of the loose soft soils and thereby increase their stability. The depths to which the ejector wells were installed ranged from about 10 m to 38.5 m. The ejector pumping was continuous on the deepest line for about 2 years. The dam height was 20 m above valley floor level.

9.6 USE OF PUMPING AND EXCLUSION METHODS IN COMBINATION

Pumped groundwater control systems are frequently used in combination with exclusion methods. The methods might be combined for a number of reasons:

   i. To reduce pumped flows in high-permeability soils
   ii. To reduce external groundwater impacts
   iii. To dewater cofferdams formed by cut-off walls
   iv. To reduce loading on cut-off structures

The decision to combine pumping and exclusion methods may be part of a strategy to 'optimize' groundwater control systems – this approach is discussed in more detail in Chapter 26.

9.6.1 Exclusion Methods to Reduce Pumped Flows in High-Permeability Soils

In high-permeability soils, where the pumped flow will be large, cut-off methods (e.g. sheet-pile walls, slurry walls, grout barriers, etc.) can be used to reduce the potential inflows to an excavation. If the abstraction points (sumps or wells) are located within the cut-off walls, the pumped flow rate will be reduced to a lesser or greater degree depending on the effectiveness of the exclusion method (Figure 9.12).

9.6.2 Exclusion Methods to Reduce External Groundwater Impacts

There are some hydrogeological and geotechnical conditions when groundwater lowering is perfectly possible from a practical point of view, but when there is a risk of detrimental impacts on neighbouring structures or natural features such as wetlands. Chapter 21 describes the range of environmental impacts potentially associated with groundwater lowering, but some of the more common impacts include settlement damage of neighbouring structures, the depletion of groundwater-dependent features such as rivers and wetlands, or the reduction in yield of nearby water supply wells.

Depending on the geological stratification at the site, exclusion methods can be used to fully or partially isolate the excavation from the wider groundwater regime. This will reduce the drawdown of groundwater levels external to the excavation and will generally reduce the corresponding external impacts.

Figure 9.12 Sheet-pile cofferdam used to reduce groundwater flow to a dewatering system located within the area enclosed by the cut-off wall. Deep wells pumped by submersible pumps are located within the 'pans' of the sheet-piles. (Courtesy of Dewatering Services Limited.)

A typical dewatering problem associated with major infrastructure projects is where new infrastructure is constructed to replace or supplement an existing plant on the same site. When dewatering was carried out for the original infrastructure, the site will have effectively been 'greenfield' with few constraints on dewatering. But when the new infrastructure is constructed, the dewatering impacts on the existing infrastructure must be considered carefully. Case History 27.10 describes the groundwater control works for Sizewell B Power Station in Suffolk, United Kingdom, where a deep cut-off wall was used to reduce the risk of impacts on neighbouring infrastructure.

9.6.3 Groundwater Pumping within Areas Enclosed by Cut-Off Walls

If the soil stratification allows the cut-off wall to form a complete cofferdam (e.g. if sheet-piles are driven down to an impermeable layer), the excavation will be isolated from the surrounding groundwater regime. However, some groundwater will remain trapped inside the area enclosed by the cut-off walls and will need to be pumped away. This is sometimes done by crude sump pumping as excavation proceeds. However, this may result in loss of fines or loosening of the soils that lie beneath the formation of the structure under construction. Sometimes, wells or wellpoints are installed inside the cofferdam to pre-drain water levels prior to excavation, thus reducing detrimental effects on the soil formation (Figure 9.13).

9.6.4 Groundwater Lowering Used to Reduce Loading on Cut-Off Structures

If the sides of an excavation are temporarily supported by sheet-pile or concrete diaphragm walls or secant pile walls, the stresses on the walls will arise partly from soil and partly from groundwater loading. In some circumstances, groundwater lowering can be used to

Figure 9.13 Pumped deep wells used to lower groundwater levels in a stratum of fine sand in an area enclosed by diaphragm walls acting as a groundwater cut-off wall. (Courtesy of Ferrer SL, Museros, Spain.)

reduce the groundwater loading on a wall so as to minimize temporary propping requirements. This can be a highly effective expedient; fewer props in the construction area can allow work to proceed more quickly and with less obstruction. One example where this was done was for the construction of cut-and-cover tunnels on the HS1 railway line (formerly known as the Channel Tunnel Rail Link [CRTL]) at Ashford, United Kingdom. Ejector wells were used to reduce pore water pressures outside a bored pile retaining wall in a very low-permeability fissured clay. The temporary depressurization achieved by the ejector wells decreased the load on the walls and allowed the number of levels of temporary props to be reduced (Roscoe and Twine, 2001; Roberts *et al.*, 2007).

However, a note of caution should be sounded. If pumping is interrupted (e.g. because of power or pump failure), the recovery of groundwater levels will increase loading on the walls and props, leading to overstressing, distortion and in the worst case, collapse. Adequate standby facilities (perhaps arranged for automatic start-up) are essential. Alternatively, pressure relief holes could be formed through the wall above formation level; if water levels rose, these would relieve the pressure behind the wall, albeit at the inconvenience of allowing the excavation to be flooded.

9.6.5 Example of Optioneering of Groundwater Pumping and Exclusion Scheme

The interaction between different combinations of pumping and cut-off wall geometries on a given site can be complex and will be influenced by the project-specific requirements for using the techniques together – for example, is a cut-off wall used to reduce pumping rates or to minimize external lowering of groundwater levels?

Figure 9.14 shows schematic sketches of some combinations of techniques that might be used when planning the dewatering of a circular shaft to be excavated into relatively competent but fractured (and therefore permeable) bedrock.

Figure 9.14a shows the simplest dewatering option, whereby external deep wells are used to lower groundwater levels, and no cut-off wall is used. This option is likely to have

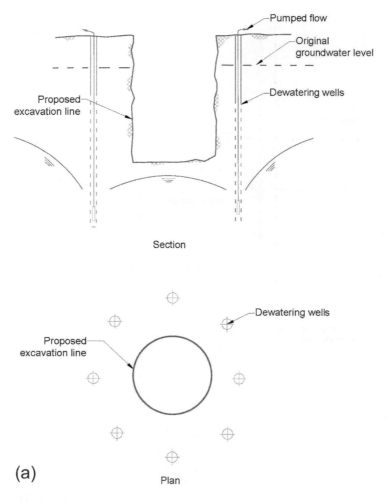

Figure 9.14 Optioneering for groundwater exclusion methods used in combination with pumping. (a) External pumping boreholes (no cut-off wall). (b) Vertical grout curtain with sump pumping. (c) Vertical grout curtain with relief wells. (d) Vertical grout curtain with internal dewatering wells. (e) Vertical grout curtain and grout basal seal.

a relatively low groundwater control cost but will potentially create an extensive zone of groundwater lowering around the shaft, which may impact neighbouring structures or groundwater-dependent features.

If there is a concern about external drawdown impacts, then a cut-off wall (in this case a grout barrier) could be installed around the perimeter of the shaft (Figure 9.14b). In this case, the base of the shaft is ungrouted, so water will still enter the base of the excavation and will be collected in sumps and pumped away. The key challenges with this option are to manage the potentially sediment-laden water pumped from the sumps and to ensure that groundwater pressures at depth are not so large that there is a risk of base instability (see Section 8.6). The addition of relief wells (Figure 9.14c) within the excavation will provide preferential vertical pathways for flow and will reduce the risk of base instability. If disposal of sediment-laden water from sumps is likely to be a problem, deep wells could be installed within the area enclosed by the cut-off wall (Figure 9.14d). The wells are used to lower groundwater levels within the excavation and, provided they have suitable filters and are

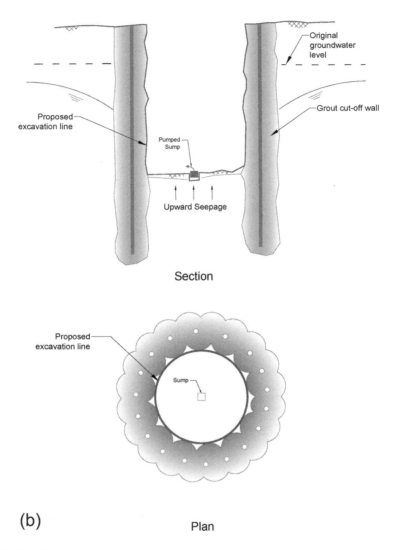

Section

Plan

(b)

Figure 9.14 (Continued)

appropriately developed following drilling, should provide discharge water with much lower sediment loads than would be derived from sump pumping.

A vertical cut-off wall of the type shown in Figures 9.14b–d will reduce the lowering of groundwater levels outside the shaft, but unless the cut-off wall either keys into a very low permeability stratum or extends to great depth, significant drawdowns may still occur around the shaft. If it is desired to reduce external drawdown to an absolute minimum, and there is no suitable very low-permeability stratum for the cut-off wall to seal into, then it may be appropriate to install a basal grout seal (see Section 18.3.1) to link with the cut-off walls and close off the bottom of the excavation (Figure 9.14e). This is likely to be the most expensive groundwater control option, but if correctly implemented, it will ensure that lowering of groundwater level outside the shaft is negligible.

These various options illustrate that pumping and exclusion methods can be used in a wide range of combinations. There is no single 'best' combination of systems. The most appropriate choice for a given site and project will depend on ground conditions, economics

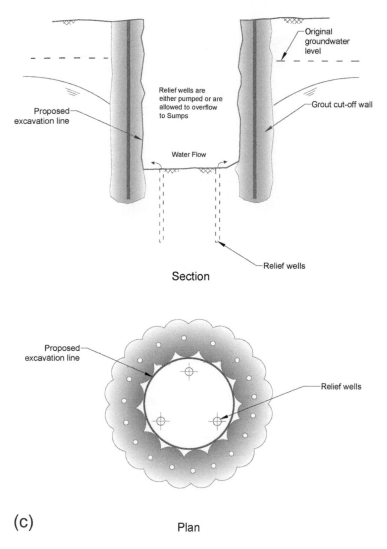

Figure 9.14 (Continued)

and the sensitivity of surrounding areas to drawdown impact, as well as the technologies available at the site locality. Optimization of groundwater control systems is discussed in more detail in Chapter 26.

9.7 GROUNDWATER CONTROL FOR SURFACE MINES

The primary focus of this book is groundwater control methods for construction projects. Groundwater control is also required in the mining industry where surface mines (often called 'open pit' mines) extend below groundwater level. The same principles of groundwater control by exclusion or pumping are used, but because of the particular constraints and requirements, there are some differences in application. This section briefly discusses groundwater control for open pit mines; further details can be found in Brawner (1982), Norton (1987), Beale and Read (2013) and Preene (2015).

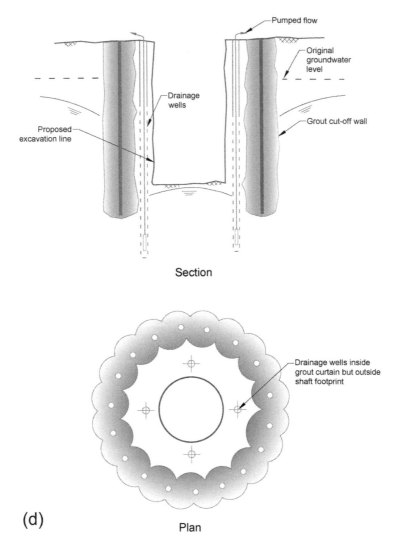

Section

(d)

Plan

Figure 9.14 (Continued)

Most open pit mines are excavated in rock (the principal exception being shallow open pits to extract sand and gravel). Groundwater will enter an open pit mine below groundwater level wherever it intercepts permeable strata or features (such as fracture zones or faults) – groundwater problems in rock are described in Chapter 8. Open pit mines are typically much deeper and larger than construction excavations; it is not unusual for open pit mines to be more than a kilometre in length and several hundred metres in depth. The period for which a mine may require dewatering is typically several years, again much longer than required for most construction excavations.

Typically, groundwater control will have the objectives of providing benefits to mining operations, which can include:

- To improve geotechnical slope stability and safety: lowering of groundwater levels and reducing pore water pressures can allow steeper slope angles to be used in pit walls (a process known as 'pit wall depressurization') and reduce the risk

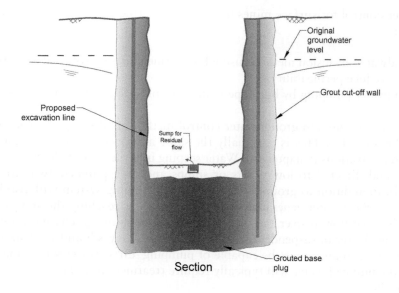

Original
groundwater
level

Grout cut-off wall

Proposed
excavation line

Sump for
Residual
flow

Section

Grouted base
plug

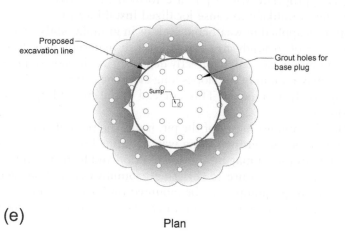

Proposed
excavation line

Grout holes for
base plug

Sump

(e)

Plan

Figure 9.14 (Continued)

of hydraulic failure of the base of the pit where confined aquifers exist below the working level.

- To provide more efficient working conditions: better trafficking and diggability within the pit, reduced downtime due to pit flooding.
- To reduce blasting costs: lowering of groundwater levels in advance of working can provide dry blast holes, reducing the need for more costly emulsion explosives.
- To reduce haulage costs: Dry product/ore and waste rock weigh less than wet material, so dewatering of rock provides a haulage cost saving.
- To reduce environmental impacts: Dewatering wells can be targeted to pump from specific geologic horizons (and low permeability cut-off walls or grout barriers can be used to exclude groundwater from key layers), potentially making use of aquitards and low-permeability layers to reduce external drawdowns that may affect shallow groundwater-dependent features such as wetlands or wells providing water supplies to nearby communities.

Groundwater control for surface mines uses the same principles as applied in construction excavations:

i. Groundwater exclusion methods using low-permeability cut-off walls or ground treatment to reduce permeability
ii. Groundwater pumping by both open pumping and pre-drainage methods

Probably the most common groundwater control method used in open pit mines is 'in-pit pumping' (Figure 9.15). This is essentially the same approach as sump pumping used for construction excavations (Chapter 14). Water seeping into the pit is collected in open drains or channels and directed to low points or sumps and then pumped away to the surface (Figure 9.15). In addition to groundwater, the in-pit pumping system will also be required to pump any surface water generated in the pit. The water reaching the sumps and pumps will typically have flowed over the pit floor and along drainage channels and will have picked up some degree of suspended solids (in the form of clay, silt and sand-sized particles). Accordingly, in-pit pumps must be capable of pumping 'dirty' water with some suspended solids, and the pumped water will typically require treatment to remove solids prior to discharge from site.

In-pit pumping is appropriate where pits are formed in relatively stable rock and the inflow of groundwater is unlikely to cause localized instability in the pit slopes and base. Where in-pit pumping is applied in weathered rock or in granular soils such as sand or sand and gravel, the seepage of groundwater through those materials may lead to instability. Furthermore, in-pit pumping can only indirectly depressurize the side slopes of the pit. High pore water pressures may remain in the slopes long after the main pit is dewatered, with the slopes draining only slowly into the pit. This can lead to the risk of large-scale geotechnical instability of the slopes.

Depending on the size and geometry of the pit, it may be possible to keep the pit almost entirely dry by in-pit pumping (with standing water confined only to small sump areas). In other cases, the bottom of the pit may be allowed to flood and form a pond or lagoon whose water level can fluctuate in response to differing groundwater and surface water inflow rates. In such cases, the in-pit pumps may be mounted on floating pontoons (Figure 14.7b) to allow them to rise and fall with the lagoon water level. Sump pumping for open pit mines is discussed in Section 14.4.

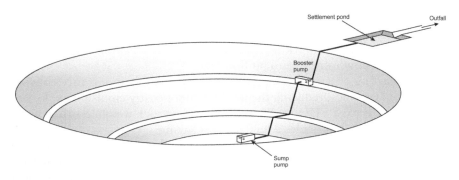

Figure 9.15 Schematic arrangement of in-pit pumping for a surface mine. (From Preene, M, Techniques and developments in quarry and surface mine dewatering, in *Proceedings of the 18th Extractive Industry Geology Conference 2014 and Technical Meeting 2015* (Hunger, E and Brown, T J, eds), EIG Conferences Ltd, London, 2015. With permission from EIG Conferences Limited, London, UK.)

Perimeter pumped wells are sometimes used as part of the dewatering of open pit mines, typically in combination with in-pit pumping (Figure 9.16). The application of deep wells is essentially the same as for construction excavations (described in Chapter 16) but on a larger scale. The deep well method has two principal advantages over in-pit pumping. First, if pumping from the dewatering wells is started long enough in advance of the pit being deepened below groundwater level, it may be possible to lower groundwater levels in advance of mining. Used in this way, the technique is a 'pre-drainage' method (sometimes known as advance dewatering), which can improve operational conditions in the mine by reducing seepages through the side slopes and base of the pit. Second, because the dewatering wells are located behind the pit slopes, in favourable geological settings they can provide a significant groundwater depressurization effect in the pit slopes. In some cases, perimeter dewatering wells may be augmented by in-pit wells, but the presence of such wells (and the associated cables and discharge pipework) in the pit may impact on mining methods and sequencing.

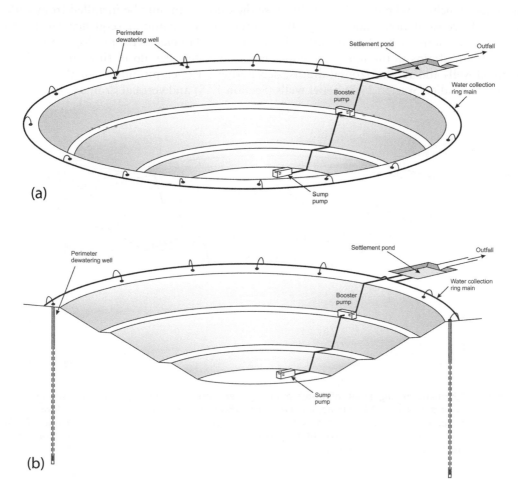

Figure 9.16 Schematic arrangement of in-pit pumping and perimeter dewatering wells for a surface mine (top) and the cut-away view (bottom). (From Preene, M, Techniques and developments in quarry and surface mine dewatering, in *Proceedings of the 18th Extractive Industry Geology Conference 2014 and Technical Meeting 2015* (Hunger, E and Brown, T J, eds), EIG Conferences Ltd, London, 2015. With permission from EIG Conferences Limited, London, UK.)

If there is good lateral hydraulic connectivity in the area of the pit slopes, pumping from perimeter deep wells can have a significant depressurizing effect on the pit slopes. However, if the rock penetrated by the dewatering wells contains low-permeability faults or other large-scale geological structures, there may be 'compartmentalization' of groundwater flow (Section 8.3.3). In these conditions, pumping from perimeter dewatering wells alone may not have a significant depressurization effect in all areas of the pit slopes. This could compromise the geotechnical stability of the pit slopes. One possible option is to install sub-horizontal drains out from the pit slopes to create additional permeable pathways for water to enter the pit (Figure 9.17 and Figure 8.16).

Typically, sub-horizontal drains are installed in lines at regular spacings, drilled out horizontally, or with a slight upward or downward inclination, from benches in the pit slopes, with a drilled diameter of 75 to 125 mm (see Section 17.5). The drains are 'passive' and are not pumped directly. The presence of the drains provides a preferential pathway for groundwater flow from the slopes into the pit. The water pressures in the slope drive water along the drain, so that water bleeds from the open end of the drain and is typically collected in drainage trenches and pumped away. Because the drains can only be installed from within the pit, their installation sequence must be integrated with the mining sequence. Such drains often flow copiously when first installed, but the flow will reduce as the pit slopes are depressurized, and drains in the upper pit slopes may completely dry up in time. In addition to pumped wells and sub-horizontal drains, dewatering for open pit mines can also make use of drainage adits (Section 17.7), relief wells (Section 17.8) and vertical drains (Section 17.9).

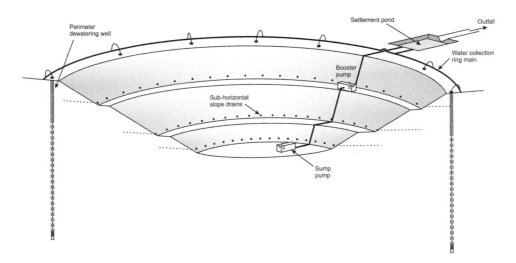

Figure 9.17 Schematic arrangement of in-pit pumping, perimeter dewatering wells and sub-horizontal slope drains for surface mine, cut-away view. (From Preene, M, Techniques and developments in quarry and surface mine dewatering, in *Proceedings of the 18th Extractive Industry Geology Conference 2014 and Technical Meeting 2015* (Hunger, E and Brown, T J, eds), EIG Conferences Ltd, London, 2015. With permission from EIG Conferences Limited, London, UK.)

Chapter 10

Groundwater Control for Tunnelling Projects

> Tunnel work is like trying to dress for dinner under the bed clothes
>
> Sir Harold Harding

10.1 INTRODUCTION

There is some merit in considering groundwater control for tunnels, shafts and associated works separately from methods used for surface excavations. While the groundwater control methods used on tunnelling projects follow the same principles – dealing with groundwater by either exclusion or pumping – the geometry and access constraints mean that different methods are sometimes used on tunnelling projects.

This chapter briefly introduces the different types of tunnelling works (including shafts and cross passages) commonly constructed and discusses the various groundwater control methods that can be deployed.

10.2 TYPES OF TUNNELLING WORKS

When considering groundwater problems during tunnelling work, it is natural to focus on the tunnels themselves – linear sub-horizontal features, often but not always cylindrical. However, there are a wide range of different geometries of tunnel works, including shafts, cross passages and caverns, constructed in a wide range of soils and rocks. As well as tunnelling for construction projects, in modern mining applications, it is not unheard of for shafts and tunnelling works to be carried out at depths in excess of 1000 m. One common factor across all these applications is that groundwater must be controlled in terms of both water inflow and instability of soil or rock exposed in the excavation (groundwater problems during tunnelling in soil and rock are described in Sections 7.7 and 8.8, respectively).

When considering the groundwater control requirements, it is useful to have a basic understanding of common methods of tunnel construction. In this chapter, the different types of excavations are categorized into five groups, each briefly described in the following sub-sections:

    i. Tunnels, which are often arranged as twin parallel tunnels on transportation projects
   ii. Shafts and vertical excavations connecting tunnel works to the surface
  iii. Cross passages linking between twin parallel tunnels
  iv. Other tunnel applications, such as adits, connector tunnels, caverns, etc.
   v. Recovery works applied after there has been a tunnel collapse or inundation

Further information on tunnelling methods can be found in Chapman *et al.* (2017). Groundwater control methods applicable to each category are discussed later in the chapter.

This chapter does not address cut-and-cover tunnels or tunnel portals and approach ramps, which are effectively conventional excavations, to which the methods of Chapter 9 can be applied. Neither are immersed tube tunnels covered here; again, the portal and casting basin excavations for these works are essentially conventional excavations. Groundwater control is often needed for the excavations associated with immersed tube projects, such as for the Conwy Crossing (Powrie and Roberts, 1990) and the Medway Crossing (Leiper *et al.*, 2000; Lunnis, 2000) in the United Kingdom and the River Lee Tunnel (White and Capps, 1997) in Ireland.

10.2.1 Tunnels

When tunnelling in soils and weak rocks, instability at the excavated face is a significant risk. To mitigate this risk, most tunnelling is carried out within a 'shield'. This is a strong cylindrical sheath pushed forwards into the ground while material is excavated from within. The shield provides circumferential support around the working area where the tunnel is being excavated and ensures that only a small area of soil/rock (the face) is exposed at any one time. Typically, in shield tunnelling, a 'primary lining' is placed immediately behind the shield to support the ground and exclude water. The primary lining is often formed of pre-cast concrete 'segments' assembled to form rings; each time the shield advances a few metres, a new ring is constructed immediately behind it.

A shield is a type of open face tunnel boring machine (TBM) that has been in use since the nineteenth century, when Sir Marc Brunel developed it for the first tunnel beneath the River Thames in London (Skempton and Chrimes, 1994). In the simplest case, the tunnel is at atmospheric pressure (known as the 'free air' condition), and groundwater can enter via the face (Figure 10.1a). Excavation can be by hand, by mechanical excavators or by a rotary cutting head that attacks the entire face simultaneously. In all these cases, groundwater inflow at the face can cause problems. A long-established variation on shield tunnelling is the use of compressed air to pressurize the tunnel, with the aim of the compressed air providing a counter pressure, thereby reducing groundwater inflow and stabilizing the ground exposed in the face (Figure 10.1b). Further details of this method are given in Section 10.4.1.

From the mid-twentieth century onwards, a new variant of shield tunnelling emerged – closed face TBMs. This method uses a pressure bulkhead close behind the face to create an isolated zone filled with a fluid, the pressure from which is used to apply a counter pressure to the face, controlling groundwater inflow and stability at the face (Figure 10.1c). Depending on the type of TBM, the counter pressure can be provided by an earth paste (earth pressure balance [EPB] TBMs) or a bentonite or polymer slurry (slurry TBMs). Further details of this type of tunnelling are given in Section 10.4.2.

Tunnelling in hard, stable rocks is sometimes done without the protection of a shield or using a partial shield to protect the miners from blocks falling from the roof of the tunnel. This means that a much larger area of unsupported ground is temporarily exposed during excavation (Figure 10.1d). Depending on rock strength and occurrence of fracturing, excavation may be by 'drill and blast' methods using explosives (typically in very strong rocks) or by mechanized excavation using 'road headers' or 'continuous miners' (boom mounted rotary cutters). Typically, a primary lining is installed behind the area of excavation. Lining methods may include sprayed concrete lining (SCL), rock bolts, rock nails, mesh support and steel arches.

In soils and rocks where relatively small excavated faces can be 'self-supporting' for a limited time, the 'sequential excavation method' (SEM) method can be used (this is also known as the new Austrian tunnelling method or NATM). This method, developed in Austria in

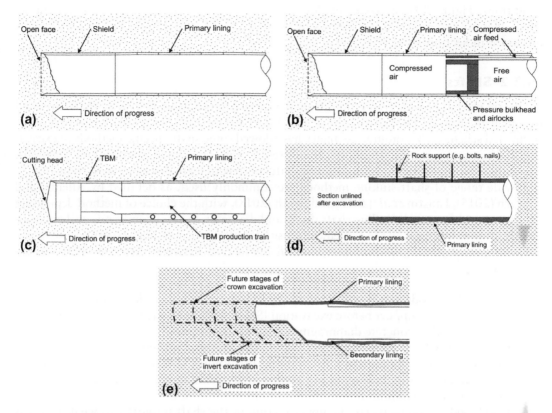

Figure 10.1 Principal tunnelling methods. (a) Open face tunnelling with shield in free air. (b) Open face tunnelling with shield using compressed air. (c) Closed face tunnel boring machine (TBM) with fluid balance at the cutting head. (d) Tunnelling in hard rock without shield. (e) Sequential excavation method (SEM) of tunnelling.

the mid-twentieth century, is essentially tunnelling without a shield, where excavation is sequenced to take advantage of the ability of stable soil and rock to 'arch' and temporarily transfer stress around a small zone of excavation. This behaviour allows the material to stand without collapse in the period between excavation and the placement of the primary lining. A typical sequence is shown in Figure 10.1e, where the sections of the upper part of the tunnel profile are excavated first and lower sections later. The key factor in the method is the rapid application of a primary lining to the excavated face to provide initial support – this normally involves SCL, sometimes called 'shotcrete'. A more permanent secondary lining is installed following excavation of the complete tunnel profile. Because of the need to ensure that the excavated zones are stable and that SCL can bond to the face, groundwater control is often required for SEM tunnelling; further details on SCL methods are given in Thomas (2019).

Tunnel projects can be constructed over great distances – for example, the Channel Tunnel between England and France (Harris *et al.*, 1996) is more than 30 km long. Tunnel projects of between 1 and 10 km total length are commonplace. It is highly likely that a given tunnel project will encounter different geological and hydrogeological conditions along the route. Furthermore, access constraints (as well as cost considerations) mean that investigation boreholes may only be widely and irregularly spaced along the tunnel route. While uncertainty over ground conditions affects all below-ground projects, it is routine in tunnel projects to have sections of the route that are poorly characterized, particularly in relation to groundwater conditions. Many tunnels meet unexpectedly challenging ground conditions; Figure 8.18 shows examples of changes in ground conditions along a tunnel route. Planning to reduce water inflows by changing the vertical alignment of the tunnel is discussed in Section 10.3.

10.2.2 Shafts

Shafts are used to provide connections from the surface to tunnel works. Some shafts are used to launch and recover TBMs, while others may be needed for access or ventilation as part of the completed scheme. The common geometry has allowed some shaft-specific methods of excavation and construction to be developed. These are sometimes known as 'shaft sinking' methods. Groundwater control methods used during shaft construction are described in Section 10.5.

A wide range of shaft sinking methods are commonly used, as detailed by Allenby and Kilburn (2015), Faustin *et al.* (2017) and Smith (2018), with the choice of method depending on a variety of factors, including shaft depth and diameter, ground conditions and environmental constraints such as the settlement sensitivity of nearby existing structures.

Faustin *et al.* (2017, 2018) make a useful distinction between different shaft-sinking methods:

1. Pre-lined methods, where a wall forming the shaft lining is installed prior to excavation (also known as support before excavation [SBE]). Examples include shafts enclosed by steel sheet-piles, concrete diaphragm walls and secant piles.
2. Concurrently lined methods, where installation of the shaft lining occurs at the same time as, or immediately after, excavation (also known as excavation before support [EBS]). Examples include caisson methods, underpinning with segmental linings and excavations with SCL.
3. Combined methods, where the upper section of the shaft (typically in weaker or less stable soil/rock) is supported by a pre-lining system and the lower portion is concurrently lined.

Different shaft sinking methods are shown schematically in Figure 10.2. In most civil engineering applications, when the shaft excavation has reached full depth, a concrete plug or slab will be placed in the base of the excavation to provide (in combination with the shaft lining) a watertight structure.

Figure 10.2a shows a shaft installed by the pre-lined method. This approach uses a physical wall (such as could be formed by steel sheet-piles, concrete diaphragm walls or secant pile walls) to support the sides of the shaft (Figure 10.3). Typically, the support wall extends below the final dig level. If the shaft is sunk through weaker strata down into stronger rock, the wall may be terminated a suitable distance below the top of the rock, and the deeper portion of the tunnel can be excavated by concurrent lining methods (an example is shown in Figure 8.7). From a groundwater control perspective, if the pre-lining system is of very low permeability, it can be used as part of a groundwater exclusion strategy. It is important to note that if contiguous concrete bored piles are used as the pre-lining system, they cannot be relied upon to exclude groundwater – this is clearly illustrated in Figure 9.7.

Figure 10.2b shows a concurrently lined shaft formed by underpinning. This method involves excavating downwards at the base of the shaft in a series of incremental stages (typically 0.75 to 2 m per increment, depending on the stability of the excavated face). A section of lining system is then used to support the exposed ground at the perimeter of the shaft (an example of this method is shown in Figure 6.3). The lining system is most commonly formed from segmental rings (of pre-cast concrete or spheroidal graphite iron [SGI]) bolted together with watertight gaskets; the gap between the extrados of each ring and the excavated ground

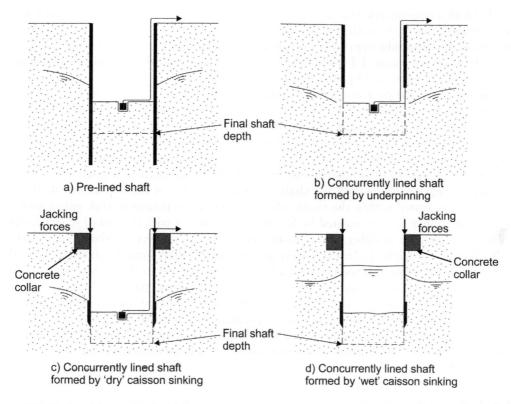

Figure 10.2 Shaft sinking methods. (a) Pre-lined shaft. (b) Concurrently lined shaft formed by underpinning. (c) Concurrently lined shaft formed by 'dry' caisson sinking. (d) Concurrently lined shaft formed by 'wet' caisson sinking.

Figure 10.3 View down a shaft where the sides are supported by concrete diaphragm walls. (Courtesy of MVB Joint Venture, Lee Tunnel Project, London, UK.) Note four relief wells installed near the centre of the shaft.

is sealed with a cement-based grout. The excavation and lining sequence is repeated to deepen the shaft until the shaft reaches final depth. This method requires relatively dry and stable conditions, and groundwater control methods are often used. A variation on this method is where the shaft is excavated downwards incrementally but with the excavated ground in the sides of the shaft supported by SCL rather than bolted segmental linings.

Figures 10.2c and d show concurrently lined shafts formed by caisson sinking. The generic civil engineering definition of 'caisson' is a watertight retaining structure. In the current discussion, we will use the more specific shaft sinking definition, where a caisson is a shaft structure constructed (as a reinforced concrete monolith) or assembled (from pre-cast segments) at ground level. The caisson is then progressively sunk into the ground in a controlled manner. In contrast to the underpinning method (where the shaft lining is extended at the bottom), in the caisson method the shaft lining is extended at the top as the shaft sinks. Material is excavated within the shaft, which allows the structure to sink either under its own weight (sometimes augmented by kentledge weights) or with the aid of special jacks.

Shafts can be sunk as either 'dry caissons' (Figure 10.2c) or 'wet caissons' (Figure 10.2d). In the dry caisson method, the excavation area inside the shaft must be dewatered to allow conventional excavation methods to be used. In the wet caisson method, the excavation area is kept flooded with water, and excavation is made underwater, typically by grab excavator (Figure 10.19). The water pressure acting on the excavated ground in the base of the shaft has a stabilizing effect, analogous to the fluid counter pressure used in closed face TBMs. Groundwater issues associated with 'wet caisson' methods are discussed further in Section 10.5.4.

Shafts can also be sunk using compressed air to exclude groundwater. This is outlined in Section 10.5.4.4.

10.2.3 Cross Passages

Many transportation projects (road or rail) use twin 'running' tunnels that are constructed parallel to each other, typically located a few tunnel diameters apart horizontally. The operational requirements of many transportation systems mean that there must be access between the tunnels at regular intervals along the route. This is necessary to meet regulatory requirements for rescue and evacuation in the event of an emergency.

The connection between running tunnels is achieved by the construction of 'cross passages' (Figure 10.4). The most common geometry is for each cross passage to be of slightly smaller diameter than the running tunnels and to be at right angles to the main tunnels. The spacing of cross passages along the route is usually controlled by operational safety, but spacings of 100 to 500 m between cross passages are typical. Effective control of groundwater is a fundamental part of any cross passage scheme; suitable methods are described in Section 10.6.

10.2.4 Other Tunnel Applications

In addition to tunnels, shafts and cross passages, there are various other situations where excavation may be made for tunnel projects. Examples include:

(i) The entry or exit of TBMs into or out of shafts, portals, outfalls and other structures, including the construction of 'soft eyes' in shaft linings to allow the TBM to enter or exit the shaft, and 'headshunt' and 'backshunt' tunnels used to facilitate the assembly/dis-assembly of TBMs at launch and recovery shafts, respectively.

(ii) Construction of station caverns, escalator chambers, tunnel enlargements, step plate junctions, cross adits or other connections where the tunnel lining has to be breached temporarily before being re-instated to a new geometry.

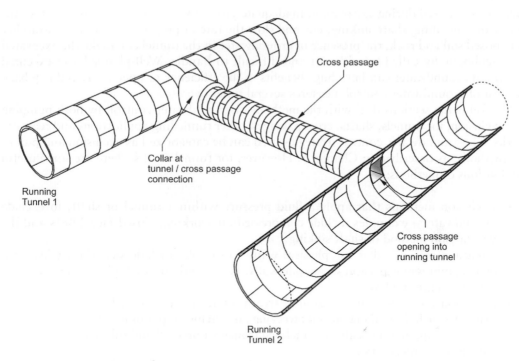

Figure 10.4 Schematic layout of cross passage between twin running tunnels.

(iii) Reduction of pore water pressures in fine-grained soils such as silts or very silty sands to reduce the risk of the soils liquefying as a result of vibration from the machinery in the TBM.

(iv) Control of groundwater where access is required to the pressurized sections of the closed face TBMs (for example to replace worn cutters or for bearing maintenance). This is often known as a 'face intervention' and can be achieved by one or more of the following: pumping; exclusion by compressed air; and ground treatment.

(v) To control groundwater velocities in order to allow the use of ground treatment methods (such as artificial ground freezing or grouting) in problematic conditions.

Groundwater control for some of these applications is discussed in Section 10.7.

10.2.5 Recovery Works

Unfortunately, tunnel projects occasionally encounter serious geotechnical and hydrogeological problems, and collapse and inundation can result; Heuer (1976) and Clay and Takacs (1997) give several examples. Groundwater is often a significant factor in tunnel collapses, and recovery works can include a significant element of groundwater control by both pumping and exclusion methods, as discussed in Section 10.8.

10.3 WATER MANAGEMENT IN TUNNELLING WORKS

As has been described in Section 7.7 (for tunnelling in soils) and Section 8.8 (for tunnelling in rocks), groundwater must be managed to avoid tunnel works being flooded out (or at least inconvenienced) by water inflows and to prevent groundwater-induced instability of soil and rock faces exposed during excavation. Inadequate groundwater control can reduce the efficiency of tunnelling, shaft sinking, etc. and slow the rate of progress. As well as instability of exposed soil and rock, the presence of groundwater in the tunnel can make the excavated spoil (colloquially called 'muck') wet and difficult to handle. Well-planned and executed control of groundwater can have huge benefits to the efficiency of operations and pay back the cost of groundwater control measures several times over.

The following sections deal with the methods of groundwater control, by both pumping and exclusion, for tunnels, shafts, cross passages, other tunnelling applications and recovery works. The groundwater control strategies used can be categorized as exclusion and pumping methods, as described in Chapter 9. However, for tunnel works, there are some useful subdivisions of methods:

- Exclusion methods that apply a fluid pressure within a tunnel or shaft, to exclude groundwater – examples include compressed air working, closed face TBMs and the wet caisson method of shaft sinking.
- Exclusion methods that use physical walls – examples include steel sheet-piles, concrete diaphragm walls and secant pile walls. These methods are typically applied vertically from ground level.
- Exclusion methods that use ground treatment techniques to reduce permeability – examples include artificial ground freezing and various types of grouting. These methods can be applied vertically and inclined from ground level and sub-horizontally from within the tunnel works.
- Pumping methods where wells are drilled from ground level.
- Pumping methods where pumping is from within tunnel works – this is known as in-tunnel dewatering. This can involve direct pumping from within the tunnel and/or pumping from small-diameter wells drilled outwards from the tunnel works.

The discussion in this chapter is mainly around the technology solutions that can be used – what techniques can be used to keep the water out, or how can we pump it away? But it is important to recognize that the overall design of the tunnel can affect the groundwater control requirements.

If the geology and hydrogeology of the tunnel route are thoroughly investigated, both by intrusive investigations and by desk studies using data sources such as geological maps, it may be possible to identify strata or beds that are more favourable to tunnelling. From a groundwater control perspective, that could involve identifying layers that are of lower permeability – such as more clayey soils or less fractured beds of rock – where groundwater inflows and associated risk of instability will be lower. If such conditions are identified at an early stage of the project, consideration can be given to changing the depth of the tunnel (known as the 'vertical alignment') to maximize the length of tunnel that will be constructed through favourable geology (Figure 10.5a). On some projects, it may not be possible to vary the vertical alignment in this way, but in fact, this approach has been used as a cost-saving and risk-reduction measure on many tunnels.

One of the largest-scale examples of this approach is the 30 km long Channel Tunnel between England and France, constructed in the 1980s/1990s. The tunnel had to traverse below the open water of the English Channel, beneath which the strata are dominated by the Chalk, a variable and fractured limestone. Tunnels in Chalk were known to experience large water inflows when open fissures were encountered. The eminent tunnel engineer Sir Harold Harding said: 'A wise tunnelling engineer should view Chalk with deep suspicion. It looks so safe when standing on 100 feet high cliffs or Chalk pits, but it can be deceptive and can vary widely and suddenly in its condition' (Harding, 1981). The consequences of

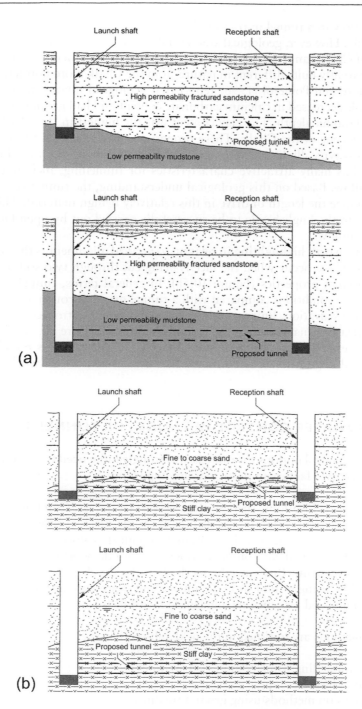

Figure 10.5 Effect of changes in tunnel vertical alignment on groundwater control requirements. (a) Tunnel alignment varied to avoid water-bearing stratum. Even though deepening the tunnel alignment will take it further below groundwater level, a deeper tunnel will experience less groundwater inflow, as it will be driven through a lower-permeability stratum. (b)Tunnel alignment varied to avoid mixed face conditions. The shallower tunnel alignment will have a mixed face of sand and clay. This can cause difficulties for tunnelling and groundwater control. The deeper tunnel alignment will have a full face of clay, improving tunnelling conditions and reducing groundwater inflows.

a large water inflow to a tunnel in fractured Chalk below the English Channel would have been catastrophic. However, geological studies indicated that a continuous layer of a subtly different type of Chalk ran from one side of the Channel to the other. This material – Chalk Marl – was apparently quite different in nature from the Chalk more familiar from onshore works. Mortimore and Pomerol (1996) report that geological investigations revealed that the beds of Chalk decreased in calcium carbonate content and had greater clay content with increasing stratigraphic depth. The clay content was the highest in the Chalk Marl, which is the lowermost zone of the Chalk. These properties mean that as a tunnelling medium, the Chalk Marl is typically less brittle and of lower permeability than the higher zones of the Chalk and has many attractive characteristics for tunnelling, including low rates of groundwater inflow. Based on this geological understanding, the tunnel vertical alignment was set to maximize the length of drive in this relatively benign material. This allowed the tunnel drives from the English side to be successfully carried out by open face TBMs with relatively minor inflows.

One other issue worth highlighting in relation to vertical alignment is the effect of 'mixed face' conditions. These occur when the face is being cut through two soil or rock types of significantly different properties (Figure 10.5b). In general terms, tunnelling is more difficult in mixed face conditions compared with a tunnel driven through a single type of soil or rock, as many excavation, lining and indeed groundwater control techniques only work effectively in a relatively narrow range of ground conditions (Figure 10.10d shows an example of groundwater control problems in mixed face conditions). If it is possible to arrange the tunnel vertical alignment to avoid mixed face conditions, tunnel works may progress more efficiently.

10.4 GROUNDWATER CONTROL METHODS FOR TUNNELS

The groundwater control options for tunnels are strongly influenced by the typical geometry of tunnel projects. Two factors in particular are important:

i. Away from shaft locations, there is often no surface access. Combined with the depth of many tunnels, this often makes drilling from the surface, either for pumping wells or for ground treatment, very difficult, uneconomical or both.

ii. The relatively confined environment associated with the working space inside the tunnel has allowed tunnelling-specific methods to be developed, whereby groundwater is excluded by using a fluid to maintain a counter pressure at the tunnel face. These include compressed air working and closed face TBMs.

The following methods are outlined in the next sub-sections:

- Open face tunnelling using compressed air working
- Closed face TBMs that apply a counter pressure to the face
- Tunnel exclusion methods using ground treatment
- Pumping methods in tunnels

10.4.1 Open Face Tunnelling Using Compressed Air Working

The longest-established method of applying a fluid pressure to the face to exclude groundwater from a tunnel is compressed air working. This method, in use since the nineteenth century (Glossop, 1976), uses a bulkhead within the tunnel (containing an airlock system capable of passing workers, materials and tunnel spoil) to isolate the tunnel face and a working area behind the face. A continuous supply of pressurized air is fed from compressors to the working area between the tunnel face and the airlock (Figure 10.1b). The tunnel is advanced by mechanized or hand excavation by miners working within a shield in a compressed air environment. In contrast to compressed air working, 'free air' is the term used to describe conventional methods, where the tunnel is at atmospheric pressure.

In the mid- to late twentieth century, the compressed air method was used widely in tunnelling in soils and weak rocks, but modern practice recognizes the health and safety risks when miners work in air pressures above atmospheric; the method is now much less common and is often limited in scope, and must be carried out under specialist medical supervision. For example, it is used to allow maintenance interventions during TBM drives rather than being the primary method of groundwater control during tunnelling. Further information on compressed air tunnelling can be found in Slocombe *et al.* (2003), Placzek (2009) and British Tunnelling Society (2012). Compressed air working can also be used for shaft construction, when an air deck and airlocks are used to seal the top of the shaft (Section 10.5.4).

One bar (100 kPa) of air pressure at the tunnel face will balance approximately 10 m head of groundwater. When the tunnel face is within a typical unconfined aquifer, the groundwater pressure will vary hydrostatically across the vertical dimension of the face, being lowest at the crown and highest at the invert of the tunnel (for a 3 m diameter tunnel, the difference in groundwater pressure between the crown and invert will be around 0.3 bar). In contrast, the air pressure is constant over the face. It is never possible to get a perfect air pressure in the tunnel to balance water pressures across the whole face. For example, if the air pressure is set to balance the water pressure at the mid height of the tunnel, some air loss may occur in the crown, and water may enter the tunnel above invert (Figure 10.6).

Technical and health and safety limitations limit the use of compressed air to around 3.5 bar (equivalent to approximately 35 m head of water). However, such high pressures are rarely used. Most compressed air working is carried out at pressures of less than 0.75 bar above atmospheric (known as low-pressure compressed air); costs rise considerably above 0.75 bar (high-pressure compressed air) due to the additional medical constraints. Groundwater

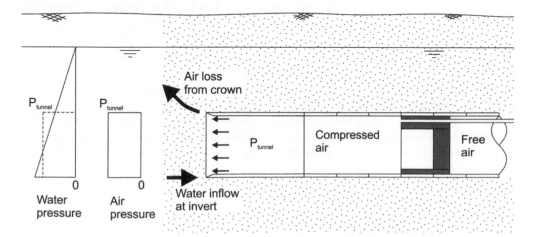

Figure 10.6 Compressed air tunnelling. A continuous supply of air is used to pressurize the working space of the tunnel (the air pressure in the tunnel is P_{tunnel}). Due to the different density of air and water, the air pressure distribution cannot precisely match the groundwater pressure distribution. Typically, air may be lost from the crown of the tunnel.

control by pumped surface wells has occasionally been used to lower groundwater levels to allow compressed air working at reduced pressures (ideally less than 0.75 bar), for example to allow access to the working head of a TBM for maintenance purposes – this is known as a 'face intervention'. The dewatering wells must be located with care to avoid compressed air escaping through the ground to the wells. An example of low-pressure compressed air, used in combination with dewatering wells, to allow TBM maintenance interventions is described by Machon and Stevens (2004).

The loss of air to the ground, plus the losses from operation of the airlock, is the reason why a continuous supply of air is needed to maintain the required pressure in the tunnel. The loss of air to the ground is greater in coarse-grained soils and permeable fractured rocks (where the larger size of the soil pores and fracture openings allows air to pass relatively freely). The escaping air will find its way to the surface and can cause various problems. Groundwater levels can rise under the influence of the upward flow of air. Air can also emerge at the surface, not only directly above the tunnel but in some cases at considerable distances. Warren *et al.* (2018) reported a case of tunnelling in Chalk (a fractured limestone) where high air losses were reported when compressed air was applied during a 36-hour long face intervention, when the tunnel crown was 26 m below ground level. The paper states:

> The escaping air made its way into the overlying floodplain sand/gravels which was confined beneath alluvial clays and as such travelled laterally to emerge either both in the river and as a 5 m spout of water/sand through a standpipe located 94 m further landward. An associated 3 m diameter, 1.5 m deep hole formed and eventually appeared at ground surface above the location of the face intervention. The hole was quickly backfilled and monitored for ongoing settlement, but the incident illustrates the need to carefully consider the geology in terms of air loss pathways.

Another potential problem is the risk of a 'blow out' – a sudden and potentially violent loss of large volumes of compressed air from the tunnel to the surface. This can occur when the air pressure in the tunnel is too high in relation to the depth of overburden above the tunnel. This is analogous to the 'buoyancy uplift' failures of the base of excavations described in Section 7.5. But in this case, it is excess air pressure causing the instability, not groundwater pressures. In 1998, a compressed air blowout occurred in London during construction of a section of the Docklands Light Railway, resulting in a large crater appearing in the grounds of a school. Investigations concluded that the cause was the use of air pressures in a shallow section of tunnel where uplift pressures could not be resisted by the overburden above the tunnel at that location (*New Civil Engineer*, 1998, 2004).

Compressed air essentially acts to repel groundwater from the tunnel face – it is important to realize that the groundwater must have a pathway to move away from the tunnel. There have been cases where compressed air has not been effective in controlling groundwater when tunnelling through generally low-permeability soil (e.g. a low-permeability Glacial Till) that contains localized more permeable sandy zones. In these cases, when groundwater is encountered, the initial application of compressed air does stop water ingress. But water inflows reappear after a short time, and further increases in air pressure stop inflows only temporarily. The problem is that the permeable zones are of limited lateral extent, and the compressed air pushes the water from the face, but there is no outlet, and the water becomes more pressurized within the sandy zones surrounded by low-permeability material. An unhelpful feedback loop develops, with each increase in the applied air pressure raising the water pressure in the permeable zone by a corresponding amount. This problem can be avoided by the installation of wells from the surface to penetrate the permeable zone and provide an outlet for the air. An alternative approach is to install sub-horizontal wellpoints into the sand zone exposed in the tunnel face and connect them to discharge to free air in the tunnel outside the airlock. This provides a route for water to drain from the permeable zone as air pressure is applied to the face.

10.4.2 Closed Face Tunnel Boring Machines (TBMs)

An alternative method of balancing ground and groundwater pressures at the tunnel face was developed in the mid-twentieth century – closed face TBMs. These machines have a shield with a rotating cutter, but the key element is that the space immediately behind the cutting head can be maintained at a controlled pressure, separated from the rest of the tunnel by a pressure bulkhead. This allows a counter pressure to be applied to the face (to exclude groundwater and support the ground at the face), while the TBM operator and miners work in 'free air' (i.e., at atmospheric pressures) conditions behind the pressure bulkhead, avoiding the health risks of working in a compressed air environment. In fact, the working area in the tunnel behind a closed face TBM is often a very dry 'shirt sleeve' working environment.

There are two principal types of closed face TBM:

1. Earth pressure balance (EPB) TBMs
2. Slurry TBMs

An EPB TBM (Figure 10.7a) works by using the spoil excavated by the rotating cutting head to form an 'earth paste' to provide support to the face. The spoil cut at the head passes into a plenum chamber behind the face. Typically, in cohesive soils such as clays, the paste formed by the spoil and groundwater is sufficiently plastic to provide adequate support and pressure to the face. In coarse-grained soils such as sands and gravels, such support diminishes unless conditioning agents (polymers, foams or bentonite) are used to generate a more cohesive paste. The spoil is removed from the plenum chamber at a controlled rate by a screw conveyor, from where the spoil passes into the tunnel, to be removed by belt conveyor or rail car. The screw conveyor is used to dissipate the pressure differential between the face and the tunnel working area. By regulating the rate at which the paste is extruded from the plenum chamber as the TBM excavates and moves forward, the screw conveyor can manipulate the support pressure to provide a suitable pressure balance at the face.

A slurry TBM (Figure 10.7b) supports the face with a bentonite or polymer slurry supplied from a surface reservoir and slurry treatment plant. The slurry can be visualized as much more fluid than the paste that provides face support in EPB TBMs, and plays a role analogous to the slurry support during diaphragm wall trenching. The spoil excavated at the cutting head is entrained in the slurry, which is pumped away through pipework along the tunnel to the surface, where the spoil is separated from the slurry. The 'clean' slurry is then circulated back to the face in a continuous process. In many slurry TBMs, a compressed air cushion or bubble in the TBM excavation chamber is used to control the slurry support pressure in order to allow the face pressure to be adjusted to balance groundwater pressures.

Modern TBMs can be viewed as underground tunnel factories, costing millions of pounds, slowly moving forwards, excavating the ground at the tunnel face and forming a lined tunnel behind. The overall production train of a TBM can be up to 150 m long, including the cutter head and shield, and equipment to erect the lining and handle the spoil.

The permanent tunnel lining is constructed within the rear section of the TBM (known as the tailskin). Most commonly, a segmental lining is used, where the segments are erected to form a complete ring. Watertightness is achieved by each segment being sealed against its neighbours by special gaskets and each completed ring being surrounded by cement-based grout (typically injected through grout holes in the segments) to backfill the overcut resulting

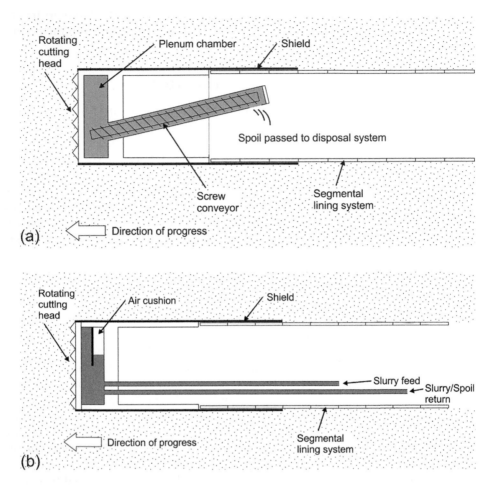

Figure 10.7 Tunnelling using closed face TBM. (a) Earth pressure balance (EPB) TBM. (b) Slurry TBM.

from the slightly larger external diameter of the TBM shield compared with the segmental lining. As the TBM excavates, it is jacked forwards from the existing lining, allowing the next ring to be added to the lining. This cycle is repeated as the tunnel progresses.

Determination of required face pressures is a complex issue and is dependent on both ground conditions and groundwater pressures. It is discussed in Shirlaw (2012). Pressures should be selected carefully: too low, and the face may be unstable and cause excessive surface settlement; too high, and the ground above the tunnel may heave.

In closed face TBMs, in normal operating conditions, the cutting head (under pressure) is not accessible from the tunnel (at atmospheric pressure). However, on long tunnel drives, maintenance access (a face intervention) may be required to the cutting head; the cutting tools may wear out and require replacement, or obstructions may need to be removed. Planned maintenance interventions can be facilitated by driving the TBM into a zone of groundwater lowering or ground treatment at a designated location on the route. This can stabilize the face and allow access for operatives (via a hatch in the pressure bulkhead) in free air to change the cutters. Alternatively, for both planned and unplanned interventions, many TBMs are configured to allow compressed air to be applied to the face in front of the pressure bulkhead – access to the face is via an airlock in the pressure bulkhead. Cutter maintenance is then carried out under carefully controlled compressed air working.

10.4.3 Tunnel Exclusion Methods Using Ground Treatment

Closed face TBMs and compressed air tunnelling take the approach of using internal tunnel pressures to exclude groundwater. However, many tunnel projects use open face TBMs, where groundwater can enter the tunnel via the soil/rock exposed in the front of the shield. For tunnelling in low-permeability soils and rocks where inflows are small and do not destabilize the face, this approach can be effective, with groundwater controlled by pumping from within the tunnel, as described in Section 10.4.4. However, groundwater can cause problems for open face tunnelling in soils and weak rocks, where face instability may occur, and in highly fractured hard rocks, where excessive inflows may be difficult to manage within the tunnel.

Ground treatment methods can be used to reduce the permeability of the soil or rock ahead of the tunnel drive – a process sometimes known as 'pre-treatment' of the ground. Methods used include various forms of grouting and artificial ground freezing. Any residual seepage into the tunnel is dealt with by maintaining some sump pumping capacity at the tunnel face. Ground treatment is also widely implemented for shafts (Section 10.5) and cross passages (Section 10.6).

Ground treatment can be deployed from the surface (Section 10.4.3.1) or within the tunnel (Section 10.4.3.2). The selection and design of ground treatment methods for tunnelling works must follow the same principles as for surface excavations. It is important to realize that not all ground treatment methods are effective in all ground conditions; applicability is discussed in detail in Chapter 18.

10.4.3.1 Ground Treatment from the Surface

Where tunnel works are not excessively deep and, importantly, where there is surface access above the tunnel, ground treatment methods can be deployed via drill holes from surface to treat a zone of ground, into which the tunnel can be driven. This approach has been used with permeation grouting, rock grouting, jet grouting and artificial ground freezing. Depending on the ground conditions, ground treatment can be used form an arch of treated ground above the tunnel drive (Figure 10.8a) or a block of treated ground totally enclosing the path of the tunnel (Figure 10.8b). Working from the surface has the advantage that ground treatment can be carried out prior to tunnelling, and the tunnel can be driven through the zone continuously, without the need to pause for in-tunnel treatments.

10.4.3.2 Ground Treatment from within the Tunnel

In many cases, the depth of the tunnel and/or lack of surface access precludes treatment from the surface. As an alternative measure, ground treatment can be deployed using subhorizontal drill holes from the tunnel to pre-treat the soil or rock ahead of the advancing tunnel face – permeation grouting or rock grouting being most commonly used. A key limitation of this approach is that pre-treatment can only be carried out a to a maximum distance (typically 20 to 30 m) ahead of the current position of the face. Once the tunnel has advanced towards the far extent of the zone of treated ground, tunnelling must stop (while the face is still in treated ground). There must then be a pause in tunnel advance while a further phase of ground treatment ahead of the face is carried out from the new position, after which the tunnel can advance again (Figure 10.9). This means that the tunnelling must progress in a repeated stop/start cycle of ground treatment, tunnel advance, ground treatment and so on. Optimizing the speed and effectiveness of the ground treatment phase is key to making this approach effective on long tunnel drives.

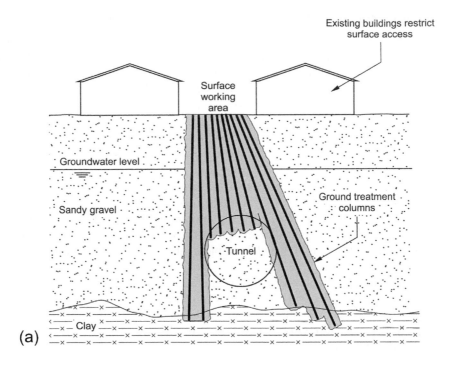

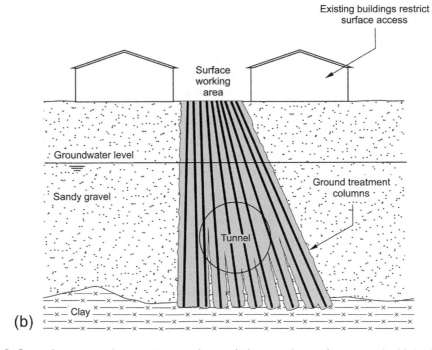

Figure 10.8 Ground treatment from surface used to exclude groundwater from tunnels. (a) Arch of treated ground above the proposed tunnel alignment. (b) Block of treated ground providing mass treatment of proposed tunnel alignment.

The most common applications in tunnelling use permeation grouting (in coarse-grained soils such as sands and gravels) or rock grouting (in permeable fractured rocks). These methods are used to form overlapping conical zones of treated ground, through which the tunnel is driven (Figure 10.9) – the process is sometimes called 'cover grouting'. This type of application for pre-grouting of tunnels is described in Daw and Pollard (1986), Garshol (2003) and Bruce (2005).

A typical sequence of working is:

1. When the tunnelling works are paused, an array of small-diameter (50 to 75 mm) holes are drilled ahead of the tunnel in a conical fan pattern, deviating away from the tunnel by 10 to 15°. Each phase of drilling and grouting is known as a 'cover' and can involve drill holes of 20 to 30 m in length. In many cases, especially in rock tunnelling, all the holes will be drilled from within the face, but occasionally, ground treatment holes will be drilled outward from inclined ports at the rear of a TBM shield. The drill holes also serve as probe holes and (depending on the methods used) can provide information on ground and groundwater conditions ahead of the tunnel.

2. The drill holes will be drilled from a stable face in pre-treated ground but will extend into untreated ground, and water may flow from them into the tunnel. The holes must be sealed by packers or by fitting a 'standpipe' (a short length of steel pipe grouted into the face and equipped with a valve).

3. Grout (of appropriate type and mix design) is then injected via the array of grout holes. Injection must be at sufficient pressure to displace the external groundwater and allow the holes to collectively create a continuous volume of treated ground of sufficiently low permeability. Cement-based grouts are most commonly used due to the low cost and wide availability of the raw materials; chemical grouts are occasionally used. Grout types are discussed in Section 18.10.

4. After a period to allow the grout to solidify, the tunnel is driven through the treated ground, stopping short of the end of the treated ground. For cover grouting in rock, typically, a plug of treated ground of minimum 5 to 6 m thickness is left beyond the tunnel face.

5. The drilling and grouting operations of Steps 1 to 4 are repeated, and the operational cycle continues.

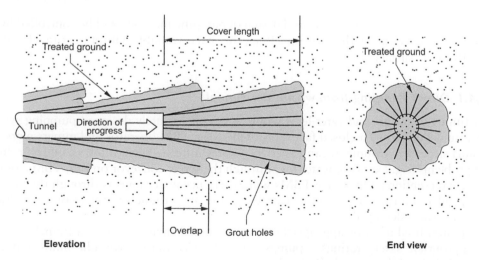

Figure 10.9 Permeation grouting or fracture grouting applied from in-tunnel to exclude groundwater. (After Daw, G P and Pollard, C A, *International Journal of Mine Water*, 5, 1–40, 1986.) This process is sometimes called 'cover grouting', with each section of grouting (in the shape of a truncated cone) termed a 'cover'.

This approach relies on the overall effectiveness of grouting to form a zone of treated ground without any significant gaps or 'windows' through which water could flow into the shaft. The consequences of a sudden inrush can be severe – ground treatment works should be designed, implemented and monitored by suitably qualified and experienced personnel.

Jet grouting has also been used for sub-horizontal ground treatment ahead of tunnels. Guatteri *et al.* (2009) describe examples where jet grouted columns were installed to form a perimeter barrier as a truncated cone around a tunnel, with a plug of treated ground at the far end of the cone. In a sequence similar to cover grouting, the tunnel advances in stages, with new zones of jet grouted treatment installed periodically, overlapping from previous treatment zones.

Artificial ground freezing has also occasionally been used for tunnel construction. However, the more complex in-tunnel requirements of freezeplants and refrigerant circulation systems make this approach cumbersome to apply in a regular cycle of treatment–tunnel advance–treatment. Where artificial ground freezing has been applied on tunnel projects, it is often in recovery works or short tunnel drives. Two examples from the United Kingdom are where a new tunnel was driven in Hull to recover a collapsed section (Brown, 2004; Munks *et al.*, 2004) and near Maidenhead, where a large box structure was jacked beneath a roadway (Wheelhouse *et al.*, 2001). The use of artificial ground freezing on shafts and cross passages is described later in this chapter.

10.4.4 Pumping Methods in Tunnelling

While many tunnels are constructed using closed face TBMs, where there is no routine requirement to remove groundwater, tunnel projects also commonly use a range of pumping techniques to control groundwater. Pumping methods used in tunnelling can be categorized as follows:

- Direct pumping from within the tunnel
- Dewatering wells from the surface
- In-tunnel dewatering wells

On any given project, the selection and design of pumping methods will be controlled by the ground conditions, the method of tunnelling, the space available within the tunnel and any access from the surface.

10.4.4.1 Direct Pumping from within the Tunnel

Every tunnel below groundwater level requires an appropriately designed pumping system. Even a tunnel driven by a closed face TBM, where the TBM operator and miners work in a dry 'shirt sleeve' environment, can experience water inflows. Such inflows can occur due to TBM problems (e.g. excessive water loss through the screw conveyor on EPB TBMs) or during face interventions. Open face TBMs will experience water ingress through the face at a rate dependent on the soil or rock permeability and any ground treatment or groundwater control measures deployed.

The pumps used must be appropriate for the confined environment of a tunnel and must have appropriate safety ratings – pumps driven by electricity, hydraulic fluids and compressed air have all been used effectively in different circumstances.

In general, the primary groundwater problem during tunnel construction is instability of soil and rock in the face, as in many cases this will occur before inflows become very large. Often, inflows will be limited to relatively low rates by external dewatering pumping or ground treatment such as grouting. The exception to this is when tunnelling in fractured hard rocks. In this case, even significant inflow rates may not destabilize the face, and tunnelling can proceed while allowing inflows at the face and using high-capacity pumps to remove the water. However, at very high inflows, direct pumping from the tunnel can become unworkable due to the the quantity of pumps and pipework required; a better solution may be to use rock grouting to reduce inflows to more manageable levels.

10.4.4.2 Dewatering Wells from the Surface

Occasionally, groundwater control for tunnel construction can be implemented by wells drilled from the surface. This approach is attractive because it avoids the need for extensive groundwater control or ground treatment from within the tunnel. Groundwater levels can be lowered in advance of construction, and the tunnel is driven through 'dewatered' ground. This method is also sometimes used on recovery projects to overcome problems such as when a TBM, designed for bedrock expected to be present over most of the tunnel length, has to traverse a short section of alluvial or glacial soils, which may be present in a buried channel or other geological feature (see Section 10.8 and Figure 8.18).

Typically, the objective is to lower groundwater levels to below tunnel invert. This should make working conditions within the tunnel more favourable, although in-tunnel pumping may still be required to deal with any residual water trapped in isolated zones within the strata.

For this approach to be practicable, surface access must be available above the tunnel, and there must be no intervening utilities or services between ground level and the tunnel. Figure 10.10 shows various possible combinations of wells used on tunnel projects. Figure 10.10d presents a problematic case of a tunnel driven in mixed face conditions, where the upper part of the face is in water-bearing sandy gravel and the lower part of the face is in stiff clay. Even extensive external dewatering with parallel lines of wells on each side of the tunnel will not be able to provide a completely 'dry' face, as some residual overbleed seepage will occur above the sandy gravel/clay interface. This could cause instability and ground loss. If this is not acceptable, then other groundwater control measures may be required, such as ground treatment of the sandy gravel by permeation grouting or using a closed face TBM.

In practice, on most urban tunnel projects, groundwater pumping systems using arrays of wells drilled from the surface are precluded by access restrictions and obstructions on the land above the tunnel route. Nevertheless, large-scale dewatering is occasionally carried out along extensive sections of tunnels. One interesting example is the High Speed 1 (HS1) railway (formerly known as the Channel Tunnel Rail Link [CTRL]) constructed in southern England in the late 1990s and early 2000s. Whitaker (2002) describes how a major dewatering exercise was carried out as part of the tunnel works with the aim of significantly lowering groundwater pressures at tunnel level to provide easier working conditions for the TBMs and reducing the ground treatment requirements for cross passages. The scheme involved abstracting up to 700 l/s from the lower aquifer beneath London (Chalk and Thanet Sand Formation) from 39 wells at 22 wellsite locations. This project faced some typical challenges for tunnel dewatering in urban areas. The project client did not own the land above the tunnel, so parcels of land close to the tunnel alignment had to be leased to provide dewatering wellsite locations. The existing sewer network did not have the capacity to handle the dewatering discharge flow, so a dedicated 3.5 km long collector pipeline (up to 600 mm in diameter) was laid beneath the streets in the area. A final interesting point about the project is that several of the dewatering wells were installed to a higher specification than is typical for dewatering wells, to the same standard as public water supply wells. These wells (and the water collector pipeline) were ultimately taken over by the local drinking water

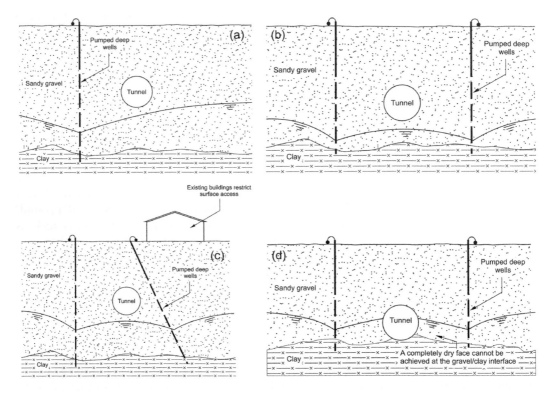

Figure 10.10 Groundwater control for tunnels using surface wells. (a) Single line of wells in deep permeable stratum. (b) Double line of wells where impermeable stratum is close below tunnel. (c) Use of inclined wells where surface obstructions are present. (d) Adequate drawdown cannot be achieved in mixed face conditions.

supply company and are now used to supply raw water to the regional water supply system (Hamilton *et al.*, 2008).

10.4.4.3 In-Tunnel Dewatering Wells

In-tunnel dewatering wells involve drilling outwards from the tunnel, either from the face or through the tunnel primary lining in the area behind the face. In general, in-tunnel dewatering can only have a local depressurization effect, so new wells would need to be continually installed to keep pace with the progress of an advancing tunnel. For this reason, this approach is not applied widely on tunnel drives but is more commonly applied for cross passage construction (Section 10.6) and other types of tunnel works (Section 10.7); further details are given in those sections. Drilling out from a tunnel below groundwater level can be difficult and challenging, especially in soils and weak rocks. This problem is discussed in detail in relation to dewatering for cross passages (Section 10.6.1.2).

One application where in-tunnel wells are used is for tunnels advanced in hard fractured rock using open face methods. Often, probe holes are drilled through the face to gather information on ground conditions ahead of the tunnel (Section 8.8.2). If the probe holes encounter significant water inflows, they can be used as in-tunnel wells (Figure 10.11).

The sequences and geometries used for probe drilling vary from project to project, but drilled diameters are typically 50 to 75 mm, with drilled lengths of 10 to 50 m per probe

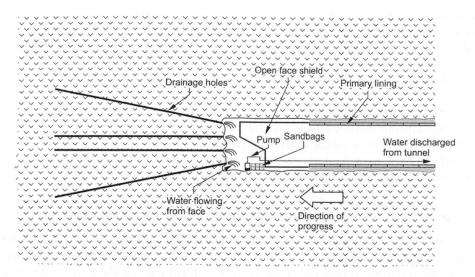

Figure 10.11 Probe holes used to drain tunnel face in rock. The number, length and geometry of probe holes will vary depending on tunnel geometry and expected ground conditions.

hole, depending on the equipment available. If flowing water is encountered, probe holes can contribute to dewatering by:

a) Allowing the probe holes to simply issue water from the face and using sump pumps to remove water from the tunnel invert (Figure 10.11). This approach is acceptable if the flow rates from the probe holes decrease rapidly when they are allowed to flow, and if the water running down the face does not cause softening, erosion or other problems.

b) Temporarily connecting the probe holes to flexible hoses to carry water directly to a sump. This has the advantage of providing a drier working environment at the face. The connection to the probe holes requires a temporary packer, equipped with a hose fitting, to be inserted in the probe holes. This approach is often used if high flows persist from probe holes and cause inefficient working conditions at the face.

As the tunnel face is advanced, new probe holes must be drilled to replace those that have been dug out. It is rare for face drainage holes in rock tunnels to be pumped; normal practice is to allow the wells to flow naturally, driven by the differential head between the tunnel and the surrounding groundwater regime.

Occasionally, dewatering during tunnel construction may be achieved by drilling of horizontal directionally drilled (HDD) wells (see Section 17.4) along or immediately parallel to the tunnel route. These drains act as preferential groundwater flow pathways to feed water in a controlled manner to the tunnel face or to a reception shaft ahead of the tunnel, from where the water is pumped away. An example of such an application is given by Peter Cowsill Limited (2001).

10.5 GROUNDWATER CONTROL METHODS FOR SHAFTS

Shafts are excavations from the surface and in principle can use the same groundwater control strategies as conventional excavations. Additionally, some groundwater control strategies are used in shafts that are not commonly applied elsewhere, including construction as flooded or 'wet' caissons and the application of compressed air.

As discussed in Section 10.2.2, the groundwater control requirements are influenced by the lining method used during shaft construction. Pre-lined methods (such as the use of steel sheet-piles, concrete diaphragm walls and secant pile walls) can allow the shaft lining system to act as a groundwater cut-off barrier.

The following methods are outlined in the next sub-sections:

- Shaft exclusion methods using physical cut-off walls
- Shaft exclusion methods using ground treatment
- Pumping methods used for shaft construction
- Methods that use a fluid pressure to stabilize the base of the shaft (including the wet caisson method and compressed air working)

10.5.1 Shaft Exclusion Methods Using Physical Cut-Off Walls

Many shafts are formed by the pre-lined construction method (described in Section 10.2.2), where a wall forming the shaft lining is installed prior to excavation (also known as SBE). Typically, the intention is that the shaft lining will form part of the permanent structure. Many of the lining methods form walls of very low permeability that can act as groundwater barriers; suitable methods include steel sheet-piles, concrete diaphragm walls and secant pile walls. Where the shaft lining wall extends below final dig level (or extends through the complete thickness of water-bearing zones) it can be used as a groundwater exclusion method. Groundwater exclusion methods will still require some form of pumping, methods of which are discussed in Section 10.5.3.

Figure 10.12 shows applications of physical cut-off walls to exclude groundwater in various ground conditions.

a) If the shaft is sunk through a water-bearing stratum (such as a sand) and terminates in an underlying low-permeability stratum (such as a clay or low-permeability bedrock), then a complete cut-off can be formed (Figure 10.12a). Groundwater pumping requirements are limited to dealing with water trapped in the soil or rock enclosed within the walls and with seepages from the low-permeability material in the base of the shaft. This arrangement can allow the shaft to be constructed with very little lowering of external groundwater levels.

b) If the shaft does not penetrate down to a low-permeability stratum, the cut-off wall will not reduce inflows to very low level, and significant pumping may be required from within the shaft (Figure 10.12b). Depending on the depth that the wall extends below final dig level, there is a risk of hydraulic failure of the soil/rock in the base of the shaft (see Sections 7.5 and 8.6). Relief wells are sometimes used in the base of shafts to reduce the risk of base instability. External groundwater levels may be significantly affected by pumping from within the shaft.

c) A base plug of treated ground (e.g. using some type of grouting) can be formed within the area enclosed by the walls, below the final dig level (Figure 10.12c). This can isolate the shaft from the wider groundwater regime and allow construction with limited pumping and little effect on external groundwater levels. Further details on ground treatment base plugs are given in Section 18.3.1.

10.5.2 Shaft Exclusion Methods Using Ground Treatment

There are many occasions when the best approach for shaft sinking is to use a groundwater exclusion method – for example, on sites where significant dewatering pumping would cause

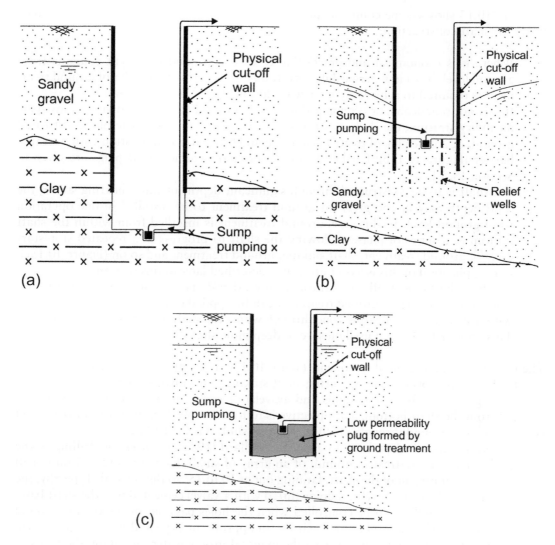

Figure 10.12 Shaft construction using physical cut-off walls for groundwater. (a) Basal cut-off into very low-permeability stratum. (b) No basal cut-off achieved; perimeter walls used to reduce total groundwater inflow. (c) No basal cut-off achieved; ground treatment base plug formed within walls.

ground settlement or would draw in contaminated groundwater. If the shaft construction method does not include a physical cut-off wall, then ground treatment methods are often used. The objective is to create a zone of low-permeability treated ground, within which the shaft can be excavated with only minimal groundwater pumping requirements.

Depending on ground conditions, it can be appropriate to use treatment methods to form a perimeter around the shaft, sealed into an underlying low-permeability stratum, or to create an approximately cylindrical block of treated ground, totally encapsulating the proposed shaft. Ground treatment methods used for shafts include permeation grouting, rock grouting, jet grouting, mix-in-place methods and artificial ground freezing. The treatment method used must be appropriate to the ground conditions; techniques are discussed in detail in Chapter 18.

Figure 10.13 shows some common applications of ground treatment to exclude groundwater for shaft construction:

a) Where it is economical and practicable to extend ground treatment columns from ground level into a low-permeability stratum, only a perimeter barrier of ground treatment is required to achieve a complete exclusion effect (Figure 10.13a).
b) If it is not practicable or economical to extend ground treatment down to a low-permeability stratum, the shaft could be constructed within a complete block of treated ground, or a base plug of treated material could be used in combination with a treated barrier (Figure 10.13b). Further details on ground treatment base plugs are given in Section 18.3.1.
c) If a shaft is to be constructed through significant depths of water-bearing strata, the total depth may be too great for ground treatment to be installed in one operation from ground level. In this case, ground treatment is installed from within the shaft bottom in a series of phases, allowing the shaft to progress in a repeated stop/start cycle of ground treatment, shaft sinking, ground treatment and so on (Figure 10.13c). One application of this is 'cover grouting', described later in this section.
d) If a deep shaft is generally sunk through strata that do not require groundwater exclusion for the full depth, ground treatment can be applied from within the shaft to deal with a specific water-bearing zone through which the shaft must pass (Figure 10.13d). This approach has been used on several deep mine shafts.

The use of sequential ground treatment (Figure 10.13c) is worth further discussion in relation to the use of 'cover grouting' during shaft sinking. This involves permeation grouting (in coarse-grained soils such as sands and gravels) or rock grouting (in permeable rocks) injected from the shaft bottom. The objective is to form an overlapping series of truncated cones of treated ground through which the shaft is sunk (Figure 10.14).

The method, described in detail by Daw and Pollard (1986), involves the drilling in the base of the shaft (in pre-treated ground) of an array of small-diameter (50 to 75 mm) drill holes across the footprint of the shaft, fanning out at 10 to 15° to the vertical. Typically, the holes are drilled through a steel standpipe (fitted with a valve) cemented into the shaft base. This allows water inflows to be controlled when the holes pass out of the treated ground into the water-bearing natural ground. Grout (of appropriate type and mix design) is then injected at sufficient pressure to displace the external groundwater and allow the holes to collectively create a continuous volume of treated ground of sufficiently low permeability. Grouting in each hole may be in stages, or by using packers to isolate particular injection zones, and 'water tests' – a form of packer permeability test – are typically carried out to assess the effectiveness of the grouting. Cement-based grouts are most commonly used due to the low cost and wide availability of the raw materials; chemical grouts are sometimes used. Grout types are discussed in Section 18.10.

After a period to allow the grout to solidify, the shaft is sunk through the treated ground, stopping short of the base of the treated ground. For cover grouting in rock, typically a plug of treated ground of minimum 5 to 6 m thickness is left below the shaft bottom. Shaft sinking then pauses, and the drilling and grouting cycle is repeated to form a further overlapping zone of treated ground, into which shaft sinking can continue.

Each stage of drilling and grouting is known as a 'cover'. Daw and Pollard (1986) indicate that a cover length of up to 20 or 30 m is typical. Where grouting is targeting sub-vertical fractures in rock, the fan of grout holes in each cover is often drilled in a 'spun' array to increase the likelihood of intercepting such fractures (Figure 10.14).

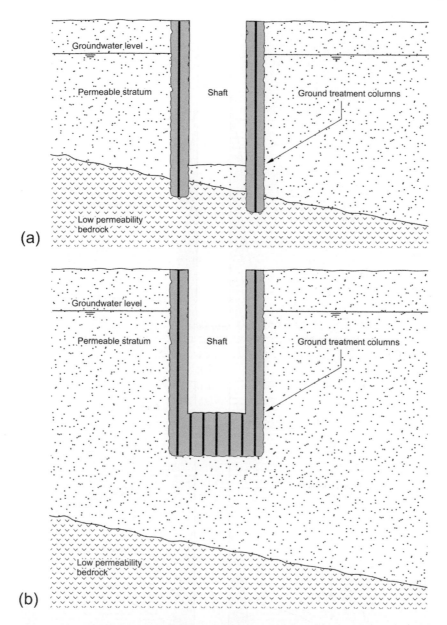

Figure 10.13 Shaft construction using ground treatment. (a) Ground treatment from ground level down to low-permeability stratum. (b) Ground treatment from ground level with base plug or fully treated zone. (c) Deep shaft advanced by sequential ground treatment from within the shaft. (d) Ground treatment targeted at deep permeable zone from within the shaft.

The method relies on the overall effectiveness of grouting to form a zone of treated ground without any significant gaps or 'windows' through which water could flow into the shaft; see Garshol (2003) and Bruce (2005). The consequences of a sudden inrush to a shaft can be severe – ground treatment works should be designed, implemented and monitored by suitably qualified and experienced personnel.

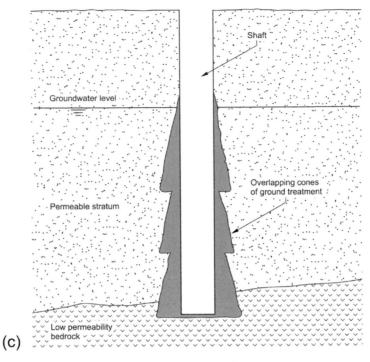

(c)

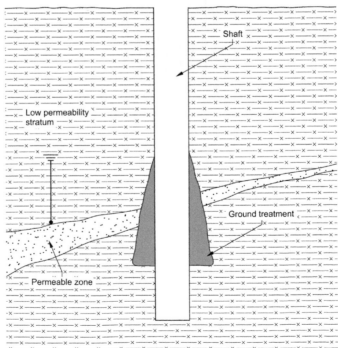

(d)

Figure 10.13 (Continued)

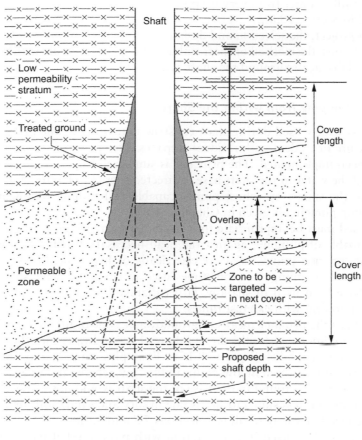

Elevation

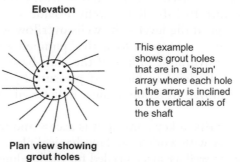

This example shows grout holes that are in a 'spun' array where each hole in the array is inclined to the vertical axis of the shaft

Plan view showing grout holes

Figure 10.14 The use of cover grouting during shaft construction. (After Daw, G P and Pollard, C A, *International Journal of Mine Water*, 5, 1–40, 1986.)

10.5.3 Pumping Methods Used for Shaft Construction

Pumping methods are used routinely during shaft construction. Pumping methods can be categorized as follows:

- Sump pumping from within the shaft (including the use of relief wells)
- Dewatering wells from the surface
- In-shaft dewatering wells

The method of bailing water from very deep shafts, occasionally used in the mining industry, is described in Section 19.8.2.

On any given project, the selection and design of pumping methods will be controlled by the ground conditions, the method of shaft sinking, the space available within the shaft bottom and the available access around the shaft at ground level.

10.5.3.1 Sump Pumping from within the Shaft

Sump pumping is widely used during shaft construction, either as the primary groundwater control measure or to deal with residual seepages after external dewatering or ground treatment has been implemented. The principle is simple: a deeper area is dug in one part of the shaft, and the water in the shaft base is directed there by shallow ditches. The water is pumped from the sumps by robust pumps capable of handling sediment-laden water. Guidance on sump pumping is given in Chapter 14.

The pumps used must be appropriate for the confined environment of the shaft bottom and must be capable of pumping against the head necessary to lift the water out of the shaft. The most common pumps used are electrically or hydraulically driven submersible pumps. Compressed air pumps are sometimes used in small-diameter shafts, and diesel-driven pumps have been used in shallow, larger-diameter shafts where there is adequate forced ventilation to deal with exhaust fumes.

Sump pumping as the primary means of groundwater control for shaft sinking is appropriate in ground conditions where the direct inflow of groundwater into the excavation will not cause significant instability (Figure 10.15a). Examples where sump pumping may be appropriate are shafts excavated in low-permeability clay, where inflows are low, or in hard fractured rock, where the ground in the base of the shaft would not be destabilized by the inflow. Sump pumping is less suited to shafts dug in silty and sandy soils or weak and erodible rock, where instability may result (see Sections 7.5 and 8.6).

Sump pumping is often used in combination with passive relief wells installed in the base of the shaft (Figure 10.15b). Relief wells provide a preferential pathway for upward flow of water to emerge at dig level – the wells overflow without pumping, which is why they are termed 'passive' in contrast to 'active' pumped wells. The use of relief wells can help reduce the risk of hydraulic failure of the shaft base. This allows any deeper water-bearing layers below dig level to bleed off pressure by overflowing into the dig, from where the water is removed by sump pumps. Relief wells are described in more detail in Section 17.8.

When working in shafts, a key constraint is access and working area. The shaft bottom will be a hive of activity with workers and mechanical plant (for both excavation and placement of shaft lining) as well as space needed for the landing of crane skips to remove spoil and provide materials. The space available for a sump and the pumps may be very limited. Furthermore, as the shaft is deepened, the sump and the pumping equipment will need to be deepened and possibly relocated during each stage of excavation. It is essential that the pumping arrangements, including discharge hoses and power cables/hydraulic hoses connecting up to ground level, allow flexibility in positioning and depth of operation.

If sump pumping is the primary means of groundwater control, then pumping from the shaft will need to be continued until the concrete base slab has been cast and has gained sufficient strength to resist groundwater uplift pressures. This is sometimes achieved by casting a vertical pipe or duct through the base slab (Figure 17.19c) and using that to continue sump pumping until structural calculations indicate that pumping can be discontinued and groundwater pressures can safely recover. The pipe through the slab can then be capped off and decommissioned (see Section 24.4).

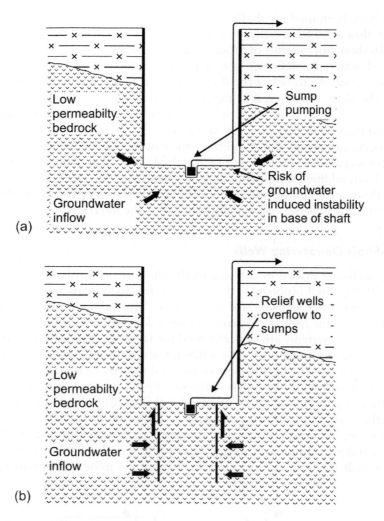

Figure 10.15 Shaft construction using groundwater pumping methods. (a) Sump pumping from within shaft. (b) Internal relief wells combined with sump pumping.

10.5.3.2 Dewatering Wells from the Surface

Where shafts are sunk through water-bearing coarse-grained soils (such as sands and gravels) or fractured rock, a common strategy is to install pumped wells around the perimeter of the shaft at ground level (Figure 10.16). The objective is to adopt a 'pre-drainage' approach and lower groundwater levels in advance of construction (Section 9.5.1). This will give much more efficient and stable working conditions in the shaft and dramatically reduce (or even eliminate) the need for sump pumping. Eliminating sump pumping means there is more working space available in the shaft bottom but also has an interesting potential knock-on effect in making disposal of dewatering water easier. Water from sump pumping water often has a high sediment load and will require extensive treatment to remove solids prior to discharge (see Section 21.9). Conversely, water discharged from appropriately designed and installed pumped wells will have very low levels of suspended solids and often requires little or no treatment prior to discharge.

Typically, well locations are set back a few metres from the edge of the shaft. Most commonly, wells are pumped by borehole submersible pumps (Chapter 16), but wellpoint systems

(Chapter 15) have been used for shallow shafts, and ejector wells (Section 17.2) have been utilized where flow rates from wells are expected to be low.

Figure 10.16 shows an example of ground conditions where external wells can potentially provide fully 'dewatered' conditions in a shaft. This is possible where a permeable stratum extends to significant depth below the deepest shaft level and the wells can extend significantly below the shaft. However, it is important to realize that if the shaft is sunk through a permeable layer (such as a sandy gravel) into a much lower permeability layer (such as a clay), it will not be possible to completely 'dewater' the gravels, and some residual 'overbleed' seepage will occur at the gravel/clay interface and can cause local instability (Figure 10.17). If groundwater seepage is expected to cause instability problems when the shaft is excavated through this level, additional measures may be required. These could include groundwater exclusion methods such as steel sheet-pile walls or permeation grouting of the gravel, or a ring of in-shaft wells to better control groundwater at the interface.

10.5.3.3 In-Shaft Dewatering Wells

Occasionally, wells are drilled in the base of the shaft or outwards through the shaft lining to deal with localized problems – this section refers to systems intended to be pumped rather than passive relief wells, which overflow without pumping. In-shaft wells can be installed vertically in the base of the shaft (Figure 6.3) or at inclinations out from the shaft (Figure 10.18). These systems are typically used for targeted drainage of localized groundwater problems and are often pumped by wellpoint or ejector well systems, with water either discharged to a sump in the shaft bottom or pumped directly to surface.

The presence of wells within the shaft will generally be an impediment to shaft excavation and construction. Furthermore, the wells are often drilled from above the final shaft bottom level; within the shaft footprint, the wells must be cut down as excavation proceeds, and at the shaft perimeter, pumps and pipework may need to be suspended from the shaft wall, making access and maintenance difficult (Figure 10.18). For these practical reasons, in-shaft wells are often only used if other measures are deemed not to be appropriate or effective.

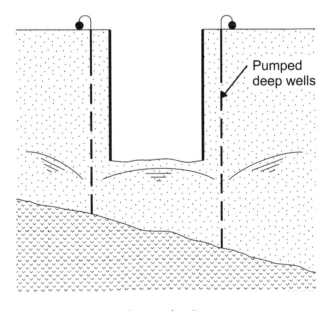

Figure 10.16 Shaft construction using external pumped wells.

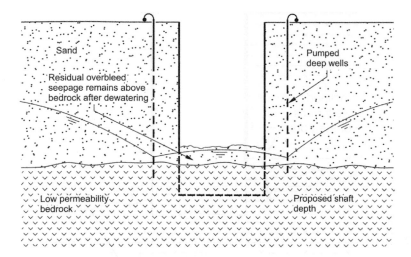

Figure 10.17 Overbleed seepage during shaft construction that passes from higher-permeability layer into an underlying lower-permeability layer. Even after dewatering wells are deployed, residual overbleed seepage may remain in the sand, trapped above the low-permeability bedrock. This can cause instability when the shaft is sunk through this zone.

10.5.4 Use of Fluid Pressure to Stabilize the Base of the Shaft

Earlier in this chapter, tunnelling methods were discussed whereby fluid pressure is used to exclude groundwater and stabilize the soil and rock in the tunnel face without the need for other groundwater control measures. For tunnels, this can be achieved by closed face TBMs of the EPB or slurry type or by the use of compressed air working. There are some shaft sinking methods that take an analogous approach and apply a fluid pressure to the base of the shaft to avoid seepage instability and hydraulic failure. Methods include:

- Shaft sinking by the wet caisson method
- Shaft boring machines in flooded shafts
- Underwater excavation within cofferdams
- Shaft sinking under compressed air working

10.5.4.1 Shaft Sinking by the Wet Caisson Method

The concept of shaft sinking by the caisson method was introduced in Section 10.2.2. In essence, part or all of the shaft lining system is constructed at ground level and is progressively sunk into the ground in a controlled manner (further details are given in Allenby and Kilburn, 2015 and Smith, 2018). Material is excavated from within the shaft, which allows the structure to sink either under its own weight (sometimes augmented by kentledge weights) or with the aid of caisson jacks reacting against a substantial concrete collar at ground level. The bottom of the shaft structure is fitted with a cutting edge and thicker section of lining (known as a 'choker ring'). The cutting edge and choker ring cut a vertical hole slightly larger than the main shaft lining above. This 'overcut' is typically kept topped up with a bentonite or polymer fluid to reduce friction and aid sinking. Once the shaft is completed to full depth, the overcut annulus is backfilled with grout.

In a common form of caisson sinking, the structure of circular shafts is formed from rings of pre-cast concrete segments. As the shaft structure sinks, new rings are added at

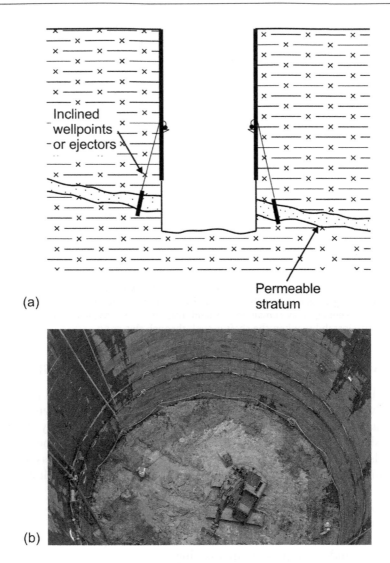

(a)

(b)

Figure 10.18 Internal drainage wells drilled out from shaft. (a) Schematic cross section through a shaft with steeply inclined wells used to target a specific permeable horizon. (b) Two-stage wellpoint system with inclined wellpoints drilled out through the shaft lining. (Courtesy of WJ Groundwater Limited, Kings Langley, UK.) The shaft was being sunk by underpinning methods down to Chalk bedrock, which had been dewatered by external pumped wells. The Chalk was overlain by fine sands of the Thanet Sand Formation, which did not fully drain down into the Chalk. Two stages of steeply inclined wellpoints drilled out through the shaft lining (with pumps located in the shaft bottom) were used to control the water in the sands and allow excavation and shaft lining by the underpinning method.

the surface to extend the lining system upwards. An alternative method is to construct the caisson by monolithic reinforced concrete methods and sink the structure in one continuous operation.

When sinking in potentially unstable ground, a key factor is to ensure that excavation and sinking of the shaft structure are coordinated, so that the caisson cutting edge is sufficiently embedded in the soil at shaft dig level. If the dig level is too close to the cutting edge, there is a risk of excessive ground loss from outside the shaft. Conversely, if the dig level is too far above the cutting edge, then the jacking forces required to sink the shaft may become excessive.

Shafts can be sunk by the dry caisson method, where the shaft is dewatered by sump pumping and/or external dewatering wells, as described in Section 10.5.3. Excavation is by conventional excavating plant, and it is straightforward to accurately control excavation levels in relation to the caisson cutting edge. The disadvantage of the dry caisson method is that the pumping will lower external groundwater levels, which can lead to ground settlements if soft compressible soils are present (Section 21.4.4).

The 'wet' or 'flooded' caisson method can allow a shaft to be sunk without other forms of groundwater control. Typically, the shaft is sunk by dry caisson methods down to groundwater level (or to a safe level above the top of a confined aquifer if hydraulic failure of the base is a concern). At that level, shaft sinking is paused, and then water is added to flood the caisson to slightly above groundwater level. Excavation of the base of the shaft then continues within the flooded caisson – this is termed 'underwater excavation'. For civil engineering shafts, excavation is normally done by a surface-mounted hydraulic excavator equipped with a telescopic grab, with the (wet) spoil deposited into dump trucks or a belt conveyor for disposal (Figure 10.19).

From a groundwater control perspective, the principle of the wet caisson method is that the head of water in the caisson (held slightly above groundwater level, by pumping water into the shaft if necessary) applies a small excess pressure to the base of the shaft excavation. In coarse-grained soils, this will prevent groundwater inflow (avoiding the risk of instability due to 'boiling' or 'quicksand' conditions). In low-permeability clays, the downward water pressure will help counteract the upward pressure from any confined aquifers below dig level. Once the shaft is excavated to the final depth, and the base of the dig has been cleaned out to remove as much sediment as possible (fine particles from the spoil tend to be washed out of the grab and settle out during excavation), a mass concrete base plug is used to seal the base of the shaft. Tremie concreting methods are used to place a thick plug (which may be several metres in thickness to resist groundwater pressures when the shaft is pumped out). Typically, the overcut annulus around the shaft will also be sealed, using cement-based grouts to displace the lubricating slurry. When the concrete plug has gained sufficient strength, the water can be pumped out, and the inside of the shaft can be accessed.

The wet caisson method requires that excavation and concrete placement be carried out underwater, to some extent working 'blind'. This creates practical difficulties:

i. It can be difficult to control excavation levels when grabbing through considerable depths of water, and problems have occurred due to over-excavation (leading to ground loss and external settlements) or when boulders are present and obstruct the downward penetration of the cutting edge.

ii. Care needs to be taken so that the tremie concrete plug is correctly placed, including ensuring that there is a good seal with the internal faces of the caisson at the level of the plug. If sediment or soft soil is left against the sides of the caisson when the base plug is cast, this can be a path of weakness for groundwater to enter the shaft when it is pumped out. In some cases, specially trained teams of divers descend to the base of the shaft and in conditions of almost zero visibility, inspect the excavation and carry out any necessary clearing and trimming of the base prior to concreting (an example of such operations is given in Machon and Stevens, 2004).

Despite these difficulties, in the hands of experienced designers and contractors, the wet caisson method can be very effective and is widely used.

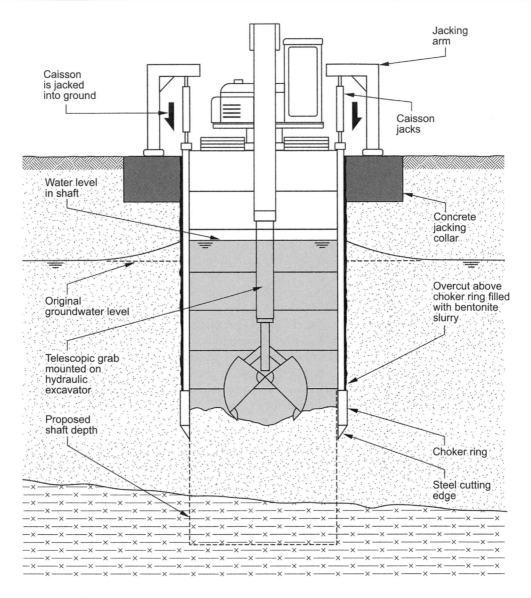

Figure 10.19 Shaft construction by the wet caisson method. The water inside the caisson applies a small excess water pressure to the base of the shaft, preventing water inflow and reducing the risk of basal instability. The base of the shaft is excavated underwater as the shaft is jacked downwards. New rings of shaft lining are added periodically at ground level as the caisson sinks.

10.5.4.2 Shaft Boring Machines in Flooded Shafts

Because of the difficulty of excavating in wet or flooded caissons, especially as the depth increases, the concept of shaft boring machines – essentially vertical TBMs – has developed. This involves a remote-controlled excavating device, working underwater, fixed to the side of the caisson a short distance above the cutting edge (Figure 10.20). The excavating device is linked by an umbilical to powerpacks and control equipment located at ground level. The cutters of the excavating device can be articulated and rotated to excavate the entire base of the shaft. As excavation proceeds, the shaft structure sinks in the same manner as any other

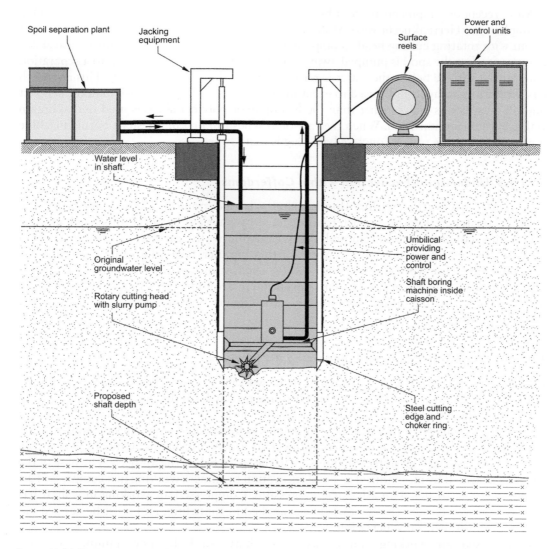

Figure 10.20 Shaft boring machine in flooded caisson. The shaft boring machine excavates underwater in a flooded caisson. An articulated cutting head excavates inside the caisson, and spoil is pumped away by a slurry pump to a spoil separation plant. After the spoil is removed, the water is returned to the caisson. As excavation proceeds, the shaft is jacked downwards. New rings of shaft lining are added periodically at ground level as the caisson sinks.

caisson, with the flexible umbilical paying out from surface reels to power and control the excavating device.

The development of shaft boring machines for civil engineering shafts started around the beginning of the twenty-first century. A very early application was in the mid-1990s by AMEC Tunnelling on the Fylde Coastal Waters Improvement Scheme in the United Kingdom (Bennett Associates, 1996; Cole, 1996). The machine was designed to sink 6 m diameter caissons to a maximum depth of 35 m and was successfully used to sink several shafts on the project. The hydraulic cutting tools excavated underwater, and the spoil accumulated in the base was removed by crane grab from the surface via an access space in the submerged excavator equipment.

More recent developments in shaft boring machines are described in Schmäh (2007). One example is the Herrenknecht vertical shaft boring machine (VSM). This uses an articulated boom with rotating cutting heads (comparable to a roadheader used for tunnel excavation) to excavate, but the spoil is pumped away as a slurry by a submersible pump to a separation plant at the surface before the water is returned to the shaft (Figure 10.20). This approach has similarities to hydromills used to excavate diaphragm wall panels, although the support fluid in a caisson is water rather than the bentonite or polymer slurries used in diaphragm wall trenches (Section 18.8). When the caisson reaches full depth, the shaft boring machine is removed, and the tremie concrete base plug is cast.

10.5.4.3 Underwater Excavation within Cofferdams

Underwater excavation within flooded shafts is occasionally carried out when the shaft is constructed within a cofferdam formed by physical cut-off walls (such as steel sheet-piles, concrete diaphragm walls or secant pile walls) rather than by the caisson method. Excavating a shaft underwater in this manner faces the same challenges as were described for wet caissons. Underwater excavation within cofferdams is usually only considered as an option when conventional groundwater control measures are deemed impracticable or if the environmental impacts from pumping are expected to be unacceptable.

10.5.4.4 Shaft Sinking under Compressed Air Working

Earlier in this chapter, the use of compressed air working to exclude groundwater from tunnel faces was described (Section 10.4.1). The same principle can be used to exclude groundwater from the base of a shaft below groundwater level. An 'air deck' is placed across the shaft. This is equipped with airlocks to allow access for workers (man locks) and to remove spoil and allow in construction materials (muck locks). A continuous supply of pressurized air is fed from surface-mounted compressors to the space below the air deck. The base of the shaft is excavated by mechanized or hand excavation by miners working in a compressed air environment at greater than atmospheric pressure. Compressed air working in a shaft has similar health and safety risks as when it is applied in tunnelling and must be carried out under specialist medical supervision.

Compressed air working can be used for shafts sunk by underpinning methods (Figure 10.21) or by caisson methods (Allenby and Kilburn, 2015). In the caisson method, the air deck is constructed on top on a monolithic concrete structure and sinks with the caisson. This approach is sometimes known as a 'pneumatic caisson' – application of the method on the Cairo Wastewater Project in Egypt is described by Coe et al. (1989).

10.6 GROUNDWATER CONTROL METHODS FOR CROSS PASSAGES

Many tunnelling schemes for transportation purposes comprise twin, parallel running tunnels to carry the road or railway. Typically, to allow emergency access between tunnels, connecting passageways must be constructed every few hundred metres. These connecting tunnels are known as 'cross passages' (Section 10.2.3).

Controlling groundwater to allow cross passage construction in workably dry and stable conditions is a significant element of many tunnelling projects, and the location and nature of cross passages present some particular challenges.

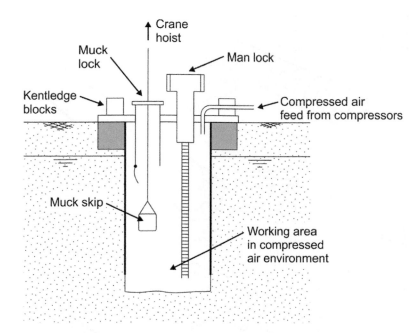

Figure 10.21 Shaft construction by underpinning under compressed air. (After Allenby, D and Kilburn, D, *Proceedings of the Institution of Civil Engineers, Geotechnical Engineering*, 168, 3–15, 2015.)

The following sections describe the general requirements of groundwater control for cross passages and then discuss the methods outlined here:

- Cross passages constructed by compressed air working
- Cross passages constructed using ground treatment to exclude groundwater
- Cross passages constructed using groundwater depressurization

10.6.1 General Principles of Groundwater Control for Cross Passages

It is stating the obvious to say that each cross passage will be different and will need its own, tailored groundwater control scheme. However, there are certain common features shared by many cross passages, which influence how groundwater control methods can be deployed. These include:

- Geometry. The most common arrangement is shown in Figure 10.22, where the cross passage is a smaller-diameter tunnel linking two running tunnels. Typically, the two running tunnels are spaced a few diameters apart. This corresponds to most cross passage lengths being in the range of 15 to 30 m. There is usually very little elevation change along a cross passage; however, some cross passages incorporate chambers for drainage sumps, dug several metres below invert.
- Excavation methods. A common feature of cross passages is that they are used to link running tunnels that were constructed by TBM methods, but they themselves are constructed without the use of a TBM or shield. Due to the small diameter and short length, typical excavation methods are hand mining or mechanized excavation using

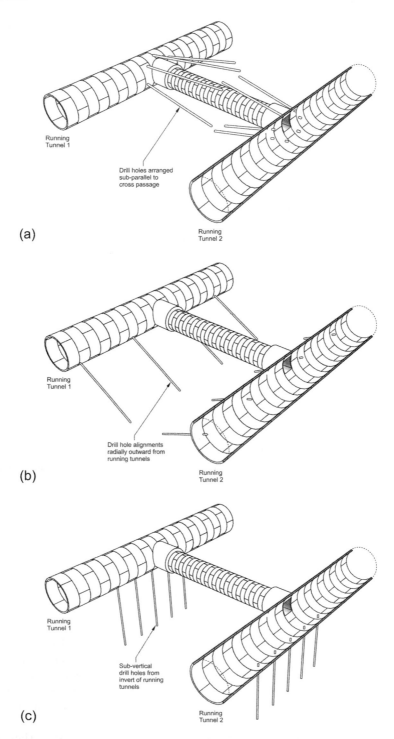

Figure 10.22 Geometry of typical drill hole arrays for ground treatment and groundwater depressurization (dewatering and ground treatment equipment not shown for clarity; well arrangements and spacings are illustrative only). (a) Drill holes sub-parallel to cross passage. (b) Drill holes radially out from running tunnels. (c) Sub-vertical drill holes downwards from the running tunnels.

small backactors or roadheaders with a primary lining (bolted segments or SCL) following close behind. Temporary support such as timber propping may be used before the primary lining is placed. These methods require that the soil/rock through which the cross passage is excavated is in a stable condition and that groundwater inflows are manageable.

- Little or no surface access. In many cases, surface access is not available above cross passage locations. Indeed, for tunnels below rivers and open water, the 'surface' above a cross passage may be under many metres of water. Accordingly, in-tunnel methods are widely used.
- Working constraints within the running tunnels. The construction of cross passages (and associated groundwater control measures) must be done from the running tunnels. This causes logistical problems, as the tunnels must continue to be used for access along the tunnel; if the tunnels are still being driven, there is a need to feed materials to, and remove spoil from, the TBM. Therefore, cross passage works must be designed and scheduled to inconvenience other works as little as possible. This can require fixed equipment such as in-tunnel pumps and freezeplants to be mounted so that they do not block access along the running tunnels. The installation equipment, such as drilling rigs, must be mobile (on either rail cars or vehicles) to allow rapid movement into and out of the tunnel. Cross passage works are sometimes planned to work from one running tunnel only or to carry out drilling and installation only on certain shifts.
- Limited ground investigation information. Lack of surface access means that often there is little pre-construction ground Investigation information at cross passage locations. It is important that as much additional information as possible is gathered during tunnelling and cross passage works (Section 10.6.1.1).
- Requirement to drill out from TBM tunnels. Every cross passage will require drilling out from the running tunnels for various purposes: probing, ground treatment and groundwater depressurization (Section 10.6.1.2).

If significant groundwater inflow or ground instability occurs during cross passage construction, there can be major delays to tunnelling programmes, and in extreme cases, it could result in the running tunnels being inundated. It is essential that cross passage works are meticulously planned, executed and supervised by experienced personnel.

A notable example of the application of groundwater control for cross passages in challenging conditions was for the Storebaelt Eastern Railway Tunnel in Denmark in the early 1990s (Biggart and Sternath, 1996). This subsea tunnel, with twin running tunnels driven by TBM, required the construction of 29 cross passages, where groundwater pressures at tunnel level were up to approximately 4.5 bar (45 m head of water). Ground conditions in different sections of the tunnel were Till deposits comprising silts and fine sand and weak rock (Marl). The methods applied to the cross passages included groundwater depressurization (in-tunnel wells and electro-osmosis, supported by the sea bed dewatering wells of Project MOSES – see Section 10.8) and ground treatment (grouting and artificial ground freezing), often used in combination (detailed descriptions of the works are given in papers by Doran, Hartwell, Kofoed, *et al.* (1995) and Doran, Hartwell, Roberti, *et al.* (1995)).

The methods applied at Storebaelt in the 1990s are very similar to those widely used in modern practice around the world. It is interesting to note that the contracting consortium and consultants on that project included companies from Scandinavia, Europe and North America. It is possible that these companies took the experience from Storebaelt and applied and developed it on their later projects in regions around the globe.

10.6.1.1 Investigation to Gather Information on Ground Conditions at Cross Passage Locations

It is often the case that there are no ground investigation boreholes close to cross passage locations. This can result in high levels of uncertainty in relation to ground and groundwater conditions at cross passage locations. It is therefore essential that all potentially relevant information is collected and reviewed, including:

i. Conceptual geological and hydrogeological models. As discussed elsewhere (Sections 5.4 and 13.4), conceptual models are fundamental to effective design and implementation of groundwater control. For cross passages, conceptual models can allow a pre-construction assessment of ground conditions, which can aid the planning of the works.

ii. Feedback from driving of running tunnels. Depending on the type of TBM, it may be possible to interrogate the parameters recorded during tunnelling in order to provide some information on ground conditions and groundwater pressures.

iii. Information from face interventions. Long TBM tunnel drives commonly have to pause periodically for 'face interventions' when the front of the cutting head (which is pressurized and inaccessible during normal operations) is accessed for inspection and maintenance. A strategy that is sometimes adopted is to plan the interventions to coincide with proposed cross passage locations. If the TBM support fluid is lowered to tunnel axis level (or lower) and compressed air applied to exclude groundwater, it can be possible to inspect the soil or rock at the face. This approach was adopted in the United Kingdom for railway tunnels through Chalk (a type of fractured limestone) beneath the River Thames for the HS1 project (Warren et al., 2003) and the Crossrail project (Kinnear, Nicholls, et al., 2018).

iv. Probe drilling. Most cross passage works include provision for probe drilling out from the running tunnels in advance of ground treatment and groundwater depressurization works. The data gathered can include samples from the drilling flush or drilling tools and information from instrumented drilling equipment, which can allow some ground conditions to be approximately correlated with parameters collected during drilling – this approach is sometimes known as measurement while drilling (MWD).

v. Piezometers. The holes formed by probe drilling out from the running tunnels can be completed as piezometers, sealed through the tunnel lining. Typically, piezometers are installed as fully grouted vibrating wire piezometers (VWPs), which monitor pore water pressures relative to the elevation of the VWP sensor (see Sections 11.7.6.3 and 22.6.3).

10.6.1.2 Drilling out from Existing Tunnels

Drilling out from tunnels is difficult and inconvenient in all circumstances. It is axiomatic that if surface access for drilling is available, it is preferable for as much ground treatment and groundwater depressurization as possible to be done from ground level. However, for many projects, much or all of the work will have to be done from within the running tunnels via holes drilled out through the lining. Even if all the treatment is done from the surface, some in-tunnel probe drilling will be required to prove ground conditions before the cross passage opening is broken out through the tunnel lining.

For any cross passage scheme, there will be a 'design' arrangement of probe holes and treatment/depressurization holes to be drilled out from the running tunnel. However, practical and logistical constraints often mean that drilling patterns need to be modified for various reasons:

- Reinforcement and joints in the tunnel lining. Most drilling will be from TBM driven tunnels lined with pre-cast concrete segments, bolted together with gaskets used to seal between adjacent segments. The presence of the joints and reinforcement within the concrete segments will restrict drill hole locations and may require small adjustments in drilling locations.
- Temporary support around cross passage openings. Forming an opening through the lining of the running tunnel for the cross passage will change the stresses on the tunnel lining, and temporary propping and strengthening is normally needed around the opening. This may affect the choice of drill hole locations.
- In-tunnel infrastructure. Various services and infrastructure are usually attached to the tunnel lining – examples include belt conveyors, ventilation ducts and utilities feeding the TBMs. The location of these may limit drill hole locations around the proposed cross passages.

While the geometry of the array of drill holes will vary from cross passage to cross passage, there are some arrangements that are used again and again (these geometries are often used in combination):

a) Drill holes aligned sub-parallel to cross passage, fanning outwards slightly (Figure 10.22a).
   i.   Drill holes from each running tunnel can form a conical array of holes around the entire circumference of the cross passage.
   ii. Drill holes can be concentrated above cross passage axis level. This approach is sometimes used to create an arch of treated ground above the cross passage.
   iii. Drill holes can be concentrated below cross passage axis level. This approach is sometimes used to create an inverted arch of treated ground below the cross passage.
b) Drill holes installed radially outwards from running tunnels (Figure 10.22b). This approach is often used in groundwater depressurization systems where the soil/rock is sufficiently permeable to allow wells relatively distant from the cross passage to generate drawdowns over a wide area.
c) Sub-vertical drill holes directed downwards (either vertically or at a slight inclination to the vertical) from the lower half of the running tunnels (Figure 10.22c). This geometry is often used to target pressurized strata below tunnel invert or as part of groundwater control for cross passages that involve deeper sumps.

The range of drill hole geometries has led to the development of specialist drilling rigs with sufficient articulation of the mast to allow drilling at various inclinations and locations within a tunnel (Figure 10.23). The rigs are typically highly mobile (on vehicles or rail cars) to allow them to be moved quickly into and out of the tunnel.

The drilling method must be selected based on the expected ground conditions and groundwater pressure outside the tunnel. When drilling out through a tunnel lining into pressurized, saturated soil or weathered rock, there is a risk that the drill hole will allow inflow of water and soil/rock particles, leading to loss of ground and collapse of the drill hole before any liner can be installed. To avoid these problems, each drill hole should be made through a steel standpipe attached and sealed into the tunnel lining.

Figure 10.23 Specialist in-tunnel drilling rig used for cross passage works. (Courtesy of WJ Groundwater Limited, Kings Langley, UK.)

Typically, the internal (tunnel) end of the standpipe is threaded or flanged to enable various equipment to be connected to allow the bore to be controlled (Figure 10.24). The whole arrangement is termed a 'blow out preventer' (BOP). Different BOP set-ups may be used, but common BOP elements include:

- A guillotine valve that can be closed to seal the hole in the event of problems.
- A relief valve (tee piece and valved outlet) that can be used to relieve high pressures and flows during drilling.

Figure 10.24 Well head arrangement (with standpipe and BOP) for drilling through tunnel and shaft linings. The hole shown is drilled out sub-horizontally from a shaft with a segmental lining.

- A 'stuffing box' (which compresses flexible packing around the rotating drilling rods/casing) or proprietary 'preventer' to seal around the rods/casing. This allows drilling without excessive water passing into the tunnel around the rotating rods/casing. The relief valve can be used to release water from the bore and direct it away from the drilling crews.

A typical drilling sequence requires the standpipe to be attached to and sealed into the tunnel lining. The guillotine valve and the stuffing box or preventer are then attached to form the BOP assembly. Drilling is carried out through the BOP, initially to penetrate the tunnel lining and then into the ground outside the tunnel. Installation of well materials (VWP piezometers, well screens, tube à manchettes, freezehole liner, temperature sensors) is also carried out through the BOP. The BOP is no longer needed (and can be disconnected from the standpipe) when the well installation is complete and water inflow is sealed off.

Various drilling methods have been used successfully on cross passage works. The choice of the most appropriate method will depend on the ground conditions and groundwater pressures around the tunnels. Roberts, Linde *et al.* (2015) describe the range of drilling techniques commonly used, and rank them in terms of ability to control bore stability.

1. Auger drilling without BOP: This method provides little bore control if pressurized sands or silts are encountered. However, it can be used in low-permeability materials such as clay, where significant water inflow is not anticipated. When it is used for probe holes, good information can be gathered on strata type. The risk with this method is that significant water inflows can be problematic, leading to bore instability and loss of ground. Contingency measures for the eventuality include leaving the auger in the bore or withdrawing and sealing the bore at the standpipe.
2. Rotary water/polymer flush drilling through a guillotine valve, without BOP: The drill string passes through the open valve. This allows the bore to be sealed with a valve as soon as the drill string is removed. This method is used when drilling in unweathered rock, where the risk of loss of ground is low, and the primary concern is high rates of water inflow if significant fractures are encountered.
3. Cased rotary water/polymer drilling through guillotine valve and BOP: The valve and BOP are attached to the standpipe and provide full control of the borehole annulus during drilling. Drilling is by rotating casing fitted with a shoe bit.
4. As Technique 3, but with lost bit drilling: A bit is attached to the end of the casing. When the hole is drilled to its final length, the well materials are installed. The bit is then detached by water pressure and the casing withdrawn, leaving the installation in place. This provides complete control of the bore throughout drilling and subsequent well or piezometer installation.

Where pressurized sand and silt horizons are encountered, the speed of installation, bore length and quality of information on the ground conditions all reduce with increased pore pressures and the corresponding increased level of bore control that is required. Even when these BOP measures are deployed, experience from Storebaelt (Hartwell *et al.*, 1994) and elsewhere indicates that maintaining bore control when drilling out from a tunnel becomes very difficult if pore water pressures are more than approximately 100 kPa (10 m head of water) above tunnel level.

10.6.2 Compressed Air Methods

Cross passages are an obvious case where compressed air working (see Section 10.4.1) can be used to exclude groundwater. An airlock fitted against the internal face of a running tunnel would allow the tunnel lining to be broken out and the cross passage excavated using compressed air to exclude groundwater. This method has been used on several projects as the primary means of cross passage construction. However, compressed air working has some disadvantages and limitations:

- On projects with smaller-diameter running tunnels, it can be difficult to install the airlocks for the cross passages and still allow access along the running tunnel.
- On deeper tunnels, the external groundwater pressures may be too high to be balanced by compressed air pressures, given the health and safety requirements of compressed air working. External groundwater pressures of up to 4.5 bar (too high for compressed air working) were one of the main reasons why the Storebaelt project adopted ground treatment and groundwater depressurization for cross passage construction (Doran, Hartwell, Kofoed, *et al.* (1995).

10.6.3 Ground Treatment Used to Exclude Groundwater

Ground treatment methods that reduce permeability are widely used to exclude groundwater during cross passage construction. Depending on depth and surface access, treatment can be done from the surface or by drilling out from the running tunnels. In most cases, ground treatment will be preceded by probe drilling from the tunnels to gather further information on ground conditions and groundwater pressures. Probe drilling can also be used after treatment to confirm that conditions are suitable for excavation.

Ground treatment techniques are described in Chapter 18. Methods that have been successfully used in cross passage construction include:

- Permeation grouting and rock grouting (Doran, Hartwell, Roberti *et al.*, 1995; Warren *et al.*, 2003)
- Jet grouting (Kinnear, Davis, Wiegand *et al.*, 2018)
- Artificial ground freezing (Kofoed and Doran, 1996; Crippa and Manassero, 2006; Ding et al., 2015; Sopko, 2019)

Ground treatment for cross passages can be used in two principal ways:

i. To reduce the permeability of the soil or rock around the cross passage to exclude groundwater from the works (any increase in strength of the treated soil/rock is of benefit but is not the primary objective) and/or
ii. To create a zone of strengthened soil or rock that can form a structural member increasing the stability of the excavation. This might be in the form of an arch or 'hood' of treated ground above the cross passage or an inverted arch below it.

In-tunnel pumps (Section 10.4.4.1) are also required to deal with any residual seepage, including any water encountered immediately outside the lining of the running tunnel when cross passage excavation commences.

Figure 10.25a shows an example where permeation or rock grouting from drill holes aligned sub-parallel to the cross passage is used to form a complete block of treated ground. The grout type (ordinary Portland cement [OPC], microfine cements or chemical grouts) and injection method (stage grouting or tube à manchettes) must be selected carefully to match the ground conditions. Typically, the grout holes are drilled and grouted in primary, secondary and tertiary patterns, with 'water testing' (a variation on packer permeability tests; see Section 12.8.3) employed periodically to assess the permeability reduction that has

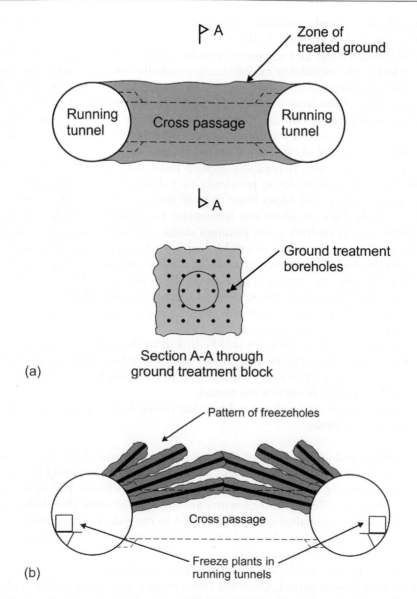

Figure 10.25 Cross-passage construction using ground treatment from within the running tunnels. (a) Cross-passage formed through block of treated ground. (b) Cross-passage constructed under arch of frozen ground. (Panel (a) after Warren, C D, et al., *Proceedings of Underground Construction 2003*, British Tunnelling Society, London, 253–265, 2003; panel (b) after Doran, S R, et al., *Proceedings of the 11th European Conference on Soil Mechanics and Foundation Engineering*, Copenhagen, Denmark, 1995.)

been achieved. The results of the water testing, together with post-treatment probe holes, are used to determine whether cross passage excavation can commence.

Figure 10.25b shows an example of ground treatment by artificial ground freezing. In this case, drill holes sub-parallel to the cross passage are used to form an arch of treated ground above the proposed excavation. Alternatively, freezeholes may be installed all around the proposed cross passage to entirely enclose the excavation within frozen ground.

Artificial ground freezing applications are typically designed using specialist software to model three-dimensional heat flow in response to the chilling effect of freezeholes. One factor that affects the application of artificial ground freezing to cross passages is 'frost expansion' of the frozen ground. As the groundwater freezes, it will expand by around 9 per cent, and this can cause an increase in pressure on the existing running tunnels (Sopko, 2019). In practice, the expansion mainly occurs radially outwards from the freezeholes, which are typically perpendicular to the running tunnels (Figure 10.25b). Therefore, only a small component of the frost pressures is applied to the running tunnels; nevertheless, assessment of frost pressures on existing tunnels should be part of the design process. A volume reduction process (and possibly an increase in permeability) will occur when the ground thaws after the end of cross passage construction.

Freezing for cross passages is typically applied using brine circulation systems (Section 18.12.1). If the cross passages are in proximity to a shaft, it may be possible to locate the freezeplant at the surface and pump brine to/from the cross passage via insulated pipes in the running tunnels. This can allow one freezeplant to supply freezeholes in both running tunnels (and to support multiple cross passages simultaneously). For cross passages distant from shafts, small freezeplants can be established in each running tunnel close by the cross passage (Figure 10.26a). The alternative freezing method of liquid nitrogen (LN) systems is rarely used for cross passage works, in part due to the health and safety risks of using LN in tunnels. Crippa and Manassero (2006) report a project in Holland where LN freezing was used on four cross passages, but in that case shaft access was available at each cross passage location, which may have partly mitigated the risks.

At the cross passage, the freezeholes will be linked to the brine circulation system via a manifold of insulated pipes (Figure 10.26b). In addition to freezeholes drilled through the tunnel lining, surface freezepipes may be attached to the inner face of the tunnel lining to chill the ground immediately outside the tunnel.

A typical sequence for artificial ground freezing (using brine circulation) deployed from within the tunnels comprises:

1. Probe drilling and pre-grouting of any disturbed zone around the running tunnels.
2. Drilling and installation of freezeholes and temperature monitoring holes. The freezeholes are typically steel tubes (sealed at the far end) 60 to 100 mm in diameter, grouted into place. Temperature sensors are installed in some drill holes, with single or multiple sensors installed in each hole. Due to the risk of large gaps between freezeholes caused by deviation during drilling, survey methods are often used to confirm that drill hole alignment is acceptable.
3. Installation of brine circulation pipework and freezeplant.
4. Freeze establishment period. Once the freezeplant is commissioned, the chilled brine (typically at temperatures of –25 to –35 °C) is circulated through the freezeholes and surface freezepipes. Temperatures are monitored via monitoring drill holes and compared against predictions from the freeze design. Typical periods to establish a complete freeze around a cross passage are 2 to 4 weeks.
5. Excavation phase and freeze maintenance period. Once the freeze has been established and cross passage excavation has begun, circulation of chilled brine continues (at a reduced thermal load), to ensure the frozen zone is maintained.
6. End of freeze and decommissioning. Following the completion of the cross passage works, brine circulation is discontinued, and the freezeplant and pipework are removed. Typically, the freezeholes are decommissioned by backfilling with grout, removing the standpipe and making good the tunnel lining. The frozen zone will thaw slowly, over several weeks, with external temperatures monitored via the temperature sensors.

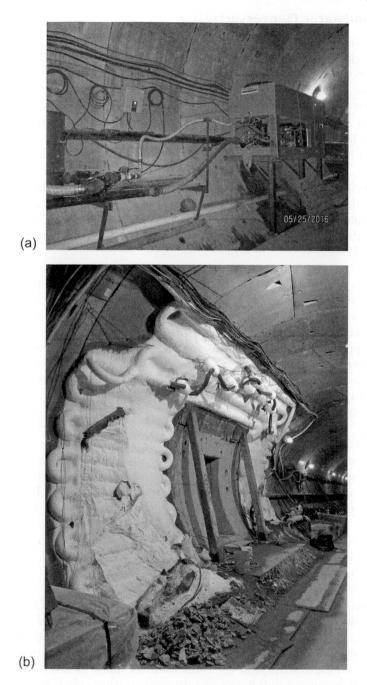

Figure 10.26 Artificial ground freezing used in cross passage construction. (a) In-tunnel freezeplant for brine freezing system. The unit shown is air-cooled, with a 5.9 ton (17,840 kcal/h) chilling capacity. (b) Array of freezeholes around cross passage opening in running tunnel. Surface freezepipes, in contact with tunnel lining, are used in this application. (Courtesy of SoilFreeze Inc., Woodinville, WA.)

10.6.4 Groundwater Depressurization

Pumped groundwater control methods are widely applied to cross passage construction, using surface wells, wells drilled out from the running tunnels or a combination of the two approaches. Strictly, these pumped methods do not typically lower groundwater levels to below excavation level, as might be done for a surface excavation. Rather, localized pumping from around the tunnel creates a depressurized zone, which means that cross passage excavations will experience much lower seepage gradients and rates of water ingress. This can allow tunnelling for cross passages using appropriate methods of temporary support.

Methods that have been used for cross passage depressurization include:

- Surface wells pumped by submersible pumps or ejectors (Linde-Arias *et al.*, 2015)
- In-tunnel dewatering wells, either flowing as passive relief wells or pumped by well-point pumps or ejector wells (Roberts, Linde *et al.*, 2015)
- In-tunnel drainage wells linked to electro-osmosis systems (Doran, Hartwell, Roberti *et al.*, 1995)

In-tunnel pumps (Section 10.4.4.1) are also required to deal with any residual seepage, including any water encountered immediately outside the lining of the running tunnel.

Due to the complex geometry of most cross passage depressurization systems, design is typically by three-dimensional numerical groundwater modelling, often combined with an observational approach to modify the system in response to field observations.

If surface dewatering methods are used, the wells can be installed and pumped significantly in advance of cross passage construction. Where in-tunnel dewatering wells are used, installation must be carefully integrated into the tunnel programme. A typical in-tunnel sequence comprises:

1. Probing and installation of piezometers
   Once the running tunnels are complete sufficiently far past the cross passage location, sub-horizontal probe holes can be drilled from each end of the cross passage. The primary objective of the probe holes is to confirm ground conditions between the tunnels. Often, the probe holes serve a secondary purpose and are installed as depressurization wells. This allows the probe holes to have VWP piezometer sensors installed to monitor water pressures before and during cross passage works or allows them to be used as passive drains to reduce water pressures during later drilling.

2. Drain drilling and passive drainage phase
   Depressurization wells are drilled outwards from the running tunnels to form the design well array. The wells are typically installed with 50 to 75 mm diameter well liner and slotted well screen, comparable to wellpoints. To avoid loss of ground into the tunnel, grout or a packer must be used to seal around the well liner at the tunnel end. As soon as each well is completed, they will flow into the tunnel (unless shut off by a valve on the standpipe at the tunnel lining). This 'passive' flow (Figure 10.27a) is of benefit because it will have some depressurization effect, making the drilling of further wells easier and reducing the time needed for active pumping to lower groundwater pressures. The water flowing from the drains should be collected by the tunnel drainage system.

3. Active drainage (pumping) phase
   The most common pumping method used for in-tunnel wells is wellpoint pumping that applies suction to the wells (Section 19.5). Electrically driven duty and standby wellpoint pumps are located in the running tunnels (Figure 10.27b) and are connected to the dewatering wells by a suction manifold attached to the tunnel lining. To reduce

Figure 10.27 Cross-passage groundwater depressurization methods using in-tunnel wells. (a) In-tunnel dewatering wells (drilled out from running tunnels) used as passive relief wells for cross passage construction. (b) In-tunnel wellpoint system used for active pumping of wells for cross passage construction. (Courtesy of WJ Groundwater Limited, Kings Langley, UK.)

the head against which the pumps have to work, the pumps usually discharge to the tunnel drainage system, from where other pumps will remove the water up to the surface. VWP pressure sensors installed in the probe holes at the cross passages are used to monitor pore water pressures.

4. Excavation phase

When pore water pressures are reduced to target levels, the lining of the running tunnel is broken out, and excavation commences. Post depressurization, there may be small water inflows when the lining is first penetrated if any trapped water is present in a disturbed zone outside the tunnel. It is not uncommon for minor overbleed seepage to occur at clay/sand interfaces or other changes in strata.

5. Decommissioning

On completion of works, dewatering wells and piezometers are typically decommissioned by backfilling with grout, removing the standpipe and making good the tunnel lining.

Various geometries of well arrays are used for cross passages (Figure 10.28):

i. In uniform deposits of moderate permeability, such as fine sands, vertical or steeply inclined wells installed downwards from the running tunnels (Figure 10.28a) may be able to reduce pore water pressures sufficiently to allow cross passage construction. Vertical wells are also used where the cross passage includes a deeper drainage sump or where a potentially pressurized zone is present a short distance below tunnel invert.

ii. In uniform deposits of low-permeability materials such as silts, wells are typically installed at close spacings sub-parallel to the cross passage. The intention is to form a fan of wells around the space where the cross passage is to be constructed (Figure 10.28b).

iii. In deposits of variable permeability (for example clay-dominated strata that contain discrete sand beds), inclined wells are installed from the running tunnels at a level and orientation where they are likely to intercept the permeable layers that require depressurization (Figure 10.28c).

One of the difficulties of installation of in-tunnel wells is placing suitable filters around the slotted well screen. Ensuring that well filters are effective is especially important in fine-grained uniformly graded soils such as silts and sands, where there is a risk of drawing fine particles out of the ground.

Filter installation must be done through the standpipe and BOP arrangement and must be suitable for use even where wells are installed sub-horizontally or at upward inclinations. Methods that have been used successfully include:

1. Well screens with preformed resin-bonded filter packs. Specific granular filter media are bonded directly to the slotted well screen in a factory-based process (Figure 16.3c). When the well screen is installed, the natural ground is allowed to collapse against the well screen when the drill casing is removed.

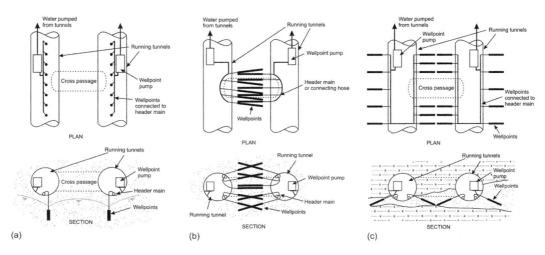

Figure 10.28 Arrangements of in-tunnel wells used for cross passage construction. (a) Vertical or steeply inclined wells installed downwards from the running tunnels. (b) Wells installed at close spacings sub-parallel to the cross passage. (c) In deposits of variable permeability, inclined wells are installed from the running tunnels at a level and orientation where they are likely to intercept the permeable layers that require depressurization.

2. Placing filter sand by pumping. This approach was developed on the Storebaelt project (Doran, Hartwell, Roberti *et al.*, 1995). When the drilling is complete to the target length, a suspension of filter sand in polymer slurry is pumped into the bore, and the slotted well screen and liner are then inserted. When fully inserted, the well screen will be surrounded by the filter slurry; the drill casing can then be removed. Once the polymer has broken down, the sand around the well screen should form a permeable filter.

10.7 GROUNDWATER CONTROL METHODS FOR OTHER TUNNELLING APPLICATIONS

In addition to the three common geometries – tunnel, shaft and cross passage – that occur at various scales on many tunnelling projects, there is a range of other tunnelling applications that are sometimes used depending on the nature of the individual projects. Different examples of this type of tunnelling works are listed in Section 10.2.4.

The type of groundwater control that is required (or indeed is practicable) will be largely controlled by the access available and the geometry of the proposed excavation as well as the ground conditions. Cases with similar geometries to shafts or cross passages are often dealt with by techniques appropriate to those applications, as outlined in earlier sections. If access at ground level is available, drilling for pumped wells or for ground treatment may be done from the surface. In-tunnel drainage and ground treatment are also widely used.

When constructing metro or underground railway systems, there is often the need to excavate 'caverns' – large underground spaces – for stations and for crossovers between lines. These are often formed by enlarging pilot tunnels driven previously through the cavern footprint. Linde-Arias *et al.* (2015) describe an example of groundwater control for station cavern construction on the Elizabeth Line (formerly Crossrail) project in London. The Stepney Green Cavern is over 50 m long, 17 m wide and 15 m high and was one of the largest caverns ever built using the SEM and SCL techniques in central London at that time. The cavern is approximately 20 m below ground level and was to be excavated through the very low-permeability London Clay Formation and the more variable Lambeth Group, which contains significant beds of clay but also includes beds and lenses of pressurized sand and silt. To reduce in-tunnel pumping (and the corresponding interference with excavation and lining), the opportunities to use surface access were maximized, and 45 ejector wells were installed from the surface. Pumping from the surface wells achieved sufficient depressurization of the sand layers to allow the eastbound section of the cavern to be constructed without the use of in-tunnel wells. However, the deeper westbound section of the cavern required further depressurization to augment the surface system. Approximately 105 in-tunnel wellpoints were installed in 90 mm diameter auger holes and pumped by a suction system, which achieved the necessary depressurization of the water-bearing zones.

One case that occurs regularly is small-scale tunnelling works to connect new tunnels into existing infrastructure. Examples include newly constructed sewers connecting into existing pumping stations and treatment works, or access passageways to connect new metro lines at interchange stations. Often, the new tunnels must be threaded between existing tunnels, foundations and buried services. Pumped groundwater control systems may not be fully effective, as water pathways may remain below or immediately around existing structures. This can allow groundwater to bypass arrays of dewatering wells. Similarly, access may preclude the use of vertical cut-off walls to completely enclose the proposed excavation.

An approach often used is ground treatment both vertically and sub-horizontally to form a zone of low-permeability treated ground around the excavations, including blocking flow paths around and below structures. Suitable techniques include permeation grouting, jet grouting and artificial ground freezing. Packer *et al.* (2018) describe the application of permeation grouting at London Underground's Victoria Station upgrade project. Pedestrian tunnel Pa16 was to be formed close beneath a Victorian-era masonry tunnel carrying subway trains. The Pa16 tunnel was to be driven by hand and machine excavation using SEM methods constructed through mixed face ground conditions. The upper section of the face comprised water-bearing sands and gravels of River Terrace Deposits, above very low-permeability stiff clay of the London Clay Formation in the lower section of the face. Groundwater control by pumped wells would not have been effective in fully dewatering the sands and gravels – some residual overbleed seepage would have been unavoidable in the zone above the top of the clay. Unstable 'running sand' conditions would likely have developed above the gravel/clay interface, leading to ground loss and settlement damage to existing tunnels and structures. Packer *et al.* (2018) describe how careful design and grout selection allowed the sands and gravels to be targeted by permeation grouting via tube à manchettes, many installed sub-horizontally. This approach was successful and allowed the mixed face conditions to be excavated in dry and stable conditions.

Artificial ground freezing is used relatively rarely but can deal with some complex geometries and challenging ground conditions. During metro construction in Copenhagen, Denmark, a connecting subway was required in a confined area between the new metro tunnels and the existing rail terminus with existing structures and tunnels present. Shafts that were part of the permanent works were constructed within freezewalls and were then used to allow the installation of sub-horizontal freezepipes to stabilize the tunnels for the subway (Hayward, 2000).

10.8 GROUNDWATER CONTROL METHODS FOR RECOVERY WORKS

If a tunnel project experiences a major collapse or inundation, then some type of recovery works will be needed to allow work to re-start. Recovery works can include reconstruction of damaged tunnels or shafts, or the construction of additional shafts or tunnels to access the failed area. Groundwater plays a key role in many tunnel 'failures'; groundwater control is an essential element in most recovery works. One factor in recovery works is that movements and ground loss can result in changes to the properties of the soil and rock in the affected areas. The strength of the soil/rock is often reduced, and the permeability is also affected – often, but not always, being higher in disturbed zones. These factors make tunnel recovery works one of the most challenging potential applications of groundwater control.

Both pumping and exclusion methods can be used. Case History 27.6 describes a system of pumped wells used to depressurize a confined aquifer after a shaft flooded due to the hydraulic failure of its base. When exclusion methods are used, the method may be selected so that in addition to acting as a barrier to groundwater, the treated ground has increased soil/rock strength and can therefore provide some structural support to a shaft or tunnel. In the United Kingdom, artificial ground freezing was used to enable the completion of two water tunnels that encountered problems during construction. On both the London Water Ring Main, as described in Section 7.7.2 (Clarke and Mackenzie, 1994), and a sewer project in Hull (Brown, 2004; Munks *et al.*, 2004), a vertical freezewall was formed around a shaft, and horizontal freezeholes were drilled around the tunnel to allow a cylinder of frozen ground to be formed around the damaged section of tunnel.

Sopko (2017) describes the application of artificial ground freezing on recovery works on two projects, where the shafts were enclosed within concrete diaphragm walls intended to exclude groundwater. In both cases, significant inflows of water and ground halted excavation above the required final dig level (subsequent investigations indicated that the inflows were due to gaps between panels in the diaphragm walls). Prior to the use of artificial ground freezing, various methods of grouting were tried but were unsuccessful in controlling inflows. The ground freezing approach allowed the shafts to be completed. On one shaft in New York, inspection of the diaphragm wall after final excavation revealed a 15 mm gap between adjacent panels in the wall, which had allowed an estimated 700 m³ of fine silty sand to flow into the shaft. Sopko (2017) states:

> The gap was basically filled with the cement-bentonite grout. However, at a depth of 40 m, it was observed that the gap was filled with frozen soil, indicating that the hydrostatic pressures blew out the grout that was temporarily providing a seal.

Problems sometimes occur when ground conditions change along the tunnel route. Hartwell (2001) describes a tunnel in Scotland that experienced a problem similar to that shown in Figure 8.18b, where a rock tunnel drive encounters a significant zone of water-bearing coarse-grained soil. The tunnel was expected to be in competent rock throughout, and construction was planned accordingly. Hartwell (2001) states:

> a 1.8 metre diameter drill and blast tunnel was being driven through coal measure sandstones and mudstones for a new sewer. After one particular blast the tunnel collapsed and started to fill with gravels, boulders and heavy water flows. The tunnel had encountered a buried glacial valley with water pressures above 3 bars [30 m head of water]. Immediate investigations confirmed the presence of the valley and variable nature of the infill material encountered by the tunnel.

The recovery works involved driving a section of tunnel in the rock in the opposite direction until the buried valley was approached. Ground treatment was then applied ahead of the new drive by permeation grouting into the glacial soils via probe holes drilled from the face. Hartwell (2001) reports: 'Some surface dewatering was used to reduce the pressures in the area but problems inevitably occurred with grout flowing to wells and causing them to fail (despite planned efforts to turn off those closest to the grouting operation).'

Holes drilled in the face for grouting purposes also had a beneficial drainage effect, and some were left ungrouted, despite some resistance from the mining crews and foremen, and were left to operate as horizontal drains to help depressurize the face. This approach allowed the tunnel to be completed across the problem area.

Perhaps the largest example of groundwater control as part of recovery works was 'Project MOSES', implemented on the Storebaelt Eastern Railway Tunnel project in Denmark, described in Biggart and Sternath (1996) and Steenfelt and Hansen (1995). Recovery works were needed after the tunnel bores were inundated when a hydraulic connection was formed between the tunnel and the sea bed above. The full face TBM was being maintained at the time and was not able to exclude the water.

Monitoring data from deep well dewatering at the tunnel portals indicated that drawdown of groundwater pressures (in the strata below the sea bed) extended out to 3–4 km from the portals. It was conceived that depressurizing the Glacial Deposits (Till containing significant sandy and silty zones) and the underlying Marl Bedrock would be able to reduce the pore water pressure at the TBM level by 3–4 bar (30 to 40 m head of water). This would mean that the pore water pressure at the TBM would be approximately 3 bar (30 m head of water), which would improve soil/rock stability at the TBM cutter head and allow regular face interventions to be carried out under compressed air. These measures would dramatically reduce the risk of further tunnel instability.

The solution was Project MOSES (Method of Obtaining Safety by Emptying Storebaelt), as shown in Figure 10.29a. This involved 43 dewatering wells (with well yields in the range

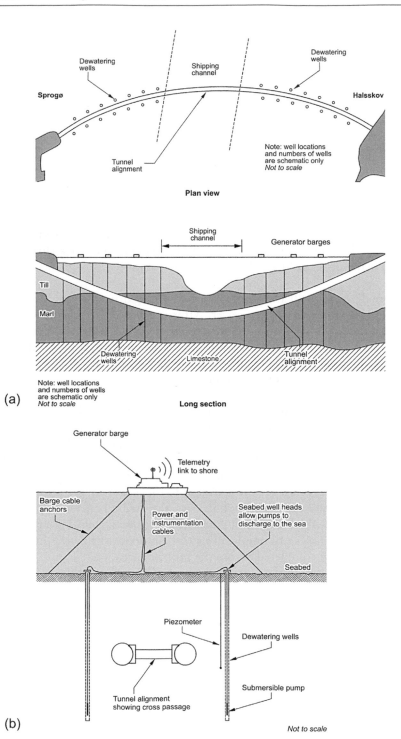

Figure 10.29 Project MOSES – subsea dewatering scheme for construction of Storebaelt Eastern Railway Tunnel. (a) Schematic plan and long section. (b) Schematic cross section. (Panel (a) after Steenfelt, J S and Hansen, S K, *Key Note Address: The Storebaelt Link – A Geotechnical View.* Aalborg: Geotechnical Engineering Group, 1995; panel (b) after Biggart, A R and Sternath, R, *Proceedings of the Institution of Civil Engineers*, 114, Special Issue 1, pp. 20–39, 1996.)

of 15 to 120 m³/h) and 12 piezometers drilled in the sea bed that operated for 30 months and pumped 45 million m³ of water from the Till and the Marl.

The dewatering wells and piezometers were drilled overwater from jack-up platforms. After completion of each well, an electric submersible pump was installed on rising main, topped by a special sea bed well head that allowed the pumps to discharge directly to the sea while submerged. The well set-up incorporated a flow meter and a pressure transducer. Power and instrumentation cables were linked to a series of six generator barges moored above the tunnel alignment (Figure 10.29b). Each barge had a 455 kW power rating and was linked to up to eight wells and three piezometers. Telemetry links were connected to monitoring and control facilities based onshore.

Project MOSES was a great success. It achieved the primary objectives of reducing tunnelling risks and allowing more rapid TBM progress. A very useful side effect was to significantly reduce pore water pressures at the cross passage locations, making the groundwater depressurization and ground treatment works easier to implement.

Chapter 11

Site Investigation for Groundwater Lowering

If you do not know what you are looking for in an investigation, you are not likely to find much of value.

Rudolph Glossop

11.1 INTRODUCTION

Site investigation was defined by Clayton and Smith (2013) as

The process by which ground engineering professionals such as geotechnical engineers, engineering geologists, and geoenvironmental engineers and scientists obtain information on the ground beneath sites that are to be developed or remediated.

This chapter describes the information that needs to be gathered to allow the design of groundwater lowering systems. Good practice in the planning and execution of site investigations is outlined, together with brief details of methods used for boring, drilling, probing, testing, etc. The analysis of the data gathered during site investigation is described in relation to design methods in Chapter 13.

This chapter deals specifically with the site investigation requirements for groundwater control projects. The successful design of such projects is highly dependent on obtaining realistic estimates of the permeability of the various soil and rock strata present beneath the site. Methods of permeability testing, their range of applicability, and their relative merits and limitations are discussed in detail in Chapter 12.

11.2 THE OBJECTIVES OF SITE INVESTIGATION

Site investigation (also known as geotechnical investigation) is the essential starting point, without which the design of any construction or geotechnical process cannot progress. Groundwater lowering is no exception to this rule. To quote the fictional detective Sherlock Holmes:

It is a capital mistake to theorize before one has data

(Sir Arthur Conan Doyle, *A Scandal in Bohemia*)

The best site investigations are deliberately planned and executed processes of discovery, carefully matched to the characteristics of the site and to the work to be carried out. Sadly, many investigations in the past have not met this challenging standard. The problems of poor or inadequate investigations were highlighted in two reports by the Institution of Civil Engineers: *Inadequate Site Investigation* (Institution of Civil Engineers, 1991) and *Without Investigation Ground Is a Hazard* (Site Investigation Steering Group, 1993).

A particular problem for the designer of groundwater lowering works is that in many cases, investigations are designed primarily to provide information for the design of the permanent works. The information needed for the design of temporary works (including groundwater lowering) is sometimes neglected. This problem can arise when the persons designing the site investigation do not have appropriate expertise or experience. Alternatively, poor communication may result in them not being informed of the likely need for dewatering, so they will not plan to gather the relevant information.

There is a wide and useful literature on site investigation, including Clayton *et al.* (1995), the *British Standard Code of Practice for Ground Investigations* (BS 5930: 2015); Eurocode 7 (BS EN 1997-1: 2004; BS EN 1997-2: 2007); and Clayton and Smith (2013), and the reader should consult these for background on the subject. The remainder of this chapter will concentrate on the particular site investigation needs for projects where temporary works groundwater lowering schemes are to be employed.

11.3 THE ROLE OF THE CONCEPTUAL MODEL

The importance of conceptual models was introduced in Section 5.4 and is further discussed in relation to design in Section 13.4. A conceptual model is a non-mathematical representation of ground and groundwater conditions that is used to help understand and communicate geotechnical and hydrogeological conditions at a site.

The conceptual model should be integral to the planning and interpretation of any investigation, especially where groundwater conditions are important. The conceptual model should identify the likely presence of aquifers and aquitards, possible ranges of groundwater levels, possible permeability of key strata, etc. The model can be used to aid the selection of investigation methods and design of monitoring wells, piezometers and other groundwater monitoring measures. The initial conceptual model is typically based on the findings of the desk study and site reconnaissance (Section 11.6) and can be used to help plan the early phases of investigation. The conceptual model should be updated as more information is gathered and interpreted. This can enable later phases of investigation to be better targeted or adapted for the ground conditions actually encountered.

11.4 PLANNING OF SITE INVESTIGATIONS

All site investigations need to be planned and designed in order that they will provide the information needed by the various designers, estimators and construction managers. There is no such thing as a 'standard' site investigation, purely because 'standard' ground conditions have yet to be discovered, no matter how much we might wish for them!

According to the Institution of Civil Engineers (1991), the designers and planners of site investigations should attempt to answer the following questions:

a) What is known about the site?
b) What is not known about the site?
c) What needs to be known?

One of the key objectives of site investigation should be the identification, as early as possible in the process, of significant potential hazards and risks associated with the ground conditions and the proposed form of construction.

On all but the smallest investigations, these questions cannot be answered by one person and may require input from specialists in soil and rock mechanics, engineering geology, geophysics, geomorphology, archaeology and hydrogeology. For groundwater lowering projects, advice from a dewatering specialist can also aid the planning of investigations.

The planning, design and ultimately, procurement of site investigations are highly specialized. It is essential that this work is guided by suitably qualified and experienced professionals, who should be associated with the site investigation from conception to completion.

Effective communication between all parties involved in the construction process is vital. This includes the site investigation as well, because without accurate and up-to-date information, how can an investigation be designed to answer the questions listed earlier? In the United Kingdom, the *Construction (Design and Management) Regulations 2015* (known as the CDM Regulations; see Chapter 25) formalize this ethos and require clients and designers to work with all parties from an early stage so that safe methods of work can be planned and adopted. This means that clients and designers must provide the designer and manager of the site investigation with details of the proposed works (e.g. location, depth and size of excavation, and support methods). Without these details, it is difficult to plan an investigation on a rational basis.

11.5 STAGES OF SITE INVESTIGATION

A site investigation includes all the activities required to gather the necessary data about the site and should consist of a number of stages:

   (i) Desk study
  (ii) Site reconnaissance
 (iii) Ground investigation
  (iv) Reporting

To many non-geotechnical specialists, the ground investigation is perceived to be the 'essence' of a site investigation. In fact, the ground investigation (which can involve trial pits, boring, drilling and probing, in situ testing, laboratory testing and the installation of monitoring wells) aims only to determine ground and groundwater conditions at the site. If other stages (especially the desk study) are neglected, inadequate investigations may result.

11.6 DESK STUDY AND SITE RECONNAISSANCE

The desk study and site reconnaissance (sometimes called a walk-over survey) are essential in any investigation – their importance cannot be over-estimated. Unfortunately, they are sometimes overlooked or considered irrelevant.

The desk study is a review of all available information relevant to the proposed project, including geological and hydrogeological maps; topographical maps; flood risk assessments; aerial photographs (if available); records of construction on nearby sites or those where similar strata and groundwater conditions were encountered; records of nearby wells, boreholes and springs; and records of mining or previous use of the site. Sources of information for desk studies are discussed by Dumbleton and West (1976), AGS (2006) and in Appendix E of Clayton and Smith (2013). Many useful data sources (such as geological mapping) can be accessed via the World Wide Web from the public domain and commercial sources.

The site reconnaissance is usually carried out after the desk study has been largely completed. It comprises a walk-over survey of the site to gather further information on the site surface conditions and access for the ground investigation, any exposed geological, groundwater or surface water features, and the nature of the areas surrounding the site.

Clayton *et al.* (1995) point out that both the desk study and site reconnaissance can provide large amounts of useful information relatively quickly and at low cost – they are by far the most cost-effective stages of site investigation. They are essential to allow efficient design of the ground investigation; failure to anticipate any predictable groundwater problems can result in poor or inadequate investigations, leading to potential problems during construction.

The desk study and site reconnaissance can also be used to gather information about the surroundings of the site, perhaps in areas where access for ground investigation cannot be obtained. For groundwater lowering projects, this is particularly useful to help determine whether there will be any adverse side effects of dewatering (such as ground settlement or derogation of water supplies; see Chapter 21) outside the site. Brassington (1986) recommends that a survey of nearby groundwater supplies be carried out to aid the assessment of the risk of derogation to water users.

The problems that may result if the desk study and site reconnaissance are neglected can be illustrated by two case histories.

a) On a project through variable glacial soils in northern England, a new sewer was laid in an open-cut trench to replace an existing sewer constructed a few decades earlier. The new sewer was generally laid parallel to the old one, apart from one section where the old sewer took a circuitous 'dog-leg' route between manholes. The new sewer was to take the obvious straight-line route between the manholes. During construction of this section, severe groundwater problems were encountered, including a flowing artesian aquifer (see Section 3.4.2) immediately beneath the base of the excavation. Work was further hampered when old abandoned sections of steel sheet-pile trench supports (probably dating from the construction of the old sewer) were encountered. With hindsight, it is fairly obvious that the old sewer was originally intended to take the direct route between the manholes, but that a very localized groundwater problem forced them to abandon work and re-route the sewer. The new sewer was eventually completed, but a desk study of old construction records might have allowed different methods to be adopted from the start, thus avoiding cost and time delays.

b) A small sewerage project involved a wellpoint system to lower groundwater levels for the construction of a manhole. The manhole itself was excavated in sands and gravels and was successfully dewatered. However, significant damage occurred to several neighbouring structures as a result of ground settlements. Subsequent investigations showed that the damaged buildings were founded on an extensive deposit of compressible peat. Although ground investigation at the manhole site did not encounter any peat, a desk study review of geological maps would have revealed the presence of peat beneath surrounding areas. Awareness of the peat stratum would have indicated the risk of damaging settlements and would have allowed appropriate mitigation measures to reduce the impact of dewatering on nearby buildings (see Section 21.4.4), to be included in the project.

11.7 GROUND INVESTIGATION

There are numerous techniques available to physically investigate a site. It is rarely obvious at the start of an investigation which methods will be suitable to gather the information needed by the project designers. It is preferable, therefore, that all but the smallest ground investigations be carried out in phases.

The first phase should be planned based on the conceptual model developed from the desk study and site reconnaissance. Typically, this phase may consist of an initial pattern of boreholes and testing across the whole site as well as the installation of monitoring wells. Consideration could also be given to the use of non-intrusive surface geophysical profiling techniques (Section 11.7.4.1), which can quickly acquire ground and groundwater information over large areas for relatively little cost. These methods are particularly suited to long linear developments such as pipelines and tunnels. Preliminary results from the first phase will allow the conceptual model to be updated, and second and subsequent phases of investigation to be designed; such work may include more closely spaced boreholes in areas where ground conditions are unclear, or perhaps specialist testing such as a pumping test.

Ground investigation usually involves the use of some combination of the following methods: boring, drilling and probing; trial pitting; in situ testing; geophysics; and laboratory testing. Additionally, the installation of monitoring wells and piezometers is an important part of any investigation where groundwater is of interest. These methods are briefly outlined in the following sections mainly in relation to British practice. Factors particularly relevant to investigations for dewatering projects are highlighted.

11.7.1 Boring, Drilling and Probing

Many methods are used to form investigation boreholes with the objective of determining ground conditions and recovering samples of soil and rock. Methods can be characterized as:

- Boring: These methods are typically used to form boreholes in soil and include cable percussion drilling, wash boring and auger boring.
- Drilling: These methods typically involve rotating drill strings (and often the use of drilling fluids) and were originally developed for core drilling in rock. However, rotary drilling methods are now also widely used to obtain high-quality geotechnical samples in some types of soil.
- Probing: This group of methods drives or pushes a small-diameter probe (which may be instrumented) and records parameters related to the penetration resistance recorded.

Boring, drilling and probing methods are described separately in the following, and their relative merits are outlined in Table 11.1. However, for simplicity, in the rest of this chapter, drilling refers to the formation, by any method, of a ground investigation borehole.

11.7.1.1 Boring Methods Used in Investigations

In Britain, one of the most common forms of boring is still light cable percussion drilling, colloquially known as 'shell and auger' drilling (Figure 11.1a). In this method, a winch is used to repeatedly raise and drop tools to excavate the base of the borehole. Soil samples may be recovered from the borehole (for description of soil type and laboratory testing), or certain types of in situ test may be performed within the borehole. A key point to note is that

Table 11.1 Advantages and Disadvantages of Methods of Boring, Drilling, Probing and Trial Pitting

Method	Advantages	Disadvantages
Boring	Suitable for use in a wide range of soils Allows soil and groundwater samples to be obtained Can allow in situ permeability tests to be carried out Allows monitoring wells (standpipes and standpipe piezometers) to be installed	Progress can be difficult if cobbles or boulders are present, or if hard obstructions are encountered in Made Ground or fill deposits Some methods require specialist equipment
Drilling	Suitable for use in rock and some soil types Allows soil and rock samples to be obtained Can allow in situ packer permeability tests to be carried out Allows monitoring wells (standpipes and standpipe piezometers) to be installed	Obtaining representative groundwater level readings can be difficult Progress can be difficult in some soils and highly weathered rock without the use of casing or specialist equipment
Probing	Provides information on soil profile CPT piezocone can provide information on soil permeability Can allow simple standpipe monitoring wells to be installed at a shallow depth Some methods allow soil samples to be obtained	Some methods do not allow soil or groundwater samples to be taken Some methods require specialist equipment Ideally should be used in conjunction with boreholes to enable correlation of soil types Penetration depth is limited in stiff materials and coarse granular soils
Trial pitting	Suitable for a wide range of strata, including very coarse soils and weak rock Allows stability and ease of excavation of strata to be directly observed Requires no specialist equipment Allows soil and groundwater samples to be obtained Can allow simple standpipe monitoring wells to be installed at a shallow depth	Safety issues related to working near excavations Depth limited to around 5 m May be difficult to progress below groundwater level in unstable soils and weathered rock Does not allow installation of standpipe piezometers

at various stages during drilling, water may be added to the borehole by the driller or may be removed by the action of the boring tools. This can lead to natural groundwater levels and inflows being masked during boring operations.

The older technique of wash boring is rarely used in Europe but is still employed in countries where labour is cheap. The basic rig is a winch tripod. The associated equipment consists of an outer pipe with a chisel bit at the lower end and a swivel head at the upper end of the wash pipe, incorporating a water pressure hose connection with a weight for driving the casing into the ground. A pump passes water down the wash pipe to slurrify the soil at the bottom of the outer casing. The return washings are not regarded as reliable for identification of soil types – though recordings of wash-water colour changes should be noted. This method is suitable for use in sands and silts, but progress in clayey soils is likely to be slow. Groundwater levels and inflows are masked in a similar way to rotary drilling.

Hollow stem continuous flight augers are suitable for use in cohesive soils but are of limited use in water-bearing coarse-grained soils; indeed, in such soils, this technique is often unworkable. The drilling spoil brought to ground surface gives only an approximate indication of soil types and horizons. Drive-in samplers can be inserted through the hollow stem to obtain strata samples at convenient depth intervals.

Figure 11.1 Drilling and boring methods used in site investigation. (a) Light cable percussion boring rig. (b) Rotary drilling rig. (Courtesy of ESG Soil Mechanics.)

11.7.1.2 Drilling Methods Used in Investigations

Rotary drilling (utilizing a rotating drill string, typically circulating a water- or air-based flush medium with polymer or foam additives) is also widely used (Binns, 1998), particularly in relatively intact rock strata but also in certain soil types. Rigs are usually mounted on four-wheel-drive trucks or tracked base machines (Figure 11.1 b). Typically, for investigation drilling, the bottom of the drill string is equipped with a specialist 'core barrel' that allows core samples to be recovered from the boreholes (for soil and rock description). Additionally, certain types of in situ tests can be carried out within the borehole. In some circumstances where high-quality characterization of strata is not required (such as drilling of holes for monitoring wells or specialist in situ tests), 'open hole' drilling is used, where core is not recovered and geological conditions are assessed from cuttings brought to the surface by the drilling fluid.

Most investigation drilling is by direct circulation methods, where drilling fluid is passed down the drill string and returns to the surface via the annulus between the drill string and the borehole walls. Investigation techniques on mining projects sometimes use reverse circulation techniques, where drilling fluid flows down the annulus and returns up the drill string. Where drilling fluids are used, groundwater levels and inflows can be difficult to determine during drilling:

- Where water-based drilling fluids are used, the net addition of drilling fluid will tend to mask any water inflows during drilling and can mean that water levels in the borehole are not representative of the surrounding ground.
- Where air-based drilling fluids are used, water entering the borehole will tend to be blown out of the hole and emerge at the surface with the compressed air. This can sometimes allow significant water inflows to be identified. However, the drill flush creates an airlift effect, so the drilling action is removing water from the ground, locally reducing groundwater levels. This can mean that water levels in the borehole are not representative of the surrounding strata.

Investigation drilling is also sometimes carried out by sonic methods (also known as rotasonic drilling). This is a relatively new technique that uses high-frequency vibration of a rotating drill string to aid penetration (Riley et al., 2018). This method can allow boreholes to be drilled without the use of drilling fluid, which can be an advantage when trying to observe groundwater inflows during drilling in relatively low-permeability strata.

11.7.1.3 Probing Methods Used in Investigations

Probing methods are sometimes used as an alternative to boring or drilling. A wide range of equipment exists and is used, but all have the principal objective of determining a profile of penetration resistance with depth. Most methods were developed as a low-cost and rapid alternative to boring and drilling. Two of the most commonly used methods are dynamic probing and static probing by the 'cone penetration test' (commonly known as the CPT).

Dynamic probing involves using relatively lightweight equipment to create a percussive action to drive the probe into the ground, producing output in the form of blows per unit depth of penetration (see, for example, Card and Roche, 1989). Window sampling is a variant of the dynamic probing method that allows soil samples to be obtained via sampling tubes driven into the ground.

Static probing by CPT (also known as 'Dutch cone' testing after the country where the method was developed) is more sophisticated than dynamic probing. The cone is pushed continuously into the ground, using reaction from the test truck, producing an output of resistance against depth (see Meigh, 1987; Lunne *et al.*, 1997). Piezocone testing and the Hydraulic Profiling Tool (HPT-CPT) are variants of the CPT method and can allow estimates of permeability to be obtained in relatively low-permeability soils (see Section 12.8.4).

11.7.2 Trial Pitting

Trial pitting is a simple and widely used method for investigation of soils and weak rocks at shallow depths. A pit is dug, exposing the strata for inspection and sampling, and often providing a much better understanding of larger-scale geological features (such as soil fabric or fracture networks in rock) than can be discerned from investigation boreholes. The depth and rate of groundwater inflows and seepages can normally be clearly identified. During excavation, it may be possible to form an opinion as to appropriate methods of full-scale excavation, if relevant.

Trial pits are normally dug by mechanical excavator (Figure 11.2). Small backhoe loaders can normally excavate to a depth of around 3 m, and larger 360° excavators may be able to work to a maximum depth of around 5 m when working from ground level.

Trial pitting is a potentially hazardous exercise and must be carried out using safe systems of work based on suitable risk assessments. Trial pits of greater than 1.2 m depth should only be entered if adequately shored and supported. Even pits of less than 1.2 m depth may be unstable, and collapses at such shallow depths can result in death and serious injury. Each pit should be assessed before entry, and if any doubt exists, the pit should not be entered. To reduce risks and avoid the need to enter trial pits, exposed soil and rock in the pit faces can be described from the surface and samples taken from the spoil in the excavator bucket. In difficult locations or where ground disturbance must be kept to a minimum, it

Figure 11.2 Trial pitting using a mechanical excavator.

may be possible to excavate pits by hand excavation. However, this method is very slow, and such excavations must employ timbering or some form of proprietary side support system. Safety in pits and trenches is discussed by Irvine and Smith (2001).

11.7.3 In Situ Testing

Various methods of in situ testing can be carried out as part of boring, drilling, probing and trial pitting with the objective of measuring various geotechnical properties of soils and rocks. The most relevant of these to groundwater lowering works are permeability tests; these are discussed in Section 11.11 and Chapter 12.

11.7.4 Geophysics

Geophysical methods involve the use of non-intrusive survey methods, typically based on sensors using acoustic energy or various frequency ranges within the electromagnetic spectrum. This approach is sometimes used in investigations for civil engineering works, but these methods are much more widely used in the oil & gas, mining and water resource fields. Geophysical methods can broadly be divided into surface methods and borehole surveys, as described in later sub-sections.

On a number of occasions, when geophysical methods have been used in investigations related to groundwater control problems, the results have been perceived to be disappointing or inconclusive. This is probably a reflection on an inappropriate choice and specification of method rather than a systematic drawback with the use of geophysics. To get the most out of geophysical surveying, it is essential that engineering geophysicists are involved at an early stage of planning and thereafter; otherwise, the method will not achieve its potential.

11.7.4.1 Surface Geophysics

Surface geophysical methods involve arrays of sensors either traversed or periodically relocated across the ground surface of an area of land, or along the route of a proposed linear feature such as a road or tunnel. Commonly applied methods include:

- Seismic refraction and reflection
- Electrical resistivity
- Magnetometry
- Ground-penetrating radar (GPR)

The commonality between the methods is that a signal is transmitted from the sensor array, and the resulting response – reflected, refracted or absorbed by the strata – is interpreted to assess some aspects of ground conditions.

In general, surface geophysical methods are used to provide information on changes in particular properties of strata beneath a site, such as depth to bedrock, and can provide information between widely spaced boreholes. This can be a useful approach along the proposed route of roads, tunnels and pipelines, where there is a large spacing between boreholes. The use of boreholes in combination with geophysics is important, because the borehole data can be used to 'correlate' or 'calibrate' the geophysical results for the site in question. Geophysical methods used for civil engineering investigations are described in chapter 4 of Clayton *et al.*, (1995) and McCann *et al.* (1997).

11.7.4.2 Borehole Geophysics

Borehole geophysical methods used in groundwater investigations are largely based on methods originally developed for the oil & gas industry. Various sensors (termed 'sondes') are slowly lowered into a borehole using a cable and winch (Figure 11.3). The sondes can measure various properties of the strata immediately around the borehole (known as formation logging) and/or of the fluid within the borehole (fluid logging). One of the challenges in applying borehole geophysical methods in investigations for groundwater lowering projects is that the borehole must not be at risk of collapse at the time of the survey (otherwise, the sonde may be trapped in the bore). In permeable coarse-grained soils, where boreholes typically need to be cased or lined to prevent collapse, the attenuating effect of the casing restricts the type of sensors that can be used. However, in fractured rock, geophysical logging has proved very useful in 'open hole' sections of boreholes that are stable without casing.

The following borehole geophysical methods have been used during the investigation for groundwater control projects:

i. Cameras and optical imaging. These methods (using downhole cameras and lights) can be used to provide images of conditions in the borehole, such as location of major fractures or level of well casings and screens. Camera surveys of wells in long-term operation can also provide useful information on clogging and encrustation of wells (Section 23.5). The usefulness of this method is limited if the water in the borehole is cloudy or if visibility through the water is restricted for other reasons.

Figure 11.3 Borehole geophysical surveys equipment. (Courtesy of European Geophysical Surveys Limited, Shrewsbury, UK.) Sondes are being prepared to be lowered down a borehole. The vehicle houses the winch and data recording.

ii. Acoustic imaging. This method (sometimes known as an acoustic televiewer) produces an image of the borehole walls from an acoustic signal transmitted and received by a rotating ultrasonic sensor in the sonde. The acoustic response can allow the nature of fractures, fissures, veins, bedding planes and lithology changes to be determined. Unlike optical imaging, this method is not affected by the water clarity.

iii. Calliper logging. This method uses a sonde with expanding measurement arms to determine the diameter of the unlined sections of boreholes in rock. In combination with other geophysical logs, this can help identify significant fractures or enlarged zones within the borehole.

iv. Natural gamma logs. The sonde detects natural gamma radiation emitted by soils and rocks. Very fine-grained (clay-sized) sediments typically emit higher rates of gamma radiation, and natural gamma logs can help detect lower-permeability layers in soils and rocks that may indicate small-scale aquitard zones within thicker aquifers.

v. Borehole magnetic resonance (BMR). The sonde uses a magnetic field to cause resonance of hydrogen nuclei in the surrounding strata (Behroozmand *et al.*, 2014). In combination with other data sources, this can be useful in assessing changes of porosity and permeability with depth (Section 12.8.7.1).

vi. Fluid logging (temperature, conductivity). The sonde measures the temperature and specific conductivity of the water in the borehole. In unlined boreholes in rock, small changes in temperature and conductivity can indicate fracture zones where water is flowing into or out of the borehole under natural hydraulic gradients (Michalski, 1989). Fluid logging can also be carried out under pumped conditions where water is pumped out to induce groundwater flow into the borehole, which gives stronger temperature and conductivity contrasts to help identify hydraulically significant fractures.

vii. Flow logging. Flowmeter sondes are used to determine vertical fluid velocity within a borehole. This can help determine areas of inflow and outflow into a borehole and can be used to give estimates of permeability (Section 12.8.7.2).

Geophysical methods used in hydrogeological and water resource investigations are described by Barker (1986), Beesley (1986) and in Chapter 4 of Sterrett (2008).

11.7.5 Laboratory Testing

Samples of soil or rock obtained from boreholes or trial pits may be tested in the laboratory. The purpose of testing can be to aid soil/rock description and classification or to determine geotechnical properties for engineering design. Properties routinely tested for include particle size distribution, strength, compressibility, permeability and chemical characteristics. Permeability is particularly relevant to groundwater control projects; laboratory testing can provide estimates from permeameter testing of samples (Section 12.7) and by empirical correlation with particle size analysis of soil samples (Section 12.6.1). Comprehensive details of geotechnical laboratory testing are given in Head (2006) and Head and Epps (2011a, 2011b). Samples of groundwater recovered during investigation may also be chemically tested in the laboratory (Section 11.10.3).

11.7.6 Installation of Monitoring Wells and Piezometers

A monitoring well (also known as an observation well) is an instrument installed in the ground at a specific location to allow measurement of the groundwater level or pore water pressure. Typically, a monitoring well is installed in a borehole drilled as part of a ground investigation to provide a means of observing groundwater levels after drilling is complete. This allows groundwater level data to be gathered:

a) After groundwater levels have equilibrated to natural conditions, following the dissipation of any disturbance to groundwater caused by drilling
b) Over an extended period following the drilling phase of an investigation, to allow seasonal and other variations in groundwater level to be observed

This section deals with the design and installation of various forms of monitoring wells. Obtaining groundwater level data from monitoring wells is discussed in Sections 11.10.2 and 22.6.

The two most commonly used designs of monitoring wells used in permeable soils and rocks are standpipes and standpipe piezometers. These provide a small-diameter well, within which the water level can be measured using a dipmeter or pressure transducer linked to a datalogger. These types of monitoring wells are most commonly installed in boreholes but are occasionally used in trial pits, where the monitoring pipe is set vertically within the backfilled pit.

11.7.6.1 Standpipes

The simplest form of monitoring well is the standpipe (Figure 11.4). This consists of a small-diameter pipe, of which the bottom section (usually at least 1 m in length) is perforated or slotted, with the base of the pipe sealed with a bottom cap. The pipe is installed in the centre of a borehole, and sand or gravel is placed around the pipe, if necessary tamped into place. Backfilling should cease at a depth of about 0.5 m below ground level, and the remainder of the hole should be sealed using puddled clay or bentonite pellets and capped off with concrete to prevent surface water or rainwater entering the borehole. It is advantageous to haunch the concrete to help to shed the surface water. Unless a special protective cover is required, the pipe should project about 0.5 m above ground level and be provided with a suitable locking cap or threaded plug. In urban areas, it is essential that the cover or capping arrangement is secure enough to resist vandalism. Some designs of covers (known as 'stopcock' covers) can be installed flush with the ground surface. These covers are sometimes

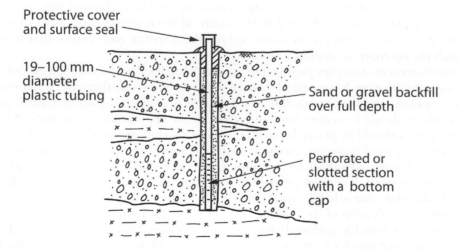

Figure 11.4 Typical standpipe installation.

preferable in vandal-prone areas because they are unobtrusive and may not attract the attention of vandals.

Plastic tubing such as uPVC or high-density polyethylene (HDPE) is an ideal material for standpipe tubing. Typically supplied in 3 m, 5 m or 6 m long pieces with threaded connections, it can readily be sawn to the desired length if necessary and joined using couplings and solvent cement. The perforated lengths of pipe are usually supplied in short lengths, such as 1.0 m or 1.5 m, and are pre-drilled or pre-slotted. If necessary, the plain pipe can be slotted on site using a hacksaw, but it should be noted that the total area of perforations should be at least twice the cross-sectional area of the standpipe. The preferred internal diameter for standpipe tubing is approximately 50 mm; this enables water samples to be taken and allows a small airline to be used to flush out the standpipe if it becomes blocked. Smaller-diameter tubing is sometimes used, but the minimum acceptable internal diameter is usually 19 mm, because this is the smallest size down which many commercial dipmeters can pass.

A standpipe is simple and cheap to install, but it is only a basic instrument. The standpipe will respond to pore water pressures in water-bearing strata along its entire depth. This is acceptable if the standpipe is used in a simple unstratified unconfined aquifer, where groundwater conditions are hydrostatic (and therefore the total head is constant with depth, so the water level will be the same in any monitoring well, irrespective of depth; see Section 3.3.2). However, if a standpipe is installed in a layered aquifer system, water can enter from more than one water-bearing layer. If the groundwater levels are different in each layer (e.g. a main water table and a perched water table), the standpipe will show a 'hybrid' water level between the two true water levels.

11.7.6.2 Standpipe Piezometers

Standpipes are not suited for use in layered or complex groundwater regimes. In such cases, it is necessary to use a 'piezometer', where the permeable section of the well is sealed into the ground so that it responds to groundwater levels and pore water pressures over a limited, defined depth only. The most common type of piezometer is the standpipe piezometer.

Figure 11.5 shows typical construction details for standpipe piezometers. The aim is to produce a 'response zone' of sand or fine gravel at the level of the stratum in which the groundwater level is to be observed. Rigid uPVC or HDPE tubing is installed in the borehole in a similar way to a standpipe, with either a short section of perforated pipe or a porous or perforated 'piezometer tip' located in the centre of the response zone. Seals of grout or bentonite pellets above (and, if necessary, below) the response zone ensure that water can only reach the tip from the desired stratum.

The installation of standpipe piezometers is more complex than for standpipes and should be carried out with care. It is essential that the seals are effective; otherwise, water may leak into the response zone from strata above or below. Where grout is used to backfill parts of the borehole, it should be cement-bentonite grout of the appropriate consistency. A layer of bentonite pellets should be placed between the grout seal and the sand filter in the response zone to avoid the sand becoming contaminated with grout (if pellets are not available, bentonite balls will have to be made up by hand). Once the lower bentonite seal is in place and has had time to swell, it is good practice to flush out the dirty water in the borehole and replace it with clean water before installing the sand filter.

A 'generic' specification of the grading of sand filters for monitoring wells for every circumstance is not possible (although Baptiste and Chapuis (2015) give some guidance in relation to ensuring that filter sand is sufficiently permeable to avoid influencing the results of in situ permeability tests in boreholes). However, in general, a filter consisting of a clean

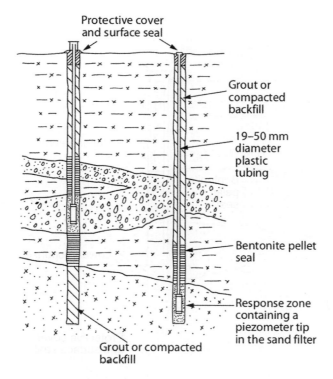

Figure 11.5 Typical standpipe piezometer installations. Two piezometers are shown, each with its response zone and piezometer tip in a different water-bearing stratum.

well-graded sand and gravel with only a small proportion of fine to medium sand is suitable for soils with some clay or silt content. For a fine sand soil, the filter should consist of coarse sand or coarse sand and gravel with not more than a few per cent medium sand. Local material may have to be used, but it is essential that the filter material is free from clay and silt. Bentonite pellets and grout seals are installed above the sand filter. The tubing should be capped off at ground level with a secure cover or headworks.

If a 'piezometer tip' is used instead of a short section of perforated pipe, this typically consists of a porous plastic element or a porous ceramic element (sometimes known as a 'casagrande element'); tips are generally 150–600 mm in length. It is good practice to soak the filter sand and ceramic element (if used) in water prior to installation – this helps avoid any air being trapped in the system and speeds up the process of equilibration between the piezometer and the natural groundwater level.

Where groundwater conditions are complex (e.g. if multiple aquifers are present), it may be necessary to install piezometer response zones at different depths to help determine whether groundwater levels vary with depth. This can be achieved in two principal ways:

a) Multiple response zones in the same borehole. It is possible to install two or more piezometer tips (each in its own response zone and separated by grout seals) in one borehole (Figure 11.6 a). If this is being contemplated, it is essential that the piezometer installations are designed and executed by experienced personnel and are supervised carefully. This is awkward work in all but the largest-diameter boreholes, and there is always the risk that the installation of the second (and subsequent) response zones and seals will affect the piezometer already installed.

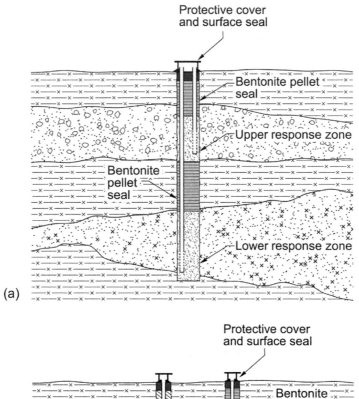

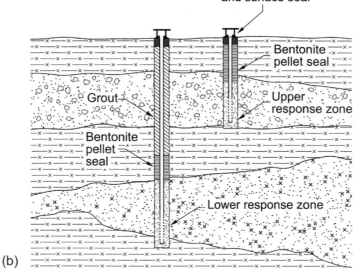

Figure 11.6 Standpipe piezometers used to monitor groundwater levels at multiple depths. (a) Multiple response zones in the same borehole. (b) An array of neighbouring piezometers with response zones at different depths.

b) An array of neighbouring piezometers with response zones at different depths. Multiple boreholes, each with a single piezometer installation, are used to provide the required range of response zone depths (Figure 11.6b). If there is any concern over the quality of workmanship of the drilling contractors and crews, this approach is often preferable, purely because the water level readings will be easier to interpret, with no worry of water leaking through poorly placed seals between response zones.

In many investigations for groundwater lowering projects, piezometers will be installed in relatively permeable water-bearing soils and rocks, where the water level inside the piezometer will respond rapidly to changes in the pore water pressure in the surrounding strata. However, piezometers in soil/rock of moderate to very low permeability may respond slowly to changes in pore water pressure. This is because a finite volume of water must flow into or out of the piezometer to register the change in pressure. This leads to a 'time lag' between changes in pore water pressure in the soil/rock and the registering of that change in the piezometer. The time lag is greater in strata of lower permeability and is greater for piezometers where larger volume flows are needed to register pressure changes.

In a standpipe piezometer, the prime factor controlling the equilibration rate is the internal diameter of the tubing; the smaller the diameter, the shorter the time lag, and the more quickly the piezometer will respond to pressure changes. In soil/rock of low to moderate permeability, it is normal to specify the internal diameter of the tubing as small as possible (19 mm is the lower practicable limit to allow monitoring by dipmeter). However, in permeable strata such as sand and gravels or highly fractured rock, the equilibration rate will tend to be rapid, and 50 mm diameter tubing can be used, allowing greater flexibility for sampling or flushing out of the piezometer.

11.7.6.3 Specialist Piezometer Installations

There may be occasions, especially in soils or rocks of low or very low permeability, when, because of the finite volume of water that must flow in order to register a change in water level, the time lag is so great that the equilibration rate of a standpipe piezometer is too slow to give useful readings. In such cases, it may be appropriate to use specialist 'rapid response' piezometers. Such instruments are characterized by the very small volume of water that must flow into or out of the sensor in order to record a change in pressure; they have been used successfully to observe pore water pressure changes in soils such as silts/clays and in unfractured rocks of very low permeability.

Prior to the 1990s, the most common forms of rapid response piezometers were hydraulic and pneumatic piezometers, where the movement of small volumes of fluid (water or air, respectively) in closed systems inside the instrument is used to balance external water pressure. Measurement of the internal pressure in the sensor system linked to the instrument allowed the external groundwater pressure to be determined.

Since the start of the twenty-first century, the vibrating wire pressure transducer (often known as a vibrating wire piezometer or VWP) has become the most widely used instrument type in rapid response piezometers. These instruments contain a metal diaphragm in hydraulic connection with the groundwater. Inside the instrument, a taut wire is stretched between the diaphragm and a stable datum. When the instrument is read, the wire is 'plucked' by passing a controlled-frequency electrical pulse along it. The taut wire resonates at a frequency related to its tension, which can be related to the deflection of the diaphragm and hence, the water pressure on the diaphragm. The VWP instruments are linked to the surface by a small-diameter data cable, which allows the instruments to be linked to dataloggers (see Section 22.4) or to specialist portable readout units. Such specialist instruments must be specified, installed and calibrated with care; instrument manufacturers can often provide useful advice.

Traditionally, rapid response instruments were installed in a borehole at a specified level in a discrete filter sand response zone in a similar way to a standpipe piezometer, but instead of an open pipe, these instruments are connected to ground level by a data cable (Figure 11.7a). This approach is effective but time consuming. An alternative approach, known as the 'fully grouted method', was developed, where the VWP instruments are installed in the borehole

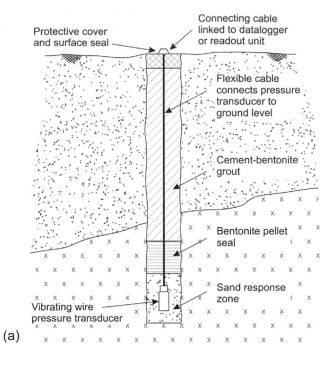

(a)

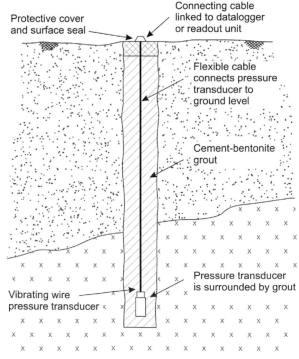

(b)

Figure 11.7 Installation of vibrating wire piezometers. (a) Installation in sand response zone. (b) Installation in fully grouted borehole. (c) Installation of multiple piezometers in fully grouted borehole.

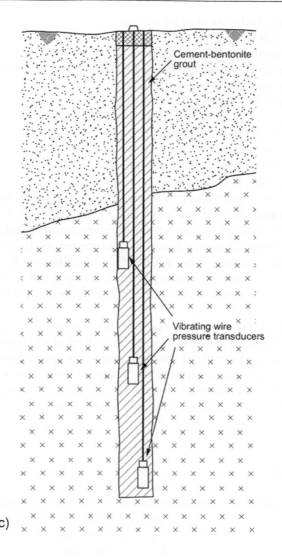

Figure 11.7 (Continued)

surrounded by cement-bentonite grout with no sand response zone (McKenna, 1995; Contreras *et al.*, 2007; Yungwirth *et al.*, 2013; Marefat *et al.*, 2019).

The fully grouted method of installation of VWP piezometers takes advantage of the fact that these instruments require only a very small volume equalization (10^{-6} to 10^{-7} m^3) to respond to water pressure changes. This means that an appropriate cement-bentonite grout is able to transmit this small volume over the short distance that separates the VWP instrument from the ground around the borehole. Studies have shown that provided the grout mix is no more than two orders of magnitude more permeable than the surrounding ground, inaccuracies in the measured pressure caused by the presence of the grout will not be significant.

Installation involves lowering the VWP instrument into the borehole, often attached to sacrificial grout pipe. The instrument is set at a specified level in the borehole (Figure 11.7b). This must be done accurately so that the pressures measured by the instrument can be related to groundwater heads (see Section 22.6.3 and Figure 22.3). The borehole is then backfilled with a carefully controlled cement-bentonite grout mix. Care should be exercised to ensure that air bubbles are not trapped in the grout around the VWP sensor – such bubbles would affect the pressure response of the instrument. To avoid air being trapped against the VWP diaphragm, the instruments are sometimes installed with the diaphragm facing vertically upwards or with the space in front of the diaphragm pre-filled with a flexible void filler. A detailed description of installation procedures used for fully grouted VWP piezometers on mining projects is given in appendix 2 of Beale and Reade (2013).

The fully grouted method also has the advantage that it can be used for the installation of multiple or 'nested' piezometers within a single borehole (Figure 11.7 c). This can be useful when there is a requirement to measure groundwater heads at different depths, such as when vertical groundwater flow is of concern.

At ground level, the data cables can be terminated with waterproof connectors (to be read later using a portable read-out unit) or are connected to a datalogger system (see Section 22.4) to allow regular readings at pre-programmed intervals. The datalogger unit can be located some distance from the piezometer itself, being connected by data cable or wireless transmission. This allows these instruments to be used where there is no permanent surface access for reading; with suitable grout seals, they have been used in boreholes beneath the open waters of rivers and lagoons, with the datalogger located on the shore.

These instruments do not record groundwater levels or total head; they record pore water pressure (above the level of the instrument). To interpret readings from the instruments correctly, the precise level at which the instrument was placed must be recorded during installation. The groundwater total head is the sum of the elevation of the instrument and the pressure head in the instrument (Figure 22.3). Determination of total head or piezometric level from pore water pressure readings is described in more detail in Sections 3.3.2 and 22.6.3.

11.8 REPORTING

To be useful to the designers and managers of the project, the site investigation must be reported in an organized, concise and intelligible manner. Traditionally, ground investigation reports were physical, hard copy documents. In modern practice, the 'deliverables' of the reporting element are produced as electronic files, formatted so that they can be printed as a conventional hard copy report if desired. However, the reports are commonly disseminated by email or web-based data sharing portals and viewed on screen.

Ideally, reporting should be carried out by geotechnical specialists who have been involved with the investigation since its inception. The minimum reporting requirement is for a 'factual report', which presents data, including the borehole logs, trial pits logs, test results and groundwater monitoring data from the ground investigation. Ideally, there should also be a separate report on the desk study and site reconnaissance, Such reports do not usually comment directly on the implications of the data gathered.

Eurocode 7 (BS EN 1997-1: 2004) sets out the requirements for a *Ground Investigation Report* (GIR), which is analogous to a factual report, perhaps with a slightly wider scope to include some basic interpretation of data. Eurocode 7 specifies that the GIR should include:

- A presentation of the available geotechnical information, including desk study information
- A factual account of all field and laboratory investigations

- A geotechnical evaluation of the information, stating the assumptions made in the interpretation of the test results
- A statement of methods adopted (citing the relevant standards)
- All relevant information on how any derived values of geotechnical parameters were determined, including any correlations used
- Any known limitations in the results

In many cases, particularly on large or complex projects, an 'interpretative report' is produced in addition to the factual report. Again, written by geotechnical specialists, this should review the ground and groundwater conditions at a site. It should include discussion of the effect of the anticipated conditions on the proposed design and construction methods. At the time the interpretative report is produced, the project design may not be finalized, but the report should discuss the geotechnical aspects of the full range of design options current at that stage. If particular potential problems in design and construction are highlighted, one of the report's conclusions may be to recommend further or specialist investigations.

Eurocode 7 (BS EN 1997-1: 2004) sets out the requirements for a *Geotechnical Design Report* (GDR), which is a form of interpretative report. Eurocode 7 specifies that the GDR should include:

- A description of the site and surroundings
- A description of the ground conditions
- A description of the proposed construction
- Design values of soil and rock properties, including justification
- Statement of the codes and standards applied
- Statements of the suitability of the site with respect to the proposed construction and the level of acceptable risks
- Geotechnical design calculations and drawings
- Recommendations for foundations and ground treatments
- Recommendations for supervision and monitoring during construction

11.9 DETERMINATION OF GROUND PROFILE

Any investigation will need to identify the nature, depth, extent and orientation of the strata beneath the site. Collectively, these parameters describe the 'ground profile', normally based on the results of boreholes, trial pits and geophysics. Determination of the ground profile is an essential part of developing the groundwater conceptual model (see Sections 5.4 and 13.4) needed to enable dewatering design.

It is essential that boreholes penetrate to adequate depth. The presence of confined aquifers or of localized zones of high permeability beneath excavations is a significant risk for many excavations. Boreholes should penetrate to a depth of 1.5 to 2 times the depth of the excavation. There have been several cases where excavations failed due to base heave (see Section 7.5 and Case History 27.6) after investigation boreholes were not taken to adequate depths but were terminated a few metres below the proposed formation level. These failures were often caused by high groundwater pressures in confined aquifers below the formation, undetected during investigation. Such problems are frustrating, because if the boreholes had been deeper and had detected the aquifer, groundwater control (for example using relief wells) would have been simple and cost-effective. As it was, major cost and time delays resulted. The only case when boreholes shallower than the recommendation can be tolerated is if the desk study clearly indicates that strata of impermeable soil/rock are present to considerable depth below formation level.

In addition to soil and rock descriptions, groundwater level information and permeability data (Section 11.11) should help identify which strata are water-bearing (and may act as aquifers) and which strata are of low permeability (and may act as aquitards and aquicludes). Compressibility test results may help identify any strata that may give rise to significant groundwater-lowering-induced settlements.

11.10 DETERMINATION OF GROUNDWATER CONDITIONS

In many investigations, the observations of groundwater conditions are totally inadequate, providing little concrete information on groundwater levels. Accurate knowledge of the likely range of groundwater levels and pore water pressures in the various strata is essential for the design of dewatering systems. This section will describe the types of groundwater observations that may be taken during investigation and will discuss their various merits and limitations. The installation of monitoring wells and piezometers is discussed in Section 11.7.6.

Table 11.2 summarizes the advantages and disadvantages of methods of determining groundwater levels. It is also useful to note that some of the observations of groundwater level inflows from trial pits and boreholes can be useful 'non-quantitative' indicators of relative permeability of the strata encountered. This is discussed further in Section 12.5.

11.10.1 Groundwater Level Observations in Trial Pits and Boreholes

The easiest and most common form of groundwater level observations is those taken in trial pits and boreholes. Unfortunately, there are two important limitations to the accuracy of readings obtained in this way:

(i) The natural groundwater inflows and levels may be masked or hidden by the excavation or boring/drilling method, particularly if water is added or removed from the pit or borehole, or if inflows of water are sealed off by drilling casing.

(ii) For a pit or borehole to show a representative groundwater level, sufficient water must flow into the pit or borehole to fill it up to the natural groundwater level. In soil/rock of moderate or low permeability, it can take a long time for the water level in the pit or borehole to come into equilibrium with the natural groundwater level. Most observations do not allow sufficient time for equilibration and so may report unrepresentative water levels.

11.10.1.1 Groundwater Levels Observed in Trial Pits

Trial pits offer a simple way to observe shallow groundwater conditions. The large size of the pit allows direct visual observation of inflows and seepages (it may be possible to categorize these as 'slow', 'medium' or 'fast' seepages on a subjective basis). The location of the seepages in relation to the soil fabric and layering in soil, or the fracture network in rock, can provide information on the relative permeability of various strata – this can be a

Table 11.2 Advantages and Disadvantages of Methods of Determining Groundwater Levels

Method	Advantages	Disadvantages
Observations in trial pits	Commonly carried out Allows seepage into excavation to be observed directly May allow perched water tables to be identified	Limited to shallow depth Standing water levels may be unrepresentative unless pit is left open long enough for equilibration to occur
Observations during boring	Commonly carried out Normally sufficient to identify water inflows from major water-bearing strata	True groundwater seepages and levels may be masked by drilling action and by addition and removal of water Seepages in soils of low to moderate permeability may not be identified Often, insufficient time is allowed for water levels to equilibrate to natural levels
Observations during rotary drilling	Commonly carried out	Groundwater seepages and levels tend to be masked by the presence of flush medium (when water-based drilling fluids are used) or by removal of water during drilling (when air-based drilling fluids are used)
Monitoring wells – standpipes	Cheap and simple to install in boreholes Useful in simple unconfined aquifers Allow monitoring of groundwater levels in the long term after boring is completed	Not appropriate for sites with confined aquifers, multiple aquifers or perched water tables May need to be purged or developed before use May need to be protected from vandalism and damage
Monitoring wells – standpipe piezometers	Relatively straightforward to install in boreholes Allow groundwater levels in specific strata to be observed Can be used on sites with complex groundwater conditions Allow monitoring of groundwater levels in the long term after boring is completed	Closer supervision of installation is needed than with standpipes May need to be purged or developed before use May need to be protected from vandalism and damage
Specialist instruments – pneumatic and electronic piezometers	Can allow accurate pore water pressure measurements in low- and very low-permeability soils Can be read remotely Allow monitoring of groundwater levels in the long term after boring is completed	Installation and calibration in boreholes can be complex; close supervision is advisable Readings may not be straightforward to interpret May need to be protected from vandalism and damage

useful way of identifying perched water tables or localized zones of higher permeability, and of observing any groundwater-related instability that may affect shallow excavations. The disadvantage of trial pits is that because of their relatively large size, a significant volume of water has to enter the pit to fill the pit up to the natural groundwater level. This process of equilibration may take several days in strata of moderate and low permeability. The pit could be left open and monitored daily, but for safety reasons, trial pits are rarely left open for long periods of time. In general, groundwater levels in trial pits may not be representative of natural groundwater levels.

11.10.1.2 Groundwater Levels Observed during Cable Percussion Boring

In British site investigation practice, a common type of groundwater observation is recorded by the drilling foreman during light cable percussion boring. Each time groundwater is encountered, these records should comprise:

a) The depth at which water is first encountered (known as a 'water strike')
b) A description of the speed of inflow (e.g. slow, medium or fast). Boring is normally suspended following a water strike and groundwater levels observed to record the rise (if any). Ideally, monitoring should continue until the water level in the borehole stabilizes, but often the rise in water level is recorded for a fixed period (typically 20 minutes) only before boring recommences.
c) The depth at which the groundwater inflow is sealed off by the temporary drilling casings.

Additionally, the drilling foreman should record whether water was added to the borehole during drilling and the water level in the borehole at the start and end of each drilling shift (together with the corresponding depth of borehole and casing at that time). All groundwater details are recorded on the driller's daily record sheet and should appear on the final borehole log, which will ultimately form part of the geotechnical factual report.

When reviewing water levels recorded during cable percussion boring, the following points must be noted:

a) In all but the most permeable strata, observing the water level rise for a short period following a water strike may not allow sufficient time for the natural groundwater level to be apparent. In many cases, the water level recorded following a water strike will be lower than the actual groundwater level in that stratum.
b) A significant rise in borehole water level following a water strike may indicate the presence of a confined aquifer, particularly if the water strike occurred at the approximate depth where the borehole passed from a low-permeability stratum into a high-permeability one. However, smaller rises in water levels are sometimes observed, even in unconfined aquifers. This occurs when the driller has drilled a short distance below the water table before noticing the inflow. The drilling action will have removed some water from the borehole, and when drilling stops, the water level will rise up to the natural level.
c) Because of the speed of boring, when drilling through soils of very low to moderate permeability (such as clays, silts and silty sands), a water strike may not be noticed at all by the driller. The spoil from the borehole will be damp or moist, but there will not be time for free water to enter the borehole. There have been cases when groundwater inflow was not recorded in investigation borings through strata of silty sand, yet subsequent excavation for the construction works encountered groundwater and needed significant dewatering. This error could have been avoided by installing monitoring wells, which allow time for equilibration of water levels.
d) The boring process will inevitably alter the water level in the borehole, so it is not representative of that in the surrounding soil. The action of the drilling tool will remove water, and the driller may be deliberately adding water as part of the drilling process. The water level in the borehole at the end of the shift is likely to be very unreliable, being highly influenced by the recent drilling activities. If drilling work is done on a day shift basis, the water level at the start of shift (next morning) may be more reliable, as the borehole water level will have begun to equilibrate overnight. Even so, the start of shift level may still be unrepresentative, especially in low-permeability strata, where longer periods may be necessary for full equilibration.

11.10.1.3 Groundwater Levels Observed during Rotary Drilling

Groundwater level observations during rotary drilling are generally unsatisfactory, because during drilling:

- If water-based drilling fluids are used, the borehole is kept topped up with water, which will mask any inflows and may also raise groundwater levels around the borehole.
- If air-based drilling fluids are used, water is continually blown out of the borehole. By observing the quantity of water blown from the borehole, it may be possible to identify significant water inflows, but it may be more difficult to spot small inflows. Furthermore, the continued removal of water from the drill hole may lower groundwater levels around the borehole.

As with boring methods, the water levels at the start of a shift tend to be more reliable than those observed at the end of the shift.

11.10.2 Groundwater Level Observations in Monitoring Wells and Piezometers

The principal drawback that affects all groundwater observations in trial pits and boreholes is that at best, they only give 'spot' readings of conditions at the time of investigation and may be affected by drilling and other activities. Furthermore, as discussed in Section 3.9, in general, groundwater levels are not constant but will vary with the seasons and in response to long-term trends (such as drought) or external influences. Observations in trial pits and boreholes cannot give information on potential variations, and (depending on the time of year when they were drilled) may not indicate the highest groundwater level that can occur at a site. This type of information can only be obtained via monitoring wells of some sort.

A defining feature of monitoring wells is that (provided they are adequately protected from damage or vandalism) they can be used to observe groundwater levels long after the main ground investigation is complete. This can allow natural changes in groundwater levels to be determined. However, this requires the instruments to be monitored for extended periods, and the practicalities of this are sometimes overlooked. If readings are to be taken manually, this will have to be included in the site investigation plan and associated costings. In remote or inaccessible sites, it may be appropriate to use datalogging systems (see Section 22.4) to record groundwater levels in order to reduce the cost associated with regular visits by personnel. The multiple readings stored in the datalogger systems can be accessed by personnel visiting the site to connect to the datalogger and download the data. Alternatively, many datalogging systems can be accessed remotely via mobile data networks – this can allow the groundwater level data to be downloaded without the need for a site visit.

A monitoring well is an instrument installed in the ground at a specific location to allow measurement of the groundwater level or pore water pressure. For determining groundwater levels, monitoring wells have a number of advantages over observations during drilling:

1. Because they are long-term installations, appropriately designed monitoring wells can allow observation of equilibrium groundwater levels, even in very low-permeability soils or rocks.
2. Monitoring wells can be installed as piezometers to record water levels or pore water pressures in a specific stratum or at defined depths.
3. Monitoring wells can be monitored for long periods of time to observe seasonal, tidal or long-term variations in groundwater level at a site.

Monitoring of groundwater levels is described in detail in Section 22.6. In summary, for the different designs of monitoring wells described in Section 11.7.6, monitoring methods used include:

- Manual water level readings in standpipes and standpipe piezometers using a 'water level indicator', commonly known as a dipmeter (Figure 22.1). The dipmeter consists of a graduated cable or tape with a probe at its tip. The probe is lowered down the well until it touches the water surface; the water completes a low-voltage circuit, and a buzzer or light is activated on the dipmeter reel.
- Automated readings in standpipes and standpipe piezometers using
  a) VWP water pressure sensors connected to datalogging systems (Figure 22.2). The sensor records the water pressure above its own level; thus, an on-site calibration is required to allow pressure readings to be converted to water levels.
  b) Combined fully submersible pressure transducer and datalogger for a single well. This is housed in a compact waterproof cylindrical casing (around 30 mm in diameter) designed to be placed below water level with no surface equipment. The unit is suspended in the well by a support wire. An on-site calibration is required to allow pressure readings to be converted to groundwater levels

  Pressure sensors should be installed and calibrated in accordance with the manufacturer's guidelines. Methods of converting water pressures to groundwater levels are described in Section 22.6.3 and Figure 22.3.
- Automated readings in rapid response piezometers, including VWP instruments, connected to datalogging systems. Typically, these instruments record the water pressure above the sensor; thus, an on-site calibration is required to allow pressure readings to be converted to water levels (Section 22.6.3 and Figure 22.3). The sensors and datalogging equipment should be installed and calibrated in accordance with the manufacturer's guidelines.

11.10.3 Groundwater Sampling and Testing

Knowledge of groundwater chemistry may be required for a variety of reasons (see Section 3.7). In particular, groundwater chemistry can influence the risk of corrosion and encrustation of equipment (Section 23.5) or how discharge water can be disposed of (Section 25.4), or it may be relevant to some potential side effects of dewatering (Chapter 21). Groundwater samples should be obtained and tested as part of the site investigation.

Where groundwater is encountered during cable percussion boring, obtaining a sample is relatively straightforward. The water sample should be taken using a clean sampling bailer as soon as possible after the seepage has begun. If water has been added during boring, this should be bailed out prior to taking the sample. It is not normal practice to take groundwater samples during rotary drilling, because the drilling fluids will tend to mix with any groundwater entering the bore; water sampled from the borehole is unlikely to be representative of groundwater.

Groundwater samples can also be taken from monitoring wells of the standpipe or standpipe piezometer type by use of a bailer or sampling pump. The water standing in the monitoring well has been exposed to the atmosphere and is unlikely to represent the true aquifer water chemistry. Therefore, it is vital to fully 'purge' the well before taking a sample.

Conventional purging methods involve pumping or bailing water from the monitoring well at a fairly steady rate until at least three 'well volumes' of water have been removed (a 'well volume' is the volume of water originally contained inside the well liner). Specialist sampling pumps should be used in preference to airlifting, since the latter method may aerate the sample, increasing the risk of oxidation of trace metals and other substances. Obtaining a water sample during a pumping test is described in Appendix 4.

Low flow purging (also known as micro purging) is sometimes used as an alternative to conventional purging. In this approach, a small sampling pump is suspended in the well opposite a specific zone in the well screen and adjusted to a very low flow rate – the objective is to generate only a very small drawdown of groundwater level: 0.1 to 0.2 m is typical. The pumped water is monitored by a portable water quality meter, and well head chemistry parameters are measured (see Section 3.7.2). Commonly measured water quality parameters include water temperature, pH, specific conductivity (EC) and dissolved oxygen (DO). The objective is to purge the well until these parameters have stabilized. Once a minimum volume has been pumped (typically the volume of sampling tubing, pump and water monitoring cell), purging is continued until the water quality indicator parameters have stabilized, and the sample is then taken. This method typically generates significantly smaller volumes of purge water compared with conventional methods.

In general, water samples should be at least 1 litre in volume. The bottles used for sampling should be clean with a good seal and should be filled to the brim with water to avoid air bubbles in the sample. The sample bottle should be washed out three times (with the water being sampled) before taking the final sample.

Groundwater samples may degrade following sampling and should be tested as soon as possible after they are taken. Ideally, they should be refrigerated in the meantime. However, even then, the sample may degrade in the bottle (for example by trace metals oxidizing and precipitating out of solution). Specialists may be able to advise on the addition of suitable preservatives to prevent this from occurring. The choice of sample bottle (glass or plastic) should also be discussed with the laboratory, since some test results can be influenced by the material of the sample bottle. Above all, it is vital to use an accredited, experienced laboratory for the chemical testing of water samples.

II.II DETERMINATION OF PERMEABILITY

Permeability (also known as hydraulic conductivity) is an essential parameter to be determined for the design of groundwater lowering systems (see Chapter 4). Even if groundwater lowering is not planned, permeability is relevant to the geotechnical design of all engineering projects that involve below-ground excavation. Site investigations for such projects should include some methods to assess the permeability of the relevant strata. A very wide range of methods are available, most of which have various advantages and limitations, which vary depending on the geology and hydrogeology of the site in question. The selection and application of methods to determine permeability are discussed in some detail in Chapter 12.

Chapter 12

Assessment of Permeability

> I often say that when you can measure what you are speaking about, and express it in numbers, you know something about it; but when you cannot measure it, when you cannot express it in numbers, your knowledge is of a meagre and unsatisfactory kind.
>
> William Thomson, Lord Kelvin, *Lecture on 'Electrical Units of Measurement'*, 1883

12.1 INTRODUCTION

Permeability (also known as hydraulic conductivity) is an important parameter for groundwater control projects. As discussed in Chapter 4, the permeability of soils and rocks can vary over a huge range, from 10^{-1} to 10^{-12} m/s – a difference of more than 10 orders of magnitude. The successful design of groundwater control projects is highly dependent on obtaining realistic estimates of the permeability of the various soil and rock strata present beneath a site. However, permeability is difficult to assess by in situ or laboratory tests, with many methods having significant limitations. Furthermore, different methods assess permeability at various scales, and values from different sources may not be directly comparable. This chapter presents commonly used methods of permeability testing, their range of applicability, and their relative merits and limitations.

12.2 THE ROLE OF PERMEABILITY IN GROUNDWATER CONTROL PROBLEMS

In geotechnical practice, permeability is a widely used parameter, which represents the ease or otherwise with which water passes through a volume of soil or rock. Geotechnical engineers typically use and report permeability in metres per second – this allows permeability values to be used directly in calculations using SI units. Hydrogeologists often report permeability in metres per day, but these values are not SI units and will give errors if not combined correctly with other units in calculations – conversion factors between different permeability units are included at the end of the book.

Permeability is a key indicator of potential groundwater problems in soils and rocks below groundwater level:

i. Strata of relatively high permeability (e.g. coarse-grained soils such as sands and gravels, or highly fractured rocks) allow groundwater to pass freely and are considered 'water-bearing'; such strata might be referred to as aquifers (Section 3.4). Excavations in such materials have the potential for significant groundwater inflows and high pumping rates if dewatering measures are deployed.

ii. Strata of intermediate to low permeability (e.g. clayey silts or mudstone bedrock with limited fracture networks) allow some limited groundwater flow under hydraulic gradients; such strata might be referred to as aquitards (Section 3.4.4). Excavations in these conditions typically do not experience large rates of groundwater inflow, but the ground is sufficiently permeable for groundwater-induced instability to occur once undrained conditions have dissipated (Section 6.6).

iii. Strata of low to very low permeability (e.g. very fine-grained soils such as silty clay that does not contain any permeable fabric, or unfractured igneous bedrock) allow only very small rates of groundwater flow under hydraulic gradients; such strata might be referred to as aquicludes (Section 3.4.3). Excavations in these conditions do not experience any significant groundwater inflow and typically do not experience groundwater-induced instability, as undrained conditions prevail around the excavation (Section 6.6).

While there might be some discussion as to the values of permeability at the boundaries between the three cases outlined here, there is no dispute as to the importance of permeability to groundwater control problems. Permeability is key to the selection of groundwater exclusion and groundwater pumping methods, as shown in Figures 9.10 and 9.11, respectively.

12.3 SOME OF THE DIFFICULTIES OF ASSESSING PERMEABILITY

Although a wide range of permeability assessment methods are available, obtaining realistic values of permeability is far from straightforward. The key problem is that, as was described in Chapter 4, soils and rocks are not homogeneous isotropic masses. Permeability is likely to vary from place to place and to vary for different directions of measurement. Further potential problems include:

a) Tests and samples may be in the wrong stratum, or in a zone of soil or rock where the permeability is not representative of the wider mass. This could be due to limitations with investigation methods or access (e.g. not being able to drill to a suitable depth/location) or could be due to an imperfect conceptual model, which does not identify the location and nature of key zones of the strata in question (e.g. testing in the wrong place or using inappropriate test methods). In either case, the permeability values may paint a misleading picture of actual conditions.

b) The act of drilling a borehole can disturb the soil/rock and affect the permeability values determined from in situ tests.

c) Samples taken from boreholes for testing in the laboratory (including particle size testing) may be affected by the drilling and sampling process and thus be unrepresentative of the ground being sampled.

d) Permeability is not measured directly. In reality, observable parameters (such as water levels or flow rates) are measured directly, and permeability is then calculated or interpreted. This introduces two types of potential error – errors in measurement of the original observation and errors in calculation of permeability, especially if a method of analysis is used that is not appropriate to the test or sample conditions.

e) Measured values of permeability will vary with the scale of the test – in effect, with the size of the sample or the volume of soil or rock within which water flow is induced. The effect of scale on assessed permeability values is discussed in detail in Section 4.6.1 and illustrated in Figure 12.1. Dewatering and groundwater control systems tend

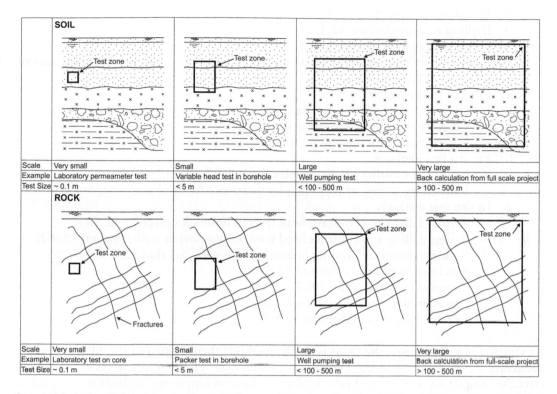

Figure 12.1 Scale effects on measurement of permeability in soil and rock. (From Preene, M and Powrie, W, *Proceedings of the XVII ECSMGE, Geotechnical Engineering: Foundation of the Future*, Reykjavik, September 2019. With permission.)

to interact with very significant volumes of the ground, and the most useful permeability estimates are large and very large-scale. Unfortunately, such data are often not available. In this chapter, the scale of different permeability assessment techniques is discussed for the various methods.

Overall, it is useful to remember the statement made by Preene *et al.* (2016): 'Even if it could be obtained, there is no single value of permeability in the ground waiting to be measured.'

If it is accepted that it can be difficult to determine meaningful values of permeability, that should not deter those involved in site investigation and dewatering design from putting their best efforts into obtaining the most useful values practicable. The following sections outline the characteristics and limitations of commonly used methods of determining permeability in geotechnical practice. The final section in this chapter discusses the relative reliability of the various techniques.

12.4 METHODS OF ASSESSING PERMEABILITY

A range of methods are available to assess permeability and can be grouped as:

1. Non-quantitative assessment methods, which do not produce numerical values of permeability but can be useful to validate (or indeed, refute) data from quantitative sources. These include:

- Visual assessment (Section 12.5.1)
- Drilling records (Section 12.5.2)
- Borehole geophysics (Section 12.5.3)

2. Quantitative methods, whereby tests or correlations are used to produce estimates of permeability. These include:
   a. Empirical assessment
      - Correlations with particle size distribution (PSD) of soil (Section 12.6.1)
      - Correlations with descriptions of rock quality (Section 12.6.2)
      - Inverse numerical modelling (Section 12.6.3)
   b. Ex situ test methods
      - Laboratory testing of soil samples (Section 12.7.1)
      - Laboratory testing of rock samples (Section 12.7.2).
   c. In situ test methods
      - Rising, falling and constant head tests in boreholes (Section 12.8.1)
      - Rising, falling and constant head tests in monitoring wells (Section 12.8.2)
      - Packer permeability testing in boreholes in rock (Section 12.8.3)
      - Specialist in situ tests (Section 12.8.4)
      - Pumping tests (Section 12.8.5)
      - Groundwater control trials (Section 12.8.6)
      - Borehole geophysics (Section 12.8.7)

As this is a practical book, the focus is on the methods themselves and the advantages and disadvantages of each approach. Theoretical background is limited to that necessary to discuss the limitations of each method – further background on some methods is provided in the Appendices. At the end of the chapter, Section 12.9 compares the different methods for assessing permeability.

12.5 ASSESSING PERMEABILITY BY NON-QUANTITATIVE METHODS

The investigation and design of groundwater control systems are carried out by professionals from a range of backgrounds, including civil engineering, geotechnical engineering, hydrogeology and engineering geology. While this represents a relatively diverse set of training and skills, these are all numerate specialisms. While we cannot claim that everybody loves 'numbers', most designers will be comfortable with handling them. This perhaps explains a common shortcoming when assessing available permeability data – there is often an undue focus on quantitative methods, those that produce numerical 'values' of permeability from tests. As discussed earlier, many test methods have limitations and may give different scales of permeability values. There have been many cases where the reported permeability values from tests at investigation stage have differed by several orders of magnitude from values back-calculated from full-scale dewatering pumping. Often, with the benefit of hindsight, there were non-quantitative indicators of potential problems with reported permeability values.

The authors believe strongly that the permeability datasets developed at the investigation stage should be validated against non-quantitative data sources such as those outlined in the following sub-sections. If there are significant discrepancies between the quantitative permeability values and the corresponding non-quantitative data, then this should be investigated, including a review of the relevant conceptual models.

12.5.1 Visual Assessment

A simple validation that is often overlooked is to assess permeability values against published ranges based on visual descriptions of strata. Every site investigation carried out to modern standards will contain, as part of the borehole logs or trial pit logs, detailed descriptions of the soils or rocks encountered. The description is normally carried out in accordance with closely defined methodologies set out in geotechnical standards. In the United Kingdom, the protocols for description of soils and rocks are described in BS EN 14688-1: 2002 and BS EN 14689-1: 2003, respectively. Similar protocols exist in the geotechnical standards applicable in other countries.

The purpose of requiring soils and rocks to be described to consistent standards is to allow an experienced engineer or geologist, reading a borehole log, to be able to glean information on the physical nature of the soil or rock encountered in the borehole. Therefore, the description of soil or rock contains a lot of information, from which it may be possible to infer approximate values for some physical properties of the material, including permeability.

Visual assessment of permeability of a soil sample is the process of assessing the soil type or PSD and, based on experience or published values (such as Table 4.1), estimating a very approximate range of permeability. This method is essential to allow corroboration of permeability test results. On *every* project, the soil descriptions from borehole or trial pit logs should be reviewed to give a crude permeability range, against which later test results can be judged. Information gathered by the desk study, such as experience from nearby projects, can be useful in this regard.

Permeability values from such sources must necessarily be generic and approximate, but experience shows that there is a strong correlation between permeability and the nature (including stress state and weathering) of the soil or rock in question. Such correlations can be used to exclude unrepresentative test results from the permeability dataset.

For example, a coarse-grained soil described as a medium sand might typically be expected to have a permeability of the order of 10^{-4} m/s; certainly, such a soil would be unlikely to have a permeability greater than 10^{-3} m/s or less than 10^{-5} m/s. If variable head permeability tests in borehole give results of 10^{-8} m/s, there is clearly some discrepancy. Either the soil description is misleading or the test results are in error or unrepresentative (or both). If this is recognized while testing is still going on, there may be a chance to modify test types or procedures to get better results. Discrepancies of this magnitude are not rare, and visual assessment of permeability can often be a more useful guide to permeability than test results, especially if the latter are limited in scope and questionable in quality.

12.5.2 Drilling Records

A wide range of data are collected during drilling and boring, both recorded manually by the drilling crews and also, where used, from instrumented drilling equipment (an approach sometimes known as measurement while drilling [MWD]). Some of this information (Table 12.1) can help to identify the presence of localized high-permeability zones at specific depths in boreholes, which may not be apparent from small-scale tests or from large-scale tests that report average permeability (e.g. a well pumping test).

For example, packer permeability tests in bedrock comprising fractured mudstone might give permeability values in the range from 10^{-6} to 10^{-7} m/s. This would imply that excavations would have relatively modest rates of groundwater inflow. However, packer permeability tests give an average permeability value over the length of the test section isolated

Table 12.1 Potential Non-Quantitative Indicators of Permeability from Drilling Records

Data from drilling records	Notes
Water strike (inflow) records from boreholes	For drilling methods (such as cable percussion boring, or rotary drilling with air flush) where added drilling fluids do not mask water inflows, records of water inflows, or changes to inflow rates, can identify zones of relatively high permeability within the borehole column.
Drilling fluid (flush) loss records, voids or tool drops in boreholes	Zones where loss of drilling fluid is noted, or where voids are identified (including when drilling tools drop suddenly), can indicate highly permeable zones.
Water levels at start and end of shift	For drilling methods that remove water (cable percussion and rotary air flush), changes in water level between start and end of shift can indicate water inflow/outflow.

between packers (see Section 12.8.3). If there are indicators of strong localized inflow, this could suggest that water flow is concentrated in a small number of fractures and that the average values from the packer test do not represent the maximum permeabilities that might be encountered locally.

12.5.3 Geophysics

12.5.3.1 Surface Geophysics

Surface geophysics was introduced in Section 11.7.4.1. Surface geophysical methods include seismic refraction methods, gravity surveying, electromagnetic surveying and resistivity soundings (Barker, 1986). These methods measure the variation in specific physical properties of the sub-surface environment and apply theoretical and empirical correlations to infer the structure of the ground and groundwater regime. Typical applications include mapping the extent of gravel deposits within extensive clay strata or locating buried channel features within drift deposits overlying bedrock (Macdonald *et al.*, 1999).

12.5.3.2 Borehole Geophysics

Borehole geophysics was introduced in Section 11.7.4.2. Downhole geophysical methods involve surveying previously constructed boreholes by slowly lowering various sensing devices (known as 'sondes') into the borehole using a winch and cable (Figure 11.3). Some types of sonde investigate the properties of the ground around the borehole (this is known as formation logging), while others measure the properties of the water in the borehole (fluid logging). The various methods available are described in Beesley (1986), chapter 4 of Sterrett (2008) and BS 7022:1988. These methods are generally applicable in boreholes drilled into rock, when more permeable fractured zones can sometimes be identified.

Some borehole geophysical methods, including flowmeter logging and borehole magnetic resonance (BMR) methods, can give quantitative values of permeability – see Section 12.8.7. Non-quantitative information on permeability variations with depth can be obtained for unlined boreholes using fluid logging to record temperature and specific conductivity (Michalski, 1989). The sondes are lowered into the borehole and provide a continuous record of fluid temperature and conductivity along the entire depth of a borehole. When applied in unlined sections of boreholes in rock, changes in gradient of the plot of these parameters can indicate localized inflow zones associated with fractures or groups of fractures. In some cases, fluid logs are carried out while the borehole is pumped; this will induce flow into the borehole and give stronger temperature and conductivity contrasts at the permeable fractures.

Borehole geophysics is a specialized technique, and results are sensitive to background hydrogeological conditions and measurement techniques used. Such investigations should be designed, executed and interpreted by suitably experienced specialists.

12.6 QUANTITATIVE ASSESSMENT OF PERMEABILITY – EMPIRICAL METHODS

There is a long history of permeability being assessed from empirical correlations with certain properties of soil or rock. Empirical correlations are not derived from theory or analysis of the physical laws of a problem; rather, they are based on observation and previous experience (Section 5.5.1 discusses empirical methods in relation to groundwater models).

Empirical methods used to derive permeability for groundwater control projects include:

   (i) Correlations with PSD of soil (Section 12.6.1)
 (ii) Correlations with descriptions of rock quality (Section 12.6.2)
(iii) Inverse numerical modelling (Section 12.6.3)

A fundamental aspect of the application of empirical methods is that the correlations are based on a finite underlying dataset and/or series of assumptions and can only be used with confidence within the limits of the observations upon which they are based. A key risk is that these methods may be applied (knowingly or unknowingly) in cases where the correlations are not valid. To avoid this, it is important that the designer has a good understanding of the basis of the empirical relationship and its limitations. These aspects are highlighted for the various methods in the following sub-sections.

12.6.1 Correlations with Particle Size Distribution (PSD) of Soil

The permeability of a soil is largely controlled by the nature of the soil pores (the viscosity of water, which varies with temperature, also is a factor, but for typical groundwater temperature ranges, this effect is small compared with the soil type). When trying to determine permeability, an attractive approach is to develop correlations between permeability and the properties of soil pores. However, in routine geotechnical analyses, it is very difficult to measure the size and properties of the soil pores. Luckily, it is well established that the properties of the pore space are strongly influenced by the size, shape, roughness and other properties of the soil particles, so the next logical step is to attempt to relate permeability to the PSD as well as other properties of soil. This allows groundwater control designers to make use of the PSD analyses routinely carried out as part of site investigations.

Empirical correlations between PSD and permeability have been in use for more than 100 years. Several different correlations are available, each of which has different ranges of application. In the following sections, the concepts and limitations of empirical correlations are discussed, and three commonly applied methods are presented:

   (i) Hazen's method (Section 12.6.1.2)
 (ii) Kozeny–Carman method (Section 12.6.1.3)
(iii) Prugh method (Section 12.6.1.4)

In the main text that follows, each method is briefly described and presented in formats that can be enumerated directly. Further details on the Hazen and Kozeny–Carman correlations are given in Appendix 1.

12.6.1.1 Background to Particle Size Correlations

Bricker and Bloomfield (2014) summarize a generic form of correlation between PSD and permeability as

$$k = \left(\frac{\gamma_w}{\mu_w} \right) C f(n) D_e^2 \tag{12.1}$$

where
  γ_w is the unit weight of water
  μ_w is the dynamic viscosity of the fluid (water in this case)
  C is a sorting coefficient
  n is porosity (which is the ratio of voids to the total volume of soil – see Section 4.3)
  $f(n)$ is a porosity function
  D_e is effective particle size

As discussed in Section 4.3, the finer fraction of particles has a disproportionate effect on the permeability of a soil mass, and in these correlations, the effective particle size is often taken as D_{10} or D_{15}.

A variety of correlations have been published. Hazen's method (Hazen, 1892, 1900, 1911) is one of the oldest and perhaps the best known, but there are many others, such as Slichter, Terzaghi, Rose (all reported in Loudon, 1952), Masch and Denny (reported in Trenter, 1999), and Kozeny–Carman (reported in Carrier, 2003). Some methods, such as Hazen's formula, conflate multiple aspects of the correlation to produce a simpler equation requiring only limited input data. Other methods, such as the Kozeny–Carman formula, apply the separate aspects of Equation 12.1 and therefore require estimates of porosity and other factors.

It is clear that different PSD correlations will produce different estimates of permeability for a given sample. Furthermore, the process of sampling and PSD testing may render the samples unrepresentative of in situ conditions, introducing further error into the permeability estimates.

A key element of the raw data for these correlations is the PSD curve of the soil sample (Figure 12.2) – methods used to determine a PSD curve are described briefly in Appendix 1. The PSD curve represents the cumulative percentage (by weight) of soil mass retained on each sieve size and allows estimation of particle size parameters such as D_{10}, D_{15}, D_{50}, D_{60} (and so on), which are used in later design methods, where D_n is the sieve aperture through which n per cent of a soil sample will pass. D_n is determined by the intercept between the PSD curve and the horizontal line of n per cent passing (Figure 12.2 shows the D_{10} intercept as an example).

Irrespective of the particular correlation used, these approaches all use data from samples recovered from boreholes (and occasionally from trial pits and other sources) rather than tested in situ. As a result, this approach has several limitations.

The sampling process may mean that the material obtained for particle size testing is unrepresentative.

(i) If bulk or disturbed samples are recovered from below the water level in a borehole, there is a risk that finer particles will be washed from the sample as it is removed from the borehole. This is known as 'loss of fines'. Samples affected in this way will tend to

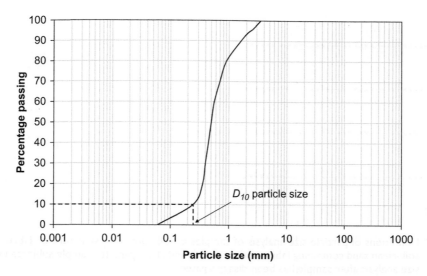

Figure 12.2 Particle size distribution (PSD) curve for a soil sample. D_{10} particle size is shown.

give over-estimates of permeability. Loss of fines is particularly prevalent in samples taken from the drilling tools during light cable percussion boring. This can be mini-mized by placing the whole contents (water and soil) of the tool into a tank or tray and allowing the fines to settle before decanting clean water. Unfortunately, in practice, this is rarely done. Loss of fines is usually less severe for tube samples or core samples; these methods may give more representative samples in fine sands.

(ii) There is a risk that fine particles from the drilling fluid (either from drilling addi-tives or from silty material suspended in water in the borehole) may contaminate the external faces of tube or core samples in coarse-grained soil. If these fine particles are included in the particle size analysis, the results of correlations will tend to under-estimate permeability. It may be possible to avoid these problems if there is sufficient material in the sample to allow the material from the edges to be discarded. Samples from rota-sonic drilling are less prone to this problem provided that the borehole can be drilled without addition of water or drilling fluids.

Before any sample is subjected to PSD analysis, it will have undergone pre-treatment. The pre-treatment and PSD analysis process is described in more detail in Appendix 1, but essentially, the sample will be disaggregated (and, if necessary, washed to separate fine and coarse frac-tions of soil) before analysis. The net effect is that any soil structure, fabric or cementation present in the in situ soil will be destroyed. The PSD analysis is carried out on an effectively homogenized sample, and the results may not represent the material as present in the ground. Consider the stratum shown in Figure 12.3a, which in situ comprises a clean sand deposit containing some laminations of silt and silty clay; the horizontal permeability will be domi-nated by the properties of the sand, and vertical permeability by the silt/clay. During the PSD analysis process, these two materials will become mixed into the mass of the sample during preparation, the PSD curve will indicate a clayey or silty sand (Figure 12.3b), and permeabil-ity correlations may give unrepresentative results. For the case shown in Figure 12.3, where the soil is anisotropic, the permeability estimate derived from PSD correlations will probably more closely represent the vertical permeability k_v than the horizontal permeability k_h.

Because of the sampling and PSD analysis process, the material as tested will bear little relation to the stress state, and hence the porosity, of the in situ soil. In all the discussions

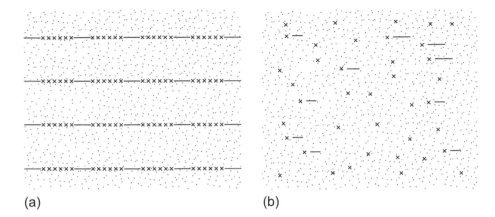

(a)   (b)

Figure 12.3 Limitations of particle size analysis of samples in soil containing structure or fabric. (a) In situ soil: clean sand containing fabric of discrete silt and clay layers. (b) Sample subjected to particle size analysis after sample has been disaggregated.

of particle size analysis, it is easy to forget that permeability is controlled by the size and nature of the soil pores, and particle size is being used as a simple (and imperfect) surrogate. While some correlations do include factors that represent stress state or porosity, the necessary input for such terms are often uncertain (for example, porosity, required for the Kozeny–Carman method, is rarely measured in the field). There are methods to estimate porosity and related parameters from other data, but this introduces further potential errors in the assessed permeability.

It is also important to consider the scale of the permeability values provided by this method. PSD analysis is typically carried out on only relatively small samples (100 to 500 g in mass) with the physical dimension of the sampled zone being some few hundred millimetres or less. Therefore, the values of permeability obtained will be very small-scale (see Figure 12.1). Figure 12.4 shows that in soil with a structure or fabric of different material types, the small scale of sampling means that different sampling locations can capture

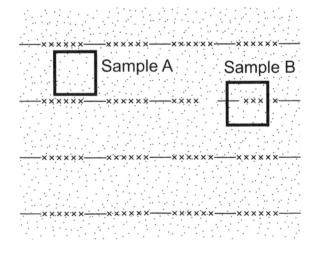

Figure 12.4 Sample variations in soil containing structure or fabric. The small scale of sampling means that different sampling locations A and B may capture different mixtures of finer or coarser material.

different mixtures of material and hence can give significantly different permeability estimates; values from PSD correlations may not be representative of the mass permeability of the strata. Large-scale tests (such as pumping tests) may give better results.

Each different empirical correlation will have limits to the particle size ranges where it is applicable; these limits should be understood before the methods are used. Furthermore, many empirical correlations are specific to certain combinations of units, and applying input parameters in the wrong units can result in further errors.

Given all the foregoing limitations, it would be tempting to think that permeability estimates from PSD correlations are of no value to the designers of groundwater control systems. However, in practice, the majority of site investigations include PSD analyses, and when used appropriately, these can form a useful part in the wider dataset of permeability values (most of which will themselves have various limitations and shortcomings). The best approach is to assess design permeability values from the full range of data sources, as discussed in Sections 12.9 and 13.4.

12.6.1.2 Hazen's Method

An American water works and sanitary engineer from New England, Allen Hazen (1892, 1900, 1911), was one of the first to propose an empirical correlation for the permeability of a sand from its PSD curve (further details are given in Appendix 1). Probably due to the simplicity of his formula, the Hazen method is still widely used by many of today's geotechnical practitioners, often without due regard to the limitations that Hazen himself stated. His objective was to determine guidelines for suitable sand gradings for water supply filtration. He determined that two important factors were the D_{10} particle size (which he called the 'effective particle size') and uniformity coefficient U, defined as

$$U = \frac{D_{60}}{D_{10}} \tag{12.2}$$

where D_{10} and D_{60} are the sieve apertures through which 10 per cent and 60 per cent, respectively, of a soil sample will pass.

Hazen included allowances for variations in the temperature of the water (see Section A1.4.1 in Appendix 1). However, the temperature of the relatively shallow groundwater in the United Kingdom varies little between about 5 °C and 15 °C; so Hazen's formula used to estimate permeability k is commonly stated as

$$k = C(D_{10})^2 \tag{12.3}$$

where
    C is a calibration factor
    D_{10} is the 10 per cent particle size taken from the PSD curves

Hazen stated in his work that his rule was applicable over the range of D_{10} particle size from 0.1 to 3.0 mm and for soils having a uniformity coefficient lower than 5. He also stated that (when k is in metres per second and D_{10} is in millimetres) his calibration factor C could vary between about 0.007 and 0.014. In practice, presumably for reasons of simplicity, C is normally taken to be 0.01, so Hazen's formula becomes

$$k = 0.01(D_{10})^2 \tag{12.4}$$

where k is in metres per second and D_{10} is in millimetres. This equation is generally taken to apply at a temperature of 10 °C, which is similar to shallow groundwater temperatures in the United Kingdom.

It cannot be stressed too strongly that, even within its range of application, Hazen's formula gives *approximate* permeability estimates only. Further background on the Hazen formula, including temperature corrections, is given in Appendix 1.

12.6.1.3 Kozeny–Carman Method

The Kozeny–Carman method (Kozeny, 1927; Carman, 1937, 1956) relates permeability to porosity and various indicators of particle size and shape. Interestingly, unlike the Hazen and Prugh methods, which were originally derived for granular filter media and soils, respectively, the Kozeny–Carman method was originally developed for a solid medium containing closely spaced pipe conduits rather than for a granular medium. Several different equations based on the method are available in the published literature (such as Loudon, 1952 and Carrier, 2003), with the differences explained by the use of different particle size indicators and different units.

Expressed in terms of soil porosity, the formulation for the Kozeny–Carman equation to estimate permeability k in metres per second for water at 10 °C is

$$k = 4.2 \times 10^{-2} \left(\frac{D_e^2}{f^2} \right) \left(\frac{n^3}{(1-n)^2} \right) \tag{12.5}$$

where
  D_e is the effective particle size in millimetres
  f is the angularity factor (values of which can be taken from Table A1.1 in Appendix 1)
  n is the soil porosity expressed as a fraction, not a percentage

D_e is the effective particle size, which (in contrast to Hazen's formula) is based on an assessment of the entire PSD curve rather than a single value. D_e can be assessed using the method given by Carrier (2003), where the fractions of the sample retained between adjacent sieve sizes are used to assess an effective particle size.

$$D_e = \frac{1}{\left[{x_1}/{D_1} + {x_2}/{D_2} + \ldots + {x_i}/{D_i} + \ldots + {x_n}/{D_n} \right]} \tag{12.6}$$

where
  $x_1 \ldots x_n$ are the fractions (not percentages) of the total mass (i.e. $x_1 + x_2 + \ldots + x_i + \ldots + x_n = 1$) retained on each sieve
  $D_1, D_2, \ldots D_i, \ldots D_n$ are the mean particle size between the respective adjacent sieves

The mean particle size D_i between the nominal sizes D_x and D_y of adjacent sieves is taken as the geometric mean of the two sizes:

$$D_i = \sqrt{(D_x D_y)} \tag{12.7}$$

Where raw data for a PSD curve is available electronically, it is a simple matter to evaluate D_e by means of a spreadsheet calculation.

One of the problems with applying the Kozeny–Carman method is that porosity n is difficult to determine, and site-specific values are rarely available for the sand-dominated soils that are often of interest to the designers of groundwater control systems. While the porosity of a sample can be determined in the laboratory, at that point, a sample of a sandy soil will have been significantly disturbed, and its porosity is likely to be very different from in situ conditions. In practice, porosity is sometimes estimated from published values (such as from Table A1.2 in Appendix 1), appropriate to the soil description, or from empirical correlations such as that given by Bricker and Bloomfield (2014), which quotes a formula from Vukovic and Soro (1992) whereby porosity can be approximately related to the PSD using the uniformity coefficient U:

$$n = 0.255\left(1 + 0.83^U\right) \tag{12.8}$$

Permeability estimates can then be obtained by combining the porosity values with D_e values into the Kozeny–Carman formula in Equation 12.5. The use of porosity values that are themselves estimates or are derived empirically is a limitation on the usefulness of Kozeny–Carman and other similar formulations and is an explanation for the somewhat erratic results that they sometimes give.

Carrier (2003) summarized some specific limits on the applicability of the Kozeny–Carman method:

1. The method is not applicable in soils that behave as clays. Carrier indicates that the method can be used in non-plastic silts.
2. The method assumes Darcian conditions (laminar flow). This is appropriate in silts, sands and even gravelly sands, but in very coarse soils, the method is invalid. An upper limit on the effective particle size D_e is around 3 mm.
3. The method is not appropriate for use in soils containing platy particles such as mica.
4. The method is not appropriate if the PSD has a long, flat tail in the fine fraction, with a significant non-zero percentage shown at the smallest recorded size. For D_e to be calculated accurately the smallest particle size present D_o must be measured or assumed.

It is important to remember that the Kozeny–Carman method is an empirical formulation and, even within its range of application, can give *approximate* permeability estimates only. Further background on the method, including temperature corrections, is given in Appendix 1.

12.6.1.4 Prugh Method

In the 1950s in America, Professor Byron Prugh (as reported in Powers *et al.*, 2007) researched and developed an empirical method for estimating the permeability of soil based on the use of particle size data together with in situ density field measurements. He checked his predictions against field measurements of permeability. Prugh's approach represents a return to the pragmatic co-ordination of academic and field observations.

Prugh plotted curves correlating permeability with D_{50} particle size for various uniformity coefficients (D_{60}/D_{10}). The D_{50} particle sizes are plotted on the horizontal axis to a log scale, and permeability is plotted on the vertical axis, also to a log scale.

Three separate sets of uniformity coefficient curves are available for:

(a) Dense soils (Figure 12.5a)
(b) Medium dense soils (Figure 12.5b)
(c) Loose soils (Figure 12.5c).

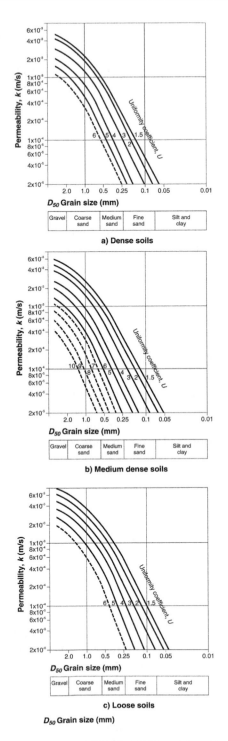

Figure 12.5 Prugh method of estimating permeability of soils. (Reproduced from Preene, M, *et al.*, *Groundwater Control – Design and Practice*, 2nd edition, Construction Industry Research and Information Association, CIRIA Report C750, London, 2016. With permission from CIRIA: www.ciria.org)

The use of these relative density indicators is effectively an inverse indicator of porosity; for a given soil grading, higher densities are predicted to result in lower permeability.

To apply the Prugh method, the first step is to assess whether the soil sample was obtained from a zone that is dense, medium dense or loose (based on Standard Penetration Test [SPT] N values from borehole logs). Relative density can be assessed from SPT N values as follows (Clayton, 1995):

- Dense soils: SPT N values greater than 30
- Medium dense soils: SPT N values between 10 and 30
- Loose soils: SPT N values less than 10

This identifies which graph should be used from Figure 12.5a to 12.5c. On the relevant plot, the D_{50} of the sample is identified on the horizontal axis and then projected upwards to the appropriate uniformity coefficient curve; from the intercept with the uniformity coefficient curve, a horizontal projection indicates the permeability value.

Prugh's data indicate that as the uniformity coefficient increases (i.e. the sample becomes less and less a single-size material), the permeability decreases noticeably. The significance of the Prugh curves, apart from their usefulness, is that they greatly help in the understanding of the inter-relationship of various factors (other than D_{10} used in Hazen's formula) affecting soil permeability. His work has been published by Powers *et al.* (2007) and in CIRIA Reports (Preene *et al.*, 2016). Like others, Prugh did not claim his method to be relevant to soils other than 'a wide range of sands', although his correlation graphs extend up to maximum uniformity coefficients of 6 to 10 and D_{50} of less than around 2 mm.

As with other particle size correlations, it is important to remember that the Prugh method is an empirical formulation, and, even within its range of application, can give *approximate* permeability estimates only.

12.6.2 Correlations with Descriptions of Rock Quality

It is common practice for the 'rock quality' of core recovered from investigation boreholes to be described in a standardized manner using one of several 'rock mass classification' systems. Classification systems include Rock Quality Designation (RQD) (Deere *et al.*, 1966), Rock Mass Rating (RMR) (Bieniawski, 1973), Q Classification (Barton *et al.*, 1974) and the Geological Strength Index (GSI) (Hoek, 1994); these are briefly discussed in Section 8.3.1.

The common element of rock mass classification systems is that they are essentially measures of the degree or intensity of fracturing of the examined core, with highly fractured rock being allocated low values and more intact rock with few fractures being allocated higher values under the classification system. Conceptually, it is obvious that in the same geological conditions, more fractured rock will be of higher permeability than less fractured rock. Therefore, we can say in mathematical terms that there should be an inverse correlation between permeability and rock quality classification – for a given system, as the classification value increases, the permeability will reduce.

While it is clear there is potential for rock quality to be empirically related to permeability, there are no generic relationships in common use. There are examples of empirical correlations developed on specific projects and in specific geological settings (Hsu *et al.*, 2011; Herrera and Garfias, 2013; Akbarimehr and Aflaki, 2019; Kolpakov and Zhdanov, 2019), but these are not currently suitable for generic use.

12.6.3 Inverse Numerical Modelling

Numerical modelling (described in Section 5.5.3) is not in itself an empirical method. However, when used to assess permeability based on data from large-scale groundwater observations, it effectively becomes an empirical assessment based on observation and previous experience.

This approach can only be used if extensive groundwater monitoring data (e.g. data from a number of monitoring wells over a long time period) are available. It also requires a thorough understanding of the geological structure and extent of the aquifers, aquitards, etc. in the area. The method involves setting up a numerical model (see Section 13.12) of the groundwater system in the vicinity of the site and running the model with a variety of permeability and other parameters until an acceptable match is obtained between the model output and the groundwater monitoring data.

This approach is not straightforward and is carried out only rarely. Since a number of parameters, in addition to permeability, will be varied during the modelling, a 'non-unique' solution may result. In other words, a number of permeability values may give an acceptable fit with the data depending on what other parameter values are used. Bevan *et al.* (2010) describe an interesting example of numerical analysis used to investigate the variation of permeability in different areas of a large excavation in Chalk that formed the tunnel portals where the HS1 railway (formerly the Channel Tunnel Rail Link [CTRL]) passed beneath the River Thames in the United Kingdom.

12.7 QUANTITATIVE ASSESSMENT OF PERMEABILITY – EX SITU TEST METHODS

An obvious and logical approach to assessing permeability is to obtain a sample of soil or rock from a borehole or trial pit and then subject it to tests in the laboratory. In relation to assessing permeability for the design of groundwater control schemes, these methods have the following limitations:

 i. The act of sampling will inevitably cause some disturbance and stress relief of the soil or rock to be tested. This will affect the measured permeability and can result in very significant errors in permeability.
 ii. Samples that can be tested in the laboratory are necessarily of modest dimensions (up to a few hundred millimetres in length and width). Therefore, the values of permeability obtained will be very small-scale (see Figure 12.1) and are unlikely to fully represent any permeable fabric in soil samples or fracture network in rock samples. The permeability values may not be directly relevant to groundwater control designs.

Despite these limitations, permeability data from these methods may be available; details and limitations are discussed in the following.

A specific limitation on laboratory tests is the effect of temperature. Laboratory tests are typically carried out at 20 °C, rather higher than shallow groundwater temperatures in the United Kingdom, which are typically 10–15 °C. The viscosity of water at 20 °C is lower than at 10 °C, and in this case, permeability recorded in the laboratory would be around 30 per cent higher than under comparable conditions in the field. However, in most cases, the effect of temperature is small compared with the other limitations of the tests, and temperature corrections are rarely applied.

12.7.1 Laboratory Testing of Soil Samples

There are a number of techniques for the direct determination of the permeability in the laboratory by inducing a flow of water through a soil sample – this approach is known as 'permeameter' testing. According to Head and Epps (2011a), there are two main types of permeameter testing:

1. Constant head test (Figure 12.6a). A flow is induced through the sample at a constant head. By measuring the flow rate, cross-sectional area of flow and induced head, the permeability can be calculated using Darcy's law. This method is only suitable for relatively permeable soils such as sands or gravels ($k > 1 \times 10^{-4}$ m/s); at lower permeabilities, the flow rate is difficult to measure accurately.
2. Falling head test (Figure 12.6b). An excess head of water is applied to the sample, and the rate at which the head dissipates into the sample is monitored. Permeability is determined from the test results in a similar way to a falling head test in a borehole or monitoring well. These tests are suitable for soils of lower permeability ($k < 1 \times 10^{-4}$ m/s), when the rate of fall in head is easily measurable.

Tests may be carried out in special permeameters, oedometer consolidation cells, triaxial cells and Rowe consolidation cells. Samples are typically cylindrical in shape, with diameters of oedometer and triaxial tests typically between 38 and 100 mm. Rowe consolidation cells are typically 250 mm in diameter. Methods of testing are described in Head and Epps (2011a).

While these tests are theoretically valid, in practice they are rarely used because of the difficulty of obtaining representative 'undisturbed' samples of soils of high or intermediate permeability (silt, sand or gravel) that are of interest in dewatering design. Even if the sample is representative of the PSD of the soil, without specialized equipment, it is very difficult to measure the density, and hence the porosity, of granular soils in situ, especially below the water table. This means that the in situ condition of the soil cannot be reproduced reliably. Similarly, any soil fabric or layering in the sample will have a profound effect on the in situ permeability but cannot be replicated in the laboratory. There is also the question of direction of flow: many laboratory permeameter tests impose vertical flow through a sample and will therefore derive vertical permeability, while in practice it is often horizontal permeability that is of interest to dewatering designers.

The only time such tests should be considered on groundwater control projects is when the permeability of very low-permeability soils (which are typically dominated by clay-sized particles) needs to be determined during investigations of potential consolidation settlements. Even then, results must be interpreted with care, as in clays with permeable fabric the size of the test sample may result in scale effects distorting the measured permeability (see Rowe, 1972). Large (250 mm diameter) samples may give more representative results than the 76 mm diameter samples routinely tested, but such large samples are rarely available.

12.7.2 Laboratory Testing of Rock Samples

Rock cores can be tested in specialist laboratory permeameters in a similar way to soil samples, where flow of water is induced through core samples recovered from boreholes. The approach has similar limitations to laboratory testing in soils, in that the rock core may be disturbed or disrupted by sampling and stress relief, and the permeability values are very small-scale and are not especially relevant to the design of groundwater control systems.

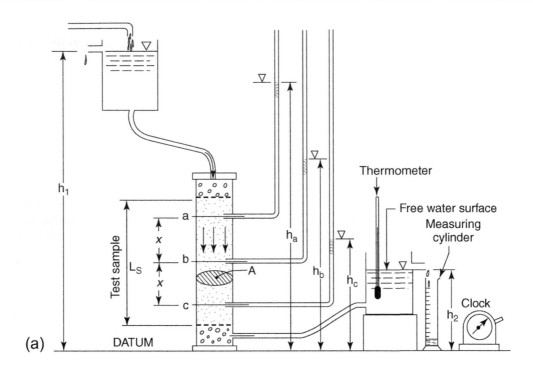

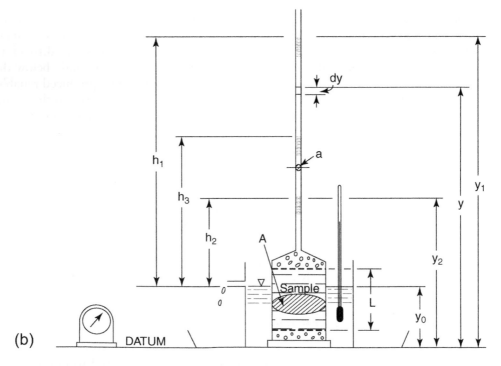

Figure 12.6 Examples of laboratory permeameter test. (a) Constant head permeameter test: downward flow. (b) Falling head permeameter test. (From Head, K H and Epps, R J, *Manual of Soil Laboratory Testing. Volume II: Permeability, Shear Strength and Compressibility Tests*, 3rd edition, Whittles Publishing, Caithness, UK, 2011. With permission.)

A further complication is that the permeability of rock samples is often dominated by flow through fractures, which tend to be highly directional. Depending on the orientation of the core axis relative to the significant fracture directions, permeability values from testing of the core may not represent the hydraulically significant fractures that are important to the design of groundwater control systems.

12.8 QUANTITATIVE ASSESSMENT OF PERMEABILITY – IN SITU TEST METHODS

In situ tests are commonly used to assess values of permeability. Common forms of in situ permeability tests in boreholes include:

(i)   Rising, falling and constant head tests in boreholes (Section 12.8.1)
(ii) In situ tests in monitoring wells (Section 12.8.2)
(iii) In situ tests in boreholes in rock (Section 12.8.3)

These tests are carried out either in boreholes (during pauses in the drilling process) or in monitoring wells installed in boreholes following completion of drilling and are based on a common principle – if the water level in a borehole is raised or lowered in a controlled manner, this will create an 'excess head' relative to background groundwater level. Depending on whether water levels are raised or lowered, this will induce flow out of or flow into (respectively) the borehole. Provided that the geometry of the 'test section' is known – this controls how water flows from the borehole – then the water level observations can be analysed to estimate permeability. These tests can only influence the soil or rock locally around the borehole. Therefore, these tests can, at best, produce 'small-scale' values of permeability representative of conditions around the borehole (see Figure 12.1). Such tests may be unduly influenced by any effects of soil/rock disturbance caused by drilling of the borehole or by local variations in geology close to the borehole.

Pumping tests, where water is pumped from a well for an extended period (typically several days), will typically influence a much larger volume of soil or rock and give more representative 'large-scale' permeability values but are more time consuming and expensive to carry out (Section 12.8.5). Occasionally, multiple wells are pumped in the form of a groundwater control trial to glean information on large-scale permeability (Section 12.8.6).

Specialist in situ tests are occasionally carried out as part of certain types of probing techniques, such as dissipation tests carried out in piezocone testing (Section 12.8.4). Borehole geophysical methods can also be used to provide permeability data (Section 12.8.7).

The test methods described in the following sections are those that are applicable to conventional civil engineering and groundwater control projects. More sophisticated methods of permeability testing have been developed in related fields, including shaft sinking for deep mining (Daw, 1984), investigations for deep geological disposal of nuclear waste (Sutton, 1996), and carbon sequestration and storage (Wiese et al., 2010). These methods are not addressed here but use the same principles as the tests described here, carried out at greater depths, at higher background pore water pressures and in lower-permeability rocks than is common for conventional tests.

12.8.1 Rising, Falling and Constant Head Tests in Boreholes

This group of tests includes:

(i) Rising and falling head tests (collectively known as variable head tests)
(ii) Constant head tests

These tests are carried out in the field on the soil/rock in situ around the borehole. They therefore avoid the problems of obtaining representative undisturbed samples that limit the usefulness of laboratory testing. Tests in boreholes are those carried out during pauses in the drilling or boring process. When the test is complete, drilling recommences – this allows several tests at different depths to be carried out in one borehole. If boreholes are drilled into rock, where the completed borehole can be stable without lining or casing, it may be possible to carry out a series of permeability tests after the end of drilling by isolating sections of the borehole between inflatable packers – this type of testing is described in Section 12.8.3.

These tests in boreholes are distinct from tests carried out in monitoring wells (Section 12.8.2) following completion of the borehole, where tests can be carried out only at the fixed level of the response zone.

Execution of variable head tests is straightforward and requires only basic equipment. The borehole is advanced to the proposed depth of test, and the original groundwater level is noted. It is essential that a representative ground water level is obtained. If necessary, the start of the test should be delayed until readings show that the pre-test groundwater level has stabilized.

Typically, the upper portion of the borehole is supported by temporary casing (which should be sealed into the upper strata to exclude groundwater from those levels). The 'test section' of exposed strata is between the bottom of the casing and the base of the borehole. The dimensions and geometry of the 'test section' have a key influence on flow conditions during the test and hence on the assessed permeability.

For a falling head test (Figure 12.7a), water is rapidly added to raise the water level in the borehole. Once the water has been added, the water level in the borehole is recorded regularly to see how the level falls with time as water flows out of the borehole into the soil. The necessary equipment comprises a dipmeter (to measure water level in the borehole), a bucket, a stopwatch and a supply of clean water (perhaps from a tank or bowser). Alternatively, if a pressure transducer linked to a stand-alone datalogger is available (Section 22.6), this can be suspended below the water level in the borehole and used to record water levels more frequently than can be achieved by manual dipmeter readings. It is essential that any water added is absolutely clean; otherwise, any suspended solids in the water will clog the base of the borehole test section and significantly affect results. Particular attention should be given to the cleanliness of tanks, buckets, etc, so that the water does not become contaminated by those means. It can be difficult to carry out falling head tests in very permeable soils (greater than about 5×10^{-3} m/s), because water cannot be added quickly enough to raise the water level in the borehole. If the natural groundwater level is close to ground surface, it may be necessary to extend the borehole casing above ground level to allow water to be added.

A rising head test (Figure 12.7b) is the converse of a falling head test. It involves rapidly removing water from the borehole and observing the rate at which water rises in the borehole. The test does not need a water supply (which can be an advantage in remote locations) but does require a means of removing water rapidly from the borehole. The most obvious way to do this is by using a bailer, which is adequate in soils of moderate permeability, but it can be surprisingly difficult to significantly lower water levels if soils are highly permeable. Alternatives are to use airlift equipment (Howell, 2013) or suction pumps or submersible pumps.

An alternate form of variable head test is the 'slug test'. Again, this involves applying rapid changes to the water level in a borehole and then observing the rate at which the water level returns to the background or natural water level. However, in a slug test, no water is added to or removed from the borehole. Instead, a heavy rod (termed a 'slug') is quickly lowered below the water level in the borehole to displace water and hence rapidly raise water levels – analogous to a falling head test (Figure 12.8a). At the end of the

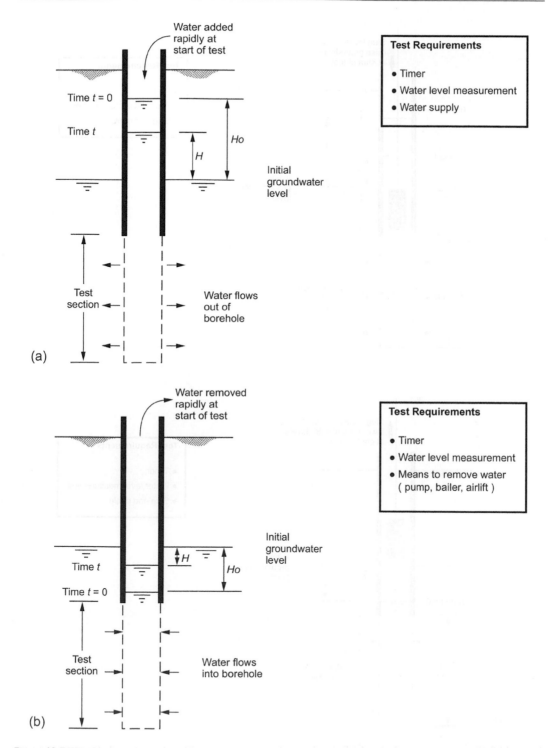

Figure 12.7 Variable head tests in boreholes. (a) Falling head test. Water is added to the borehole to raise water levels, inducing flow from the borehole into the surrounding strata. (b) Rising head test. Water is removed to the borehole to lower water levels, inducing flow from the surrounding strata into the borehole.

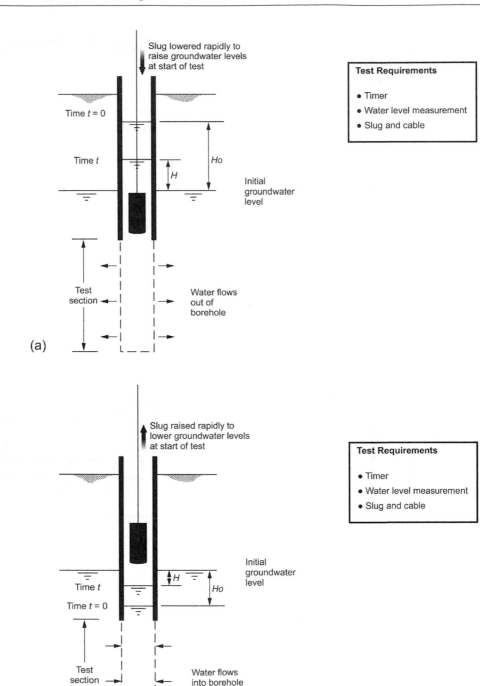

Slug lowered rapidly to raise groundwater levels at start of test

Time $t = 0$

Time t

Ho

H

Initial groundwater level

Test section

Water flows out of borehole

(a)

Test Requirements

• Timer
• Water level measurement
• Slug and cable

Slug raised rapidly to lower groundwater levels at start of test

Initial groundwater level

Time t

H

Ho

Time $t = 0$

Test section

Water flows into borehole

(b)

Test Requirements

• Timer
• Water level measurement
• Slug and cable

Figure 12.8 Slug tests in boreholes. (a) Falling head phase of slug test. The slug is rapidly lowered and submerged to raise water levels, inducing flow from the borehole into the surrounding strata. (b) Rising head phase of slug test. The slug is rapidly raised out of the water to lower water levels, inducing flow from the surrounding strata into the borehole.

falling head stage, when water levels have equilibrated, rapid removal of the slug from the water level will cause a sudden lowering of water level – analogous to a rising head test (Figure 12.8b). Slug tests have the advantage that no water supply or equipment to pump or bail water is needed.

For the relatively permeable soils and rocks of interest in groundwater lowering problems, variable head tests can be analysed using the work of Hvorslev (1951), which is one of the methods included in the British Standards on geotechnical testing (BS EN ISO 22282-1:2012 and BS EN ISO 22282-2:2012). Hvorslev assumed that the effect of soil compressibility on the permeability of the soil was negligible during the test, and this is a tolerable assumption for most water-bearing soils. If in situ permeability tests are carried out in relatively compressible silts and clays, different test procedures and analysis may be required (see Brand and Premchitt, 1982).

The Hvorslev analysis is based on the following assumptions:

- Flow conditions are laminar, so that Darcy's law is valid.
- The soil/rock and groundwater are incompressible.
- The soil or rock around the test section is saturated.
- The water inflow or outflow during the test does not change background groundwater level around the test section.

A further implicit assumption is that the borehole is 100 per cent hydraulically efficient – i.e. the borehole itself does not introduce any hydraulic resistance in addition to that from the properties of the soil or rock mass.

For the Hvorslev analysis, permeability k is calculated from variable head tests using

$$k = \frac{A}{FT} \tag{12.9}$$

where
   A is the cross-sectional area of the borehole casing (at the water levels during the test)
   T is the basic time lag
   F is a shape factor dependent on the geometry of the test section

T is determined graphically from a semi-logarithmic plot of H/H_o versus elapsed time as shown in Figure 12.9. H_o is the excess head in the borehole at time $t = 0$, and H is the excess head at time t (the excess heads H and H_o are measured relative to the original groundwater level). Additional notes on the analysis of variable head tests are given in Appendix 2.

Values of shape factor F for some commonly occurring borehole test section geometries were prepared by Hvorslev (1951) and are shown in Figure 12.10. It should be noted that the shape factor is not unitless and has the dimension of length; care should be taken to use appropriate units in calculations. Where the intention is to calculate permeability in metres per second, F should be calculated in metres by applying L and D in metres into the equations shown in Figure 12.10.

The simplest test section is when the temporary casing is flush with the base of the borehole, allowing water to enter or leave the borehole through the base only. If soil will stand unsupported, it may be possible to extend the borehole ahead of the casing to provide a longer test section. If the soil is not stable, the borehole could be advanced to the test depth and the test section be backfilled with filter sand or gravel as the casing is withdrawn to the top of the test section.

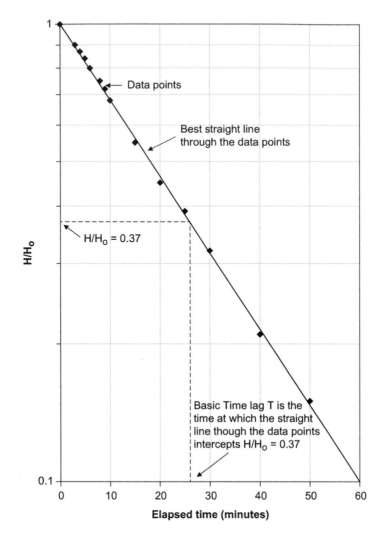

Figure 12.9 Analysis of variable head tests.

For the generic case of a cylindrical test zone of length L and diameter D (Figure 12.11), where $L/D > 10$ (as stated in BS EN ISO 22282-2:2012), permeability from variable head tests can be calculated as

$$k = \frac{d^2 \ln\left(2L/D\right)}{8LT} \tag{12.10}$$

where d is the diameter of the section of tubing within which the water level rises and falls.

If the soil or rock has anisotropic permeability – k_h and k_v in the horizontal and vertical directions, respectively – Hvorslev (1951) states that permeability can be assessed for the geometry of Figure 12.11 using Equation 12.11, provided that $mL/D > 4$:

$$k_h = \frac{d^2 \ln\left(2mL/D\right)}{8LT} \tag{12.11}$$

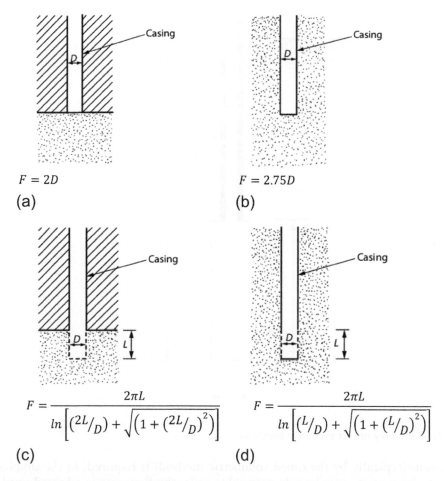

$F = 2D$

(a)

$F = 2.75D$

(b)

$$F = \frac{2\pi L}{\ln\left[(2L/_D) + \sqrt{\left(1 + (2L/_D)^2\right)}\right]}$$

(c)

$$F = \frac{2\pi L}{\ln\left[(L/_D) + \sqrt{\left(1 + (L/_D)^2\right)}\right]}$$

(d)

Figure 12.10 Shape factors for permeability tests in boreholes. (a) Soil flush with the bottom of the casing at the impermeable boundary. (b) Soil flush with the bottom of the casing in uniform soil. (c) Open section of the borehole that extended beyond the casing at the impermeable boundary. (d) Open section of the borehole that extended beyond the casing in uniform soil (After Hvorslev, M J, *Time Lag and Soil Permeability in Groundwater Observations*, Waterways Experimental Station, Corps of Engineers, Bulletin No. 36, Vicksburg, Mississippi, 1951.)

where m is the permeability transformation ratio

$$m = \left(k_h / k_v\right)^{0.5} \tag{12.12}$$

Constant head tests (Figure 12.12) involve adding or removing water from a borehole to maintain a constant excess head while the flow rate is recorded. Constant head tests are most often carried out as inflow tests (where water is added to the borehole), but outflow tests (where water is pumped out of the borehole) can also be carried out. The equipment required is rather more complex than for variable head tests, as some form of flow

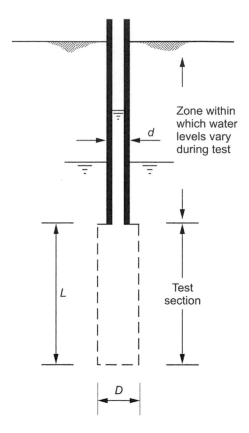

Figure 12.11 Geometry of test section in borehole.

measurement (typically by the timed volumetric method) is required. In the simplest form of the test, appropriate to relatively permeable soils, the flow rate is adjusted until a suitable constant head is achieved, and the test is allowed to continue until a steady flow rate is established. A consistent supply of clean water is required for inflow tests, and this can be a disadvantage in remote locations.

Permeability k is calculated for constant head tests using

$$k = \frac{q}{FH_c} \tag{12.13}$$

where

    q is the constant rate of flow
    H_c is the constant excess head (measured relative to original groundwater level)
    F is the shape factor (from Figure 12.10)

As with the other equations in this chapter, care must be taken to use consistent units. Where the intention is to calculate permeability in metres per second, q should be in cubic metres per second, not litres per second or litres per minute, as might be given in field records.

It is well known (see for example Black, 2010) that variable and constant head tests in boreholes have a number of limitations and may be subject to a number of errors. When

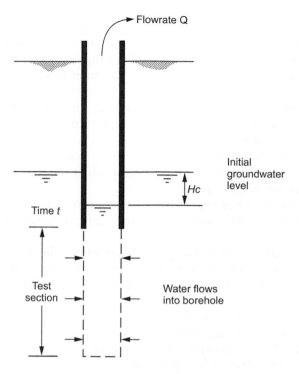

Flowrate Q

Test Requirements

- Timer
- Water level measurement
- Water supply (for inflow test)
- Means to remove water
  (for outflow test)

Initial groundwater level

Hc

Time t

Test section

Water flows into borehole

Figure 12.12 Constant head test in borehole. Example shown is a test where water is pumped out from the borehole.

carrying out these tests (and when reviewing the results), it is essential that these factors are considered.

1. Tests in boreholes only involve a relatively small volume of soil or rock around the test section – these tests provide 'small-scale' values of permeability. If the soil/rock is heterogeneous or has significant fabric, such tests may not be representative of the mass permeability of the strata. Large-scale tests (such as pumping tests) may give better results.

2. One of the most common causes of errors in test analysis is the use of unrepresentative background groundwater levels, for example if the pre-test water level had been affected by drilling activities. If this occurs, the calculated excess heads H will be wrong, and the calculated permeability value may be grossly in error. When analysing the test, it is important to validate the reported pre-test and post-test water levels against the conceptual model (and levels from other boreholes) to determine whether the background water levels have been correctly assessed. Ideally, where tests are carried out during pauses in drilling, testing should not start until monitoring shows that water levels in the boreholes have stabilized.

3. For variable head tests, the conventional Hvorslev method of analysis is based on the assumption that the initial increase/decrease in water level in the borehole occurs instantaneously. Obviously, in practice, the change in water level will not be instantaneous but will require a finite time period. All possible steps should be taken to keep to a minimum the period of adding/removing water at the start of the test. Black (2010) suggests that test results will be compromised if the test initiation phase (when water

levels are changed) exceeds 10 percent of the duration of the subsequent equilibration period. If this criterion is not achieved, for example where a hose pipe is used to continuously add water to a well over several minutes, the water level response will be different from the Hvorslev assumptions, and erroneous permeability values may result.

4. Results of tests that add water to the borehole (falling head and constant head tests) can be significantly affected by clogging or silting up of the test section as water is added. It is vital that only totally clean water is added, but even then, silt already in suspension may block flow out of the borehole. It is not uncommon for inflow tests to under-estimate permeability by several orders of magnitude.

5. In loose granular soils, tests that induce flow into the borehole (rising head and constant head tests that remove water) may cause piping or boiling of soil at the base of the borehole. This could lead to over-estimates of permeability.

6. The drilling of the borehole may have disturbed the soil in the test section, changing the permeability. Potential effects include particle loosening, compaction, or smearing of silt and clay layers.

7. If the natural groundwater level varies during the test (due to tidal or other influences), the results may be difficult to analyse. If significant groundwater level fluctuations are anticipated during a test of, say, 1 or 2 hours' duration, tests in boreholes are unlikely to be useful.

8. If the drilling casing does not provide an effective seal to isolate the test section, then leakage of water into or out of the test section may occur from other strata. This will affect the water level response during the test and may lead to erroneous results. Black (2010) indicates that for tests in strata of relatively low permeability (less than around 10^{-7} m/s), leakage is a particular concern, because the theoretical water flow rates into the strata are very low. Therefore, even very low rates of leakage can significantly affect the test response.

Although these tests have a number of limitations, they are inexpensive to execute and are widely used and included in many standards and textbooks. It is good practice to carry out both rising and falling head tests in the same borehole to allow results to be compared. In any event, results from in situ tests in boreholes should be reviewed against the anticipated conceptual model (Section 13.4) for the site and treated with caution until supported by permeability estimates from other sources.

12.8.2 Rising, Falling and Constant Head Tests in Monitoring Wells

Variable and constant head tests can be carried out in monitoring wells (standpipes and standpipe piezometers) following completion of boring (Figure 12.13). Installation techniques for monitoring wells are discussed in Section 11.7.6.

While these tests can only be carried out at the depth of the monitoring well response zone, they have the advantage that they can be executed (and repeated if necessary) after boring has been completed without the time pressures associated with working in pauses in the drilling process. Tests are analysed using the same methods as for tests in boreholes, and permeability can be estimated from Equations 12.9 through 12.13. In those equations, d is the diameter of the section of tubing within which the water level rises and falls, and D is the diameter of the test section, usually taken as the diameter of the borehole within which the monitoring well is installed.

The limitations on variable and constant head tests in boreholes also apply to tests in monitoring wells. Additional problems may result from the nature of the monitoring well

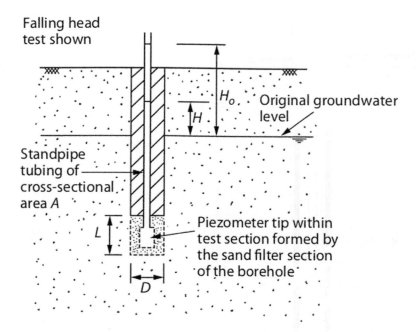

Figure 12.13 Variable head tests in monitoring wells.

itself. If the standpipe or piezometer has not been installed to the highest standards, it may be partially clogged and therefore can present some restriction to the flow of water into or out of the monitoring well. In such cases, any tests will merely determine the permeability of the piezometer rather than the permeability of the soil/rock beyond. To reduce this problem, all monitoring wells should be purged or developed (by pumping or airlifting) prior to testing. Testing should not commence until the water level has recovered to its equilibrium level.

12.8.3 Packer Permeability Testing in Boreholes in Rock

The in situ testing techniques described in Section 12.8.1 are commonly used as part of site investigations in soil and can also be applied to boreholes in rock. However, in practice, a different approach is often taken for permeability testing in boreholes drilled through rock strata. This is because the boreholes drilled in rock may be stable without the use of casing, and this can provide long sections of boreholes that can be incrementally tested. This approach can be useful, because groundwater flow in rock is typically dominated by flow along fractures (see Section 8.3). A borehole drilled through a stratum of rock may pass through both relatively unfractured zones (which will be of low permeability) and more fractured zones (of higher permeability). Packer permeability testing can be carried out in different zones of a borehole. Borehole geophysical surveys (Sections 12.5.3.2 and 12.8.7) can provide useful data to identify the levels at which permeability testing should be carried out.

Packer permeability tests can be carried out in sections of rock borehole that are stable without casing (Figure 12.14). A discrete 'test section' is isolated from the rest of the borehole using inflatable packers (Figure 12.15). After the packers are inflated and the pre-test water level measured, water is pumped into or out of the test section with the flow rate and pressure head monitored. If test sections are set at different depths in the borehole, multiple tests may be carried out sequentially along the borehole length after drilling is completed or can be executed incrementally, during pauses in drilling, as the borehole is deepened.

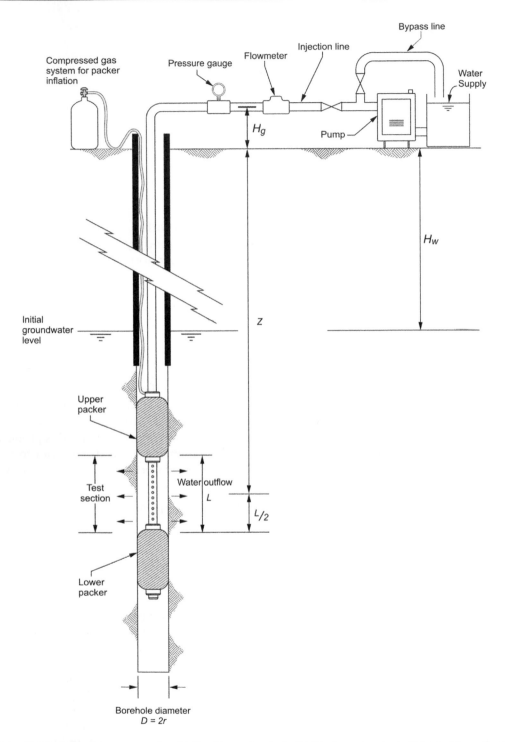

Figure 12.14 Packer test geometries. (a) Double packer test. (b) Single packer test. (Adapted from Preene, M, *Quarterly Journal of Engineering Geology and Hydrogeology*, 52, 182–200, 2019.) Note: aboveground pressure measurement system shown.

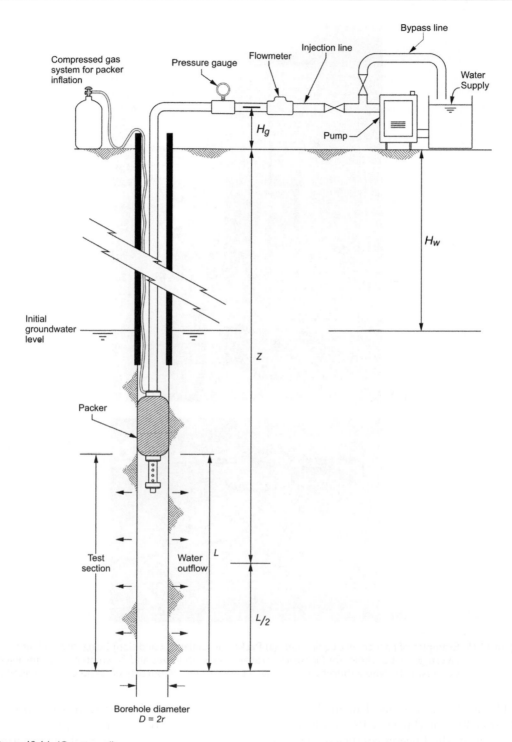

Figure 12.14 (Continued)

Figure 12.15 Examples of packer test equipment. (a) Packer (in deflated condition) being inserted into site investigation borehole. (b) Pipework arrangement with valves and flowmeter for water injection test in borehole (borehole is visible in background). (Courtesy of Geotechnics Limited.)

The packer test as used in modern geotechnical practice is derived from the work of Maurice Lugeon (Lugeon, 1933), who defined a standard test protocol (the Lugeon test) and a new unit, the Lugeon coefficient (Lu), for water injection testing as part of the grouting programmes for dam foundations being constructed in the early twentieth century. Lugeon's innovation was to define standard test parameters to give an empirical measure of water take (the Lugeon coefficient) calculated on a common basis for each test. This allowed a rational comparison of water take between boreholes and between tests at different levels in the same borehole. The Lugeon coefficient *Lu* is defined as water absorption measured in litres/minute into a 1 m test section at an excess pressure of 10 bar (1000 kPa; 1 MPa; 102 m head of water); excess pressure is defined as the pressure above the ambient ground-water pressure at the midpoint of the test section:

$$Lu = \frac{Q_{1000}}{L}$$

(12.14)

where

Q_{1000} is the water take of the borehole (in litres per minute) at an excess pressure of 1000 kPa

L is the length of the test section

Equation 12.14 requires the use of the stated units to obtain the correct values; equivalent equations for Imperial units are used in North American practice.

Published literature typically uses a correlation between the Lugeon coefficient and permeability of 1 Lu $\approx$ 1 × 10^{-7} m/s. Preene (2019) discusses the basis of correlation between Lu and permeability units and shows that the precise correlation depends on the borehole diameter and the length of the test section. For common test geometries, the correlation is in the range 1 Lu $\approx$ 0.9 × 10^{-7} to 1.5 × 10^{-7} m/s.

A strict interpretation of the definitions of the Lugeon test requires that high excess pressures be used (ideally 1000 kPa), based on its origin as a test to mimic the injection of grout into fractured rock. In most groundwater control applications, the use of such high excess pressures is neither necessary nor advisable (due to the risk of the excess water pressures causing fracture dilation or hydrojacking). Guidance on suitable excess water pressures is given later in this section and in Appendix 3.

The question of terminology is important. The authors suggest that in the context of groundwater control projects, the most appropriate terminology is to view these tests as packer permeability tests (rather than Lugeon tests), with the objective of determining permeability in metres per second rather than reporting Lugeon coefficient values. When applied to grouting projects, tests adhering more closely to the Lugeon test template may be appropriate (Paisley *et al.*, 2017).

The packer test method is described in Walthall (1990), BS EN ISO 22282-3:2012 and Preene (2019) – further details are given in Appendix 3. The test section can be isolated above and below by packers (double packer test; Figure 12.14a) or between a packer and the base of the borehole (single packer test; Figure 12.14b).

The most common type of packer test is an inflow test, where water is injected into the test section while the flow rate and injection pressure are recorded. Outflow tests, where water is pumped from the borehole, can also be carried out, although the test equipment is more complicated (Price and Williams, 1993); a number of studies (including Brassington and Walthall, 1985) have concluded that outflow tests are preferable to inflow tests. However, in relatively small-diameter investigation boreholes, it is much easier to inject water than to remove it (which requires a downhole pump and more sophisticated equipment). In routine investigations for groundwater control projects, injection tests are used almost exclusively and are the focus of the following section (and Appendix 3).

A common format of packer permeability tests involves five consecutive injection phases (each phase is of the same duration – typically 10 or 15 minutes; longer durations are sometimes used). Injection pressures are varied rapidly at the end of each phase, giving a stepwise transition of pressure between phases in three ascending and descending test pressures, A, B, C in the sequence A1–B1–C–B2–A2. Example test sequences are given in Appendix 3.

The overall water injection phase is of short duration (50 to 75 minutes), and therefore the test is 'small-scale' (see Figure 12.1) and can only influence a modest volume of rock around the borehole. Bliss and Rushton (1984) used numerical modelling to investigate packer testing and concluded that the zone of rock where groundwater flow is affected significantly during a packer test is limited to a radius of around 10 m from the borehole.

A packer test is typically analysed by treating each phase of the test as a steady-state constant head injection test, with each phase independent of the others; the test will generate five pairs of data points for injection flow rate Q and applied excess head H, which is calculated from the pressure P_t in the test section. As described in Appendix 3, plots of Q vs. H can be useful to help interpret different flow conditions in the test.

For a test section of length L in a vertical borehole of diameter D in a uniform isotropic aquifer, provided that L/D greater is than 4, the permeability k can be determined for each test phase by

$$k = \frac{Q}{2\pi HL} \ln\left(\frac{2L}{D}\right) \tag{12.15}$$

The same equation is used for double packer tests and single packer tests (where there is potentially some 'end effect' outflow directly from the base of the test section). For $L/D > 4$, the flow from the end effect is a very small proportion of the total flow and does not have a major effect on the calculated permeability.

If permeability conditions are anisotropic, the horizontal permeability k_h can be determined (for tests with $L/D > 4$) from

$$k_h = \frac{Q}{2\pi HL} \ln\left(\frac{2mL}{D}\right) \tag{12.16}$$

where

$$m = \left(\frac{k_h}{k_v}\right)^{0.5} \tag{12.17}$$

and k_v is the vertical permeability.

It is important to remember that these equations calculate the average permeability $k_{average}$, assuming that the injected water leaves the test section uniformly across the cylindrical boundary. The test results may not be representative of more fractured and highly permeable zones within the test section.

The design of packer test should consider the length and depth of the test section, the excess head applied to the test section, and the maximum permeability that can be determined with the available equipment:

Length and depth of test section. The geometry of the test section is defined by test length L and depth Z of the midpoint of the test. Different approaches include testing the whole borehole length, incrementally by multiple test sections, to give a vertical profile of permeability vs. depth, or targeting testing at specific horizons inferred to be more or less permeable than the typical host rock. Examples of test lengths commonly used are:

i.   General characterization of the vertical distribution of permeability can use consecutive test sections from groundwater level down to the base of the borehole. Test lengths of between 3 and 10 m are typical.

ii. If permeability is expected to vary significantly with depth (e.g. due to weathering or fracturing), consecutive test sections with shorter lengths may be carried out over the relevant depths. Test lengths of between 2 and 6 m are typical.

iii. Where potentially permeable faults or zones of multiple discontinuities may be present in the rock, tests can be carried out at the depth of the permeable zone and, for comparison purposes, at other depths. The length of test section should allow the packers to be located at suitable levels to seal against relatively intact rock.

iv. If discrete permeable horizons are expected. then test sections can be targeted to these zones. Test lengths of between 1 and 3 m are typical. The same approach can be used if potential permeable features are identified from core logs or borehole geophysics. For comparison, some tests should be done at other depths where the features are not indicated to be present.

Packer tests are rarely attempted in the unsaturated zone above groundwater level. The assumptions of the Hvorslev method of analysis are invalid in unsaturated conditions, and the results will be different from tests in the saturated zone due to the injected water filling voids and changing rock saturation. It is also difficult to assess the initial groundwater head and therefore the applied excess head (the test section would be reported as 'dry' before the start of the test).

Applied excess head. The maximum test excess pressure P_{max} should be selected with care. High injection pressures can give a risk of hydraulic dilation of fractures/joints within the rock around the borehole. Preene (2019) indicates that for geotechnical investigations, tests with applied excess heads in the range 5 to 25 m (50 kPa < test excess pressure < 250 kPa) can give good results. As noted in Appendix 3, to reduce the risk of hydraulic dilation of fractures/joints, the maximum excess test pressure P_{max} should not exceed the vertical effective stress σ'_v in the test section, and this check should be made if high test pressures are contemplated.

Upper permeability limit of equipment. A given set of packer testing equipment will have a maximum injection flow rate Q_{max}. This limits the maximum permeability that can be determined for a given excess head in the test section. Preene (2019) indicates that for Q_{max} of 150 l/min (a common equipment configuration) and a target excess head of 10 m, the maximum permeability that can be determined is around 1×10^{-4} m/s.

Like other forms of in situ permeability tests, packer tests have a number of limitations and may be subject to a number of errors. When carrying out these tests (and when reviewing the results), it is essential that these factors are considered:

1. Due to the relatively short duration of the injection phase, packer tests in boreholes only involve a relatively small volume of rock (and associated fracture network) around the test section – these tests provide 'small-scale' values of permeability. Such tests may not be representative of the mass permeability of the rock. Large-scale tests (such as pumping tests) may give better results.

2. Measured permeability values will be affected if the drilling process has blocked or enlarged natural fractures. Walthall (1990) states that prior to a packer test, the borehole should be cleaned out to remove all drilling debris and also recommends that the borehole be developed by airlifting. Even without drilling-related effects, the mere presence of the borehole may lead to stress relief and stress re-distribution around the borehole, changing the local permeability.

3. A problem that sometimes occurs is when packers do not form an effective seal with the borehole walls. During injection, water will leak from the test section and can give completely misleading results. It is good practice to try to locate packers in sections that are likely to give good seating for packers. If obtaining high-quality test results is important, it may be worthwhile carrying out a borehole geophysics

survey (including a calliper survey) prior to packer testing and selecting packer set-ting depths on that basis.

4. A common error in the analysis of packer tests is the use of an incorrect initial groundwater level when analysing the test results. When analysing the test, it is important to validate the reported pre-test and post-test water levels against the conceptual model (and levels from other boreholes) to determine whether the back-ground water levels have been correctly assessed.

5. Problems with the test execution can also affect results. Examples include insuf-ficient water supply; restrictions in pumps or pipework systems limiting applied heads or flow rates or causing fluctuations during test phases (when steady-state conditions should apply); and use of dirty or sediment-laden water for injection (which can result in clogging of the test section).

6. Errors can result from inappropriate measurement and processing of field observations:

   a) Errors in the field can include misreadings of flowmeters or pressure gauges, or the use of out-of-calibration equipment (flowmeters and pressure sensors).

   b) Errors in the processing of field observations can include misestimation of the excess head in the test section (which is not measured directly but has to be calculated from other observations, as described in Appendix 3).

   c) Errors in the calculation of permeability values for each test phase can include arithmetical errors or using inappropriate units (e.g. using flow rates in litres per minute when attempting to calculate permeability in metres per second).

12.8.4 Specialist In Situ Tests

Specialist in situ tests are occasionally carried out as part of certain types of probing tech-niques. These include dissipation tests carried out in piezocone testing (Lunne *et al.*, 1997) or in situ permeameters (Chandler *et al.*, 1990). These tests are generally used only in fine-grained soils (such as silt and clays) of low or very low permeability, and are not generally used in investigations for groundwater lowering except to investigate the permeability of aquitards, where the risk of settlement is of concern.

The most common of these specialist tests is the dissipation test, carried out as part of cone penetration test (CPT) piezocone probing. When the piezocone is advanced during probing in low-permeability soils, an excess pore water pressure will build up at the cone, greater than the natural groundwater pressure, as a result of penetration. If penetration is stopped, the excess pore water pressure will dissipate at a rate controlled by the coefficient of consolidation with horizontal drainage c_h, which is related to the stiffness E'_o and hori-zontal permeability k_h of the soil by

$$c_h = \frac{k_h E'_o}{\gamma_w} \tag{12.18}$$

where γ_w is the unit weight of water. Analysis of the test involves determining the degree of dissipation R at a given time t, where

$$R = \frac{u_t - u_o}{u_i - u_o} \tag{12.19}$$

u_t = pore water pressure at time t
u_o = equilibrium pore water pressure at the level of the test
u_i = pore water pressure recorded at the start of the test

In general, the dissipation test should be continued at least until R is 0.50 or lower. A typical dissipation test plot is shown in Figure 12.16. To estimate c_h, the time t_{50} for 0.5 degree of dissipation is determined and applied to the following formula:

$$c_h = \frac{T_{50}}{t_{50}} r_o^2 \qquad\qquad (12.20)$$

where

the time factor T_{50} is obtained from published theoretical solutions

r_o is the equivalent penetrometer radius

The theoretical solution to be used to estimate c_h should be chosen with care; it will be influenced by the strength characteristics of the soil and by the location of the pore water pressure measuring point on the piezocone tip; Lunne *et al.* (1997) present a number of possible solutions that may be used to obtain T_{50} and r_o.

Once c_h has been determined, k_h can be estimated using Equation 12.18 provided that the soil stiffness is known. The determination of permeability from piezocone dissipation tests involves a lot of assumptions and uncertainties; values determined in this way should be treated as a general indicator only. The penetration of the piezocone itself may disturb the soil and may locally reduce the permeability around the cone. This can be particularly acute if the soil has a laminated structure, when the penetration of the cone through clay layers may smear clay over the more permeable sand layers. These factors may result in piezocone results under-estimating the in situ permeability.

Specialist versions of CPT, known as the Hydraulic Profiling Tool (HPT-CPT), can be used to assess the variation of soil permeability with depth (Reiffsteck *et al.*, 2010). During the HPT-CPT push, water is injected into the formation, and the rate of flow and water pressure are recorded. The results give an indication of relative permeability. By means of

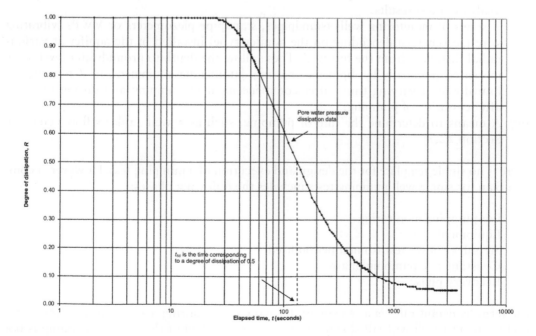

Figure 12.16 Analysis of pore water pressure data from piezocone dissipation test.

several static 'slug' tests, the site-specific hydraulic behaviour can be validated and, under favourable conditions, translated into absolute values of permeability.

12.8.5 Pumping Tests

A correctly planned, executed and analysed pumping test is often the most reliable method of determining the mass permeability of water-bearing soils and rocks. This is principally because the volume of soil/rock through which flow of water is induced is significantly greater than in the cases of variable and constant head tests in boreholes and monitoring wells. Under the classification shown in Figure 12.1, pumping tests can provide large-scale values of permeability. Unfortunately, due to the relatively high cost of a pumping test compared with these methods, it is used less frequently than is desirable.

The conventional form of pumping test involves controlled pumping from a single well and monitoring the flow rate from the well and the drawdown of groundwater levels in monitoring wells at varying radial distances away. This type of test is essentially the same as that used by hydrogeologists when investigating the potential for groundwater supply. There is therefore an extensive body of literature providing methods of analysis of such tests in a variety of hydrogeological settings.

This section mainly discusses single well tests, but alternative types of pumping test are outlined in Section 12.8.5.2. Groundwater control trials (which have similarities to multiple well pumping tests) are described in Section 12.8.6.

12.8.5.1 Single Well Pumping Test

The essentials required for a single well pumping test (Figure 12.17) are:

 (i) A central water abstraction point, usually a single deep well, which forms the 'test well'. Ideally, this should penetrate the full thickness of the aquifer, as this simplifies analysis of test results.
 (ii) A series of monitoring wells (standpipes, standpipe piezometers or VWPs (vibrating wire piezometers) depending on the aquifer type) installed in the aquifer at various radial distances from the test well. These allow the depth to groundwater level to be measured (manually by dipmeter or using datalogging equipment) to determine the drawdown at varying times after commencement of pumping and during recovery after cessation of pumping.
(iii) A means to determine the rate of pumping, such as a weir tank or flowmeter (see Section 22.7).

There is no single template for the design and execution of a pumping test. However, a common plan for single-well pumping tests includes the following phases:

1. Pre-pumping monitoring
2. Equipment test
3. Step-drawdown test (sometimes omitted)
4. Constant rate pumping phase
5. Recovery phase

The term 'constant rate' is a misnomer. In many cases, during a pumping test, the water level in the test well will fall slowly during the later period of the test. If the pump is not adjusted, the pumped flow rate will tend to reduce slightly as the increased drawdown in the

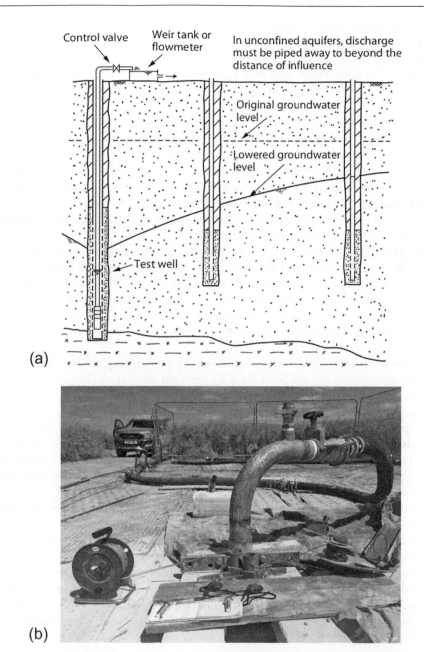

Figure 12.17 Well pumping tests. (a) Components of a pumping test. For clarity, only two monitoring wells are shown, but several are normally required. (b) Well head arrangement for test well. Test equipment visible includes well head with sampling point and control valve, with the discharge pipe leading to a flowmeter. A dipmeter is being used to record water levels in the well. (Panel (b) courtesy of Stuart Wells Limited.)

well changes the discharge head at the pump. In some hydrogeological settings, for example where the aquifer is bounded and large drawdowns are generated, the pumped flow rate may reduce significantly. Where the test is for the design of groundwater control systems, it is normal practice to allow the flow rate to reduce and not to attempt to adjust the pump to keep the flow rate constant.

To obtain the most relevant data for groundwater control schemes, the 'constant rate' phases of pumping test are typically carried out at the maximum achievable flow. The test well is commonly 'over pumped' with drawdown in the well lowered to close above the pump intake. As a result, there may be a significant seepage face, so the water level in the pumped well often does not provide a useful guide to conditions in the aquifer outside the well.

In addition to monitoring of groundwater levels and the discharge flow rate, a test is also a useful opportunity to obtain groundwater samples from chemical testing. Further details on the execution of pumping tests is given in Appendix 4. Test methods are defined in BS ISO 14686: 2003 and BS EN ISO 22282-4:2012. The drillers' borehole log of the test well itself may provide additional information about ground conditions. Geophysical logging of the well may also provide useful information (see Sections 12.5.3.2 and 12.8.7). Such methods can be especially useful for wells installed into rock aquifers to allow identification of fractured zones where groundwater is entering the well.

A pumping test is a relatively expensive way of determining permeability. The cost of a pumping test is seldom justified for a small project or for routine shallow excavations. However, for any large project or deep excavation, or where groundwater lowering is likely to have a major impact on the construction cost and/or schedule, one or more pumping tests should be carried out. It is essential that a pumping test is planned to provide suitable data for dewatering design. Issues to be considered include:

- The drawdown that will be observed in nearby monitoring wells (which at the end of constant rate phase should be at least 10 per cent of that in the proposed dewatering system). The drawdown should also be significantly greater than any natural variations in groundwater level due to tidal or other effects.
- The duration of the constant rate phase. For pumping tests used for the design of groundwater lowering systems, typical durations of pumping are between 2 and 7 days. Longer tests (with durations up to 60 days) are occasionally carried out, especially where the environmental impact of drawdown on distant groundwater-dependent features (see Chapter 21) is a concern.
- The number and position of monitoring wells (which should allow the drawdown pattern around the test well to be fully identified).
- The design of the test well (which, ideally, should be of similar design to, and be installed by the same methods as, the proposed dewatering wells).
- The requirement (or benefits) of additional monitoring over and above monitoring of pumped flow rates and groundwater levels in monitoring wells. Additional monitoring may include more detailed water quality monitoring or observations of groundwater levels in more distant groundwater-dependent features (e.g. water supply wells, wetlands, springs or streams).

To meet these aims, a pumping test is generally carried out in the second or subsequent phases of ground investigation, when ground conditions have been determined to some degree. This allows the depth of the test well screen and monitoring well response zones to be selected on the basis of the data already gathered.

A fundamental requirement for a pumping test is the need to dispose of the pumped water. For most pumping tests, the volume of water generated is too large to store on site. The water is typically discharged to surface watercourses or to sewers (pumping and re-injection tests are discussed in Section 12.8.5.2). When a pumping test is planned, checks must be made that the disposal route has adequate capacity, and the necessary permissions must be obtained from the regulatory bodies (such as Environment Agency in England) or from the utility companies responsible for the sewer network. Regulatory permissions associated with pumping of groundwater are discussed in Section 25.4.

Analysis of the test results can provide information that is useful in a number of ways:

a) Data from the step-drawdown test can be used to analyse the hydraulic performance of the well in order to determine well losses and efficiency. This approach is widely used in the testing of water supply wells (see Clark, 1977, for methods of analysis), but is less widely relevant to temporary works wells for groundwater lowering purposes. It can still be useful when trying to optimize the performance of deep well systems and may allow comparison of the performance of wells drilled and developed by different methods.

b) Data from the constant rate pumping phase can be used to estimate the permeability and storage coefficient of the volume of aquifer influenced by the test. The permeability values estimated from the test results are used as an input parameter in the design methods described in Chapter 13. Permeability is estimated by conventional hydrogeological analyses (described later), which can also provide some information about the aquifer boundary conditions.

c) Observations of the way drawdown reduces with distance from the test well can be used to construct a distance–drawdown plot, which can then be used to design groundwater lowering systems by the cumulative drawdown method (see Section 13.8.2). This method is interesting because the design does not need a permeability value, since that is implicit in the distance–drawdown plot.

Most pumping tests are analysed by 'non-steady-state' techniques, which are relatively flexible methods and can be applied to data even as the test is continuing. This allows data to be analysed almost in 'real time'. 'Steady-state' methods of analysis can be used but may require much longer periods of pumping than are necessary with non-steady-state methods. In general, analysis by non-steady-state techniques is to be preferred.

There are a wide variety of methods of analysis that can be used to analyse the results of the constant rate pumping phase, many of which are usefully summarized in Kruseman and De Ridder (1990). Each of the methods is based on a particular set of assumptions about the aquifer system (unconfined, confined or leaky), the well (fully penetrating or partially penetrating) and the discharge flow rate (generally assumed to be constant). Methods suitable for analysis of data from the recovery phase are also available.

These methods should be viewed as a 'tool kit' providing a range of possible analysis methods. Provided that the basic details of the aquifer and well are known, it is normally straightforward to select one (or more than one, if uncertainty exists over aquifer conditions) method appropriate to the case in hand. Commonly used methods of analysis fall into two main types:

(i) Curve fitting methods: These typically involve displaying on a log–log graph, for each monitoring well, drawdown against elapsed time. The data will generally form a characteristic shape. The data curve is then overlain with a theoretical 'type curve' and the relative positions of the two curves adjusted until the best match of the shape of the two curves is obtained. Once a match is achieved, the permeability and storage coefficient can be determined by comparing values from each curve. These methods were developed from the work of Theis on simple confined aquifers, but variations are available for various other cases (see Kruseman and De Ridder, 1990). The curve fitting process can be done manually but can be rather tedious. However, in recent years, commercial software packages are being increasingly used for pumping test analysis, which speeds up the process.

(ii) Straight line methods: This approach involves plotting sets of data so that characteristic straight lines are produced, allowing the permeability and storage coefficient to be determined from the slope and position of the line. These methods are a special case of the Theis solution based on the work of Cooper and Jacob (1946) and are often called the Cooper–Jacob methods. Two approaches are possible and can be used on the same drawdown dataset from a test:

    a. Time–drawdown diagrams involve plotting the drawdown data from one monitoring well against elapsed time since pumping began (Figure 12.18); this process is repeated for all monitoring wells being analysed.

    b. Distance–drawdown diagrams plot the drawdown recorded (at a specific elapsed time) in all monitoring wells against the distance of each monitoring well from the test well (Figure 12.19).

The Cooper–Jacob straight line method is a widely used method of analysis on groundwater control projects, mainly due to its relative simplicity. The original Cooper–Jacob method was based on horizontal flow to fully penetrating wells in confined aquifers, but it can also be used in unconfined aquifers where the drawdown is a small proportion (less than 20 per cent) of the original aquifer saturated thickness.

For the time–drawdown data from a single monitoring well, aquifer permeability k and storage coefficient S are determined as follows. From the semi-log graph (Figure 12.18), draw a straight line through the main portion of the data (the data will then deviate from the straight line at early times and possibly at later times). From the graph, obtain the slope of the straight line, expressed as Δs, which is the change in drawdown s per log cycle of time. Also determine t_o, the time at which the straight line intercepts the zero drawdown line – k and S are then obtained from

$$k = \frac{2.3q}{4\pi\Delta sD} \qquad\qquad (12.21)$$

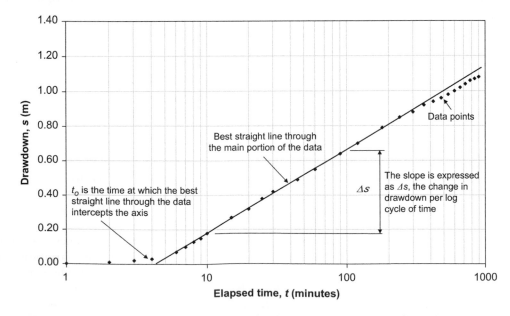

Figure 12.18 Analysis of pumping test data: Cooper–Jacob method for time–drawdown data.

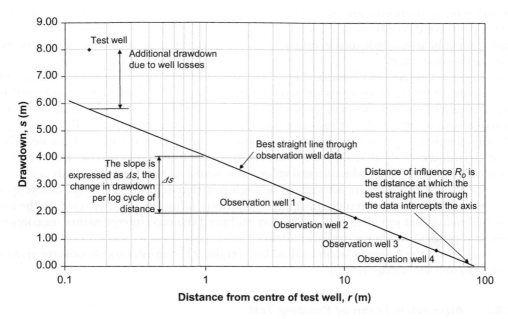

Figure 12.19 Analysis of pumping test data: Cooper–Jacob method for distance–drawdown data.

$$S = \frac{2.25kDt_o}{r^2} \tag{12.22}$$

where

  q is the constant flow rate from the test well
  D is the aquifer thickness
  r is the distance from the centre of the test well to the monitoring well

This process can be repeated for each monitoring well.

For the distance–drawdown approach, data from all monitoring wells, at elapsed time t after pumping started, are plotted on the semi-log graph (Figure 12.19). A straight line is drawn through the monitoring well data. If the test well has a very much larger drawdown than the monitoring wells, this may be the result of well losses. In such cases, the straight line should be based on the monitoring wells only and should not include the test well drawdown. From the graph, obtain the slope of the straight line, expressed as Δs, which is the change in drawdown s per log cycle of distance. Also determine R_o, the distance at which the straight line intercepts the zero drawdown line – k and S are then obtained from

$$k = \frac{2.3q}{2\pi\Delta sD} \tag{12.23}$$

$$S = \frac{2.25kDt}{R_o^2} \tag{12.24}$$

The analysis can be repeated for various elapsed times. This approach also allows the distance of influence R_o to be estimated, which can be a useful check on values used in later dewatering designs.

Care must be taken to ensure that consistent units are used in these equations. To obtain permeability k in metres per second (its usual form in dewatering calculations), the following conventions are used: Δs, D, R_o and r must be in metres; q must be in cubic metres per second (not litres per second – this is a common cause of numerical errors); and t and t_o must be in seconds (not in minutes, which is often the most convenient way to plot drawdown–time data).

Because the Cooper–Jacob method is only a special case of the more generic Theis solution, a check must be made to ensure that the method is valid. The Cooper–Jacob method can be used without significant error provided that $(r^2 S)/(4kDt) \leq 0.05$. This means that the approach is valid provided that t is sufficiently large and r is sufficiently small.

Comprehensive analysis of a pumping test may produce several values of permeability. Time–drawdown analysis will produce one result for each monitoring well, and distance–drawdown analysis can produce several values depending on how many times are analysed. These permeability values are all likely to be slightly different either due to uncertainties in analysis or in response to changes in aquifer conditions across the zone affected by the test. This highlights that pumping test results may still need detailed interpretation; in complex cases, it is prudent to obtain specialist advice.

Variations on the Cooper–Jacob method for certain other aquifer conditions are given in Kruseman and De Ridder (1990).

12.8.5.2 Alternative Forms of Pumping Test

Although the single well pumping test is the most commonly applied test, other types can be relevant in some circumstances. Different test types are summarized in Table 12.2, which also includes the phases commonly included in conventional single-well tests.

Where artificial recharge of groundwater (see Section 17.13) is planned, special pumping test programmes involving re-injection of water may be required. Typically, the test is started with a conventional pumping phase, whereby water is pumped from well(s) and discharged to surface water or a sewer. In later stages, the water from the pumped well(s) is diverted to recharge well(s), usually located at a significant distance from the pumping wells. This allows data to be gathered on both the capacity of the recharge wells and the interaction between the pumping and recharge wells. A test using two pumping wells and three recharge wells is described by Roberts and Holmes (2011).

12.8.6 Groundwater Control Trials

Groundwater control trials are an extension of the ethos of pumping tests in that they are large-scale in situ tests to determine the hydraulic properties of the ground. Furthermore, a carefully planned trial can provide other useful information.

Instead of a single test well, a groundwater control trial involves pumping from a line or ring of wells of some sort (wellpoints, deep wells or ejector wells). Obviously, to specify a suitable trial, the main dewatering design must be reasonably advanced to allow selection of the appropriate pumping method, well spacing, well screen depth, etc. As with a pumping test, the trial is pumped continuously for a suitable period (typically 1 to 4 weeks) while monitoring discharge flow rate and water levels in monitoring wells. The results of such tests are normally analysed using the design methods of Chapter 13 to 'back-calculate' an equivalent soil permeability. Powrie and Roberts (1990) describe an example of a trial using ejector wells.

In addition to determining the permeability, trials provide opportunities to investigate other issues relevant to the proposed works:

(i) Because several wells have to be installed, the relative performance of proposed installation methods (e.g. different types of drilling or jetting) can be assessed.

Table 12.2 Characteristics of Different Phases of Pumping Tests

Test type	Typical duration	Outline of typical test and test objective
Phases of conventional single well pumping test		
Equipment test	15–60 minutes	Short period of continuous pumping. Objective is to confirm pumps, pipework, etc. are functioning and to inform setting of pumping rates for later phases.
Yield test	1–8 hours	Single well pumped at nominal constant rate. Objective is to estimate well yield. Test duration is too short to determine hydrogeological conditions as reliably as can be achieved by longer constant rate test phases.
Step test	8–12 hours	Single well pumped in stepwise fashion with increasing flow rates (typically 60 to 100 min per step). Objective is to estimate well performance, including well yield.
Constant rate pumping test	1–60 days	Single well pumped at nominally constant rate. Objective is to assess drawdown in aquifer over a wide area and allow derivation of hydrogeological parameters and boundary conditions. Typical test durations up to 7 days for groundwater control design. Longer duration tests (>7 days) are more relevant to projects where external environmental impacts are a concern.
Other types of pumping test		
Pumping and re-injection test	2–28 days	Water pumped from well(s) and re-injected to other well(s). Typically, re-injection is via recharge wells located a significant distance from the pumped wells. Objectives are same as for constant rate test plus estimating recharge well capacity and assessing interaction between abstraction and recharge wells.
Dewatering trial/ pilot test	5–28 days	A group of wells, typically forming all or a sub-section of the proposed dewatering system, is pumped. Objective is to investigate the effectiveness of the full system (see Section 12.8.6).
Cut-off wall pumping test	1–5 days	A well or sump within the area enclosed by a cut-off wall is pumped. The objective is to investigate the effectiveness of the cut-off wall. Typically, groundwater levels are observed in monitoring wells outside the cut-off walls to observe any external drawdown (see Section 18.14).

Adapted from Preene, M, et al., Pumping tests for construction dewatering in Chalk, in *Engineering in Chalk: Proceedings of the Chalk 2018 Conference* (Lawrence, J A, Preene, M, Lawrence, U L and Buckley, R, eds), ICE Publishing, London, pp. 631–636, 2018.

(ii) If the trial consists of a ring of wells, a trial excavation can be made inside the ring during the trial. This excavation can provide data about ease of excavation of the soil; stability of dewatered excavations; and trafficability of plant across the dewatered soil. There have been cases where trial excavations have been combined with large-scale compaction tests to assess the suitability of the excavated material as backfill elsewhere on site.

Figure 12.20 shows examples of groundwater control trials on sites where rings of well-points were installed and the area within the ring excavated during pumping. This type of trial can not only provide information on the feasibility of dewatering at the site; it can also allow battered side slopes to be cut at various trial slopes and provide useful information on the handling characteristics of the dewatered soil.

Groundwater control trials are probably the most expensive form of in situ permeability test, but for large projects in difficult ground conditions, they can reduce risks of problems during the main works, and the cost may be justified on that basis. If the dewatering system is to be designed by the observational method (see Section 13.3.2), a trial can be a key part of the scheme.

(a)

(b)

Figure 12.20 Wellpoint dewatering trial. (a) Ring of wellpoints installed around trial area. A 20 by 20 m rectangular ring of wellpoints has been installed around the trial area. Local excavations are made within the dewatered area to assess the behaviour of the dewatered soil when excavated. (b) Battered excavation inside the wellpoint ring with sides at various trial slope angles. A trial excavation is made within a 40 by 40 m rectangular ring of wellpoints in a silty sand. Trial batter slopes are formed to assess the stability of the soil following groundwater lowering. (Courtesy of WJ Groundwater Limited, Kings Langley, UK.)

12.8.7 Borehole Geophysics

Borehole geophysical methods (Section 11.7.4.2) have traditionally been used extensively in hydrogeological studies for the development of groundwater resources but are used less commonly in investigations for construction projects. These methods do not generally give *direct* estimates of permeability values; as discussed in Section 12.5.3, they are typically used to provide non-quantitative data to allow indirect estimation of permeability and identification of zones of relatively higher or lower permeability.

However, there are some borehole geophysical techniques that can provide quantitative values of permeability. In this approach, special sensors (termed 'sondes') are slowly lowered into a borehole using a cable and winch (Figure 11.3). Two methods are discussed in the following sub-sections: borehole magnetic resonance (BMR) and borehole flowmeter logging.

Borehole geophysics is a specialized technique, and results are sensitive to background hydrogeological conditions and the measurement techniques used. Such investigations should be designed, executed and interpreted by suitably experienced specialists.

12.8.7.1 Borehole Magnetic Resonance (BMR)

Borehole magnetic resonance (BMR) is a technique that has been widely used in the oil & gas industry for more than two decades to evaluate reservoir properties. This technique is now sometimes used for investigations for construction and mining projects. As the sonde is lowered down the borehole, a magnetic field is used to cause resonance of hydrogen nuclei in the surrounding strata (Behroozmand *et al.*, 2014). This allows the sensing method to determine porosity and the proportion of water in the pores that is potentially mobile and immobile (bound to clay particles or held in small pores by capillary forces). Processing of the data can allow direct estimates of permeability. The continuous nature of the logging process means that a continuous trace of permeability with depth is generated. In combination with other data sources, this can be useful in assessing changes of permeability with depth.

12.8.7.2 Borehole Flowmeter Logging

Flowmeter logging uses a sonde to measure the vertical velocity of water flow in a borehole, which can identify zones of inflow and outflow. The most common technique uses a flowmeter sonde with an impellor that rotates to record vertical water velocity. The sonde is lowered into and out of the well at a constant winch speed, providing a continuous record of vertical water velocity; an example log is shown in Figure 12.21a. An alternative approach is the heat pulse flowmeter, where a sonde is lowered to a specific depth and held there for a few minutes (to allow water flow to stabilize). The sonde then generates a heat pulse, and vertical water velocity is determined from the time taken for the pulse to reach temperature sensors along the length of the sonde. The sonde is then moved to a different depth and the process repeated.

A vertical trace on a flowmeter log indicates constant velocity, which implies no inflow or outflow to that section of the borehole. Changes in the velocity trace indicate inflow at that section of borehole, with the inflow rate proportional to the slope of the velocity line. As the sonde is lowered down the borehole, the velocity will reduce progressively as each inflow zone is passed.

Flowmeter logging is normally carried out after at least 24 hours have elapsed since the most recent work on the borehole (either drilling or development) – this is to allow the establishment of stable groundwater flow conditions. Logging is then done under unpumped (static) conditions, when the vertical water flow is due to natural movement only. Unfortunately, the small velocities typically observed can make inflow zones difficult to differentiate. A more useful approach is to log under pumped conditions. This requires the installation of a temporary pump in the upper section of a borehole with the flowmeter logs run beneath the pump. This gives a greater velocity contrast along the borehole and makes it much easier to identify inflow zones, as illustrated in Parker *et al.* (2010, 2019). The vertical flow velocities can be combined with the borehole diameter (measured by calliper log) to provide values of vertical flow rate. This can allow the borehole inflow to be assessed from the difference in vertical flow rate between discrete depths. If the overall transmissivity (permeability multiplied by aquifer thickness) has been determined (e.g. from a pumping

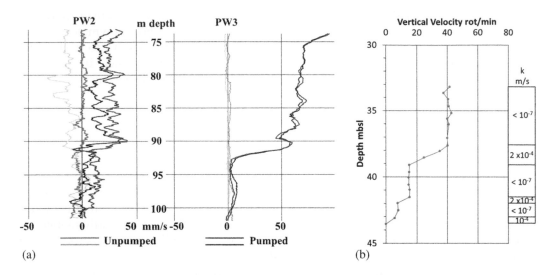

Figure 12.21 Examples of geophysical logs using impellor flowmeter to record vertical water velocity in boreholes. (a) Impellor flowmeter log from two wells in fractured Chalk bedrock tested under unpumped and pumped conditions, Beckton, East London, United Kingdom. The change in velocity under pumped conditions indicates significant inflow at approximately 90 to 91 m depth. (b) Annotated impeller flowmeter log with assessed permeability based on velocity data, Storebaelt, Denmark. (From Roberts, T O L and Hartwell, D J, Chalk permeability, in *Engineering in Chalk: Proceedings of the Chalk 2018 Conference* (Lawrence, J A, Preene, M, Lawrence, U L and Buckley, R, eds)., ICE Publishing, London, pp. 425–430, 2018. With permission.)

test), the vertical flow rate profile can be used to allocate the transmissivity proportionally to different depth zones of the borehole. The thickness of each zone can then be used to derive a permeability for each zone of transmissivity.

Figure 12.21b shows an example of a flowmeter log assessed in this way. Roberts and Hartwell (2018) state that analysis of a pumping test gave a transmissivity that implied an average permeability of 10^{-5} m/s over the whole length of the borehole. The velocity recorded by a flowmeter log indicated a small number of distinct flow zones with little inflow in between. The velocity log allowed inflow to be allocated to different depth zones, and transmissivity to be proportioned and permeability assigned for each vertical section. The interpreted log in Figure 12.21b shows that assessed permeabilities are high (of the order of 10^{-4} m/s) in the flow zones and much lower ($<10^{-7}$ m/s) in the other zones; there is no zone with a permeability of 10^{-5} m/s, as implied by the pumping test.

12.9 COMPARISON OF PERMEABILITY ASSESSMENT METHODS

The foregoing sections have described the methods commonly used to estimate permeability. Meaningful and representative values of permeability are essential for the design of groundwater lowering projects, but the designer may be presented with a very wide range of values (perhaps two to three orders of magnitude), with a much greater variation than expected based on the anticipated heterogeneity of the soil or rock. Unfortunately, limitations in the permeability testing and analysis methods often introduce some 'apparent' variation (independent of the properties of the aquifer) in the results. Some of the variation may be due to permeability being assessed at differing scales by different methods (Figure 12.1). Groundwater control systems influence large volumes of the ground, and ideally, large-scale

Table 12.3 Tentative Guide to the Reliability of Permeability Estimates

Method	Characteristics and limitations	Typical scale (Figure 12.1)	Relative cost	Reliability[1]
Non-quantitative methods				
Visual assessment	The description or classification of the soil or rock is compared with published permeability values of similar strata – can be very useful to validate quantitative data from other sources. For soils, this approach can allow the order of magnitude of permeability to be estimated. Less applicable to fractured rock.	Very small to small	Very low	Good
Drilling records	Observations of water inflow during drilling can indicate zones of higher permeability. Some drilling methods may mask water inflows, reducing the reliability of the information.	Very small to small	Very low	Poor to good
Borehole geophysics	Sensors (sondes) are lowered into the borehole to record properties of the strata and the fluid in the borehole. Can give indirect estimates of permeability and allow identification of zones of higher and lower permeability. Typically applicable in rock to identify fractured horizons within boreholes.	Very small to small	Moderate to high	Poor to good
Quantitative methods				
Empirical methods				
Correlations with particle size distribution (PSD)	Elements of soil grading determined from PSD curves of soil samples are used in empirical correlations. Each correlation method (e.g. Hazen; Kozeny–Carman; Prugh) has a limited range of PSDs to which it should be applied; use outside those ranges can result in gross errors. Samples may be unrepresentative because a) sampling and the PSD testing process results in homogenization of samples and loss of structure and fabric, and b) fine particles may be lost from the sample (typically flushed out into the borehole fluid). Gives very poor estimates in laminated or structured soils or where silt and clay content is significant.	Very small	Very low	Very poor to moderate
Correlations with description of rock quality	Permeability can be correlated with rock mass classification systems such as RQD, etc. It is possible to develop project-specific correlations, but no generic correlations are available.	Very small	Very low	Not widely applicable
				(Continued)

Table 12.3 (Continued) Tentative Guide to the Reliability of Permeability Estimates

Method	Characteristics and limitations	Typical scale (Figure 12.1)	Relative cost	Reliability[1]
Inverse numerical modelling	Pumping rate and groundwater level data are used to assess permeability from full-scale dewatering using numerical modelling tools. Can only be applied during or after significant phases of a project are in progress provided that adequate monitoring is in place. Useful to refine designs for later phases. Only of value if the hydrogeology of the aquifers/aquitards at the site is defined reasonably clearly and if adequate monitoring data are available.	Large to very large	High (when including cost of pumping tests and monitoring)	Good
Ex situ test methods				
Laboratory testing of soil samples	Water flow is induced through a cylindrical specimen of soil. Sample size is limited to a few hundred millimetres. Disturbance (disruption of structure and fabric) and changes in void ratio and stress state due to sampling mean that permeability values may not be representative. Such tests are of little use in granular sand or gravels, as representative samples cannot be obtained without unacceptable disturbance. Can produce good results in clays and some silts where minimally disturbed samples can be obtained. In soils with fabric and structure, permeability estimates are affected by sample size; smaller samples may under-estimate permeability.	Very small	Low to moderate	Very poor to moderate
Laboratory testing of rock samples	Water flow is induced through a cylindrical core of rock. Sample size is limited to a few hundred millimetres. Disturbance and changes in stress state due to sampling mean that permeability values may not be representative. Permeability in rock is controlled by the orientation and connectivity of fracture networks. Permeability estimates are affected by sample size; smaller samples may under-estimate permeability.	Very small	Low to moderate	Very poor to moderate

(Continued)

Table 12.3 (Continued) Tentative Guide to the Reliability of Permeability Estimates

Method	Characteristics and limitations	Typical scale (Figure 12.1)	Relative cost	Reliability[1]
In situ test methods				
In situ tests in boreholes: falling head, rising head and constant head tests	Common method (also known as slug tests) for testing discrete zones during borehole drilling, especially in soils. Falling head tests cause water to flow out of the borehole, and rising head tests induce flow into the borehole. In high-permeability strata, water levels change rapidly during the test, and the rate of change can be difficult to record manually (the problem can be overcome by the use of VWP pressure sensors linked to dataloggers). In low-permeability soils, a test may need to last several hours, as excess heads slowly dissipate. The presence of a disturbed zone (caused by drilling) around the borehole can result in unrepresentative responses. Can test only a small volume of soil at the base of the borehole. • Falling head tests and constant head tests where water is added to the borehole may be prone to clogging of the test section. • Rising head tests and constant head tests where water is removed to the borehole may be prone to boiling or loosening of the soil in the test section.	Small	Low	Very poor to moderate
In situ tests in monitoring wells: falling head, rising head and constant head tests	Common method (also known as slug tests) for carrying out permeability tests in monitoring wells. Falling head tests cause water to flow out of the well, and rising head tests induce flow into the well. In high-permeability strata, water levels change rapidly during the test, and the rate of change can be difficult to record manually (the problem can be overcome by the use of VWP pressure sensors linked to dataloggers). In low-permeability soils, a test may need to last several hours, as excess heads slowly dissipate. The presence of a disturbed zone (caused by drilling) around the monitoring well response zone can result in unrepresentative responses. Can test only a small volume of soil or rock around the response zone of the monitoring well. • Falling head tests and constant head tests where water is added to the well may be prone to clogging of the test section, especially if the added water is not completely clear.	Small	Low	Very poor to good
				(Continued)

Table 12.3 (Continued) Tentative Guide to the Reliability of Permeability Estimates

Method	Characteristics and limitations	Typical scale (Figure 12.1)	Relative cost	Reliability[1]
Packer permeability testing in boreholes in rock	Widely used for testing of boreholes in fractured rock, typically by injecting water to create an excess head in a section of borehole isolated by inflatable packers, to induce flow out of a borehole. Produces an average permeability for the test section; it can be difficult to determine whether the permeability is associated with many distributed fractures or a smaller number of discrete fractures.	Small	Low to moderate	Very poor to good
Specialist in situ tests	Pore water pressure changes are recorded by piezocones and other probing methods. Appropriate to silts, clays and some sands. Can test only a small volume of soil near the probe but can be used to profile permeability with depth. If the tests involve penetrometers, disturbance or smear of soil layering may occur, affecting results.	Very small to small	Moderate	Moderate to good
Pumping tests	Water is pumped from a well while pumped flow rate and drawdown in neighbouring monitoring wells are monitored. Provided that the pumped flow rate is sufficient to create a large drawdown in the strata around the well, and if pumping continues for an extended period (typically several days), a large volume of the ground can be influenced. The large-scale permeability values from this method are a good match for the needs of dewatering designers. May also provide information on hydraulic boundary conditions. Appropriate to soils and rocks of moderate to high permeability. Rarely appropriate for low-permeability strata (such as clays and silts or unfractured rock) unless pumped by ejector wells and monitored by appropriate types of rapid response monitoring wells (such as VWP piezometers). Results can be difficult to interpret if multiple aquifer systems are present beneath the site.	Large	Moderate to high	Good to very good
Groundwater control trials	Similar to well pumping tests but typically pumping from multiple wells. Potentially affects even greater volumes of the ground. Appropriate for large-scale projects in difficult ground conditions or where the observational method is being used. Requires careful planning.	Large to very large	High to very high	Very good
Borehole geophysics	Flowmeter logging can be used to determine a profile of vertical flow velocity within the borehole that can be used to determine vertical variations in permeability. Typically applicable in rock to identify relative permeability of fractured horizons within boreholes.	Very small to small	Moderate to high	Poor to good

1 When used for groundwater control purposes and applied in soil or rock conditions suitable to the method, and when analysed appropriately.

permeability values (such as can be obtained from well pumping tests) should be used in design. Small-scale and very small-scale permeability values are less useful, as they may not include the effect of more permeable pathways (coarser fabric in soils and fracture networks in rock).

Permeability is a difficult parameter to determine accurately, and when selecting values to be used in design, some uncertainty is unavoidable. During design, this can be addressed by:

• Reviewing the permeability values against the conceptual model
• Validating test results against non-quantitative data sources to exclude likely unrepresentative values
• Carrying out sensitivity analyses to evaluate the effect of differing permeability on drawdown and flow rate

When presented with values from a range of methods, deciding on the range of permeability to be used in design calculations is made easier if the relative reliability of each method is known; some guidance is given in Table 12.3. The selection of appropriate permeability values for the design of groundwater control systems is discussed in relation to conceptual models in Section 13.4.

Chapter 13

Design of Groundwater Lowering Systems

Any design that relies for its success on a precise calculation is a bad design.

Hugh Golder

13.1 INTRODUCTION

The philosophy and basic methods for the design of groundwater lowering systems are outlined in this chapter. The main emphasis of this chapter, and indeed of most of this whole book, is to deal with field-proven, simplistic but practical methods applicable to many common situations while providing advice on the approach to more complex problems.

The uncertainty inherent in any ground engineering process requires a 'questioning' or 'testing' approach to be adopted in design, whereby nothing is taken for granted. Sensitivity or parametric analyses can be used to assess how the design could cope with differing conditions. Alternatively, the observational method might be used to vary the design based on records taken during construction. Both these approaches are outlined, and the vital importance of developing a realistic conceptual model is stressed. The important effect that geometry and geological structure have on design is also described.

The basic designers' 'tool kit' – the formulae and concepts used in routine designs – is presented and its application discussed. Methods for estimation of steady-state discharge flow rate and for selection of well yield and spacing are described in detail. Other design issues, including assessing the base stability of excavations, are also outlined. The basic tenets of groundwater modelling are discussed in relation to more complex problems. Several simple design examples are presented in Appendix 5.

13.2 WHAT IS DESIGN?

In the context of dewatering, design is the process of developing a workable and economic solution to a groundwater problem that will affect (or is already affecting, if dewatering is applied as a remedial measure) an excavation. In some cases, design starts with a clearly defined and understood groundwater problem, and the main design objective is to develop an appropriate solution. In other situations, the nature of the groundwater problem may not be well understood, and the initial stages of design can involve investigation and definition of the problem before specific solutions can be developed.

In essence, there are three elements to dewatering design: modelling, analysis and judgement.

Modelling: To develop a potentially viable dewatering solution, it is necessary to formulate and record the various elements of the problem – in other words, to develop a model that allows the problem to be described and communicated. Every dewatering problem should have a *conceptual model* developed at an early stage (see Sections 5.4 and 13.4). For some complex or important dewatering designs, numerical models may be developed as part of later analysis (Section 13.12). Development of the conceptual model is the foundation of good dewatering design. If this stage is neglected, then the chances of successful outcomes are reduced significantly.

Analysis: This is the stage of design at which numerical calculations or analyses are carried out, for example to estimate the required pumping flow rate or to assess potential environmental impacts. Analysis may be by the conventional closed-form analytical equations described in this chapter or may be by numerical modelling. It is essential that the analysis methods used are appropriate to the conceptual model developed earlier. Analytical models are discussed in Section 5.5.

Judgement: Modelling and analysis alone are not sufficient to ensure the design of effective groundwater control systems. An element of judgement is required to ensure that the proposed design is realistic and practicable. Judgement cannot be learned from a book such as this but is developed through accumulated experience. But as a guide, judgement might involve comparing the proposed dewatering design with established empirical guidance or with examples of similar dewatering systems in comparable ground conditions. Some specific examples of the use of judgement are given in the empirical checks applied to the design examples in Appendix 5.

Successful design will typically require all three elements to be applied in some combination. Where dewatering design produces results that are perceived to be unsuccessful, perhaps where the target drawdown is not achieved or where it becomes clear that excessive pumping capacity has been provided, it is often found that one of the elements of design has been neglected.

In the authors' experience, the design of groundwater control systems is typically not governed in detail by national or international design standards. This is unlike some forms of geotechnical engineering design (for example related to the design of piles or earth retaining structures), where requirements may be set out more formally in design standards. For example, while Eurocode 7 (BS EN 1997-1:2004) includes a section on dewatering, it limits its requirements to relatively general comments that the design should be based on the results of ground investigations, that the objective of design should be to achieve stable excavations, and that dewatering systems should be reliable and subject to monitoring. This approach of setting 'high level' requirements without setting prescriptive limits on possible design methods is common with design standards in some other parts of the world.

13.3 DESIGN APPROACH

There are two main philosophical views of the design of groundwater lowering or dewatering systems:

1. That the design process is essentially a seepage calculation problem, which can be assessed using application of groundwater and hydrogeological theory. This approach implies that the major problem is estimating the discharge flow rate, and that selection and design of equipment are a secondary matter, carried out once the flow rate has been determined. We could call this the 'theoretical' approach, where the analysis element of design dominates. Alternatively,

2. That the design process must concentrate on selecting the appropriate type of well, well spacing and pump size for the ground conditions. Direct estimation of the flow rate is a less important issue (and indeed, may never actually be calculated precisely). This can be thought of as the 'empirical' approach using case history experience, where the judgement element of design dominates.

The authors consider that both the theoretical and the empirical have their advantages and disadvantages depending on ground conditions and the nature of the project.

For example, there are a number of cases which are sufficiently common that, once site investigation has confirmed there are no unusual complications, they can be designed purely empirically – almost by rule of thumb, based on the established capabilities of standard dewatering equipment. For example, shallow trench excavations in homogeneous sand deposits of moderate permeability can almost always be dewatered by wellpoint systems with a spacing of 1 to 2 m between wellpoints. This has been practically proved over many decades. The empirical method is not applicable in more complex (or less clearly identified) geometries and ground conditions. If applied in such circumstances, it can lead to considerable difficulties. Some of the pros and cons of empirical methods are discussed in Section 5.5.1.

The theoretical approach can be used whether or not there is empirical experience of the case in hand. The approach may involve fairly simple closed-form equations or analyses, or it may require more complex numerical modelling. The results of the calculations or modelling are used to specify the number and type of wells, pumps, etc. that will be required. Problems often arise when the theoretical approach does not take into account the limitations and advantages of the various dewatering techniques. If these issues are not considered, impractical or uneconomic system designs may result. The challenge of theoretical methods is that they require a solid conceptual understanding to be able to select and apply an appropriate theoretical model. If this is not done correctly, the results can be disappointing. This leads to the well-known saying by Professor Victor de Mello (cited in Burland, 2008b) that 'Water has an unfortunate habit of seeping through every theory.'

The best design approaches incorporate elements of both the theoretical and empirical methods. The theoretical method requires a 'conceptual model' of the ground and groundwater regime to be developed, following which calculations are carried out. Simple and fairly basic calculations are perfectly acceptable and may be preferred in many cases, provided that they are compared with an empirical approach. The empirical method should be used as a 'sanity check' to ensure that the proposed groundwater lowering system is realistic and practicable. For example, if the output of a theoretical design recommends a single stage of wellpoints to achieve a drawdown of 10 m, this is clearly not going to work. More subtly, if a wellpoint spacing of, say, 15 m is recommended, this should be looked into more closely, since this is outside the normal range of wellpoint spacings – there may be problems with the conceptual model, methods of analysis or selection of the dewatering method.

This chapter presents the methods that can be used in combined theoretical and empirical approaches. The main emphasis is to deal with field-proven practical means pertinent to the majority of groundwater lowering projects. The various methods presented form a 'tool kit' of techniques, which if selected with care, can deal with a wide range of real problems.

One defining feature of the design of any geotechnical process (including groundwater lowering) is that there will be some uncertainties in the ground. These uncertainties may result from the site investigation being of limited or inappropriate scope. Alternatively, even following a comprehensive site investigation, the sheer variability and complexity of the revealed ground conditions may give rise to uncertainty in design.

Uncertainty will affect the way the design is progressed. In principle, there are two basic approaches to design:

(i) Pre-defined designs. This might be thought of as the 'traditional' design process, whereby one geological profile and set of parameters are selected and used to produce a single set of design predictions. The design is implemented with fairly basic monitoring, limited to checking that the design is 'effective' in the gross sense. Design and construction methods are not reviewed unless the original design is 'ineffective' (e.g. not achieving the target drawdown in a dewatering design). In such cases, corrective action (alternative design and construction methods) would be taken.

(ii) The observational method. This contrasting design process begins with more than one geological profile and set of parameters and more than one configuration for the required dewatering system. These might range from 'most probable' to 'most unfavourable', perhaps with a number of intervening conditions as well. More than one set of design predictions are produced, together with measurable 'trigger values' (see Section 22.3.1), to allow the detailed effectiveness of the design to be observed during construction. The data taken during construction are continually reviewed to allow design and construction methods to be altered incrementally to match the behaviour of the ground and groundwater.

13.3.1 Pre-Defined Designs for Groundwater Lowering

The pre-defined approach to dewatering design involves the selection of a geological profile (particularly the depth and sequence of aquitard layers) and important parameters (primarily permeability) and then applying these to an appropriate method of seepage analysis (such as one of those presented in Section 13.8). The result of this calculation is the 'estimated' or 'design' discharge flow rate and drawdown distribution. Such an approach sometimes has an air of certainty or finality, especially if carried out by civil engineers more used to relatively consistent and reliable materials like steel and concrete.

In reality, few groundwater lowering systems can be designed successfully in such a regimented way. As was discussed in Chapter 3, permeability and aquifer/aquitard boundaries can have a dramatic effect on dewatering systems; the difficulty of accurately determining permeability was described in Chapter 12. These factors mean that it is unrealistic to carry out a single seepage analysis and to then expect the dewatering system to have a high likelihood of performing adequately. The eminent geotechnical engineer Dr Hugh Golder made the pertinent point (cited in Burland, 2008a) that 'Any design that relies for its success on a precise calculation is a bad design.'

The best groundwater lowering systems are flexible and robust in nature, able to cope with ground conditions slightly different from those anticipated with few, if any, minor modifications. Such systems are also easy to modify or upgrade if ground conditions are substantially different from those expected.

The pre-defined approach to design can still be used, but normally more than one set of calculations is carried out as part of 'sensitivity' or 'parametric' analyses.

- A sensitivity analysis is a set of repeated calculations using varying values of a single parameter in each calculation. Often, for dewatering designs, permeability is a key parameter, so seepage calculations are carried out for a range of possible values. The question being addressed is 'can the groundwater lowering system, as designed, cope with the range of discharge flow rates corresponding to the possible range of permeability?' If the answer is yes, all is well. If the answer is no, but the system could be easily (and quickly) modified to handle flow rates at either extreme end of the range, then the system may still be acceptably robust. If the answer is no, and the system cannot be readily modified, there is a risk that the system will not be able to handle all the possible flow rates. In such cases, it may be prudent to try to develop an alternative, more flexible design and test that against the results of the sensitivity analysis.

- Parametric analyses are a rather broader version of sensitivity calculations. The question addressed is 'what parameters are influential to the design?' A number of parameters or conditions are systematically varied in calculations to investigate which have the greatest effect on the design. This may allow the design to be varied or may prompt additional site investigation work to refine estimates of certain parameters or clarify key issues. In addition to permeability, aquifer boundary conditions can have an important effect on dewatering designs. A parametric study might look at the effect of a close source of recharge affecting the distance of influence or the base of the aquifer being deeper than expected. Again, the design process should include an assessment of the ability of the proposed dewatering system to cope with such possible conditions.

13.3.2 The Observational Method for Groundwater Lowering

The observational method allows a rational approach to construction where there is uncertainty over ground conditions or over the most suitable and economical dewatering or ground treatment options. The method uses construction observations to gather information about the behaviour of the ground and groundwater, and then modifies construction in response. The method is much more than carrying out parametric or sensitivity analyses. It should be a continuous and deliberately planned and managed process of design, construction control, monitoring and review that allows previously defined modifications to be incorporated into the construction process when necessary. The object is to achieve a robust process that provides for economical construction without compromising safety. The background, methods and case histories of the observational method are described in CIRIA Report C185 (Nicholson *et al.*, 1999) and in Hardy *et al.* (2018). Two main variants are normally identified:

a) The observational method applied from the inception of the project (sometimes known as the *ab initio* approach)

b) The observational method applied during construction (variously known as the best way out method or the *ipso tempore* approach, depending on the circumstances). In this variant, the observational method is used to allow progress when unexpected problems occur on site or to allow unplanned modifications to be made on a rational basis in response to field observations. It is often applied when a 'pre-defined' design has proved unsatisfactory and modifications are required.

The observational method can be applied to many groundwater lowering systems. This is because they can be easily modified (by the addition of extra wells or by using pumps of different capacity) and because easily observable parameters (such as drawdown and discharge flow rate) can be used to interpret how the system is performing. Examples of the observational method applied to groundwater lowering systems are given in Roberts and Preene (1994b), Nicholson *et al.* (1999) and Preene *et al.* (2016).

Probably the most common application of the observational method to groundwater control systems is the best way out method, used when a pre-defined method does not work and an alternative must be developed. It has been used when dewatering systems have had to be uprated or modified when the pre-defined design failed to achieve the design aims.

Effectively, the initial system (which was not installed with the observational method in mind) is monitored and used as a large-scale pumping test or trial to allow remedial measures to be selected.

The *ab initio* approach is less widely applied to groundwater control projects but has been used on larger projects where ground conditions are known to be complex or where the project design is not going to be finalized until construction is well advanced. The number of wells and pumping capacity can be optimized, based on the data gathered during the groundwater lowering works themselves. Optimization of the systems should be carefully considered to avoid the temptation to use the bare minimum number of wells. A truly optimized system will have adequate standby plant and alarm systems, plus some additional wells (over and above the minimum) as an allowance against loss of wells or performance due to construction damage, clogging or biofouling, and so on. Optimization of groundwater control systems is discussed in detail in Chapter 26.

One possible reason why the *ab initio* application of the observational method has mainly been applied to larger projects is the need for clear management of the design and construction process. This is necessary to allow construction feedback to be obtained and linked into the ongoing design process in a timely manner. If smaller contracts are managed on this basis, there is no reason why the benefits of the observational method could not be applied to a wider range of projects.

The methods presented in the remainder of this chapter are based largely on pre-defined methods of design but applied to give robust and flexible systems. The observational method will not be considered further, but it is an approach that the dewatering designer should be familiar with, since it is another part of the 'tool kit' to be applied when appropriate.

13.4 DEVELOPMENT OF CONCEPTUAL MODEL

The essential first step in the design process is the development of a conceptual model of the ground and groundwater conditions. Conceptual models were introduced in Section 5.4 and are a non-mathematical representation of ground and groundwater conditions that are used to help understand and communicate geotechnical and hydrogeological conditions at a site.

Designers should use the conceptual model as the foundation for their subsequent choices and decisions about which of the methods in the designer's 'tool kit' is appropriate for use for a particular project or site. If a 'poor' conceptual model – i.e. one that is not a good match for actual conditions – is used, then the outcome of the design process may be seriously flawed. The value of the conceptual model to dewatering design cannot be overstated.

The conceptual model is based on the available data on ground and groundwater conditions, the proposed construction project and the environmental sensitivities of the surrounding area. An advantage of developing the conceptual model at an early stage is that the designer will have to critically review the available data. This will help identify any significant gaps or uncertainties in the dataset.

Uncertainty is unavoidable when working below ground. Even if time and money were unlimited, it is not possible fully characterize ground and groundwater conditions. One common problem is that groundwater levels are often only monitored for a finite period, and the data available may not identify true maximum and minimum groundwater levels. The best conceptual models identify and highlight key areas of uncertainty that could affect groundwater control designs. This can help justify the need for further site investigation or allow designers to plan more robust or flexible dewatering systems to address the uncertainty.

The conceptual groundwater model depends on a very wide range of factors that are relevant to hydrogeological conceptual models, as described in Section 5.4, and listed here:

   (i) Aquifer type(s) and properties
  (ii) Aquifer depth and thickness
 (iii) Presence of aquitards and aquicludes
 (iv) Distance of influence and aquifer boundaries
  (v) Initial groundwater level and pore water pressure profile
 (vi) Presence of compressible strata
(vii) Geometry of the proposed works
(viii) Groundwater lowering technique
 (ix) Period for which groundwater lowering is required
  (x) Depth of proposed wells
 (xi) Environmental constraints

If information is available to address most, if not all, of these factors, then a conceptual model can normally be developed. Conceptual models should not be unnecessarily complex (complexity in modelling is discussed in Section 5.3). On smaller projects in straightforward ground conditions, a conceptual model may simply be a list of the expected conditions. Figure 13.1 shows an example of a simple pro-forma to record key data. Conceptual models are outlined in each of the design examples given in Appendix 5.

Any consideration of groundwater flow in general, or of the equations presented later in this chapter, highlights that aquifer permeability is a critically important parameter. Chapter 12 has described the plethora of techniques available to estimate permeability, from the simple to the very complex. When assessing permeability values to be used in calculation, the designer should not visualize a single permeability value to be used in the conceptual model but rather, a range of realistic values to be used in sensitivity analyses. The range of permeability values may represent uncertainty due to natural variations in permeability, or limitations in the permeability test methods or results.

It is difficult to give simple, useful guidelines on the selection of realistic permeability ranges. There will always be some reliance on judgement and experience, but the following advice is relevant:

- Be aware that different methods of assessing permeability produce results of greater or lesser reliability. Table 12.3 provides some guidance. Consider the relative merits of each method when assessing permeability from the available results.
- Always compare the permeability results with 'typical' values of permeability from published correlations with soil types (such as Table 4.1) or, even better, from experience at nearby sites. It is also good practice to compare permeability values used in design with 'non-quantitative' indicators of permeability (see Section 12.5). This approach is vital in excluding unrealistically high or low permeability results. For example, few experienced engineers would expect a slightly silty sandy gravel to have a permeability of 1×10^{-8} m/s, yet falling head tests in such soils often produce results of that order.
- If permeability estimates from various differing techniques produce broadly similar results, in agreement with typical values for those soil types, then the design range of permeability could be assessed from the full range of data. If there are large discrepancies in the data from various methods, some of the data may have to be excluded from the assessment process. Again, Table 12.3 and Section 12.5 may be of help in assessing the reliability of data.

PMC DEWATERING	
Site:	Eastern resewerage, phase 1
Location:	Anytown
Prepared by:	MP
Basic Information	
Aquifer type and properties:	Unconfined aquifer. Generally described as fine to medium sand. PSD results indicate k range of 1 to 5×10^{-4} m/s. Falling head tests give lower results.
Initial groundwater level:	1.5 mbgl at northern end, varying steadily to 1.8 mbgl at southern end. No long-term monitoring carried out, so background or seasonal variations unknown.
Depth to base of aquifer:	Not proved. Deepest borehole penetrated to 18 m depth and was still in sand.
Aquitards/Aquicludes present:	Some thin clay layers at 3 to 5 m depth in southern section; could result in overbleed seepage.
Recharge/Barrier boundary:	None apparent
Excavation depth and geometry:	Trench works, mainly 2.5 to 3.5 m deep. Trench width less than 1.5 m.
Maximum drawdown:	Maximum drawdown is to 0.5 m below deepest dig = 4 m depth. This implies a maximum drawdown of 2.5 m.
Dewatering period:	4 weeks per 100 m section.
Compressible strata:	None indicated in site investigation data.
Possible dewatering technique:	Wellpoints parallel to trench. Wellpoints 6 m deep, so will be partially penetrating. Risk of overbleed in south.
Sketch:	
Other Notes	

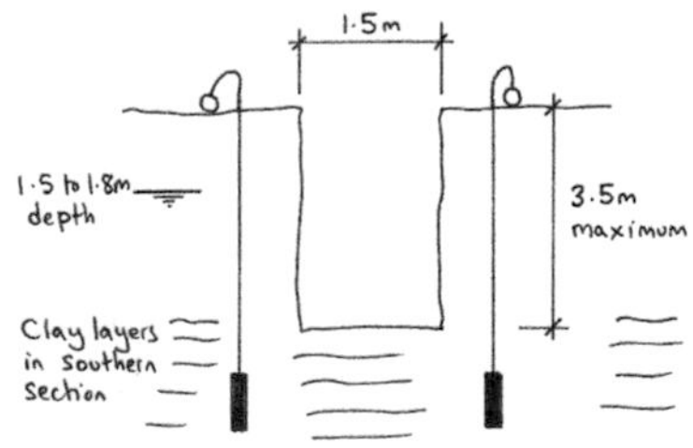

Figure 13.1 Example of simple conceptual model for groundwater lowering system.

- Always consider the important aspects of the design when selecting the permeability range to be used in calculations. Mistakes have been made when designers have focused too much on estimating the highest likely permeability to ensure a pumping capacity sufficient for the maximum possible flow rate. This is not always the most appropriate approach. It is true that in high-permeability aquifers, the total flow rate may be critical, and assessing permeability at the upper end of the possible range may be a robust approach. But alternatively, in soil and rock of low to moderate permeability, a critical case in design may be if the permeability is at the lower end of the possible range, when yields may be very low, necessitating unfeasibly large numbers of wells.

Once the conceptual model exists, an initial view must be taken of the dewatering method to be used and the likely geometry of the system.

13.5 EXPECTATIONS OF ACCURACY

It is important to be realistic with expectations of accuracy. Dewatering design should not be viewed as a precise analytical process resulting in single numerical values for key design outcomes such as pumped flow rates. The preceding parts of this chapter have repeatedly highlighted the likely uncertainty in design parameters and boundary conditions, and the resulting need to consider a range of design scenarios by means of sensitivity and parametric studies. In most circumstances, a dewatering design should not report a single value for calculated quantities such as pumped flow rate, time to achieve drawdown, estimated ground settlement or lowering of water levels at nearby groundwater-dependent features. The use of single values in design reports and calculation summaries may give the reader a false sense of precision; it is preferable in calculations and design reports that a range of values be quoted, for example 'the predicted pumped flow rate will be between 10 and 15 l/s, based on the assumptions stated in the calculations'.

Figure 13.2 presents some interesting data from a study of around 20 pumped groundwater control systems in soils of low to moderate permeability, where dewatering was carried out by wellpoints, deep wells or ejector wells. For each case, the pumped flow rate (q_c) was predicted by standard closed-form analytical methods based on a conceptual model and parameter values carefully derived from the available ground investigation information. These values were then compared with the pumped steady-state flow rate recorded in the field (q_r). This study showed that even with high standards of conceptual modelling, design and parameter selection, the best that could be achieved was to predict the pumped flow rate within a factor of three times greater than or less than the flow rate observed in the field. In a small number of cases shown in Figure 13.2, the observed pumped flow rate varied from

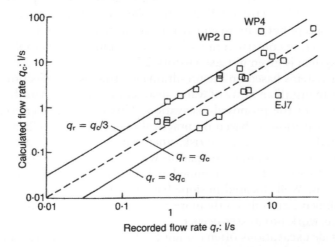

Figure 13.2 Comparison of calculated and recorded flow rates from dewatering systems. (From Preene, M and Powrie, W, *Géotechnique*, 43, 191–206, 1993. With permission.)

the observed value by a factor greater than three. In these cases, retrospective analysis indicated that the actual ground conditions differed significantly from those indicated in the site investigation information.

The data in Figure 13.2 should not be taken to indicate that methodical and rational dewatering design is futile. Rather, they show that the designer should avoid assigning undue precision to calculated values. The design process should identify the most likely value for calculated values but should also indicate the likely maximum and minimum values. These maximum and minimum values are important, as they need to be compared with the capabilities of the proposed dewatering system. For example, if the proposed dewatering system can cope with the predicted maximum and minimum values of flow rate, either in its nominal layout or with easily applicable modifications, then a robust dewatering design is likely to result. However, if the maximum and minimum predictions could not be handled by the dewatering system or would require expensive and time-consuming modifications, it may appropriate to re-visit the selection of the dewatering method and equipment and to develop a more flexible solution.

13.6 SELECTION OF METHOD AND GEOMETRY

There are a number of pumped groundwater lowering methods available (see Section 9.5), and part of the design process is to select an appropriate technique that will satisfy the various constraints on the project in hand.

A useful starting point when selecting a technique is Figure 9.11. When the required drawdown and estimated soil permeability are known from the conceptual model, the appropriate method can be chosen. Where more than one method is feasible, the choice between them may be made on cost grounds, local availability of equipment, or expertise of those carrying out the works.

13.6.1 Equivalent Wells and Slots

In practice, the required drawdown for an excavation can seldom be achieved by a single well; most dewatering systems rely on several wells acting in concert, the rare exceptions being, in a high-permeability stratum, the use of a large central sump pumping installation with radial drains (see Case History 27.2). The collector well system (see Section 17.6) is a variation on this concept but is more widely used for agricultural or land drainage purposes than for civil engineering construction (although an example of an application to a permanent dewatering system is given in Case History 27.12).

For dewatering systems based on the pre-drainage principle (see Section 9.5.1), the established approach is to install an array of water abstraction points, generally sited immediately adjacent to the area of the proposed excavation. This preferred location of wells (at the perimeter of the excavation) is based on practical experience. If wells are sited within the area to be excavated, they are greatly at risk, since maintaining wellheads, riser pipes, power cables and discharge mains is rarely compatible with the activities of heavy excavation plant. Where possible, wells are best located immediately outside the excavation area or within the batter of excavations. Wells located in slope batters, while presenting an initial excavation and access impediment, are inherently more secure (in comparison with wells in the main excavation area) as work progresses deeper.

Very large or wide excavations often cannot be dealt with adequately or economically by wells sited only around the perimeter. As highlighted in the previous paragraph, the practicalities of maintaining and/or reinstalling damaged wells in the middle of a 'live' area of excavation are far from trivial. In many such cases, unpumped relief wells (see Case History 27.4) could be used inside the excavation area. These act as vertical drainage paths, and their usefulness is not affected by progressive shortening as the excavation is deepened. The disadvantage of relief wells is their continuous water discharge, which must be disposed of by an effective drain and sump system. The disposal problem is readily controlled where the rates of flow are small (i.e. the underlying stratum is of low permeability, say lower than 10^{-6} to 10^{-5} m/s). Relief wells are described in detail in Section 17.8.

In general, the multiple well systems used for groundwater lowering can be classified as either 'linear' systems (for installation alongside trench excavations) or 'ring' systems (for installation around circular or rectangular excavations). Figure 13.3 shows definitions of these geometries.

This leads to the practical question: 'how can we model the flow to such systems?' It would be very tedious to consider the flow to each individual well and the complex interaction between them. A useful approach is to consider the groups of wells as large 'equivalent wells' or 'equivalent slots', thereby allowing simple and accessible formulae to be used to estimate the flow rate.

We can define an equivalent well as a groundwater lowering system where, on a gross scale, flow of groundwater to the system is radial. Radial flow implies that flow lines converge to the well from a distant, diffuse source of water. The concept of the 'zone of influence' of a well (discussed in detail in Section 5.8) describes the area of the aquifer affected by pumping from a well (Figure 13.4a). The radius of influence R_o is a theoretical concept describing the zone of influence. R_o is defined as the distance from the centre of the well to the edge of an idealized circular zone of influence.

An equivalent slot is a line of closely spaced wells forming a groundwater lowering system. If the line of wells is very long, the flow of groundwater to the system is planar (although there will be some radial flow to the ends of the slot). Under planar conditions, flow lines do not converge but are parallel, resulting in quite different flow conditions from radial flow. The edge of the theoretical zone of influence will be parallel to the line of wells (Figure 13.4b). The distance of influence L_o is the distance from the line of wells to the idealized edge of the zone of influence.

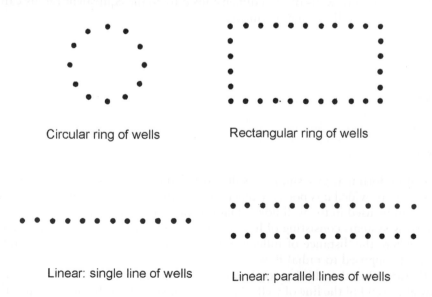

Circular ring of wells   Rectangular ring of wells

Linear: single line of wells   Linear: parallel lines of wells

Figure 13.3 Plan layouts of groundwater control systems: linear and ring arrays.

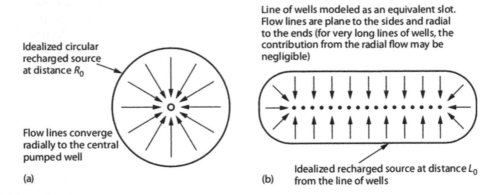

Figure 13.4 Zone of influence. (a) Radial flow. (b) Planar flow.

Of course, the equivalent well and slot concepts are approximations, introduced purely to make the estimation of discharge flow rate more amenable to simple solutions. Nevertheless, there is considerable justification for using these simplifications in appropriate conditions. The equivalent well concept was proposed by Forchheimer (1886). He based his work on that of Dupuit and analysed radial flow towards a group of wells. By means of correlation with field data, he demonstrated the acceptability of the concept for most practical purposes, provided that the wells are spaced in a regular pattern. The equivalent well concept was later endorsed by Weber (1928), based on extensive field data. The equivalent slot approach, where a long line of closely spaced wells is treated as one continuous water abstraction slot, is implicit in the work of Chapman (1959), who studied flow to wellpoint systems. The equivalent well and slot simplification is an established practical method, used in the design sections of Powers *et al.* (2007) and CIRIA Report C750 (Preene *et al.*, 2016).

For radial flow to rings of wells, this approach requires estimation of the equivalent radius r_e of the system (Figure 13.5). For a circular ring of wells, r_e is simply the radius of the ring. For a rectangular ring of wells of plan dimensions a by b, the equivalent radius can be estimated by assuming a well of equal perimeter:

$$r_e = \frac{(a+b)}{\pi} \tag{13.1}$$

or equal area:

$$r_e = \sqrt{\frac{ab}{\pi}} \tag{13.2}$$

In practice, both formulae give similar results provided that the ring of wells is not very long and narrow (i.e. provided a is not very much greater than b). The estimate of r_e determined in this way can be used in the well flow equations presented in Section 13.8.

Long narrow systems consisting of lines of closely-spaced wells (where a is much greater than b), or where the distance of influence is small, are likely to operate in conditions of planar flow (as opposed to radial flow). These systems may be better simplified to equivalent slots (Figure 13.6). In addition to planar flow to the sides, there will be a component of radial flow at the end of the line of wells (Figure 13.4). For relatively long systems, the radial flow component is likely to be relatively minor and is sometimes neglected. For shorter

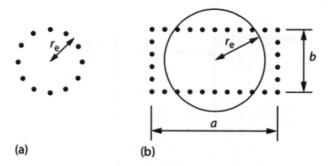

(a)   (b)

Figure 13.5 Equivalent radius of arrays of wells. (a) Circular system of radius *re*. (b) Rectangular system

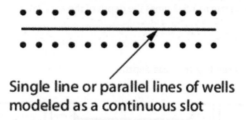

**Single line or parallel lines of wells
modeled as a continuous slot**

Figure 13.6 Equivalent slots.

systems, the radial flow to the end may be a significant proportion of the total discharge and should be incorporated into calculations.

13.6.2 Geological Structure, Well Depth and Underdrainage

Geological layering and structure may have a controlling effect on the geometry of groundwater lowering systems, in particular the well depth and the level of well screens. There are situations where it may be possible to use the geological structure to advantage to enhance the performance of the dewatering system. The potential to do this should have been identified in the conceptual model. Some options are discussed in the following.

In a homogeneous permeable aquifer, the wells must penetrate to sufficient depth to achieve the required drawdown. As a rule of thumb, widely spaced wells should penetrate to one and a half to two times the depth of the excavation. This is relevant to two aspects of dewatering design:

i. To ensure that the wells have adequate 'wetted screen length' (see Section 13.9) even after drawdown
ii. To reduce the risk of base instability in excavations due to the presence of unrelieved groundwater pressures in deeper layers (see Section 13.7) (Figure 13.7)

In an aquifer that extends to great depth, the wells may not need to penetrate to the base of the aquifer but may be designed to be 'partially penetrating' with a depth controlled by the need for adequate wetted screen length; if the aquifer does not extend to great depth below excavation formation level and is underlain by an aquiclude or aquitard, the wells will have to fully penetrate the aquifer. In practice, obtaining the required drawdown for excavation can be very problematic if the residual aquifer thickness below the excavation is much less than around one-third of the original saturated thickness (Figure 13.8).

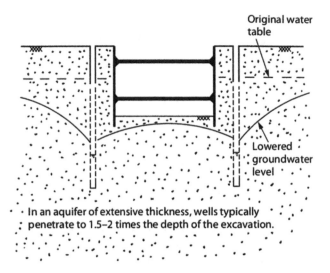

Figure 13.7 Wells in aquifers extending to great depth

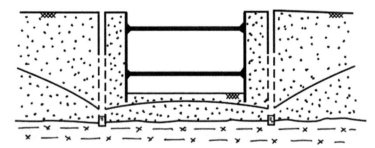

Figure 13.8 Wells in aquifers of limited thickness.

Obtaining the final part of the drawdown is difficult because the presence of the low permeability layer restricts the wetted screen length of each well. This reduces the yield of each well and its corresponding ability to generate further drawdown. There are two possible approaches in this case. One is to install the wells at much closer spacings than normal to obtain adequate wetted screen length by having a large number of wells. This approach can be economic with the wellpoint or ejector well method, but less so with deep wells, due to the greater cost per well. The second approach is to install an economical number of wells and to then try to manage the residual 'overbleed seepage' (see Section 7.6.2) by protecting the faces of the excavation with sandbags, geotextile filters or other erosion prevention measures (Figure 17.24). This allows water to enter the excavation without causing damaging loss of fines; the seepage water must then be removed by sump pumping.

If the aquifer is not homogeneous but consists of layers of greater and lesser permeability, well depth and screen level must be dictated by the layering. The basic requirement is to abstract water *directly* from the most permeable layers (or more strictly, the layers of the highest transmissivity) in preference to intermediate- or lower-permeability layers. This will ensure that well yields are maximized, because the permeable layers will readily feed water to the well screens.

If the most permeable layer (such as a gravel stratum) is beneath layers of moderate permeability (such as a silty sand), the most efficient system will involve wells pumping directly from the gravel. This will promote downward drainage of water from the sand, to the gravel, and thence to the wells (Figure 13.9). This important case is known as the 'underdrainage' approach and is widely used where ground conditions allow. It is nearly always the best option, even if the wells have to be slightly deeper than first thought in order to intercept the permeable layer.

In contrast, where a permeable layer overlies a less permeable layer, it is likely that a dewatering system abstracting purely from the lower layer may not achieve the required groundwater lowering. The rate of recharge to the more permeable stratum will exceed the rate at which water can be abstracted from the underlying lower-permeability layer. Hence, if an excavation has to penetrate into the underlying stratum of lower permeability, it may be necessary to provide two abstraction systems. One system must pump from the overlying, more permeable layer. This allows the second pumping installation, screened in the underlying and less permeable stratum, to operate without being overwhelmed by seepage from the overlying layer.

13.7 ASSESSMENT OF BASE STABILITY

In addition to estimation of the required pumped flow rate, an important part of the design process is an assessment of base stability for an excavation to avoid the risk of 'hydraulic failure'. The wider issues around hydraulic failure are discussed in Sections 7.5 and 8.6 for excavations in soil and rock, respectively.

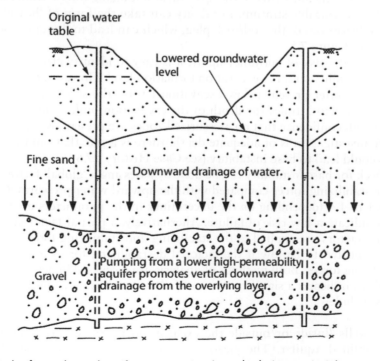

Figure 13.9 Pumping from a layered aquifer system using the underdrainage principle.

The two types of ground conditions relevant to hydraulic failure of the base of an excavation are:

i. Excavations where the formation level is in a low-permeability stratum that acts as an aquitard. In this case, the risk of instability is related to buoyancy uplift, where, if a pressurized confined aquifer is present below dig level, there can be upward movement of the base of the excavation, which can result in 'base heave' or a 'blow' (see Sections 7.5.1 and 8.6.1).

ii. Excavations where the formation level is in permeable soil or highly weathered rock. The principal risk is that the base of the excavation fluidizes due to upward seepage gradients, resulting in running sand or 'quicksand' conditions (see Section 7.5.2).

13.7.1 Assessment of Risk of Buoyancy Uplift

Instability by this mechanism is a risk when:

a) The excavation formation level bottoms out in a layer of very low-permeability soil (e.g. clay) or rock (e.g. relatively unfractured mudstone) that acts as an aquitard.
b) A permeable stratum forming a confined aquifer or water-bearing zone is present beneath the aquitard at a shallow depth below excavation formation.
c) The piezometric level in the confined aquifer is significantly above dig level.

The presence of the aquitard prevents significant vertical seepage into the excavation, meaning that the groundwater pressures in the deep aquifer will not be reduced.

There is a risk of instability by buoyancy uplift if the 'stabilizing forces' – comprising the weight of the plug of soil/rock below dig level plus any contribution from soil/rock shear strength – do not significantly exceed the upward 'destabilizing forces' from the water pressures in the deep permeable stratum. Instability can take the form of 'heave' (i.e. upward movement and distortion) of the soil/rock plug, which can lead to flooding and inundation of the excavation (Figure 7.5).

When assessing the risk of buoyancy uplift, it is easy to identify criterion (a) – the presence of low-permeability strata at excavation formation level. However, to identify criterion (b) – a confined aquifer present below excavation level – the site investigation boreholes should extend to 1.5 to 2 times the depth of the excavation (unless the conceptual model indicates with high confidence that no pressurized zones are present over that depth range). Unless the site investigation is of adequate depth, there is a risk that unidentified confined aquifer zones could lead to base instability (see Case History 27.6).

A design check for base instability can be done by assessing a 'global' factor of safety F in relation to the stabilizing forces compared with the upward water pressures. Global factors of safety of 1.1 to 1.4 (depending on the uncertainty in the parameters used) are typically considered to indicate a low risk of failure. Alternatively, a partial factor approach can be taken, whereby factors are applied to the stabilizing and stabilizing forces, and the risk of instability is indicated to be low if the factored destabilizing forces are greater than the factored stabilizing forces. Both methods of analysis are discussed in detail in Section 7.5.1.

Where analysis indicates a significant risk of failure by buoyancy uplift, possible mitigation measures include:

1. Pumped wells (inside or outside the excavation) to reduce groundwater pressures in the confined aquifer. The piezometric level in the aquifer should be lowered

sufficiently that the stabilizing forces exceed the destabilizing forces by a suitable factor of safety.

2. Relief wells (Section 17.8) within the excavation to provide a preferential pathway for upward groundwater flow in order to reduce groundwater pressures in the confined aquifer. As excavation progresses below the initial piezometric level, the relief wells will overflow, and the water can be pumped away by sump pumps.

3. Very low-permeability physical cut-off walls to isolate the confined aquifer immediately below the excavation. Provided that the cut-off walls are watertight, this will prevent further recharge, thus leaving only the water pressure contained in the aquifer within the cut-off walls to be dealt with by relief wells.

4. Increasing the downward pressure on the base of the excavation by excavating underwater, with a tremie concrete plug used to seal the base on completion (this is the basis of the 'wet caisson' method used for shaft construction; see Section 10.5.4).

13.7.2 Assessment of Risk of Fluidization Due to Upward Seepage Gradients

Instability by this mechanism is a risk when:

a) The excavation base bottoms out in a layer of soil or highly weathered rock of moderate or high permeability (e.g. silts, sands or gravels).

b) Dewatering pumping is carried out from within the excavation, either by sump pumping or from pumped wells of limited depth.

This problem is commonly of concern where the sides of the excavation are formed by low-permeability cut-off walls and where groundwater control pumping is concentrated within the excavation.

There is a risk of instability if the upward hydraulic gradients caused by pumping reduce effective stresses in the soil to very low levels (Figure 7.11). In extreme cases, effective stresses will approach zero, and the soil will fluidize and lose all strength – so-called 'quicksand' or 'running sand' conditions. As well as destabilizing the base of the excavation, this can lead to loss of passive support to excavation shoring systems, which can result in the catastrophic collapse of excavations.

The magnitude of the upward hydraulic gradient is a key indicator of the risk of this type of failure. Fluidization will theoretically occur when the upward hydraulic gradient exceeds a critical value i_{crit}. Different conditions that can lead to fluidizations of the base of an excavation due to large upward seepage gradients include:

a) Unconfined aquifer comprising relatively uniform permeable stratum (Figure 7.10a).

b) Layered aquifer system with shallow high-permeability stratum (Figure 7.10b).

c) Layered aquifer system with deeper high-permeability stratum (Figure 7.10c)

The mean hydraulic gradient i_{crit} for upward seepage is defined as the head difference dh divided by the flow path length dl (see Figure 7.9). Theoretically, fluidization will occur when $i_{crit} \approx 1$, so it is important that this value is not approached. Typically, the predicted upward hydraulic gradients should not exceed i_{crit}/F. F is a factor of safety with typical values used in design ranging from 4 to 6. Lower values of F (lower than 4 to 6) are sometimes used but can only be justified where site conditions (permeability and seepage boundary conditions) are known with a high degree of confidence, hence reducing uncertainty.

Methods of analysis are discussed in detail in Section 7.5.2. Equations to estimate upward hydraulic gradients in the base of excavations enclosed by vertical cut-off walls are given in Section 13.8.4.

Where analysis indicates a significant risk of fluidization due to upward seepage gradients, possible mitigation measures include:

1. Increasing the flow path by ensuring that cut-off walls penetrate a sufficient depth below formation level.
2. Providing preferential, engineered flow paths for groundwater to enter the excavation. This can be achieved by installing an array of relief wells (Section 17.8) within the excavation.
3. Reducing the driving head by lowering groundwater levels outside the area of excavation enclosed within the cut-off walls. This could be done by a system of dewatering wells outside the excavation while still sump pumping from within the excavation.
4. Reducing the driving head by keeping the excavation partly or fully topped up with water and excavating underwater, and then placing a tremie concrete plug in the base. This method is sometimes used for the construction of shafts and cofferdams and is often known as the 'wet caisson' method (see Section 10.5.4).
5. Changing to a dewatering approach based on external wells pumped to lower groundwater levels to below formation level, thereby avoiding the need for sump pumping.

13.8 ESTIMATION OF STEADY-STATE DISCHARGE FLOW RATE

Two key unknowns to be determined during design are the steady-state discharge flow rate and the yield, number and design of wells necessary to achieve that flow rate. Some commonly used methods to estimate discharge flow rate are described in this section. Estimation of well yields is described in Section 13.9.

13.8.1 Steady-State Well and Slot Formulae

This section presents simple closed-form solutions that can be used to estimate the steady-state discharge flow rate from systems treated as equivalent wells or slots. The emphasis here is on simple methods to be used when the conceptual model indicates conditions not too different from the idealizations and assumptions discussed in the following. For more complex cases, or where conditions differ dramatically from those discussed here, analysis by numerical modelling may be appropriate (see Section 13.12).

The simple formulae for radial flow to wells are generally based on the work of Dupuit (1863), which used certain simplifications and assumptions about the aquifer properties and geometry:

- Darcy's law is valid everywhere in the aquifer.
- The aquifer is isotropic and homogeneous – thus, the permeability is the same at all locations and in all directions.
- The aquifer extends horizontally with uniform thickness in all directions without encountering intermediate recharge or barrier boundaries within the radius of influence.
- Water is released from storage instantly when the head is reduced.
- The pumping well is frictionless and fully penetrates the aquifer.

- The pumping well is very small in diameter compared with the radius of influence, which is an infinite source of water forming a cylindrical boundary to the aquifer at distance R_o.

In reality, several of these assumptions are unlikely to be fully satisfied. For example, soils are usually stratified and generally exhibit anisotropic permeability – in soils, horizontal permeabilities are often greater than those in the vertical direction; often by more than one order of magnitude. Similarly, in rock, the permeability may be dominated by fracture flow and may vary greatly from point to point.

Dupuit made a further, important assumption. This was that the groundwater flow to the well was horizontal. This is a valid assumption for fully penetrating wells in confined aquifers but is invalid (at least close to the pumping well) in unconfined aquifers (Figure 5.8) or if the well is only partially penetrating. Dupuit's analysis was purely for the radial flow case, but Muskat (1935) did analogous studies for planar flow to slots using similar idealized assumptions.

Nevertheless, despite the idealizations and simplifications inherent in the formula, experience has demonstrated that the Dupuit-based formulae can be successfully used to estimate the steady-state pumping requirements for relatively short-term dewatering purposes. These methods are used in the design sections of Mansur and Kaufman (1962) and Powers *et al.* (2007).

The empirical evidence that the Dupuit methods give reasonable estimates of flow rate is supported by a number of theoretical studies. Hantush (1964) stated: 'The Dupuit-Forchheimer well discharge formulae, despite the shortcomings of some of the assumptions, predict the well discharges within a high degree of accuracy commensurate with experimental errors.' The assumptions have a more significant effect on the accuracy of the lowered groundwater level profile around a well, but even then, it is generally accepted that the Dupuit approach can predict drawdowns to acceptable accuracy at distances from the well of more than one and a half times the aquifer thickness.

The commonly used formulae for estimation of the steady-state discharge flow rate are listed in Table 13.1 together with diagrams of the idealized geometry. The formulae are categorized by whether the aquifer is confined or unconfined, whether flow is radial or planar, and whether the well or slot is fully or partially penetrating. All these conditions must be clarified during development of the conceptual model before the formulae can be applied.

A significant qualification on the use of these formulae is that the results will only be as valid (or invalid!) as the parameters used in them. Previous sections have discussed the selection of permeability values for design purposes and the need for sensitivity and parametric analyses. A similar approach should be applied when using the steady-state formula. Additionally, other parameters should be selected with care, including:

a) Equivalent radius (r_e) of the system. For radial flow cases, this can be estimated from Equation 13.1 or 13.2.
b) Radius of influence (R_o) for radial flow cases. The radial flow cases assume a circular recharge boundary at radius R_o. This is a theoretical concept representing the complex behaviour of real aquifers (see Section 5.8); the distance of influence is not a constant on a site but is initially zero and increases with time. However, the simplification of an empirical R_o value is a useful one. The most reliable way of determining R_o is from pumping test analyses presented as a Cooper–Jacob straight-line plot of distance–drawdown data (see Section 12.8.5). If no pumping test data are available, approximate values of R_o (in metres) can be obtained from Sichardt's formula (which is actually based on earlier work by Weber):

Table 13.1 Simple Formulae for Estimation of Steady-State Flow Rate

Case	Schematic diagram	Formula for steady-state flow rate Q		Notes
Radial flow to wells				
Fully penetrating well, confined aquifer, circular source at distance R_o (Theim equation)		$$Q = \dfrac{2\pi kD\left(H - h_w\right)}{\ln\left[R_o/r_e\right]}$$	(13.3)	k = permeability D = thickness of confined aquifer H = initial piezometric level in aquifer h_w = lowered water level in equivalent well r_e = equivalent radius of well R_o = radius of influence
Fully penetrating well, confined aquifer, line source at distance L_o (method of images)		$$Q = \dfrac{2\pi kD\left(H - h_w\right)}{\ln\left[2L_o/r_e\right]}$$	(13.4)	k = permeability D = thickness of confined aquifer H = initial piezometric level in aquifer h_w = lowered water level in equivalent well r_e = equivalent radius of well L_o = distance to line source
Fully penetrating well, unconfined aquifer, circular source at distance R_o (Dupuit–Forchheimer equation)		$$Q = \dfrac{\pi k\left(H^2 - h_w^2\right)}{\ln\left[R_o/r_e\right]}$$	(13.5)	k = permeability H = initial water table level in aquifer h_w = lowered water level in equivalent well r_e = equivalent radius of well R_o = radius of influence
Fully penetrating well, unconfined aquifer, line source at distance L_o (method of images)		$$Q = \dfrac{\pi k\left(H^2 - h_w^2\right)}{\ln\left[2L_o/r_e\right]}$$	(13.6)	k = permeability H = initial water table level in aquifer h_w = lowered water level in equivalent well r_e = equivalent radius of well L_o = distance to line source

(Continued)

Table 13.1 (Continued) Simple Formulae for Estimation of Steady-State Flow Rate

Case	Schematic diagram	Formula for steady-state flow rate Q	Notes
Partially penetrating well, confined aquifer		$$Q_{pp} = BQ_{fp}$$ (13.7)	Q_{pp} = flow rate from partially penetrating well Q_{fp} = flow rate from fully penetrating well B = partial penetration factor for radial flow (obtained from Figure 13.10a)
Partially penetrating well, unconfined aquifer		$$Q_{pp} = BQ_{fp}$$ (13.8)	Q_{pp} = flow rate from partially penetrating well Q_{fp} = flow rate from fully penetrating well B = partial penetration factor for radial flow (obtained from Figure 13.10b)
Plane flow to slots Fully penetrating slots, confined aquifer, flow from lines sources on both sides of slot		$$Q = \frac{2kDx\,(H - h_w)}{L_o}$$ (13.9)	x = linear length of slot k = permeability D = thickness of confined aquifer H = initial piezometric level in aquifer h_w = lowered water level in equivalent slot L_o = distance of influence
Partially penetrating slots, confined aquifer, flow from lines sources on both sides of slot		$$Q_{pp} = \frac{2kDx\,(H - h_w)}{(L_o + \lambda D)}$$ (13.10)	x = linear length of slot k = permeability D = thickness of confined aquifer H = initial piezometric level in aquifer h_w = lowered water level in equivalent slot L_o = distance of influence λ = partial penetration factor (obtained from Figure 13.10c)

(Continued)

Table 13.1 (Continued) Simple Formulae for Estimation of Steady-State Flow Rate

Case	Schematic diagram	Formula for steady-state flow rate Q		Notes
Fully penetrating slots, unconfined aquifer, flow from lines sources on both sides of slot		$$Q = \frac{kx\left(H^2 - h_w^2\right)}{L_o}$$	(13.11)	x = linear length of slot k = permeability H = initial water table level in aquifer h_w = lowered water level in equivalent slot L_o = distance of influence
Partially penetrating slots, unconfined aquifer, flow from lines sources on both sides of slot		$$Q = \left[0.73 + 0.23\left(P/H\right)\right]\frac{kx\left(H^2 - h_w^2\right)}{L_o}$$	(13.12)	x = linear length of slot k = permeability H = initial water table level in aquifer h_w = lowered water level in equivalent slot L_o = distance of influence P = depth of penetration of slot below original water table
Plane and radial flow Rectangular systems, confined aquifer		$$Q = kD\left(H - h_w\right)G$$	(13.13)	k = permeability D = thickness of confined aquifer H = initial piezometric level in aquifer h_w = lowered water level in equivalent well G = geometry shape factor (obtained from Figure 13.11)

$$R_o = 3000(H - h_w)\sqrt{k} \tag{13.14}$$

where $(H - h_w)$ is the drawdown (in metres) and k is the permeability (in metres per second). This formula needs to be modified when used to analyse large equivalent wells. Dupuit assumed that the radius of the well was small in comparison to the radius of influence, but often the radius r_e may be large in comparison to R_o. In such cases, the following equation can be used:

$$R_o = r_e + 3000(H - h_w)\sqrt{k} \tag{13.15}$$

When estimating R_o, it is important to review the calculated distance of influence to avoid using wildly unrealistic values. In the authors' experience, values of less than around 30 m or more than 5000 m are rare and should be viewed with caution. It may be appropriate to carry out sensitivity analyses using a range of distance of influence values to see the effect on calculated flow rates. For the radial flow case, R_o appears in a log term, so small errors do not have a significant effect on calculated flow rates, but the possibility of gross error exists if a very large or very small R_o is used.

c) Distance of influence (L_o) for planar flow cases. L_o (in metres) can be estimated from Sichardt's formula, but a different calibration factor must be used:

$$L_o = 1750(H - h_w)\sqrt{k} \tag{13.16}$$

where k is in metres per second and $(H - h_w)$ is in metres. The distance of influence appears as a linear term in the planar flow equations – the estimated flow rate is inversely proportional to L_o. The distance of influence must be chosen with care, and sensitivity analyses are strongly recommended.

d) Lowered water level (h_w) inside the equivalent well or slot. The equivalent well or slot method requires that the lowered water level (inside the well or the slot) used in equations is the groundwater level in the excavation area itself. Obviously, the water level in each individual well will be lower (perhaps considerably so), but this drawdown would not be representative of the drawdown in the equivalent well or slot.

Table 13.1 includes equations to estimate flow rates from partially penetrating wells. Such a well penetrates only the upper part of the aquifer, and its yield will generally be less than that of a fully penetrating well of similar design and construction. Equations 13.7 and 13.8 allow the flow rate Q_{pp} from a partially penetrating well to be estimated by multiplying the flow rate Q_{fp} from a fully penetrating well by a partial penetration factor B, which varies between 1 (for a fully penetrating well) and zero.

Figure 13.10 shows values of B for individual wells (of radius r_w less than approximately 0.5 m). These are based on equations developed by Kozeny for confined aquifers (Figure 13.10a) and by Borelli for unconfined conditions (Figure 13.10b); in both cases, isotropic permeability conditions are assumed in the analysis. In theory, these factors cannot be directly applied to equivalent wells of large diameter, but they can be used to assess the approximate relationship between flow rate and penetration. If partial penetration is a major factor on a project, for example when dewatering in very thick aquifers, numerical modelling (Section 13.12) should be considered.

The flow to partially penetrating slots, used to model long lines of closely spaced wells or wellpoints, can be estimated from Equation 13.10 and Figure 13.10c for confined aquifers and Equation 13.11 for unconfined aquifers.

Figure 13.11 shows shape factors to be used with equation 13.13 for mixed radial and planar flow to rectangular excavations in confined aquifers.

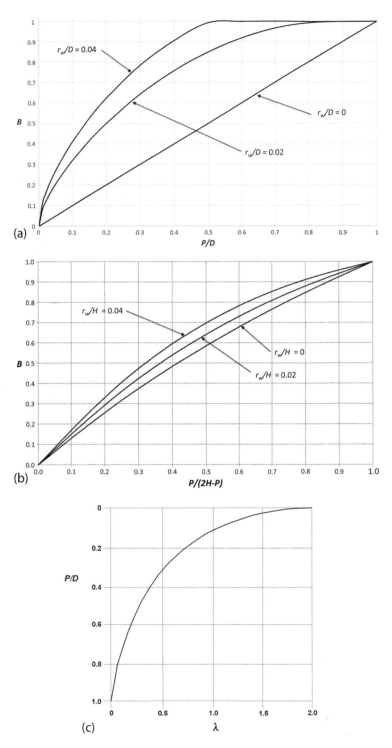

Figure 13.10 Partial penetration factors for wells and slots. (a) Radial flow to wells in confined aquifers. (b) Radial flow to wells in unconfined aquifers. (c) Planar flow to slots in confined aquifers. (After Mansur, C I and Kaufman R I, *Foundation Engineering* (Leonards, G A, ed.), McGraw-Hill, New York, pp. 241–350, 1962.)

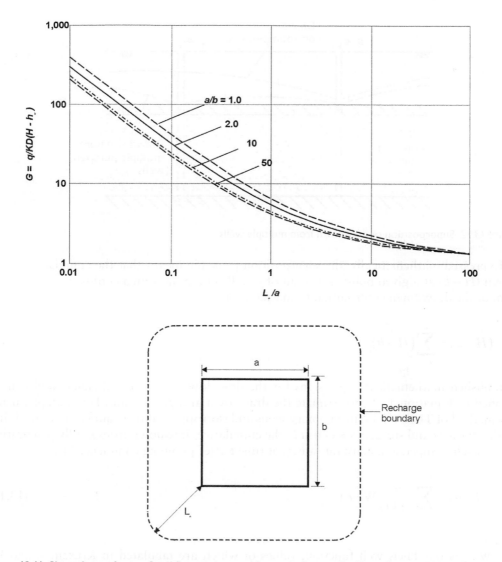

Figure 13.11 Shape factor for confined flow to rectangular equivalent wells. (From Powrie, W and Preene, M, *Géotechnique*, 42, 635–639, 1992. With permission.)

13.8.2 Cumulative Drawdown Analysis – Theoretical Method

The formulae described in the previous section are used to analyse systems of closely spaced wells, modelled as equivalent wells or slots. Such an approach is less satisfactory if the wells are widely spaced; in those cases, a cumulative drawdown (or superposition) method may be more suitable.

This method takes advantage of the mathematical property of superposition applied to drawdowns in confined aquifers. In essence, the total (or cumulative) drawdown at a given point in the aquifer, resulting from the action of several pumped wells, is obtained by adding together (or superimposing) the drawdown from each well taken individually (Figure 13.12). This approach is theoretically correct in confined aquifers but is invalid in unconfined aquifers, where the changes in saturated thickness that occur during drawdown complicate the interaction of drawdowns.

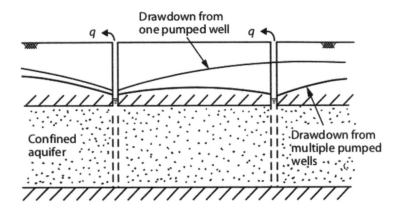

Figure 13.12 Superposition of drawdown from multiple wells.

Expressed mathematically, the superposition principle means that the cumulative drawdown $(H - h)$ at a given point as a result of n wells pumping from a confined aquifer is the sum of the drawdown contribution from each well:

$$(H - h) = \sum_{i=1}^{n} (H - h)_i \tag{13.17}$$

Established mathematical expressions for the drawdown from an individual well can be applied to Equation 13.17 to estimate the drawdown at a given point. For example, using the method of Theis (1935) in a homogeneous and isotropic confined aquifer of permeability k, thickness D and storage coefficient S, the cumulative drawdown from n fully penetrating wells, each pumped at a constant rate q_i, at time t after pumping commenced is:

$$(H - h) = \sum_{i=1}^{n} \frac{q_i}{4\pi kD} W(u_i) \tag{13.18}$$

where

　　$W(u)$ is the Theis well function, values of which are tabulated in Kruseman and De Ridder (1990)

　　$u = (r^2 S)/(4kDt)$

　　r is the distance from each well to the point under consideration

For values of u less than about 0.05, the simplification of Cooper and Jacob (1946) can be applied, giving

$$(H - h) = \sum_{i=1}^{n} \frac{q_i}{4\pi kD} \left\{ -0.5772 - \ln \left[\frac{r_i^2 S}{4kDt} \right] \right\} \tag{13.19}$$

This can also be expressed in decimal logarithms as

$$(H - h) = \sum_{i=1}^{n} \frac{2.3 q_i}{4\pi kD} \log \left[\frac{2.25 kDt}{r_i^2 S} \right] \tag{13.20}$$

In many aquifers, the condition of $u<0.05$ is satisfied after only a few hours' pumping, which means that Equations 13.19 and 13.20 can generally be used for the analysis of groundwater control systems in confined aquifers.

If the target drawdown $(H - h)$ in the excavation area is known, these equations can be solved to determine the number, location and yield of wells necessary to achieve the required drawdown. This also allows the total discharge flow rate (the sum of flow from all the wells) to be determined. This method is most suitable for systems of relatively widely spaced wells. It is mainly used for deep wells and occasionally for ejector well systems; it is rarely used for wellpoint systems.

The following points should be considered when applying the method:

(i) The method has been reliably applied to the estimation of drawdown within the area of excavation, away from the pumped wells themselves. Estimating the cumulative drawdown inside each well is more difficult, because well losses may not be accurately known. If large well losses occur, the method is less reliable, because the drawdown contribution becomes uncertain.

(ii) Application of the method requires that the aquifer parameters and well yields be estimated. In practice, the most reliable way to obtain suitable estimates is from analysis of a pumping test. If pumping test data are not available, the estimated cumulative drawdowns should be treated with caution unless there is a high degree of confidence in the parameter values used in calculations. If a pumping test has been carried out, the graphical cumulative drawdown method (described in the subsequent section) may be a more appropriate method of analysis.

(iii) It may be possible to obtain the required drawdowns in the proposed excavation using a few wells pumped at high flow rates or a larger number of wells of lower yield. Similarly, varying the well locations around (or within) the excavation may produce significantly different drawdowns in the area of interest. In years gone by, investigating the effect of the various options was a tedious process. However, using personal computers, it is possible to write routines or macros for spreadsheet programs to evaluate Equation 13.20, allowing many options to be rapidly considered. When evaluating the various options, it is vital that realistic well yields are used (see Section 13.9); otherwise, too many or too few wells will be specified.

(iv) Equations 13.18 through 13.20 include a term for the time since pumping began, so each cumulative drawdown calculation is for a discrete time t. The time used in calculation will depend on the construction programme. If the programme shows that a 2 week period is available for drawdown (between the installation of the dewatering system and the commencement of excavation below the original groundwater level), then that case should be analysed. However, in reality, there may be problems with the installation of a few of the wells and pumps, so not all n wells will be pumping for the full 2 week period. It may be prudent to design with the objective of obtaining the target drawdown in a rather shorter time.

(v) The assumptions inherent in Equations 13.18 through 13.20 (isotropic confined aquifer, fully penetrating wells and constant flow rate from each well) obviously will not apply in all cases. Provided that the basic aquifer conditions are confined or leaky, it may be possible to use Equation 13.18 for other conditions by substituting an alternative expression in place of the Theis well function $W(u)$. Kruseman and De Ridder (1990) give well functions for a number of cases, including leaky aquifers, anisotropic permeability, partially penetrating wells and variable pumping rates.

The cumulative drawdown method assumes that individual wells do not interfere significantly with each other's yield. For wells at wide spacings (greater than around 20 m) in confined aquifers (where the aquifer thickness does not change with drawdown), interference is usually low. In such cases, the observed drawdowns are likely to be close to those predicted directly from the cumulative drawdown method. However, in general, observed drawdowns will be slightly lower than predicted. It is not unusual for observed drawdowns to be between 80 and 95 per cent of the calculated values when applied using reliable parameters derived from pumping tests. To allow for this, the total well yield (or the number of wells to be installed) should be increased (by dividing by an empirical superposition factor J of 0.8 to 0.95). For example, the total system flow rate Q is determined from the sum of the individual flow rates q_i from n wells:

$$Q = \frac{1}{J} \sum_{i=1}^{n} q_i \tag{13.21}$$

The cumulative drawdown method is invalid in unconfined aquifers (or confined aquifers where the drawdown is so large that local unconfined conditions develop). This is because the saturated thickness decreases as drawdown increases, making each additional well less effective compared with the initial wells. Although the method is theoretically invalid in unconfined conditions, where drawdowns are small (less than 20 per cent of the initial saturated aquifer thickness), the method has been successfully applied using an empirical superposition factor J of 0.8 to 0.95. For greater drawdowns in unconfined aquifers, the cumulative drawdown method has been applied using empirical superposition factors of 0.6 to 0.8.

13.8.3 Cumulative Drawdown Analysis – Graphical Method

If distance–drawdown data are available describing the aquifer response to the pumping of a single well, a graphical cumulative drawdown method can be used. This approach is based on the Cooper–Jacob straight line method of pumping test analysis (see Section 12.8.5), which uses Equation 13.19 expressed as

$$(H-h) = \sum_{i=1}^{n} \frac{q_i}{2\pi k D} \ln\left(\frac{R_o}{r_i}\right) \tag{13.22}$$

where all terms are as described previously, apart from R_o, which is the radius of influence at time t. The equation is evaluated graphically and is used to obtain the total drawdown $(H - h)$ at the selected location, resulting from a given array of wells, without the need to evaluate the aquifer parameters.

The method is described in detail by Preene and Roberts (1994) and involves the following steps:

1. Determine the target drawdown level in critical points of the excavation. Typically, critical points where drawdown is checked include the centre and corners of the excavation. Normally, the target drawdown level is a short distance (0.5 to 1 m) below the excavation formation level.
2. From the pumping test data, construct a drawdown–distance plot on semi-logarithmic axes. Drawdowns recorded in monitoring wells at a given time after pumping commenced are plotted, and a best straight line is drawn through the data (Figure 13.13a). For short-duration pumping tests, the data used are normally from the end of the test. The drawdown in the pumped well is typically ignored, as it may be affected by well losses.

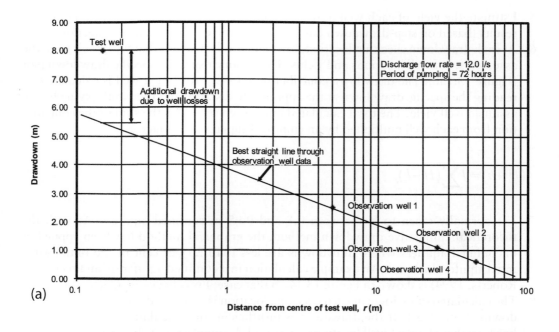

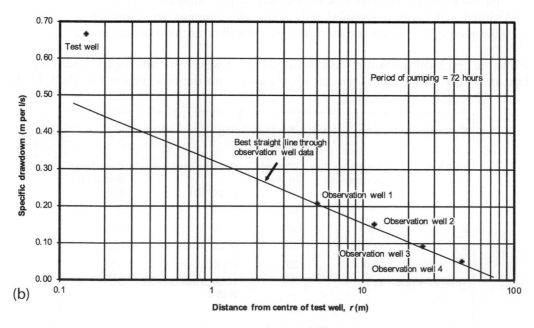

Figure 13.13 Cumulative drawdown analysis: graphical method. (a) Distance–drawdown plot. (b) Specific drawdown plot.

3. Convert each drawdown data point to specific drawdown by dividing by the discharge flow rate recorded during the test. A straight line is then drawn through the monitoring well data to obtain the design-specific drawdown plot (Figure 13.13b); this plot shows the drawdown that results from a well pumped at a unit flow rate.

4. Draw a plan of the excavation and groundwater lowering system, marking on the well locations and the points where drawdown is to be checked. Measure and record the distances from each well to each drawdown checking point.

5. Estimate the yield of each well in the system. This may be based on the pumping test results (based on step-drawdown data) or may involve the guidelines of Section 13.9.

6. At each drawdown checking location, calculate the drawdown that will result from the assumed set of well locations and yields. This is done using the specific drawdown plot (Figure 13.13b). The drawdown contribution $(H - h)_i$ from each well is calculated by reading the specific drawdown at the appropriate distance and then multiplying by the assumed well yield. The total calculated drawdown $(H - h)$ is the sum of the contribution from each well multiplied by an empirical superposition factor J:

$$(H-h) = J\sum_{i=1}^{n}(H-h)_i \qquad (13.23)$$

7. J is normally taken to be between 0.8 and 0.95 in confined aquifers. As with the theoretical cumulative drawdown method, the graphical method has been applied in unconfined aquifers where drawdowns are less than 20 per cent of the initial saturated aquifer thickness. An example calculation in a confined aquifer (from Preene and Roberts, 1994) is shown in Figure 13.14; in that case, J was back-calculated to be 0.92.

8. The calculated drawdown at each checking location is compared with the target drawdowns from Step 1. If the drawdown is insufficient, the calculation is repeated after having either changed well locations; increased the number of wells; or increased individual well yields. It is vital that the well yields assumed are achievable in the field. If the assumed yields are too large, the system will not achieve its target drawdowns.

While the graphical method is most commonly used where site investigation pumping test data are available, the technique can also be used with the observational method. In this

---

Estimation of drawdown at Well 8 due to pumping on four other wells

Well No.	Discharge flow rate per well (l/s)	Distance to Well 8 (m)	Specific drawdown (m per l/s)	Calculated drawdown (m)
1	8.5	82	0.079	0.67
2	8.5	100	0.072	0.60
6	11.0	50	0.082	0.91
7	11.0	20	0.103	1.13
Total flow rate	**39.0**		**Total drawdown at Well 8**	**3.31**

Actual drawdown recorded at Well 8 after 44 hours = 3.06 m

Therefore, drawdown achieved is 3.06/3.31 = 92 per cent of calculated cumulative drawdown

Figure 13 14 Case history of cumulative drawdown calculation. (Data from Preene, M and Roberts, T O L, The application of pumping tests to the design of construction dewatering systems, in *Groundwater Problems in Urban Areas* (Wilkinson, W B, ed.), Thomas Telford, London, pp. 121–133, 1994. With permission.)

approach, one of the first wells in the groundwater lowering system is pumped on its own in a crude form of pumping test. Drawdowns are observed in the other dewatering wells (which are unpumped at that time); these data allow distance–drawdown plots to be produced. The cumulative drawdown calculations are then used to help with the decision-making progress to decide whether to install additional wells. As each new well is installed and pumped, further drawdown data are collected and the predicted drawdowns compared with the actual. In these cases, it is often found that as each additional well becomes operational, the empirical superposition factor J reduces further as interference between wells increases.

13.8.4 Estimation of Flow Rates Where Cut-Off Walls Are Present

The analytical methods presented in the foregoing sections (13.8.1 through 13.8.3) are applicable to groundwater flowing directly to excavations where water can enter both the sides and the base of the excavation. However, in some situations, a low-permeability cut-off wall may be used either as part of the excavation support system or to reduce groundwater inflows to the excavation (see Section 9.4). An example of this geometry is shown in Figure 9.3. For these cases, the widely used 'equivalent well' model of groundwater flow to an excavation is invalid. It is necessary to use analytical methods that take account of vertical flow. Alternatively, these geometries can be analysed using numerical groundwater models (see Section 13.12).

Kavvadas *et al.* (1992) developed an analytical solution for groundwater flow into excavations where cut-off walls are present at the sides of an excavation for the geometry shown in Figure 13.15a. The solution is based on the following assumptions:

- The width of the excavation is b, with identical cut-off walls on both sides.
- The excavation is of length a and is long in comparison to its width (which means that radial effects at the end of the excavation can be neglected).
- The soil or rock is assumed to be homogeneous and to have an isotropic permeability of k.
- The cut-off wall is impermeable and penetrates to depth d below the lowered groundwater level in the excavation (assuming that groundwater level is lowered below formation level).
- The aquifer is underlain by an impermeable stratum at depth. The base of the cut-off wall is a distance s above the impermeable base of the aquifer.
- Pumping from the excavation does not lower the external groundwater level.

The solutions for flow rate q per unit length of excavation perimeter and maximum upward hydraulic exit gradient i_{max} in the base of the excavation are:

$$q = 0.85k\left(H - h_w\right)\left[1 - (0.2)^{s/0.5b}\right]\left(\frac{d}{0.5b}\right)^{-0.5}\left(\frac{d_1}{0.5b}\right)^{-0.125} \tag{13.24}$$

$$i_{max} = 0.5\left(\frac{H - h_w}{d}\right)\left[1 - (0.2)^{s/0.5b}\right]\left(\frac{d}{0.5b}\right)^{-1/3}\left(\frac{d_1}{0.5b}\right)^{-0.125} \tag{13.25}$$

where H, h_w and d_1 are as shown in Figure 13.15a and are:
   H = initial water table level in aquifer
   h_w = lowered water table level in aquifer
   $(H - h_w) = d_1$ = drawdown in excavation

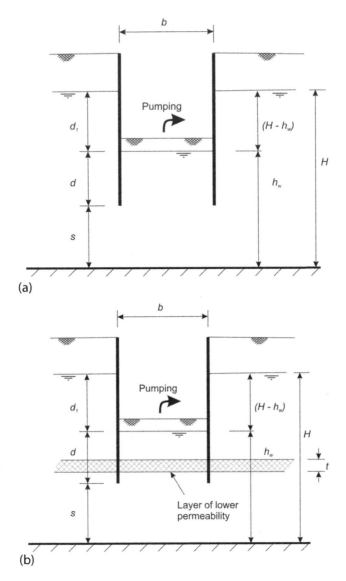

Figure 13.15 Groundwater inflow to excavation where cut-off walls are present. (a) Uniform soil conditions. (b) Where a lower permeability layer is present.

Kavvadas *et al.* (1992) give an equation to calculate the total flow rate Q from a rectangular excavation of length a and width b, taking into account corner effects. This can be expressed as

$$Q = q\left\{(2[a+b])(0.70+0.30[1-b/a])\right\}$$ (13.26)

where q is flow rate per unit length of excavation perimeter for very long excavations (Equation 13.24).

Kavvadas *et al.* (1992) also note that for rectangular excavations, the maximum hydraulic gradient will occur in the corner of the excavation and will be larger than that calculated from Equation 13.25. For square excavations, the maximum hydraulic gradient may be 1.7 times that calculated from Equation 13.25.

The analytical methods presented in Equations 13.24 through 13.26 are applicable to cases where cut-off walls are used in unconfined aquifers to reduce inflows to the excavation. Because the method assumes that pumping does not cause any external drawdowns, the method is probably conservative and is likely to give estimates of flow rate more applicable to the early phases of pumping, before storage release (see Section 13.8.6) has occurred, rather than long-term pumping rates.

Kavvadas *et al.* (1992) also provide a solution for the case where a continuous horizontal layer (of thickness *t*) of lower-permeability material is present beneath excavation formation level and above the toe of the cut-off wall (Figure 13.15b). In that case, the solutions for flow rate q_r per unit length of excavation perimeter and maximum upward hydraulic exit gradient i_{maxr} in the base of the excavation are:

$$q_r = \frac{q}{\mu} \tag{13.27}$$

$$i_{maxr} = \frac{i_{max}}{\mu} \tag{13.28}$$

$$\mu = 1 + 0.25\left(\frac{t}{d}\right)\left(\frac{k}{k_r} - 1\right) \tag{13.29}$$

where

q_r = flow rate per unit length of excavation perimeter for the case when low-permeability layer is present

i_{maxr} = maximum hydraulic exit gradient for the case when low-permeability layer is present

q = flow rate per unit length of excavation perimeter for uniform permeability case (Equation 13.24)

i_{max} = maximum hydraulic exist gradient for uniform permeability case (Equation 13.25)

t = thickness of the low-permeability layer

k_r = permeability of the low-permeability layer

k = permeability of the remainder of the aquifer

13.8.5 Estimation of Flow Rates into Tunnels

The analytical methods presented in the foregoing sections (13.8.1 through 13.8.4) are applicable to the typical geometries of dewatering systems around trenches or excavations. They are not directly applicable for estimation of groundwater inflow to tunnels. The following section discusses some aspects of groundwater inflow to tunnels.

The potential for groundwater inflows exists whenever tunnels are driven below groundwater level through permeable soil or rock strata. The tunnel acts as a drain, with water flowing into the tunnel by gravity. The rate of water inflow may have an effect on the methods and equipment employed during construction. Relatively small inflows may simply be a cause of discomfort and inefficiency in excavation and tunnel lining operations. More substantial inflows may have a huge influence on the works, in extreme cases giving a real risk of inundation of the tunnel. Groundwater problems when tunnelling in soil and rock are discussed in Sections 7.7 and 8.8, respectively.

Inflows can be reduced or managed by various methods, including the use of compressed air, tunnel boring machines (TBMs), grouting and other forms of ground treatment, or by pumping or other methods (see Chapter 10). During the planning of tunnelling projects, it is often necessary to estimate potential water inflows as part of the decision-making process.

Only then can appropriate ground treatment and tunnelling methods be specified. However, groundwater flow conditions around a tunnel are complex. Accurate prediction of inflows is difficult for a number of reasons, including:

- The natural inhomogeneity of the permeability of soils (due to layering and change in material type) or of rocks (due to variations in fracture spacing, opening and orientation, presence of faults and so on).
- The complex time-dependent interaction between the tunnel and the surrounding ground. For example, even the rate of tunnel advance will affect the inflows, since a slowly advancing tunnel may drain the ground ahead of it, reducing inflows.
- The effect of tunnel construction methods. If tunnels are lined immediately after excavation, inflows may be restricted to areas locally around the working face. Some tunnelling methods, such as compressed air working or closed face TBMs, may almost totally exclude groundwater from the tunnel. If the tunnel is to remain unlined, or is lined in one operation late in the project, inflows are possible over a much wider area.

These factors mean that even when comprehensive site investigation data are available, experience of the local geology and construction experience will be useful when estimating inflows. This section presents an analytical approach that can be used in support of engineering judgement.

The use of analytical methods requires permeability estimates for the soil or rock through which the tunnel is to be driven. In addition to the results of permeability tests, the description of the soil type or rock type, rock quality and rock fracture descriptors are vital to allow engineering judgement to be applied and to highlight any 'rogue' readings within the permeability data (see Section 12.9).

Groundwater inflow to tunnels can take two principal forms:

- Flow from a porous medium, such as a granular soil (e.g. sand or gravel), and from porous rock in which the flow through multiple discrete fractures means that the flow of water is widely distributed through the rock mass as a whole. The medium may be isotropic and homogeneous or anisotropic and heterogeneous. In most such cases, the flow of groundwater will be laminar, and Darcy's law will apply. The permeability k of the medium and groundwater recharge conditions are important controlling parameters of inflows.
- Flow from discrete fissures, fractures, discontinuities or solution features, as might be found when tunnelling in competent rock. Such flow of water may be turbulent, and Darcy's law does not apply. Inflows are controlled by the discontinuity aperture size and spacing, the extent of the fracture system and any recharge conditions. Any inflow calculations require the definition of likely discontinuity conditions.

Laminar flow of groundwater through porous media is described by Darcy's law (Equation 3.2). The formulae in this section are derived from Darcy's law. Darcy's law can apply to groundwater flow through fractured rocks provided that the fracture-induced permeability is assumed to give the same effect as a homogeneous porous medium. This might be the case if the fractures are approximately uniformly closely spaced. In both soils and rocks, if very high inflows occur, turbulent flow may develop just outside the edge of the tunnel excavation, and Darcy's law will be invalid in that area.

Whatever method of analysis is used, some form of sensitivity or parametric analysis may be appropriate to assess the likely range of inflows based on the available data. These estimates may help determine whether ground treatment to reduce permeability (such as grouting) will be essential or should be included only as a contingency. It is useful to consider the sensitivity of calculated inflows to selected parameters.

Tunnel diameter: A number of studies (including Fitzpatrick *et al.*, 1981) have highlighted that inflow rates are not especially sensitive to tunnel diameter, and inflows are not dissimilar for small and large tunnels. Wrench and Stacy (1993) state that this is supported by the observation that inflow to pilot bores is often as large as it is into the final tunnel.

Depth below groundwater level: This has an important effect on calculated inflows, which increase rapidly as the depth is increased.

Soil or rock permeability: The permeability of the soil or rock is critical to the calculation of inflows. This inflow comes not only from the level of the tunnel itself but also from the soil or rock above and below the tunnel – this should influence the planning of site investigation permeability testing. In many fractured rocks, the overall permeability reduces with depth, and this will affect tunnel inflows. Wrench and Stacy (1993) suggest that simple calculations may over-estimate inflows if the permeability values used are based solely on the level of the tunnel.

Groundwater recharge and storage: If the tunnel is driven rapidly through a stratum where there is sufficient groundwater recharge and storage (from either precipitation or flow from other strata) to prevent groundwater levels being lowered, the tunnel will not significantly disturb drawdown water levels. If the water table is not lowered, inflow to the tunnel will remain steady and will not decline with time (Figure 13.16a). Conversely, if there is limited groundwater recharge and storage, and especially if the tunnel is driven slowly, allowing more time for water to drain into the tunnel, groundwater levels will be lowered (Figure 13.16b). Inflows will decline from their initial values as a 'distance of influence' develops away from the tunnel. The long-term water inflows are unlikely to be zero but will stabilize at some flow rate in balance with the available storage and recharge.

A further complication occurs when considering groundwater inflow to tunnels in rock, where inflow is via fractures. It is well known that the act of forming a tunnel will increase the effective stress in the rock around the tunnel. This is due to a combination of redistribution of stress around the tunnel and pore water pressure reduction due to inflows to the tunnel opening. This can result in a 'tightening' of rock around the tunnel, which will close fracture openings and reduce permeability locally. This zone of reduced permeability around the tunnel is sometimes known as a 'skin' zone.

Steady-state groundwater inflow Q to tunnels in conditions where Darcy's law is valid (i.e. where the flow is not concentrated in a small number of discrete features) can be estimated from equations analogous to those for vertical wells. Palmström and Stille (2015) present a generic equation for water inflow Q to a tunnel or cavern as

$$Q = \frac{2\pi k H x}{G} \tag{13.30}$$

where
k = permeability of the soil or rock
D = diameter of the tunnel
x = length of the tunnel being assessed
H = head difference between the groundwater level and the tunnel axis
G = geometry factor

A widely applied version of this formulation is based on the work of Goodman *et al.* (1965) for the geometry shown in Figure 13.17 and the assumptions stated in the following

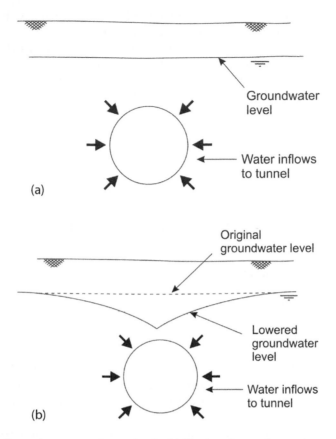

Figure 13.16 Effect of tunnels on groundwater levels. (a) No drawdown of groundwater levels. The ground-water storage and/or recharge is sufficiently large to prevent any significant lowering of ground-water levels as a result of water draining into the tunnel. (b) With drawdown of groundwater levels. Water draining into the tunnel results in lowering of groundwater levels.

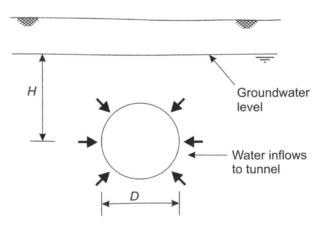

Figure 13.17 Groundwater inflow to tunnels.

list. In the Goodman equation, the geometry factor is ln(4H/D), where D is the diameter of the tunnel, so inflow can be calculated from

$$Q = \frac{2\pi k H x}{\ln\left(\frac{4H}{D}\right)} \tag{13.31}$$

Goodman's equation is based on the following assumptions:

- A tunnel of infinite length.
- Steady-state conditions.
- Soil or rock of homogeneous and isotropic permeability.
- No drawdown of the groundwater level.
- Flow occurs to the cylindrical surface (sometimes known as the extrados) of the tunnel only; no allowance is made for the flow to the end of tunnel drive (i.e. the working face).

For tunnels in rock, Palmström and Stille (2015) present an equation that takes account of the skin zone of reduced permeability as

$$Q = \frac{2\pi k H x}{\left[\ln\left(\frac{4H}{D}\right) + \varepsilon\right]} \tag{13.32}$$

where ε is a skin factor. Palmström and Stille (2015) suggest that the skin factor should be between 2 and 5.

Because of the limitations in the simplifying assumptions, equations such as that of Goodman should be used with caution but can provide order of magnitude estimates of likely tunnel inflows. The Goodman equation (13.31) often over-estimates inflows, because it does not account for drawdown of groundwater levels, nor does it allow for a reduced permeability zone around the tunnel. Moon and Fernandez (2010) present factors that can be used to reduce the calculated inflows from the Goodman equation to allow for both these phenomena.

Methods for estimating groundwater inflows to rock tunnels are discussed in more detail in Heuer (1995, 2005). Due to the complex geometry of many tunnel construction projects, numerical groundwater modelling (see Section 13.12) can be a suitable tool to develop groundwater inflow estimates for different stages of tunnel construction.

13.8.6 Storage Release and Uprating of Pumping Capacity

Many of the methods used to calculate discharge flow rates assume that 'dewatered' conditions have developed and that a zone of influence of drawdown exists in the aquifer around the groundwater lowering system. To reach this condition, water must be released from storage in the aquifer within the zone of influence (see Section 3.4). This means that during the initial period of pumping, before the steady state is approached, an additional volume of water must be pumped.

In confined aquifers, the quantity of water from storage release is small and is only significant during the first few hours of pumping. Its effect on the necessary pumping capacity is often neglected. In contrast, in unconfined aquifers, the water from storage may be significant and may persist for several weeks or more, dependent on the aquifer permeability, the aquifer recharge boundaries and the pumping rate. Powers et al. (2007) suggest that storage release is likely to be a major issue in permeable unconfined aquifers where the proposed discharge rate is higher than 60–70 l/s.

Storage release means that either:

(i) A system designed with a capacity equal to the steady-state flow rate will take longer than anticipated to achieve the target drawdown

or

(ii) The design system flow rate should be increased above the steady-state estimate to deal with water from storage and to ensure that drawdown is achieved within a reasonable time period.

The release of water from storage has the same effect as reducing the distance of influence used in calculation. If the 'long-term' distance of influence is used in design, drawdown may only be achieved slowly. Alternatively, if the designer uses a 'short-term' distance of influence in design, drawdown may be achieved rapidly, but the system may be over-designed in the long term. This is only likely to be an issue in high-permeability unconfined aquifers.

If the distance of influence used in design is taken from analysis of short-term pumping test data (see Section 12.8.5), this will include for the effects of water released from storage within the aquifer, and calculated flow rates are likely to achieve drawdowns fairly rapidly. This also applies to systems designed by the graphical cumulative drawdown method, where pumping test data are used directly.

If no pumping test data are available, and the distance of influence is estimated from empirical formulae such as Equations 13.14 through 13.16, then judgement must be used where distances of influence of several hundred metres are predicted in high-permeability soil or rock. Systems with pumping capacities designed on that basis will be able to cope with steady-state inflows once the zone of influence has developed. However, they may be overwhelmed by storage release during the early stages of pumping, and drawdown may take a long time to be achieved, leading to delays in the construction programme. Designers sometimes overcome this problem by using rather smaller distances of influence, which predict higher flow rates; this helps ensure that drawdown is achieved in a reasonable time.

If the distance of influence used in the design has been estimated from Equations 13.14 through 13.16, the following equations can be used to crudely estimate the time t, which would be required for this distance of influence to develop. The equations are for radial flow (from Cooper and Jacob, 1946)

$$R_o = \sqrt{\frac{2.25kDt}{S}} \tag{13.33}$$

and for planar flow (from Powrie and Preene, 1994a)

$$L_o = \sqrt{\frac{12kDt}{S}} \tag{13.34}$$

where
  D is the aquifer thickness
  k is the permeability
  S is the storage coefficient

Strictly, Equations 13.33 and 13.34 are only valid in confined aquifers, but they can be used in unconfined aquifers where the drawdown is not a large proportion of the original saturated thickness. Typical values of specific yield (approximately equal to S in an unconfined aquifer) are given in Table 3.1.

If the design distance of influence will take a long time (more than a few days) to develop, it is possible that the system should be designed assuming a smaller R_o or L_o (values typically used in high-permeability soils are in the range 200–500 m). Changing the design in this way would increase the required pumping capacity of the system. This would allow the water released from storage to be handled by the system and would result in the drawdown within the excavation area being achieved within a reasonable length of time.

Storage depletion is perhaps of most significance for high-capacity dewatering systems operating for long periods of time, as might occur with a dewatering system for an open pit mine (Section 9.7), or for a permanent dewatering system (Chapter 20). For such schemes, if hydrogeological analysis indicates that the pumped flow rates will initially be high but will reduce after several weeks or months of pumping as aquifer storage depletion occurs, one option is to use large pumps for initial pumping and then swap them for smaller pumps. The labour cost of the pump changeout operation will be recovered in reduced energy usage over future years. This approach is described in more detail by Rea and Monaghan (2009).

13.8.7 Other Methods

Occasionally, other methods are used to estimate steady-state flow rates.

Flow net analyses (Section 5.6.1) are sometimes used to model flow patterns not amenable to simplification as equivalent wells or slots. A flow net is a graphical representation of a given two-dimensional groundwater flow problem and its associated boundary conditions. Flow nets are one of the common forms of output of numerical groundwater models (see Section 13.12). However, as described by Cedergren (1989), hand sketching of flow nets can be used to obtain solutions to certain flow problems, considered in either plan or cross section, for isotropic or anisotropic conditions. Typical problems where flow nets are used include seepage into excavations or cofferdams where the presence of partial cut-off walls alters the groundwater flow paths (Figure 5.2).

Very rarely, physical models or electrical resistance or resistance-capacitance analogues (see Rushton and Redshaw, 1979) are used to analyse groundwater flows (see Sections 5.6.2 and 5.6.3). In the past, they were used more commonly to analyse complex problems, but in recent times, advances in numerical modelling methods have made these techniques largely obsolete. A rare recent application of the use of electrical analogues is described by Knight *et al.* (1996).

13.9 SPECIFICATION OF WELL YIELD AND SPACING

Having determined the total required pumping rate, the next step is to determine the yield, spacing and depth of wells (be they wellpoints, deep wells or ejector wells). The design of well filters is covered in Section 13.10.

13.9.1 Well Yield

Each well must be able to yield sufficient water so that all the wells in concert can achieve the required flow rate and hence the required drawdown.

Water enters a well where the well screen penetrates below the lowered water table in an unconfined aquifer, or where it penetrates a saturated confined aquifer. This depth of penetration is known as the 'wetted screen length' and is an important factor in the selection of well depths to achieve adequate yield (Figure 13.18).

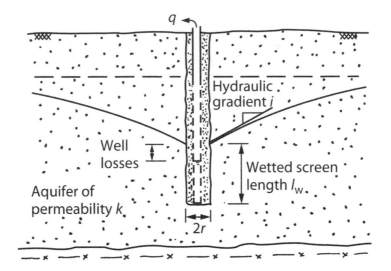

Figure 13.18 Wetted screen length of wells.

In theoretical terms, the yield q into a well can be described (from Darcy's law) by

$$q = 2\pi r l_w k i \tag{13.35}$$

where
  r = the radius of the well borehole (not the diameter of the well screen). This assumes that the well filter media are of significantly greater permeability than the aquifer
  l_w = the wetted screen length below the lowered groundwater level
  k = aquifer permeability
  i = hydraulic gradient at entry to the well

  The designer has no control over aquifer permeability but can vary the length and diameter of the well within limits determined by the geology of the site and the availability and cost of well drilling equipment. Experience suggests that any well will have a maximum well yield, beyond which the well will 'run dry', i.e. the water level in the well will reach the pump inlet or suction level, preventing further increases in flow rate.
  In 1928, Sichardt published a paper entitled 'Drainage capacity of wellpoints and its relation to the lowering of the groundwater level', which examined the yield of pumped wells based on the records of numerous groundwater lowering projects. He determined empirically that the maximum well yield is limited by a maximum hydraulic gradient i_{max} that can be generated in the aquifer at the face of a well. The Sichardt limiting gradient is generally taken as relating the aquifer permeability k (in metres per second) to the maximum hydraulic gradient at the face of the well by

$$i_{max} = \frac{1}{15\sqrt{k}} \tag{13.36}$$

The flow rate based on the limiting hydraulic gradient is the theoretical maximum amount of water that a *well* can yield – in very high-permeability aquifers, the potential well yield may be so large that the actual flow rate is controlled by the *pump* rather than the well.

The Sichardt gradient is probably a reasonable way to estimate the maximum well yield in a relatively high-permeability aquifer (k greater than about 1×10^{-4} m/s). However, work by Preene and Powrie (1993), who analysed well yields in a large number of dewatering systems in fine-grained soils, has suggested that Equation 13.36 may not be appropriate for lower-permeability aquifers. The Sichardt gradient may over-estimate hydraulic gradients (and hence well yields) in soils of permeability lower than about 1×10^{-4} m/s. The work of Preene and Powrie indicated that in lower-permeabilty soils the hydraulic gradients were generally less than 10, and that an average of six was not unreasonable. These two approaches are combined in Figure 13.19 to give the maximum well yield per unit wetted screen length for wells of various diameters of bored hole. Yield per unit length for other diameters can be calculated using Equation 13.36 and a limiting hydraulic gradient appropriate to the aquifer permeability.

Figure 13.19 can be used to design deep well or ejector systems. Wellpoint systems (or systems of ejector wells with short screens at very close spacings) are analysed slightly differently, as will be described later.

For wells at wide spacings in an aquifer that extends to some depth below the excavation, once a well diameter is assumed, Figure 13.19 can be used to determine the minimum wetted screen length (below the lowered water level) needed to obtain the total flow rate from the deep well system. This would then allow the number of wells and wetted depth per well to be estimated. For example, if 120 m of wetted screen length was estimated, this could be equivalent to 12 wells with 10 m wetted screen each, or eight wells with 15 m wetted screen each, and so forth. Obviously, a check then needs to be made that the originally assumed well diameter is large enough to accommodate well screens and pumping capacity to produce the design yield. If necessary, the well diameter must be increased. Guidance on the size of deep wells and ejector wells required to produce given discharge rates per well is given in Section 16.3 and Section 17.2, respectively.

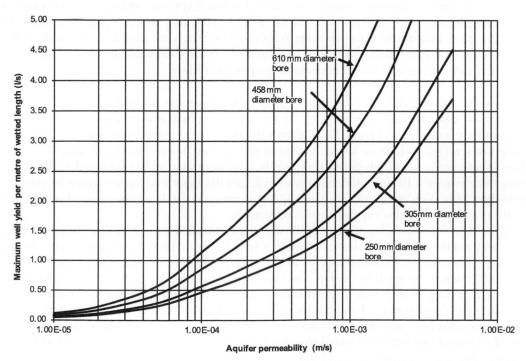

Figure 13.19 Maximum yield per unit wetted length of wells.

Estimation of well yield is a point in design where judgement and experience can be vital, so the designer should consider the following:

1. The wetted screen length l_w will be rather less than the penetration of the well below the 'general' lowered water level. This is because of the additional drawdown around each well due to each individual cone of depression (Figure 13.18). The difference between the wetted depth and the excavation drawdown level can be estimated from pumping test results. If no pumping test data are available, l_w must be estimated based on engineering judgement.
2. The yields calculated by the methods described earlier are theoretical maximums and may not be achieved in practice. Well yields may be reduced by use of inappropriate filter material or poor development. The method of drilling will also influence yield, with jetted boreholes generally being more efficient than rotary or cable tool percussion drilled holes. Experience also suggests that two competent drillers using the same methods and equipment on holes a few metres apart can produce wells with wildly differing yields. There is no conclusive explanation for this phenomenon, but it is likely to be related to the precise way each driller uses the bits, casing and drilling fluids and how that affects a thin layer of aquifer just outside the well.
3. If the geology does not consist of one aquifer that extends to great depth, this may affect well depth and will restrict the flexibility of the designer in specifying l_w. If the aquifer is relatively thin, screen lengths will be limited by the aquifer thickness, and a greater number of wells will be required. If there is a deeper permeable stratum that, if pumped, could act to underdrain the soils above, it may be worth deepening the wells (beyond the minimum required) in order to intercept the deep layer (see Section 13.6.2).
4. It is always prudent to allow for a few extra wells in the system over and above the theoretically calculated number. Typically, for small systems, at least one extra well is provided, or for larger systems, the number of wells may be increased by around 20 per cent. This allows for some margin for error in design or ground conditions but also means that the system will be able to achieve the desired drawdown if one or two wells are non-operational due to maintenance or pump failure.

In a similar manner, the potential yield of a wellpoint or an ejector well installed in a jetted hole can be assessed using Equation 13.36 and a limiting hydraulic gradient appropriate to the aquifer permeability. Figure 13.20 shows the theoretical maximum yield of 0.7 m long wellpoint screens. Yield from longer or shorter screens can be estimated pro rata. The figure shows that even in very high-permeability soils, a conventional wellpoint is unlikely to yield more than around 1 l/s (special installations of larger diameter and longer screen length may yield more, but such applications are rare). A maximum possible wellpoint yield of 1 l/s is a useful practical figure for the designer to remember.

Figure 13.20 is based on a disposable wellpoint installed in a 200 mm diameter sand filter and a self-jetting wellpoint installed in a jetted hole of 100–150 mm diameter. When dealing with jetted wells, the diameter of the jetted hole may be uncertain, requiring further judgement to be used. In theory, a wellpoint system could be installed in an analogous manner to a deep well system by determining the necessary total wetted screen length and the corresponding number of wellpoints and then adding additional wells as a contingency. In practice, this process is rarely carried out. Wellpoint equipment is almost always used in one of a limited number of standard spacings (Section 15.5). Methods of selecting wellpoint spacings are described in Section 13.9.3.

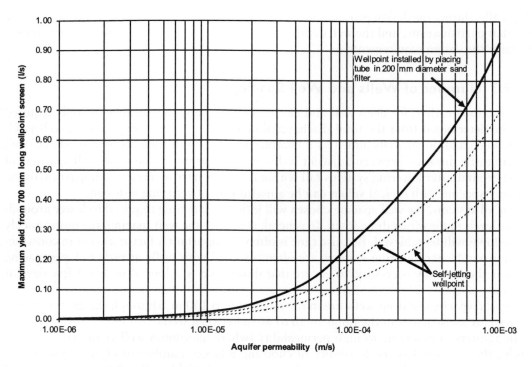

Figure 13.20 Maximum yield of wellpoints.

The estimated well yield can be used to select the capacity of the pumping equipment. For deep wells and ejector wells, the pumping equipment is located in each well and is sized directly from the well yield. For wellpoint systems, one pump acts on many wellpoints in concert. The designer has the choice of a few larger pumps or a greater number of smaller units (see Section 15.7). When estimating the required pump capacity from the calculated steady-state flow rate, an additional allowance must be made for the greater flow rate from storage release during the initial period of pumping (see Section 13.8.6).

13.9.2 Depth of Wells

The depth of wells used in groundwater control systems must be selected based on consideration of several different criteria, including:

a) Requirements to ensure adequate well yield (Section 13.9.1). Typically, in design, calculations are made to ensure that the wells penetrate sufficiently far into the target permeable stratum that adequate flow rates can be yielded.

b) Requirements to ensure that base stability is addressed (Section 13.7). Well depth may be influenced by the need to ensure that deep permeable layers are depressurized. As a rule of thumb, if pumped wells or relief wells penetrate to at least one and a half to two times the depth of the excavation, they are likely to penetrate sufficiently deeply to intercept any deep problematic layers. If the wells are rather shallower, high groundwater pressures could remain at depth and lead to hydraulic failure of the excavation base.

c) Geological structure (Section 13.6.2). The well depth may be selected in design to target particular strata, either as part of an underdrainage strategy or to ensure that multiple aquifers are depressurized.

Typically, during the design process, one of these criteria will be identified as the critical one for that particular site, and the well depth should be checked to ensure that it is deep enough to meet the design requirements.

13.9.3 Number of Wells and Well Spacing

As described earlier, for deep well and ejector well systems, the number of wells required can be determined from the total discharge flow rate divided by the predicted well yield, with some additional wells added as a contingency. This will then allow the average well spacing (the distance between adjacent wells) to be determined. Because well spacings of most groundwater lowering systems fall within relatively narrow ranges, comparison of the 'design' spacing with typical values can be a useful way of verifying a design.

For all systems, the well spacing chosen will influence the time required to lower groundwater levels to the target drawdown. In general, the closer the well spacing, the more quickly drawdown will be achieved. Because time is often as important a factor as cost in construction, groundwater lowering systems are often installed with wells at rather closer spacing than is theoretically necessary, to ensure that drawdown is achieved within a few days or weeks.

Typical spacings of deep wells are in the range of 5 to 100 m, although the great majority of systems use spacings in the range of 10 to 60 m between wells.

In aquifers of moderate to high permeability (where maximum well yields are relatively high), the designer has the flexibility of choosing a larger number of closely spaced low-yield wells or a smaller number of widely spaced high-yield wells. When potential well yields are large, high-capacity pumps may not be available, and the output from each well may be controlled by the pump performance; this should be considered when estimating well numbers and spacings. If the aquifer extends to great depth, it may be possible to use relatively few very high-capacity wells of very great depth at spacings of several hundred metres. This approach would require extensive pump test data, backed up by numerical modelling.

In low-permeability strata, the well yield will be relatively low, and the option of a small number of widely spaced high-yield wells is not available to the designer. To achieve the total discharge flow rate, the wells will be at close spacings. If a well spacing of less than around 5–10 m is suggested by the design, or if the well yield is lower than around 0.7 l/s, an ejector well system could be considered as an alternative to deep wells.

Ejector well systems can be used in two ways:

(i) They can be used in soil/rock of moderate permeability as an alternative to low-yield deep wells or as 'deep wellpoints' to achieve drawdowns in excess of 5–6 m. When used in this way, well spacings are similar to those used for deep wells (5 to 10 m) or wellpoints (1.5 to 3 m).

(ii) In strata of low permeability, they can be used as a vacuum-assisted pore water pressure control method (see Section 9.5.4). Because of the limited area influenced by each well, ejector wells tend to be installed at close spacings (1.5 to 5 m) and operate at only a fraction of their pumping capacity.

In contrast to deep well and ejector well systems, wellpoint spacings are rarely selected on a yield per well basis. The number of wellpoints is determined by first selecting the wellpoint spacing from a fairly narrow range; the number of wellpoints is then determined from the length of the line or ring of wellpoints. Typical wellpoint spacings are in the range of 0.5 to 3.0 m (see Table 15.1 for spacings categorized by soil type). Closer spacings tend to reduce the time to achieve drawdown compared with wider spacings. Wellpoint spacing is controlled by different factors in soils of high and low permeability.

a) In high-permeability soils (such as coarse gravels), the total discharge flow rate will be large. Each wellpoint may be operating at high yields. A rule of thumb is that a conventional wellpoint cannot yield more than about 1 l/s, and the spacing is sometimes based on reducing the average wellpoint yield below 1 l/s. Wellpoints tend to be installed at close spacings in high-permeability soils. If a spacing in the range of 0.5 to 1.0 m is suggested by design calculations, wellpoint dewatering may not be the most appropriate method. High-capacity suction wells (Section 16.10) at close spacings might be considered as an alternative.

b) In low-permeability soils (such as silty sands), the total flow rate will be low. Consideration of the maximum wellpoint yield may suggest that fairly wide (5 to 10 m) spacings may be possible. In reality, because of the limited area influenced by each wellpoint, such a system is likely to perform poorly, and a very long time may be required to achieve drawdown. The solution is to install wellpoints at closer spacings of 1.5 to 3.0 m.

13.10 SPECIFICATION OF WELL SCREENS AND FILTERS

Pumped wells of all types (wellpoints, deep wells and ejector wells) suitable for use in soils require a 'well screen'. Most commonly, this comprises a perforated vertical pipe, within which the pump or suction pipe is placed. The well screen must:

a) Allow water to freely enter the well to ensure an adequate yield
b) Retain the soil around the well to prevent the well collapsing and avoid fine particles being continuously drawn from the ground with the pumped water

The latter criterion requires that a filter medium, often termed a 'filter pack', be present in the annulus between the screen and the wall of the well or sump.

Filter packs can be formed by placing, at the time of well installation, specially selected granular material (typically of sand or gravel size) in the annulus around the well screen – these are 'artificial' gravel packs (Section 13.10.2). In some coarse-grained soils, it is possible to develop a 'natural' filter pack from the soil immediately around the well screen (Section 13.10.3). Wells installed in competent rock can sometimes be installed without a well screen or filter pack (Section 13.10.5).

As an alternative to conventional granular filter packs, placed during well installation, pre-formed filters (fitted when the well screens are manufactured) are also available (see Section 16.3.3).

Wellpoints have the same requirements for filter packs as other types of dewatering wells, but filter design is rarely done for such projects. Instead, the use of standard wellpoint screen designs (Section 15.3.1) and 'sanding in' techniques (Section 15.4.8) usually ensure that adequate filters are in place. Sanding in may not be required where wellpoints are installed into well-graded gravelly sands and sandy gravels. In such cases, an effective natural filter pack can be formed by the washing action of the jetting process.

13.10.1 Requirements for Well Screens and Filter Packs

The well liner consists of plastic or steel pipe set vertically in the well. The perforated section of the well liner is slotted or perforated to allow water to enter; this is termed the 'well screen' and is typically located at depths within the well where the principal water-bearing zones are expected. At other depths, the well liner is unperforated and is often known as the 'well casing'.

When selecting a well screen and casing, care must be taken to ensure that the wall thickness of the base pipe forming the screen and casing is adequate to withstand collapse pressures from soil and groundwater loadings. There have been occasions when thin-walled plastic well screens have collapsed during development and pumping. Collapses have also occurred during placement of grout seals, when fluid pressure from the grout can be high. In most of these cases, further problems were avoided by installing additional wells with thicker-walled screens – the screen manufacturer should be able to provide guidance.

For wells installed in soil, the filter media must be selected based on the particle size distribution (PSD) of the aquifer material to meet the following two conditions:

(i) To be sufficiently coarse that the filter pack is significantly more permeable than the aquifer in order to allow water to enter the well freely
(ii) To be sufficiently fine that the finer particles are not continually withdrawn from the aquifer

The selection of any filter media has to be a compromise between these two conflicting conditions. Concentrating on condition (i) will give a high-yield well but with an increased risk of continuously pulling sand or fines into the well. Concentrating on condition (ii) will prevent movement of fine particles but may restrict well yield.

An additional requirement is that the material chosen must be suitable for placement in wells (see Section 16.6) with minimum segregation – filter media of uniform grading are preferred for this reason.

As noted earlier, well filters are designed against two conflicting criteria – the need to be highly permeable and the need to prevent continuous movement of fines. Add to this the problems that aquifers are variable and heterogeneous (a filter that works on one well may not work on one at the other end of the site) and that suitable filter materials may not be available locally, and it is obvious that experienced judgement is needed for filter design. The problem of aquifer variability can be addressed by carrying out a thorough site investigation. The problem of limited availability of suitable filter material (especially on remote sites) may require less than ideal filters to be used. In such cases, a series of trial wells using local materials may be appropriate. If the locally available material does not give acceptable well performance, the use of pre-formed filters (e.g. geotextile mesh or resin-bonded gravel; see Section 16.3.3) might be considered, since shipping costs for these materials may be lower than for bulky filter gravels.

13.10.2 Design Rules for Artificial Filter Packs and Slot Size

There have been numerous theoretical and practical studies of design methods for granular filters in relation to water supply wells and for dams. CIRIA Report C750 (Preene *et al.*, 2016) summarized suitable criteria for design of granular filters for dewatering purposes as follows:

1. $D_{15filter} > 4 \times D_{15aquifer}$. This satisfies condition (i) in the previous sub-section to ensure that the filter material is of sufficiently high permeability. For widely graded materials, this should be applied to the finer side of the filter grading envelope and the coarser side of the aquifer grading envelope. Additionally, the filter material should contain no more than 5 per cent of particles finer than 63 μm.

2. $D_{15filter} \leq 5 \times D_{85aquifer}$. This satisfies condition (ii) (to retain the aquifer material around the filter) and is known as Terzaghi's filter criterion. For widely graded materials, this should be applied to the coarser side of the filter grading envelope and the finer side of the aquifer grading envelope.

3. $U_{filter} < 3$. This allows the filter to be placed without risk of segregation. U is the uniformity coefficient ($U = D_{60}/D_{10}$) – very uniform materials (consisting of only a small range of grain sizes) have a low U. If U is greater than 3, there is a risk of segregation during placement, and the filter material should be placed carefully by tremie pipe (see Section 16.3.4).

Application of these criteria to an aquifer grading would produce a filter grading envelope as shown in Figure 13.21. A filter material that falls within the envelope is then selected from those available.

Application of these criteria to real aquifers results in a relatively narrow range of materials that are suitable for use as granular filter media. At one end of the range, a relatively low-permeability silty sand aquifer might need a 0.5–1.0 mm filter sand, while at the other end, a very high-permeability coarse gravel might need a 10–20 mm pea gravel. Materials outside these ranges are rarely used. Rounded uncrushed aggregates are generally considered to have higher permeabilities than the same grading of angular crushed material. For this reason, rounded materials are preferred for use as filter media.

The perforations in the well screen should be chosen to match the filter pack and avoid large percentages of the filter material being able to pass into the well. The most common type of perforation is a 'slot' – a linear opening of constant width (Figures 16.3a and 16.3b). In general, the slot size should be approximately equal to $D_{10filter}$ (in other words, the slot should retain 90% of the filter material). There is another condition that must be considered when specifying the slot size on the well screen – there must be a large enough total area of slots (known as the 'open area') to allow the desired flow to pass through the screen. Open area is defined as the total area of slots or apertures expressed as a percentage of the total area of well screen. In general terms, the open area is much greater for well screens

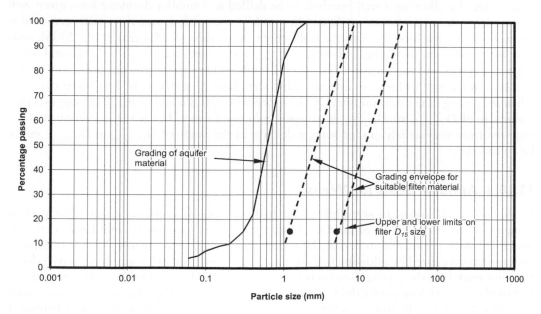

Figure 13.21 Aquifer and filter grading envelopes.

with larger slot sizes than with smaller ones. The slotted well screens used for temporary works wells typically have open areas of between 5 and 20 per cent for fine and coarse slots, respectively. The manufacturer's catalogues should indicate the actual percentage open area for the slot size and screen type in use.

Sterrett (2008) recommends that sufficient open area be provided to ensure that the average screen entrance velocity (well discharge divided by total area of screen apertures) is lower than 0.03 m/s. Parsons (1994) and Houben (2015) argue that this 'entrance velocity' approach has little theoretical basis. Nevertheless, it is an established rule of thumb that produces wells which perform at least adequately in many situations. Houben (2015) highlights that if analysis indicates that the screen entrance velocity is excessively high, then, because the velocity is relatively insensitive to the diameter of the well screen, the most practical design change is to increase the screen length.

When the filter is placed, it may be relatively loose, and it is likely to compact a little during development due to the flow of water re-arranging particles and also due to the finest filter particles being drawn into the well. Accordingly, the filter pack when placed must extend to some height (normally at least 0.5–1 m) above the top of the slotted well screen to ensure that aquifer material does not come into direct contact with the well screen if the filter pack compacts. Sometimes, permanent tremie tubes are left in place to allow the filter pack to be topped up by the addition of more material.

13.10.3 Design Rules for Natural Filter Packs and Slot Size

There are some types of granular soils for which it is not necessary to introduce artificial filter media, but where it is possible to directly develop the aquifer, remove the finer particles and form a 'natural' filter pack in the aquifer immediately outside the well screen. The term 'natural' is somewhat of a misnomer, since the intention is to use the development process to modify the in situ material to form a permeable and stable filter pack around the well. The key distinction is that no new 'artificial' granular filter media are added to the well annulus.

This method can be employed in coarse well-graded soils such as sandy gravels. It can give cost savings by allowing a well borehole to be drilled at a smaller diameter for a given well screen size – the space for an artificial filter pack is not needed, and the aquifer material is allowed to collapse directly onto the well screen. According to Misstear *et al.* (2006), soils may be appropriate for natural filters if $D_{40aquifer} > 0.5$ mm and $U_{aquifer} > 3$. Natural filters are not appropriate for uniform fine-grained soils, where there are not sufficient coarse particles in the soil to form an effective filter structure. Slot size must be chosen carefully when proposing to use natural filters. CIRIA Report C750 (Preene *et al.*, 2016) suggests that a slot size of $D_{40aquifer}$ to $D_{50aquifer}$ is acceptable in most cases, but if the maximum yield is required from very widely graded soils, a slot size in the range of $D_{60aquifer}$ to $D_{70aquifer}$ might be considered.

13.10.4 Annular Seals for Wells

In shallow applications in straightforward hydrogeological conditions, the filter pack is sometimes continued up to ground level (i.e. the annulus around the well liner is entirely backfilled with permeable material). However, this well design creates the risk of aquifer contamination caused by allowing surface water (or water from other aquifers) to pass down the filter pack into the aquifer (see Section 21.6.1). To avoid these problems, it is good practice to place a very low-permeability seal in the annulus above the filter pack. Annular seals are often formed of bentonite pellets or grout of either neat cement or cement-bentonite (materials and methods for such seals are given in Section 16.6). The objectives of such seals can include:

- Preventing surface water from flowing down the well annulus
- Preventing vertical flow of water between different aquifers
- Avoiding water being drawn directly from compressible aquitards
- Allowing effective decommissioning of wells (Section 24.3)

13.10.5 Dewatering Wells in Rock

If wells are drilled in harder fractured rock, it may be possible to install the well with an unslotted well casing in the top of the well but with the lower section installed as an unlined 'open hole' section, without well screen and filter pack. This design relies on the unlined section through the rock being stable with little risk of collapse.

Where wells are installed in weak or weathered rock, a perforated well screen is sometimes installed with the aim of reducing the risk of the pump or suction pipe being trapped or damaged by blocks of rock falling into the well. In such cases, the inflow to the well is predominantly from fractures, and (provided that the well has been adequately developed) a filter pack is not needed to prevent movement of fine particles. However, it is good practice to fill the annulus between the well screen and the borehole wall with a coarse permeable filter gravel, which acts as a formation stabilizer. The formation stabilizer prevents weaker blocks of aquifer rock from collapsing against and distorting the well screen. Formation stabilizers should be highly permeable to allow free passage of water; 10–20 mm pea gravel is often used.

13.11 OTHER CONSIDERATIONS

Although the prime concern of most groundwater lowering designs is to estimate the steady-state flow rate and the corresponding number and yield of wells that will be required, other issues sometimes need to be addressed during design.

13.11.1 Estimation of Drawdown Distribution around Well

As pumping continues, the zone of influence around a dewatering system will increase with time. This means that the area affected by drawdown will expand, initially rapidly and then progressively more slowly as time passes (Figure 5.9). This is clearly a complex problem, and complete solution probably requires a suitably calibrated numerical model. However, there are some simpler methods that can be used to estimate the drawdown pattern around an excavation at a given time. These methods are outlined in this section.

In dewatering design, the drawdown distribution may be required for the following reasons:

a) To determine how far the zone of significant settlements will extend during the period of pumping. This is useful if the impact of potential environmental impacts (see Chapter 21) is being assessed. This normally involves producing a plot of the drawdown versus distance at time t after pumping began. Successive plots at greater values of t can show the development of the zone of influence. Each plot of drawdown versus distance at a given time is known as an isochrone.

b) To determine the time required to achieve the target drawdown in a particular part of the excavation. This is normally only an issue in low-permeability strata, where it may influence the construction programme. In moderate- to high-permeability soil/rock, experience has shown that most appropriately designed systems should achieve the target drawdown within 1 to 10 days' pumping.

In many cases, these calculations are unnecessary because the time to achieve drawdown and the risk of side effects are not of major concern. In other cases, the methods presented here can be used to *approximately* determine the time-dependent drawdown distribution. All these analyses assume that the aquifer is homogeneous and that no recharge or barrier boundaries are present within the zone of influence (i.e. all pumped water is derived purely from storage release). Numerical modelling should be considered where aquifer boundaries are likely to be present.

If the groundwater lowering system consists of relatively widely spaced wells, then cumulative drawdown methods can be used. The theoretical approach (Equations 13.17 through 13.20) can allow the predicted drawdown at a selected point to be plotted against time. Alternatively, the distance–drawdown profile (or isochrone) can be determined at a given time after pumping commences. By repeating the calculation for different times, a series of isochrones showing the drawdown pattern at each time interval can be produced.

If the system consists of closely spaced wells in a regular pattern, the system can be analysed as an equivalent well or slot. Analyses that assume constant drawdown in the well or slot can then be used. The constant drawdown assumption is a reasonable one for most dewatering systems (apart from during the first few hours of pumping) and has given acceptably accurate results in practice.

For horizontal planar flow to an equivalent slot in a confined aquifer, the drawdown curve can be expressed as a parabola (Powrie and Preene, 1994a), shown in dimensionless form in Figure 13.22. The drawdown s at distance x from the slot can be determined from this figure provided that the drawdown s_w at the slot and the distance of influence L_o are known. The drawdown at the slot is normally taken as the same as the drawdown inside the excavation (not the drawdown in individual wells), and L_o is estimated from

$$L_o = \sqrt{\frac{12kDt}{S}} \tag{13.37}$$

where
  D is the aquifer thickness
  k is the permeability
  S is the storage coefficient
  t is the time since pumping began

For horizontal radial flow to an equivalent well in a confined aquifer, a time-dependent solution was developed by Rao (1973); this has been plotted in Figure 13.23. The drawdown s at radius r from the centre of the well can be determined from this figure that provided the drawdown s_w at the slot and the time factor T_r are known.

$$T_t = \frac{kDt}{Sr_e^2} \tag{13.38}$$

where r_e is the radius of the equivalent well, and all other terms are as defined previously.

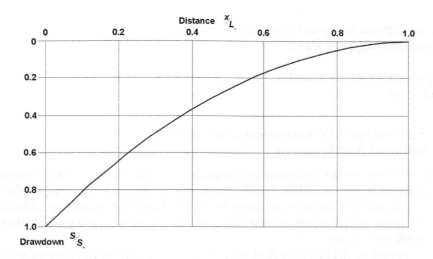

Figure 13.22 Dimensionless drawdown curve for horizontal planar flow to an equivalent slot.

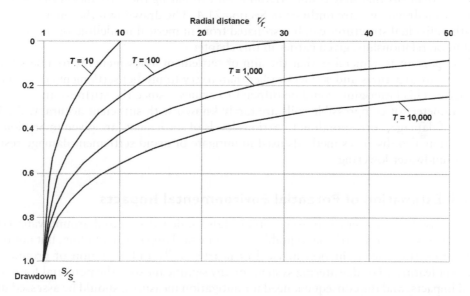

Figure 13.23 Dimensionless drawdown curve for horizontal planar flow to an equivalent well. (After Powrie, W and Preene, M, *Géotechnique*, 44, 83–100 1994.)

These methods are only theoretically valid in confined aquifers but can be used without significant error in unconfined aquifers where the drawdown is not a large proportion of the original saturated thickness.

Equations 13.37 and 13.38 use the storage coefficient S, which is appropriate for soil/rock of moderate to high permeability. These methods can also be used where pore water pressure control systems are employed in low-permeability soils, but the drainage characteristics of the soils may be expressed in terms of c_v, the coefficient of consolidation:

$$c_v = \frac{kE_o'}{\gamma_w} = \frac{kD}{S} \qquad (13.39)$$

where
   E'_o is the stiffness of the soil in one-dimensional compression
   γ_w is the unit weight of water

For applications in low-permeability soil, Equation 13.39 can be substituted into Equations 13.37 and 13.38.

13.11.2 Estimation of Groundwater-Lowering-Induced Settlements

Ground settlements are an unavoidable consequence of the effective stress increases that result from groundwater lowering. In most cases, the settlements are so small that there is little risk of damage or distortion to nearby buildings. However, if compressible soils (such as peat or normally consolidated alluvial clays and silt) are present, it is possible that damaging settlements may occur.

If the conceptual model has identified the presence of significant thicknesses of compressible strata within the zone of influence, it will be necessary to consider the magnitude of ground settlements that may result. Methods for estimating the settlements that will result from a given drawdown are outlined in Section 21.4. The drawdown that may be expected beneath individual structures can be estimated from numerical modelling or the drawdown distribution relationships given earlier in this chapter.

It is important to remember that the aim of settlement assessments is to assess the risk of damage to structures and services rather than to try to predict settlements to the nearest millimetre. This latter aim would be difficult to achieve, since the stiffness and consolidation parameters of compressible soils are rarely known with sufficient accuracy. Predicted degrees of damage, corresponding to various levels of settlement, are given in Section 21.4. That section also discusses methods used to mitigate or avoid settlement damage resulting from groundwater lowering.

13.11.3 Estimation of Potential Environmental Impacts

As is discussed in Chapter 21, in some circumstances, dewatering and groundwater control works may result in a risk of unacceptable environmental impacts, including, but not limited to, ground settlement, reduction in yield of nearby wells and depletion of groundwater-dependent features. For dewatering systems of any significant size, the potential for environmental impacts, and the consequent need for mitigation measures, should be assessed during design. Assessment of environmental impacts is discussed in Section 21.10.

13.12 NUMERICAL MODELLING

Numerical modelling, in various forms, is a routine method in modern geotechnical analysis. It can enable the analysis of complex groundwater problems that cannot readily be resolved by closed-form analytical solutions such as the Dupuit–Forchheimer equations and similar methods. The background to numerical modelling is discussed in Section 5.5.3, and further detail on the numerical modelling process can be found in Anderson *et al.* (2015).

This section principally discusses the use of proprietary numerical groundwater modelling packages to solve complex groundwater flow problems not normally amenable to solution by other means. However, computing power can help designers in other ways. This includes spreadsheet programs used to evaluate multiple versions of design equations that might previously have been carried out by hand. It is fairly easy to write routines for a spreadsheet to allow repeated sets of calculations to be performed as part of sensitivity or parametric analyses. Automating this process on a personal computer or other device dramatically speeds up this process compared with hand calculation.

- The software divides the overall problem, its geometry and boundary conditions into a number of discrete smaller mathematical problems to be solved individually. Therefore, the geometry of the problem must be defined as part of the conceptual model before it can be modelled.
- The software can only follow instructions given to it by the user. If there are errors in the input data, or more importantly, if the conceptual model (on which the groundwater model is based) is unrealistic or is poorly translated into the numerical model, then gross errors are likely to result.

The numerical modelling process can be divided into a number of stages:

a) Development of the conceptual model (see Section 13.4). Even though it does not involve touching a computer, this is the most important stage of the numerical modelling exercise. The conceptual model must quantify the geometry, aquifer parameters and boundary conditions that will be used to define the numerical model. If the conceptual model does not adequately reflect actual conditions, the modelling results are unlikely to be realistic or useful.

b) Selection of software and setting up of numerical model. Once the conceptual model is defined, software capable of modelling those conditions can be selected. The software is then used to create the groundwater model (the set of instructions defining the relevant geometry, properties and boundary conditions) for the problem. Any errors or omissions in the input data and instructions will have an effect on the results produced by the model.

c) Verification and calibration. These activities are essential to allow any errors in the input data or model formulation to be identified, and for the user to develop some degree of confidence in the validity of the output. The aim of verification is to answer the question 'has the model done what we intended it to do?' To answer this question, the input data must be scrutinized for errors and the output of the model must be compared with known analytical solutions. It is unlikely that an analytical solution will be available for the whole model, but it may be possible to simplify all or part of the model and compare it with results calculated by, say, flow net or equivalent well methods. If errors are detected, these should be corrected and verification repeated until an acceptable agreement is obtained. Following verification, the model should be calibrated against field data such as monitoring well readings in various parts of the modelled area. Calibration is a trial and error procedure whereby the model parameters and boundary conditions are varied (within realistic ranges chosen from the conceptual model) until there is acceptable agreement between the field data and the model output.

d) Prediction and refinement. A verified and calibrated model can be used to predict the results of interest (flow rates, drawdowns, settlements, etc). Parametric and sensitivity analyses can be carried out to assess the effect on the results of different well arrays or aquifer conditions. For larger or longer-term projects, it may be possible to refine predictions by further validation and calibration against results from the monitoring of the dewatering system – this can be used as part of the observational method.

13.12.1 Potential Applications of Numerical Modelling

There are a number of ways that numerical models can be used in the design of groundwater lowering systems. Perhaps the most obvious is to use the model, following a comprehensive site investigation, as a predictive design tool to finalize the dewatering system. The conceptual model is developed from the site investigation results, and the numerical model is run repeatedly, adding, removing or relocating wells and pumping capacity until the target drawdown is obtained at specified points or other design requirements are satisfied.

There is another way that numerical modelling can be used as an aid to both design and site investigation. If a groundwater model is created at an early stage of the project (perhaps using a conceptual model based on the site investigation desk study), it can be used as a preliminary design tool to crudely model the effect of various scenarios, including possible ground conditions and construction options. Outputs from the modelled scenarios may highlight particular issues to be addressed by the ground investigation. Similarly, the effect of changing the size of excavation, depth of cut-off walls and so on can be investigated; this may be useful information for designers. Any potential side effects of groundwater lowering can also be quantified. The model is then developed, recalibrated and refined as additional data is gathered, and continues to provide information for designers on the effect of various options. This approach has been successfully adopted on a number of larger projects, including those where the observational method was used.

Numerical groundwater models are not needed for every dewatering project. Numerical modelling is of most value in cases where traditional closed-form analytical solutions are either unavailable or too cumbersome for practical use. Situations where numerical models should be considered include:

- *Complex geometry and geology.* Where the conceptual model and/or dewatering system is complex, perhaps with multiple aquifers and intermediate aquitards, nearby recharge or barrier boundaries, or artificial recharge systems, suitable closed-form solutions may not be available, or their use may require such simplification of the conceptual model that the resultant errors or uncertainties are unacceptable. Numerical models have the potential to allow a design to represent the conceptual model with fewer simplifications, thereby giving more realistic and relevant results.
- *Anisotropy and non-uniform permeability conditions.* Analytical methods are available that can address anisotropic permeability and zones of different permeability, but these are applicable to only a narrow range of cases. Numerical models have the potential to allow the effect of these conditions to be assessed either as part of the main design process or as part of sensitivity or parametric studies. Bevan *et al.* (2010) describe an example where numerical modelling was used to investigate the impact on a major dewatering system of large-scale variations in permeability in the area of the excavation.
- *Transient analyses.* Where non-steady-state conditions, such as rate of drawdown or rate of water level recovery following the end of pumping, are of interest, numerical modelling can be very useful. This is particularly the case for multiple well systems with non-uniform geometries. Transient numerical models are often significantly more time-intensive and require better calibration data than steady-state models.
- *Environmental impacts.* Numerical modelling is widely used on major dewatering projects to assess the potential environmental impacts on the groundwater environment (see Section 21.4). Where the potential impacts of concern include the movement of contaminated groundwater, it is essential that the modelling software used has the capability to simulate the transport of contamination in groundwater flow. Numerical

modelling is typically also used to aid the design of mitigation measures. For example, where dewatering-related settlements are a concern, modelling scenarios may be run to assess the effectiveness of different locations and geometries of cut-off walls and artificial recharge wells. Crompton and Heathcote (1993) and Edwards (1997) describe a case where numerical modelling was used to predict long-term groundwater level rises (and the corresponding requirement for long-term dewatering) when a tidal barrage was constructed across Cardiff Bay, UK.

- *Multiple scenarios.* A huge advantage of numerical modelling is the ability to establish base numerical models and then vary key parameters or aspects in order to allow multiple scenarios to be modelled. This can be of great value in helping the project team understand the risks if ground conditions or boundary conditions found in practice vary from the initial conceptual model.

13.12.2 Potential Pitfalls of Numerical Modelling

Fundamentally, numerical modelling is no better or worse than analyses using closed-form solutions for dewatering design. Done well, by experienced design teams, both methods can give great results. Conversely, applied unwisely, both approaches can result in poor estimates of discharge flow rates and drawdowns, inappropriate designs and consequent problems during construction.

Any designer needs to be realistic that the outputs of any model are not a perfect simulation or prediction of reality; however, that should not stop us from using models in design. We should aspire to develop useful models even if they are imperfect. The statistician George Box said: 'Remember that all models are wrong; the practical question is how wrong do they have to be to not be useful' (Box and Draper, 1987). Therefore, it is essential that the designer 'challenges' or attempts to validate the outputs of the design process with the aim of identifying potential errors or unrealistic results that could reduce the validity of design outcomes. With numerical modelling, the review and validation of design outputs is especially important.

Some of the potential pitfalls that can affect numerical modelling exercises are outlined in the following list:

- *Treating the model as a 'black box'.* There is a real temptation, especially among those new to numerical modelling, to consider the modelling software packages, and the numerical models constructed from them, as 'black boxes', the workings of which are a mystery. This approach is a mistake. All software packages have their limitations and simplifications. The user must be aware of the limitations of the software to ensure that it is being applied appropriately. Similarly, as a result of software constraints, the models formed using the software may not be a perfect match for the conceptual model. It is essential that any simplifications in the numerical model are identified and understood.

- *Divergence between the conceptual and the numerical model.* Any conceptual model will, unavoidably, be an imperfect representation of reality. Similarly, a numerical model will almost always involve some simplification of the conceptual model to allow a workable model to be formulated. There is a risk, if the modelling specialist is not working closely with the dewatering designer or other team member who developed the conceptual model, that the numerical model will diverge unacceptably from reality. It is essential that numerical modelling is not carried out in isolation but is fully integrated with the rest of the dewatering design process.

- *Expecting a unique solution.* It is important that dewatering designers recognize that numerical models may, in some circumstances, give non-unique solutions. This can occur where complex geometries and multiple boundary conditions exist, and when the available calibration data might not allow the effect of a particular feature – such as a surface water body potentially linked to the aquifer – to be uniquely quantified. Such uncertainty of model outputs might require the design process to include sensitivity analysis and/or gathering of more data to allow more detailed calibration of the model.
- *Lack of empirical checks of modelling results.* Where a dewatering system is designed based on a numerical model, it is essential that empirical checks are carried out on the proposed dewatering system. For example, well yields, well spacing and well depths should be compared with typical values given elsewhere in this book. If this is not done, then the resulting dewatering system may be unworkable.
- *Presenting a false sense of precision.* Most numerical modelling software packages provide results in the form of tabulated values of water levels, flow rates, etc. that can be easily converted to graphs and contour plots, which can then readily be incorporated into design reports. There is a real risk that people reading the results will be impressed by the presentation of results and forget about the underlying simplifications and uncertainties. It is essential that those carrying out and reporting on numerical modelling work communicate the uncertainties and do not give a false sense of precision to the results. For example, care should be taken in selecting the number of significant figures used when quoting results, and it may be appropriate to quote key results (such as discharge flow rates) as a range of values rather than a single figure.

13.13 DESIGN EXAMPLES

Some design examples are presented in Appendix 5 to illustrate the application of the methods presented in this chapter. For ease of reference, the relevant equation and figure numbers are noted. The examples do not merely cover the numerical aspects of design but also discuss some of the issues over which 'engineering judgement' must be exercised.

Design example 1: Ring of relatively closely-spaced deep wells in a confined aquifer. This case is modelled as an equivalent well using radial flow equations.

Design example 1a: This analyses the case of example 1 using the alternative method of using shape factors for flow to equivalent wells in confined aquifers.

Design example 2: A line of partially penetrating wellpoints alongside a trench excavation in an unconfined aquifer. This case is analysed by modelling the line of wells as an equivalent slot under planar flow conditions. The effects of assuming different aquifer depths and of including the contribution from radial flow to the end of the slot are assessed.

Design example 3: Ring of widely spaced deep wells around a large excavation in a confined aquifer. The cumulative drawdown method is used to design the system.

SECTION 3

CONSTRUCTION

The value of an engineering science is determined by what it can accomplish as a tool in the hands of a practising engineer.

Karl Terzaghi

Chapter 14

Sump Pumping

[Sump pumping] is not the sort of thing one can learn from a book; it is learned down in the mud, preferably while equipped with boots of some height.

J Patrick Powers

14.1 INTRODUCTION

Sump pumping is the simplest, cheapest and probably most widely used method of groundwater control. It can be a very effective expedient to allow construction work below groundwater level, but if used inappropriately can cause significant problems, including ground instability and environmental problems due to the discharge of dirty water. This chapter addresses the formation of pumping sumps and associated gravity drainage channels for the control of surface water and groundwater. Sump pumping during tunnelling and shaft construction is discussed in Chapter 10. The pumps suitable for sump pumping applications are dealt with in Chapter 19.

Chapter 27 includes two Case Histories (27.2 and 27.7), which illustrate some applications and pitfalls of sump pumping.

14.2 APPLICATIONS AND LIMITATIONS OF SUMP PUMPING

Sump pumping is the most basic of the dewatering methods. The method is a commonly used type of 'open pumping' – see Section 9.5.1. In essence, it involves allowing groundwater to seep into the excavation, collecting it in locally deeper areas known as sumps and then pumping the water away for disposal. Sumps are provided for two separate purposes, though the form of a sump may be similar for either requirement:

a) To collect surface water run-off channelled to it by means of collector ditches or channels for discharge to a disposal point or area (see Section 14.3).
b) To collect and discharge pumped water to aid lowering of the groundwater for a shallow excavation (Section 14.4); also, sump pumping may be required for drainage of toe drains for battered slopes (see Section 14.5) and to deal with perched water and overbleed seepage in excavations that are dewatered by pre-drainage methods such as wellpoints or deep wells.

The preferred method of disposal of water collected by the sumps is by means of pumping. Typically, each sump is equipped with a robust, simple, pump – a 'sump pump' (see Section 19.4).

Sump pumping can be a very effective and economic method to achieve modest draw-downs in well-graded coarse soils (such as gravelly sands, sandy gravels and coarse gravels) or in hard fractured rock. However, it does have significant limitations, some of which are related to the 'open pumping' nature of the method, whereby groundwater is allowed to seep into the excavation before it is pumped away.

In Chapters 7 and 8, we saw how groundwater flow *into* an excavation can have a destabilizing effect on fine-grained soils or highly weathered rocks. Unfortunately, in some ground conditions, the seepage of water into the excavation during sump pumping can cause major problems. These problems can include fine particles being washed out of the soil/rock with the water – this is known as 'loss of fines'. Loss of fines can lead to ground movements and settlements, because material is being removed from the ground, giving the potential for the formation and collapse of sub-surface voids (see Section 21.4.3). Disposal of the water can also create problems, because if loss of fines occurs, the discharge water will have a high sediment load, which can cause environmental problems at the disposal point (see Section 21.9). These issues are discussed further in Section 14.8 and Section 14.9.

Powers (1985) lists soil types where the use of sump pumping has a significant risk of causing loss of fines. These include:

(i) Uniform fine sands
(ii) Soft non-cohesive silts and soft clays
(iii) Soft and weathered rocks, where fractures can erode and enlarge due to high water velocities
(iv) Rocks where fractures are filled with silt, sand or soft clay, which may be eroded
(v) Sandstone with uncemented layers that may be washed out

In these soil and rock types, even the best-engineered sump pumping systems may encounter problems. Potential problems can be avoided by employing groundwater lowering using a pre-drainage method such as an array of wells (wellpoints, deep wells or ejector wells) with correctly designed and installed filters. Provided that the wells are located outside the main excavation area, these methods have the advantage that they draw water *away* from the excavation, improving stability – avoiding the destabilizing flows into the excavation that are associated with sump pumping.

The initial discharge water from any pumping operation will be discoloured due to the presence of fines. Continuation of the discolouration of the water can be tolerated only if the pumped water is entirely derived from surface run-off. It cannot be tolerated if the discolouration continues when the flow is derived from groundwater – this is a warning of potential danger, for it indicates continuing withdrawal of fines from the formation.

14.3 SUMPS FOR SURFACE WATER RUN-OFF

Some surface water in the construction area is unavoidable, for example as the result of rainfall into the excavation. Even in very arid climates, surface water may be a problem due to wash water from concreting operations, seepage through defects in cut-off walls, etc. Sump pumping is often part of the strategy to deal with surface water by collecting it and conveying it to a disposal point. Strategies for control of surface water in excavations are discussed in Section 9.2, and in Section 17.10 in relation to drainage of slopes.

When an excavation exposes low-permeability soil or rock (such as clay or mudstone), all surface water (from precipitation, water from surrounding areas running down ramps, and other sources) will tend to pond in low points in the excavation. Movement of construction plant through surface water ponding will lead to deterioration of the surface, especially on clays and mudstones. This will inhibit efficient use of plant, and any significant excavation should have a plan to minimize such ponding of surface water.

Surface water can also enter the excavation from the surrounding ground-level areas.

- It is good practice to dig a collector ditch around the ground-level perimeter of an excavation to prevent surface run-off from entering the excavation (Figure 14.1). Collector drains should be lined with an impermeable membrane to avoid potential upstream recharge that might cause a rotational slip of the excavation slope.
- Measures should also be taken to prevent water running down any access ramps, and in multi-level excavations, to prevent surface water in the dig from running from shallow levels into the deeper areas.

Within the excavation, it is good practice to form the surfaces of the construction areas so that they are not level; they should be gently sloped, so that all surface water is shed to suitably sited and constructed collector channels or drains. The open ditch method (Figure 14.1a) should be used within the excavation only where the presence of open ditches does not inhibit construction work. As an alternative, agricultural type drains can be used

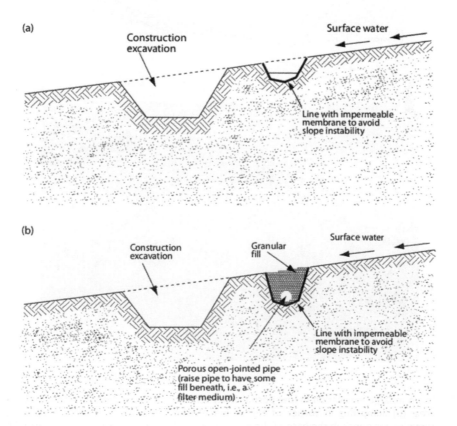

Figure 14.1 Drains for collection of surface water run-off. (a) Open ditch. (b) Agricultural drain. (After Somerville, S H, *Control of Groundwater for Temporary Works*, Construction Industry Research and Information Association, CIRIA Report 113, London, 1986.)

(see Figure 14.1 b), but the surface must be regularly scarified to reduce the effect of clogging by suspended particles in the surface water. This cleaning dictum should be applied to both the open ditch and the agricultural type drains.

The fall on the bed of a collector drain should be sufficient to minimize silting up but not so steep as to cause erosion. Near the sump, it may be prudent to increase the width of the drain to allow a flow velocity low enough to prevent erosion. The alternative is to provide check weirs at intervals along the line of the drain.

The surface water run-off within an excavation should be channelled to a conveniently located pumping sump as shown in Figure 14.2. The forms of construction of the sump(s) are described in the following.

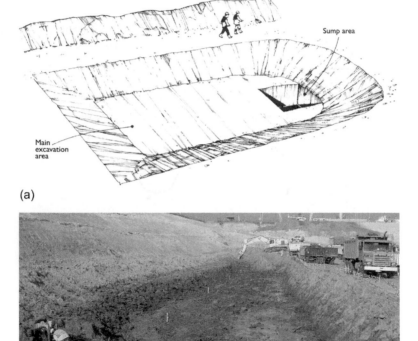

(a)

(b)

Figure 14.2 Sump in large excavation in low-permeability stratum. (a) Typical sump within main excavation area. (b) Sump dug in corner of excavation. Without drainage ditches feeding water to the sump, water ponds in the base of the excavation, and a wet excavation results. (Panel (a) reproduced from Somerville, S H, *Control of Groundwater for Temporary Works*, Construction Industry Research and Information Association, CIRIA Report 113, London, 1986. With permission from CIRIA: www.ciria.org; panel (b) courtesy of WJ Groundwater Limited, Kings Langley, UK.)

14.4 SUMPS FOR GROUNDWATER PUMPING

In order to be effective, a sump for the purposes of pumping groundwater within an excavation for construction or mining purposes must satisfy some basic criteria:

i. The sump must be deep enough. At all times, the depth of a sump must be a generous amount deeper than the formation level of the excavation at that time and also of the bed of any collector drain(s) leading into it. Ideally, there should be an allowance for some sediment build-up in the base of the sump to reduce the risk of the pump or suction hose becoming stuck or blocked.

ii. The sump must be large enough to accommodate the intended size of pump or suction hose.

iii. The sump must be physically stable or must be provided with means to support the sides. If the sump is not stable, pumps or suction hoses may become stuck in the sump due to collapses or debris.

iv. It must be possible to construct the sump safely, including the sections of the sump below groundwater level (where the ground may be unstable). For deep excavations, it is common practice to form the sump(s) to the maximum depth to which the ground is stable, and when the excavation has been dug down to the new, lowered groundwater level, the sumps are further deepened. This process can be repeated to allow the excavation and sumps to be deepened incrementally. However, this can limit the rate at which the excavation can be deepened, and for deep excavations, a better approach may be to use a pre-drainage method such as wellpoints or deep wells.

v. The sump should be hydraulically stable and efficient. This means that where the sump includes a structure or support system, it should have some form of filter that does not pass significant quantities of fine particles and also does not provide excessive resistance to water flowing into the sump.

vi. The sump should not be an unacceptable impediment to construction. This can be a problem in small excavations and shafts, where a sump (and associated pumping equipment) may take up a significant proportion of the available working area. Particular aspects of sump pumping for shaft sinking are discussed in Section 10.5.3.1.

Ideally, sumps should be located at the edge of the excavation so that they are outside the final structure that is being constructed. In this arrangement, it is straightforward to continue pumping from the sumps until the construction work is sufficiently complete that groundwater control is no longer needed. On sites where space is limited and the structure fills the entire excavation (for example when an excavation is made inside a concrete diaphragm wall or secant pile wall), the sumps may have to be located within the footprint of the structure. In this case, the chamber of the sump will have to be extended up to pass through the base slab of the structure with pumps relocated to slab level. When dewatering is no longer required, the sumps will have to be sealed though the slab (see Section 24.4 in the chapter on decommissioning).

Sump pumping can be carried out using different pump types, as described in Section 19.4. A very common approach to sump pumping uses suction pumps of the vacuum-assisted self-priming centrifugal type or reciprocating diaphragm type, typically powered by diesel engines. In shallow excavations, the sump pump can be located outside the excavation at ground level with a suction hose laid from the pump down into the sump (Figure 14.3a). However, suction lift limitations mean that the maximum effective depth to which a well-maintained suction pump will operate is about 6 m below the level of the pump. For deeper

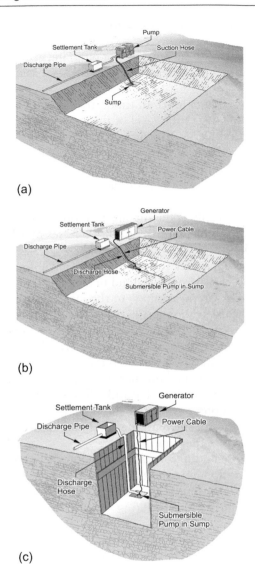

Figure 14.3 Sump pumping. (a) Sump pumped by a suction pump located at ground level. (b) Sump pumped by a submersible pump. (c) Sump pumping in small excavation with vertical sides supported by shoring. A submersible pump is shown in this application.

excavations of greater depths, it will be necessary to re-locate the pumps at a lower level within the excavation as work progresses. When diesel-driven pumps are used in deep excavations, the safe systems of work should include adequate forced ventilation to deal with exhaust fumes.

For deeper excavations, it is common to use submersible pumps (either electrically or hydraulically driven) suspended within the sump itself (Figure 14.3b and c)

In highly permeable ground conditions (sands and gravels and highly fractured rock), the need for sufficient pumping capacity, is paramount since a greater pumping capacity is needed to initially dewater an excavation than is required to maintain the water level at a steady state in its finally lowered position. Where possible, the pumping plant should be installed in multiple units so that the additional units required to give the increased capacity for the initial pumping load can be shut down as the required levels are reached. The spare pumpsets should be left in position to act as a standby in case of breakdown or other emergency. Intermediate pumping may also be required in a subsequent backfilling state.

Some of the water flowing to the sumps will have flowed over exposed soil or rock and is likely to transport fine particles with the water. These fine particles are likely to be abrasive and capable of causing wear and damage to the pumping equipment. A sump of a generous size will allow some settlement of the larger (and probably more abrasive) fine particles. Adequate provision should be made for periodic servicing of the pumps and removal of accumulated sediment. The pump should be suspended so that the bottom of the unit is about 300 mm above the bottom of the sump to allow for some build-up of sediment. This does not apply to smaller shallow sumps where the suction hose only is in the sump.

A common arrangement for a sump in granular soil is to suspend the pump in a permeable chamber formed by a 200 litre drum or similar (Figure 14.4a) with many holes punched through the sides of the drum. Outside the drum is placed an annulus of fine gravel to act as a crude filter. Other forms of sump construction are shown in Figure 14.4. If the excavation is in relatively stable rock, then it may be possible to simply form a relatively basic pit with steep sides, without the need for a chamber or filter gravel (Figure 14.5). Where any sump forms an open body of water, there should be suitable fencing and edge protection to reduce the risk of personnel or the public falling into the water.

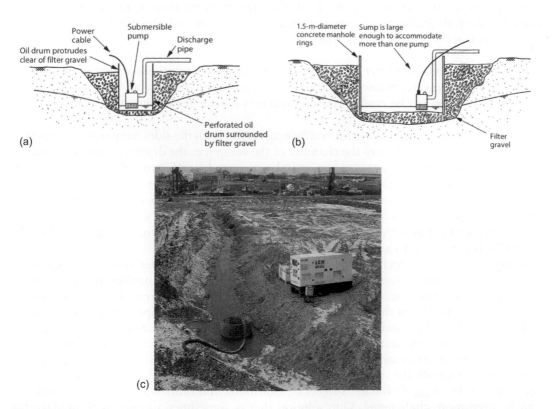

Figure 14.4 Typical forms of sump construction for excavations in soil. (a) Small sump formed using a perforated oil drum. (b) Large sump formed using concrete manhole rings. (c) Concrete manhole ring sump fed by drainage ditch. (Panel (c) courtesy of WJ Groundwater Limited, Kings Langley, UK.)

Figure 14.5 Example of a large sump dug in relatively stable fractured rock. The sump is pumped by a suction pump with the suction hose below groundwater level.

Sumps should be dug to a greater depth than the main excavation and should be maintained in their original form throughout the construction period, though deepened if necessary as excavation proceeds. This will:

a) Allow placement of filter media that may be necessary to minimize loss of ground
b) Keep groundwater below excavation level at all stages of the work
c) Allow changes to be made in the construction scheme for the main excavation

Most sumps in granular soils are formed by excavation, with the sides temporarily supported by sheeting for stability before the chamber of the sump (i.e. the drum or similar) is placed. In certain circumstances, jetted sumps may be preferred (see Figure 14.6). A suitable-size placing tube is jetted into the ground to the required depth. A disposable intake strainer and flexible suction hose are lowered into the placing tube in a manner similar to the positioning of a disposable wellpoint and riser pipe. Filter media is placed within the placing tube, around the strainer and riser pipe, as the placing tube is withdrawn. The upper end of the riser pipe is connected to a suitable pump.

Sump pumping also forms an important part of the groundwater control strategies for many open pit mines (Section 9.7), although in that industry, the method is referred to as 'in-pit pumping'. The objective is to achieve a general lowering of groundwater levels within the pit by pumping from the lowest point (the sump), which is equipped with the necessary pumps to transfer the water out of the pit (Figure 9.15). Often, the sumps are very large and can effectively become water storage lagoons (Figure 14.7a). Typically, the sumps receive surface water run-off as well as groundwater. Due to the large size of typical open pit mines, during heavy rainfall, the surface water flow rates to the sumps can be large, even in temperate climates. In tropical climates subject to monsoons and rainy seasons, the peak inflows can be even more challenging. One solution that is sometimes adopted is to manage the very bottom of the open pit as a large sump, within which water levels are kept low for the majority of the time, but after storm events the sump is allowed to fill up with water, acting

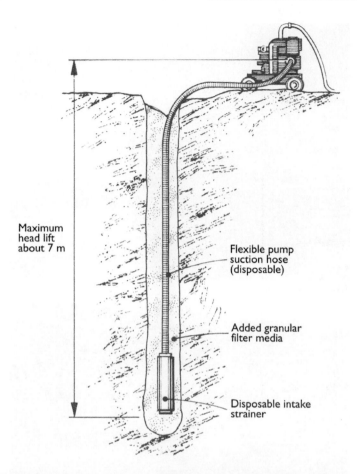

Maximum
head lift
about 7 m

Flexible pump
suction hose
(disposable)

Added granular
filter media

Disposable intake
strainer

Figure 14.6 A jetted sump. (Reproduced from Somerville, S H, *Control of Groundwater for Temporary Works*, Construction Industry Research and Information Association, CIRIA Report 113, London, 1986. With permission from CIRIA: www.ciria.org)

as a storage buffer to reduce the peak pumping rate necessary to prevent the main body of the pit being flooded. Pumps mounted on floating pontoons (Figure 14.7b) are able to move up and down with the water level, and pump over several days or weeks to lower the water level in the sumps back down to normal operating level.

14.5 DRAINAGE OF SIDE SLOPES OF AN EXCAVATION

In addition to dealing with groundwater seepages into the base of an excavation and run-off from rainfall, sump pumping has a role to play in dealing with seepages of water from side slopes. The destabilizing effect of seepages through slopes was discussed in Section 7.6.1, and methods for draining slopes are described in Section 17.10. Typically, slope drainage systems are intended to provide a relatively free pathway for water to emerge from the slopes. Sump pumping is commonly used to manage this water.

A toe drain (Figure 17.24) linked to pumped sumps is commonly used as part of slope drainage strategies. The water pumped from the sumps fed by the toe drain should be clear – i.e. no fines. The presence of fine particles in the pumped seepage water would indicate that the

Figure 14.7 Examples of in-pit pumping used in open pit mines. (a) Large sump excavated in rock in the base of an open pit mine. (b) Pontoon-mounted submersible pump used in sump where water levels fluctuate significantly due to storm water flows.

filters around the drain(s) are inadequate; fines are being continually removed, and eventually the slope will become unstable. The sumps of the seepage collection system should be as described in Section 14.4. The expected seepage rates of flow should indicate the dimensions of the sumps that will be required.

14.6 SUMP PUMPING OF SMALL EXCAVATIONS

A common application for sump pumping is in small excavations or trenches (see Irvine and Smith, 2001) where the sides are supported by some form of shoring or trench sheeting

(Figure 14.3c). In this case, the shoring systems are often not intended to be 'watertight' and so do not form a fully effective groundwater cut-off wall, and water will seep through the shoring system. In these conditions, it is acceptable to use sump pumping in stable fractured rocks and in permeable coarse-grained soils such as gravels and clean sand/gravel mixtures with low sand and silt content. Even though the shoring system is not completely impermeable, often the presence of trench sheets or trench boxes will limit the inflow to be pumped (see Figure 14.8).

The essential point is that the discharge water must be clear. If there is evidence of significant sediment in the pumped water, then sump pumping must be stopped. When dealing with finer-grained soils (such as silty sand), a system of wellpoints adjacent to the side(s) of the trench (Section 15.8) should be used instead of sump pumping. Case History 27.7 describes an example where sump pumping during shaft sinking caused settlement problems when fines were pumped continuously.

14.7 SUMP PUMPING FROM WITHIN CUT-OFF WALLS

Control of groundwater by exclusion (Section 9.4) involves forming a notionally impermeable physical cut-off wall or barrier around the perimeter of the excavation to prevent groundwater from entering the working area (Figure 9.2). Typically, the cut-off is vertical and penetrates down to a very low-permeability stratum that forms a basal seal for the excavation.

In this arrangement, some groundwater will be trapped in the area enclosed by the cut-off walls. Sump pumping is commonly used to remove such trapped water as the excavation is dug, in addition to dealing with water from rainfall and any leakage through the walls. The same principles and limitations of the sump pumping method apply as for applications where cut-off walls are not used.

14.8 SUMP PUMPING PROBLEMS

In practice, sometimes a deep well or wellpoint installation fails to establish the total groundwater lowering required – perhaps because the wellpoints were not placed deep enough or

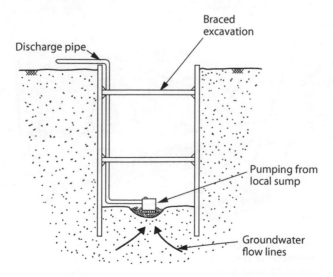

Figure 14.8 Sump pumping from within trench.

were spaced too far apart. In such circumstances, the safe, correct procedure is to install further deep wells or a second stage of wellpoints (see Section 15.10.2).

There have been occasions where this correct procedure was not followed, but instead, efforts were made to achieve that final lowering – say another 0.5 m of drawdown– by additional sump pumping. Figure 14.9 shows a typical result of the messy conditions created by this incorrect application. In this case, the foundation slab for a valve chamber was required to be constructed at a modest depth below the water table. The proposed excavation into a very silty sand stratum was ringed by a wellpoint installation – the correct approach. Unfortunately, the requisite amount of groundwater lowering was not achieved. The reason for this shortfall in lowering is not known. Perhaps the wellpoints were not installed deep enough; perhaps there were many air leaks in the pipework, so a good vacuum was not established; perhaps the permeability of the soil was lower than predicted, so the potential achievable lowering was less. There would appear to have been no close observation of what was happening while sump pumping was continued. The sump pumping resulted in the continuous removal of fines from the sides of the excavation. It can be seen that the wellpoints and their risers (originally vertical) were moved towards the sump pumped excavation by the considerable movement of the surrounding ground.

Another example of an effect of sump pumping is shown in Figure 14.10. A bund of predominantly granular material was placed to protect an excavation sited beside a tidal estuary. A cut-off wall was formed to penetrate to an underlying stratum of low permeability and so exclude water in the estuary as the tide level rose. Prior to closure of the cut-off, it was judged to be acceptable to carry out limited sump pumping of the excavation so that work inside the bund could be continuous even around the times of high tides. There came

Figure 14.9 Significant ground movement caused by inappropriate sump pumping. The wellpoint risers around the perimeter of the excavation had originally been installed vertically. Loss of fines due to poorly controlled sump pumping resulted in ground movements, distorting the wellpoints from the vertical.

Figure 14.10 Outwash fans due to sump pumping. The suction hoses for the sump pumps were mounted on the crude pontoon shown in the left of the photograph. This helped pumping continue as the water level within the excavation varied with the tide.

a day when accidentally, pumping was continued longer than necessary. The photograph shows a series of outwash fans due to the transport of fine sand in prolonged seepage flows.

If the transport of fines cannot be controlled by the construction of sumps with adequate filters, and if the discharge water cannot be adequately treated, a change in dewatering method should be considered. The use of a well system (wellpoints, deep wells or ejector wells) with adequate filters is normally a viable alternative to prevent loss of fines occurring.

14.9 DISPOSAL OF WATER FROM SUMP PUMPING OPERATIONS

If the water flowing to the sumps is removing fines from the soil, the pumped water will have a significant sediment load of sand, silt and clay-size particles. Sump pumps are normally tolerant of some sediment in the water and are likely to continue to operate unless the sediment load is exceptionally high, when they can become choked with sand and silt. However, the discharge of the sediment-laden water at the disposal point is likely to cause environmental problems.

Increasingly, there are situations where sump pumping may be technically feasible but where environmental concerns associated with the discharge of 'dirty' water from the site preclude the use of sump pumping. Discharge of water with a significant sediment load can cause a range of potential problems (see Section 21.9 for more detailed discussion of this issue). If the water is discharged to a sewer, the sediment may build up in the sewer, reducing capacity and causing the sewer to back up. If the water is discharged to a watercourse, the sediment will have a harmful effect on aquatic plant, fish and insect life.

It can be difficult to economically remove silt and clay-size particles from discharge water. However, in recent years, the use of mobile 'silt traps' based on the principal of lamella plate and other settlement processes has become common; when appropriately deployed, these units can significantly reduce the sediment load in discharge water. Water treatment options for suspended solids are discussed in Section 21.9.

The use of water treatment to remove suspended solids from water derived from sump pumping should only be considered where investigations have confirmed that the loss of fines from the soil (see Section 14.2) will not cause the creation of voids or localized instability in the excavation. If any risk exists of instability caused by loss of fines, then sump pumping should be curtailed and replaced with an alternative method such as the use of a well system (wellpoints, deep wells or ejectors) with adequate filters, where loss of fines should not occur. Case History 27.7 describes an example where sump pumping during shaft sinking caused settlement problems when fines were pumped continuously.

Chapter 15

Wellpoint Systems

Dewatering is regarded by the civil engineering profession in general as something of an art, the practice of which is best left to the cognoscenti

William Powrie

15.1 INTRODUCTION

For small and medium-sized construction projects of limited depth, wellpointing is a frequently used pumping method for the control of groundwater. It is much used for shallow pipeline trenching and the like. This chapter describes the wellpoint pumping method, including good practices, installation procedures and practical limitations. Variations to the wellpoint system to cope with differing soil conditions are considered.

The deep well system is usually more appropriate for deep excavations and is addressed in Chapter 16. A variation on the conventional wellpoint system applicable to pipeline trenching in open country, often referred to as horizontal wellpointing, is addressed in Chapter 17 together with other less commonly used groundwater lowering techniques, including ejector wells.

Chapter 27 includes two Case Histories (27.3 and 27.8), which illustrate some applications of wellpoint dewatering.

15.2 WHICH SYSTEM: WELLPOINTS OR DEEP WELLS?

There is a certain amount of confusion, probably due to some looseness of terminology, concerning the precise differences between wellpoint and deep well systems. Essentially, a wellpoint pumping system uses a partial vacuum to suck groundwater up from a group of small-diameter wells (wellpoints) to the intake of a pump unit, which then pumps it to a disposal area. Its application is constrained by the physical limits of suction lift. In contrast, a deep well system consists of a group of wells, each having a submersible pump near the bottom of the well, which likewise pump to a disposal area; the method is not constrained by suction lift limitations. This is the key practical distinction between the two systems.

From the consideration of both economic and technical criteria, it is likely that wellpointing will not be the most appropriate technique to use if the following conditions exist:

- Large excavations or where excavation depths are greater than 12 to 15 m
- Where there is a pressure head in a confined aquifer below an excavation, which should be reduced to preserve stability at the formation level.

For such conditions, a deep well system, sometimes supplemented by a system of relief wells, might be considered.

15.3 WHAT IS A WELLPOINT SYSTEM?

A wellpoint system consists essentially of a series of closely spaced small-diameter water abstraction points (the wellpoints) connected via a manifold to the suction side of a suitable pump. The wellpoint method in its modern form is around 100 years old (see Section 2.7) and is employed worldwide. There are some variations in equipment and terminology from country to county (for example, in some localities, wellpoints are known as spearpoints or sandpoints), but the basic principles of the method are universal.

The wellpoint technique is the pumping system most often used for modest-depth excavations, especially for trenching excavations and the like (see Figure 15.1). In appropriate ground conditions, a wellpoint system can be installed speedily and made operational rapidly. The level of expertise needed to install and operate a wellpoint system is not greatly sophisticated and can be readily acquired. However, as with any ground engineering process, having experienced personnel to plan and supervise the works can be crucial in identifying and dealing with any change in expected ground conditions.

A wellpoint is a small-diameter water abstraction point (the well screen), sometimes referred to as a 'strainer' (so-called because of the wire mesh or other strainer of the self-jetting wellpoint), through which the groundwater passes to enter the wellpoint. They are installed into the ground at close centres to form a line alongside (Figure 15.1), or a ring around (Figure 15.2), an excavation. The perforated wellpoint is typically about 0.7–1.0 m in length and 40–50 mm nominal diameter. Each is secured to the bottom end of an unperforated pipe (the riser pipe) of slightly smaller diameter; 38 mm diameter pipe is commonly used. However, where the 'wetted' depth is limited due to the proximity of an impermeable surface (see Figure 15.17), it is preferable that the length of the wellpoint should be shorter (usually 0.3–0.5 m) to restrict the risk of air intake at maximum drawdown. Often, it will be necessary to install wellpoints at closer centres to compensate for the lower screened length of each short wellpoint.

Each wellpoint is connected to a header main (typically of 150 mm diameter) that is placed under vacuum by a wellpoint pump (Figure 15.2). The header main is normally made of high

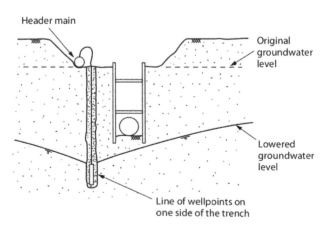

Figure 15.1 Single-sided wellpoint system. Section through trench with line of wellpoints close to one side of trench.

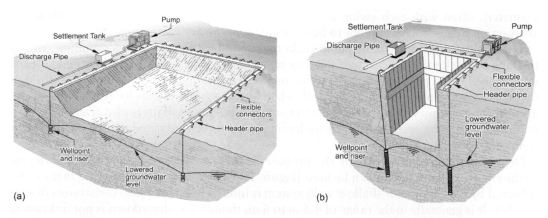

(a)   (b)

Figure 15.2 Wellpoint systems. (a) Wellpoint system around a large excavation. (b) Wellpoint system around a small excavation with vertical sides supported by shoring. Note: standby pumps not shown.

impact plastic (which is light and easy to handle), although steel pipe is sometimes used, especially when there is a risk of damage from construction activities. The pipe is typically supplied in 6 m lengths and is joined on site by simple couplings that allow a certain degree of skew to allow the header main to be laid around gentle curves if necessary. For sharper curves, 90 and 45 degree bends are available, as well as tee pieces, blanking ends, etc.

The header main has connections for wellpoints at regular intervals, perhaps every metre. The individual wellpoint riser pipes are connected to the header main via 'swing connectors', sometimes known simply as 'swings' (Figure 15.3). The swing connectors provide some flexibility in connecting the wellpoints (which are unlikely to be installed in a precise straight line) to the relatively rigid header main. Before the advent of readily available flexible plastic hose, the connectors were made from a series of metal pipework bends, which, when swung around, gave the necessary articulation (see Figure 2.12) – hence 'swing connectors'. Nowadays, swings are typically formed from flexible plastic hose (of 32–50 mm

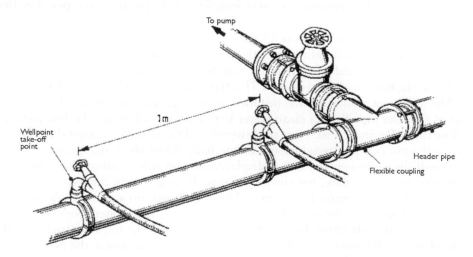

Figure 15.3 Flexible connection from wellpoint riser to suction manifold via trim valve. (Reproduced from Somerville, S H, *Control of Groundwater for Temporary Works*, Construction Industry Research and Information Association, CIRIA Report 113, London, 1986. With permission from CIRIA: www.ciria.org)

diameter), often with each swing incorporating a trim valve (Figure 15.14) to allow the abstraction rate of each wellpoint to be regulated if necessary (see Section 15.7).

The applied vacuum from the pump sucks the groundwater from the surrounding ground through the wellpoint screens, into the riser pipes, through the swing connectors and thence up into the header pipe and so to the pump intake. The pump then forces the water through the discharge main for ultimate disposal. Thereby, the system effects local lowering of the groundwater within the area that it encompasses. The header main is effectively used as a manifold to allow a small number of wellpoint pumps to be connected to a large number of wellpoints (Figure 15.4).

Depending on the geometry of the proposed excavation, there is a range of different wellpoint arrangements that can be used (Figure 15.5). The amount of lowering that can be achieved by a wellpoint (or shallow well) system is limited by the physical constraints of suction lift. It is generally in the range of 4.5 m to 6 m, though 7 m drawdown is not unknown. It depends very much on the efficiency of the water/air separation device on the pump and the vacuum efficiency of the total installation (i.e. the airtightness of the system pipework). In addition, it depends on the geological structure and the permeability of the soil mass. The width of the excavation is also pertinent to achieving the required amount of lowering at its centre. Ideally, the header main should be just above the static groundwater level to minimize the amount of suction lift. This may entail excavating down to groundwater level prior to installation of the wellpoints, followed by the installation of wellpoints and header main at that level (Figure 15.5d and Figure 15.6). Furthermore, to maximize the efficiency of a wellpoint system, the suction intake of a wellpoint pump should ideally be at the same level as the header main. Often, this will require that a small pit be dug to lower the pump body to the requisite level.

Where the land area available for construction is not constrained, the wellpoint system can be used for deeper excavations by means of a multi-stage wellpoint installation (Figure 15.7). Further details on multi-stage wellpoint systems are given in Section 15.10.

15.3.1 Types of Wellpoint

The wellpoints are vital components of every installation. There are two types of wellpoint:

(i) The self-jetting wellpoint (Figure 15.8a)

These are known as self-jetting since they can be installed without the use of a placing tube. The wellpoint and riser are metal and therefore are rigid, with the wellpoints connected to the bottom of the riser pipe. This type may be recoverable for subsequent reuse.

The wellpoint and riser are installed using high-pressure water supplied to the top of the riser pipe from a clean water jetting pump. There is a hollow jetting shoe below the wellpoint screen. Near the lower end of the shoe, a horizontal pin is located; above the pin is a lightweight loose-fitting ball. When the water pressure is applied to install the wellpoint, the ball is displaced downwards to allow the passage of the high-pressure water flow, but the pin retains the ball within the jetting shoe. When pumping starts, the applied vacuum sucks the ball up onto a shaped spherical seating and thereby seals the lower end of the wellpoint, so that water from the surrounding ground can only enter through the screen section. The unperforated riser should extend to near the lower end of the wellpoint screen to minimize the potential for air intake at maximum drawdown.

While the self-jetting wellpoint can be subsequently extracted for reuse, it is not uncommon for the riser pipes to be damaged during extraction and so to need straightening or even replacement before reuse. In addition, the wellpoint screens may need 'desanding' before being suitable for reuse.

(a)

(b)

Figure 15.4 Wellpoint pumps connected to header main. (a) A wellpoint pump is connected to the header main laid around an excavation. In this case, the sides of the excavation are supported by steel sheet-piles; at the time of the photo, partial excavation has been made to allow the installation of a propping system. (b) A set of three wellpoint pumps is connected to the header main with valves to isolate each pump. The centre pump is a standby pump to be used if either of the other pumps is out of service. (Panel (a) Courtesy of Dewatering Services Limited, Sandbach, UK; panel (b) courtesy of WJ Groundwater Limited, Kings Langley, UK.)

(ii) The disposable wellpoint (Figure 15.8b)

The wellpoint and riser are usually of plastic materials and therefore inert to corrosion. They are installed by means of a placing tube or holepuncher using a similar high-pressure water jetting technique for installation as for the self-jetting wellpoint.

While many purpose-manufactured plastic wellpoints have been marketed for some time, common practice is to adapt low-cost thin-walled convoluted uPVC perforated land drainage pipe to form disposable wellpoints with woven mesh stocking or 'coco' wrapping. The later forms a good filter and in conjunction with the normal sanding-in is very effective, even in difficult silty soils. As with the self-jetting wellpoint, the riser

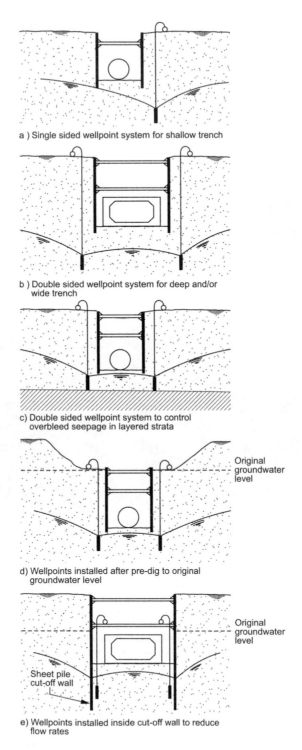

a) Single sided wellpoint system for shallow trench

b) Double sided wellpoint system for deep and/or
      wide trench

c) Double sided wellpoint system to control
     overbleed seepage in layered strata

d) Wellpoints installed after pre-dig to original
     groundwater level

Original
groundwater
level

e) Wellpoints installed inside cut-off wall to reduce
     flow rates

Original
groundwater
level

Sheet pile
cut-off wall

Figure 15.5 Possible arrangements of wellpoint dewatering systems. (a) Single-sided wellpoint system for shallow trench. (b) Double-sided wellpoint system for deep and/or wide trench. (c) Double-sided wellpoint system to control overbleed seepage in layered strata. (d) Wellpoints installed after pre-dig to original groundwater level. (e) Wellpoints installed inside cut-off wall to reduce flow rates.

Figure 15.6 Wellpoint header main installed below original ground level. Prior to dewatering, local excavations were made down to natural groundwater level along the line of wellpoints, and the header main and pumps installed at that level. (Courtesy of WJ Groundwater Limited, Kings Langley, UK.)

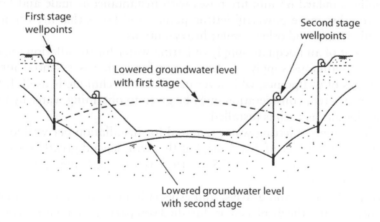

Figure 15.7 Two-stage lowering using wellpoints.

(typically made of uPVC ducting) should extend to near the bottom of the screened length. The bottom end of the disposable wellpoint is sealed. Generally, the riser pipe is a nominal 6 m long, but it can be longer.

The disposable wellpoints are not recoverable, but some of the plastic riser pipes can sometimes be recovered for future reuse. The disposable wellpoint is very appropriate to use for long-duration pumping duty.

15.4 WELLPOINT INSTALLATION TECHNIQUES

Self-jetting wellpoints and placing tubes used with disposable wellpoints are installed using high-pressure water supplied from a high-pressure jetting pump.

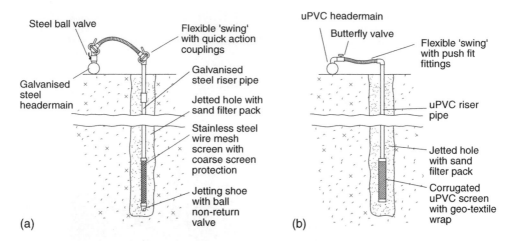

Figure 15.8 Disposable and reusable self-jetting wellpoints. (a) Self-jetting wellpoint. (b) Disposable well-point. (Reproduced from Preene, M, *et al.*, *Groundwater Control – Design and Practice*, 2nd edition, Construction Industry Research and Information Association, CIRIA Report C750, London, 2016. With permission from CIRIA: www.ciria.org)

The capacity of a typical jetting pump (see Section 19.5.5) allows it to supply water via the jetting hoses at a rate of about 20 l/s and at a pressure of 6–8 bar. The flexible jetting hoses are usually standard 63 mm fire hoses with instantaneous male and female connections. There are other more powerful jetting pumps and hoses that are often used for the holepuncher, pile jetting and other similar heavy-duty uses.

The procurement of an adequate supply of jetting water for installation must be resolved for each site. A continuous supply of clean water for jetting is essential for the efficient placing of all types of wellpoints, whichever installation technique is used. If the source of jetting water is restricted, consideration must be given to obtaining water by pumping from the first few wellpoints as each is installed.

Generally, the volume of jetting water required for installation per wellpoint with a 6 m long riser will be of the order of 1–1.5 m³, but this will depend greatly on soil conditions. However, it can be in the range of 0.5–35 m³, the latter figure being applicable to jetting in a very permeable river gravel.

Jetting in compact sands and gravels, and especially in an open gravel, may be difficult and slow, mainly because the displaced or slurrified soil particles tend not to be washed from the jetted hole to ground surface due to rapid dissipation of pressurized water into the open permeable formation. However, there is some compensation, because in such permeable soils, sanding-in is unlikely to be required.

The wellpoint system is very flexible. For instance, subsequent to the initial installation, extra wellpoints can be placed speedily to deal with localized trouble spots. It is especially useful for shallow-depth trench excavations, where very often the pumping period is expected to be of short duration.

15.4.1 Installation of Self-Jetting Wellpoints

The wellpoint and its steel riser pipe are assembled to the required length, and a flexible jetting hose is attached to the top of the riser pipe via a jetting adapter. The other end of the flexible jetting hose is connected to the supply outlet from the high-pressure jetting pump.

It is prudent to form a small starter hole into which to position the bottom end of the wellpoint, or, for a line of closely spaced wellpoints, to form a trench and so restrict general flooding of the site area. Prior to turning on the supply of jetting water, at the required wellpoint position, the jetting crew must upend the wellpoint and riser with jetting adapter and jetting hose all connected (Figure 15.9). There is considerable weight in the riser and hoses – this is a strenuous and tricky task. Such operations should be carefully planned to reduce the risk of injury to operatives; it is essential that there is adequate manpower available and that the working area is uncluttered and stable underfoot.

The jetting pump forces high-pressure water down the metal riser pipe, past the ball at the lower end of the wellpoint. A strong jet of water emerges from the base of the wellpoint. The high-pressure water slurrifies the granular soil immediately below the bottom of the

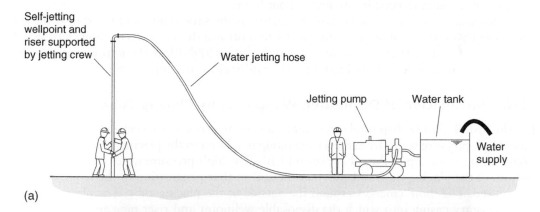

(a)

(b)

Figure 15.9 Installation of self-jetting wellpoint. (a) Installation crew and equipment for self-jetting wellpoint. (b) Self-jetting wellpoint being installed for basement excavation. (Panel (a) reproduced from Preene, M, et al., Groundwater Control – Design and Practice, 2nd edition, Construction Industry Research and Information Association, CIRIA Report C750, London, 2016. With permission from CIRIA: www.ciria.org; panel (b) courtesy of WJ Groundwater Limited, Kings Langley, UK.)

wellpoint, enabling downward penetration to be made until the required depth is achieved. The slurrified soil is washed to the surface with the jetting water and emerges in the annulus around the riser pipe. This turbulent flow of water and soil is colloquially known as the 'boil'. On reaching the required depth, the output from the jetting pump is throttled back until sanding-in is completed. Then, and only then, should the water be turned off completely and the jetting adapter and hose disconnected.

In suitable soils, such as clean sands, the installation of self-jetting wellpoints is speedy – in some cases, it may be only a matter of a few minutes per wellpoint for the actual jetting to depth. In such circumstances, it is, therefore, very cost-effective.

As a rough guide, where the Standard Penetration Test (SPT) N values are less than about N = 25–30, both self-jetting wellpoints and lightweight manhandle-able placing tubes (for installing disposable wellpoints) can be used. Either technique requires minimal mechanical plant but has a greater need for manual labour force.

In denser gravels, one of the controlling factors is the sand content of the soil. If the soil is sufficiently sandy to allow the jetting water to return to the surface as a 'boil', jetting will probably be feasible. If gravels have little sand content and allow the jetting water to dissipate, the 'boil' will be lost, and jetting will be difficult, if not impossible.

15.4.2 Installation of Disposable Wellpoints by Placing Tube

For the installation of disposable wellpoints, a similar process is mobilized using a placing tube. High-pressure water from the jetting pump is applied to the placing tube via the appropriate fitting at the top of the tube (Figure 15.10). The high-pressure water emerges from the bottom of the placing tube and slurrifies the soil so that penetration to the required depth is achieved. The 'boil' emerges around the outside of the placing tube. The placing tube is the temporary casing into which the disposable wellpoint and riser pipe are then installed centrally and sanded-in.

Modest cranage will be needed to handle the placing tube (e.g. a backhoe loader or a small 360° hydraulic excavator). However, if access for plant is restricted, where jetting conditions are easy, a lightweight 100 mm placing tube that can be manhandled may be used.

(a)   (b)   (c)

Figure 15.10 Installation of wellpoints using placing tube. (a) Steel placing tube is suspended from excavator. (b) Placing tube is jetted into the ground. (c) Plastic disposable wellpoint is installed. (Courtesy of Dewatering Services Limited, Sandbach, UK.)

The placing tube itself is a robust open-ended steel jetting tube of nominal 100 or 150 mm casing having a wall thickness of 4–5 mm. The fittings at the top of the tube are arranged such that not only can high-pressure water be applied from a jetting pump, but there is in addition the facility to apply compressed air to assist downward penetration through difficult ground conditions.

When an excavator is used to handle the placing tube, the top of the tube should be fitted with a platform or anvil, such that if difficult ground conditions are encountered (e.g. a thin layer of stiff clay or random cobbles), the bucket of the machine can be used to press down on the platform when installation progress is slow. The anvil also serves to protect the cap and various fittings at the top of the tube.

On reaching the required depth, the jetting water is turned off, the cap at the top of the tube is removed, the wellpoint and its riser are inserted to depth centrally inside the placing tube, and the placing tube is withdrawn as sanding-in progresses.

As a rough guide, the placing tube technique for wellpoint installation is appropriate where the SPT N values are about N = 40 or below. Where many random cobbles are expected and the SPT values are about N = 35–40, and where SPT values are above N = 40, the holepuncher technique should be considered.

In recent years, there has been a tendency to extend the use of the placing tube technique to replace the more labour-intensive use needed for the self-jetting wellpoint and the light-weight placing tube techniques.

15.4.3 Installation Using the Holepuncher and Heavy-Duty Placing Tube

The holepuncher technique has been used successfully to progress through soils having SPT N values below N = 30 and up to about N = 65. While it is often used to install wellpoints in hard or difficult ground conditions, it is also used for the installation of deep wells. For wellpoint applications, the holepuncher is typically suitable for installations up to 15 m deep.

The holepuncher (often called a 'sputnik') is a simple robust form of wash boring equipment having in addition a drop hammer driving facility. It consists of an outer heavy-duty casing (usually 200–250 mm nominal bore but can be up to 450 mm bore or even 600 mm for deep well installations) and an inner wash pipe that has a weighted head and can be used also as a drop hammer. Usually, the top of the wash pipe has two intake ports for the supply of high-pressure jetting water and also has a facility for a compressed air supply to be connected. The 'boil' of washings rises to ground level in the annulus between the inner wash pipe and the outer casing (Figure 15.11).

Cranage will be needed for the use of a holepuncher. The crane must have free fall on two hoist lines. If compressed air is used to assist installation, the minimum compressor capacity needed is of the order of 55 l/s (120 cfm) at about 10 bar.

The forerunner to the holepuncher was the heavy-duty placing tube, which is still used by some organizations. This equipment similarly consists of a thick-walled casing (usually 150–250 mm bore) with a box at the head containing lead weights. There are two jetting water connections and a compressed air connection – again similar to the holepuncher – provided just below the top. The displaced soil spoil is backwashed to ground level outside the casing tube. Similarly, a crane with free fall hoist lines is required to handle the heavy-duty placing tube.

The wellpoint and its riser pipe are installed and sanded-in using the same procedures as for the placing tube method.

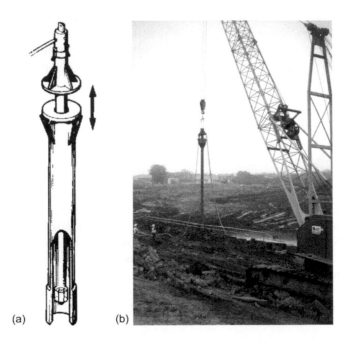

(a)   (b)

Figure 15.11 Holepuncher for installation. (a) Schematic view of holepuncher. The inner wash pipe is raised or lowered during jetting to allow the weighted head to be used as a drop hammer. (b) Installation in progress. The holepuncher is suspended from a crane. (Panel (a) reproduced from Somerville, S H, *Control of Groundwater for Temporary Works*, Construction Industry Research and Information Association, CIRIA Report 113, London. With permission from CIRIA: www. ciria.org; panel (b) courtesy of WJ Groundwater Limited, Kings Langley, UK.)

There are particular safety hazards associated with using a holepuncher or heavy-duty placing tube. The tube itself is a heavy and unwieldy device, not easily operated by crane drivers who have never used one before. The volume of jetting water and compressed air applied inputs a lot of energy into the ground; this can result in gravel and cobble fragments being ejected from the top of the tube with great velocity, to land some distance away. Especially in urban areas or on confined sites, the risks to operatives and the public should be carefully assessed when considering using the holepuncher technique.

15.4.4 Installation by Rotary Jet Drilling

The rotary jet drilling technique (Figure 15.12) is a modern adaptation of jetting by placing tube and is now widely used. An excavator-mounted drill mast with hydraulic rotary head and swivel allows the placing tube to be rotated as it is jetted in. The rotary head can also be used to apply downward force to the placing tube to aid penetration. This system has been used in a range of conditions, including clays, sands, sandy gravels and weak rock such as weathered sandstones. It can be used as an alternative to a holepuncher to avoid some of the safety hazards associated with the latter method.

15.4.5 Installation through Clay Strata

Where cohesive soils overlie water-bearing granular soils, it may be expedient to form a pre-bore through the cohesive soil layers using a flight auger attachment fitted to a hydraulic excavator (Figure 15.13) and then, having penetrated the clay soils, change to a water jetting technique.

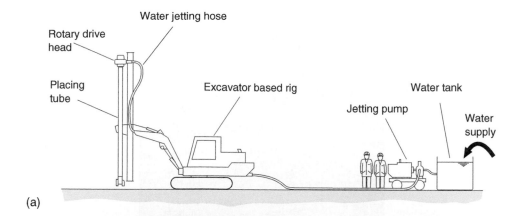

(a)

(b)

Figure 15.12 Rotary jet drilling rig. (a) Schematic view of rig. (b) Installation in progress. (Panel (a) reproduced from Preene, M, *et al.*, *Groundwater Control – Design and Practice*, 2nd edition, Construction Industry Research and Information Association, CIRIA Report C750, London, 2016. With permission from CIRIA: www.ciria.org; panel (b) courtesy of Hölscher Wasserbau GmbH, Haren, Germany.)

Installation by jetting is likely to be difficult if clay layers have to be penetrated, since due to the cohesiveness of clays, the size of the jetted annulus outside the wellpoint and riser or placing tube is likely to be very restricted (or almost non-existent) and so constrain the backwash of soil particles to the surface.

A technique used in North America to install self-jetting wellpoints in these conditions is 'chain jetting'. A chain is attached to the lower end of the wellpoint and wound around the riser to ground level prior to jetting in. The lifting and surging of the wellpoint and riser are effected using the chain, not the riser pipe. The chain will increase the size of the hole made through the clay and thereby create an annulus through which granular soils from below the clay layer can be washed to the surface and into which subsequently the sanding-in filter media can be installed from the base of the wellpoint to ground level.

Figure 15.13 Hydraulic auger attachment for pre-boring through cohesive strata. (Courtesy of WJ Groundwater Limited, Kings Langley, UK.)

When jetting in mixed soils (such as glacial till), the clay impediment may be further exacerbated by the presence of cobbles and boulders. If it is necessary to install wellpoints in such soil conditions, a rotary jet drilling rig or holepuncher is likely to be preferred.

Where it is difficult to install wellpoints by jetting, a site investigation type drilling rig may be appropriate. Rotary or cable percussion methods are used to drill a borehole, and the wellpoint and riser are installed centrally in the borehole, with the annulus backfilled with a sand filter. This onerous variation on installation techniques is more likely to be appropriate to a pore water pressure reduction project but may be needed for a water lowering requirement in a highly stratified soil formation.

15.4.6 The Merits of Jetted Hole Installations

A water jetting technique using a self-jetting wellpoint, a placing tube, a holepuncher or rotary jet drilling establishes holes without side smear and so provides a more efficient drainage hole than a hole made using a site investigation drilling technique or a continuous flight auger technique. Hence, from the superiority of the resulting water abstraction properties apart from cost considerations, the water jetting technique is to be preferred wherever practicable.

15.4.7 The Need for Filter Media

During pumping, apart from the initial pumping period, fines should not be continuously withdrawn from the surrounding ground – so the discharge water should be clear. If this is not so, the installation is faulty, for assuredly, continuous withdrawal of fines will lead to instability.

Prevention of continuous movement of fines can be achieved by means of a column of filter media around each wellpoint and its riser pipe by sanding-in. This is very similar to the provision of a filter pack around the screen of a deep well (see Section 13.10), but the grading of the filter media for sanding-in of a wellpoint is not as critical as it is to a deep well installation.

The wellpoints must have sufficient flow capacity through the wellpoint screen to provide adequate water abstraction from the surrounding soil. This should not be a problem when using adequately designed wellpoints. Where there is the need to draw down to a level close to an impermeable interface, it is prudent to use short (say 0.3–0.5 m length) wellpoint screens to reduce the risk of drawing in air and the consequent need for repetitive adjustments to the trim valves. In a high-permeability soil, the reduced length of wellpoint will restrict its potential water passing capacity, and a compensating reduction in spacing between wellpoints will be needed. Thus, the number of wellpoints required will be greater.

The conventional coconut fibre or woven mesh stocking wrapping to the disposable wellpoints provides a filter additional to that of the sanded-in media.

15.4.8 Sanding-In

Efficient sanding-in of wellpoints and risers helps prevent the removal of fines. The dewatering efficiency of individual wellpoints will be improved by providing preferential down drainage paths from any overlying perched water tables that may be due to variations in vertical permeability in heterogeneous strata. Generally, a washed sharp sand, similar to a medium-coarse concreting sand, will be a satisfactory filter or formation stabilizer. Sanding-in around the wellpoint and its riser pipe is essential in silty soils and fine sands but may be omitted when installing wellpoints in well-graded gravelly sands and sandy gravels. In such soils, an effective natural filter pack (see Section 13.10.3) can be formed by the washing action of the jetting process.

The procedure for sanding-in a self-jetting wellpoint and its riser is as follows. Upon reaching the required depth, throttle back on the jetting pump so that the water emerging in the 'boil' from the wellpoint hole at ground level just gently 'bubbles'. This upward flow of water should maintain the oversized jetted hole temporarily while sanding-in is progressed. If this oversized hole is not maintained because the jetting pump has stopped, the soil may collapse around the wellpoint, and the integrity of the filter column will be impaired. In general, the technique for the placing of the filter media is very basic, shovelling sand into the annulus, which is being kept open by the 'bubbling' rising water, until the annulus is filled to ground level.

Sanding-in is vital to achieve an efficient installation in layered or other non-homogeneous soils. It has been known for only two or three shovels of sand to be placed in a wellpoint annulus – this is totally inadequate!

Disposable wellpoints installed by placing tube are sanded-in in a similar manner. After the water is turned off and the top cap removed from the placing tube on reaching the required depth, the wellpoint and riser are installed centrally within the placing tube. The filter sand is placed around the wellpoint and riser within the placing tube as it is withdrawn. The level of the top of the filter sand should always be above the level of the bottom of the placing tube as it is being withdrawn. Some installation crews sand-in by adding sand around the outside of the placing tube as it is withdrawn.

The soil structure is important – is the soil mass nearly homogeneous, or is it anisotropic? If the soil structure is markedly anisotropic, there will be potential for encountering a series of perched water tables or local trouble spots. Adequate sanding-in is vital to the successful drainage of anisotropic soils. Also, where a stratum of soft clay is penetrated by a self-jetting wellpoint, the clay may squeeze in before the vertical column of filter sand is placed – the filter sand may bridge at the top of the clay layer.

15.5 SPACING OF WELLPOINTS AND DRAWDOWN TIMES

The theoretical number of wellpoints required for a particular project, and the associated spacings, will be indicated by the calculations outlined in Chapter 13. However, the spacing of wellpoints for simple trenching excavations is often determined from past experience of working in similar soils.

In practice, the spacing between wellpoints tends to be influenced by the spacing of the take-off points on the suction header main supplied by the manufacturer. These are mostly at 1 m centres, so usually, actual spacings will be at 1 m, 1.5 m and 2 m centres. A spacing of 0.5 m can be achieved with two parallel header mains, each with 1 m take-off spacings. It can be difficult to actually install wellpoints at 0.5 m spacings in a single line; jetting in one wellpoint may blow the adjacent wellpoints out of the ground. In such cases, it may be appropriate to install the number of wellpoints equivalent to one line at 0.5 m spacings but actually lay the wellpoints out in two parallel lines, each at 1.0 m spacing.

While the spacing selected to effect drawdown will depend on permeability and soil structure, programme time requirements are often of primary importance. If rapid drawdown is needed, wellpoints should be installed at closer centres. Typical spacings and approximate drawdown times for a single-stage installation are given in Table 15.1.

15.6 SEALED VACUUM WELLPOINT SYSTEMS

Gravity drainage to a wellpoint installation in low-permeability fine-grained soils having permeability lower than about 1×10^{-5} to 5×10^{-5} m/s is usually very slow. The rate of drainage can be improved by sealing the top of the annulus around the wellpoint riser at the topsoil zone with bentonite pellets, puddled clay or a cement/bentonite mix having a putty-like consistency. This allows the suction action of the pump to generate a partial vacuum in the entire filter column, increasing the hydraulic gradient between the soil and the wellpoint. This is often described as vacuum wellpointing and is one of the methods of pore water pressure control used in fine-grained soils (see Section 9.5.4).

This technique can be quite effective in stratified soils provided that proper sanding-in has been achieved from the bottom of the wellpoint to the underside of the clay plug. On occasions, it has been used effectively to reduce the moisture content of low-strength clay soils and thereby increase slope stability. The vacuum tank unit pumpset (see Section 19.5.3) would be the appropriate type to use for this particular duty. Ejector well systems (see Section 17.2) might be considered for use in place of a vacuum wellpoint system.

The wellpoints should be closely spaced because of the limited effect of individual wells in low-permeability soils. If this technique is applied to low-permeability soils in a loose condition, the side slopes of the excavation should be protected from sudden disturbances – such as the use of any vibration techniques – as these might cause liquefaction.

15.7 WELLPOINT PUMPING EQUIPMENT

Different types of pumps used for wellpoint applications are described in Chapter 19 – the most commonly used types are double-acting piston pumps or vacuum-assisted centrifugal pumps. The following section discusses how the pumps are deployed on site. A typical

Table 15.1 Typical Wellpoint Spacings and Drawdown Times

Soil type	Typical spacings (m)	Drawdown (days)
Fine to coarse gravel	0.5 to 1.0	1 to 3
Clean fine to coarse sand and sandy gravel	1 to 2.0	2 to 7
Silty sands	1.5 to 3.0	7 to 21

Note: If there is a risk of residual overbleed seepage, wellpoint spacings should be at the closer end of the range.

medium-sized wellpoint installation will incorporate a 100 mm or 150 mm duty pump plus a standby pump connected to a 150 mm sized header main (which acts as a suction manifold) linked to the flexible swing connectors of each individual wellpoint via its riser. The individual component parts have been described earlier.

15.7.1 Duty or Running Pumps

The total connected pump capacity must be sufficient to deal with the greater rate of abstraction required during initial drawdown. In unconfined aquifers of moderate to high permeability (such as sands and gravels), the calculated steady-state pumping rate is unlikely to be adequate to achieve acceptably rapid drawdown. In such circumstances, the initial rate of pumping may be up to twice the calculated equilibrium rate of pumping required to maintain drawdown. It is common practice on start-up to operate both the running and the standby pumpsets to achieve a fast drawdown.

When dealing with high flow rates or a large installation, it is prudent to provide multiple pumpsets in the system rather than a single large output running pump, since this affords greater flexibility. This often results in better fuel economy and is economical on the provision of standby pumps.

For systems running with a single duty pump, the ideal position for the pump is at the middle of the suction header main so that water is being abstracted from an equal number of wellpoints to either side of the pump station. Where two duty pumps are used, they may be positioned adjacent to each other – thus at the end of each set's header line – and a standby pump may be positioned and connected so as to pump from either set (Figure 15.4b). This is acceptable only if the pumping load required of either running pump is significantly below its rated capacity.

Generally, wellpoint installations are operated using diesel-driven units. However, if the running pump is to be electrically powered (perhaps in order to reduce noise levels), it is often acceptable and economical that the standby pump be diesel-powered. Since groundwater lowering pumpsets operate 24 hours per day, fuel is a significant cost factor. Fuel consumption should be highlighted in the build-up of cost estimates. The costs of alternative sources of power (such as mains electric power) should be compared if their provision is practicable. Energy use during pumping is discussed in Section 19.11.4.

The fuel consumption of the double-acting piston pump (Section 19.5.1) is more modest than that of the centrifugal pump; also, the wear and tear is less because of the slower rate of movement. However, the maximum amount of lowering that can be achieved using a piston pump is generally limited to about 4.5–5.5 m. An efficient system operated with a vacuum-assisted self-priming centrifugal pumpset could achieve an additional 1 m of drawdown in similar soil conditions.

These are general indications for guidance only. Actual achievements will depend upon the permeability of the soils and the mechanical efficiency of the individual pumpsets and of all the pipework connections of the installed system. Air leaks (at the header pipe and wellpoint connections) must be minimized, because they can have a very significant effect on the amount of suction available at the wellpoints and so affect the lowering that is achievable.

15.7.2 Standby Pumps

In general, groundwater lowering systems should operate continuously 24 hours per day, 7 days per week. Hence, it is imperative that the installed system incorporates facilities to ensure that pumping is indeed continuous. Generally, this can be provided by connecting additional (standby) wellpoint pump(s) to the suction header and discharge mains with suitably placed valves for a swift changeover of operation from the running pump to the standby unit as and when required. This might be necessary in the event of an individual running pump failure or maintenance stoppage to check oil levels, etc.

A judgement should be made, before commencing site work, of the likely effects of a cessation of pumping. The decision should be based on the answers to the following two questions:

(i) Is there a significant risk that a cessation of pumping will cause instability of the works? If the answer to this question is 'yes', the provision of standby pumping and/or power facilities is essential.

(ii) Will a cessation of pumping create only a relatively minor mess that can be cleaned up afterwards at an acceptable cost in terms of both inconvenience and delay? If the answer to this question is 'yes' and there is no safety risk, a judgement has to be made on whether or not to accept the cost savings by not providing standby pumps while recognizing that a risk is being taken.

15.7.3 Operation of a Wellpoint System and 'Trimming'

The amount of lowering of the water level that can be achieved by a wellpoint system is governed by:

1. First and foremost – the physical bounds of suction lift. Elevation above sea level and, to a lesser extent, ambient temperatures have some input into this limitation. As ground elevation above sea level increases, available suction decreases. Also, ambient temperatures and ground elevation are of significant relevance to the rating of the engine required to drive the pump to achieve the necessary rate of pumping.

2. The hydraulic efficiency of the total wellpoint installation – this includes the pipework sizing above and below ground as well as the adequacy of the wellpoint sanding-in procedures and the air/water separation unit on the pumpset.

3. Air leaks. These can drastically reduce the amount of vacuum that is available to withdraw water from the soil via the abstraction points.

As suction is applied to self-jetting wellpoints, their ball valves are seated, and the groundwater is sucked through the wellpoint screen only. The bottom of the riser pipe terminates near the base of the wellpoint screen so as to minimize or delay the intake of air when maximum drawdown is being achieved.

The bottom end of a disposable wellpoint is sealed, and so, similarly, the groundwater is sucked through the screen to the bottom of the riser pipe and thence to the suction header main.

As the water level at each wellpoint is drawn down to near the level of the top of the screen, there will be a risk of entraining air with water and thereby reducing the amount of available vacuum. The trim valves (Figure 15.3) at each individual wellpoint's flexible connection (known as a 'swing' connector) to the header main enable the experienced operator to adjust the amount of suction such that the intake is predominantly water and the amount of air intake is minimized. This is necessary to ensure that the wellpoint system operates at the maximum achievable drawdown.

The adjustment of the valves on each wellpoint is known as 'trimming' or 'tuning' and is an important part of operating a wellpoint system. A poorly trimmed system may achieve significantly less drawdown than a system that has been trimmed correctly. Trimming is probably more of an art than a science, but it can normally be mastered with experience.

As the system is trimmed (i.e. the wellpoint valves are closed or throttled), the vacuum shown on gauges on the pump and header main should increase. This is because the amount of air entering the system via the wellpoint screens is reducing, allowing the pump to generate more vacuum. When trimming a system, it can be very satisfying to see the vacuum increase as a result of your efforts. However, it must be remembered that the aim is not to maximize vacuum but to maximize drawdown and flow rate. It is important not to become obsessed with obtaining greater and greater vacuum. An overzealously trimmed system (where all the wellpoint valves are almost closed) will have a very impressive vacuum but will pump little water and generate little drawdown – this is a classic mistake made by novices when first attempting trimming!

A wellpoint that is pumping problematical volumes of air will probably be producing 'slugs' or 'gulps' of air and water. The momentum of each slug of water will make the wellpoint swing connector jump up and down, perhaps quite violently – this is known as 'bumping'. A 'bumper' is trimmed by slowly closing the valve until the flow is smooth and then re-opened slightly – the last action is vital to avoid over-trimming. A small steady flow of air along the swing connector is acceptable (provided the rest of the system is functioning well); it is probably counterproductive to attempt to trim a system so that no air at all enters the wellpoints.

Wellpoint equipment sets supplied by British, North American and Australian suppliers usually include trim valves to regulate the rate of water/air flow to each wellpoint (Figure 15.14a). However, suppliers based on the continent of Europe generally do not supply trim valves on flexible swing connectors (Figure 15.14b), unless specifically asked to do so. The reason for this is not clear. South-East Asian suppliers also tend to omit trim valves and to follow continental European philosophy. The inclusion of trim valves involves added cost, but the authors consider that the potential for increased lowering by adjustment of trim valves, especially for a multi-stage wellpoint installation, is cost-effective.

15.8 WELLPOINT INSTALLATIONS FOR TRENCH EXCAVATIONS

A considerable amount of shallow depth trenching for pipelaying is carried out worldwide. Irvine and Smith (2001) give overall guidance to good trenching practices and include many specific recommendations concerning support of the excavations and other safety guidelines, both with and without the need for groundwater control. This book does not address the many methods of trench support. However, it is important that for each and every trench pumping installation, the basic concepts described should be observed, though depending

(a)

(b)

Figure 15.14 Examples of flexible connectors between wellpoint and header main. (a) Flexible connection with trim valve. (b) Flexible connections without trim valve. (Panel (a) courtesy of Dewatering Services Limited, Sandbach, UK; panel (b) courtesy of Loots Grondwatertechniek, Amsterdam, The Netherlands.)

upon local conditions, some adjustments may be appropriate. The method of 'horizontal wellpoint' using drains laid by special trenching machines is covered in Section 17.3.

15.8.1 Single-Sided Wellpoint Installations

A pipelaying contractor will much prefer having the header main and associated wellpoints only on one side of the trench, because this will allow uninterrupted access for his plant and equipment on the other side of the spread (Figure 15.15). The basic question is how to assess whether a single-sided installation will be adequate. Water levels must be reduced below formation level. The critical point is the bottom of the trench on the side remote from the line of wellpoints of Figure 15.15. If the drawdown phreatic surface is below that point, a single-sided installation is adequate.

A single-sided system will be suitable only if a sufficient depth of permeable soil exists beneath formation level and if there are no significant layers or lenses of impermeable material in the water-bearing soils, especially above the level of the base of the wellpoints.

The slope and shape of the drawdown surface are very relevant and primarily depend upon the soil permeability. The lower the permeability, the steeper the slope of the drawdown cones that individual wellpoints will establish. The spacing between wellpoints must be close enough to ensure that their individual cones of depression interact and so produce a general and fairly uniform lowering.

The dewatering contractor has no control over soil conditions on site but does have control over the spacing between wellpoints. If the proposed excavation depth is modest, and the depth of the unconfined water-bearing stratum is considerable, the wellpoints can be installed at wide spacings on long riser pipes. The use of long riser pipes has the merit of allowing the top of the wellpoint screen to be set at greater depth, minimizing the possibility of a wellpoint sucking air and reducing the available vacuum. This may be offset by increased difficulty in the handling of long risers during installation. Generally, this variation is more suitable for an installation that is to be pumped for a significant period of time than for a progressive type installation, when the pumping period will be of short duration.

Experienced engineering judgement is needed when deciding on wellpoint spacings and whether a single-sided installation will be adequate. The conditions favourable to the use of a single-sided wellpoint installation for trench excavation are:

a) A narrow trench.
b) Effectively homogeneous, isotropic permeable soil conditions that persist to an adequate depth below formation level.

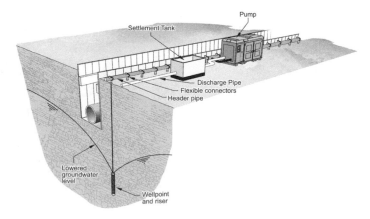

Figure 15.15 Single-sided wellpoint system for pipeline construction.

c) Trench formation level not more than about 5 m below standing groundwater level. The actual depth achievable depends upon the permeability of the water-bearing soil beneath the formation and the overall pumping efficiency of the installation.

For trenching in excess of about 5 m deep, it is possible to use a second stage of wellpointing, as described in Section 15.10.2, provided that there is sufficient depth of permeable soil beneath final formation level and the width of the wayleave is sufficient. The sides of the trench excavation must be safely supported throughout.

Single-sided systems work well in isotropic soil conditions. In practice, soils are often heterogeneous and anisotropic. In such conditions, double-sided systems, where wellpoints are installed on both sides of an excavation, may be more effective.

15.8.2 Double-Sided Wellpoint Installations

As the name implies, a double-sided system has wellpoints along both sides of the trench (Figure 15.16). This layout naturally restricts working access to the edges of the trench and can make life irksome for the pipelaying contractor.

The conditions likely to require the use of a double-sided wellpointing for trench excavation are:

a) A relatively wide trench
b) Trench formation level 4.5 to 6 m or more below standing groundwater level
c) Impermeable stratum close to formation level, or low-permeability layers or lenses present above formation level

If there is an impermeable stratum (such as a clay layer) at or close to formation level (Figure 15.17b), even if a double-sided wellpoint system is employed, there may be some residual seepage or overbleed near the interface between the overlying permeable soil and the underlying impermeable stratum. In this situation, particular thought must be given to the wellpoint depth and spacing. The wellpoints must be 'toed into' the impermeable stratum – in effect, they are installed to penetrate a few hundred millimetres into the clay, forming a sump of sorts around each wellpoint to maximize drawdown. This requires careful installation – overzealous 'toeing in' will result in the wellpoint screen being installed

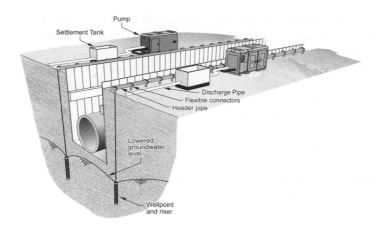

Figure 15.16 Double-sided wellpoint system for pipeline construction.

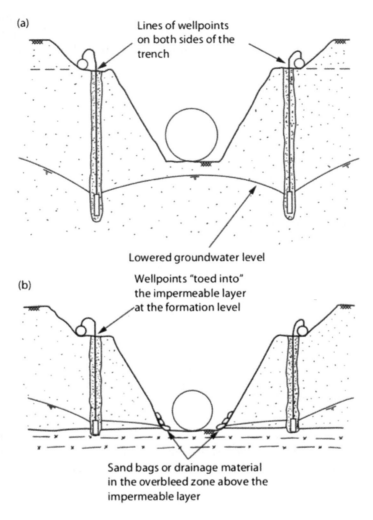

Figure 15.17 Double-sided wellpoint installation. (a) Permeable stratum extends below formation level. (b) Impermeable stratum present near formation level.

too deeply into the clay, thereby becoming clogged and ineffective. The wellpoints should be installed at rather closer spacings than normal to try to intercept as much of the overbleed seepage as possible.

Even if these measures are adopted, some overbleed seepage is likely to pass between the wellpoints and enter the trench. It is essential to control the overbleed so that pore water pressures do not build up and the water flow does not continuously transport fines and risk instability of the sides of the trench. Some form of sand-filled permeable bags, granular drainage blanket or geotextile mesh should be placed as indicated in Figure 15.17b. This will allow the water to flow into the excavation without the build-up of pore water pressures and movement of fines. Thus, the stability of the trench will be preserved. The water flowing to the trench must be continuously removed by conventional sump pumping to prevent standing water building up in the working area.

There is a hybrid of single- and double-sided systems that can be used when trenching in extensive permeable strata where an impermeable layer exists above formation level (Figure 15.18). These conditions could be dealt with by means of a double-sided wellpoint installation, but this could be an encumbrance to the contractor's excavation and pipelaying activities. The alternative is to have a single-sided wellpoint installation on one side of the trench excavation with vertical sand drains on the other side. The vertical drains (see Section 17.9) consist of a series of holes jetted at diameters and spacings similar to the wellpoints and penetrating below the low-permeability layer. These are backfilled with sand of the same grading as that used for the sanding-in of the wellpoints. These will provide downward drainage of the perched water on the other side of the trench. This should reduce troublesome overbleed on the side of the trench remote from the wellpoints.

15.8.3 Progressive Installation for Trench Works

The same principles that are applicable to wellpoint systems for static excavations are also pertinent to pipeline trench excavations where the working area moves progressively as each section of pipe is placed. There are additional complications because of the need to advance the groundwater lowering system to keep pace with the trench works. The initial lowering syndrome to establish drawdown is always at the head of the progressively advancing installation.

Generally, the equipment length for a wellpoint system for a rolling trench excavation should be about three to four times as long as the planned weekly advancement of the trench (Figure 15.19). This allows time for the extraction of wellpoints, risers and manifolds behind completed work, their progressive installation ahead of the work, and – the least certain factor – the time to establish drawdown of the groundwater ahead of trenching excavation and subsequent pipelaying and backfilling.

Where the soil is of relatively low permeability (such as a silty sand), the time for drawdown will be protracted, so in such soils, it is prudent to allow for a greater length of operating equipment to be installed and operating ahead of the length of open excavation.

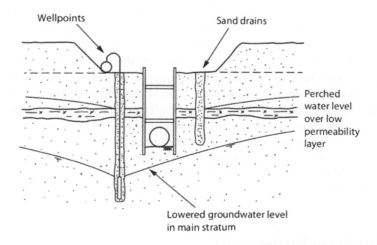

Figure 15.18 Single-sided wellpoint installation with sand drains to aid control of perched water table.

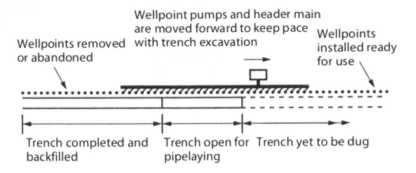

Figure 15.19 Method of wellpoint system progression for trench works.

15.9 WELLPOINTING FOR WIDE EXCAVATIONS

Where a wide excavation is required, a perimeter wellpoint installation alone may not be sufficient to establish adequate lowering at the middle of the excavation. The lower the permeability, the more likely it is that the amount of lowering at the centre of the excavation will be insufficient. It might be prudent, even necessary, to install additional wellpoints within the excavation area.

15.10 WELLPOINTING FOR DEEPER EXCAVATIONS

15.10.1 Long Risers and Lowered Header Mains

If the excavation width is modest, but the required drawdown is greater than achievable by a single stage, and the thickness of permeable soils beneath formation level is significant, drawdown can be increased by installing wellpoints on longer than normal 6 m risers to an adequate depth below formation level.

The additional drawdown is achieved by initially pumping the wellpoints using pumps and header mains laid at about standing groundwater level. Pumping of this initial installation will establish some lowering of the water level. Excavation is then made down to the lowered water level, and a second suction header main is installed at the lower level. This lower main is pumped while the upper suction main is still active. Progressively, each individual wellpoint (having shortened its length of riser) is disconnected from the upper header main and connected to the lower header main. Pumping should be continued from the original upper main until all wellpoint risers have been shortened and connected into the lower main. Using the standby pump from the upper header main to begin pumping on the lower header main can reduce the number of pumps required for this method. When all the wellpoints have been connected to the lower main, the upper pump can be moved down to act as a standby unit for the lower main. The upper header main is then dismantled and removed.

This process could be described as a two-stage installation having only one stage of installation of the wellpoints and risers (see Case History 27.3 for an example of where this procedure was used).

15.10.2 Multi-Stage Wellpointing

Multi-stage wellpoint installations (Figure 15.20) can be used for deep excavations as an alternative to deep wells (see Chapter 16) or ejector wells (see Section 17.2). If a multi-stage

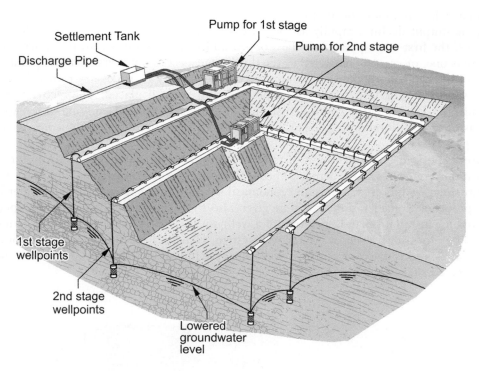

Figure 15.20 Two-stage wellpoint system.

wellpoint system is used, it is necessary to make sufficient allowance for side slopes or batters to excavate safely to formation depth and berms to support header mains of lower stages.

The recommended site procedure for the installation of multi-stage sets is as follows:

a) Excavate to about the standing water level; it may be possible to excavate to around 0.5 m below the water level.
b) Install and connect first-stage wellpoint system around the perimeter, making due allowance for subsequent slope batters and berms.
c) Pump continuously on the first-stage system.
d) Excavate to about 0.5 m above the lowered water level.
e) Install and connect the second-stage wellpoint system around the perimeter, likewise allowing for excavation batters and berms for subsequent stages.
f) Pump continuously on the second-stage system as well as continuing to pump on the first stage.
g) Again, excavate to about 0.5 m above the level to which water has been lowered by pumping on both first and second stages.
h) Continue the sequence of excavation, wellpoint installation and pumping until formation level is reached.

This sequence of operations will entail short halts in excavation as each further intermediate-stage level for wellpoints and header main, etc. is installed for the further lowering of the groundwater.

At each stage, in addition to installation of the wellpoint system, a number of monitoring wells should be installed to monitor the lowering being achieved and to indicate the depth to which the next stage of excavation can be taken.

Often, when pumping on multi-stage systems of more than two stages, the first-stage pumping output declines rapidly as the third stage is brought into operation. When this happens, the first-stage pumps can be stopped and transferred down to the fourth stage, if there is one, or can be connected to the third stage as standby units. The wellpoints and risers of the top stage should not be extracted, since it may be necessary to reactivate the upper stage(s) as the structure is being built up and backfill placed. The recovery of the groundwater level must be controlled so that at every stage of the building of the works, there is no risk of flotation of the partially completed structure due to an uncontrolled rise of the groundwater.

Three- and four-stage wellpoint installations have been operated satisfactorily for deep excavations (Figure 15.21). There have been, and will be in the future, many projects where a multi-stage wellpoint system is an economical solution to a particular groundwater lowering problem. Consideration should be given also to the fact that the degree of expertise required for installation and operation of a wellpoint system is less sophisticated than that appropriate for a deep well or ejector well system. Hence, there may be occasions or geographical locations where the use of a wellpoint system will be the more appropriate choice for a deep construction project.

Figure 15.21 Four-stage wellpoint system. The sand being dewatered is slightly cemented, which allows the side slopes to stand very steeply in the short term. The cementing is sufficiently weak that the ground unravels in the event of water inflow but in a 'dewatered' state can stand almost vertically. (Courtesy of WJ Groundwater Limited, Kings Langley, UK.)

Chapter 16

Deep Well Systems

That local drainage by shafts will be effectual seems placed beyond any reasonable doubt by the complete sympathy existing between the levels of the water in the shafts at a considerable distance from each other, and by the simultaneous effect of one pump in different shafts.

Robert Stephenson, *Reports on the construction of the Kilsby Tunnel*, 1836

16.1 INTRODUCTION

A deep well system consists of an array of bored wells pumped by submersible pumps. The wells act in concert – the interaction between the cones of drawdown created by each well results in groundwater lowering over a wide area. Because the technique does not operate on a suction principle, greater drawdowns can be achieved than with a single-stage wellpoint system. This chapter addresses the temporary works deep well groundwater lowering system and describes good practices to be used during installation and operation.

The principles and applications of deep wells are discussed. The methods used for well drilling, installation, development and operation are outlined, and some practical problems with the operation of deep well systems are presented. The vacuum deep well and bored shallow well systems, which are variants on the deep well method, are described briefly.

Chapter 27 includes two Case Histories (27.4 and 27.7), which illustrate some applications of deep well dewatering.

16.2 DEEP WELL INSTALLATIONS

A deep well system consists of an array of wells, each well pumped by a submersible pump. Each well consists of a bored hole (typically formed by a drilling rig) into which a special well liner is inserted. The liner consists of plastic or steel pipe, of which a section is slotted or perforated to form a well screen to allow water to enter; other sections consist of unperforated pipe (the well casing). Typically, the annulus between the borehole and the well screen/casing is backfilled with filter media or formation stabilizer to form what is known as the filter pack.

The wells are generally sited just outside the area of proposed excavation (although for very large excavations, wells may be required within the main excavation area as well as around the perimeter). A deep well system has individual pumps positioned near the bottom of each well; usually, the pumps are borehole electro-submersibles (see Section 19.6). The well screen and casing provide a vertical hole into which a submersible pump attached to its riser pipes can be installed (and also recovered as and when required).

A typical deep well system (Figure 16.1) consists of several wells acting in concert. Each well creates a cone of depression or drawdown around itself, which in a high-permeability aquifer may extend for several hundred metres. The interaction between the cones of drawdown from the wells produces the cumulative drawdown required for excavation over a wide area. Apart from pumping tests, deep wells are rarely used in isolation or individually; the method relies on the interaction of drawdowns between multiple wells.

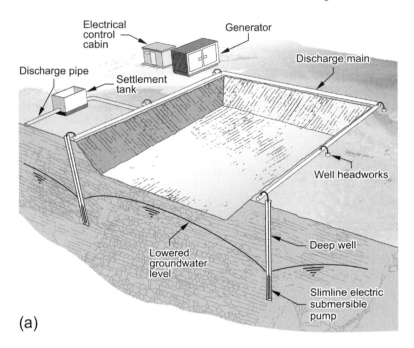

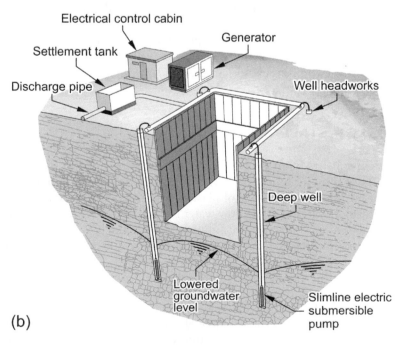

Figure 16.1 Deep well system. (a) Deep well system around a large excavation. (b) Deep well system around a small excavation with vertical sides supported by shoring.

The components making up an individual temporary works well are shown in Figure 16.2. Generally, for most temporary works requirements:

1. The well screen and casing sizes will be in the range of 150 to 300 mm diameter. The well screen and casing are typically plastic, with steel being used only rarely.
2. The drilled borehole sizes will be in the range of 250 to 450 mm diameter.
3. The well depths will be in the range of 10 to 50 m. Occasionally, wells are drilled to greater depths, especially for shaft or tunnel construction projects and for open pit mines.
4. When deep wells are installed through granular soils and weak or weathered rock, perforated well screens and filter gravel are required, as shown in Figure 16.2. If wells are drilled in harder fractured rock, it may be possible to install the well with an unslotted well casing in the top of the well but with the lower section installed as an unlined 'open hole' section without well screen and filter pack.

The vital feature of the deep well system compared with the single-stage wellpoint method is that the theoretical drawdown that can be achieved is limited only by the depth of well and soil/rock stratification. The wellpoint method (see Chapter 15) is limited by the physical

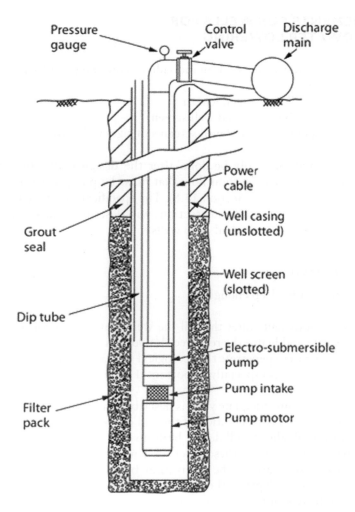

Figure 16.2 Schematic section through a deep well.

bounds of suction lift. In contrast, the drawdown of a deep well installation is constrained only by the depth/level of the intake of the pump(s) – provided, of course, that the power of the pump is adequate to cope with the total head from all causes. Hence, the rated output of the installed pumps should match the anticipated well yield and drawdown in the well. The energy costs of operating a deep well installation are likely to be competitive due to the greater efficiency of borehole pumps compared with the total system efficiency of a multi-staged wellpoint installation.

The well screens, pumps and other materials are similar to those used for water supply wells. However, since the working life of a temporary works well will almost always be significantly shorter than the life of a water supply well, temporary works wells can be constructed by rather cheaper and simpler methods. Also, the onerous health regulations to control the risk of water contamination during construction of water supply well installations are mostly inappropriate to temporary works wells.

The initial cost of installing a deep well system is significant. A high standard of expertise in the design and control of installation procedures is required to ensure that the appropriate good practices are implemented throughout and to promote optimum and economic performance.

16.3 REQUIREMENTS FOR WELLS FOR GROUNDWATER LOWERING

There are three major requirements for an individual temporary works well:

1. The depth of the well and screen length
2. The diameter of the borehole and well screen
3. The filter media (and associated well screen slot size) to be used

The design of a well will generally be done after the design of the overall groundwater lowering system has estimated the total flow that must be pumped to obtain the required drawdown for the excavation (see Section 13.8). The specification of well yield and spacing is outlined in Section 13.9. The following sections will describe some of the practical issues associated with the design of individual deep wells.

16.3.1 Depth of Well

In essence, the well must be deep enough to:

1. Be able to yield sufficient water that all the wells pumped in concert can achieve the required flow rate and hence the required drawdown
2. Be of sufficient depth to penetrate the geological strata in which groundwater pressures are to be lowered (this is especially important if deep confined aquifers are present)

The design of a well to ensure adequate yield is described in Section 13.9. The depth requirements are generally met by a consideration of the soil or rock stratification at the site. There are certain rules of thumb about well depth, built up over many years of practice, and well designs should be compared with these to check for gross errors. If the aquifer extends for some depth below the base of the excavation being dewatered, the wells should be between 1.5 and 2 times the depth of the excavation. Wells significantly shallower than this are unlikely to be effective unless they are at very close spacing (analogous to a wellpoint system). Design issues regarding the depths of wells to reduce the risk of base failure of excavations are covered in Section 13.7.

If the geology does not consist of one aquifer extending to great depth, this may affect well depth. If the aquifer is relatively thin, screen lengths will be limited, and a greater number of wells will be required. If there is a deeper permeable stratum that, if pumped, could act to 'underdrain' the soils above (see Section 13.6.2), it may be worth deepening the wells (beyond the minimum required) in order to intercept the deep layer.

It is always prudent to allow for a few extra wells in the system over and above the theoretically calculated number. This not only allows for some margin for error in design but also means that the system will be able to achieve the desired drawdown if one or two wells are non-operational due to maintenance or pump failure.

16.3.2 Diameter of Well

In general, the diameter of the well will be chosen to ensure that the borehole electro-submersible pump (see Section 19.6) to be used will fit inside the well screen and casing, and that any necessary filter media can be placed around the well screen. This will allow the necessary drilled diameter of the borehole to be determined. Sometimes, when working in remote locations or in developing countries, the selection has to be made in reverse, beginning with the borehole diameter that can be drilled by locally available equipment and then working backwards to the size of pump that can be accommodated.

The starting point for determining the size of the bore is the diameter of the pump to be installed in the well screen. Once the expected individual well yield (see Section 13.9) at the steady-state rate of pumping has been determined, a pump should be selected from a manufacturer's or hirer's catalogue that has a maximum rated performance of between about 110 and 150 per cent of the steady-state flow rate at the anticipated working head. This allows for some additional pumping capacity to help establish the drawdown.

The pump manufacturer's catalogue will list the minimum internal diameter of well screen necessary to accommodate the pump to be used, assuming that the wells are perfectly straight and plumb. In practice, most wells deviate from the ideal alignment, and using a slightly larger screen diameter reduces the risk of a pump getting stuck down a well. Some general guidance on well screen diameters is given in Table 16.1. The recommended minimum well screen diameters are generally larger than those quoted by the pump manufacturers. Even so, if a well has a large amount of deviation, even a very small pump may become jammed at the tight points in the well.

Table 16.1 Recommended Well Screen and Casing Diameters

Maximum submersible pump discharge rate (l/s)	Recommended minimum internal diameter of well screen and casing[a] (mm)	Recommended minimum diameter of boring[b] (mm)
5	125–152	250–275
10	152–203	300–325
15	165–250	300–375
20	180–250	300–375
25	203–300	325–425
44	250–350	375–475

[a] Diameter will depend on external dimensions of pump used.
[b] Minimum diameter of boring is based on nominal filter pack thickness of 50 mm. Slightly smaller drilled diameters may be feasible if a natural filter pack can be developed in the aquifer.

Knowing the required minimum internal diameter, select an appropriate well casing and screen from the manufacturer's catalogues. The standard sizes of well screens are unlikely to exactly match the internal diameter needed for the pump, so the next available size up of well screen and casing is used.

Typically, the wells used for groundwater lowering in soils will have a 'filter pack' of sand or gravel placed in the annulus between the borehole wall and the well screen (see Figure 16.2). Preferably, the annular thickness of the filter pack should be about 75 mm (but never lower than about 50 mm or greater than 100–150 mm). Thus, the minimum size of the borehole to form the well should be the external diameter of the well screen and casing (including at joints where the diameter may be greatest) plus twice the thickness of the filter pack. This diameter is unlikely to exactly match the standard sizes of drilling equipment, so the borehole should typically be drilled using the next size up of drill bits, taking care to check that this does not result in filter pack thicknesses in excess of 150 mm.

16.3.3 Types of Well Screen

In most cases, the well screens used on dewatering projects will be proprietary products manufactured specifically for the purpose. Around the world, there are many companies specializing in the production of well screens and casings in various diameters, materials and specifications, which can provide materials for dewatering projects. The well casing (the unperforated sections) and well screen (perforated sections) are typically provided in sections of between 1 and 6 m in length. During installation into the well, the sections will be joined together by either threaded joints or glued joints (for plastic screens) or welding (for steel screens). Various fittings are available for casing and screens, including base caps, lifting heads and adaptors.

The most common type of well screen used on dewatering projects is manufactured from plastic: either uPVC or high-density polyethylene (HDPE). The base pipe is perforated with regular arrays of linear slots of a specified size (Figure 16.3a). The most common commercially available slot sizes are from 0.5 to 3 mm.

Continuous slot well screens have a single continuous slot (of constant width) running in a spiral round the pipe diameter. The best-known configuration is steel screens where the slot is formed by a single triangular or trapezoidal shaped wire welded in a spiral around a base pipe or a cylindrical structure of rods. The slot is formed between the loops of wire (Figure 16.3b). These steel screens are typically very strong and are hydraulically very efficient due to the large total slot length on each section of screen. The disadvantage of these steel screens is their relatively high cost compared with conventional plastic well screens. Continuous slot screens are available in stainless steel, galvanized steel, carbon steel and also plastic materials.

Well screens are also available with pre-fitted or pre-formed filter packs. These are fitted with gravel filter media of a specified particle size, avoiding the need for a conventional filter pack around the well screen. This can be useful in situations where there may be practical difficulties in installing conventional filter packs. The two main types of pre-fitted filter packs are:

1. Resin-bonded screens. Specific granular filter media are bonded directly to the slotted well screen in a factory-based process (Figure 16.3c).
2. Pre-packed screens. The gravel filter media is held in place (packed) between two specially matched concentric screens.

In some cases, pre-formed filters are formed by wrapping geotextile mesh around a slotted base pipe. The most hydraulically efficient designs provide a high open area by placing a

Figure 16.3 Types of well screen. (a) Perforated plastic screen. The screen is fitted with threads for joint-ing. (b) Stainless steel continuous slot screen. The screen is fitted with weld rings for jointing. (c) Perforated plastic screen with pre-formed filter pack. This is an example of a pre-formed filter pack with the filter media bonded in place with resin. The screen is fitted with threads for jointing. (Panels (a) and (c) courtesy of Boode Water Well Systems; panel (b) courtesy of Johnson Screens.)

very coarse spacer mesh under the filter mesh itself to allow water to flow around the outside of the base pipe beneath the filter mesh. Guidance for selection of the size of openings in the filter mesh can be obtained from the manufacturers. Once selected, the mesh is normally applied to the screen in the factory. Granular filters or formation stabilizers may be used in conjunction with geotextile screens.

The wall thickness of the base pipe forming the well screen and well casing must be adequate to withstand collapse pressures from soil and groundwater loadings. If the screen and casing are not strong enough, the well liner may collapse or 'squeeze' – reduce in diameter – when differential pressures are greatest. High pressures on the liner can occur during the placement of grout seals, due to fluid pressure from the grout, or if the well is rapidly emptied during well development or the early stages of pumping. Most cases of collapse are associated with a relatively thin-walled plastic base pipe. In most of these cases, further problems were avoided by installing additional wells with thicker-walled screens – the screen manufacturer should be able to provide guidance. Collapse of steel well screens is rare, but these are not commonly used for temporary works wells on cost grounds.

16.3.4 Filter Media and Slot Size

A temporary works groundwater lowering well (Figure 16.2) typically consists of a well screen and casing installed centrally inside a borehole formed by a drilling rig. The annulus between the screen and the borehole wall is filled with granular filter media to form a 'filter pack'.

The filter media must be selected (based on the particle size distribution [PSD] of the aquifer material) to meet the following two conditions:

(i) To be sufficiently coarse that the filter pack is significantly more permeable than the aquifer in order to allow water to enter the well freely
(ii) To be sufficiently fine that the finer particles are not continually withdrawn from the aquifer

The selection of any filter media has to be a compromise between these two conflicting conditions. Concentrating on condition (i) will give a high-yield well but with an increased risk of continuously pulling sand or fines into the well. Concentrating on condition (ii) will prevent movement of fine particles but may restrict well yield. In practice, only a relatively narrow range of materials are suitable for use as granular filter media. At one end of the range, a relatively low-permeability silty sand aquifer might need a 0.5–1.0 mm filter sand, while at the other end, a very high-permeability coarse gravel might need a 10–20 mm pea gravel. Materials outside these ranges are rarely used.

The slot size (the minimum width of the slot) of the well screen should be chosen to match the filter and avoid large percentages of the filter material being able to pass into the well. The well screen slot sizes used in practice also fall within a relatively narrow range – slot widths of 0.5 to 3 mm are typical.

The methods used to determine acceptable gradings for well filters and associated well screen slot size are discussed in Section 13.10.

The filter media are placed in the annulus around the well screen by being fed down a tremie pipe or in some circumstances by being poured in from the surface without a tremie pipe (see Section 16.6). When the filter is placed, it may be relatively loose, and it is likely to compact a little during development. Accordingly, the filter pack when placed must extend to some height (normally at least 0.5–1 m) above the top of the slotted well screen to ensure that aquifer material does not come into direct contact with the well screen if the filter pack compacts. Sometimes, permanent tremie tubes are left in place to allow the filter pack to be topped up by the addition of more material.

There are some types of coarse-grained soils for which it is not necessary to introduce artificial filter media, but where it is possible to directly develop the aquifer, remove the finer particles and (by the process of well development) form a natural filter pack in the aquifer immediately outside the well screen. This method can be employed in coarse well-graded soils such as sandy gravels. It can give cost savings by allowing a well borehole to be drilled at a smaller diameter for a given well screen size – the space for a filter pack is not needed, and the aquifer material is allowed to collapse directly onto the well screen.

In addition to conventional granular filter media placed during well installation, pre-formed filters (fitted at manufacture) are also available (see Section 16.3.3).

16.4 CONSTRUCTING DEEP WELLS

The methods used to construct wells for temporary works groundwater lowering purposes have much in common with those used to form water supply wells, but there are some differences in techniques and equipment. A general appreciation of well drilling methods can be gained from some of the publications in the water supply field (Stow, 1962, 1963; Cruse, 1986; Roscoe Moss Company, 1990; Rowles, 1995; Sterrett, 2008).

There are four main stages in forming a groundwater lowering well:

1. Drilling of borehole (Section 16.5)
2. Installation of materials: screen, casing, filter gravel, etc. (Section 16.6)
3. Development of well (Section 16.7)
4. Installation and operation of pump (Section 16.8)

These are described in the following sections.

16.5 DRILLING OF WELL BOREHOLES

Many methods are available for the formation of well boreholes. The technique selected will depend upon the type of equipment available in the territory; the expertise of the well boring organization selected; the soil/rock at the site; and the groundwater conditions anticipated.

Several different methods are used commonly to form well boreholes:

i. Cable tool percussion drilling
ii. Wash boring or water jetting
iii. Rotary drilling by direct circulation
iv. Rotary drilling by reverse circulation
v. Rota-sonic drilling

Further details are given in the subsequent sections of text.

For temporary works wells, cable tool percussion drilling is limited to well depths of about 20 to 35 m, but it has been used down to 50 m depth on occasion – for water supply wells, the method has been used to considerably greater depths. Wash boring and hole-puncher equipment have been used to install wells to about 35 m depths. Rotary drilling rigs can cope with significantly greater depths if necessary.

The jetting, wash boring and rotary drilling techniques, when used appropriately, generally provide more productive wells and require less development. The effectiveness (i.e. potential yield) of a rotary drilled hole depends greatly on the properties of the drilling fluid used and the adequacy of the removal of the drilling residues on completion of the hole – this often gives rise to much debate between groundwater lowering specialists, drilling contractors and drilling additive suppliers about the relative merits of the various marketed drilling slurries. The merits of various types of drilling fluids and additives are discussed by Roscoe Moss Company (1990) and Sterrett (2008). Adequate development of the well on completion is vital to ensure its best performance (see Section 16.7).

Throughout the well drilling operation, the arisings from the drill hole should be observed and logged to determine whether soil/rock conditions are as expected. If not, it may be prudent to determine whether the design of subsequent deep wells needs to be varied. For instance, the arisings might reveal an unexpected low-permeability layer within the wetted depth. However, do not ignore the fact that information gleaned from well boreholes is rarely of as high quality as that obtained from dedicated site investigation holes. Any judgements made on the basis of observing the well borehole arisings can only be somewhat speculative.

Whatever drilling technique is used, it is essential that wells be relatively straight and plumb; otherwise, there will be severe operational problems with the submersible pumps. Water supply wells are typically required to be drilled to a verticality tolerance of 1 in 300. This tolerance is probably unnecessary for temporary works wells. Verticality requirements are often not explicitly specified for temporary works wells, but if they are, then 1 in 100 appears to be a more reasonable requirement.

16.5.1 Cable Tool Percussion Drilling

The original cable percussion drilling rigs made use of a reciprocating mechanism known as a 'spudding arm' or 'walking beam' to repeatedly lift and lower a bit (or chisel) suspended on the end of a wire rope inside telescoping sizes of temporary boring casings. The action of the bit breaks up the soil or rock at the base of the borehole to form a slurry with the groundwater entering the well (Figure 16.4a). As the slurry builds up, it deadens the percussive action of the bit; the bore has to be bailed periodically to remove the slurry. The largest rigs were powerful, and progress could be made, albeit slowly, through even boulder and bedrock formations. Inevitably, the arisings are not much better than a mashed-up slurry of mainly indistinguishable consistency. This type of rig is still in use by some drillers and is especially prevalent in developing countries.

Some time later, there came a variation on this technique, the early tripod bored piling rig, which also made use of temporary boring casings of telescoping sizes. The cable passes over a pulley at the top of the tripod and is raised and lowered by a winch linked by a clutch to a small donkey engine. To make progress through granular formations, a sand pump or shell is used, and to penetrate a clay stratum, a clay cutter is used; chisels are used if cobbles or boulders are encountered. Tripod rigs became more mobile and towable, and self-erecting tripod rigs are still commonly used for site investigation in the United Kingdom, where there are colloquially known as 'shell and auger' rigs, although they are more correctly known as light cable percussion rigs.

While light cable percussion rigs are most commonly used for 150–200 mm diameter site investigation boreholes, the larger models (of 2 and 3 tonne winch capacity) are widely used to drill temporary works wells. Figure 16.4b shows a 2 tonne rig being used to drill a 300 mm diameter well bore.

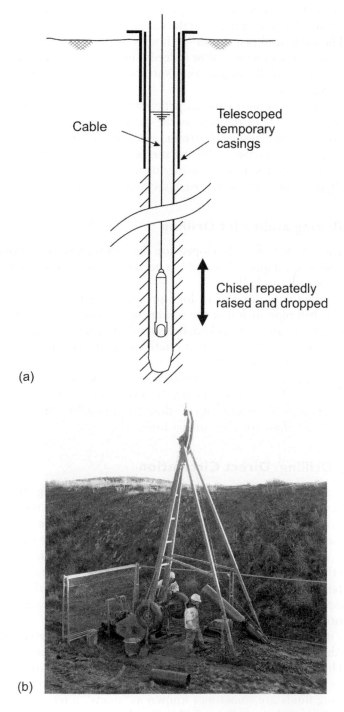

Figure 16.4 Light cable percussion boring. (a) Cable percussion boring method. (b) Light cable percussion boring rig.

Whatever type of cable tool percussion rig is used, drilling will generally be slower than by the rotary method. This is mainly due to the time required to install and remove the temporary casings. The cable tool percussion method is most suitable for soils such as sands, gravels and clays. Progress can be very slow indeed if used to drill rock formations, although the method has been successfully used in some soft rocks such as weathered Chalk or highly weathered Triassic Sandstone.

When using temporary boring casing to penetrate a multi-layered soil structure, there is risk of smearing of the side of the hole as the casing is being installed and withdrawn. This could mask or block the permeable strata, especially if the individual soil/rock layers are thin. Also, when using percussion boring methods, there is some risk that the boring cuttings and slurry will clog the water-bearing formation, so subsequent well development should be carefully monitored and supervised (see Section 16.7).

16.5.2 Wash Boring and/or Jet Drilling

The technique of jetting was first developed for the installation of wellpoints – relatively small-diameter holes (see Chapter 15) – and has been extended to the making of larger holes for the installation of wells and piles.

Essentially, it entails the use of a jet of high-pressure water being applied at the bottom of the wash pipe or drill pipe such that the soil at the level of the water jet is slurrified (i.e. put into an almost liquid state). The soil particles tend to be flushed back up to ground level, leaving a bored hole that has been washed relatively clean with little side smear, such as might result from cable percussion drilling.

One of the most common applications of this method is the use of a holepuncher or heavy-duty placing tube to jet wells into place. Figure 16.5 shows a 300 mm diameter holepuncher in use. Holepunchers have been used up to diameters of 600 mm and depths of 36 m, although depths of more than 20 m are uncommon.

16.5.3 Rotary Drilling: Direct Circulation

This method uses a drill bit attached to the bottom end of a rotating string of hollow drill pipe (sometimes known as drill 'rods'). Bits are typically a tricone rock-roller bit consisting of moveable cutters, or fixed drag bits; the rotating action of the bit breaks up the soil or rock in the base of the bore into small cuttings. A continuous supply of drilling fluid is pumped down the drill pipe to cool the rotating bit. The fluid rises back to the surface in the annulus between the drill pipe and the borehole walls and flushes the cuttings to the surface (Figure 16.6a). Hence, the drill fluid must have sufficient 'body' and rising velocity to retain the cuttings in suspension throughout their upward travel to the surface and also to provide a degree of support to the borehole to prevent collapse.

The bit and the drilling fluid are each vital components of the process, and an understanding of their interdependent functions is necessary for successful usage of the process (see Sterrett, 2008). The drilling fluid may be water-based with long-chain polymer compounds added to the water to form a fluid with better support and cutting transportation properties. Water-based drilling fluids are sometimes known as 'muds', a term that dates from when bentonite slurries were used routinely to drill wells – bentonite is now rarely used in drilling fluids for water wells. If air (supplied by a powerful compressor) is the drilling fluid, small amounts of additive may be used to form a 'foam' or 'mist' to carry the cuttings out of the bore. Many modern drilling additives are naturally biodegradable, so that any residues left in the ground after development will decay and are less likely to act as an impediment to flow of water into the well.

Figure 16.5 Holepuncher used for installation of deep wells. (Courtesy of T O L Roberts.)

On reaching the required depth of bore, the drill string and bit are withdrawn, with the borehole left filled with drilling fluid prior to installation of the well screen and filter pack. If possible, it is good practice prior to installation to try to displace the fluid in the borehole and replace with a clean, thin solution of fluid.

Rotary drilling rigs are available in a variety of sizes, from small site investigation units that can be towed behind a four-wheel drive vehicle to rigs suitable for shallow oil & gas wells that need to be delivered on several articulated trucks and have to be assembled by crane. For groundwater lowering deep wells, self-contained rotary rigs mounted on trucks or on crawler bases or four-wheel drive tractors (Figure 16.6b) are most commonly used. Such rigs have drilled wells to depths in excess of 100 m.

16.5.4 Rotary Drilling: Reverse Circulation

This is a variation on rotary drilling often used to drill large-diameter water supply wells but used only rarely to construct temporary works deep wells. The principle of operation is the same as for direct circulation rotary drilling, but the direction of flow of drilling fluid is reversed (Figure 16.6c).

The mixture of drilling fluid and cuttings in suspension flows up the hollow drill pipe with a fast rising velocity because the internal diameter of the drill pipe is limited. The borehole is

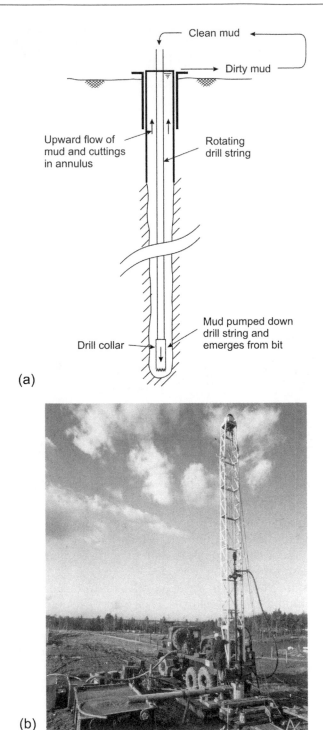

(a)

(b)

Figure 16.6 Rotary drilling. (a) Direct circulation rotary drilling method. (b) Truck-mounted direct circulation rotary drilling rig. (c) Reverse circulation rotary drilling method. (Panel (b) courtesy of
British Drilling and Freezing Company Limited.)

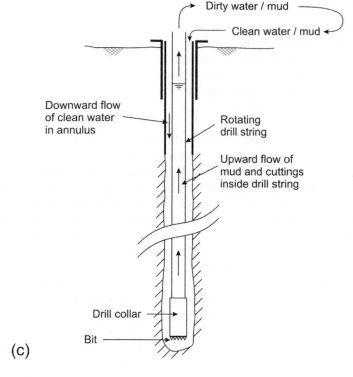

Figure 16.6 (Continued)

kept topped up with drilling fluid, which allows the return flow of fluid to descend slowly in the annulus – of greater cross-sectional area than the drill pipe – between the drill pipe and the borehole wall. The level of the drilling fluid must always be a few metres above the level of the groundwater to help maintain a stable bore. As with the direct circulation method, on reaching the required depth of bore, the drill string and bit are withdrawn, and the borehole is kept topped up with fluid.

16.5.5 Rota-Sonic Drilling

A relatively recent development is sonic drilling (also known as rota-sonic drilling). In this method, a top-drive drill head generates vibratory energy via a pair of eccentric counter-rotating weights. The vibration is directed to the drill string to cause it to resonate, creating displacement, shearing and fracturing of the material through which the drilling progresses (Riley *et al.*, 2018). The drill string is rotated to evenly distribute resonance at the drill bit face. This reduces friction between the drill string and the ground, allowing very rapid rates of penetration. In some cases, this method can allow boreholes to be drilled without the use of drilling fluid, significantly reducing the volume of fluid waste generated during drilling. This can be an advantage when drilling on contaminated sites (e.g. for groundwater remediation schemes).

The rota-sonic drilling method uses an outer casing and an inner drilling tool. Typically, the inner tool is advanced first, and then the outer casing is advanced over the inner tool to support the bore while the inner tool (and spoil) is removed. The final hole is cased, so the well materials (well screen and filter) are installed inside the bore as the outer casing is incrementally extracted.

16.6 INSTALLATION OF WELL MATERIALS

Once the bore is complete, the well casing and well screen are placed in the hole. Typically, the casing and screen consist of threaded pipe of between 150 and 300 mm diameter, supplied in lengths of between 1 and 6 m. The diameter of casing and screen required is determined as described in Section 16.3. The casing sections are plain (unperforated) and the screen sections are perforated (generally by slotting), and for temporary works wells are typically made from uPVC or HDPE – steel screen and casing are used only rarely (see Section 16.3.3).

Prior to installation, the drilling crew should be instructed on the number and sequence of casing and screen lengths to be installed to ensure that the screens are located within the water-bearing horizons. The screen and casing are installed in sections by lowering into the borehole using the drilling rig's winch. A bottom cap is fitted to the first length of screen or casing that is lowered into the bore; further sections are then added until the string of screen and casing is installed to the required depth. It is often a good idea to place a few hundred millimetres of filter gravel in the base of the bore before commencing the installation of the screen and casing. This prevents the bottom length sinking into any soft sediment or drilling residue remaining in the base of the hole.

When the screen and casing are in position, the filter pack (see Sections 13.10 and 16.3.4) must be placed in the annulus between the well screen and the borehole wall. The filter pack is formed from filter media generally consisting of uniformly sized gravels, although coarse sands are sometimes used. The filter material may be supplied in bags or in bulk loads. Ideally, to ensure correct placement, the gravel should be installed using one or more tremie pipes. The tremie pipe would typically be a uPVC or steel sectional pipe of 50 mm internal bore (although sizes down to 32 mm bore have been used to place uniform sands). The filter media are poured slowly into a hopper at the top of the tremie, perhaps washed down by a gentle flow of water. Patience is vital in this operation – adding the filter media too quickly will cause the tremie pipe to block or 'bridge'. If this occurs, the tube will have to be removed and flushed clear, causing delays and inconvenience. The tremie tube is raised slowly, keeping pace with the rising level of filter media.

It is sometimes acceptable to place the filter media without a tremie pipe by pouring into the annulus from the surface. This is only acceptable provided that the filter media is very uniform and will not segregate as it falls down the bore and if the wall of the borehole is very stable (e.g. if supported by temporary casing). Again, care must be taken to ensure that the filter media are added slowly. If too much is added at once, a blockage or 'bridge' may occur; this can be difficult to clear and may result in the screen and casing having to be removed and the complete installation recommenced.

If temporary drill casing is used to support the borehole, this must be removed in sections as the filter media is added. The level of the filter media should not be allowed to rise significantly above the base of the temporary casing; otherwise, a 'sand lock' may occur between the temporary drill casing and the well screen. This can cause the well screen to be pulled out when the temporary casing is withdrawn. The best practice is to alternately add some filter media, pull the temporary drill casing out a little, add more filter media, pull the drill casing further, and so on, monitoring the level of the filter media continuously. Boreholes drilled by rotary methods without temporary casing do not have this constraint, and filter media can be placed in one continuous, steady operation.

The filter pack is sometimes brought up to ground level, but it is good practice to place a very low-permeability seal above the filter pack (see Section 13.10.4 and Figure 16.2). This reduces the risk of aquifer contamination by surface water (or water from other aquifers) passing down the filter pack into the aquifer (see Section 21.6.1). The seal is often formed of bentonite pellets or grout of either neat cement or cement-bentonite (see Sterrett, 2008 for guidance on grout mixes and placement). If the grout were to be placed directly on top of the filter pack, some grout might be lost into the filter media, compromising its permeability. To avoid this, a 1–2 m thick layer of bentonite pellets should be placed on top of the filter pack and allowed to swell before the grout is placed.

16.7 WELL DEVELOPMENT

Development is a process, carried out between completion of the well and installation of the pumps, with the aim of removing any drilling residue or debris from the well and maximizing the yield of clean sand-free water. If development is not carried out, not only might the yield be low, but also the borehole electro-submersible pump will be damaged as a result of pumping sand-laden water.

The most commonly used forms of development in granular aquifers are intended to induce two-way groundwater flow between the well and the aquifer to remove any loose particles from the filter pack and aquifer immediately around the well. This will increase the aquifer permeability locally and will remove any potentially mobile soil particles that might damage the operational pumps. It is most important that the development generates two-way flow (alternately into and out of the well) to help dislodge mobile soil particles that may be loosely wedged in soil pores – this is much more effective than continuous pumping. Development can only be effective in wells that have appropriately designed and installed filter packs (including natural filter packs). No amount of development can correct a well with a filter pack that is too coarse or is discontinuous due to installation problems.

There are a wide range of development techniques (see Sterrett, 2008) used on water supply wells, but in practice, most temporary works wells are developed by one of three techniques:

1. Airlift pumping. Air from a compressor is used to lift water from the well up an eductor tube formed from plastic or steel pipe of 75–150 mm diameter (Figure 16.7). This is a robust method of pumping and, with suitable equipment, can transport significant volumes of sand with the water. Reversal of flow is achieved by lowering the airline past the bottom of the eductor tube and delivering a short blast of air into the well. Alternatively, airlift pumping can be used for a few seconds to raise a 'slug' of water to just below ground level, at which point the air is turned off, causing the water to fall back down the well, inducing flow out of the well and into the aquifer. The airlift method is most effective when the water level in the well is near the surface and becomes less efficient for deeper water levels – further details on airlift pumping methods are given in Section 19.8.1.

2. Surge block. A tight-fitting block is lowered into the well screen (Figure 16.8) and pulled sharply upwards using a tripod and winch (such as a light cable tool percussion rig). As the block moves upwards, water will be forced out of the well above the block and drawn in below – thereby achieving reversal of flow. There are a number of ways that the block can be surged, using either many short strokes or fewer long ones (see Sterrett, 2008); some methods of surging are known as 'swabbing the well'. If a special surge block is not available, a weighted shell or bailer of slightly smaller diameter than the well screen is sometimes used. The sediment and debris that builds up in the well will need to be removed by airlift pumping or bailing.

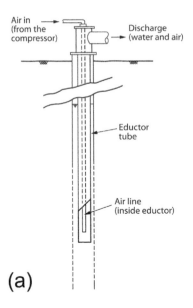

Figure 16.7 Well development by airlift pumping. (a) Airlift with eductor tube. (b) Water pumped from a well during well development.

3. Jetting. Less commonly used, this method involves lowering a jetting head (mounted on drill pipe) down inside the well screen. The drill pipe and jetting head are slowly rotated, and high-pressure water is pumped down the drill pipe and jets horizontally at the screen via small nozzles in the head. The jetting generally forces flow into the aquifer and so may need to be alternated with airlift pumping to get flow reversal and also to remove any sediment or debris generated by jetting.

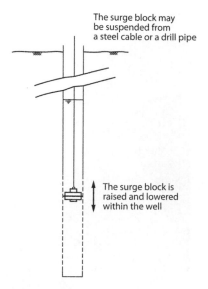

The surge block may
be suspended from
a steel cable or a drill pipe

The surge block is
raised and lowered
within the well

Figure 16.8 Well development by surge block.

16.7.1 Monitoring of Pumped Water for Suspended Solids

Development is usually discontinued when the well no longer yields sand or fine particles when pumped by the airlift. This is normally monitored by observing the discharge water, which initially will appear very dirty or discoloured but should become clear as development proceeds. However, a note of caution should be sounded. Sometimes, the flow of discharge water may appear visually clear but still contains small but significant amounts of fine sand – enough to cause problems for the pump and to create voids around the well if sand pumping continues in long-term pumping.

A simple way to check for suspended sand in pumped water is to take a sample of water in a clean white plastic tub (of the sort used to hold soil samples). Any sand will be clearly visible in the bottom of the tub. Specialist sediment sampling equipment can also be used:

i. A 1 litre conical container called an 'Imhoff cone' (Figure 16.9a) can be used to check for sediment in the water. The sediment settles in the bottom of the cone, and the quantity is read off in millilitres per litre.

ii. Where more accurate assessment of sand content is required, continuous sand samplers can be attached to the well head via a threaded tapping and a small but continuous flow rate passed through the sampler. The Rossum sand sampler (Rossum, 1954), shown in Figure 16.9b, is one example. This captures sand in the glass vial at the base of the sampler. Combined with records of the flow rate through the sampler, this allows the sand concentration (in millilitres per litre) to be estimated.

The question is often asked – how long should it take to develop a well? There is no simple answer to this, but experience suggests that many wells in granular aquifers will take between 6 and 12 hours to be effectively developed. Certainly, if a well still yields copious amounts of sand after more than 2 to 3 days (say 20 to 30 hours' development), the well is unlikely to improve, and a replacement well (possibly with a different filter media) should be considered.

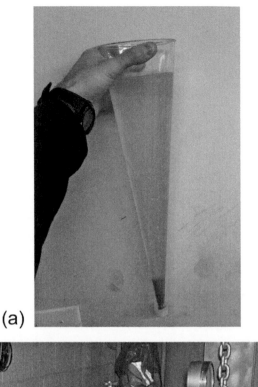

(a)

(b)

Figure 16.9 Specialist methods used to determine sediment content in pumped water. (a) Imhoff cone. (b) Rossum sand sampler.

16.7.2 Well Development by Acidization

For wells drilled in carbonate rocks, such as Chalk, wells may be developed by acidization. This is a technique commonly used in water supply wells (see Banks *et al.*, 1993), where acid is introduced into the well to dissolve any drilling slurry and to improve the well–aquifer connection by dissolving aquifer material in fractures locally around the well. In practice, significant quantities of concentrated acid (up to several tonnes in some cases) are introduced into the well. The reaction between the acid and the carbonate rock generates large volumes of carbon dioxide gas, which must be carefully controlled by fitting a gas-tight head plate (equipped with specialist valves and pressure relief devices) to the top of the well. Acidization is not a straightforward procedure and should be planned and undertaken by experienced personnel, with particular emphasis on ensuring that the necessary health and safety measures are in place. Acidization is not commonly used to develop temporary works wells, but the method has been used on some deep groundwater lowering wells into the Chalk beneath London, including for the Jubilee Line Extension tunnel project in the 1990s.

16.7.3 Re-Development of Wells during Long-Term Dewatering Pumping

If wells are operated for long periods of time (several months or years) and are affected by encrustation or clogging, well performance may deteriorate. In such cases, well performance may be improved by periodically taking individual wells out of service for redevelopment by one of the methods described earlier (including acidization, which has been used on heavily encrusted wells). Remedial methods to address the clogging and encrustation of wells are discussed in Section 23.5.

16.8 INSTALLATION AND OPERATION OF DEEP WELL PUMPS

Once the wells have been developed, the pumps can be installed. This section deals primarily with the routine installation and operation of borehole electro-submersible pumps – by far the most commonly used type in deep well systems. More specialist pumping aspects, including operational problems, are discussed in Section 19.6.

For groundwater lowering applications, the pump is normally suspended vertically near the base of the well so that potential drawdown is maximized. The base of the pump should be at least 1 m above the base of the well to avoid the pump becoming stuck in any sediment that may build up at the bottom of the well.

The pump is connected to surface by the riser pipe, which is a steel or plastic pipe that carries the discharge from the pump up to ground level (see Figure 16.2). The riser pipe is typically between 50 and 150 mm in diameter for the smallest and largest pumps used for temporary works. If the riser pipe and joints are appropriately rated, it may be used to support the weight of the pump, riser pipe and water column. Otherwise, the weight must be supported by straining cables or ropes from the pump, tied off at the well head. It is good practice to install a dip tube (of 19–50 mm diameter) alongside the riser pipe to allow access for a dipmeter to be used to monitor the water level in the well.

Most commonly, the riser is supplied in sections (typically 3 to 6 m in length) with threaded, flanged or specialist joints. A crane or hoist is used to lower the pump and string of riser pipe into the well; extra sections are added until the pump has been lowered to the design level (see Figure 19.9). The electrical power cable connected to the pump is also paid out as the pump is lowered into position. The cable should be kept reasonably taut and be taped or tied to the riser pipe every few metres. If this is not done, then any slack in the cable may form 'loops' down the well, interfering with the subsequent installation of dip tubes.

Where pumps are installed at significant depths (greater than 50 m), the installation of sectional riser pipes can be a slow and tedious process. An alternative is to use flexible riser pipe (effectively a high-specification layflat hose designed for that purpose) to suspend the well in the pump. The flexible riser is connected to the submersible pump in a single length and (where space allows) can be quickly lowered into the well by passing over a roller pulley at the well head, using a vehicle to control the rate of descent into the well (Figure 16.10).

At ground level, the riser pipe is capped off with a headworks arrangement – typically including a control valve and pressure gauge – attached to the top of the riser pipe and

(a)

(b)

Figure 16.10 Installation of deep well submersible pump on flexible riser pipe for a well as part of a mine dewatering system (a) Borehole submersible pump (in foreground) connected to flexible riser pipe ready for installation. (b) Flexible riser pipe being lowered into well using roller pulley and vehicle traversing toward the well head. (Courtesy of Angus Flexible Pipelines Limited, Lancaster, UK.)

connected to the discharge pipework (Figure 16.11). The best sort of valve for use at the well head is the gate valve. It is preferable because its slide plate is at right angles to the water flow and its position can be finely adjusted by its screw mechanism to regulate the flow. Its sensitivity of opening adjustment is good, and when fully open, it offers less resistance to flow than any other type of valve. The other type of valve sometimes used for well output control is the butterfly valve. It is less costly than the gate valve (size for size), but the sensitivity of opening control is less than that of the gate valve.

(a)

(b)

Figure 16.11 Examples of well head arrangements for dewatering wells. (a) Well head with valve for a high-capacity deep well. (b) Well head for a low-capacity deep well, with valve, flowmeter and electrical control panel. (Panel (a) courtesy of WJ Groundwater Limited, Kings Langley, UK.)

The pump power cable is connected to an electrical control unit located either alongside the well head or in a central control house. Most borehole electro-submersible pumps are powered by a three-phase supply; if the phases are connected in the wrong order, the pumps can run (or rotate) backwards (i.e. anticlockwise rather than clockwise). Pumps running backwards move very little water, so on installation, it is important to 'check the direction' of each pump in turn. One method is to start the pump with the control valve fully closed and observe the pressure gauge. If the needle indicates a high pressure (consult the pump manufacturer's data sheets for the expected value), the motor is running in the right direction. But if the indicated pressure is low, the motor may be running backwards. Turn the pump off, isolate the pump switchgear; have an electrically competent person change over two of the phase wire connections, and re-start the pump against a closed valve. The pressure gauge should register full pressure, indicating that the motor is now running in the right direction.

For groundwater lowering projects, the pumps generally run continuously (24 hours per day, 7 days per week), with the operating water level in individual wells only a little above the level of each pump intake. Generally, the flow from each pump is adjusted to achieve this condition by manipulation of the control valve on each well head. The valve should be partially closed to impose just sufficient back pressure (extra artificial head) to constrain the well output and maintain the optimum operating level. In some installations, especially where well yields are low, the pumps are operated intermittently using sensors in the well to trigger the pump to start when the water level is high and stop when the water level is too low. The control of deep well pumps is discussed in Section 19.9.

If the well operating level is too close to the level of the pump intake, the pump will tend to 'go on air' from time to time – this is when the water level in the well reaches the pump intake, and air is drawn into the pump. In this condition, the pump will race when 'on air', and the riser pipe will tend to vibrate. This must not be allowed to continue in the long term, because under such conditions, the pump motor will be damaged severely. The symptoms of 'going on air' can be detected by observing the needle of the pressure gauge at the well head sited upstream of the control valve (see Figure 16.2) and by observing vibration of the riser pipe. When the discharge flow contains some air (rather than water only), the gauge needle will repeatedly drop to zero. If these erratic variations in pressure gauge readings are observed, close the control valve gradually until the pressure gauge needle remains steady – a very slight fluctuation in the needle position is acceptable. Alternatively, if the gauge needle remains steady, the well operating level may be above the design level. If this condition is suspected, gently open the control valve until the needle starts to flicker, and then close the control valve by a very small amount.

16.8.1 Encrustation and Corrosion

Deep wells operating for long periods may sometimes be affected by encrustation due to chemical precipitation or bacterial growth in the well screen, filter pack, pump or pipework. Encrustation can reduce the efficiency of the well (reducing yield and increasing well losses) and may increase the stress on the pump, leading to more severe wear and tear and, ultimately, failure.

In certain types of groundwater, corrosion of metal components in pumps and pipework may be severe. Recognition of such conditions is vital when specifying equipment – for example, maximizing the use of plastic pipework can reduce the problems, leaving only the submersible pumps made from metal.

The nature of these problems, and appropriate avoidance and mitigation measures, are discussed in Section 23.5.

16.9 VACUUM DEEP WELL INSTALLATIONS

If deep wells are installed in a low-permeability aquifer and well yields are low, well performance can be enhanced somewhat by sealing the top of the well casing and applying a vacuum. The vacuum is generated by an exhauster unit (a small vacuum pump) located at ground level. Typically, the exhauster unit would be connected to a manifold, allowing it to apply vacuum to several wells (Figure 16.12). Because air flow to the exhauster will be low (once vacuum is established), the pipework connecting the exhauster and the wells need only be of small diameter, perhaps 50 mm or less.

For the method to be effective, the top of the well filter pack must have a bentonite or grout seal; otherwise, air may be drawn through the gravel pack into the well, preventing the establishment of a vacuum. A typical vacuum deep well is shown in Figure 16.13. The

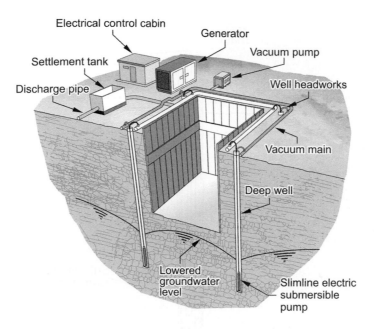

Figure 16.12 Deep well system with vacuum applied to the wells.

top of the well needs to be sealed airtight between the well casing and the pump riser pipe, power cable and vacuum supply pipe. Purpose-built well head seals are available, but on-site fabrication of seals using plywood disks and copious amounts of sealing tape has also proved effective.

The application of vacuum to deep wells is unlikely to be a panacea for a poorly performing system, especially if poor yields are associated with inappropriate filter gravel gradings or inadequate well development. The yield from the wells will not be transformed wholesale, but, at best, will increase by 10–15 per cent, occasionally more. Even if a partial vacuum equivalent to 8 m head of water can be maintained in the wells, do not expect the drawdown outside the wells to increase by a similar amount. It may be that the vacuum is mainly used to overcome well losses at the face of the well. The application of vacuum can cause operational problems, including increased risk of collapse of well screens or sand being pulled through the filter pack.

16.10 SHALLOW SUCTION WELL INSTALLATIONS

The bored shallow well system is a synthesis of the deep well and wellpoint systems. Bored shallow wells are constructed in the same way as a deep well system but use the wellpoint suction pumping technique to abstract the water (Figure 16.14). Hence, the amount of lowering that can be achieved is subject to the same limitations with a wellpoint system, namely that drawdowns in excess of 6 m below pump level are difficult to achieve. The method is most useful on congested urban sites and where the soil permeability is high. Because the wells are of larger diameter than wellpoints, they generally have a greater yield and so can be at greater centres. This means that the wells create fewer constraints on the activities of the steel fixers, shutter erectors and other trades needing to carry out work within the dewatered excavation. The shallow well method is known by some as 'jumbo' wellpointing, since the wells can be thought of as widely spaced, grossly oversized wellpoints.

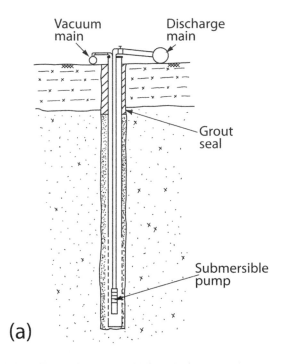

(a)

(b)

Figure 16.13 Vacuum deep well. (a) Schematic section through a vacuum deep well. (b) Well head of vacuum deep well. The wellhead is sealed, and the exhauster unit (next to the well) is used to generate a vacuum in the well. (Panel (b) courtesy of Hölscher Wasserbau GmbH, Haren, Germany.)

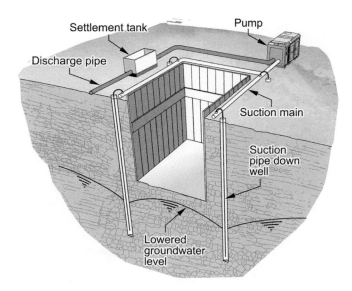

Figure 16.14 Shallow suction well system.

Chapter 17

Other Dewatering Systems

Make the plan to suit the ground and not the ground to suit the plan.

Sir Marc Brunel

17.1 INTRODUCTION

This chapter describes some less commonly used dewatering and groundwater control systems, and outlines conditions appropriate to their use.

The systems described in this chapter are:

(i) Ejector well systems, appropriate to low-permeability soils such as very silty sands or silts
(ii) Horizontal wellpointing, mainly suitable for dealing with large trenching or pipeline projects of limited depth
(iii) Horizontal wells and drains
(iv) Collector wells, drainage tunnels and adits
(v) Relief wells and vertical drains
(vi) Slope drainage and siphon drains
(vii) Electro-osmosis, applicable to pore water pressure control in very low- permeability soils or for increasing the shear strength of very soft soils.
(viii) Artificial recharge systems.
(ix) Applications of dewatering and groundwater control technologies for the control or remediation of contaminated groundwater.

Some of the techniques described are specialized and perhaps rather esoteric in nature. An engineer or geologist working in groundworks and excavations might spend a whole career in the field and never have to apply any of these techniques. Nevertheless, it is important to be aware of the specialist methods that may be of help when faced with difficult conditions.

17.2 EJECTOR WELLS

The ejector well system (also known as the eductor or jet-eductor system) is suitable for pore water pressure reduction projects in low-permeability soils such as very silty sands, silts or clays with permeable fabric. In such soils, the total flow rate will be small, and some form of vacuum assistance to aid drainage is beneficial (see Section 9.5.4); the characteristics of ejector wells are an ideal match for these requirements. Chapter 27 includes Case History 27.5, which describes an application of ejector wells for dewatering during shaft construction.

Essentially, the method involves an array of wells (which may be closely spaced like well-points or widely spaced like deep wells) with each well pumped by a jet pump known as an ejector. Ejector well dewatering was developed in North America in the 1950s and 1960s, when jet pumps used in domestic supply wells were first applied to groundwater lowering problems. Since then, the technique has been applied in Europe, the Far East and the former Soviet Union. Ejector wells were not widely employed in the United Kingdom before the late 1980s; the A55 Conwy Crossing project was one of the first UK projects to make large-scale use of ejector wells (Powrie and Roberts, 1990). Some applications of the ejector well method in the United Kingdom are described by Preene and Powrie (1994) and Preene (1996).

17.2.1 Merits of Ejector Well Systems

The ejector well system works by circulating high-pressure water (from a tank and supply pumps at ground level) down riser pipes and through a small-diameter nozzle and venturi located in the ejector in each well. The constriction of flow through the nozzle converts pressure head to velocity head (the water passes through the nozzle at high velocity), thereby creating a zone of low pressure and generating a vacuum of up to 9.5 m of water at the level of the ejector. The vacuum draws groundwater into the well through the well screen, where it joins the water passing through the nozzle and is piped back to ground level via a return riser pipe and thence back to the supply pump for recirculation. A schematic ejector well system is shown in Figure 17.1. Two header mains are needed. A supply main feeds high-pressure water to each ejector well, and a return main collects the water (consisting of the supply water plus the groundwater drawn into the well) coming out of the wells. This large amount of pipework is needed to allow the recirculation process to continue. Pumping equipment for ejector wells is discussed in Section 19.7.

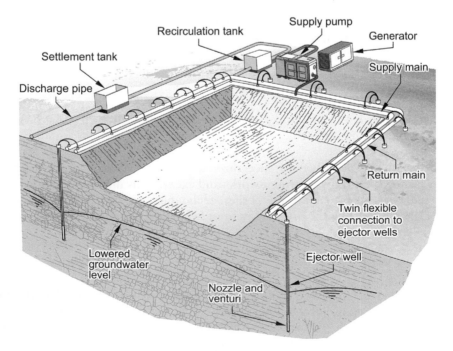

Figure 17.1 Ejector well system.

The most obvious practical advantage of an ejector well system is that ejectors will pump both air and water; as a result, if the ejectors are installed in a sealed well in low-permeability soil, a vacuum can be developed in the well. This is one of the main reasons why ejectors are suited to use in low-permeability soils, where the vacuum is needed to enhance drainage of soils into the wells. Another advantage is that the method is not constrained by the same suction lift limit as a wellpoint system (see Chapter 15). Drawdowns of 20–30 m below pump level can be achieved with commonly available equipment, and drawdowns in excess of 50 m have been achieved with systems capable of operating at higher supply pressures.

These characteristics mean that ejector well systems are generally applied in one of two ways:

(i) As a vacuum-assisted pore water pressure control method in low-permeability soils
(ii) As a form of 'deep wellpoint' in soil/rock of moderate permeability as an alternative to a two-stage wellpoint system or a low-flow-rate deep well system

It is also important to be aware of some of the practical limitations and drawbacks of the ejector well system. Perhaps the most significant drawback is the low mechanical (or energy) efficiency of ejector systems. In low- to moderate-permeability soils, where flow rates are small, this may not be a major issue, but in higher-permeability soils, the power consumption and energy costs may be huge in comparison to other methods. This is probably the main reason why the ejector well system is rarely used in strata of high permeability. Another potential problem is that due to the high water velocities through the nozzle, ejector systems may be prone to gradual loss of performance due to nozzle wear or clogging. This can often be mitigated by regular monitoring and maintenance, but it may make long-term operation less straightforward.

17.2.2 Types of Ejectors

An ejector is a type of jet pump (see Section 19.7). Several different designs of ejectors are available, each having different characteristics. These different ejector designs can be categorized into two types based on the arrangement of the supply and return riser pipes:

(i) Single (or concentric) pipe (Figure 17.2). This design has the supply and return risers arranged concentrically with the return riser inside the supply riser. The supply flow passes down the annulus between the pipes and through the ejector and then returns up the central pipe.
(ii) Twin (or dual) pipe (Figure 17.3). Here, the supply and return risers are separate, typically being installed parallel to each other.

In both cases, a non-return valve is present within the ejector to prevent any injected water flowing into the well screen and thence into the ground.

Both types are used in dewatering systems. In the simplest configuration, the ejector is lowered into the well, suspended on riser pipes, in a similar fashion to submersible pumps in the deep well method.

- Twin-pipe ejectors can be accommodated in well screens and well casings of approximately 100 mm internal diameter and larger.
- Single-pipe ejectors can be accommodated in slightly smaller well screens and well casings of approximately 75 mm internal diameter and larger.

Single-pipe ejectors can also be deployed in an alternative arrangement where the outer riser pipe is also used as the well casing, provided that it has sufficient pressure rating. This

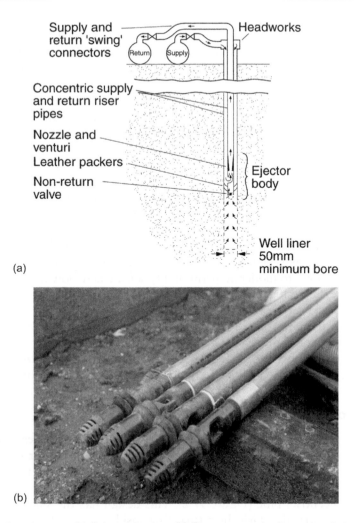

Supply and return 'swing' connectors

Headworks

Concentric supply and return riser pipes

Nozzle and venturi

Leather packers

Non-return valve

Ejector body

Well liner 50mm minimum bore

(a)

(b)

Figure 17.2 Single-pipe ejectors. (a) Schematic view. (b) Ejectors attached to central riser pipes ready for installation into well. (Panel (a) reproduced from Preene, M, *et al.*, *Groundwater Control – Design and Practice*, 2nd edition, Construction Industry Research and Information Association, CIRIA Report C750, London, 2016. With permission from CIRIA: www.ciria.org; panel (b) courtesy of WJ Groundwater Limited, Kings Langley, UK.)

allows ejectors to be installed in well casings of 50 mm internal diameter. The ejector is set at a level within the unperforated casing, above the well screen.

The installation and connection of twin-pipe ejectors suspended in a well involves rather simpler plumbing than for single-pipe designs; this can make the twin-pipe type more suitable for use in localities where skilled labour is scarce, and it is desired to keep dewatering equipment as simple as is practicable.

17.2.3 Installation Techniques

The ejectors themselves (the jet pumps) are installed in wells, which are generally installed by similar methods to deep wells: cable percussion boring, rotary drilling, jetting or auger boring (see Chapter 16). Well casings and screens and filter packs are then installed in the borehole. If single-pipe ejectors are to be used, the smaller-diameter casings are sometimes installed by methods more akin to wellpointing (see Chapter 15) than deep wells. Ejector

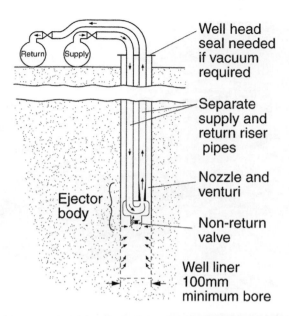

Well head
seal needed
if vacuum
required

Separate
supply and
return riser
pipes

Nozzle and
venturi

Ejector
body

Non-return
valve

Well liner
100mm
minimum bore

Figure 17.3 Twin-pipe ejectors – schematic view. (Reproduced from Preene, M, *et al.*, *Groundwater Control – Design and Practice*, 2nd edition, Construction Industry Research and Information Association, CIRIA Report C750, London, 2016. With permission from CIRIA: www.ciria.org)

systems are very intolerant of any fine particles in the system, and the wells should be developed (see Section 16.7) following installation and prior to placement of the ejectors and risers.

The ejectors are connected to the supply and return riser pipes (twin or single-pipe depending on design) and are lowered down to the intended level, typically near the base of the well. The headworks are fitted to seal the top of the riser pipes to the well liner, and the flexible connections are made to the supply and return mains (Figure 17.1 and Figure 17.5). This is needed to connect each ejector to the pumping station(s), which supply the high-pressure water that is the driving force behind the pumping system.

17.2.4 Ejector Pumping Equipment

The pumping equipment in an ejector system consists of three main elements: the ejector (and associated riser pipes and headworks); the supply pumps; and the supply and return pipework. The characteristics of ejector pumping equipment are discussed in Section 19.7. The net pumping rate (known as the induced flow rate) of each ejector should be designed based on the anticipated well inflow. Ejector performance curves can then be used to determine the required supply flow per ejector (at the specified pressure).

The supply pumps should be chosen to supply the required total supply flow at the necessary pressure, taking into account friction losses in the system. Adequate standby pumping capacity should be provided. Supply pumps are typically high-speed single or multi-stage centrifugal pumps; the pump and motor may be either horizontally or vertically coupled (Figure 17.4). If the system consists of relatively few ejector wells, one pump may be used to supply the whole system. However, large systems (more than 15 to 25 ejector wells) could be supplied by one large duty pump (Figure 17.4a) or a bank of several smaller duty pumps connected in parallel (Figure 17.4b). The use of several smaller pumps can allow a more flexible approach, with additional pumping capacity being easy to add as needed, and allows more economical use of standby pumps.

(a)

(b)

Figure 17.4 Ejector supply pumps. (a) Horizontally coupled pumps. (b) Vertically coupled pumps. (Courtesy of WJ Groundwater Limited, Kings Langley, UK.)

The supply pipework is normally made of steel with the pipes and joints rated to withstand the supply pressure. The return pipework does not need such a large pressure rating but may still be under considerable positive pressure, especially if the wells are drawing air into the system. Supply header pipes are typically 100 or 150 mm in diameter. Return header pipes are commonly of a slightly larger diameter to deal with higher flow rates and entrained air; 150 to 200 mm diameter is typical. Air elimination valves may need to be included in the return header to prevent airlocking. Figure 17.5 shows the pipework arrangement for a large-scale ejector well system.

The components of an ejector set-up form a complex hydraulic system. For all but the smallest of systems, the design of the pumping systems should be carried out with care. Work by Miller (1988), Powrie and Preene (1994b) and Powers *et al.* (2007) is recommended as further reading to those faced with such a problem.

Figure 17.5 Ejector pipework. An ejector headworks is shown in the foreground. The supply and return header pipes can be seen in the background, linked to the ejector by flexible hoses. (Courtesy of WJ Groundwater Limited, Kings Langley, UK.)

17.2.5 Operation of an Ejector Well System and Potential Imperfections

The operation of an ejector well system is relatively straightforward. In general, individual ejectors do not need adjusting or trimming, as might be required for wellpoints. The system should be very stable in operation, and the supply pressure (displayed on gauges at the supply pump) should hardly vary once the recirculation tank and supply/return pipework are primed and running. Any changes in supply pressure may indicate a problem with the system and should be investigated.

Successful long-term operation of ejector well systems relies on ensuring that the circulation water is not contaminated with suspended fine particles of soil or other detritus (and that sand or silt is not being drawn in through improperly designed, or inadequately developed, well screens). Significant amounts of suspended solids will damage the supply pumps (which are generally intended for clean water only) and may build up in the ejector risers and bodies, restricting flow. However, the most serious effect of suspended solids is excessive wear of the nozzle and venturi in the ejector due to the abrasive action of the particles as these pass through the nozzle at high velocity. Even low levels of suspended solids can cause nozzle wear over weeks or months of pumping. As the nozzle wears, its opening enlarges, the supply pressure falls and system performance deteriorates. Figure 17.6 shows examples of pristine, moderately worn and severely worn nozzles.

Suspended solids may enter the circulation water in a number of ways:

1. The material may have been present when the system was assembled. If appropriate supervision and workmanship are not employed, it is not unknown for tanks and pipework to contain sand, dead leaves and other extraneous matter left over from previous use or storage. If these are not cleaned out prior to commissioning the system, the system will probably either clog up after a few minutes or suffer severe nozzle wear over the next few days – neither is particularly satisfactory. When installing a system, the

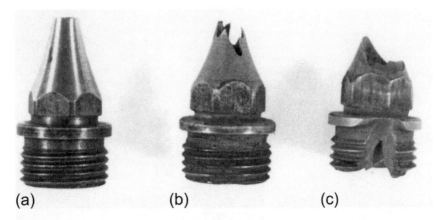

(a)   (b)   (c)

Figure 17.6 Examples of ejector nozzle wear. (a) New nozzle. (b) Moderately worn nozzle: after 6 months in a slightly silty system. (c) Severely worn nozzle: after 1 month in a moderately silty system. (Courtesy of W Powrie.)

    pipes should be as clean as is practicable. Prior to start-up, the system must be primed with *clean* water. The water must be run to waste to flush the system clean; flushing should only stop when the water runs clear. It is important to allow for this flushing out when estimating the volume of water needed for priming.

2. Silt or sand particles may be drawn from the wells into the system either during the initial stages of pumping (due to inadequate well development) or continually (due to ineffective well filters). These problems should be avoided by ensuring that appropriate well filters are in place and that development is not neglected.

3. Silt or sand may be drawn in from an individual well that may have a poorly installed filter or where the well screen has been fractured by ground movement. If this well is not identified and switched off, it will continually feed particles into the system. The circulating action means that the sand from one well can damage all the ejectors in a system.

4. Soil or debris may have inadvertently been added to the circulation tank as a result of construction operations. This has occurred where spoil skips have been craned over the tank location and small amounts of spoil have fallen into the tank. This can be avoided by fitting a lid to the tank.

5. The growth of biofouling bacteria (see Section 23.5) can generate suspended solids in the form of iron-related compounds associated with the bacteria's life cycle. This is less straightforward to deal with, but one of the simplest solutions is to periodically dispose of the circulation water and flush out the system with clean water.

If significant nozzle wear does occur, once the cause of the problem has been identified and dealt with, the ejectors will need to be removed from the wells. The worn nozzles and venturi should be replaced with pristine items before the ejectors are replaced in the wells. If there are many ejectors in the system or they are particularly deep, this can be quite an undertaking. Nozzle wear is definitely one case where prevention is better than cure.

    Biofouling (item 5 in the list) may also cause problems by allowing encrustation to build up in and around the ejectors, causing clogging rather than wear. In general, if the level of dissolved iron in the groundwater is more than a few milligrams per litre, the potential for clogging should be considered (see Section 23.5).

17.3 HORIZONTAL WELLPOINTS

The horizontal wellpoint method uses a horizontal flexible perforated pipe, pumped by a suction pump, to effect lowering of water levels. Typically, the perforated pipe (the horizontal wellpoint) is installed by a special land-drain trenching machine (Figure 17.7). One end of the pipe is unperforated and is brought to the surface and connected to a wellpoint suction pump. The method is also known as horizontal sock dewatering or sock dewatering (after the 'sock' of geotextile filter mesh sometimes used to cover the flexible pipe).

A horizontal wellpoint is very efficient hydraulically because it has a very large screen area, and horizontal flow will be planar to the sides of the perforated pipe. This contrasts with flow local to vertical wellpoint systems, where flow lines converge radially to each wellpoint, and where the screened area is limited by the short length of the wellpoint screens.

The main use of the method is for large-scale shallow cross-country pipelines as an alternative to single-stage wellpointing (Section 15.8), when rapid rates of installation and progression are required (Figure 17.8). The principal restriction on the use of the method is local availability of the specialist trenching machines to lay the perforated pipe at adequate depth. Trenching machines capable of installing drains to 6–8 m depth are relatively commonly available in Holland and North America but are less readily available in the United Kingdom. Large trenching machines were used on British pipeline and motorway cutting projects in the 1960s and 1970s, but nowadays, most of the trenching machines in use in the United Kingdom are limited to installation depths of 3–4 m, although a small number of machines with deeper capability (6–7 m) are in use.

In addition to pipeline work, horizontal wellpoints are occasionally used instead of vertical wellpoints to form perimeter dewatering systems around large excavations for dry docks and the like (Anon, 1976). If large drawdowns are required, multiple stages of horizontal wellpoints can be installed in a similar way to vertical wellpoint systems. Horizontal wellpoints have also been used to provide groundwater abstraction systems for water use in shallow sand aquifers (Brassington and Preene, 2003).

Although the horizontal method is best suited to the long straight runs associated with pipelines or the perimeter of large excavations, it can also be used to form a grid or herringbone

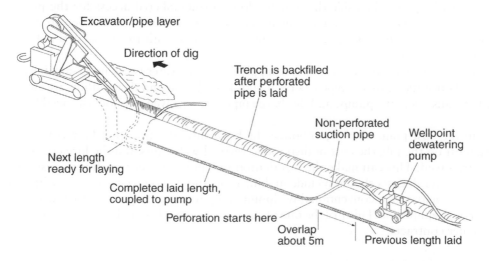

Figure 17.7 Horizontal wellpoint installation using a land drain trenching machine. (Reproduced from Preene, M, et al., *Groundwater Control – Design and Practice*, 2nd edition, Construction Industry Research and Information Association, CIRIA Report C750, London, 2016. With permission from CIRIA: www.ciria.org)

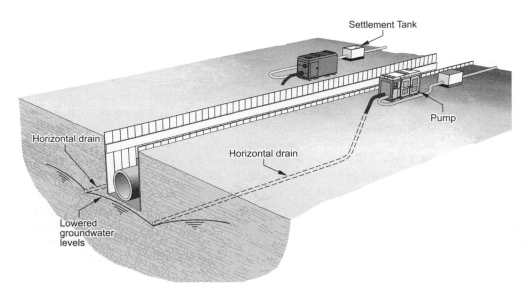

Figure 17.8 Horizontal wellpoint system for pipeline construction. In this example, two horizontal well-points are installed, one on each side of the excavation.

pattern to dewater large areas to shallow depth. The method has also been used to consolidate areas of soft soils in conjunction with vacuum pumping systems (Anon, 1998).

17.3.1 Merits of Horizontal Wellpoint Systems

There are two principal practical advantages to the use of the horizontal wellpoint method. First, very rapid rates of installation can be achieved by specialist trenching machines (up to 1000 m per day in favourable conditions); this can be vital when trying to keep ahead of the installation of cross-country pipelines. Second, the absence of vertical wellpoints and surface header pipes alongside the trench allows unencumbered access for the pipe-laying operations. This has the additional benefit that there is less above-ground dewatering equipment, which might be damaged by the contractor's plant with the associated risk of interruption of pumping.

Other practical advantages are that a supply of jetting water is not necessary for installation, and once the drainage pipe has been laid, installation and dismantling are simple and rapid, because only the pumps and discharge pipes are involved, without the need for header pipes.

From the cost point of view, although the horizontal drain cannot be recovered and is written off on the job, the cost of disposable vertical wellpoints and the hire cost of header pipes are saved. This can make the installation rate per metre of trench very cost effective. However, overall costs should include mobilization and demobilization costs, which can be high (in comparison to conventional wellpoint equipment) for large trenching machines and supporting equipment. This may be one of the reasons why this method tends not to be used on smaller contracts.

17.3.2 Installation Techniques

Horizontal drains can be installed using conventional trench excavation techniques (e.g. by hydraulic excavator). However, these methods tend to be slow and may have problems in maintaining the stability of the trench while the drain is placed; such methods will only be cost effective on relatively small projects or where rapid progression is not desired.

Horizontal wellpoint systems are more commonly installed using crawler-mounted trenching machines (Figure 17.9) equipped with a continuous digging chain, which typically cuts a vertical-sided trench of 225 mm width as the machine tracks forwards. Depths of installation are between 2 and 6 m, although deeper depths can be achieved by the very largest machines. A reel of flexible perforated drainage pipe feeds through the boom supporting the digging chain and is laid in the base of the trench. As the machine tracks forwards, either the spoil is allowed to fall back into the trench, or the trench is backfilled with filter media.

The perforated pipe used as the drain is typically uPVC or HDPE land drain of 80–100 mm diameter (although 150 mm pipe is sometimes used); the pipe is generally wrapped in a filter sock of geotextile mesh, coco matting or equivalent. The pipe comes in continuous reels, perhaps 100 m in length. One end of the pipe is sealed with an end plug, and the other end is unperforated for the first 5–10 m. When the machine starts to cut the trench, the unperforated end is fed out first and is left protruding from ground level. The machine tracks away, cutting the trench and laying the drain almost simultaneously (Figure 17.10). When the reel of drain runs out, the sealed end is left in the base of the trench. A new reel of drain is fitted to the machine, and the next section is laid in a similar way, with an overlap of around 5 m between sections (Figure 17.7). Some trenching machines have the facility for addition of filter media above the drain to improve vertical drainage in stratified soils. Filter media should be selected on the same basis as for vertical wellpoint systems (see Section 15.4.8).

When a horizontal drain is installed for pipeline dewatering, topsoil is typically stripped off prior to installation. This is normal practice in order to allow the site to be reinstated at the end of the project but has the added advantage of allowing the trenching machine to track on the firmer sub-soil. Some of the larger machines weigh up to 32 tonnes and can be difficult to operate on soft soils. It may be necessary to fit wider crawler tracks to reduce ground pressures. Many trenching machines can start trenching from ground level by rotating the digging boom into the ground while the digging chain is cutting. However, in many cases, a starter trench is dug by a conventional excavator, which allows the trenching machine to start cutting with the boom in the vertical position.

Figure 17.9 Specialist trenching machine for installation of horizontal drains. (Courtesy of Hölscher Wasserbau GmbH, Haren, Germany.)

Figure 17.10 Trenching machine cutting trench and laying horizontal drains. (Courtesy of Hölscher Wasserbau GmbH, Haren, Germany.)

For pipeline works, the drain is typically installed along the centreline of the proposed pipeline at a depth below the pipeline formation level (Figure 17.11) or as parallel drains on either side of the trench (Figure 17.8). When pumped, this allows the drains to directly dewater the area beneath the proposed pipeline. When construction is complete, the horizontal drain is left in place and abandoned. Occasionally, it is necessary to grout up the drain at the end of the project to prevent any influence on long-term groundwater conditions and avoid the creation of an artificial horizontal pathway for groundwater flow (see Section 21.6.3).

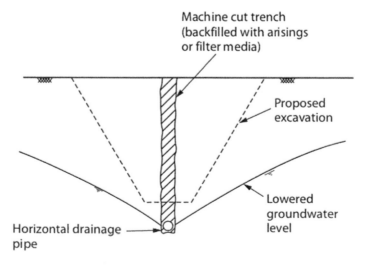

Figure 17.11 Installation of horizontal drains for pipeline trench.

The specialist trenching machines can be effective tools but are not without their problems. First, it can be difficult to detect unexpected ground conditions through which the drain has been laid. If coarse gravel, cobbles or boulders are present, progress may be slowed, and wear to the trenching machine may be excessive; in extreme cases, there is a risk of the digging chain breaking. If layers of soft clay are present in conjunction with a high water table, the clay may 'slurry up' and coat the perforated pipe, clogging it as it is laid. If such ground conditions are anticipated, judgement should be applied before committing to the horizontal wellpoint method.

17.3.3 Pumping Equipment

Horizontal wellpoints are pumped by connecting a conventional wellpoint pump to the unperforated section of the drainage pipe where it emerges from the ground. As the horizontal system is pumped on the suction principle, it is subject to drawdown limitations for similar reasons to the vertical wellpoint system. In general, the maximum achievable drawdown will be limited to between 4.5 and 6 m depending on ground conditions and the type of pump used (see Section 19.5).

17.4 HORIZONTAL DIRECTIONALLY DRILLED (HDD) WELLS

Horizontal drilled wells are a technique used occasionally as part of groundwater control on civil engineering projects. This section describes the horizontal directional drilled (HDD) method rather than sub-horizontal drilled drains used to stabilize slopes in rock (these are discussed in Section 17.5).

Horizontal wells are often considered for use where groundwater must be abstracted from beneath inaccessible areas or from areas where the disruption associated with surface drilling is undesirable. Applications have included installation of permanent dewatering systems (see Chapter 20) beneath existing built up areas (Figure 17.12) and extraction of contaminated groundwater or leachate (Cox and Powrie, 2001) without the risks of cross-contamination associated with vertical drilling. HDD wells have also been used as part of dewatering during tunnel construction. HDD wells can act as preferential groundwater flow pathways (essentially as horizontal relief wells) to feed water in a controlled manner to the tunnel face or to a reception shaft ahead of the tunnel drive, from where the water is pumped away (Peter Cowsill Limited, 2001). HDD wells have also been used as recharge wells to re-inject water as part of artificial recharge schemes and also as part of dewatering systems for open pit mines (Mansel *et al.*, 2012).

HDD techniques used for installation of dewatering wells have been developed from techniques originally developed to allow oil and gas pipelines to be drilled beneath obstacles such as roads and rivers (Kummerer, 2015). The methods are essentially based on rotary drilling, with a drilling fluid (mud) used for borehole support, cooling/lubrication of the steerable drill bit and transport of cuttings. Steering is typically achieved using either asymmetric or 'bent sub' drill bits that can drill either along the drilling axis or at a deviation of a few degrees from the axis. The drill string and bit are controlled to allow the bore to be deviated, with the position of the drill bit monitored by magnetic or inertial survey methods.

The main difference between the HDD methods used for dewatering applications and those used for pipeline placement is that the mud used to support the hole during drilling is normally based on biodegradable polymer, rather than bentonite, to minimize the reduction of permeability associated with the drilling process. HDD installation lengths in excess of 1000 m are possible.

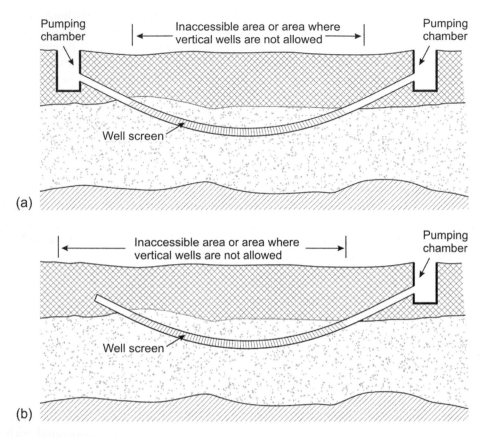

Figure 17.12 Horizontal directionally drilled (HDD) well to pump from beneath urban area. (a) Double-entry completion. (b) Single-entry completion.

The successful design and installation of HDD wells is challenging, and the unit cost of a typical HDD well will be high due to the length of the well and the complexity of the operation. It is essential that experienced designers and contractors be consulted when planning HDD well systems.

HDD wells are categorized as double-entry completions or single-entry completions based on the installation method.

Double-entry completions (Figure 17.12a): A pilot hole is drilled from a launch pit. The bore typically is angled down on entry, is then steered to run approximately horizontally beneath the target zone, and then is steered upwards to emerge at a reception pit. The pilot bit is then removed from the drill string and replaced with a larger reamer bit, which is then pulled back through the borehole to enlarge it to the required diameter. In some configurations, additional drill rods are added at the reception pit behind the reamer as it is pulled through. A series of reamers of progressively increasing size may be used to enlarge the hole to its final diameter. Double-entry completions are the most commonly used method, because better control of hole stability is possible compared with single-entry completions.

Single-entry completions (Figure 17.12b): This method (also known as a blind-ended hole) is used when there is no access for a reception pit and the well must be installed from one end only. Once the pilot hole is drilled, the bit is withdrawn, and a succession of reamers of progressively increasing size are used to enlarge the hole. Because reaming is carried out by pushing the bit (rather than pulling it as in double-ended completions), there is a risk that the enlarged hole will not follow the pilot bore. Well lengths that can be achieved with single-entry completions are significantly lower than for double-entry methods.

The installation of well screen is much more difficult in HDD wells than for conventional vertical wells, primarily due to the long length of the wells and the fact that the wells are deliberately deviated in direction. This results in large tensile stresses on the screen during installation.

The string of well screen is normally installed by being pulled through the hole by the drill rods (in single-completion wells, the well screen must be pushed in from the entry pit). Well screens are normally formed from HDPE, carbon steel or stainless steel. The percentage open area of the screen is typically lower than for vertical wells to ensure high tensile and compressive strength of the screen. It is difficult to install a conventional granular filter pack around an HDD well screen, so it is common practice instead to use pre-packed gravel filter, mesh or geotextile filter screens (see Section 16.3.3) suitably protected to survive the screen being pulled through the well during installation.

Development of HDD wells after completion of drilling is also problematic, because the small open areas of the well screen and the types of filters used mean that much of the development energy is dissipated in the screen and filter system and never reaches the ground. Typically, water jetting and intermittent pumping and flushing with pressurized water are used to develop HDD wells.

17.5 SUB-HORIZONTAL DRAINS

Sub-horizontal drains are a method of passive drainage used to reduce pore water pressures in rock slopes (and occasionally in soil slopes). The objective of the drains is to provide preferential drainage pathways to allow groundwater to flow from the face of a slope (see Sections 8.5 and 8.7.4). As described in Section 9.7, the method is widely used to aid the depressurization of rock slopes in open pit mines (Leech and McGann, 2008; Beale and Read, 2013), but it is also used in rock excavations for civil engineering purposes.

A common geometry is to install the drains in multiple lines at different levels within the excavation (Figure 17.13). Within each line, the drains are drilled either horizontally or with a slight upward inclination, from benches in the pit slopes, with a drilled diameter of 75 to 125 mm.

Different forms of drain construction are shown in Figure 17.14:

(a) In competent rock, the drain holes can be left unlined if the geological data indicates that they are unlikely to collapse (Figure 17.14a).
(b) In less competent material, a slotted well screen is installed in the bore of the drain on completion of drilling. In the simplest form, the well screen is simply brought to the slope face, with no seal, and water flows from both the screen and the annulus around it (Figure 17.14b).
(c) In some cases, uncontrolled flow of water from the end of drains may cause softening, erosion or instability of the rock immediately behind the slope. In these cases, the annulus around the well screen should be sealed by grout or with a packer (Figure 17.14c). This can allow flexible pipes to be connected to the drain to direct flow to drainage ditches without running down the slope face. If a valve is fitted to the end of the well screen, the drain can be used as a piezometer by closing the valve and observing the water pressure using a pressure gauge or vibrating wire piezometer (VWP) transducer.

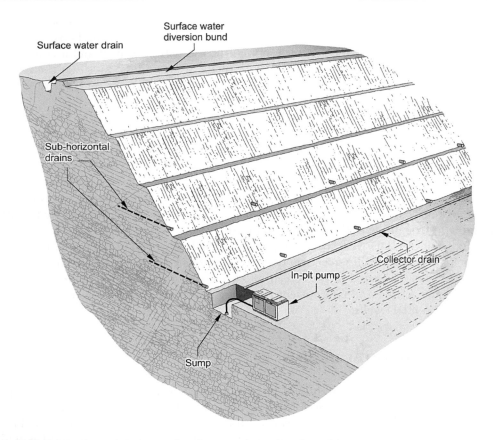

Figure 17.13 Sub-horizontal drains used to depressurize a slope in rock.

This technique is a passive drainage method, and the drains are not pumped directly. The water pressure in the slope drives water along the drain, so that water bleeds from the open end of the drain and is typically collected in drainage channels and pumped away by sump pumps. Where lines of drains are relatively high up on the slope, and the drainage channel is on a bench, there is a risk that water seeping from the base of the channel may destabilize lower sections of the slope. It may be necessary to line the channel (with concrete or geosynthetic membranes) to prevent this happening.

Because flow is driven by the hydraulic gradients, the physical gradient of the drains is not critical, and in theory, horizontal drains can be effective. In practice, drains are often drilled into the slope face with an upward inclination of 5° to 10°. This gives some self-cleaning effect to help remove any sediment drawn into the drain and reduce the risk of blockages. Drilled lengths of drains can vary widely, depending on the slope geometry, but 10 to 100 m is typical. Shorter drains can be drilled by relatively conventional drilling rigs fitted with articulating booms to allow horizontal drilling. Longer drains may be drilled by specialist HDD rigs.

Because sub-horizontal drains can only be installed from within the excavation, their installation sequence must be integrated with the excavation programme to ensure that the required slope depressurization is achieved before the excavation is taken deeper. For typical applications in fractured rock, the drains will flow copiously when first installed, but the flow will reduce as the slopes are depressurized; drains in the upper sections of the slopes may dry up completely.

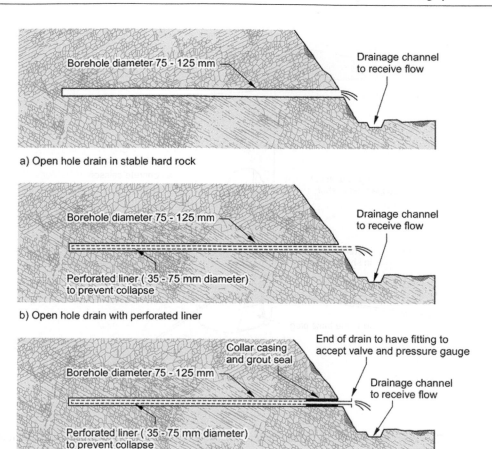

a) Open hole drain in stable hard rock

b) Open hole drain with perforated liner

c) Drain with perforated liner and collar seal

Figure 17.14 Different types of sub-horizontal drain. (a) Open hole drain in stable rock. (b) Open hole drain with perforated liner. (c) Drain with perforated liner and collar seal. The drains are shown with a horizontal alignment, but typically are drilled into the slope face with an upward inclination of 5° to 10°.

17.6 COLLECTOR WELLS

A collector well consists of a vertical shaft typically 3 to 6 m in diameter, sunk as a concrete-lined caisson, from which laterals (horizontal or sub-horizontal screened wells, typically of 100–250 mm diameter) are jacked or drilled radially outwards (Figure 17.15). Collector wells are sometimes known as Ranney wells after a proprietary system (first developed by Leo Ranney in 1933) used to form this type of well (Drinkwater, 1967). Ranney's concept was that by placing well screens horizontally into a permeable formation, a much greater length of screen would be productive, and therefore a collector well would have a much greater yield than a conventional vertical well.

A collector well is pumped by a submersible or lineshaft pump located in the shaft, which lowers the water level in the well. This creates a pressure gradient along the laterals, causing them to flow freely into the main shaft. This method is particularly suited to large-capacity permanent water supply installations in shallow sand and gravel aquifers. Water supply installations are described in Hunt (2002) and Hunt *et al.* (2002).

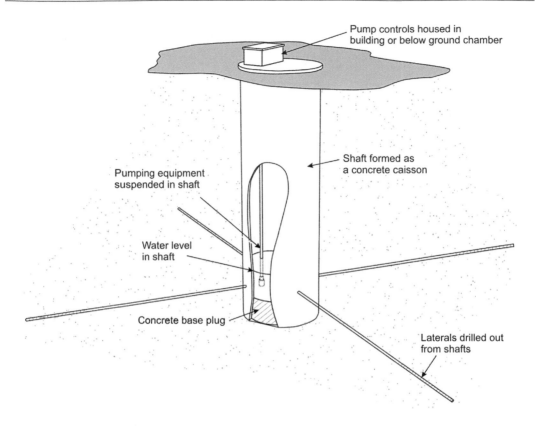

Figure 17.15 Collector well. Pumping from the shaft of the collector well creates an hydraulic gradient along the lateral wells, which draws water into the shaft.

Collector wells are rarely used for groundwater control operations. In general, the cost of constructing the central shaft and forming the laterals is likely to be prohibitive for temporary works applications (some of the challenges of drilling out from shafts and tunnels below groundwater level are discussed in Section 10.6.1.2). Nevertheless, the method has occasionally been used, for example to aid dewatering of tunnel crossings beneath roads and railways where access to install conventional wells was restricted (Harding, 1947). In the early 2000s, four collector wells were used to form a long-term dewatering system beneath a tunnel in Glasgow in Scotland (Schünmann, 2005). This project is described in Case History 27.12 in Chapter 27.

17.7 DRAINAGE TUNNELS AND ADITS

In the section on groundwater control for tunnelling in rock (Section 8.8), we saw that a rock tunnel will often act as a drain and receive groundwater inflow. The earlier sections concentrated on the problems this can cause during tunnel construction. However, in some circumstances, it may be possible to use the drainage effect of a tunnel as a groundwater lowering measure in its own right.

This leads to the approach of forming a drainage tunnel (also known as a drainage adit or drainage gallery), where a gently sloping tunnel is driven through rock beneath the zone where groundwater depressurization is required (Figure 17.16a). The construction of a

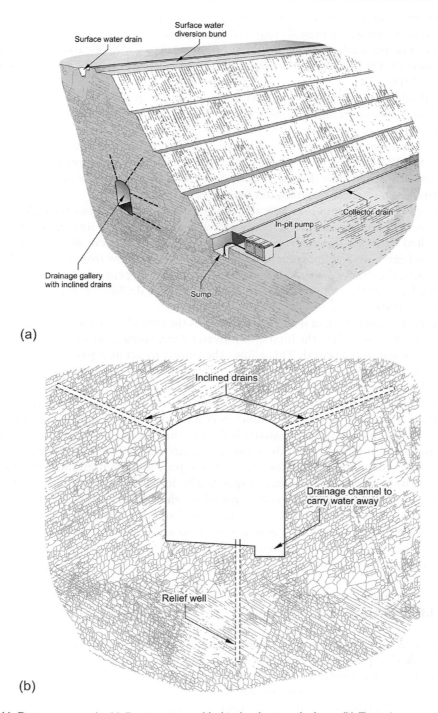

Figure 17.16 Drainage tunnels. (a) Drainage tunnel behind a deep rock slope. (b) Typical section through a drainage tunnel showing inclined drains and relief wells.

tunnel for drainage purposes is a major task and should not be undertaken lightly. Due to the high cost and long timescale required to construct a drainage tunnel, the method is rarely economic on small-scale or short-term projects. Typical applications are to stabilize landslip zones (Siddle, 1986) or to depressurize around and beneath open pit mines (Bar and Baczynski, 2018).

The method is typically applied in fractured rock and requires the construction of a tunnel (which may be several hundred metres long) behind a slope or beneath an open pit mine. Groundwater control will be required during tunnel construction, as discussed in Section 8.8. In most cases, as the tunnel face is advanced, drilling teams follow up and drill out from the tunnel to install inclined drains (similar to those described in Section 17.5 but at steeper inclinations) and relief wells (Figure 17.16b). Some of the inclined holes can be installed as VWP piezometers to allow groundwater pressures to be monitored.

During operation, the tunnel is typically at atmospheric pressure (occasionally, a vacuum is applied to the entire tunnel), so groundwater will drain passively into the tunnel. In rock that is faulted or has a dipping structure, the tunnel itself will cut across the geological structure and can have an important drainage effect by direct flow from fractures. However, in many cases, the drains drilled out from the tunnel are the primary drainage means, and the principal role of the tunnel itself is to allow access for drilling of drains and to convey the water to a disposal point.

The method is most efficient if topography allows the tunnel to be constructed with a portal at an elevation lower than the final groundwater depressurization level, so the tunnel can slope gently down to the portal. This allows the tunnel to act as a passive (i.e. unpumped) drain with water running down the tunnel by gravity to flow out of the tunnel portal. A drainage channel is sometimes cut in the base of the tunnel, with the tunnel floor sloping toward the channel to concentrate the flowing water in one area to make maintenance access easier (Figure 17.16b). If a low-level tunnel portal is not possible, drainage tunnels can be pumped via sump pumps placed in shafts linked to the tunnel. Drainage tunnels are often required to function for long periods of time, and safe access is required for maintenance; safe systems of work should allow for the confined space and the need for ventilation.

Drainage tunnels may be constructed specifically for drainage purposes, with the tunnel dimensions controlled by the access required for the rigs to drill the drains – minimum tunnel heights of 2 to 4 m are typically required to allow drain drilling. However, on some mining projects, tunnels originally constructed for access or exploration purposes have been identified as having a significant drainage effect and have subsequently been incorporated into mine depressurization programmes.

17.8 RELIEF WELLS

When an excavation is made into a low-permeability soil or rock above a confined aquifer, there is a risk that pore water pressures in the confined aquifer may cause the base of the excavation to become unstable. There is a risk of 'hydraulic failure', where the base of the excavation may 'heave' upward because the weight of soil remaining beneath the excavation is insufficient to balance the uplift force from the aquifer pore water pressure (see Sections 7.5 and 8.6). One way of avoiding this potential instability is to reduce pore water pressures in the aquifer by pumping from an array of deep wells (or for shallow excavations, wellpoints) located outside the excavation.

Such pumped well systems are termed 'active' because water is pumped directly. The contrasting system of relief wells (also known as pressure relief wells or bleed wells) offers an alternative method of reducing pore water pressures in confined aquifers. These systems are

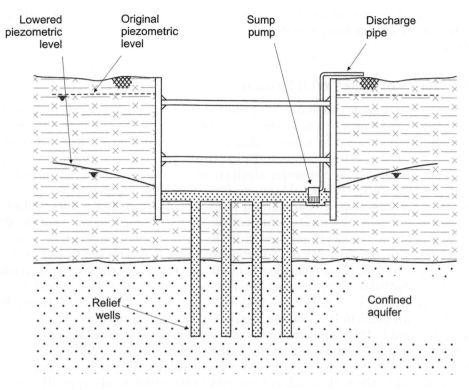

Figure 17.17 Relief well system. The relief wells overflow into a granular drainage blanket in the base of the excavation. The water is pumped away from a sump in the drainage blanket.

'passive'; this means that they are not directly pumped but merely provide preferential pathways for water from the aquifer to 'bleed' away, driven by the existing groundwater heads.

A schematic view through a typical relief well system is shown in Figure 17.17. The wells are normally drilled prior to commencement of excavation, or at least before the excavation has progressed below the piezometric level in the aquifer. As excavation continues, the wells will begin to overflow, relieving pore water pressures in the aquifer and ensuring stability. The water flowing from the relief wells is typically disposed of by sump pumping (Chapter 14).

The purpose of relief wells is to provide an engineered preferential pathway for upward groundwater flow from pressurized strata beneath an excavation or a structure. The method is applied in two principal cases:

1. For an excavation, relief wells must be located within the area of the dig. As excavation progresses, groundwater will flow up the relief wells into the excavation, thereby reducing pore water pressures below the base of the excavation. The water entering the excavation is typically disposed of by sump pumping.
2. For a below-ground structure such as a basement, relief wells can be used for long-term reduction of groundwater pressures below the structure. Relief wells should be located within the structure and overflow into a drainage system, from where the water is pumped away (long-term groundwater control systems are described in Chapter 20).

The discussion of relief wells in this section will be mainly restricted to wells of relatively large diameter (greater than 100 mm) formed by drilling or jetting and backfilled with sand or gravel. This type of relief well is distinct from the smaller-diameter vertical path-ways that can be formed using prefabricated vertical drains using the methods discussed in Section 17.9.

17.8.1 Merits of Relief Well Systems

When used in appropriate conditions, the principal advantages of relief well systems are cost and simplicity. The simplest relief wells consist of gravel-filled boreholes (Figure 17.19a); since the wells do not need to accommodate pumps or well screens, they can be of mod-est diameter, reducing drilling and installation costs. The water flowing from the wells is removed by conventional sump pumps, which are more readily available and more robust in use than borehole electro-submersible pumps used to pump from deep wells.

Relief wells are best employed in shafts or deep cofferdams where the sides of the excava-tion are supported by a structural cut-off wall or other retaining structure and the stability of the excavation base is the primary concern. The method is most appropriate to use where the excavation base is in stiff clay or weak rock (such as Chalk, soft sandstones or fractured mudstones). Because the water from the relief wells overflows onto the excavation formation, there may be a risk of the water causing softening of exposed soil or rock (especially in clays that have a permeable fabric). This can lead to difficult working conditions. It may be pos-sible to avoid this problem by installing a granular drainage blanket and network of drains (Figures 17.17 and 17.18) to direct water to the sumps and prevent ponding in the excavation.

A key question when considering a relief well system is: how many wells of a given diam-eter will be required? An initial stage is obviously to estimate the rate at which groundwater must be removed by the wells to achieve lowering to formation level – typically, this is esti-mated by treating the excavation as an equivalent well (see Section 13.6.1). Theory suggests that the capacity of a gravel-filled vertical drain can be estimated by direct application of Darcy's law (see Section 3.3):

$$Q = kiA \tag{17.1}$$

where
  Q is the vertical flow rate along a relief well (m³/s)
  k is the permeability of the gravel backfill (m/s)
  A is the cross-sectional area of the well (m²)
  i is the vertical hydraulic gradient along the well

The vertical hydraulic gradient in the relief well can be estimated from the length of the well and the head difference between the zone being pressure relieved and the excavation. This will give the theoretical maximum vertical flow capacity of a gravel filled well. These values are theoretical maximums, but field experience suggests that such flows are rarely achieved in practice. In fact, the actual capacity may be significantly less than the theoretical capacity for a variety of reasons, including smearing or clogging of the borehole wall during drilling, or segregation of the filter material during placement.

While the number of relief wells required must consider the well capacity, some thought must also be given to the spacing between wells. If small flows are predicted, only a few wells may be necessary to deal with the volume of water. However, it may be prudent to install additional relief wells to ensure that the distance between wells is not excessive (say, no greater than 5–10 m). If the wells are widely spaced and the ground conditions may be variable (especially if the wells are installed into fractured rock), there is a danger that the wells may not adequately intercept sufficient permeable zones or fractures. This could

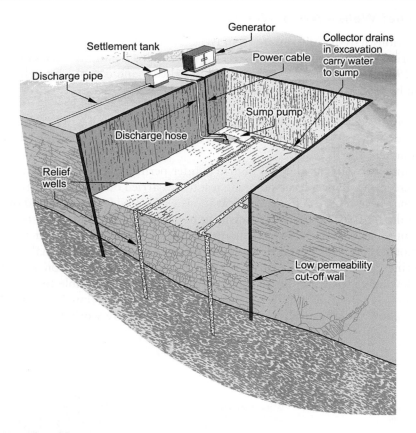

Figure 17.18 Relief wells linked to a sump pumping system.

lead to unrelieved pressures remaining, with the possibility of local base heave in the areas between the wells.

17.8.2 Installation Techniques

Relief wells are typically drilled by similar methods to deep wells: cable percussion boring, rotary drilling, jetting or auger boring (see Chapter 16). The borehole is drilled to full depth, any drilling fluids used are flushed clean, any well screen is installed, and then the filter media (sand or gravel) are added to backfill the bore up to the required level. The diameter of drilling is typically between 100 and 450 mm.

The filter media may be placed in the well via a tremie pipe or may be simply poured in from ground level. This latter approach is acceptable provided that the filter media has a very uniform grading so that there is little risk of segregation of the filter particles as they settle to the bottom of the bore. In practice, many relief wells installed in weak rocks, where the performance of the filter is not critical, are filled with uniform coarse gravel. Ideally, to permit maximum transmission of water, the gravel should be of the rounded pea shingle type of 10–20 mm nominal size. However, on remote sites, it may be necessary to use locally available material (which may consist of angular crushed aggregates) and accept some reduction in well efficiency. In certain cases, where the long-term performance of the relief wells is critical, the gravel used may need to be designed as a filter medium to match soil conditions in a similar way to deep wells (see Section 13.10). If no well screen is installed, it is not normally possible to develop the relief well.

17.8.3 Relief Wells – Are a Well Casing and Screen Needed?

Previous sections have mainly discussed gravel-filled relief wells, but there are cases when it is appropriate to install casings and screens (surrounded by a gravel pack) in relief wells (Figure 17.19b). Although they introduce additional cost and complexity, relief wells with screens can be used in the following cases:

a) If the confined aquifer is of high permeability, a screened well will have a greater vertical flow capacity than a purely gravel filled well. The use of screened wells may reduce the number of wells required.

b) If a pumping test is needed to confirm aquifer permeability and flow rate, the casing and screen can allow a basic pumping test to be carried out using a submersible pump. This may be appropriate for large excavations where a widely spaced grid of relief wells is installed, based on an initial assessment of permeability. Pumping tests are then carried out on some of the wells to estimate the actual permeability and determine whether additional relief wells are needed to fill in the gaps between the original wells.

c) If during critical stages of construction (e.g. during casting of concrete structures), overflowing water would be an inconvenience, it may be possible to install submersible pumps in the wells and temporarily lower water levels below formation level. This would allow the critical activities to be completed in more workable conditions.

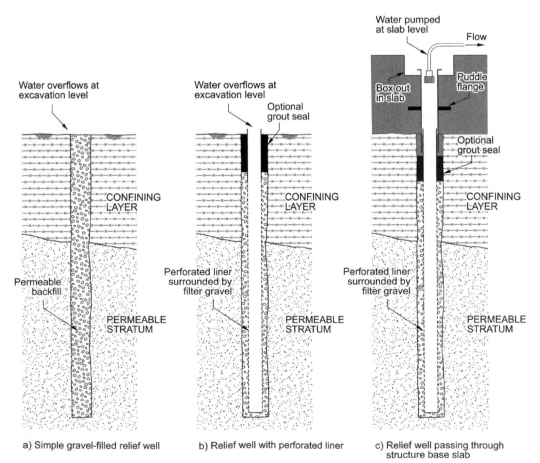

a) Simple gravel-filled relief well   b) Relief well with perforated liner   c) Relief well passing through structure base slab

Figure 17.19 Alternative designs of relief wells. (a) Simple gravel-filled relief well. (b) Relief well with perforated liner. (c) Relief well passing through structure base slab.

To be effective, relief wells must be located within the excavation. If the structure being constructed completely fills the excavation (e.g. where a shaft is formed within concrete diaphragm walls or secant pile walls), then the relief wells must be within the structure. In many cases, the relief wells need to remain in operation until the works have been completed to a stage where the weight of the structure is sufficient to resist the original (unrelieved) groundwater pressures. In this case, the relief wells must pass through the reinforced concrete base slab of the structure (Figure 17.19c). Typically, vertical steel ducts are cast into the slab, with waterproofing measures (such as puddle flanges) used to reduce risk of the ducts creating a pathway through the slab (Figure 17.20). The requirements for puddle flanges (and similar measures) where wells penetrate the base slab of a structure is discussed in Section 24.4 in relation to decommissioning of wells.

Relief wells are sometimes also used as part of the permanent works to provide long-term pressure relief for deep structures (e.g. basements or deep railway cuttings) constructed above confined aquifers (see Section 20.2). Permanent relief wells are often installed with screen and casings to allow the wells to be cleaned out and re-developed if their performance deteriorates after several years of service.

17.9 VERTICAL DRAINS

There is sometimes the need to create engineered vertical flow paths to enable downward groundwater flow. This is achieved by the installation of vertical drains – these are effectively the mirror image of relief wells. The most common application of vertical drains is to reduce drainage path lengths to accelerate the rate of consolidation when thick layers of fill are placed over compressible alluvial soils (Institution of Civil Engineers, 1982). In the context of groundwater lowering, vertical drains can be used to help drain perched aquifer

Figure 17.20 Relief wells passing through base slab of structure under construction. (Courtesy of MVB Joint Venture, Lee Tunnel, London, UK.) Four relief wells are fitted with vertical pipes used to duct the well casing through the reinforced concrete base slab; shown under construction in the photograph.

conditions where an intervening aquiclude prevents groundwater from draining down to a lower aquifer, which is being dewatered (see Figure 15.18 and Figure 17.21). The water trapped in the upper aquifer may threaten the stability of excavation slopes or cause seepage into structures unless it is drained.

The two most commonly used designs of vertical drains are:

i. Pre-fabricated vertical drains (PVDs) – Figure 17.22a: A mandrel (a stiff, small diameter casing, closed at its base), is driven into the ground, taking with it a narrow, flexible PVD, which comprises a permeable core surrounded by filter geotextiles (Figure 17.23). The mandrel is withdrawn, leaving the PVD in place.

ii. Sand drains – Figure 17.22b: A borehole is formed by drilling, jetting or driving a mandrel, which is then filled with sand or gravel of high permeability. Occasionally, a slotted well screen is installed in the drain (Figure 17.22c). This has the effect of significantly increasing the vertical flow capacity of the drain and also allows the water levels in the drains to be recorded in order to help monitor the performance of the drains.

Each drain has relatively low capacity for vertical flow. It is usually necessary to install the drains in lines or in a grid pattern at a relatively close spacing between drains (typical spacings are in the range of 0.5 to 3 m).

17.10 SLOPE DRAINAGE

Groundwater seepage from slopes can have a significant destabilizing effect in both soils and rocks. Even where groundwater levels are lowered by pre-drainage methods such as

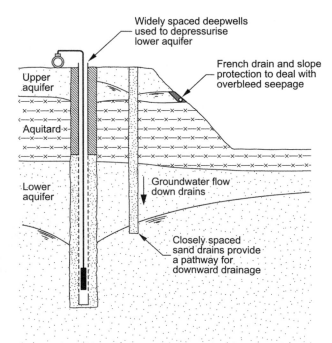

Figure 17.21 Vertical drains used to promote the drainage of perched water downwards into lower dewatered aquifer.

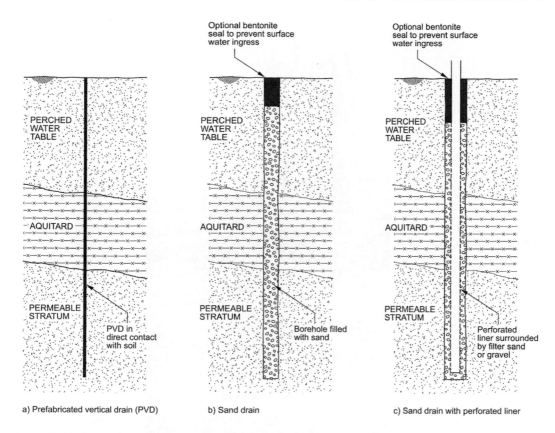

Figure 17.22 Types of vertical drains. (a) Prefabricated vertical drain (PVD). (b) Sand drain. (c) Sand drain with perforated liner.

wellpoints, deep wells or ejector wells, localized seepages and perched water tables can still cause problems (see Sections 7.6 and 8.7).

Slope drainage techniques cover a range of passive drainage methods based on the principle of constructing permeable zones or features on or within slopes to allow water to drain more easily from the ground. Typically, after emerging from the slope, the water will gravitate down the slope and will reach a drain at the toe from where the water will ultimately be discharged.

Slope drainage is applied in various circumstances:

(i) Temporary applications in low-permeability soils and rocks as the primary groundwater drainage measure to stabilize slopes for construction excavations

(ii) Temporary applications in higher-permeability soils and rocks, as an auxiliary drainage measure where groundwater is lowered by sump pumping or pre-drainage methods, when there is a need to manage residual seepage (such as overbleed or perched water) emerging from the slope

(iii) Permanent (i.e. long-term) applications in low-permeability soils and rocks, installed during construction as a planned design measure to be the primary groundwater drainage method to ensure the stability of slopes that form part of structures and infrastructure such as road and rail cuttings (see Section 20.3.4)

(iv) Permanent (i.e. long-term) applications in low-permeability soils and rocks, installed at some stage after construction as part of remedial drainage measures following slope failure or problematic ground movements

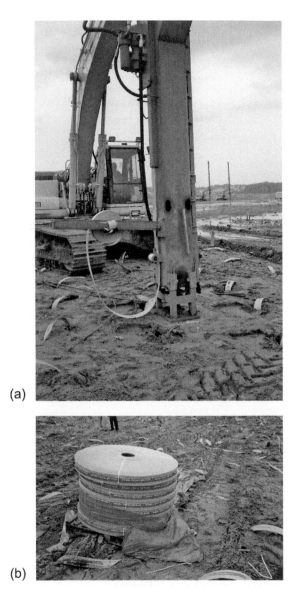

(a)

(b)

Figure 17.23 Prefabricated vertical drains. (a) Vertical drain installation rig for installation of prefabricated vertical drains (PVDs). A mandrel is used to place PVD to the required depth. The tops of several previously installed vertical drains can be seen protruding from the ground around the rig. (b) Reel of prefabricated vertical drains (PVD) prior to installation. The tops of several previously installed vertical drains can be seen in the surrounding ground surface.

The control of surface water (Section 14.3) also plays an important role in slope drainage, on the slope itself, on the ground at the toe and on the ground surface close to the crest of the slope.

Slope drainage methods are most applicable in soils and rocks of relatively low permeability (lower than around 10^{-6} to 10^{-7} m/s). If passive drainage is attempted for slopes cut in highly permeable strata, groundwater flow rates may be too high, and active drainage (e.g. by wellpoints, deep wells or ejector wells), or groundwater exclusion methods may be more appropriate. Slope drainage methods are described in more detail in Hutchinson (1977) and Carder *et al.* (2008).

17.10.1 Toe Drains

When groundwater emerges from a slope, either by direct seepage or from passive slope drains, the water will naturally gravitate down the slope. If this water is not collected and removed, it will pond at the toe. As a minimum, this will be a nuisance and will lead to wet and messy working conditions; in some circumstances, instability may occur at the toe of the slope. It is good practice to install a drain at the toe of the slope – a 'toe drain' – to collect the water and carry it to discharge either by gravity (if topography allows) or by sump pumping.

Common arrangements for toe drains are:

(i) A shallow gravel-filled trench (which may be lined with a filter geotextile and contain a perforated pipe) – this type of drain has the advantage that it can also receive some direct groundwater flow, further draining the toe of the slope.

(ii) An open channel or ditch – typically used in very low-permeability soils and rocks, where the main objective is to collect surface water and small quantities of seepage running down the slope, and significant direct groundwater seepage is not expected.

Typically, a toe drain is constructed close to the toe of the slope, with the intention of collecting water seeping from the slope and through the ground immediately below the toe (Figures 17.24 and 17.25a). It is essential that the gradient and outfall level of toe drains are selected to avoid water backing up in the drain. Water standing in the drain could result in increased infiltration into the ground and localized waterlogging and instability of the lower sections of the slope.

Toe drains used for construction purposes can vary widely in accordance with local practices and the preference of the site team. Where a trench drain is used, the basic requirement is a backfilled drainage trench containing a perforated carrier pipe. Sometimes, the carrier pipe is laid on a bed of blinding concrete to prevent any standing water in the drain infiltrating into the ground. Toe drains may be deployed in conjunction with sand-bagging of the toe of the slope or weighting with graded filter material. Alternatively, the fill can be weighted with loose-laid timbers and sacking with straw or hay padded behind them. Modern practice also allows the use of one of the various proprietary geotextile fabrics. The materials to be used will depend on local availability in the country concerned. It is advantageous to place the filter material back up the slope a short distance and so prevent seepage in that area from transporting fines (Figure 17.24). Periodic maintenance may be needed to ensure that toe drains remain efficient.

Where an open channel or ditch is used as a toe drain, there is a risk of water seeping into the ground. It is good practice to line the channel with a thin layer of concrete or a geosynthetic membrane. Often, sediment and other detritus can build up at the toe of a slope, and this may accumulate in the channel, requiring periodic maintenance cleaning to restore flow capacity.

Toe drains can form an important part of surface water control measures within excavations. Ideally, gradients of the excavated floor areas should be formed so that run-off water is directed to collector drains (including toe drains), and the risk of ponding should be assiduously avoided. The various drains should feed to sumps, usually sited at the corners of excavations below the general excavation level, and made big enough to hold sufficient water for pumping and keep the excavation floor relatively workable and free from water. A pump should be provided for each sump and connected to a discharge pipe (types of sump pump are described in Chapter 19).

The drains leading to the sumps should be so arranged as to allow drainage of the whole excavation and have sufficient fall to prevent silting up. Channels and ditches should be

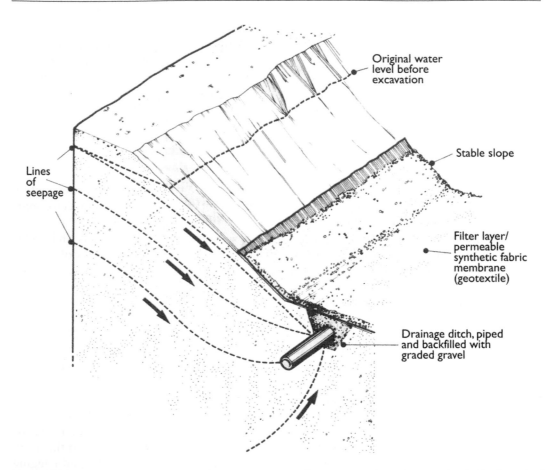

Figure 17.24 Toe drain of the gravel filled trench type. (Reproduced from Somerville, S H, *Control of Groundwater for Temporary Works*, Construction Industry Research and Information Association, CIRIA Report 113, London, 1986. With permission from CIRIA: www.ciria.org)

maintained by cleaning out from time to time. Ditches should be sufficiently wide to allow a water velocity low enough to prevent erosion. This may be achieved by constructing check weirs at intervals along the ditch. Additionally, the ditches can be improved by laying concrete blinding, paving material, porous open-jointed pipes or other agricultural land drainage piping surrounded by filter material.

The water pumped from the sumps fed by toe drains of the gravel-filled trench type should be clear – i.e. no fines. The presence of fine particles in the pumped seepage water would indicate that the filters around the drain(s) are inadequate, that fines are being continually removed, and eventually the slope will become unstable.

17.10.2 Drainage Measures at the Crest of the Slope

Water infiltration in the vicinity of the crest of a slope can cause significant stability problems. If water is allowed to pond above a slope, it may infiltrate into the ground, reducing the stability of the slope. If surface water from the surrounding area is allowed to run down the slope, this has the potential to cause erosion at the crest and on the slope itself. Furthermore, if any tension cracks are present near the crest, these can become filled with water, potentially destabilizing the slope.

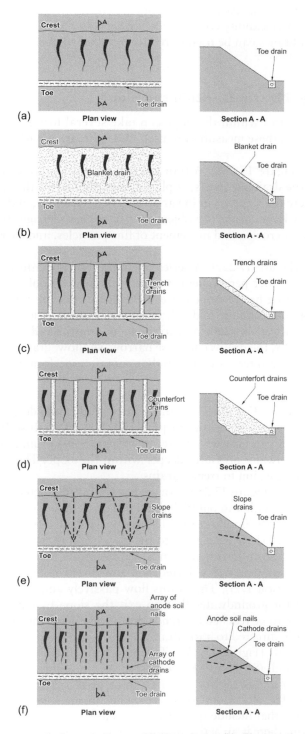

Figure 17.25 Common types of slope drainage. (a) Toe drain. (b) Blanket drain. (c) Trench drains. (d) Counterfort drains. (e) Sub-horizontal drains. (f) Electro-osmotic drains. (panel (f) after Alder, D, *et al.*, Design principles and construction insights regarding the use of electrokinetic techniques for slope stabilisation, *Proceedings of the XVI ECSMFGE, Geotechnical Engineering for Infrastructure and Development* (Winter, M C, Smith, D M, Eldred, P J L and Toll, D G, eds.), ICE Publishing, London, pp. 1531–1536, 2015.)

Lined collector trenches, set back slightly from the crest, can be used to divert water away from the slope. Low-permeability cover layers (see Section 18.13) such as blinding concrete or geosynthetic membranes can be used to reduce infiltration of surface water into the area near the crest of the slope.

17.10.3 Drainage Measures within the Slope

Passive drainage measures within the slope can take several forms, which may be used in combination. Typically, these measures are used in combination with a toe drain:

(i) Blanket drain (Figure 17.25b). A thin layer of permeable granular filter sand/gravel, and/or geotextile drainage layers, is placed over the face of the slope. The method is often used where the slope is cut in variable or layered strata, and diffuse seepages are anticipated over large areas. The objective is to collect seepages from several areas of the slope, prevent erosion and movement of fine particles, and carry the water to a toe drain.

(ii) Trench drains (Figure 17.25c). A series of linear trench drains are installed running down the slope, either perpendicular to the toe or at an oblique angle to the toe to form an intersecting herringbone pattern. Trench drains are typically shallow (0.6 m deep or less, but occasionally to depths of up to 2 m) to provide a pathway for shallow groundwater to leave the slope and to intercept run-off and control infiltration into the slope. Typically, the drains comprise narrow, shallow, gravel-filled trenches (which may be lined with a filter geotextile and/or contain a perforated pipe) and feed water to a toe drain.

Counterfort drains (Figure 17.25d). These drains are significantly deeper than trench drains with the objective of penetrating below any potential slip surface. Installed perpendicular to the toe of the slope, they can provide a degree of strengthening or mechanical buttressing as well as lowering groundwater pressures deeper within the slope. These drains discharge to a toe drain in a similar fashion to trench drains.

Sub-horizontal drains (Figure 17.25e). An array of small-diameter (50 to 100 mm) holes are drilled into the slope, either in parallel alignment at right angles to the toe or forming a radial pattern from each drilling location. The holes are drilled out horizontally or with a slight upward inclination (5 to 10 degrees to the horizontal is typical). In competent rock, the drain holes can be left unlined if geological data indicate that the bores are unlikely to collapse. In less competent material, a slotted well screen is installed in the drain on completion of drilling (see Section 17.5). The drains flow passively and are intended to provide a preferential pathway for groundwater flow to exit the slope by connecting to permeable zones that are too deep to be reached by trench drains or counterfort drains. Water bleeding from drains is typically directed to a toe drain.

Electro-osmotic drains (Figure 17.25f). This is a specialist approach that uses electro-osmotic drainage methods (Section 17.12) combined with sub-horizontal drains. The method is described in Alder *et al.* (2015) and requires an array of sub-horizontal electrodes to be drilled into the slope. The electrodes comprise anodes (formed by steel soil nails) and cathodes (formed by drains that include specialist electrically conductive geotextile filter mesh). Following installation, there is an active drainage phase where a direct current (DC) voltage is applied across the electrodes to promote drainage from low-permeability soils, with the water being drawn to the cathodes by electro-osmotic forces. Following the active drainage phase (which may last several months), the voltage is discontinued and the cathodes act as passive drains, while the anodes remain in place to reinforce the slope.

Additionally, vertical drains (Section 17.9) and siphon drains (Section 17.11) are passive methods that have been successfully applied to long-term drainage of slopes and landslide stabilization projects.

Where slope drainage is used to control groundwater in the long term for infrastructure slopes (such as for road and rail cuttings) or as part of landslide stabilization measures, the drainage systems may clog or become blocked by detritus, and performance may deteriorate. Long-term infrastructure drainage systems will require appropriate monitoring and maintenance; further details can be found in Spink *et al.* (2014).

17.11 SIPHON DRAINS

The technique of siphon drains is a passive drainage method developed in the 1980s to allow water to be pumped by gravity (without the need for external energy input) via permanently primed siphon pipes (Mrvik *et al.*, 2010). This approach has found useful applications in landslide drainage and stabilization schemes where long-term pumping at low flow rates is required, and where the sloping terrain allows the siphon system to be arranged at appropriate levels (Bomont *et al.*, 2005).

Siphon drain systems are used to pump from wells located in the zone that is to be drained or depressurized. Wells are typically at approximately 5 m spacing. The suction end of the siphon pipe is installed in the well in a container so that the end of the pipe remains submerged below water even if the water level in the well falls below the level of the siphon pipe inlet. The top of the container is set at the target drawdown level in the well (known as the 'reference level'). The siphon pipe is connected at ground level to an accumulator located down slope at the reference level (Figure 17.26). The siphon pipe is usually installed in a duct with the accumulators in an outlet manhole, all at a suitable depth below ground level for frost protection. The purpose of the accumulator is to maintain the prime in the siphon.

Once primed, when the water level in the well is above the 'reference level', water will flow through the siphon (to be discharged at the outlet). Flow will continue until the water level in the well falls to the 'reference level', when flow will slow and stop. At the top of the siphon

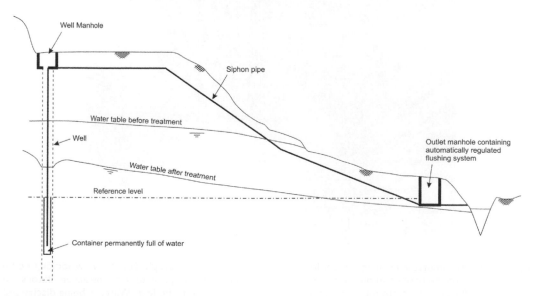

Figure 17.26 Siphon drain system.

tube, the water pressure will be low, and gas may come out of solution, creating bubbles. These gas bubbles are flushed out when the siphon is flowing, but if the flow slows and stops, the gas bubbles may build, potentially leading to loss of prime in the siphon. When the water level in the well rises, the accumulator automatically re-primes the siphon, flushing out any accumulated gas bubbles in the line, and re-starts the flow. Typically, the siphon pipes from several wells would be plumbed into a single outlet manhole. Figure 17.27 shows the accumulator flushing system from several wells in operation in an outlet manhole.

The siphon drain technique is suited to soils of relatively low permeability (lower than 1×10^{-5} m/s). Depending on the size of the equipment, flow rates of up to 15 l/min (0.25 l/s) are possible. Drawdowns of up to 8.5 m can be achieved below the top of the siphon pipe.

17.12 ELECTRO-OSMOSIS

Electro-osmosis is suitable for use in very low-permeability soils such as silts or clays with no permeable fissures or fabric, where groundwater movement under the influence of hydraulic gradients created by pumping would be excessively slow. Electro-osmosis causes groundwater movements in such soils using electrical potential gradients rather than hydraulic gradients (see Section 4.2.3). A direct current (DC) is passed through the soil between an array of anodes and cathodes installed in the ground. The potential gradient causes positively charged ions and pore water around the soil particles to migrate from the anode to the cathode, where the small volumes of water generated can be pumped away by wellpoints or ejectors (Figure 17.28). The method can reduce the moisture content of the soil, thereby increasing its strength. In many ways, electro-osmosis is not so much a groundwater control method as a ground improvement technique. One of the drawbacks of the method is that it is a decelerating process, becoming slower as the moisture content decreases.

Figure 17.27 Accumulator flushing system for siphon drains. The photograph shows a view looking down into the base of an outlet manhole for a system of several siphon drains. The accumulators are flushing the system to remove gas that may cause prime to be lost. Water is being discharged into the base of the outlet manhole. (Courtesy of TPGEO.)

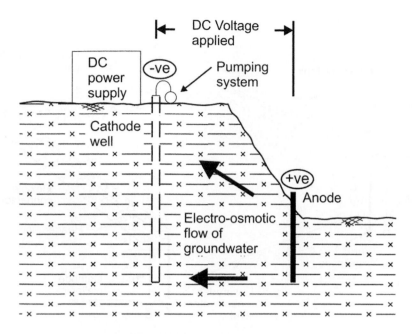

Figure 17.28 Principle of electro-osmosis applied to stabilization of an excavation.

Electro-osmosis is a very specialized technique and is used rarely, mainly when very soft clays or silts need to be increased in strength – Farmer (1975) indicates that the technique becomes relevant at permeabilities lower than around 1×10^{-7} m/s. Casagrande (1952) describes the development of the method.

Some relatively recent applications are given by Fetzer (1967), Casagrande *et al.* (1981), Doran, Hartwell, Roberti, Kofoed, and Warren (1995) and Alder *et al.* (2015). Applications have included stabilization of construction excavations, ground improvement for cross passage construction during tunnelling, acceleration of consolidation beneath embankment dams, and long-term stabilization of slopes. In some applications, electro-osmosis is used in conjunction with electrochemical stabilization (see Bell and Cashman, 1986), when chemical stabilizers are added at the anodes to permanently increase the strength of the soil.

In application, the electrode arrangements are straightforward, typically being installed in lines, with a spacing of 3 to 5 m between electrodes. In some arrangements, anodes and cathodes are placed in parallel lines (Figure 17.29). Alternatively, the electrodes can be placed in one line, in an alternating anode–cathode–anode sequence. Water is drawn to the cathodes and must be pumped away, so these can be formed from steel wellpoints or steel well liners. If it is desired to use plastic well liners at the cathodes, a metal bar or pipe (to form the electrode) needs to be installed in the sand filter around the well (Figure 17.30a). Specialist electrically conductive geotextile filter membranes (where steel conductors are woven into the mesh) are available and can be used with plastic well screens to form the cathode wells (Jones *et al.*, 2005; Alder *et al.*, 2015). The anodes are essentially metal stakes (Figure 17.30b); gas pipe, steel reinforcing bar, old railway lines or scrap sheet-piles can be used.

Applied voltages in are generally in the range of 30 to 100 V. Effectiveness can be improved if the potential gradient can be in the same direction as the hydraulic gradient. Casagrande (1952) states that the potential gradient should not exceed 50 V/m to avoid excessive energy losses due to heating of the ground. However, it might be advantageous to operate at 100 to

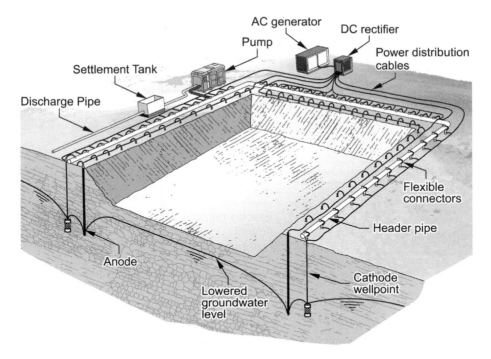

Figure 17.29 Electro-osmosis system for stabilization of excavation. (After Lomizé, G M, *et al.*, *Proceedings of the 4th International Conference on Soil Mechanics and Foundation Engineering*, London, 1, 62–67, 1957.)

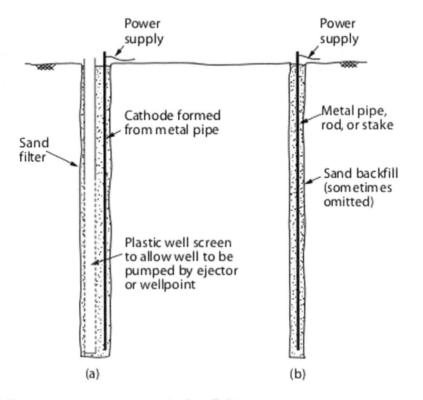

Figure 17.30 Typical electrode details. (a) Cathode well. (b) Anode.

200 V/m during the first few hours to give a faster build-up of groundwater flow. Reduction in power consumption may be possible if the system can be operated on an intermittent basis.

17.13 ARTIFICIAL RECHARGE SYSTEMS

Artificial recharge is the process of injecting (or recharging) water into the ground in a controlled way, typically by means of special recharge wells or trenches (Figure 17.31). Artificial recharge systems are not straightforward in planning or operation and are carried out on only a small minority of groundwater lowering projects. One of the most significant difficulties is the management and mitigation of clogging of recharge wells. Chapter 27 includes two Case Histories (27.8 and 27.9) that describe applications of artificial recharge.

For groundwater lowering applications, water from various sources has been used for artificial recharge:

- By far the most common case is where the recharged water is the discharge from the dewatering abstraction system, injected into the ground to control or reduce drawdowns around the main excavation area or simply as a means to dispose of the discharge water.
- Less commonly, mains water from the municipal supply is used for recharge. This is normally done to reduce the risk of clogging of recharge wells when the quality of the dewatering discharge water indicates the likelihood of severe clogging problems. The downside of this approach is the high cost of the mains water.
- On rare occasions, in coastal areas where saline intrusion of aquifers has occurred (see Section 21.4.11), it may be environmentally acceptable to recharge seawater into the ground. Bock and Markussen (2007) describe the artificial recharge system for the Copenhagen Metro project; in the harbour area, where groundwater was already saline, water from the harbour was used for recharge purposes. It was assessed that the low dissolved iron content of the harbour water relative to the pumped groundwater would reduce the severity of clogging of recharge wells.

17.13.1 Applications of Artificial Recharge Systems

When used in conjunction with groundwater lowering systems, artificial recharge can be used as a mitigation measure to control or reduce drawdowns away from the main area of

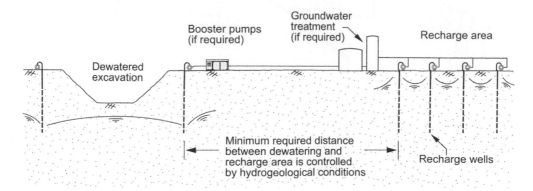

Figure 17.31 Artificial recharge to control drawdowns around a groundwater lowering system.

groundwater lowering in order to minimize the potential environmental impacts described in Chapter 21.

Artificial recharge is also occasionally used as a means to dispose of some or all of the discharge water if other disposal routes are not practicable. If this option is being considered, it must be noted that in many countries, including the United Kingdom, formal permission (see Section 25.4) must be obtained from the regulatory authorities to allow artificial recharge – there is no automatic right to recharge groundwater back into the ground. Similar regulations exist in several countries.

Recharge wells or trenches must be located with care. If the system is intended to reduce drawdowns at specific locations, then the recharge points will generally be between the groundwater lowering system and the areas at risk. The recharge locations may be quite close to the pumping system, and much of the recharge water could recirculate back to the abstraction wells, leading to an increase in the pumping and recharge rate. Unless the soil or rock stratification can be used to reduce the connection between the pumping and recharge system, a physical cut-off barrier could be used to minimize recirculation of water through the ground (Figure 17.32). If the water is being recharged to prevent derogation of water supplies, the recharge should be carried out further away to avoid excessive recirculation.

At locations where there is only a single aquifer present, the pumped water can be discharged into the same aquifer provided that there is sufficient horizontal distance between the dewatering system and the recharge location (Figure 17.31). Where multiple aquifers are present, it may be possible to recharge the water to a different aquifer than the one it was pumped from (Figure 17.33). If there is an aquitard providing some degree of hydraulic separation between the two aquifers, this can allow the recharge wells and dewatering system to be separated by a relatively small horizontal distance. The disadvantage of this approach

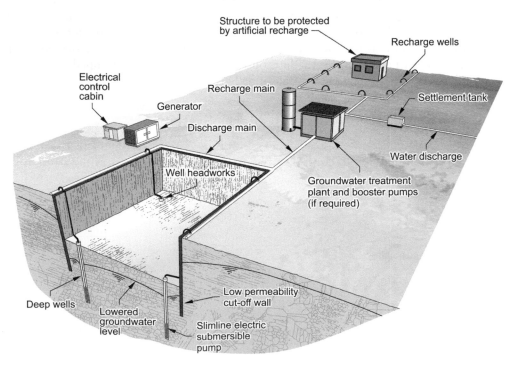

Figure 17.32 Artificial recharge system used in combination with low-permeability cut-off wall around excavation.

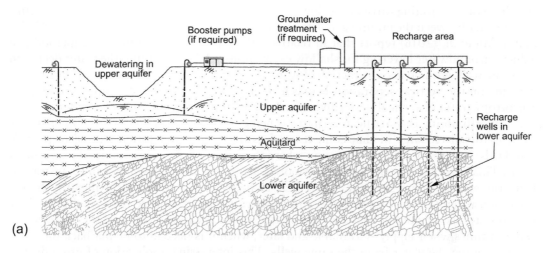

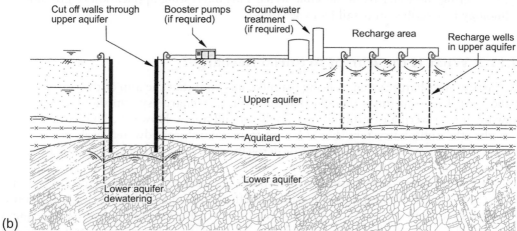

Figure 17.33 Artificial recharge in multiple aquifer systems. (a) Groundwater lowering required in upper aquifer with groundwater recharged to lower aquifer. (b) Physical cut-off wall used to exclude groundwater in upper aquifer. Groundwater lowering is required in lower aquifer with groundwater recharged to upper aquifer.

is that the water quality may be different in the two aquifers, and the mixing of water in the recharge wells may promote more severe clogging than might otherwise occur. Powrie and Roberts (1995) and Roberts and Holmes (2011) describe cases in the United Kingdom where excavations enclosed within concrete diaphragm walls were dewatered by deep wells within the dig, pumping from Chalk bedrock. The pumped water was passed to shallow recharge wells in an overlying gravel aquifer. This allowed the recharge wells to be located relatively close outside the diaphragm wall, with little risk of significant recirculation of water between the recharge wells and the dewatering system (this is the geometry shown in Figure 17.33b).

An interesting variation on artificial recharge systems that abstract from and recharge water to different aquifers is the vertical circulation well (VCW) described by Holzbecher *et al.* (2011) and Yulan *et al.* (2016). Each well in the system has a single well liner in the bore with a shallow abstraction well screen and a deeper re-injection well screen. A specialist pump and packer arrangement is used to allow pumping from the upper screen while simultaneously injecting through the lower screen. There is an obvious risk of short-circuiting of water between the pumping and re-injection well screens, but Holzbecher *et al.* (2011) and Yulan *et al.* (2016) report that in relatively thick permeable aquifers, the method can achieve groundwater lowering for shallow excavations while recharging the water deeper in the same well, with no net abstraction of water. It is reported that the success of the method depends on the vertical separation between the two well screens and is influenced by the anisotropy (ratio of vertical permeability to horizontal permeability) of the aquifer.

It is worth noting that artificial recharge is being increasingly applied in the water supply field. Artificial recharge in that context involves injecting water (often treated to potable standards) into the aquifer at a location where there is a supply at ground level (e.g. from a river). This injection effectively provides more water to the aquifer, which can be abstracted from the aquifer down gradient, where it is to be used. A variant on artificial recharge, used increasingly in the United States and Europe, is aquifer storage recovery (ASR). ASR involves storage of potable water in an aquifer at one location. This is done by injecting water at times when supply is available and later, when it is needed to be put back into supply, recovering the water from the same wells. This interesting application of groundwater technology is described in detail by Pyne (1994).

17.13.2　What Is the Aim of Artificial Recharge?

If an artificial recharge system is being planned, it is vital that the aim of the system is clear at the start. It is sometimes thought that an artificial recharge system should prevent any drawdown or lowering of groundwater levels around a site. However, Powers (1985) suggests that the aim of a system should be not to maintain groundwater levels *per se* but to prevent side effects (such as settlement or derogation of water supplies) reaching unacceptable levels.

Groundwater levels vary naturally in response to seasonal recharge (Section 3.9) and under the influence of pumping. Other natural effects include tidal groundwater responses in coastal areas (Section 3.6.3). A specification for an artificial recharge system that required 'zero drawdown' would be not practicable. It may be more appropriate to set a target such that groundwater levels in selected observation piezometers shall not fall below defined levels (which are likely to be different in different parts of the site).

Assuming that groundwater levels around a site vary naturally with time, and that this variation is not causing detrimental side effects, the lowest acceptable groundwater level in a monitoring well is often set as the seasonal minimum. Occasionally, if the groundwater lowering system is intended to generate very large drawdowns in relatively stiff aquifers, the lowest acceptable groundwater level is set rather lower, at a few metres below the seasonal minimum.

It is sometimes also useful to specify maximum groundwater levels in selected piezometers. This reduces the temptation for overzealous recharging, raising the groundwater levels above the seasonal maximum, which could lead to problems with flooded basements and the like. By defining an allowable minimum and maximum groundwater level in each monitoring well (with at least 0.5 m between maximum and minimum levels), the system operator will have some modest leeway within which to adjust the system.

17.13.3　Recharge Trenches

A simple method of recharge for shallow applications is the use of recharge trenches, sometimes known as infiltration trenches (Figure 17.34a). These are excavated from the surface, and the dewatering discharge is fed into them from where the water infiltrates into the ground.

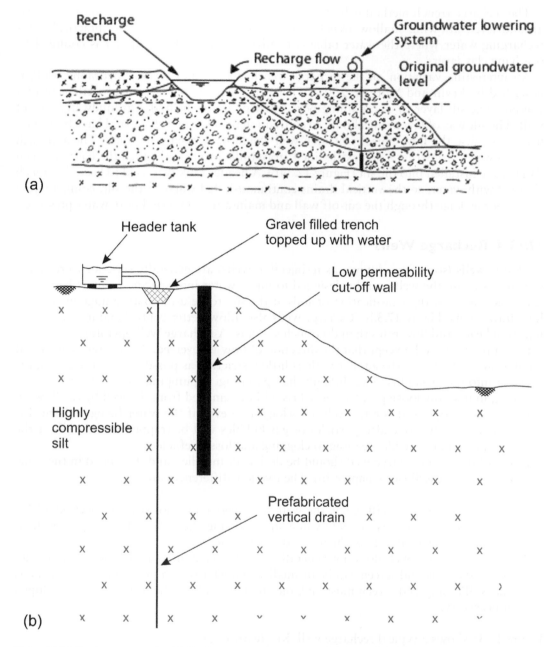

Figure 17.34 Recharge trench. (a) Recharge trench in unconfined aquifer. (b) Recharge trench used in combination with prefabricated vertical drains. (After Ervin, M C and Morgan, J R, *Canadian Geotechnical Journal*, 38, 732–740, 2001.)

Recharge trenches have been employed to good effect when pumping from mineral workings (Cliff and Smart, 1998) but are less effective in construction situations and are rarely used. Clogging of the trenches (by the growth and decay of vegetation or the build-up of sediment) reducing infiltration rates is often a problem. Trenches typically require periodic cleaning out by excavator. It is difficult to control and measure the volumes entering the trenches, making this method less easy to adjust than a system of recharge wells. An overflow channel is normally required to prevent overtopping of the trenches.

The recharge trench method is best suited to unconfined aquifers with water tables near ground level, allowing shallow trenches to be used. The method is less practicable for recharging water where the water table is relatively deep or if the aquifer is confined by overlying clay layers.

An interesting variation on the recharge trench, which they term a 'hydraulic wall', is described by Ervin and Morgan (2001). Their requirement was to control settlement of a compressible silt layer by maintaining pore water pressures behind a groundwater cut-off wall. The silt was of low permeability, so conventional recharge wells would have been inefficient, because the flow capacity of each well would have been very small. The approach adopted was to construct a geotextile-lined, gravel-filled trench, through the base of which PVDs were installed into the silt (Figure 17.34b). A header tank was used to keep the trench charged with water, so the vertical drains continuously fed water to the silt, counteracting any minor leakage through the cut-off wall and maintaining external pore water pressures.

17.13.4 Recharge Wells

Recharge wells (sometimes known as re-injection wells) are generally superior to recharge trenches because the wells can be designed to inject water into specific aquifers beneath a site; this can allow the stratification of soils at the site to be used to advantage in controlling drawdowns (Figure 17.33). Recharge wells also allow better control and monitoring of injection heads and flow rates than do trench systems. A recharge well operates in a similar way to a pumping well except that the direction of flow is reversed. A recharge well should allow water to flow into the aquifer with as little restriction as possible. However, recharge of water into the ground is more difficult than pumping. A pumping well is effectively self-cleaning, in that any loose particles or debris will be removed from the well by the flow. In contrast, a recharge well is effectively self-clogging – even if the water being recharged is of high quality, any suspended particles or gas bubbles will be trapped in the well (or the aquifer immediately outside), leading to clogging and loss of efficiency.

In broad terms, a recharge well should be designed, installed and developed in the same way as a pumping well (see Chapter 16). The two key differences are:

To maximize hydraulic efficiency, the well filter media and slot size (see Section 13.10) should be as coarse as possible while still allowing the well to be pumped without continuous removal of fines during re-development.

Because a recharge well does not generally have to accommodate a submersible pump, the well casing and screen can be of smaller diameter than a pumping well. However, the well casing and screen must be large enough to allow the well to be re-developed if necessary.

Figure 17.35 shows a typical recharge well. Key features are:

- Well casing and screen, surrounded by filter media, with a grout or concrete seal around the well casing to prevent water short-circuiting up the filter pack to ground level. If the top of the well casing is also sealed around the recharge pipework, this can allow recharge flows to be fed by header tanks raised above ground level to increase the injection rate in each well.
- A down spout to prevent the recharge water from cascading into the well and becoming aerated. Aeration of the water may promote biofouling and other clogging processes and should be avoided as much as possible.

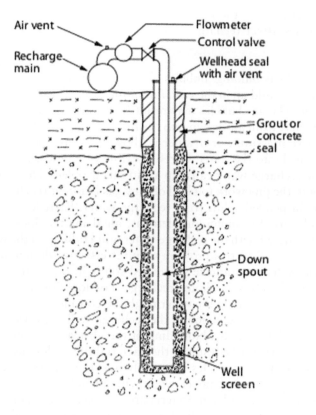

Figure 17.35 Recharge well.

- Air vents at the top of the well and pipework to purge air from the system when recharging commences and to prevent airlocks in the system.
- Control valve and flowmeter to allow monitoring and adjustment of flow to the well.
- If water is to be recharged into a shallow aquifer, a system of recharge wellpoints may be considered (an example is given in Case History 27.8). The recharge wellpoints are installed at close spacings using similar methods to conventional wellpointing (see Chapter 15).
- Almost all recharge wells suffer from clogging to some degree, and it is vital that the design used allows appropriate re-development. This is discussed further in the following sections.

17.13.5 Water Quality Problems and Clogging

Experience has shown that it is much harder to artificially recharge water into the ground than it is to abstract it. There are various rules of thumb which say that to recharge water back into the same aquifer from which it came, two or three recharge wells will be needed for every abstraction well. Many of the practical difficulties with artificial recharge arise from water quality problems leading to clogging. The designer and operator are being unrealistic if they do not expect recharge wells or trenches to clog (to a lesser or greater degree) in operation.

It is worthwhile considering the mechanisms that can lead to the clogging of recharge wells (similar processes affect recharge trenches). Pyne (1994) has outlined five clogging processes:

1. Entrained air and gas binding. Bubbles of vapour (such as air or methane) present in the recharge water build up in the well, inhibiting flow of water through the filter pack into the aquifer. The gas bubbles may result from the release of dissolved gases from the recharge water; from air drawn into the recharge pipework, where changes in flow generate negative pressures; or if water is allowed to cascade into the recharge well, from air entrained into the water. Careful sealing of pipework joints and minimizing aeration of the recharge water can help avoid this problem. Degassing equipment can be installed onto the pipework to bleed off any gas before it reaches the recharge wells.
2. Deposition of suspended solids from recharge water. Particles (colloidal or silt- and sand-sized soil particles, biofouling detritus, algal matter, loose rust or scale from pipework, etc.) carried with the recharge water will build up in the well and filter pack, blocking flow. Control of this problem requires effective design and development of the abstraction wells to minimize suspended solids in the water; the discharge from sump pumping is rarely suitable for recharge. For low-flow-rate systems, the use of sand filter systems to clean up the water might be considered.
3. Biological growth. Bacterial action in the recharge wells can result in clogging of the wells themselves. Additionally, biofouling of the abstraction wells and pipework can release colloidal detritus (the result of the bacterial life cycle) into the water. The flow of water will carry these particles inexorably into the recharge wells, leading to further clogging. The severity of any clogging will depend on several factors (see Section 23.5). Since some of the problem bacteria are aerobic, minimizing any aeration of the recharge water is advisable. Periodic dosing of the wells with a dilute chlorine solution (or other disinfecting agents) has been used to inhibit bacterial growth.
4. Geochemical reactions. The recharge water can react with the natural groundwater or with the aquifer material. Such reactions are most likely if the recharge water is not groundwater from the same aquifer (e.g. if mains water is used for recharge, or water from one aquifer is recharged back to another aquifer) or if the pumped groundwater is allowed to change chemically prior to re-injection. Typical reactions include the deposition of calcium carbonate or iron/manganese oxide hydrates. The potential significance of these reactions can only be assessed following study of the aquifer and groundwater chemistry, but they are generally reduced in severity if water pressure and temperature changes during recharge are minimized. Chemical dosing of the recharge water might be considered to inhibit particular reactions.
5. Particle re-arrangement in the aquifer. Although this is not usually a significant effect, the permeability around the well may reduce due to loose particles around the well being re-arranged by the flow of recharge water out of the well. This effect can be minimized by effective development on completion and periodic re-development.

These clogging effects should be considered when designing and operating an artificial recharge system. After all, a recharge system is not going to achieve its aims if it is unable to continue to inject water into the ground.

17.13.6 Operation of Recharge Systems

An artificial recharge scheme broadly consists of an abstraction system, a recharge system and a transfer system (between the abstraction and recharge points), which may include water treatment elements to prevent the effects of clogging. The abstraction system should be straightforward in operation and maintenance (see Chapter 23), but the transfer and recharge systems may be more problematic. The crux of the issue is to manage the recharge water quality to limit the clogging of recharge wells to acceptable levels, which will not prevent the system from achieving its targets.

As has been stated previously, most artificial recharge systems will suffer from clogging. There are two approaches to dealing with this, often used in combination:

Prevention (or reduction) of clogging – principally by treatment of the discharge water to remove suspended solids and retard clogging processes such as bacterial growth.

Mitigation of clogging – accepting that clogging will occur and then planning for it. This may involve providing spare recharge wells, to be used when efficiency is impaired, and implementing a programme of regular re-development of clogged wells.

Measures to prevent clogging must focus on ensuring that the recharge water is of high quality and does not promote clogging processes. Possible measures include:

a) Removal of suspended solids by filtration with sand filters or bag or cartridge filters. Filters will require periodic cleaning, back-washing or replacement as solids collect in them.
b) Chemical dosing to precipitate out problematic carbonates or iron and manganese compounds.
c) Intermittent or continuous chlorine dosing of the recharge water to reduce bacterial action. It is also a relatively common practice to disinfect recharge wells with chlorine or other biocides on first installation and following later rehabilitation.
d) Prevention of aeration of water to reduce the risk of gas binding.
e) Use of mains water if the pumped groundwater cannot be rendered suitable.
f) Careful management of the system on start-up to avoid slugs of sediment-laden water entering the wells (the initial surge of water may pick up sediment deposited in the pipework and should be directed to waste, not to the recharge wells).

Measures to mitigate clogging include:

- Designing recharge well screens to have ample open area to accept flow. Sterrett (2008) recommends that the average screen entrance velocity (recharge flow divided by total area of screen apertures) is less than 0.015 m/s, approximately half the maximum velocity recommended for abstraction wells.
- Re-developing the recharge wells on a regular basis to rehabilitate the wells (see Section 16.7). This is most commonly done by airlift development, sometimes in combination with chemical treatment, which dislodges any sediment, loose particles, precipitates and bacterial residues. Depending on the severity of clogging, wells in a system may be re-developed on a rolling programme, with an individual well being re-developed every few weeks or months. Even after re-development, wells may not return to their pre-clogging performance – if the system operates for very long periods, additional recharge wells may need to be installed after a few years to prevent the overall system performance from dropping to unacceptable levels.
- Providing additional recharge wells over and above the minimum number of recharge wells necessary to accept the recharge flow. In this way, the flow can still be accepted if some of the wells are badly clogged, allowing more time for re-development.
- Monitoring of artificial recharge systems follows good practice for abstraction systems (see Chapter 22). For recharge wells, the water level below ground level (by dipping) or the pressure head above ground level (by pressure gauge) should be monitored. Ideally, flowmeters should be installed on each well to record recharge flow rates, but some types may be affected by clogging. At the very least, the overall recharge rate should be monitored, perhaps by V-notch weir.

17.14 DEWATERING AND GROUNDWATER CONTROL TECHNOLOGIES USED FOR THE CONTROL OR REMEDIATION OF CONTAMINATED GROUNDWATER

Many construction projects are carried out on or near sites where there is a legacy of soil and groundwater contamination. Such contamination may result from current or former industrial uses, pre-existing pollution incidents, or landfills and waste management sites. Groundwater pumping on or in the vicinity of such sites is likely to have an impact on the pre-existing contamination, for example by causing plumes of contaminated groundwater to move in different directions than otherwise. Such impacts must be assessed during the design process; this is discussed in Section 21.4.10.

Where the existence of groundwater contamination is known or suspected, the technology commonly used for groundwater control can be applied as part of a strategy to control or remediate contaminated groundwater. Methods based on groundwater pumping (i.e. sumps, wells and artificial recharge) and groundwater exclusion (i.e. cut-off walls) can be used. Potential applications of these techniques on contaminated sites are described in the following sections. Case History 27.11 describes an application of contamination remediation using cut-off walls and groundwater abstraction.

Groundwater control technologies may be used on sites where contaminated groundwater is an issue in order to meet one or more of the following objectives:

(a) Use of pumping systems to control or manipulate the movement of contaminated groundwater in pre-existing plumes. In some circumstances, the pumping is intended to lower the groundwater levels around the contaminated area so that groundwater flow is towards the source, preventing outward migration of contaminants – this approach is sometimes termed hydraulic containment (Figure 17.36a).

(b) Use of pumping systems to extract and treat mobile contaminants either from the source zone or from the surrounding contaminant plume – this approach is sometimes termed 'pump and treat' (Figure 17.36b).

(c) Use of pumping systems located close to a contaminant source (e.g. a leaking tank or silo) to attempt to intercept leakage before it has the opportunity to migrate away from the source (Figure 17.36c).

(d) Use of pumping systems to lower groundwater levels in order to achieve secondary objectives. These objectives may include holding groundwater levels below a known contamination source zone; dewatering of excavations to allow removal of contaminated soils or structures; or lowering of groundwater levels to promote volatilization of contaminants for soil vapour extraction or increase bacterial access to oxygen as part of wider remediation strategies (Figure 17.36d).

(e) Installation of low-permeability cut-off barriers to exclude or contain zones of contaminated soil or groundwater (Figure 17.36e).

(f) Installation of buried passive treatment systems (often associated with sections of low-permeability cut-off barrier) – these systems are known as permeable reactive barriers or PRBs (Figure 17.36f).

The design and execution of groundwater control on contaminated sites, or the use of groundwater control methods as part of remediation schemes, is complex. It is essential that the team developing any such scheme include designers and professionals experienced in contamination remediation.

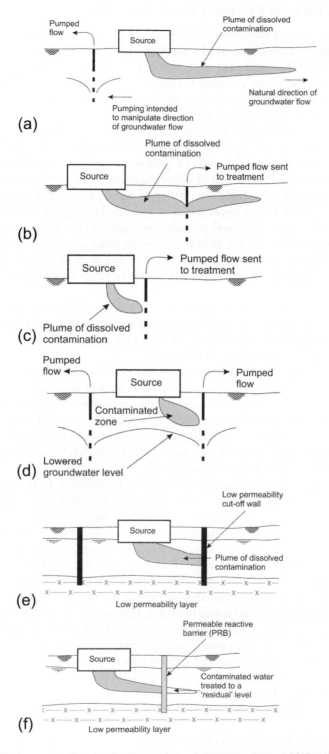

Figure 17.36 Groundwater control technologies used on contaminated sites. (a) Pumping systems to control movement of contaminated groundwater. (b) Pumping system to extract and treat mobile contaminants (pump and treat). (c) Pumping systems to intercept leakage close to source. (d) Pumping systems to lower groundwater levels to facilitate other processes. (e) Cut-off barriers to exclude or contain contaminated groundwater. (f) Permeable reactive barriers (PRBs).

17.14.1 Applications of Pumping Systems on Contaminated Sites

In most cases, there are no fundamental differences in the groundwater pumping systems applied on contaminated sites compared with sites where contamination is not an issue – the physics of groundwater flow and the mechanics of pumping are the same in both cases. The key difference is that the water pumped is expected to contain elevated levels of contaminants and therefore is likely to require some form of treatment or specialized disposal route.

One case where groundwater pumping systems on contaminated sites may need to diverge slightly from the norm is where significant quantities of LNAPL (light non-aqueous phase liquids) exist as a layer of 'free product' floating on top of the water table. This situation can arise on sites contaminated by spills of petroleum and other light hydrocarbon compounds. If conventional well designs and pumping systems are used, there is a risk that the free product will become mixed with the pumped water, producing an emulsion that can be problematic to treat. A better approach is to provide larger-diameter wells with screens that extend above the water table and provide two separate pumping systems in each well. The water pump would be set deep, below the free product level, and shallow 'skimmer' or 'scavenger' pumps would float on the water level in the well to directly draw off the free product. Methods suitable for use on sites where free product is present are given in Holden *et al.* (1998).

In general, to produce an effective, efficient and environmentally acceptable groundwater abstraction system, the design and construction of groundwater pumping systems on contaminated sites should take the following considerations into account, in addition to the general principles outlined elsewhere in this book.

The presence of suspended solids (fine particles of silt/sand/clay) in pumped water significantly increases the difficulties of treatment. It is almost unavoidable that sump pumping operations will generate water with significant suspended solids content. Therefore, to simplify treatment requirements, sump pumping should be avoided on contaminated sites whenever possible. Systems based on wells (such as wellpoints or deep wells) should be used instead, since these systems, when designed, installed and developed appropriately, tend to produce water with low suspended solids content.

Peak water flow rate requiring treatment should be minimized. In many cases, the capital cost of a groundwater treatment system is more or less proportional to the peak flow rate. Reducing the flow rate of contaminated water will significantly reduce costs. Flow rates may be minimized by the use of groundwater cut-off barriers, by ensuring that only the minimum necessary drawdowns are generated, and by segregating contaminated and 'clean' water (which can be allowed to bypass the treatment process) as far as practicable.

Assuming that the site investigation has identified the extent and location of the contaminants, it may be possible to target dewatering wells (or screened sections within wells) directly on the most contaminated zones or strata. Ultimately, a system could be designed to have 'clean' wells pumping uncontaminated water (which need not be treated) and a separately connected system of 'contaminated' wells, perhaps pumping from a different depth or stratum, producing water that is then fed to a treatment plant.

Any wells should be designed appropriately so that they do not act as potential vertical migration pathways for pollution. It may be appropriate for wells to have grout seals above and below the well screens to reduce the risk of creation of artificial vertical flow paths (see Section 21.6).

Surface water should be managed efficiently. On most contaminated sites, there is a risk that any water allowed to run or pond on top of exposed soils will itself become contaminated and therefore potentially require treatment. Good surface water management practices (see Sections 9.2 and 14.3) should be applied to keep surface water away from contaminated areas as far as possible. Any contaminated surface water should be segregated from 'clean' surface water.

The most common way that pumped water containing elevated levels of contamination is managed involves the water being passed through an on-site treatment system (see Section 17.14.3) to reduce contamination levels before being discharged to a conventional water disposal route such as sewer or surface watercourse. In some cases, the treated water is recharged back into the ground via recharge wells or trenches.

On rare occasions, when modest volumes of very highly contaminated or difficult-to-treat water are involved, the water may be transported off site by road tanker to a specialized licensed disposal facility – this approach quickly becomes impractical if the volumes are more than around 100 cubic metres per day. As an illustration. a large articulated road tanker will hold around 20 m^3 of water. Therefore, a minimum of five road tankers would be needed to transport every 100 m^3 of water for disposal. The cost and disruption of large numbers of road tanker movements militate against this approach on most projects.

Other chapters in this book give specific details on pumping systems such as sump pumping (Chapter 14), wellpoints (Chapter 15), deep wells (Chapter 16) and ejector wells (Section 17.2). Applications of groundwater pumping systems on contaminated sites are described in Holden *et al.* (1998).

17.14.2 Applications of Barrier Systems on Contaminated Sites

Barrier systems based on low-permeability ground treatment or cut-off walls are widely applied on sites where contaminated groundwater is a concern. If used to enclose a contaminated site, they have the obvious advantage of potentially isolating the contamination from the wider groundwater regime, either as a long-term strategy or to facilitate other remediation or construction activities. The most commonly used barrier methods on contaminated sites are slurry trench walls and sheet-pile walls, although other methods are used occasionally. An example of a slurry trench application on a contaminated site is given in Case History 27.11.

It is important to recognize that no groundwater cut-off barrier is truly impermeable. Leakage should be expected through joints in the wall (see Section 18.3). Indeed, even a very low-permeability material has a finite permeability and will permit some seepage in the very long term directly through the mass of the barrier. In any event, the soil or rock within the area enclosed by the barrier will contain water, which will need to be pumped out in advance of any excavation. Therefore, even extensive and high-specification dewatering barrier systems will not completely avoid the need to deal with contaminated water.

A further complication occurs when groundwater cut-off walls are used to enclose large areas. If the barrier formed by the cut-off wall is keyed into an underlying low-permeability stratum, the groundwater within the area enclosed by the walls will be trapped there – in fact, that is probably the objective of the barrier. The complication arises because precipitation falling on the ground surface within the enclosed area is likely to infiltrate into the ground. If the rate of infiltration is significant, then it is possible that over time (perhaps several years). groundwater levels will rise within the area enclosed by the wall. This may lead to problems with flooding of buried structures and services or waterlogging of the ground. If water levels rise, an outward hydraulic gradient will be created across the wall. Where leaks or imperfections exist in the wall, this may promote leakage of contaminated water out of the area enclosed by the wall. One solution is to use long-term pumping to maintain the groundwater levels in the enclosed area slightly lower than groundwater levels outside the wall. This maintains an inward hydraulic gradient and prevents outward migration of contamination via groundwater flow. This approach was adopted in Case History 27.11.

A technique based on groundwater barrier technologies that is occasionally used to remediate groundwater migrating across or out of a site is the PRB. This approach accepts that the existence of an artificial barrier in the ground will result in hydraulic gradients across the barrier, even in the absence of pumping. In PRBs, all or part of the barrier is formed from a permeable material that can react with groundwater, removing contamination as groundwater flows through it as a result of natural groundwater flow (Figure 17.36f). Since the mid-1990s, PRB systems have been applied on sites where there is a desire to reduce levels of contamination in groundwater migrating from a site while avoiding the cost and maintenance issues associated with long-term pumping.

Specific details of the techniques used to form low-permeability barriers are given in Chapter 18. Applications of low-permeability barrier systems on contaminated sites are described in Privett et al. (1996). Permeable reactive barriers are described in Carey et al. (2002).

17.14.3 Groundwater Treatment Technologies

Where groundwater is pumped on contaminated sites, there is a potential requirement to treat the water prior to disposal. The requirement for treatment may be to reduce levels of contamination to levels acceptable for disposal to sewer (Section 25.4), to meet regulatory requirements for disposal to surface waters or artificial recharge to groundwater (Section 17.13), or to reduce environmental impacts resulting from discharge of groundwater (Section 21.9).

A wide range of treatment methods is available. Often, a given contaminant could be treated by several quite different methods applied in series; the choice of method will depend on the concentration of contaminants, the discharge flow rate, the duration of pumping, and the availability of treatment equipment and technologies. Some of the available technologies are described by Nyer (2009) and Holden et al. (1998).

The most commonly used groundwater treatment technologies are:

Buffer storage: Holding tanks or lagoons are normally provided upstream of the main treatment system to help smooth out any peaks or lows in flow rate or contaminant concentration. The buffer storage is normally sized to provide sufficient storage to allow the dewatering system to continue to operate while the treatment plant is off-line (due to maintenance or breakdowns) for specified periods. The tanks or lagoons also will allow some settlement of suspended solids in the pumped water.

Oil–water separation: The presence of even limited amounts of 'free product' (LNAPLs, immiscible with water; see Section 21.4.10) can cause problems with later treatment stages. An oil–water separator is often provided as one of the first stages of treatment to remove any free product from the treatment stream (the oil will require separate disposal).

Metals removal: The most common approach to metals removal is to promote the conversion of metals from soluble to insoluble form and then to settle out and remove the resulting precipitates. Typically, the pH of the water is raised (made more alkaline) by the addition of bases to reduce the solubility of metals. Coagulation and flocculation follow by addition of specialist chemicals to promote precipitation. Finally, the precipitates are removed by a process of settling/clarification, where the settled solids are collected and removed as sludge for further processing and disposal.

Filtration: Filtration by sand filters or bag or cartridge filters is sometimes carried out to reduce suspended solids in the treatment stream. Filters will require periodic cleaning, back-washing or replacement as solids collect in them.

Granulated activated carbon (GAC) filtration: GAC has the ability to adsorb hydrocarbons and other compounds from liquid and vapour streams and is widely used to treat contaminated groundwater. The water is passed through tanks or vessels containing the GAC, and the contaminants sorb onto the surface of the GAC granules. The capacity of a given quantity of GAC to absorb contamination is large but finite and will eventually be exhausted. When that happens, the spent GAC is normally removed from site (for disposal or regeneration and later re-use) and replaced with fresh material. Typically, a treatment system would comprise two GAC vessels (primary and secondary) connected in series. Water quality is tested between the primary and secondary units. When the sorption capacity of the primary unit is exceeded, excessive levels of contaminants will be detected at the sampling point between the two units. When this occurs, the flow is redirected to the original secondary unit, while the GAC is replaced in the primary unit.

Air stripping: This technique is used primarily to remove volatile and semi-volatile hydrocarbons from water. Such compounds are relatively easy to transfer from the liquid phase (water) to the vapour phase (as an off-gas). This is achieved by pumping the contaminated water to the top of an air stripping tower and allowing the water to cascade downwards over a multitude of plastic packing elements within the tower. At the same time, a fan blows air upwards through the tower. The air–water contact, distributed over the very large surface area of the packing elements, promotes the stripping of the volatile compounds from the water, which leave the tower in the off-gas airflow. The off-gas is often treated by passing through vapour phase GAC to reduce air emissions. Air stripping is often used in series with GAC filtration. If the air stripper is located upstream of the GAC filter, the volatile hydrocarbons will be removed before the GAC filter. This reduces the contaminant loading on the GAC filter and will probably increase the life of the GAC media.

Figure 17.37 Modular groundwater treatment plant used to treat discharge from dewatering system. (Courtesy of Hölscher Wasserbau GmbH, Haren, Germany.)

For most construction-related groundwater control or remediation projects, the groundwater treatment system will be required for a period of several months, up to perhaps a maximum period of 1 to 2 years. In these applications, groundwater treatment plants commonly comprise temporary systems of modular steel tanks and vessels (Figure 17.37), delivered to site and then commissioned in suitable configurations to meet the project requirements. For very long-term remediation projects, it may be cost effective to construct groundwater treatment plants to similar standards as permanent wastewater treatment systems (this approach was taken for Case History 27.11).

Chapter 18

Methods for Exclusion of Groundwater

Water has an unfortunate habit of seeping through every theory

Victor de Mello

18.1 INTRODUCTION

On many below-ground construction projects, it may be necessary to deploy ground engineering techniques to exclude groundwater from an excavation. As has been discussed in Chapters 9 and 10, these exclusion methods may be deployed on their own or in combination with dewatering pumping techniques.

While the principal focus of this book is on groundwater control techniques based on pumping technologies, it is appropriate for the dewatering practitioner to have at least a basic understanding of exclusion methods used to form cut-off barriers or walls. This knowledge will be useful when comparing pumping and exclusion options or when developing methods to mitigate environmental impacts (see Chapter 21). This chapter provides a brief overview of the more commonly used groundwater exclusion techniques, including the factors affecting the application of each technique. The systems described in this chapter are:

   (i) Steel sheet-piling
  (ii) Vibrated beam walls
 (iii) Slurry trench walls
  (iv) Concrete diaphragm walls
   (v) Concrete secant pile walls
  (vi) Grout barriers
 (vii) Mix-in-place barriers
(viii) Artificial ground freezing
  (ix) Low-permeability surface barriers

The chapter ends with a discussion of some of the methods that can be used to verify the performance of groundwater exclusion solutions.

18.2 PRINCIPAL METHODS FOR GROUNDWATER EXCLUSION

A wide range of exclusion techniques are available to form cut-off walls or barriers around civil engineering excavations. Those commonly used for groundwater control purposes are summarized in Table 9.1. They have been categorized by the way each method interacts with the ground. Displacement barriers (such as steel sheet-piling or vibrated beam walls) are inserted into the ground, displacing soil and as a consequence generating little or no spoil. In contrast, excavated barriers (such as slurry trenching, concrete diaphragm walls and secant pile walls) involve excavating material from the ground (thereby generating spoil) and replacing it with barrier material. Injection methods (the various forms of grouting) involve injecting and/or mixing special fluids into the ground, where they solidify to block ground-water flow in pores and fractures. Another category of exclusion methods is the system of artificial ground freezing, where the ground is modified by thermal means. The ground is frozen to form a barrier of very low-permeability frozen ground and groundwater.

The selection of a given exclusion method used to form a cut-off barrier will depend on the ground conditions and other potential constraints on a given project. Primary constraints are required depth of wall; ground conditions; geometry of wall (some methods can be used horizontally or inclined to the vertical, while others are limited to vertical applications); and whether the barrier is intended to be permanent or temporary (temporary barriers can be useful in reducing long-term groundwater impacts, as discussed in Section 21.7 and Section 24.5). The various methods are described in the remainder of this chapter.

There may be occasions when different exclusion methods are used in combination. In the same way that if multiple aquifer systems are present, more than one pumping technique may be used, exclusion methods may also be combined to deal with the particular stratifica-tion on a site. Soudain (2002) describes a project where a 2.23 km long groundwater cut-off wall was retrofitted around a large waste disposal pit used to dispose of half a million animal carcasses following a foot and mouth disease outbreak in the United Kingdom. In that case, a jet grouted barrier was first formed in the fractured bedrock at the site. Then, a slurry trench wall was excavated through the overlying drift into the weathered bedrock and was keyed into the upper 1.5 m of the jet-grouted zone to provide a continuous low-perme-ability barrier to a greater depth than could have been achieved by slurry trenching alone.

18.3 GEOMETRIES OF EXCLUSION APPLICATIONS

The application of exclusion methods to provide a physical cut-off wall around an excava-tion is described in Section 9.4 and shown in Figure 18.1. In some applications, there is a requirement to provide a temporary or permanent retaining structure to form the walls of an excavation – a wall totally enclosing an excavation is often called a 'cofferdam'. The groundwater cut-off barrier may also act as a structural element of the permanent works, such as the perimeter of a basement. These requirements will affect the choice of technique used to form the cut-off wall, because not all methods are suitable for structural appli-cations. On tunnelling projects, the particular geometries of the various elements being constructed (tunnels, shafts, cross passages) can be a controlling factor in the selection of groundwater exclusion methods. Groundwater exclusion applications for tunnelling proj-ects are described in detail in Chapter 10.

The most common application is to install vertical cut-off walls around an excavation, as shown in Figure 18.1. However, some exclusion methods, including grouting and artifi-cial ground freezing, can be installed in non-vertical geometries to produce various cut-off geometries (Figure 18.2). Some methods can be used to form low-permeability plugs to seal the base of excavations (Section 18.3.1). Horizontal very low-permeability surface barriers can also be used to reduce infiltration from rainfall and other sources (Section 18.13).

One of the features of the methods used to form cut-off walls is that they are typically not installed continuously but for practical reasons, are installed in discrete elements. These ele-ments (typically panels or columns) are designed to overlap or intersect to form continuous

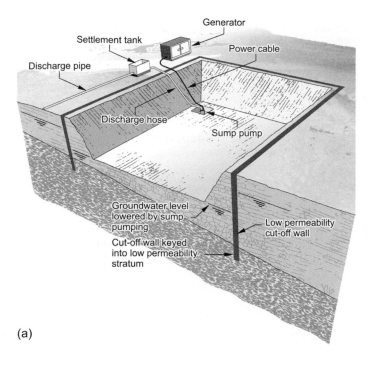

(a)

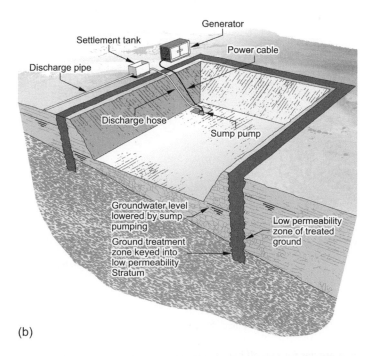

(b)

Figure 18.1 Groundwater control by exclusion where there is a shallow very low-permeability stratum. (a) Cut-off walls penetrate into very low permeability stratum. (b) Groundwater exclusion using physical cut-off walls. (c) Groundwater exclusion using ground treatment.

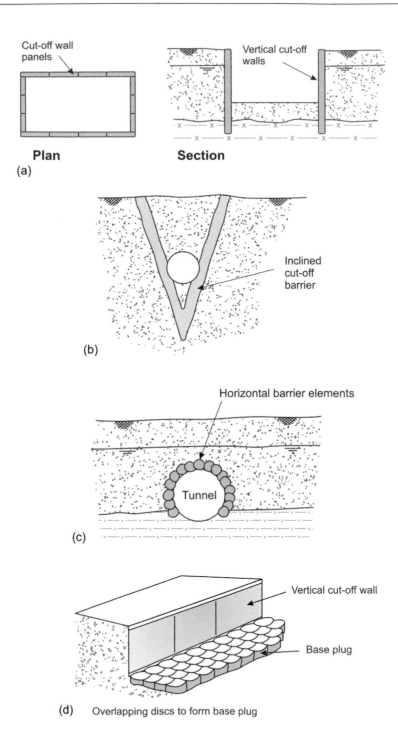

Figure 18.2 Typical geometries for cut-off barriers (a) Vertical barriers. (b) Inclined barriers. (c) Horizontal barriers arranged to exclude groundwater from a tunnel, located at the interface between higher- and lower-permeability strata (d) Base plug consists of overlapping columns of ground treatment.

barriers of very low-permeability material (Figure 18.3). Accordingly, a key element of the design and construction of a cut-off wall is ensuring the integrity of the joints between the elements of the wall. When, as occasionally happens, cut-off walls do leak significantly, the most likely cause is seepage through one or more joints. Leakage through joints is often caused by deviation of alignment of adjacent panels or columns in the cut-off wall. As the panels deviate, they may cease to intersect, leaving a gap through which water can pass. Case History 27.10 describes the extensive concrete diaphragm wall around Sizewell B Power Station in Suffolk, United Kingdom, where monitoring of groundwater levels during dewatering indicated the presence of a leak, which required sealing by grouting.

18.3.1 Basal Seals

Where there is a very low-permeability stratum (such as clay or unfractured bedrock) at a convenient depth, vertical walls from the surface will be able to penetrate down to that stratum (Figure 18.1). Typically, the bottom of the wall is 'toed into' the very low-permeability stratum – i.e. the wall penetrates a short distance into the stratum. The depth of 'toe-in' is

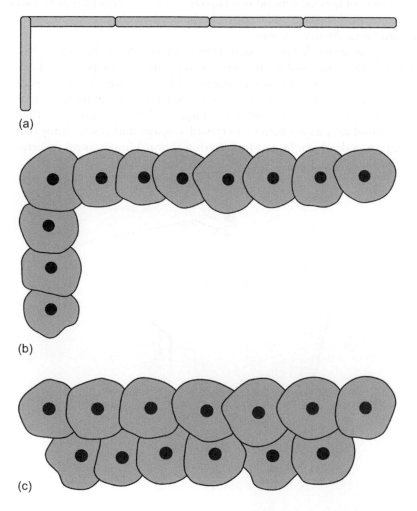

Figure 18.3 Elements used to form cut-off walls. (a) Intersecting panels or structural elements. (a) Overlapping columns or piles. (c) Multiple rows of overlapping columns.

typically between 1 and 3 m. The combination of the artificial cut-off walls and the natural stratum can act to almost totally prevent groundwater flow into the excavation.

If a suitable very low-permeability stratum does not exist or is so deep that it is not economical or technically feasible to install a cut-off wall to sufficient depth, vertical elements alone cannot form a complete groundwater exclusion barrier. If there is a need to totally exclude groundwater, a horizontal barrier can be used to form a basal seal (often called a 'base plug') below the excavation dig level to prevent upward vertical seepage (Figure 18.4). The construction of horizontal barriers is relatively rare but has been carried out using jet grouting, mix-in-place, permeation grouting and artificial ground freezing techniques.

The objective is to provide a continuous horizontal barrier to seal up against the vertical elements of the cut-off barrier. When the area enclosed by the cut-off walls is dewatered and excavated, it will be buoyant (see Section 7.5.1) and will experience unbalanced upward groundwater forces. The configuration of the basal seal may be influenced by the need to resist buoyancy forces. This is illustrated by Figure 18.5, which shows four cases:

a) Thick base plug immediately below excavation formation level (Figure 18.5a). In this case, the zone of treated ground is relatively thick. It serves two purposes – to act as a barrier to vertical seepage and to provide the deadweight to resist buoyancy forces due to upward groundwater pressures.

b) Thin base plug at depth below excavation formation level (Figure 18.5b). The zone of treated ground is thin, and its thickness is controlled by the permeability of the treated material and the required residual seepage. The resistance to uplift pressures is provided by the weight of the plug of treated ground and the untreated material above it.

c) Base plug formed as an inverted arch (Figure 18.5c). In this case, the base plug of treated ground acts as a barrier to vertical seepage and resists buoyancy forces by a combination of deadweight and transferring upward forces as an inverted arch.

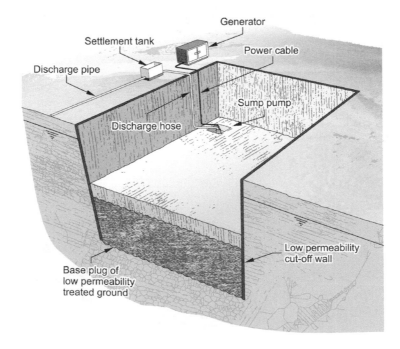

Figure 18.4 Groundwater control by exclusion using physical cut-off walls with horizontal barrier (base plug) to seal base of excavation.

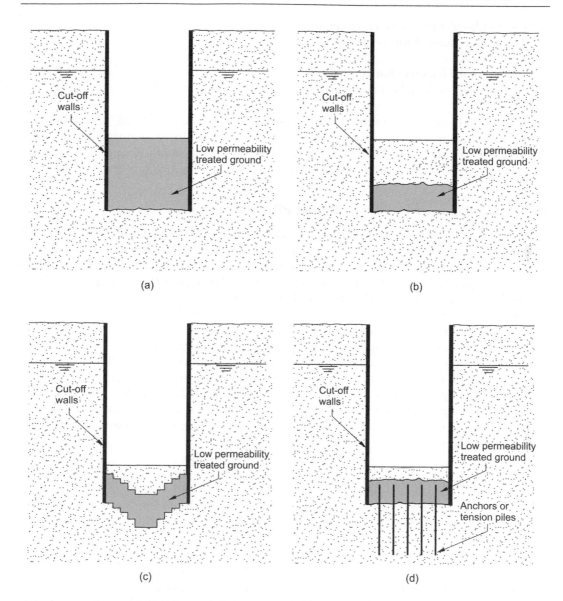

Figure 18.5 Different configurations of horizontal barrier (base plug) to resist buoyancy forces. (a) Thick base plug immediately below excavation formation level. (b) Thin base plug at depth below excavation formation level. (c) Base plug formed as an inverted arch. (d) Anchored base plug.

d) Anchored based plug (Figure 18.5d). The treated ground acts as a barrier to vertical seepage and resists buoyancy forces by a combination of deadweight and the action of tension piles or anchors.

18.4 PERMEABLE WALLS

The focus in this chapter is on methods that can form elements (including vertical walls) of very low permeability to act as barriers to groundwater flow. It is important to realize that there are other methods used to form retaining walls around excavations that are not suitable (without other measures) to be used as a groundwater exclusion method. Two commonly used methods of this type are:

- Contiguous concrete bored pile walls
- Soldier pile walls (also known as king post walls or Berlin walls)

It is important to realize that even though these walls should be considered 'permeable' and do not form groundwater exclusion barriers, the walls are of lower permeability than the surrounding soil and rock. Therefore, the presence of the wall will influence groundwater seepage patterns. For contiguous pile walls, this is discussed by Richards *et al.* (2016).

18.4.1 Contiguous Concrete Bored Pile Walls

In this method, cast in situ concrete piles (typically of 600–1200 mm diameter) are installed at spacings such that the pile casings can be installed without interfering with adjacent piles or so that the auger used to form the pile is not damaged by cutting adjacent piles. A small gap (of the order of 75–150 mm) remains between the piles (Figure 18.6). Further details on the method can be found in Robinson and Bell (2018b).

Contiguous pile walls are not watertight in their conventional form, and their primary use is as an earth retaining structure, which may be incorporated into the permanent works. Water can seep through the gaps between the piles, and if excavation is made below groundwater level in potentially unstable soils (where there is significant sand or silt content) or in highly weathered rock, then fine particles may be washed out of the ground via the gaps. In some cases, the application of grouting methods has been used to seal the gaps so that the wall can act as a groundwater cut-off barrier. However, this is not ideal, and it is normally preferable to use a form of wall construction that can exclude groundwater without such remedial measures.

18.4.2 Soldier Pile Walls

The soldier pile wall method (also known as king post walls or Berlin walls) is used to form retaining walls for excavations (Figure 18.7). The installation sequence can be summarized as:

i. In advance of excavation, vertical structural elements (the soldier piles or king posts) are installed along the line of the wall at a typical spacing of 2 to 3 m.
ii. Excavation then proceeds incrementally downwards in stages of around 1 m. The excavated face is immediately supported by horizontal elements (commonly called 'laggings') placed to bridge between the soldier piles. Packing, mortar or grout may be placed behind the lagging to increase support.
iii. The excavation and lagging sequence is repeated to allow the excavation to reach full depth.

Further details on the method are given in Fernie *et al.* (2012).

Water can seep through the gaps between the lagging, and if excavation is made below groundwater level in potentially unstable soils (where sandy or silty material is present) or in highly weathered rock, then fine particles may be washed out through the lagging. The method is mainly used when groundwater-induced instability is not a major concern – either because ground conditions are favourable or because pre-drainage methods of groundwater control have been deployed to lower groundwater levels.

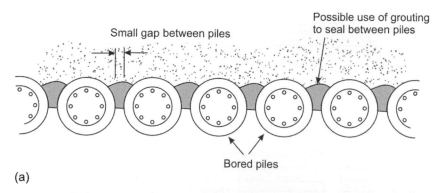

Small gap between piles

Possible use of grouting
to seal between piles

Bored piles

(a)

(b)

Figure 18.6 Contiguous concrete bored pile walls. (a) Contiguous pile wall. Adjacent piles do not intersect, and a small gap remains between piles. In this example, grouting is used to attempt to seal between piles. (b) Contiguous pile wall with capping beam installed. The gaps between adjacent piles are clearly visible. In this case, groundwater level has been lowered below excavation formation level by external dewatering pumping.

18.5 STEEL SHEET-PILING AND COMBI-PILE WALLS

Steel sheet-pile walls consist of a series of interlocked steel sections (typically of a 'Z' or 'U' profile), which are driven, vibrated or pushed into the ground to form a continuous barrier (Figure 18.8). Sheet-pile walls can be used (subject to appropriate design and/or the provision of propping) in a structural role to support the sides of an excavation as well as acting as a barrier to groundwater flow (Figure 9.12). An important characteristic of sheet-pile walls is that they can potentially be used to provide temporary cut-offs. If the sheet-pile wall does not form part of the permanent works, it may be possible to remove the sheet-piles during the later stages of construction, thereby reducing the potential groundwater impacts from a permanent groundwater cut-off barrier being left in place. Extracted sheet-piles will commonly be used on subsequent construction projects, so any given sheet-pile may be used to provide cut-off walls on several projects during its life.

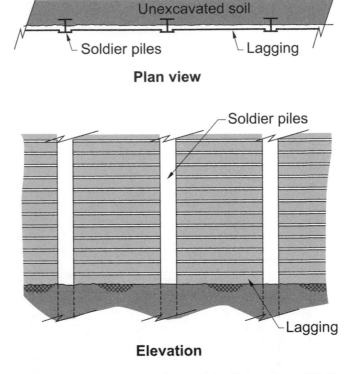

Figure 18.7 Soldier pile walls. (After Fernie, R, et al., Embedded walls, in *ICE Manual of Geotechnical Engineering, Volume II Geotechnical Design, Construction and Verification* (Burland, J, Chapman, T, Skinner, H and Brown, M, eds), ICE Publishing, London, pp. 1271–1288, 2012.)

Traditionally, sheet-pile walls are installed by driving of the piles, where a piling hammer or vibrator is suspended from a service crane and attached to each pile in turn to drive it into the ground. These methods are still in widespread use, but it is recognized that they generate levels of noise and vibration that may not be acceptable, especially in urban areas or where sensitive structures are present. In recent years, there have been great advances in so-called 'silent' methods of piling installation. These methods 'press' the piles into the ground by the application (in stages limited by the throw of the jacks) of steady pressure from hydraulic jacks rather than the dynamic forces from impact or vibration. The hydraulic pile presses used for this task use the reaction supplied either from previously installed piles (Figure 18.9) or from a large base machine. Such methods generate significantly lower levels of noise and vibration compared with traditional driving methods.

As has been discussed previously, the ability of a cut-off wall to exclude groundwater is highly dependent on the watertightness of its joints. A key issue with sheet piles is that because of the narrow width of an individual sheet-pile (typically 400 to 600 mm, measured in the direction of the wall), very many piles are needed to form a wall of any length, and as a consequence, there will be very many joints in the wall. On a typical sheet-pile section, either end will be formed into a hook profile known as the 'clutch'. When used in a cut-off wall, the clutch of each pile slides into the clutch of its neighbour on either side and, in theory, provides both a structural interlock and a relatively watertight seal. In reality, the clutches may leak significantly, and several proprietary sealants are available and can be installed in the clutches prior to driving in an attempt to reduce leakage through clutches. It should be recognized that if pile installation and alignment is not carefully controlled, or

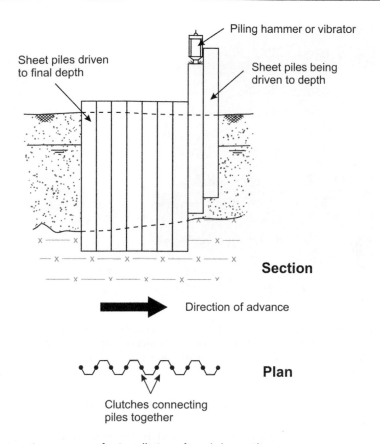

Figure 18.8 Construction sequence for installation of steel sheet-piling.

if driving conditions are difficult, it is possible that piles may 'declutch' during installation. If this occurs, part of the pile depth separates from its neighbours, which may leave a large permeable 'window' in the cut-off wall through which water can flow.

A variant on sheet-pile walls is the combined pile or 'combi-pile' wall (Figure 18.10). Such walls comprise primary elements formed of large-diameter heavy steel sections (often tubular steel piles) driven in a line at a spacing such that gaps remain between them. These gaps are filled by secondary elements typically comprising steel sheet-piles, which are linked to the primary elements by clutches. These walls are appropriate when the structural loading on a wall is high; the primary elements provide the principal structural strength, and the secondary elements complete the wall, allowing it to act as a groundwater barrier.

Further information on steel sheet-piling can be found in Williams and Waite (1993), BS EN 12063:1999 and Fernie *et al.* (2012).

18.6 VIBRATED BEAM WALLS

Vibrated beam walls involve successive overlapping insertions of a steel probe or mandrel into the ground, and the injection of grout (typically a cement-bentonite slurry) into the slot left behind by the mandrel as it is withdrawn. The mandrel used to form the wall typically comprises a heavy-duty steel 'I' beam, driven into the ground by vibrator, hence the generic term 'vibrated beam walls'. The method is known by a variety of other names, including

Figure 18.9 Installation of steel sheet-piling by the press-in method using hydraulic jacks. (Courtesy of Giken Europe BV, London.) The hydraulic jacking unit sits on top of the previously driven piles. The jacking unit grips onto the piles and presses new piles into the ground using the resistance of the neighbouring piles.

thin grouted membrane, ETF wall or vibwall, depending on the application and the company carrying out the work.

A typical installation sequence for a vibrated beam wall is shown in Figure 18.11. A vibrator is mounted atop the I beam, operated on a leader suspended from a service crane. A grout pipe is fixed to the web of the I beam to feed an injection nozzle located close to the base of the beam. The vibrator drives the beam to the target depth (grout may be injected at this stage to aid driving). In favourable conditions, depths of 20 m may be achieved, but penetration difficulties may be caused by the presence of cobbles or boulders. Pre-boring is sometimes used to loosen the ground prior to beam insertion. As the I beam is withdrawn, grout is injected down the pipe to fill the void left by the beam. Once withdrawn, the beam is re-inserted into the ground to form the next section of wall, ensuring that there is sufficient overlap to provide continuity of the wall. The whole process is repeated until the length of wall is completed. A section of vibrated beam wall exposed during excavation is shown in Figure 18.11c.

A vibrated beam wall has little structural value, and the principal groundwater control applications have been to form deep temporary cut-off walls through easily penetrated ground conditions such as sands. A typical example was for the Conwy Crossing project in the United Kingdom, where a vibrated beam wall was used to form a cut-off wall through a bund of hydraulically placed sand fill around a temporary casting basin used to construct elements for an immersed tube tunnel (Powrie and Roberts, 1990). In that application, the bund was to be dredged out in the latter stages of the project to allow the tunnel units to be floated out. The modest structural strength of the wall was an advantage, because it did not impede dredging of the bund.

Because of the relatively narrow width of the majority of the wall (which is formed by the void left by the web of the I beam), there is a risk that deviation or deflection of the beam will result in loss of overlap between adjacent beam insertions. If this occurs, water may be

Figure 18.10 Combi-pile walls. The wall comprises primary elements (large-diameter tubular steel piles) and secondary elements (steel sheet-piles), which are linked together by clutches.

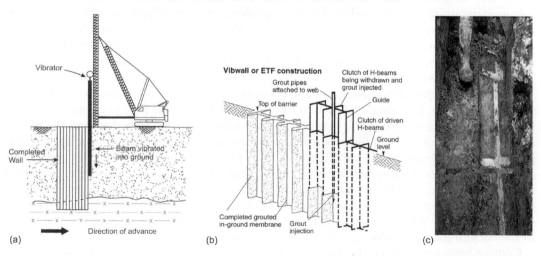

Figure 18.11 Construction sequence for vibrated beam cut-off walls. (a) Installation sequence for vibrated beam cut-off walls. (b) Schematic detail of vibrated beam cut-off walls. (c) Exposure of vibrated beam wall during excavation. (Panel (b) from Woodward, J, *An Introduction to Geotechnical Processes*, Spon, Abingdon 2005. With permission; panel (c) courtesy of Bachy Soletanche, Burscough, UK.)

able to pass through the wall. In some instances, clutches of two or three beams are driven together in an attempt to reduce the risk of deviation.

Further information on vibrated beam walls is given in Privett *et al.* (1994) and McWhirter (2000).

18.7 SLURRY TRENCH WALLS

A slurry trench is formed by the excavation of a trench supported by a slurry (most commonly, but not always, a bentonite suspension) during excavation. Depending on the type of construction, a wall formed of soil-bentonite mixture or a self-hardening cement-bentonite mixture is produced. In its common forms, a slurry trench has limited lateral strength and is not intended to act as a structural wall for either temporary or permanent works unless reinforced with support elements such as beams or steel cages. Its primary objective on civil engineering projects is to act as a barrier to groundwater flow; if the wall is to act as a retaining structure, a concrete diaphragm wall (see Section 18.8) may be more appropriate. On contaminated sites, slurry trench walls (often in combination with high-density polyethylene [HDPE] membranes set into the wall) are used to prevent sub-surface migration of contaminants (Privett *et al.*, 1996). Case History 27.11 describes a project where a slurry trench wall was used on a contaminated site.

The most common form of excavation is continuous excavation by a backactor fitted with an extended boom and dipper arm (Figure 18.12a), which is effective for depths down to 15–20 m. Specialist long-reach backactors may be able to work at depths as deep as 25–30 m (Figure 18.12b). The excavator digs a trench, which is kept topped up with bentonite or cement-bentonite slurry (Figure 18.12c). During excavation, the role of the slurry is to support the trench and prevent collapse. The level of slurry in the trench is kept topped up above groundwater level. This allows a thin 'filter cake' of bentonite to form on each face of the trench as clay particles are filtered from the slurry as it seeps into the surrounding soil. The filter cake reduces loss of slurry from the trench and allows a differential head to develop between the slurry and the groundwater outside the trench. This differential head of slurry plays a key role in maintaining trench stability during excavation, and typically, these walls can only be constructed if there is a minimum 1 to 2 m head of slurry above natural groundwater level.

Soil-bentonite walls are most commonly used in countries influenced by American construction practices, while cement-bentonite walls are common in European-influenced areas.

Soil-bentonite walls are constructed by the excavation of a trench under bentonite slurry. The backfill is prepared by mixing soil and bentonite adjacent to the trench. Typically, at the start of the trench, the initial section of backfill is placed by tremie pipe to ensure consistent filling. Once the initial backfill is in place, the remaining material is pushed into the trench from one end to form a steadily progressing shallow slope of fill material, which tracks the progress of the trench excavation (Figure 18.13). The material used for the fill is commonly spoil from the trench, although if the natural ground has limited fines content, imported fill may be used to give better wall properties. The mixing of the soil and bentonite at the surface is often done by a bulldozer tracking and blading clean bentonite slurry into small stockpiles of soil. Where some limited structural strength is needed, cement may be added to the backfill mix to produce a soil-cement-bentonite wall.

Cement-bentonite walls can be constructed by either single-stage or two-stage construction:

- In single-stage construction, the trench is excavated under the self-hardening cement-bentonite slurry, which later sets as the permanent backfill (Figure 18.14a). This approach may be satisfactory for shallow trenches where the excavation time is much lower than the slurry setting time.

(a)

(b)

(c)

Figure 18.12 Construction slurry trench wall by backactor. (a) Excavation by conventional backactor fitted with extended boom and dipper arm. (b) Excavation by specialist long reach backactor. (c) Trench kept topped up with bentonite slurry during excavation. (Panel (a) courtesy of Arup, London; panel (b) courtesy of Inquip Associates Inc., Santa Barbara, CA; panel (c) courtesy of Inquip Associates Inc, Santa Barbara, CA.)

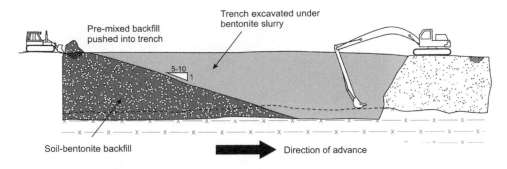

Figure 18.13 Construction sequence for soil-bentonite slurry trench wall.

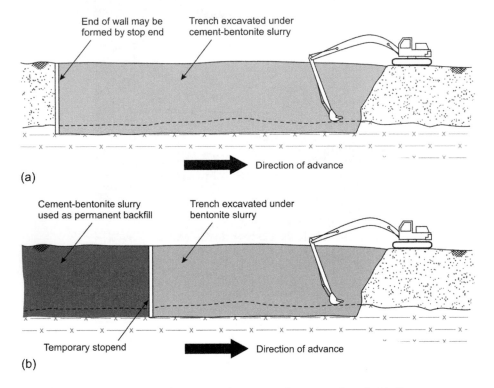

Figure 18.14 Construction sequence for cement-bentonite slurry trench wall. (a) One-stage construction for cement-bentonite wall. (b) Two-stage construction for cement-bentonite wall.

- In two-stage construction, the trench is supported by a bentonite slurry during excavation, which is then replaced with the permanent cement-bentonite mix to form the final wall (Figure 18.14b). In this approach, temporary stop ends are used to separate the bentonite slurry used in excavation from the permanent cement-bentonite mix. This is beneficial to reduce the risk of cross-contamination affecting the properties of both materials.

The very top of the slurry trench wall may experience cracking due to drying out, and it is good practice to dig out the top 500 to 1000 mm and replace with a cap of compacted clay (Figure 18.15). The cap may also incorporate a shallow depth of HDPE membrane (typically 2 m depth), particularly where the wall is to have a long design life.

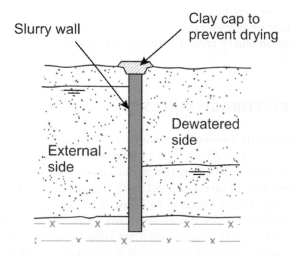

Figure 18.15 Capping of slurry trench wall with compacted clay.

While excavation by backactor is by far the most common method of construction for slurry trench walls and can achieve depths of 15–20 m using widely available equipment (and depths of 25 m or deeper using specialist backactors), other methods are also used. Draglines are occasionally used and are effective down to 25 m depth of wall. On rare occasions, specialist trenchers have been used (Brice and Woodward, 1984; Schünmann, 2004) down to 8 m depth.

Deeper walls (down to 100 m wall depth) require construction by the non-continuous panel method using diaphragm walling grabs or hydromills (see Section 18.8). In some cases, to reduce costs, the upper 25 m or so of a deep trench may be dug by the cheaper backactor method, after which grabs or hydromills are deployed to excavate the base of the trench to full depth. Slurry trench walls to such depth are used only rarely in civil engineering applications and are more relevant to deep cut-offs for dam construction or to reduce seepage into open pit mines.

Typical wall widths are in the range of 300 to 3000 mm. Wall permeabilities of 10^{-7} to 10^{-9} m/s can be achieved. Typical target permeabilities in performance specifications are in the order of 10^{-8} to 10^{-9} m/s for cement-bentonite walls (without the use of polyethylene membranes); soil-bentonite walls typically produce permeabilities in the order of 10^{-9} m/s. The required thickness of the wall is determined to ensure that the hydraulic gradient across the wall is not excessive or to provide an adequate design life in aggressive ground conditions. In groundwater control applications, the hydraulic gradient is normally limited to 10–30.

The approximate mix design for the slurry is an important part of slurry trench wall work and should be carried out by specialists. Typically, a cement-bentonite slurry will comprise a mixture of various proportions of bentonite, ordinary Portland cement [OPC] and ground granulated blast-furnace slag (GGBS). The typical slurry density will be approximately 1.1 tonne/m^3.

To ensure the stability of the trench during excavation, the head of bentonite slurry must be maintained 1 to 2 m above groundwater level. Where groundwater level is close to ground level, it may be necessary to build up ground level by the placement of fill to ensure that the slurry head can be increased to an adequate level.

Because slurry trench walls are typically constructed either by continuous excavation or by the panel method, they are not particularly suitable when forming cut-off walls of complex plan layouts. They are most suitable for forming cut-off walls where the plan layout is essentially rectangular or polygonal with relatively long runs of wall between corners.

Further information on slurry trench walls is given in Jefferis (1993) and Institution of Civil Engineers (1999).

18.8 CONCRETE DIAPHRAGM WALLS

Diaphragm walls (commonly known as D-walls) are concrete walls, typically cast in situ, formed within a trench supported by a slurry (most commonly, but not always, a bentonite suspension) during excavation. The concrete is placed into the trench, displacing the slurry and forming a concrete wall in direct contact with the ground. In typical applications, the wall is constructed of reinforced concrete to allow it to act as a structural element of the permanent or temporary works (Figure 18.16). Less commonly, the wall may be intended purely as a groundwater cut-off barrier, with no structural role, and may be formed of unreinforced plastic concrete (plastic concrete is a relatively ductile concrete formed from aggregate, cement and bentonite).

Typically, the construction sequence for a diaphragm wall involves the wall being excavated and concreted in discrete 'panels' (Figure 18.17). Alternate primary panels are constructed, with the end of each panel formed by a temporary steel section known as a 'stop end'. Later, the stop end is removed by a crane, and the secondary panels are constructed between the primary panels. Typically, wall widths are 800–1500 mm (although thicknesses of up to 2500 mm have been installed), and panel lengths are 3–7 m.

Diaphragm walls are constructed by specialist excavating equipment, normally suspended from a heavy-duty crane or base machine, which forms the trench within which each panel will be constructed. Excavation is carried out between surface 'guide walls', which ensure positional accuracy and prevent the top of the wall enlarging as a result of the insertion and removal of the digging tools. To prevent the trench collapsing during excavation, it is kept flooded with a support fluid. The support fluid is typically a bentonite suspension with some polymer admixtures to control its properties – the fluid is commonly referred to as bentonite

Figure 18.16 Concrete diaphragm walls around a rectangular excavation. The diaphragm wall is tied back by ground anchors. Dewatering wells are being used to control groundwater within the excavation.

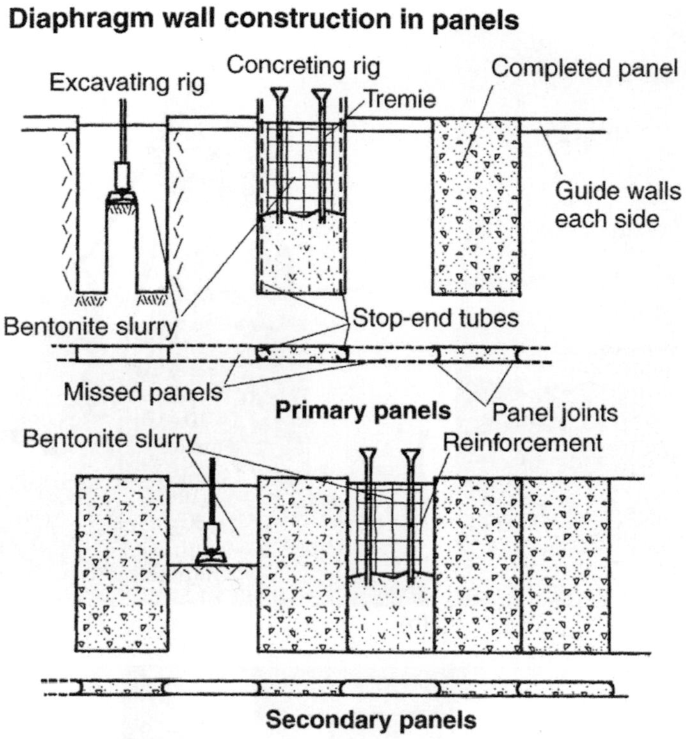

Diaphragm wall construction in panels

Figure 18.17 Construction sequence for concrete diaphragm wall. (From Woodward, J, *An Introduction to Geotechnical Processes*, Spon, Abingdon 2005. With permission.)

slurry. The trench is kept topped up with slurry to a level above external groundwater level, so the hydrostatic pressure of the slurry on the trench faces improves stability in a similar fashion to slurry trench walls (Section 18.7). The slurry will become contaminated with sediment during the excavation process and is typically pumped from the trench, passed through a desanding plant and recycled back into the trench.

When panel excavation is complete and residual sediment has been removed from the bentonite slurry in the trench, the concrete can be placed. For reinforced concrete walls, a pre-fabricated reinforcement cage is lowered into the slurry-filled trench. A specially designed concrete mix is then tremied into the trench. Concrete placement is carefully controlled to ensure continuous placement from the bottom up to displace the bentonite slurry from the trench.

Excavation for diaphragm walls is usually carried out by either grabs or milling machines with rotating cutting tools (known as hydromills). Because the trench is kept topped up with bentonite slurry, each time the digging tool is lowered into the trench, it disappears beneath the slurry; the plant operator is effectively digging 'blind'. Therefore, great care must be taken to ensure verticality and alignment of the panels in order to ensure that wall continuity is achieved over the full depth. Concrete guide walls at ground level play a key role as positional guides for the panel excavation. Most modern equipment is now fitted with sensors to allow verticality to be checked and adjusted as required during excavation.

The traditional excavation equipment for diaphragm walls is a heavy-duty rope operated grab (Figure 18.18a), suspended on either a cable or a Kelly bar from a modified crawler crane rig. More recently, hydraulic grabs, which have a much higher closing force, have become commonplace. Grabs are suitable for excavation in soils and some soft rocks. Since

(a)   (b)

(c)

Figure 18.18 Excavation equipment for concrete diaphragm walls (a) Diaphragm walling grab. (b) Hydromill diaphragm walling rig. (c) Close-up of rotating cutters on hydromill. (Panel (a) courtesy of Bachy Soletanche, Burscough, UK; panels (b) and (c) courtesy of Cementation Skanska Foundations, Maple Cross, UK.)

the 1980s, there has been considerable development in hydromills (Figure 18.18b). These units are sometimes referred to as rockmills or hydrofraises (after one of the first units, developed by Soletanche in the 1980s). In the most common configuration, drum cutters rotating on a horizontal axis (Figure 18.18c) are used to break up the soil or rock in the base of the excavation, which is then pumped with the slurry, via a reverse circulation system, to a desanding plant on site. The clean slurry is then pumped back to the panel excavation. Hydromills typically allow faster excavation rates than grabs and have been effective in very stiff soils and in rock, where excavation by grab would be slow and problematic. Diaphragm walls have been installed to depths of more than 150 m.

In order to ensure that the trench is adequately supported during excavation, the head of slurry must be maintained at least 1 to 2 m above groundwater level. Occasionally, dewatering systems such as wellpoints have been used to temporarily lower groundwater levels during wall excavation to ensure that sufficient slurry pressures are maintained in the trench.

Because diaphragm walls are typically constructed by the panel method, they are not particularly suitable for use in cut-off walls of complex plan layouts. They are most suitable for forming cut-off walls where the plan layout is essentially rectangular or polygonal with relatively long runs of wall between corners. If complex plan geometries are required, concrete secant pile walls may be more appropriate. Leakage occasionally occurs at panel joints (see Case History 27.10 for Sizewell B Power Station) and may require retrospective sealing.

Further information on diaphragm walling can be found in Puller (2003), BS EN 1538:2010 and Robinson and Bell (2018a).

18.9 CONCRETE SECANT PILE WALLS

Secant pile walls are formed from circular concrete piles installed in close proximity to form a line to support the perimeter of the excavation and act as a barrier to groundwater flow. The essence of the secant pile method is that alternate piles are bored at spacings of less than one diameter so that they will then intersect and form a continuous concrete wall (Figure 18.19). This is different from contiguous pile walls (Section 18.4.1), where a small gap remains between piles.

Secant pile walls have been used in groundwater control applications to form the support structures for conventional excavations and also for circular and elliptical shafts. Because the wall is formed of individual piles whose relative position can be carefully controlled, secant pile walls are better able to provide walls of complex plan arrangements (curved walls or complex intersections) than wall systems such as diaphragm walls, which are installed in larger panels.

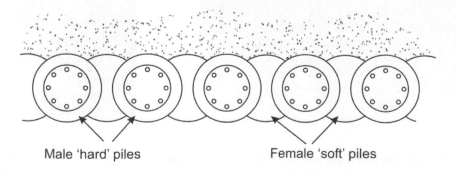

Male 'hard' piles   Female 'soft' piles

Figure 18.19 Concrete secant pile wall. Adjacent piles intersect to form a continuous wall. Female piles are installed first, and male piles are subsequently installed, cutting into the female piles.

The method involves the installation of bored cast in situ concrete piles (typically of 600–1200 mm diameter) in a specific construction sequence, where alternate piles (known as primary or 'female' piles) are installed in each line, with secondary or 'male' piles subsequently installed to fill in the gaps. The piles are spaced at less than one diameter so that the male piles cut into the female piles in a process known as 'overcutting'. The male piles cut a secant out of the female piles, thereby creating a continuous wall (Figure 18.20).

Secant pile walls are sometimes categorized based on the nature of the female piles (the male piles are almost always reinforced structural concrete):

- Hard – soft walls: The female piles ('soft' piles) are constructed of an unreinforced cement-bentonite mixture. These walls are essentially temporary works, because the female piles do not have the long-term durability to form part of the permanent works. These walls are sometimes included in permanent works by the addition of a structural concrete liner on the inner face once exposed by excavation.
- Hard – firm walls: The female piles ('firm piles') are constructed of unreinforced concrete to allow them to be incorporated into the permanent works. The concrete used in the female piles is significantly weaker than in the male piles to avoid constructability problems and make overcutting of female piles easier.
- Hard – hard walls: The female piles are formed from structural concrete of equal strength to the male piles (and may contain reinforcement of suitable geometry to avoid being damaged during overcutting). Special piling rigs are needed to perform the overcutting of the male piles into the high-strength female piles.

For pile diameters in the range of 600 to 1200 mm, typical overcut (the distance the male piles cut into the female piles) will be in the range of 100 to 250 mm depending on ground conditions and the design requirements of the wall. Construction sequence and programming are important for secant pile walls. The female piles are installed first, leaving gaps to

(a) (b)

Figure 18.20 Concrete secant pile wall (a) Secant pile wall exposed by excavation Bored pile wall is capped by a concrete beam and supported by tubular steel props. (b) Secant pile wall exposed showing unreinforced female piles and reinforced male piles. (Panel (a) courtesy of Balfour Beatty Ground Engineering, Basingstoke, UK; panel (b) courtesy of Bachy Soletanche, Burscough, UK.)

be filled in by male piles. But if the female piles are left too long before the male piles are cut into them, the concrete in the female piles will have developed high strength, which will create practical problems for male pile installation. Excessive concrete strength in female piles can lead to verticality and alignment problems with male piles. It is possible for male piles to pop out of alignment at depth, resulting in groundwater flow paths through the wall. Therefore, male pile installation should be programmed to be carried out at an appropriate time, based on the concrete mix design, after the female piles are installed.

Secant pile walls can be installed to depths of up to 30 m if suitable boring plant is available. It is important to understand that the practical depth over which an effective secant pile wall can be constructed is the depth over which pile secanting can be guaranteed. This will be determined by the piling tolerances achieved during the works and not merely the depth that can be achieved by the piling rig. Reinforced concrete guide walls (similar to those used in diaphragm walling) are typically installed at ground level as an aid to achieving good positional and verticality tolerances.

Additional information on bored pile walls is given in Puller (2003), BS EN 1536:2010 and Robinson and Bell (2018b).

18.10 GROUT BARRIERS

18.10.1 Principles of Grouting

Grouting encompasses a range of techniques used in ground engineering to modify the properties of soils and rocks by the controlled injection of special fluids (grout), which set or harden to modify the ground properties. In the context of this chapter, we are primarily interested in grouting methods that reduce the permeability of soils and rocks, although many of the methods also increase the soil and rock strength.

Rawlings *et al.* (2000) define grouting in the following terms:

Grouting in ground engineering can be defined as the process of controlled injection of material, usually in a temporary fluid phase, into soil or rock, where it stiffens to improve the physical characteristics of the ground for geotechnical engineering reasons. In such a situation, grouting is a process in which the remote placement of a pumpable material in the ground is indirectly controlled by adjusting its rheological characteristics and manipulating the placement parameters (pressure, volume and flow rate).

There are seven principal types of grouting techniques:

1. *Permeation grouting:* Filling or partially filling permeable pores within a soil by grout injection without disturbing the structure of the soil.
2. *Rock grouting:* Filling or partially filling fractures, joints and other discontinuities in the rock mass by grout injection without creating new fractures or opening existing fractures.
3. *Jet grouting:* Disruption of existing soil structure by water/grout jets and in situ mixing with, and replacement by, injected grout.
4. *Soil mixing:* Disruption of existing soil structure by mechanical tools (e.g. augers or cutters) and in situ mixing with, and replacement by, injected grout.
5. *Hydrofracture grouting:* Deliberate fracturing of the ground (soil or rock) using grout under pressure. Typically used to inject grout into otherwise accessible voids to reduce permeability or increase strength.

6. *Compaction grouting:* Injection of stiff mortar or paste-like grout into the ground to displace and compact the soil in situ.

7. *Compensation grouting:* Use of compaction, permeation or hydrofracture grouting in a controlled, responsive manner as an intervention between an existing structure and an engineering operation. A typical example is compensation grouting associated with tunnelling works, where it is applied to minimize tunnelling induced movements on existing structures (Mair *et al.*, 1995).

Permeation grouting, rock grouting and jet grouting are the main types of grouting used to form groundwater cut-off barriers to exclude water from civil engineering excavations. Each of these techniques will be briefly described in subsequent sections. Additionally, soil mixing can be used to form mix-in-place barriers, which are sometimes used for ground water exclusion. Mix-in-place barriers are described in Section 18.11.

Further background information on grouting methods is given in Rawlings *et al*, (2000), BS EN 12715:2000, Bell (2012) and Hughes *et al.* (2018).

18.10.2 Types of Grout

The nature of the grout used – its type and its specific properties – is fundamental to the success or otherwise of any grouting exercise. Essler (2012) states:

> The design of the grout is important as it must penetrate the interstitial pores with minimum disturbance. If the grout viscosity is too high or the particulate size is too large then complete uniform permeation will not be possible.

It follows that successful application of grouting methods requires some understanding of different types of grout. Grouts used for groundwater exclusion purposes on civil engineering projects fall into two main classes: suspensions and solutions.

Suspension grouts comprise solids suspended in water without being dissolved. True suspension grouts contain particles that are large enough to settle under gravity, with the associated problems of maintaining good grout quality during mixing and handling. Cement-based grouts, which consist of mixtures of cement and water, sometimes with additives such as bentonite or pulverised fuel ash (PFA), fall into this category. Colloidal suspensions (such as colloidal silica) are those in which the particles are so fine that there is no settlement under gravity, and particles are kept in suspension by Brownian motion. The finite size of the solid particles in suspension grouts limits their ability to permeate or penetrate further into small pores or fractures in soils and rocks.

Solution grouts do not contain solid particles and generally 'gel' (solidify) on setting. The most common form of solution grouts are the so-called 'chemical grouts'; these materials are most commonly silicate compounds or resins. The lack of particles in solution grouts means that they typically can permeate or penetrate further into small pores or fractures in soils and rocks compared with typical cement-based suspension grouts.

Table 18.1 summarizes the range of typical ground types that can be treated by various grouts. Characteristics of the grout types most commonly used in groundwater exclusion applications are given in the following sections.

18.10.3 Cement-Based Grouts

Cement-based suspension grouts are used widely around the world. This reflects the wide availability of the raw materials and the relatively simplicity and low cost of grout preparation.

Table 18.1 Indicative Grouts for Different Types of Ground (Based on BS EN 12715:2010)

Type of ground	Soil/rock properties	Permeation grouting	Rock or contact grouting
Granular soil	Gravel, coarse sand and sandy gravel $k > 5 \times 10^{-3}$ m/s	Pure cement suspensions Cement-based suspensions	
	Sand $5 \times 10^{-5} < k < 5 \times 10^{-3}$ m/s	Microfine cement suspensions Solutions	
	Medium to fine sand $5 \times 10^{-6} < k < 1 \times 10^{-4}$ m/s	Microfine cement suspensions Solutions Special chemicals	
Fissured rock	Faults, cracks, karst $e > 100$ mm		Cement-based mortars Cement-based suspensions (clay filler)
	Cracks, fractures 0.1 mm $< e < 100$ mm		Cement-based suspensions Microfine cement suspensions
	Microfractures $e < 0.1$ mm		Microfine cement suspensions Silicate gels Special chemicals

e = fracture width; k = permeability

In their simplest form, cement-based grouts are suspensions of cement in water, with the water–cement ratio manipulated to provide the required rheological properties. The stability (resistance to sedimentation of particles) of a grout for a given water–cement ratio may be improved by the addition of clay, bentonite or chemical additives. For filling large voids (old mine workings or karst voids), a filler such as PFA may be added as a substitute for cement and to reduce cost per unit volume of grout. Accelerators may be added to adjust setting time.

The key to understanding cement-based grouts and their limitations is to recognize that the cement particles suspended in the water are of a finite size that is comparable to the openings in many soils and rocks. Conventional OPC typically has a particle size distribution in which 95 per cent of particles (D_{95}) are finer than 100 μm. This means that in all but very permeable soils (coarse gravels with little sand or silt content) and rock with wide openings, OPC-based grouts are not very effective (see Table 18.1). The problem which occurs is that during injection, cement particles quickly become wedged in pores or fractures, preventing further grout penetration. Hence, the grout penetrates only a short radius from the point of injection, and it is very difficult to form a continuous grout barrier of treated ground between injection holes.

The penetration of cement-based grouts is improved by the use of microfine and ultrafine cements. During the manufacture of these materials, the cement is ground extensively to produce smaller particle sizes than OPC. Two particle size parameters used to describe these cements are the D_{95} particle size (95 per cent of the particles, by weight, are finer than this value) and the Blaine value (expressed in square metres per kilogram), which is a measure of the specific surface of the particles. The larger the Blaine value, the finer the particle size distribution. Microfine and ultrafine products from different manufacturers will have different properties. Some approximate guidance is given here:

- Microfine cement: $D_{95} < 20$ μm; Blaine value > 600 m²/kg
- Ultrafine cement: $D_{95} < 10$ μm; Blaine value > 600 m²/kg

Grouts using these raw materials therefore contain smaller particles and have the potential to penetrate/permeate smaller openings (soil pores/rock fractures). Microfine and ultrafine cement is significantly more costly than conventional OPC cement; grout will be correspondingly more expensive and is typically only used to grout fine openings in soil or rocks that cannot be penetrated by OPC-based grouts.

The penetrability of cement-based grouts can also be improved by decreasing the grout viscosity and gel strength (using additives to reduce viscosity by flocculation). Alternatively, penetrability can be aided by increasing the resistance to filtering effects by the addition of activators (dispersing agents).

18.10.4 Chemical Grouts

The most common types of chemical grouts are:

- Silicate-based grouts
- Resin grouts
- Colloidal silica grouts

Silicate grouts are based on sodium silicate solutions, to which an acid hardening agent is added. The viscosity of the mix changes over time, reaching a solid state as a gel. By varying the type and concentration of chemical components in the grout, a wide range of properties can be obtained. There are two principal types of gel:

- Soft gels (water-tightening gels)
- Hard gels (strengthening gels)

Soft gels have a low concentration of silicate, with gelling typically obtained by adding mineral reagents. The low viscosity (similar to that of water) means that they can be injected into fine-grained sands, where they have a 'water-tightening' effect, reducing permeability. These grouts have a short design life (around 6 months to 2 years) and are so only suitable for temporary works of relatively short duration. The permeability of treated ground may increase over time due to degradation of the grout by various processes, including washing out and erosion.

Hard gels have a higher silicate concentration and organic reagent content and produce a stronger gelled grout. The design process may need to address the conflicting issues of grout strength and durability versus penetrability of the grout into the target ground.

Resin grouts are based on solutions of organic products, either in water or in non-aqueous solvents that can form a gel under given conditions. Resin grouts can potentially have very low viscosity (and hence can penetrate very small soil and rock openings) and setting times that can be controlled between a few seconds and several hours (depending on the grout components and reagents). Resin grouts are typically much more expensive than other types of chemical grouts. They are most typically used where durable grouts are required and no other grout types are feasible.

Resin types used for permeability reduction and water exclusion purposes include acrylic, polyurethane and phenolic grouts. Some resins are of significant toxicity and should only be used by experienced designers and contractors with suitable health protection and control measures in place.

Colloidal silica grouts have been developed to allow very fine-grained sands to be successfully grouted. These grouts consist of a nanometric colloidal suspension of silica particles (typical size 0.01 to 0.1 µm), which acts like a true liquid and can penetrate into very small pores and fractures. The grout is applied using the 'two component' method, where immediately prior to injection, a second component of a simple salt solution is added to the silica suspension. This disturbs the colloidal state of the particles and causes a strong and durable gel to form. The gel time can be closely controlled by varying the type and concentration of the second component. The advantage of colloidal silica grouts is that they are permanent and do not suffer the durability problems of silicate grouts. These grouts have the further advantage that they contain neither solvents nor toxic constituents.

18.10.5 Design of Grout Barriers

It should be recognized that the successful installation of a groundwater cut-off barrier by grouting methods is a complex process and will be influenced by many factors, including ground conditions, the fabric and structure of the soil or rock, grout properties (rheology) and the method of grout delivery. The eminent civil engineer Sir Harold Harding (one of the pioneers of modern grouting techniques in British construction practice) once said that one of the challenges of grouting was 'manipulating it in the ground beyond vision or arm's length' (Harding, 1947). It is certainly true that injecting grout into the ground is one thing, but delivering it accurately to the desired location and achieving the required permeability reduction may be quite another challenge.

A grouting solution for a given site should be developed on a bespoke basis, based on a thorough ground investigation, and the expert advice of experienced grouting designers should be obtained – design principles for grouting projects are outlined in Essler (2012). On projects of any significant size, or where the success of grouting is critical, a trial section of grouting is often carried out in advance of the main cut-off works to allow the grouting design to be finalized.

A key element of the design of a grouted groundwater barrier is the selection of the grout type and properties. The choice of grout is influenced primarily by the permeability of the ground, particularly the likely size range of the interconnected voids in soil or interconnected fractures in rock. In soils, an empirical parameter known as 'groutability ratio' (defined as the ratio between the particle sizes representing 15 per cent of the ground [D_{15}] and 85 per cent of the grout particles [D_{85}]) can be used to assess the penetrability of particulate grouts. In rock, maximum grout particle size to fracture width is considered, a ratio of three being commonly used. Table 18.1 summarizes the range of typical ground types that can be treated by various grouts.

Various grouting techniques have the potential to generate significant volumes of solid and slurry spoil, which must be disposed of appropriately and in accordance with the relevant regulations. In many cases, the spoil will contain greater or lesser quantities of grout, so the practical and regulatory aspects of disposal of cement-based or chemical grouts must be considered.

In most cases, grouted groundwater barriers are effectively permanent and will remain in place long after construction works are finished. It may be necessary to assess the long-term impact of the change to groundwater flow patterns (see Section 21.7) or consider the need for the barrier to be physically breached at the end of construction (see Section 24.5). On rare occasions, where the groundwater chemistry has the potential to interact with the grout barrier, the risk of leaching of polluting chemicals from grout into the groundwater may need to be assessed. This is especially relevant for PFA-based grouts, which can leach heavy metals and are increasingly less favoured for this reason.

In addition to the usual health and safety issues on civil engineering sites, there are some risks particularly associated with the use of grouts. Cement-based grouts can create dust problems during mixing, and there is a risk of 'cement burns' if liquid grout is allowed to stay in prolonged contact with skin. Some types of chemical grouts are harmful to the skin and may give off potentially toxic or irritant fumes. Appropriate health and safety measures should be put in place during grouting operations.

18.10.6 Permeation Grouting

Permeation grouting involves the injection of liquid grout, via boreholes of some sort, at a steady injection pressure without disturbing the soil structure. It is used in soils (and very highly weathered rocks that effectively act as granular soils) to fill or partially fill the pore spaces, displacing the water and air in those spaces. The grout sets or gels in the pore spaces, reducing the potential for intergranular groundwater flow and thereby reducing permeability (Figure 18.21).

In groundwater control applications, permeation grouting is commonly carried out from lines or triangular grids of closely spaced injection holes. The intention is that the grout permeates radially from each drill hole and merges with the grouted zones around neighbouring drill holes. In this way, a more or less continuous 'grout curtain' of lower mass permeability than the surrounding ground can be formed. The injection holes can take various forms:

- Lancing. This involves the percussive driving of a simple steel tube (25–38 mm in diameter), through which the grout is injected. The lance is withdrawn by jacking. This is a relatively crude method and is limited by the capacity of the equipment to depths of 4 to 6 m.

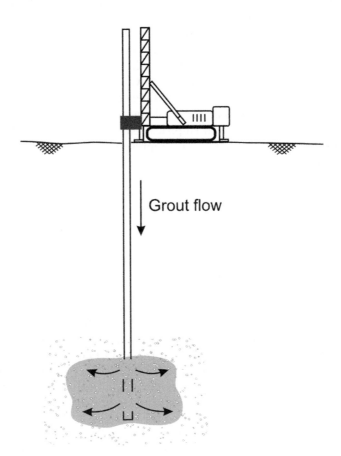

Grout flow

Figure 18.21 Permeation grouting. Liquid grout is injected at a steady pressure, without disturbing soil structure, to fill or partially fill the soil pore spaces

- End of casing (EoC) method. This is similar to the lancing method but uses a rotary drilling rig and a drilling fluid to remove cuttings. Grout injection is directly from the end of a cased borehole (typically of 75–114 mm diameter). Depth is usually limited to around 10 m. The casing is withdrawn by jacks or the action of the drilling rig.
- Tube à manchettes (TAMs). TAM pipes are specialist grouting tubes, typically steel or plastic pipes 50 to 60 mm in diameter, installed in drill holes and permanently secured into place by filling the annulus around them with weak grout (sleeve grout). The TAM pipes have multiple injection ports along their length at regular spacing, allowing multiple injections at different depths. TAM grouting is described in more detail later in this section.

The sequence for permeation grouting is typically injection into each hole of a finite, predetermined volume of grout in a sequence of primary, secondary and possibly tertiary holes. The intention is for the primary injections to create grout zones that overlap or coalesce to some degree. Final closure between the primary injections is achieved by injection into the secondary and tertiary holes. Typical final spacing between grouting holes varies from 0.8 to 1.3 m in fine-grained sands up to 2 to 4 m in more permeable gravels.

During grouting, the effectiveness of the injection programme in reducing permeability and hence, the need for additional phases of grout injection are normally investigated by a programme of 'water testing'. A water test is basically an in situ permeability test (see Section 12.8) carried out within the grouted zone. Where water tests are used as a part of the design or verification process for a grouting programme, it is essential that a sufficient number of tests are carried out to ensure that test results are not skewed by local variations in natural ground conditions (verification testing is described in Section 18.14).

More sophisticated permeation grouting programmes may use the tube à manchette technique (TAM grouting). TAM pipes are specialist grouting tubes, permanently secured into place by filling the annulus around them with weak grout (sleeve grout). The TAM pipes contain grout ports at regular vertical intervals (commonly between 330 and 1000 mm) protected by external rubber seals (Figure 18.22). Using a special double packer lowered down inside the TAM pipe, each level of grout ports can be isolated, allowing grout injection at a specific level, and injection pressures and volumes can be monitored and controlled at each level. When grout is injected, the grout pressure opens the rubber seal covering the grout ports, breaks the sleeve grout and allows grout to pass into the soil. The TAM method allows controlled grouting, because the level of grout injection can be controlled, and repeated injections can be made from any grouting port.

The appropriate selection of grout characteristics is fundamental to successful permeation grouting. For example, when cement-based suspension grouts are used, flow blockages may occur in the soil pores due to filtration of grout particles in the flow paths. This may prevent grout flowing the necessary distance from the injection point. The viscosity of grout, both on injection and later as the grout begins to gel, will also restrict grout flow. The applicability of grouts is summarized in Table 18.1. In general, effective use of low-pressure permeation grouting using conventional cement based-grouts is limited to gravels with modest sand content. The use of microfine and ultrafine cement grouts and chemical grouts may allow permeation grouting to be effective in medium-grained and possibly fine-grained sands.

The reduction in permeability of treated ground that can be achieved by permeation grouting will be affected by the grout design and placement methods, but the achievable effects will be dominated by the nature of the soil. In gravels and medium to coarse-grained sands, a treated permeability of the order of 5×10^{-7} to 1×10^{-6} m/s is achievable with good practice.

The grout used is normally designed to penetrate the finest groutable soil pores, based on the objectives of the grouting programme. Sometimes, different grout types are used in

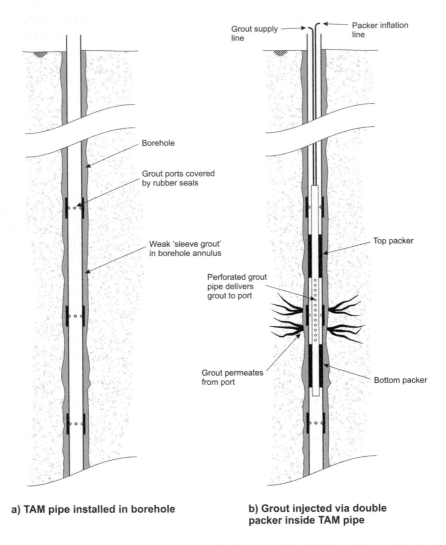

a) TAM pipe installed in borehole

b) Grout injected via double packer inside TAM pipe

Figure 18.22 Grouting using TAM. (a) TAM pipe installed in a borehole. (b) Grout injected via double packer inside a TAM pipe.

a grouting programme. Cheaper cement-based grouts may be used for a sequence of initial grouting to fill the larger void spaces, followed by a later sequence of grouting with more expensive chemical grouts to fill as many of the remaining smaller void spaces as possible.

In general, well-controlled permeation grouting will produce only modest quantities of spoil (associated with the drilling of injection holes). Low-pressure permeation grouting should not cause problems of ground heave if carefully controlled and provided that injection pressures are maintained below hydrofracture levels.

Further information on permeation grouting is given in Rawlings *et al.* (2000), BS EN 12715:2000, Bell (2012) and Hughes *et al.* (2018).

18.10.7 Rock Grouting

Rock grouting is analogous to permeation grouting but applied specifically with the objective of injecting grout to penetrate and fill or partially fill existing fractures, joints and other voids within a rock mass. The grout gels or sets, blocking water flow pathways through the rock and thereby reducing mass permeability (Figure 18.23).

Rock grouting will typically strengthen rock formations because the presence of hardened grout will restrict the potential for movement along fracture or joints in the rock mass. In civil engineering applications, rock grouting is used to reduce groundwater inflows in excavations into rock for shaft sinking, tunnel construction, dry docks, etc. Rock grouting has also been used to provide low-permeability basal seals (Figures 18.4 and 18.5) to reduce inflow into the base of excavations, as described by Davis and Horswill (2002).

Rock grouting is typically carried out in stages from small (typically 50–80 mm) diameter drill holes. For each stage, the relevant part of the borehole is isolated by packers, allowing grout injection to be targeted at specific zones:

- In descending stage grouting, a discrete section of hole (the stage length) is drilled, the drill string removed and the grout injected (water injection testing may also be carried out prior to grout injection). After a period to allow for an initial set of the grout, the grouted stage length is drilled out, and drilling continues in virgin rock to the bottom of the next and lower stage. The lower stage is then injected, and the drilling and grouting cycle is repeated for subsequent stages. In this way, the hole is grouted progressively from top to bottom of the target depth zone to be grouted.
- In ascending stage grouting, the borehole is drilled to full depth in a single pass; the method therefore relies on the rock being sufficiently stable to prevent the borehole collapsing. A packer is then used to isolate an initial stage at the base of the borehole, and grout is injected. After a period to allow an initial set of the grout, the packer is

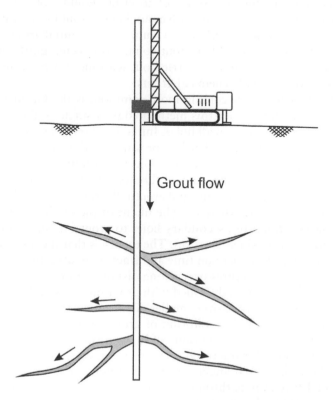

Figure 18.23 Rock grouting. Liquid grout is injected at a steady pressure to fill or partially fill existing fractures within the rock.

moved upwards in the hole and re-set to seal the stage and grout is injected. The grouting cycle is repeated for subsequent stages, moving progressively from the bottom to the top of the target depth zone to be grouted.

It is generally accepted that descending stage grouting allows better control over grouting, but the method is slower and hence more expensive than ascending stage grouting. Where boreholes are sufficiently stable to allow ascending stage grouting, that method is often used on the basis of lower cost.

The spacing between rock grouting drill holes in a completed grouting pattern is typically in the range of 1 to 4 m, usually in single lines or multiple parallel lines depending on the size of discontinuities to be penetrated and the target permeability of the treated ground. In rock grouting, the drill holes are typically drilled in a 'split-spacing' sequence in which additional holes are drilled and grouted between previously grouted holes. Primary holes are drilled at wide spacing (perhaps 5–6 m) between holes. These holes are grouted before secondary holes are drilled and grouted in the intervening spaces, halving the effective grout hole spacing. Tertiary holes may then be drilled and grouted between the primary and secondary holes, and so on. At each stage, the grout take (volume of grout injected before a pre-determined 'refusal' pressure or maximum injection volume is reached) for each hole would be recorded, and a series of 'water tests' (in situ permeability tests) would be carried out. As later stages of holes are grouted, grout takes should reduce, and the results of water tests should indicate a reduction in permeability. These data are used to determine the need for additional injections.

The primary and secondary injections are intended to fill the larger fractures and discontinuities in the rock. Tertiary and later stages of grouting should grout the finer discontinuities. Typically, the grouting is carried out with a number of grout mixes, initially using 'thin' grouts at high water–cement ratios with fixed volumes. If the initial grout volume is injected without resistance, then the next thicker grout volume is injected until a pressure build-up is seen. Later stages of grouting may be carried out with more penetrating grouts, such as microfine cements, or occasionally chemical grouts.

A further approach, which is becoming more common, is the Grout Intensity Number or GIN grouting methodology. With this methodology, a single, stable and relatively low-viscosity grout is used for all stages in all holes. Initial primary holes are drilled and grouted at wide spacing with the grout injected at low pressures and using relatively high volumes. The GIN value is the product of pressure and injection volume and is determined during initial injections or in advance, utilizing a trial (Lombardi, 2003). Generally, hole spacing is reduced using the split-spacing technique, and eventually, the tertiary holes are injected with relatively high pressures and low volumes. The intent of the method is to use the primary holes to grout the largest fractures, secondary holes to grout fractures of intermediate size, and tertiary holes to grout the finer fractures. The theory is that the distance the grout travels for a certain pressure is dependent on time and fracture width; thus, at low pressure, the grout will only fill the larger fractures over a long distance, and when drilling intermediate holes, the widest fractures are already sealed with grout but the smaller ones are not. Thus, grouting is carried out progressively over smaller distances corresponding to finer fractures.

In relation to the planning and monitoring of grouting operations, the permeability of rock masses is sometimes referred to in units of Lugeon coefficient (Lu). The Lugeon coefficient describes the potential for fractured rock to accept water as assessed by a packer permeability test (see Section 12.8.3) under specific conditions. A value of 1 Lu is defined as a water acceptance of 1 litre/minute through a 1 m length of 76 mm diameter borehole under an applied head of 10 bar (102 m head of water) above groundwater level. The Lugeon coefficient can be related to conventional SI units, as 1 Lu is approximately equal to 1×10^{-7} m/s.

Most rock grouting is carried out using cement-based grouts. Conventional cement grouts are effective at sealing rock openings of 0.5 mm or larger and can achieve a minimum permeability of around 5 Lu (approximately 5×10^{-7} m/s). With the use of microfine and ultra-fine cements or chemical grouts, it may be possible to achieve a permeability of the grouted mass of around 1 Lu (approximately 1×10^{-7} m/s).

Further information on rock grouting is given in Houlsby (1990), Rawlings *et al.* (2000), BS EN 12715:2000, Bell (2012) and Hughes *et al.* (2018).

18.10.8 Jet Grouting

In contrast to permeation grouting and rock grouting, where there should be little disruption of the soil/rock structure, jet grouting is a more aggressive method, which locally disturbs the ground. The method uses high-pressure jets of water or grout to disaggregate and erode the natural soil or rock structure by flushing away a significant proportion of the particles, which are replaced by grout to produce a mixture of grout and soil.

The method uses a jetting monitor (from which the jets emit radially) located near the base of a specialist drill string or jetting pipe. The jet-grouting process typically involves the monitor being slowly raised out of the ground while being rotated (Figure 18.24). Correctly executed, this will produce a column of treated material (a 'jet-grouted' column), which is of low permeability and potentially significantly greater strength than the untreated soil. Continuous vertical barriers are achieved by the installation of multiple overlapping jet-grouted columns. Jet grouting can also be used in inclined or sub-horizontal alignments and has also been used to form basal plugs (Figure 18.4) within excavations, formed from overlapping columns, as described by Newman *et al.* (1994). If the jetting pipe is raised without rotation, the jetting action can create a thin planar zone of jet-grouted material – termed a jet-grouted 'panel'.

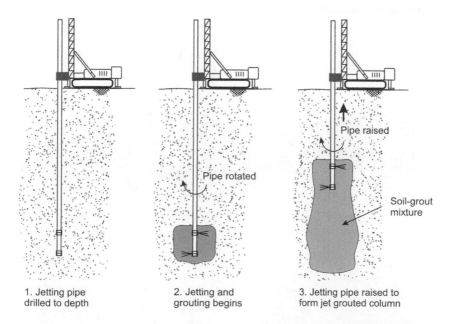

1. Jetting pipe drilled to depth

2. Jetting and grouting begins

3. Jetting pipe raised to form jet grouted column

Figure 18.24 Jet grouting. High-pressure water or grout jets are used to form a column of disturbed soil/rock around the jetting pipe, flushing away a significant proportion of the soil/rock particles, which are replaced by grout to form a column of soil-grout mixture.

Because of the potential strength of jet-grouted material, it can play a role in temporary ground support during excavation. A wide range of soils can be treated, both coarse-grained and fine-grained as well as weak and weathered rock. Permeabilities of 10^{-7} to 10^{-9} m/s have been achieved in jet-grouted barriers.

A jet-grouted column is formed by drilling in the jetting pipe (typically 89 to 114 mm diameter) to the maximum depth of the column (in difficult ground conditions, a pre-drilled hole may be used). One or more high-pressure radial jets of fluid are then emitted from the monitor at the base of the jetting pipe (Figure 18.25), while the pipe is rotated slowly. The jetting pipe is then lifted slowly while rotating to create the column of mixed soil and cement. Continuous barriers can be formed by either the 'fresh-in-fresh' or the 'fresh-in-hard' method:

- In the fresh-in-fresh sequence, adjacent columns are formed sequentially without waiting for the grout to set. This allows the new column to easily cut into the neighbouring column, thereby forming a continuous barrier.
- In the fresh-in-hard sequence, neighbouring columns are not formed until the existing columns have gained some strength. This reduces the risk of the drilling activity causing washout of the neighbouring column but means that the previously hardened columns may block the eroding effect of the jets and may create 'shadow' zones where the erosion and mixing of material is incomplete.

Jet grouting is normally carried out using cement-based grouts, either cement only or cement with bentonite or other fillers. Three types of jet-grouting systems are available:

Single fluid system (Figure 18.26a): A high-pressure grout jet (typically at 30 to 60 MPa) is used to cut the column, eroding and replacing the soil. Excess soil and grout slurry (spoil return) is forced up the annulus around the jetting pipe to the surface. The primary mechanism with the single fluid method is injection and mixing of grout, with only limited replacement of soil.

Figure 18.25 Jet-grouting rig The radial jets at the base of the drill string are shown operating above ground level. (Courtesy of Keller Geotechnique, West Yorkshire, UK.)

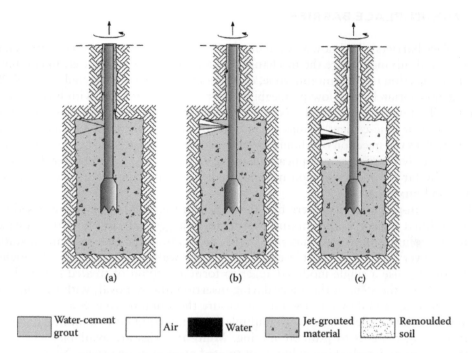

| | Water-cement grout | | Air | | Water | | Jet-grouted material | | Remoulded soil |

Figure 18.26 **Types of jet-grouting system. (a) Single fluid – grout only. (b) Double fluid – grout/compressed air. (c) Triple fluid – water/compressed air/grout. (From Croce, P, et al., *Jet Grouting: Technology, Design and Control*, CRC Press, Boca Raton, Florida, 2014. With permission.)**

Double fluid system (Figure 18.26b): A grout jet is shrouded in compressed air (typical pressure 0.2–1.5 MPa). The purpose of the air shroud is to improve the cutting action of the jet, increase the energy of the jet and hence allow a greater volume of ground to be treated by a single column insertion. The double fluid method typically involves greater soil replacement, and hence spoil volumes, than the single fluid method.

Triple fluid system (Figure 18.26c): A water jet, shrouded in air, is used to erode the soil, with a separate nozzle, lower down the monitor, used for grout injection. Typical pressures are air 0.5–1.5 MPa; water 4 MPa; grout 0.5–3 MPa. Because the only purpose of the grout is to mix with the material previously disturbed by the water jet, it can be delivered at lower pressures. The triple fluid method typically involves greater soil replacement, and hence spoil volumes, than the single fluid method.

The diameter of ground treated by a single column insertion depends on the method used, soil type, grout characteristics, injection pressures, and lift and rotation speeds. Typical column diameters of treated ground for a single jet-grouted column are 0.5–1.2 m for the single jet methods and up to 3.0 m for the double or triple jet method.

Jet-grouting operations are complex and must be carefully controlled. Three potential problems or challenges with jet grouting are:

  i. Ground heave if grout injection and spoil removal are not carefully controlled
  ii. Hydrofracturing of the ground as a result of spoil returns becoming blocked
  iii. Generation of large volumes of slurried spoil (of the order of 50 m³ of spoil per rig per shift)

Further information on jet grouting is given in Essler (1995), Lunardi (1997), BS EN 12716:2001, and Croce *et al.* (2014).

18.11 MIX-IN-PLACE BARRIERS

Mix-in-place barriers (also known as soil mix barriers) are low-permeability barriers formed in situ by 'soil mixing', where the mechanical action of an auger or cutter, in combination with grout injection into the mixing zone, creates a zone or block of treated material. When it is used appropriately, very low-permeability barriers can be formed with little or no spoil generated. Two main types of soil mixing have developed. In dry soil mixing, a dry powder (often cement or lime-cement) is injected into the in situ mixing zone. In contrast, wet soil mixing involves the injection of a liquid grout (typically a cement-based grout) into the mixing zone. For civil engineering applications to create barriers to groundwater flow, wet soil mixing is by far the most common method used. The following section deals exclusively with wet soil mixing.

When mix-in-place barriers were first developed in the 1990s, they were based on the use of modified augers as the mixing tool to form a vertical cut-off barrier formed of overlapping columns of treated material. In this application, the geometry and installation sequence are very similar to those used for secant pile walls (see Section 18.9). Single-axis installation rigs use a single modified auger to form a column of treated material. Grout is injected down the stem of the auger during insertion and removal, with the auger often being 'worked up and down' in the bore to ensure thorough mixing. Soil mixing is often carried out with modified augers with multiple grout injection points, mixing paddles and gaps in the auger flights to promote mixing. Multi-axis rigs are available, which install two, three or more interlocking columns of treated ground in one cycle. Nominal column diameters are in the range of 600 to 2500 mm. Depths of the order of 40 m are possible with land-based equipment.

Since the early 2000s, an alternative construction method for mix-in-place walls has emerged in the form of cutter soil mixing (CSM) equipment, derived from hydromill cutters used for diaphragm wall construction (Schöpf, 2004). In contrast to auger-based soil mixing, where the cutting tool rotates on a vertical axis to form a column, CSM rigs use multiple cutting wheels rotating on a horizontal axis to form panels. In a typical construction sequence, the cutter unit descends, disturbing and mixing the soil, with relatively little injection of fluid. Once the target depth is reached, the cutting tools are reversed, and as the cutter ascends, grout is injected to form the mix-in-place material. To achieve a very homogeneous zone of treated ground, the descending and ascending phases are sometimes repeated.

Wall construction is achieved in an analogous way to diaphragm walls, whereby discrete panels are formed, each cut into its neighbour to achieve an overlap. CSM units are typically mounted on a crawler-mounted drill rig (Figure 18.27). This technique can be used to construct walls to depths of 60 m and thicknesses of up to 1.2 m.

Vertical mix-in-place barriers can also be installed using modified trenching machines, as described by Hardwick (2010). Special trenchers, similar in principle to those used to install horizontal wellpoints (see Section 17.3), use a rapidly moving chain cutter/mixer to break up the ground and form a narrow slot of disturbed material while grout is injected (Figure 18.28). The direction of the cutter/mixer chain is periodically reversed to ensure adequate mixing. Trenchers are available to install barriers to depths of 6 to 7 m.

In appropriate conditions, mix-in-place methods have the potential to produce barriers with permeability in the range of 10^{-7} to 10^{-9} m/s. The treated material is typically of higher strength than the natural ground and may potentially, with appropriate design, play some role in the temporary works structural support of an excavation.

Further information on mix-in-place barriers is given in Greenwood (1989), Blackwell (1994), BS EN 14679:2005, and Denies and Huybrechts (2018).

Figure 18.27 CSM cutter soil mixing unit. (Courtesy of Golder Associates, Toronto, ON.)

Figure 18.28 Mix-in-place barrier installed by trencher. (Courtesy of Bachy Soletanche, Burscough, UK.)

18.12 ARTIFICIAL GROUND FREEZING

Artificial ground freezing involves the installation of a series of closely spaced, small-diameter boreholes (known as freezeholes), typically to form a ring around an excavation. A very low-temperature refrigerant is then circulated through the freezeholes. This chills the ground around each freezehole, freezing the water contained within the soil pores or rock fractures and causing a cylinder of frozen ground to form around each freezehole. As the circulation of the refrigerant continues, the cylinders of frozen ground will slowly increase in diameter. The objective is for the cylinders to eventually intersect and coalesce to form a continuous 'freezewall' around the excavation (Figure 18.29). The formation of a complete freezewall is termed 'closure'. Once closure has been achieved, the circulation of refrigerant must continue, albeit at a reduced thermal load.

A suitably designed freezewall will be of very low permeability and can be very effective at excluding groundwater from an excavation or tunnel. Frozen ground is also very strong, and a freezewall can form part of temporary works structural support for an excavation.

Unlike most other techniques for groundwater exclusion, artificial ground freezing is truly temporary. A freezewall can only be maintained by the continued circulation of refrigerant, although once the freezewall has been established, refrigerant flows can normally be reduced relative to levels at the start of freezing (this is sometimes known as the 'maintenance freeze' phase). On completion of a project, refrigeration is discontinued, and the freezewall will normally thaw slowly (probably over a period of months), leaving behind no permanent barrier to groundwater flow. This may be a benefit in areas where there are concerns over long-term impacts on groundwater flow (see Section 21.7).

There are two principal types of systems used for artificial ground freezing: first, the circulation of chilled brine in a closed system and second, the use of liquid nitrogen (LN).

18.12.1 Brine Circulation Systems

This configuration uses a closed circuit of low-temperature brine (typically calcium chloride) circulated through the freezeholes (Figure 18.30a). The brine is circulated to a portable freezeplant (effectively a large refrigerator) powered by electric or diesel prime movers, which dissipates the extracted heat to atmosphere via cooling towers or evaporative condensers. A large freezeplant is shown in Figure 18.30b; a smaller unit used for tunnel cross passage construction is shown in Figure 10.26a.

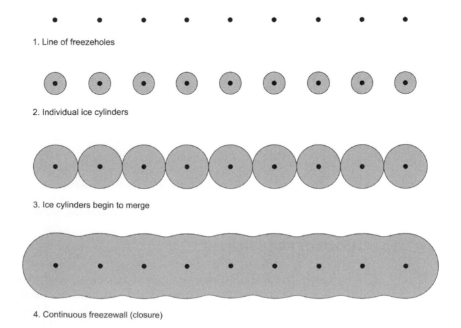

1. Line of freezeholes

2. Individual ice cylinders

3. Ice cylinders begin to merge

4. Continuous freezewall (closure)

Figure 18.29 Development of freezewall.

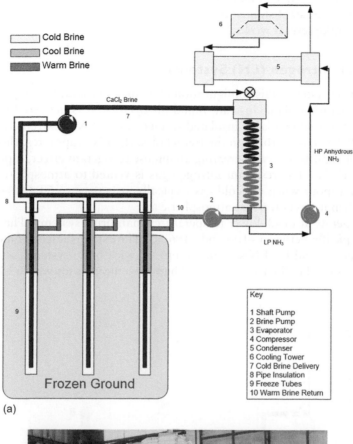

(a)

(b)

Figure 18.30 Artificial ground freezing using brine circulation. (a) Schematic view of brine circulation system (b) Portable freezeplant. This freezeplant is driven by a 180 kW electric motor. The output is 166,320 Kcal/hour when evaporating at −37.5 °C. (Panel (a) from Auld, F A, *et al.*, *Proceedings of the XVI ECSMFGE, Geotechnical Engineering for Infrastructure and Development* (Winter, M C, Smith, D M, Eldred, P J L and Toll, D G, eds), ICE Publishing, London, pp. 901–906, 2015. By permission of the authors and courtesy of British Drilling and Freezing Company Limited, Nottingham, UK; panel (b) courtesy of British Drilling and Freezing Company Limited, Nottingham, UK.)

Brine is typically circulated at temperatures of between –25 and –35 °C. At these circulation temperatures, the establishment of a complete freezewall (time for closure) is relatively gradual and may take several weeks.

18.12.2 Liquid Nitrogen (LN) Systems

In contrast to brine systems, this configuration of artificial ground freezing does not rely on mechanical refrigeration plant. Instead, liquid nitrogen (LN) is delivered to site (typically by road tanker) and held in a large insulated storage vessel. LN is extremely cold (–196 °C) in liquid form. When passed through the freezeholes, the LN evaporates, thereby absorbing the latent heat of vaporization and creating an intense refrigerant effect, rapidly chilling the ground (Figure 18.31). The resultant nitrogen gas is vented to atmosphere, often causing plumes of water vapour when the cold gas (typically at between –60 and –100 °C) causes condensation when it meets warmer atmospheric air.

LN systems operate at much lower temperatures than brine systems. The temperature in freezeholes is typically between –100 and –196 °C. The establishment of a complete freezewall is much more rapid for LN systems compared with brine systems, a complete freeze typically being established in 1 to 2 weeks. There have been cases where LN systems have

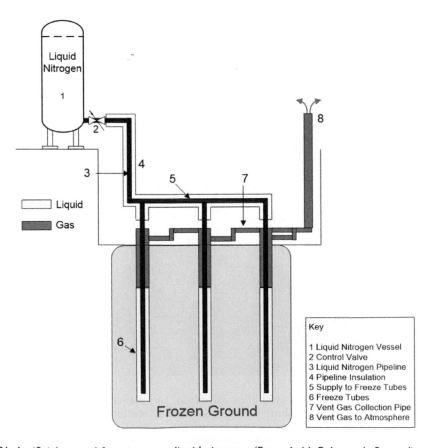

Figure 18.31 Artificial ground freezing using liquid nitrogen. (From Auld, F A, et al., Proceedings of the XVI ECSMFGE, Geotechnical Engineering for Infrastructure and Development (Winter, M C, Smith, D M, Eldred, P J L and Toll, D G, eds), ICE Publishing, London, pp. 901–906, 2015. By permission of the authors and courtesy of British Drilling and Freezing Company Limited, Nottingham, UK.)

been used to rapidly establish a freezewall, and then, to reduce costs, the maintenance freeze has been continued by brine circulation (Viggiani and Casini, 2015).

18.12.3 Typical Applications for Artificial Ground Freezing

There are two typical situations when artificial ground freezing is used:

- Planned use: This is when ground investigations have identified the potential for difficult excavation conditions, for example very unstable water-bearing strata, very deep excavations and so on. In such cases, artificial ground freezing is planned for from the start, typically with twin objectives of preventing groundwater inflow and temporarily increasing the strength of the soil or rock around the excavation. In these cases, freezing can be by either brine or LN systems depending on ground conditions and the time available to establish the freezewall (LN being used in situations when the freezewall must be established rapidly).
- Emergency (recovery) use: There have been numerous occasions when artificial ground freezing has been used as part of recovery operations following collapse, large-scale instability or inundation of an excavation or tunnel (see Section 10.8). Following such events, there is normally a need to control groundwater and stabilize ground to allow remedial works to be carried out. Because of the need for rapid establishment of freeze and the difficulties of stabilizing the disturbed ground around a collapse, LN is often the refrigerant of choice in these applications. Clarke and Mackenzie (1994) and Brown (2004) describe two case studies where artificial ground freezing was used as part of recovery works for tunnelling projects that had encountered difficulties.

Typical configurations of artificial ground freezing installations include vertical freezewalls around shafts or excavations (Figure 18.32) or horizontal freezes to provide support around tunnels and other underground spaces. Applications of ground freezing on tunnelling projects (including for tunnels, shafts and cross passages) are discussed in Chapter 10.

The most common application is to create a 'peripheral freeze' to fully enclose the excavation but minimizing the volume of frozen ground that must be excavated. The freezewall is designed to be largely outside the excavation line, but the excavated face should be slightly into frozen material. The exposed face will be frozen, and insulation is typically applied after excavation to reduce the risk of significant thawing. In some cases, a 'mass freeze' is carried out, where a significant volume of soil or rock is frozen, into which the excavation is dug. This requires a more intensive freezing effort and means that the excavation will have to be formed in the very hard frozen ground. This approach is sometimes used in tunnel recovery works after collapse or major instability has occurred (see Section 10.8).

The freezeholes are typically of relatively small diameter (100–150 mm) with a design spacing between freezehole centres of between 0.75 and 1.5 m. An unperforated tube is installed in each freezehole to allow circulation of the refrigerant. Steel tubing is commonly used, but medium-density polyethylene (MDPE) can be used if the freezeholes are located in zones that are to be excavated. It is essential that drilling is carried out within accurate tolerances of setting out, alignment and straightness so that two adjacent freezeholes do not diverge at depth. If this occurs, the time for the freezewall to 'close' may be longer than anticipated, or in extreme cases, closure may not be achieved in the area of the misaligned freezeholes, and a 'window' may remain in the wall, through which groundwater can flow. Often, freezeholes are accurately surveyed along their length prior to commencement of refrigerant circulation, so that any areas of misaligned freezeholes can be identified and additional holes drilled if necessary. In some cases, directional drilling methods are used to allow freezeholes to be 'steered' into the desired alignment.

(a)

(b)

Figure 18.32 Typical layouts of artificial ground freezing systems (a) Artificial ground freezing system (using brine as the refrigerant) prepared ready for sinking of a circular shaft. (b) Artificial ground freezing system (using liquid nitrogen as the refrigerant) as part of the recovery works on a tunnel project. The white plumes are water vapour caused by the low temperatures of the nitrogen gas exhausted to atmosphere from the four towers visible in the background. (Courtesy of British Drilling and Freezing Company Limited, Nottingham, UK.)

The development of the freezewall is commonly monitored by means of thermocouples or other temperature sensors in boreholes located close to the freezewall.

18.12.4 Effect of Groundwater Flow on Artificial Ground Freezing

High rates of groundwater flow can impede successful ground freezing. Groundwater flowing across a freezewall will carry heat away, reducing the net cooling power available to establish or maintain a freezewall. This will increase the time to achieve closure of the freezewall or may even prevent the establishment of a complete freezewall and allow 'windows' to remain, through which groundwater can flow.

The interaction between groundwater flow and artificial ground freezing systems is discussed by Schmall and Dawson (2017). Groundwater flow past a freezehole will transfer some of the cooling effect downgradient, which in plan view will result in an egg-shaped formation of frozen ground around the freezehole (rather than the perfect cylinders shown in Figure 18.29). Even where groundwater is flowing, the egg-shaped frozen zones may eventually be able to merge and close the freezewall. However, if groundwater velocity is very high, the water flowing past a freezehole will carry away sufficient cooling effect to prevent the frozen zone around each tube merging with its neighbours, and a window for groundwater flow will remain.

It is generally accepted that typical brine circulation systems may experience problems if groundwater flow velocities are greater than about 2 m/day. The lower refrigerant temperatures used in LN systems mean that these systems may potentially be effective in groundwater flow velocities up to 20 m/day. The velocities quoted above are Darcy velocities (sometimes known as Darcy flux), defined as the velocity equivalent to the flow per unit cross-sectional area of a porous medium (see Section 3.3.4). This is an apparent velocity, equivalent to the rate at which water would flow through a porous medium if it were an open conduit; it does not represent actual velocities in soil pores or rock fractures. The Darcy velocity v is related to the permeability k and the hydraulic gradient i in the direction of flow by Darcy's law:

$$v = -ki \tag{18.1}$$

Typical groundwater velocities that occur naturally are discussed in Section 3.3.4. Velocities high enough to affect artificial ground freezing systems are rare in soils or rocks under natural hydrogeological conditions. The risk of problematic velocities is more likely to be related to a combination of the following:

- Artificial hydraulic gradients created by nearby groundwater pumping or other external influences
- Concentration of water flow in localized higher-permeability zones, either natural (e.g. intensely fractured zones in bedrock) or artificial (e.g. highly permeable drainage pathways associated with existing structures or utilities, etc.)

Various measures can be used to mitigate the effect of high groundwater velocities and reduce the risk of problems with closure of the freezewall (these measures may be used in combination):

i. Increasing the intensity of the freezing effort by installing a greater number of more closely spaced freezeholes
ii. Using a lower-temperature refrigerant (LN instead of brine)

iii. Using grouting to reduce the permeability of the most transmissive zones (this approach reduces k in Equation 18.1)

iv. Using carefully controlled pumping to limit hydraulic gradients and (as shown by Equation 18.1) thereby reduce groundwater velocities

18.12.5 Other Limitations of Artificial Ground Freezing

The principle of artificial ground freezing is that the circulation of refrigerant in the freeze-eholes will chill and ultimately freeze the groundwater in pores and fractures.

If the soil or rock is unsaturated (for example above groundwater level or where 'dry' cavities exist), then there may be insufficient free groundwater to allow artificial ground freezing to be effective.

Groundwater control works are occasionally carried out where groundwater is notably brackish or saline. This is a relevant factor for artificial ground freezing, as salinity decreases the freezing point of water to below 0 °C. This will increase the thermal load required to freeze the ground and may result in slower closure of the freezewall. However, artificial ground freezing has been used successfully in brackish waters with significant salinity. Provided that the water quality is taken into account in the freeze design, elevated salinity need not be an obstacle to application of the method.

Ground movements (heave and settlement) due to freeze and thaw effects are sometimes associated with artificial ground freezing as a result of the formation and thawing of ice lenses. Such ground movements are of particular concern in clays, silts and organic soils.

Very cold pipework, which is a necessary part of artificial ground freezing systems, presents specific health and safety risks (cold burns). Similarly, if leaks of refrigerant (such as LN) occur, there will be risks to personnel due to, for example, asphyxiation due to excess of nitrogen gas in confined spaces such as tunnels and excavations. Health and safety protocols specific to artificial ground freezing systems should be used.

Further information on artificial ground freezing is given in Harris (1995) and Auld et al. (2015).

18.13 LOW-PERMEABILITY SURFACE BARRIERS

Occasionally, the infiltration of rainfall and other surface water can cause problems for excavations, and it may be necessary to prevent surface infiltration by using low-permeability cover layers to seal exposed soil/rock surfaces or faces. Examples of where this may be required include:

a) Excavations in soil or rock where perched water table conditions exist (Figure 18.33a). If precipitation and surface water are allowed to pond at surface and infiltrate into the ground, then this may feed the perched water table, causing seepage and instability problems in the excavation slopes.

b) Excavation in fractured rock, especially where the dominant fracture or bedding direction dips towards the excavation (Figure 18.33b). If there is significant infiltration into the ground behind the crest of the slope, then seepage and instability problems can occur where the fractures/bedding planes intersect the excavation slopes.

c) Excavations where artificial bodies of surface water exist behind the crest of the slope (Figure 18.33c). The surface water body could be a water storage pond or could simply be an unlined surface water ditch that contains standing water. The seepage from the pond/drain can cause seepage or instability problems in slopes.

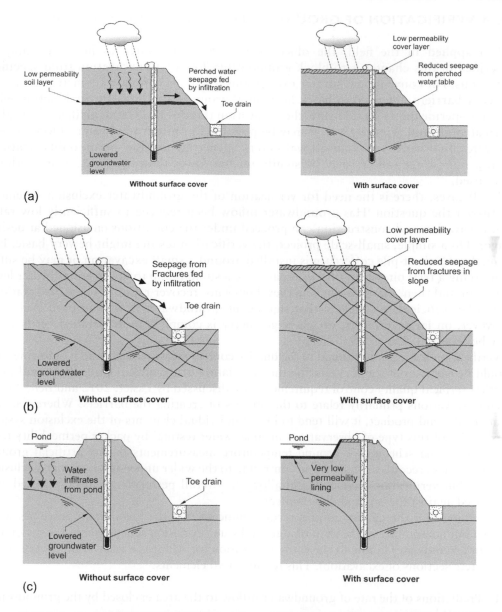

Figure 18.33 Low-permeability surface barriers used to reduce seepages from slopes. (a) Low-permeability cover layer used to reduce seepage rates from perched water table. (b) Low-permeability cover layer used to reduce seepage rates from shallow fracture networks when excavating in rock. (c) Low-permeability pond lining used to reduce recharge of groundwater close behind the slope.

These problems can be mitigated by capping the relevant surface areas (e.g. behind the crest of the slopes) with materials of low permeability to reduce infiltration and maximize run-off. Possible materials that can be used to form surface barriers include:

- Mass concrete blinding
- Sprayed concrete lining (SCL)
- Compacted clay fill
- Low permeability geosynthetic membranes

18.14 VERIFICATION OF GROUNDWATER EXCLUSION METHODS

When applied in the field, the objective of the groundwater exclusion techniques described in this chapter is to block groundwater flow to an excavation from specified directions. For conventional surface excavations, the most common geometry is vertical walls or barriers, intended to block horizontal groundwater flow. In combination with very low-permeability strata below the excavation (Figure 18.1) or an artificial base plug of treated ground (Figure 18.4), it may be possible to limit groundwater inflow to very low rates. In other cases, if cut-off walls do not penetrate to a low-permeability stratum, total groundwater inflows may be significant but lower than if no exclusion methods were used.

In all cases, there is the need for verification of the groundwater exclusion scheme – to answer the question 'Has groundwater inflow been reduced to sufficiently low rates that excavation and construction can proceed under the conditions envisioned at design stage?' On a simple, small-scale project, the verification testing might be very basic. For example, if a sheet-pile cofferdam is installed around a small excavation, it may be sufficient to dig a trial pit within the cofferdam, use a sump pump to lower groundwater level in the pit and then note that the water level does not recover rapidly. These observations would be sufficient to conclude that significant groundwater flows had been excluded. However, on larger projects, or where geometry is complex, more detailed verification may be required.

Verification testing is additional to the quality control measurements and monitoring that should be carried out routinely during the installation of the cut-off walls or groundwater barriers (typical quality control requirements are discussed by Essler, 2012 and Bell, 2012). Such observations primarily relate to the process of creating the barriers. Where they do relate to the end product, it will tend to be for individual elements of the exclusion system – examples of this type of observations include 'water testing' by packer permeability testing on grouting schemes, or ground temperature measurements during artificial ground freezing. These records are useful and contribute to the wider understanding of an exclusion scheme, but verification testing should allow the overall performance to be assessed and compared against the original design intent.

Assuming that the quality control records indicate that the installation processes were carried out acceptably, verification of a groundwater exclusion scheme should focus on the rates of groundwater inflow to the entire excavation (or where the excavation is very large, to discrete sections of excavation). This requires two elements:

1. Predictions of the rate of groundwater inflow to the area enclosed by the groundwater cut-off barriers
2. Practicable means to determine the rate of inflow in the field

The verification process involves comparing the predictions with the reality.

18.14.1 Prediction of Inflow to Groundwater Exclusion Schemes

Estimates of the inflow into an excavation enclosed by groundwater exclusion barriers will require the use of an analytical model of some kind (Section 5.5), but before any calculations can be done, a conceptual model must be developed. In Section 5.4, a conceptual model was defined as a simplified non-mathematical representation of a groundwater system to help understand and communicate groundwater conditions.

For a conceptual model of an excavation within groundwater cut-off barriers, the relevant groundwater conditions are the volume of water trapped within the soil or rock in the area enclosed by the cut-off walls, and the various potential sources of inflow.

During the initial transient phase, while groundwater levels are being pumped down, the pumped flow rate will include a component due to removal of groundwater trapped within the excavation. For an excavation of conventional layout, with vertical groundwater barriers around the perimeter enclosing plan area A_p, if the required drawdown of groundwater levels is s, the volume of water V trapped within the soil/rock can be estimated from

$$V = A_p.s.S_y \tag{18.2}$$

where S_y is the specific yield of the soil or rock. S_y is a measure of the proportion of water that can drain from an unconfined aquifer and is generally lower than the soil porosity (typical values are given in Table 3.1). If ground conditions within the excavation include layers with significantly different specific yield, Equation 18.2 can be used to estimate the water that will potentially drain from each layer within the depth range over which groundwater levels will be lowered; these volumes are then summed to estimate the total quantity of groundwater within the excavation.

Steady-state inflow (after groundwater levels are lowered within the excavation) potentially comprises several components:

1. General seepage through the perimeter groundwater barriers. This seepage is typically horizontal and is controlled by the overall mass permeability of the barrier.
2. General seepage through the base of the excavation. This seepage is vertical and is controlled by the mass permeability of the natural strata below the excavation or of any artificial base plug of treated ground.
3. Flow through any defects or imperfections in the groundwater barriers (vertical walls and base plugs). Examples of defects include leaking joints between diaphragm wall panels; flow beneath vertical barriers where insufficient toe-in is achieved into very low-permeability strata; and 'windows' of untreated or unfrozen ground within grout barriers or freezewalls, respectively.
4. Groundwater derived from direct precipitation into the excavation. The nature of the exposed ground will affect how much precipitation contributes to inflow. If surfaces are covered in low-permeability layers (e.g. blinding concrete), most rainfall will run off, and (assuming it is collected as surface water) very little will infiltrate into the ground. If permeable soil (e.g. coarse gravels) or highly fractured rock is exposed, a significant proportion of rainfall will infiltrate and contribute to groundwater pumping rates.

One approach to estimating the inflow is to develop a numerical groundwater model based on the ground conditions, excavation geometry and the anticipated properties of the groundwater barriers. Steady-state models can predict the long-term pumping rates after the trapped water has been removed. More complex transient models are required to assess the rate of drawdown during the pumping-down phase.

As an alternative to numerical modelling, for excavations with simple geometry, Darcy's law (Section 3.3.2) can be used to estimate inflow. Darcy's law relates flow rate Q to permeability k, the cross-sectional area of flow A and the hydraulic gradient:

$$Q = -k\left(\frac{\Delta h}{l}\right)A \tag{18.3}$$

When applied to seepage perpendicular to cut-off walls and groundwater barriers:

Q = flow rate through an area of cut-off wall or barrier

A = cross-sectional area of cut-off wall or barrier through which the water flows

k = mass permeability of the cut-off wall or barrier

l = length of the flow path (the thickness of the cut-off wall or barrier)

Δh = difference in groundwater level between the internal and external faces of the cut-off wall or barrier

($\Delta h/l$) is the hydraulic gradient across the barrier. The negative term is necessary in Equation 18.3 because flow occurs down the hydraulic gradient – i.e. from high head to low head.

For steady-state vertical seepage through grouted base plugs, in most cases. Δh will be constant. If the base plug is of uniform thickness, the hydraulic gradient will also be constant. For horizontal seepage through vertical cut-off walls, Δh and the hydraulic gradient will typically vary with depth. Equation 18.3 may need to be evaluated for discrete vertical depths and the total flow through the well estimated by summing the inflow over the total depth of wall.

For steady-state estimates, Δh should be based on the drawdown to final excavation level. During the pumping-down phase. the inflow can be estimated incrementally for various drawdown levels to show how the inflow rate increases as the water level inside the excavation is lowered.

For large excavations, the inflow will be dominated by the flow through the planar faces of the walls or base plug, and the flow at corners will be only a modest contribution. Darcy's law can be used to estimate the horizontal inflow through cut-off walls and vertical flow through base plugs. These can then be summed to give an estimate of total groundwater inflow (excluding defects in the barriers).

For small excavations, the rainfall component of the pumping rate will be very small and is usually not estimated. However, for very large excavations that are open for long periods, the rainfall component should be estimated to confirm that it is not a significant component of the steady-state pumping rate.

18.14.2 Methods to Determine Total Inflow to Groundwater Exclusion Schemes

Methods to measure inflow to an excavation enclosed by groundwater barriers are simple in concept but can be more complex in practice. Two types of test are commonly carried out:

1. Pumping-down tests: This involves recording how much water must be pumped from the excavation to lower groundwater levels by a given amount.
2. Recovery tests: This involves pumping to lower groundwater levels within the excavation. Pumping is then stopped, and the inflow is assessed from the rate of recovery of groundwater levels.

Pumping-down tests require at least one pumping well and several monitoring wells to be installed within the area enclosed by the groundwater barriers. In a process directly comparable to pumping tests (Section 12.8.5), water is pumped from the well for a continuous period of several hours or days while flow rate and groundwater levels (inside and outside the excavation) are monitored. Successful pumping-down tests require a reasonable horizontal hydraulic connection between the pumping well and the monitoring wells. This type of test is difficult to carry out if the wells are installed in low-permeability strata.

Recovery tests are often used when the yield from a pumping well is expected to be very low and a pumping-down test would be difficult to execute. The method has some superficial similarities with rising head permeability tests (Section 12.8.1) in that water is removed from a well and the rate of water level recovery is observed. However, unlike rising head tests, where the water removal phase should be very brief, in this context, the well should be pumped for a longer period – at least several tens of minutes – with the objective of draining as much water as possible from the area around the well. In that way the recovery of water levels should be more representative of the excavation as a whole, not just the ground around the well itself.

For an excavation of conventional layout, with vertical groundwater barriers around the perimeter enclosing plan area A_p, the rise in groundwater level Δs over time t can be converted to an inflow Q by

$$Q = \frac{A_p.\Delta s.S_y}{t} \tag{18.4}$$

and the rate of water level recovery ($\Delta s/t$) can be related to an inflow rate by

$$\frac{\Delta s}{t} = \frac{Q}{S_y A_p} \tag{18.5}$$

For both types of test, groundwater levels should also be recorded in any monitoring wells outside the excavation. The objective of this monitoring is to identify any external drawdown of groundwater, which may indicate leakage through the groundwater barriers.

18.14.3 Assessment of Acceptable Performance

The principle of verification testing is that field observations are compared with relevant predictions of groundwater conditions.

Positive indicators for verification of an effective groundwater exclusion scheme include:

a) Flow rates during pumping-down tests less than or equal to predicted values
b) Rates of water level rise during recovery tests less than or equal to predicted values
c) No significant drawdown in external monitoring wells

Point (c) merits further discussion. External monitoring wells can be very useful, and observations of drawdown of external groundwater levels during a pumping-down test can be a key indicator of cut-off wall leakage (see Case History 27.10). However, in the converse case, if no significant external drawdowns are observed during a pumping-down test, this does not always indicate that the groundwater barriers are fully effective. If the excavation is surrounded by highly permeable strata, considerable leakage into the excavation may occur before significant external drawdowns can be discerned.

Chapter 19

Pumps for Groundwater Lowering Duties

All hands to the pump.

Old English idiom

19.1 INTRODUCTION

Pumps are ubiquitous in groundwater control systems. Even schemes based on groundwater exclusion require pumps to deal with trapped groundwater and surface water. There is a wide range of pumps available for pumping water. However, many pumps are not suitable for temporary works dewatering installations, so equipment must be selected with care. The categories of pumpsets appropriate to a particular groundwater lowering site requirement will depend on the technique in use.

In this chapter, the basic principles of pumping are introduced, and different categories of pumps are discussed. The pump types appropriate to the most common groundwater lowering techniques (sump pumping, wellpoints, deep wells and ejector wells) are described together with some less commonly used pumping methods. Pumps must be combined with other elements, including controls and pipework, to form the wider pumping system, and these other aspects are also discussed. The chapter ends with a discussion of various issues associated with the selection and sizing of pumps.

19.2 PRINCIPLES OF PUMPING

On most groundwater control projects, standard or 'off the shelf' pumpsets will be used, either bought new, hired-in or re-used from a previous project. Different types of pumps have different characteristics and are available with different nominal pumping rates. A further complication is that the output of a given pump will vary depending on the 'duty' imposed on it (which is largely controlled by the total head against which the pump must work). The pumping system must be sized so that it can achieve the anticipated flow rate both once drawdown is established and also during the initial period of pumping, when flow rates may be higher.

A groundwater control designer or practitioner does not need to understand every aspect of pump construction at the nuts and bolts level, and similarly, need not be expert in the hydrodynamic theories that describe how pumps move water. However, in order to select the right pumps and pipework for the job, it is necessary to grasp the basics of pump theory. The essential elements necessary to analyse pumping systems are given in the following sub-sections.

19.2.1 What Is a Pump?

Mankind has used pumps since ancient times. In essence, a pump is a device that uses an external energy source to lift fluid (in this case, water) against gravity. The first pumps were human-powered (Figures 2.1 and 2.2); then came horse power and during the Industrial Revolution, steam power (Figure 2.4). In modern groundwater control practice, pumps are usually powered by electrical motors or diesel prime movers. This highlights that an effective pump has two principal components – the mechanism for moving water (sometimes known as the 'wet end') and the energy source (the motor or prime mover) – collectively, they form a 'pumpset'.

19.2.2 Different Pumping Methods

Different physical mechanisms can be used to lift water. Pumps that might be encountered in groundwater control schemes can be categorized into four types. Typical applications are shown in Table 19.1:

- Direct lift pumps. This method involves physically lifting water in a container or series of containers. Bailing is an example of this type of pumping.
- Displacement pumps. This type of equipment (also known as positive displacement pumps) takes advantage of the fact that water is effectively incompressible and can therefore be efficiently 'pushed' or 'displaced'. If a piston is used to displace water in a pump cylinder, the volume of water moved is equal to the displacement of the moving piston. Wellpoint piston pumps are an example of this method.
- Velocity pumps. This method is based on the principle that when water is propelled with sufficient speed, the momentum can be used either to create a flow or to generate a pressure, thereby drawing water through a pump. An example is a centrifugal pump, where a rapidly rotating impeller is used to propel water out radially at high velocity. Jet pumps, as used in ejector wells, are also classified as velocity pumps.

Table 19.1 Different Pump Types Used to Lift Groundwater in Construction and Mining Applications

Physical mechanism of pumping	Commonly used pump types	Less commonly used pump types
Direct lift pumps	–	Bailing from boreholes or wells during permeability testing Bailing from deep shafts during shaft sinking
Displacement pumps	Diaphragm type submersible sump pumps Diaphragm type suction pumps used for wellpointing and sump pumping Piston type suction pumps used for wellpointing	Progressive cavity (mono) pumps, used in deep mining applications
Velocity pumps	Centrifugal type submersible sump pumps Centrifugal type suction pumps used for wellpointing, sump pumping and wellpoint jetting Electro-submersible centrifugal turbine pumps used in deep wells Centrifugal high-pressure supply pumps for ejector well systems Jet pumps used in ejector wells	Lineshaft centrifugal turbine pumps used in deep wells
Buoyancy pumps	Airlifting used for well development	Airlifting used for continuous dewatering pumping

- Buoyancy pumps. This type of equipment uses air (or other gases) bubbled through water. This reduces the density of the affected volume of water, which then rises through the remaining body of fluid. Airlift pumping is based on this principle.

The physical mechanism of pumping has an important effect on how a pump responds to changes in pumping conditions. For example, a displacement pump, which pushes a discrete volume of water, will, provided that the pump speed is unchanged, move essentially the same flow rate against a high head as it will against a low head. In contrast, the output of a velocity pump will change much more significantly and may be much lower when pumping against a higher head.

19.2.3 Mathematics of Pumping

As defined earlier, a pump is a mechanical device used to lift water. The power required to do so is controlled by the laws of physics and can be expressed mathematically by relatively few parameters. The power requirement P (kW) to pump a water flow rate Q (m³/s) against a total head H (m) is expressed as

$$P = \frac{Q \times H \times \rho_w \times g}{\eta} \tag{19.1}$$

where
  ρ_w is the density of water (1000 kg/m³)
  g is the acceleration due to gravity (9.81 m/s²)
  η is the combined efficiency of the pump and motor (the pumpset)

The energy consumption E (kWh) of a pumpset (assuming the parameters do not vary with time) is the power P multiplied by the pump run time t (hours):

$$E = \frac{Q \times H \times \rho_w \times g \times t}{\eta} \tag{19.2}$$

A key controlling parameter affecting how much water is pumped for a given power input is the total head H (also known as the total dynamic head). Total head is a measure of the resistance against which the pump has to operate and is made up of the vertical distance the water has to be lifted plus the additional head caused by friction losses and velocity heads in the pipework and fittings. The relationship between these components for submersible and suction pumps is discussed in the following.

The breakdown of total head is different for a suction pump, where the pump sits above the water level and must raise the water up to itself before pumping it away, than for a submersible pump, where the pump mechanism sits below water level, and there is no suction lift.

For a suction pump (Figure 19.1a), total head H comprises

$$H = h_s + h_d + f_s + f_d + \left(\frac{v_s^2}{2g} \right) + \left(\frac{v_d^2}{2g} \right) \tag{19.3}$$

where
  h_s = static suction lift
  h_d = static discharge head
  f_s and f_d = friction losses in the suction and discharge pipework, respectively
  v_s and v_d = water velocity in the suction and discharge pipework, respectively

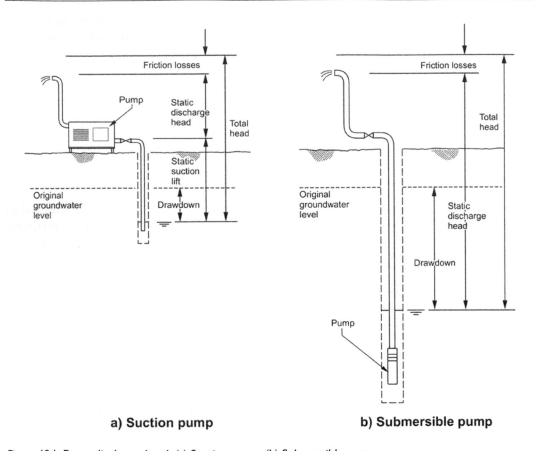

a) Suction pump   **b) Submersible pump**

Figure 19.1 Pump discharge head. (a) Suction pump. (b) Submersible pump

$v^2/2g$ is velocity head and represents the kinetic energy within the flowing water in the pipework. For most groundwater pumping systems, with pipework sized to minimize friction losses, the velocity head will be small and is usually ignored in calculations. Equation 19.3 then becomes

$$H = h_s + h_d + f_s + f_d \tag{19.4}$$

For submersible pumps (again neglecting velocity head), total head H comprises (Figure 19.1b)

$$H = h_d + f_{\text{sub}} + f_d \tag{19.5}$$

where f_{sub} is the friction loss in any submerged intake pipework (for most pumps used for groundwater lowering purposes, this is zero).

Methods of estimating friction losses in pipework are given in Section 19.11.

The flow rate produced by a given type of pumpset varies (to a greater or lesser degree) with the total head against which the pump must act. In general, the greater the total head, the lower the pumping rate. Different types of pump will respond to changes in head in different ways; to aid designers in pump selection, manufacturers produce performance charts – termed 'pump curves' – describing the idealized performance of each model of pumpset. An example pump curve is shown in Figure 19.2. This curve is for a centrifugal pump and indicates several characteristics common to that pump type:

- At zero flow (e.g. if the discharge side of the pump is closed off by a valve), there is a maximum total head that the pump can deliver. The pump cannot move water against a total head (static head plus friction losses) greater than the maximum head.
- As the pumped flow rate increases, the head the pump can deliver decreases. Initially, the pumping rate decreases by only modest amounts as the head increases, but at higher heads, the pump output reduces significantly for each additional increment in total head.
- The efficiency of the pumpset varies along the pump curve. The point on the pump curve with the highest efficiency is called the 'best efficiency point' (BEP), which is often around the midpoint of the flow rate range of the pump. Ideally, a system should be designed to allow a pump to be operated at its BEP. In practice, this can be very difficult to achieve in groundwater control applications due to the varying duty applied to the pump during drawdown and in steady-state operation. Even so, the designer should recognize that it can be inefficient to operate a pump at the very low end or the very high end of its pump curve, and it may be better to select an alternative pump of smaller or larger nominal capacity, respectively.

Figure 19.3 is an example of a 'family' of pump curves for a particular size of borehole electro-submersible turbine pump (see Section 19.6.1) used in deep well applications. This type of pump is commonly configured to have multiple stages of impellors, the discharge from one impellor feeding into the inlet of the next-stage impellor. In this way, multiple stages of impellors can be connected together in series to allow pumps to operate against higher heads. Figure 19.3 shows pump curves for units of nominal flow rate between 6 and 16 m^3/h (1.7 and 4.4 l/s) with between 5 and 26 stages of impellers.

Figures 19.2 and 19.3 also show the curve for NPSH (net positive suction head). This is sometimes called the 'required NPSH' (NPSHr). This represents the head (specified by the manufacturer) that must be maintained at the pump inlet to avoid cavitation. Cavitation occurs when the pressure in the pumped water falls to close to the vapour pressure of water (which for typical groundwater temperatures of 10 °C to 15 °C, is approximately 1.2 and 1.7 m head of water, respectively). If the vapour pressure is reached, the water will 'flash' to vapour, and bubbles will form within the flowing water. Cavitation occurs when the entrained bubbles are carried into the higher-pressure zones within the pump, where they collapse, causing the pump to run very noisily, as if it were pumping gravel. Cavitation can damage the impellers and other internal components of the pump and should be avoided.

The risk of cavitation is reduced by ensuring that the pressure at the pump inlet – termed the 'available NPSH' (NPSHa) – is greater than NPSHr. NPSHa can be estimated by starting with the head of water equivalent to atmospheric pressure (10 m for practical purposes) and then subtracting the sum of the suction lift, the friction losses in the suction pipework, and the vapour pressure of water.

- For a wellpoint pump used to lower groundwater level to 4 m below pump inlet, assuming 0.2 m friction losses in the wellpoint and header main, and assuming a water temperature of 10 °C (so the vapour pressure is 1.2 m), NPSHa = 10 m – (4.0 + 0.2 + 1.2); the NPSHa is therefore 4.6 m. If the pump used requires NPSHr greater than 4.6 m, then cavitation damage may result.

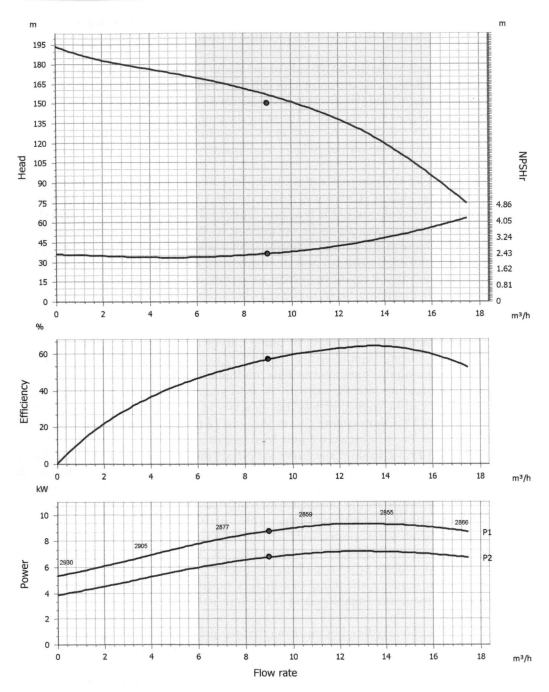

Figure 19.2 Example pump curve for centrifugal pumps In this example, the shaded area represents the manufacturer's recommended range of operation for this pump. (Courtesy of Geoquip Water Solutions, Ipswich, UK and Franklin Electric, Dueville, Italy.)

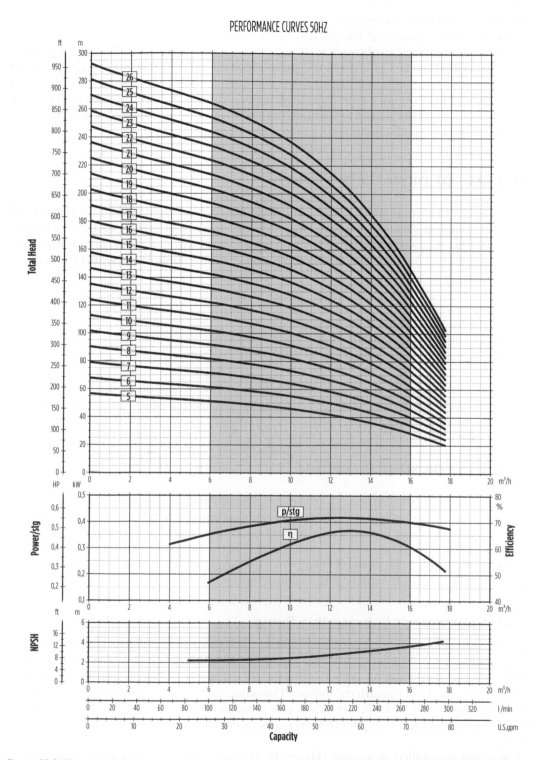

Figure 19.3 Example family of pump curves for borehole electro-submersible pumps. In this example, the grey shaded area represents the manufacturer's recommended range of operation for this family of pumps. Pump curves are shown for units with between 5 and 26 stages of impellers. (Courtesy of Geoquip Water Solutions, Ipswich, UK & Franklin Electric, Dueville, Italy.)

- For a submersible pump, there is no suction lift, and the depth of water submergence is treated as a 'negative' suction lift. For a borehole submersible pump with the inlet submerged under 2 m of water at 10 °C, and assuming zero friction losses in the pump inlet, NPSHa = 10 m − (−2.0 + 1.2); the NPSHa is therefore 10.8 m. If the pumping water level were drawn down to the pump inlet, NPSHa would be 8.8 m, which in theory would not cause problems due to suction cavitation but would allow the pump to 'draw air'. This should be avoided for borehole submersible pumps, which ideally should be operated with at least 1 m of water above inlet level.

19.3 TYPES OF DEWATERING PUMPS

It is important that appropriate pumps are selected for a given dewatering technique. Each application requires different pumping characteristics in terms of aspects such as maximum pumping rates, maximum heads, ability to handle the presence of air or suspended solids in the water and so on. It is logical to group pumps by the techniques to which they can be applied:

  i. Sump pumping (Section 19.4)
 ii. Wellpoint systems (Section 19.5)
iii. Deep well systems (Section 19.6)
 iv. Ejector well systems (Section 19.7)

Other, less widely applied pumping methods are described in Section 19.8.

The overall pumping system includes control systems and pipework. These are described in Sections 19.9 and 19.10, respectively.

19.4 UNITS FOR SUMP PUMPING

Sump pumping (Chapter 14) is one of the most common pumping applications on construction sites. Many types of pumps are used for sump pumping; some are suitable, but some others are not. The sump pumping duty is most frequently needed to deal with both groundwater and surface water run-off. Unfortunately, usually, little or no consideration is given to filtering the water to be pumped. Thus, the units used for sump pumping must be able to cope with some suspended solids in the pumped water.

19.4.1 Contractor's Submersible Pump

A pump type commonly used for sump pumping is known as the contractor's submersible pump (Figure 19.4). The most common type is an electric submersible unit with a sealed motor that usually runs in oil. This is a turbine centrifugal pump of the bottom intake type (i.e. the water intake is beneath the motor). The pump itself is installed in a sump within the excavation and is linked by power cables to a generator, which is typically located outside the excavation (Figure 14.3b), but in large excavations, the generator may be located within the dig.

The pump is installed in a sump (Figure 14.4) to allow it to remain at least partly submerged; the pumped water flows around the outside of the motor casing and thereby helps to keep the motor cool. Most units can operate on intermittent 'snore', where the water level is drawn down to the pump invert, and the pump draws both air and water. Some pumps can

Figure 19.4 Contractor's electric submersible pump. (Courtesy of Geoquip Water Solutions, Ipswich, UK.)

be controlled by 'float switch', which allows the pump to operate intermittently in response to high or low water levels in the sump (see Section 19.9.2). It is prudent to suspend the unit so that the bottom of the pump intake is about 300 mm above the bottom of the sump; this allows for some accumulation of sediment in the sump. These units are less mechanically inefficient than some other types of pump. When high flows are involved, the energy costs are significant.

The submersible pump does not operate on the suction principle, so there is no limit to the possible drawdown provided that pumps of sufficient power and head rating are used. In civil engineering applications, commonly used pumps range from 4.5 kW units with 75 mm outlets (capable of pumping around 10 l/s at 10 m head) up to 40 kW units with 200 mm outlets (of capacity 100 l/s at 25 m head). Higher-head pumps are available for deeper applications such as shaft sinking. The pumps are designed to tolerate significant quantities of suspended solids up to sand-sized (and in some cases gravel-sized) particles.

Hydraulic submersible pumps are also available, in which the submersible pump is linked by small-diameter hoses to a diesel-driven hydraulic power pack located outside the sump at ground level. Compressed-air-driven pumps (of both the centrifugal and the diaphragm type) are sometimes used on tunnelling projects to avoid the spark ignition risks associated with electrical equipment.

19.4.2 All-Purpose Self-Priming Centrifugal Suction Pump

These pumps operate on the suction principle with the pump located outside the sump and a suction hose run down into the sump (Figure 14.3a). However, suction lift limitations mean that the maximum effective depth to which a well-maintained suction pump will operate is about 6 m below the level of the pump.

These pumps are of a similar design to centrifugal pumpsets used for wellpointing duties (see Section 19.5.2) but adapted to be tolerant of some suspended solids in the pumped water. This requires the pump to be manufactured with generous clearance between the impeller and the volute casing to allow fines to pass. However, by allowing for extra clearance in the pump internals, some hydraulic efficiency is lost, and 'all-purpose' pumps normally have reduced performance compared with dedicated wellpoint units. As with units for wellpointing duties, pumps may be driven by either diesel or electrical power.

19.5 WELLPOINT DEWATERING PUMPS

The wellpoint pump operates on the suction principle and is positioned at some elevation above the level of the installed wellpoint screens (see Figure 15.2) – generally at the level of the header main. It is required to suck the groundwater into the perforated wellpoints, up the individual unperforated riser pipes, into the suction header main to the intake of the wellpoint pumpset and thence discharge the pumped groundwater via a discharge manifold to a disposal point. It follows, therefore, that an efficient wellpoint pumpset must develop:

a) Adequate vacuum to lift the groundwater from the groundwater level within the well-points and deliver the groundwater to the pump intake
b) Sufficient residual head to discharge the pumped water to the disposal area

All efficient wellpoint pumpsets are designed to pump only clean water with minimal suspended solids. There are three categories of pump types commonly used for wellpoint applications:

(i) Double-acting piston pumps
(ii) Self-priming centrifugal pumps
(iii) Vacuum tank units

19.5.1 Double-Acting Piston Pump

The reciprocating double-acting piston pump (Figure 19.5) is widely employed on wellpoint installations. It has long been used in Holland and elsewhere in continental Europe but only became widely used in the United Kingdom from the 1990s onwards. One of the main

Figure 19.5 Reciprocating double-action piston pumpset used for wellpointing. (Courtesy of WJ Groundwater Limited, Kings Langley, UK.)

advantages of this type of pump is that energy consumption is significantly lower than that of a vacuum-assisted self-priming centrifugal pumpset of comparable output.

The wellpoint piston pump operates on the displacement principle. Power is supplied by a rotating shaft driven by a diesel engine or electric motor. Cranks within the pump mechanism convert the shaft rotation into the reciprocating action of two parallel pistons aligned horizontally. The pistons are a tight fit within cylinders fitted with suction and delivery valves. The pistons are 'double acting', so each stroke of each piston simultaneously draws water into and expels water from the cylinder. The crank timing creates a pulsating pressure fluctuation in the suction system. Some manufacturers claim that this is of benefit in reducing clogging of wellpoint filters.

If there are suspended fine particles in the abstracted groundwater (perhaps because the sanding-in of wellpoints is inadequate), these will act as an abrasive that will cause wear of the pistons and cylinder liners. In time, this will lead to a reduced vacuum and associated deterioration of pumping efficiency.

Units are available with diesel or electric prime movers. A typical 100 mm unit (which refers to the nominal size of the discharge outlet) has a power requirement of 5.5 kW and can pump up to 18 l/s at 10–15 m head. The larger 125 mm unit (7.5 kW) can pump up to 26 l/s at 10–15 m head. Comparison with the capacities of self-priming centrifugal units shows that piston pumps are generally suited to lower flow rate situations and that if high flows are anticipated, a centrifugal unit may be a better choice. Because the pump is of the displacement type, the pumping capacity varies little with total head (up to the maximum head rating of the pump).

These pumps are nominally self-priming, but even for a pump in good condition, priming may be slow. Accordingly, initial priming (by filling the pump with clean water) should be considered. The amount of vacuum that it can generate when primed is slightly lower than that generated by a comparable self-priming vacuum-assisted centrifugal pumpset. As a rule of thumb, a piston pump may not be appropriate if a suction lift of more than 4.5–5.5 m is required.

19.5.2 Vacuum-Assisted Self-Priming Centrifugal Pumpset

This pump comprises a centrifugal pump of the volute type combined with a separate vacuum pump. A centrifugal pump is a rotodynamic pump where an impeller rotating at high speed pushes water radially outward. The water flung out from the impeller is directed through the tangential outlet of a curved casing (the volute) enclosing the impeller, where the kinetic energy of the water is converted to pressure. Water leaving the central eye of the impeller creates suction, drawing more water to the pump. The separate vacuum pump removes air from the pipework upstream of the impeller to allow the pump to quickly self-prime.

The pumpset has four separate components:

1. When the water enters the pump inlet, it first passes into an enclosed chamber (the float chamber) fitted with internal baffles and a float valve. This chamber serves to separate the air and water drawn into the pump, and is connected to the vacuum pump, which extracts air from the top of the float chamber. The inlet side of the float chamber is connected to the header main, and the outlet is connected to the inlet of the centrifugal pump unit. The volume of this chamber must be sufficient to ensure adequate separation of the air from the water. This is especially important when the rate of water flow from the header main is substantial. The air is extracted by the vacuum pump so that the water fed to the eye of the pump impeller has little or no air; otherwise, there will be a risk of loss of efficiency of the pump when operating. The purpose of the float

valve is to shut off the vacuum when the water level in the float chamber reaches a predetermined level, thereby preventing carry-over of water and potential damage to the vacuum pump; and to re-establish vacuum when the water level in the float chamber has fallen.

2. A vacuum pump (known as an exhauster unit) connected to the float chamber, to augment significantly the small amount of vacuum generated by the centrifugal pump unit itself. This creates a strong suction effect to draw groundwater (and air) continuously into the float chamber to flood the impeller and prime the pump. The vacuum pump is driven from the pump shaft, typically by a belt drive.

3. A 'clean water' centrifugal pump (i.e. with a minimal clearance between the impellers and the inside of the volute casing) to discharge the water from the float chamber to the discharge main. Generous clearance between impeller and volute is only appropriate if this pump type is also to be used as a contractor's 'all-purpose' pump (see Section 19.4) for dealing with solids in suspension, but the extra clearance needed for the all-purpose duty impairs efficient performance for a wellpointing duty.

4. The prime mover will generally be either diesel or electric. Its rating must be adequate to drive both the vacuum pump and the water pump to produce the expected flow with adequate allowance to cope with total head from all causes. If it is anticipated that the pump will be required for duty at a ground elevation significantly above mean sea level, and/or that the ambient temperature will be above normal, a larger engine will be needed to produce the required output.

A centrifugal wellpoint pump is operated continuously at a high vacuum (i.e. low NPSH) at the eye of the impeller and so is liable to cavitation. A cavitating pump exhibits a rattling sound similar to a great snoring noise, and the unit tends to vibrate. If the condition persists, the surfaces of the impeller will become pitted and the bearings damaged. The pump shaft may be fractured.

The most commonly used size of centrifugal pump for wellpoint installations is the 150 mm discharge pump size with a 10 l/s vacuum pump (Figure 19.6a). Such pumps typically have water flow capacities of up to 55 l/s at 10 m head. The 100 mm size is also used on some small wellpoint installations (capacity up to 40 l/s at 10 m head). Larger units are available in the 200 mm, 250 mm and 300 mm sizes but are not often required; it is generally preferable to use parallel arrangements of two or more 150 mm pumpsets in place of larger units (Figure 15.4b). Unlike piston pumps (which operate on the displacement principle), the pumping capacity of centrifugal pumps will reduce significantly as total head increases.

Wellpoint pumps are most commonly powered by diesel prime movers. These units are relatively noisy, and 'silenced' units are almost universally used in built-up areas. Silenced pumps have the entire pump and prime mover enclosed in a noise-reducing housing, as shown in the example of a 300 mm pump unit in Figure 19.6b. In fact, silenced units are far from silent; if noise is a major concern, then electrically powered pumps (running from a mains supply) should be considered, as these are the quietest units available.

The vacuum pump (exhauster unit) is an important element to assist the priming of centrifugal wellpoint pumpsets. Two types of vacuum pumps are used:

(i) The liquid ring or water recirculating vacuum pump
(ii) The flood lubricated or oil sealed vacuum pump

The vacuum units for the wellpoint application are required to operate continuously over a wide range of air flow rates from the very high to the very low. The service duty to be fulfilled is onerous. Generally, the vacuum pump is belt-driven off the drive shaft of the prime mover.

(a)

(b)

Figure 19.6 Vacuum-assisted self-priming centrifugal pumpset used for wellpointing. (a) Unsilenced wellpoint pumpset. (b) Silenced wellpoint pumpset. (Courtesy of Andrews Sykes plc, Wolverhampton, UK.)

The energy consumption of the vacuum pump is of the order of 15 per cent of the total of the motor output. The heat that may be generated can be considerable, so the cooling arrangements must be reliable; this requirement is particularly pertinent to the oil-sealed vacuum unit.

The liquid ring vacuum pump recirculates some of the pumped groundwater, which is generally cool. Thus, the cooling requirements are less demanding than for the flood lubricated vacuum pump. However, if it is expected that the pumped water will be especially corrosive, then special cooling modifications may be desirable, depending on the anticipated length of the pumping period and the chemical characteristics of the water to be pumped.

The liquid ring type is simpler than the flood lubricated vacuum pump, but the degree of vacuum that it can generate is slightly (about 5 per cent) lower). It can be more robust and be made capable of handling high air flow rates. The air handling capacity of the liquid ring pump is in the range of 1.4 to 14 m^3/min, as compared with the 1 to 3 m^3/min capacity of the flood lubricated pump – both at 0.85 bar and ambient temperature and pressure.

Adequate separation of air from water in the float chamber is very important when a flood lubricated pump is used. If the air is not adequately separated from the water, some water will be 'carried over' to the vacuum pump and cause emulsification of the oil that forms the vacuum seal; this will cause damage to the vacuum pump.

The vacuum-assisted self-priming centrifugal pump may also be used for sump pumping of clean filtered groundwater (see Chapter 14) and for operating a bored shallow well system (see Section 16.10).

19.5.3 Vacuum Tank Unit

The vacuum tank unit (Figure 19.7) is a pump that is sometimes used on wellpoint systems where the total pumping rate is very low. The pump is effectively a large-scale version of the float chamber in a vacuum-assisted centrifugal wellpoint pump. The tank has a large volume (typically 1 to 5 m^3) to give more efficient separation of air from the pumped water. A continuous vacuum is applied to the tank by one or more vacuum pumps; this draws water to the vacuum tank, which gradually fills with water. The tank is fitted with one or more water pump(s) to discharge the water. Level sensors are used to switch the water pumps on or off depending on the level of the water in the tank. Typically, the whole unit is electrically operated. The vacuum tank unit is efficient, especially for dealing with low rates of flow, because while the application of vacuum is continuous (and in most units, the amount of applied vacuum can be varied depending on the duty required), the energy for pumping the water is only mobilized as and when required to empty the tank. However, the sophistication of the total unit is considerable.

The vacuum tank unit is also useful for ground improvement of very low-permeability soils, since by reducing the moisture content, the shear strength of the soil can be increased. The rate of increase in shear strength will be slow. The rate of pumping will tend to be very low, perhaps of the order of 100 to 200 l/h. A self-priming centrifugal pump could not cope with such a duty; it would grossly overheat within a relatively short period of time.

19.5.4 Comparison of Merits of Piston Pumps vs. Vacuum-Assisted Centrifugal Wellpoint Pumps

Compared with a piston pump of the same nominal outlet size, a vacuum-assisted centrifugal pump can pump more water and achieve a slightly greater suction lift. However, the initial capital cost of a centrifugal pump will be greater than that of a piston pump. Furthermore, the fuel (or electrical power) consumption of the centrifugal unit will significantly exceed that of a piston pump. Piston pumps also have the advantage that their output reduces much less than that of a centrifugal pump when pumping against significant discharge heads.

Because of their lower capital cost and low fuel consumption, piston pumps may be the most economical solution on many low- and moderate-flow-rate wellpoint projects. Centrifugal pumps are more likely to be appropriate for high-flow-rate systems, e.g. in sands and gravels. However, the difference in fuel consumption between the piston and centrifugal wellpoint pumps can be so significant that the most economical solution, especially on longer-term projects, may be to replace centrifugal pumps with multiple units of the piston type.

(a)

(b)

Figure 19.7 Vacuum tank unit for wellpointing. (a) Skid-mounted vacuum tank unit. (b) Vacuum tank unit connected to wellpoint system. (Courtesy of Hüdig GmbH & Co. KG, Celle, Germany.)

19.5.5 Wellpoint Jetting Pumps

Wellpoint jetting units are high-pressure centrifugal clean water pumps used for the installation of wellpoints and their risers, placing tubes, etc. Their outputs are usually in the range of 12 to 200 l/s at pressures of 4 to 23 bar. The most commonly used jetting pump for the installation of wellpoints is rated to deliver about 20 l/s at a pressure of about 6–8 bar.

19.6 PUMPS FOR DEEP WELLS

The deep well method involves pumping from an array of wells with an individual pump installed in each well (Figure 16.1). Two types of pump are commonly used to pump from deep wells:

1. The borehole electro-submersible turbine pump
2. The vertical lineshaft turbine pump

The pump end is similar in configuration in both types, but the drive is different. The pump is of the centrifugal type and is submerged below the water level within the well. Within the pump, the rotodynamic action of an impeller rotating at high speed pushes water radially outwards. Unlike a wellpoint centrifugal pump, where the water is directed tangentially via the volute casing, in this arrangement, the water flung out from the impeller is directed through a ring of diffuser channels to slow down the water and raise its pressure in a similar fashion. The water flows from the diffuser channel along the axis of the pump – this configuration is known as a centrifugal turbine pump. In multi-stage pumps, the water flows into the eye of the next-stage impeller, which boosts the water pressure further. In this way, multi-stage pumps can operate against very high heads.

- In an electro-submersible unit, the pump is driven by a submersible electric motor incorporated in a common casing with the pump (the wet end). The complete unit is positioned near the bottom of the well. Typically, multiple pumps are connected to a common power source, either diesel generators or mains power, with cables feeding down into each well to supply the motor.
- In a lineshaft unit, the pump element (the wet end) is likewise positioned near the bottom of the well but is driven via a lineshaft by a motor (either electric or diesel) mounted at the surface.

19.6.1 Borehole Electro-Submersible Turbine Pumps

Generally, in developed countries, deep well pumps for temporary works projects are of the borehole electro-submersible turbine type. These are much used by water supply organizations. Hence, a great variety of these are readily available; almost off the shelf in many instances. The submersible turbine pump has a high mechanical efficiency – 70 to 80 per cent is common. Figure 19.8 is a cutaway photograph of a typical electro-submersible turbine pump.

The borehole submersible units are engineered to a slim cylindrical profile and so economize on the borehole and well screen diameter necessary to accommodate the pump. The smallest electro-submersible pumps in common use have a capacity of around 0.5 l/s and can be installed inside a well screen and casing of 110–125 mm internal diameter. Pumps of 10 l/s capacity can normally be installed inside 152 mm internal diameter (the old imperial six inch size) well screen and casing. Pumps of up to 40–80 l/s capacity are available and require well liner and casing of 250–300 mm or larger internal diameter.

The pump manufacturer's specification will normally state the minimum internal well diameter into which a given model of pump can be installed and operated. This should be used as a guide, but remember that the manufacturer's recommendation will assume the well is straight and true. In reality, the well casing may have been installed with slight twists or deviations from plumb, and these may make the pump a tight fit when, in theory, it should pass freely. Pumps do sometimes become stuck and have to be abandoned down

Figure 19.8 Borehole electro-submersible turbine pump. (Courtesy of Grundfos A/S, Bjerringbro, Denmark.) The pump shown is a 'cut-away' demonstration version to show the shaft and impeller inside the wet end at the top part of the pump.

the well (resulting in the need not only for a new pump but also for a new well), much to the chagrin of all concerned. If cost and available drilling methods allow, it is good practice to slightly increase the diameter of the well liner and casing above the minimum required for the pump.

For each pump capacity, there is a range of pumps offering the same flow rate but at increasing head. The additional head is achieved by adding additional stages of impellers to the wet end with a corresponding increase of the power requirements of the electric motor (see the family of pump curves in Figure 19.3). Pumps are typically constructed largely from stainless steel, but some plastic, cast iron or bronze components may also be used.

Most borehole submersible turbine pumps have their electric motor located in the lower part of the pump body casing with the wet end above. The wet end consists of the water intake and stator (these are both integral parts of the casing) and the multiple impeller stages, which are fixed to the drive shaft from the motor.

There is a waterproof seal on the pump shaft between the motor and the pump unit. Beneath the motor, there is a bottom bearing, which is designed to take the weight and end thrust of the whole pumpset. Most submersible turbine pump units supplied for deep wells have three-phase motors. However, single-phase units are available for low duties and outputs. They have motors up to about 5 kW.

The pump is installed centrally in the well on the end of its riser pipe (Figure 19.9); installation methods are discussed in Section 16.8. Prior to installation, the well should have been developed so that the pumped groundwater has no fine particles in suspension (see Section 16.7).

Figure 19.9 Lowering borehole electro-submersible pump and riser into well. (Courtesy of Grundfos A/S, Bjerringbro, Denmark.)

In order to prevent water in the riser pipe running back through the impellers into the well when the power is switched off, there should be a non-return valve immediately above the pump outlet. Such a backflow would be harmful if an attempt were made to restart the pump while the water in the riser pipe was still running back; under such circumstances, the additional start-up load could cause overloading. However, a non-return valve will mean that the riser pipe will remain full of water when the pump is switched off. This increases the weight to be lifted when the pump is removed from the well. It is acceptable to drill a 5 mm diameter hole in the non-return valve to allow water to drain from the riser pipe sufficiently slowly.

19.6.1.1 Use of Oversized Pumps

Occasionally, it may not be possible to equip a well with the appropriate output submersible pump, and so a pump rated at an output greater than necessary is installed in the well, and the pump will have to be operated with the well head valve 'throttled' to provide some additional resistance to reduce the flow rate. Often, in such circumstances, the lack of sensitivity on the control valve will make it difficult (sometimes even impossible) to adjust the well output so that continuous pumping is practicable. There are two alternative expedients that can be employed in such circumstances:

a) At the well head, fit an additional small bore pipe and valve to bypass the main control valve. Operation involves closing the small bypass valve completely and adjusting the main valve to achieve the best possible coarse adjustment, and then gradually opening the bypass valve to make the best fine adjustment. This mode of operation makes continuous pumping feasible.

b) Modify the pump controls for intermittent running with automatic stop/start. This requires sensing units in the pump electrical controls linked to upper and lower level sensors in the well (see Section 19.9.2).

19.6.1.2 Encrustation and Corrosion

Deep wells operating for long periods may sometimes be affected by encrustation due to chemical precipitation or bacterial growth in the well screen, filter pack, pump or pipework. Encrustation may reduce the efficiency of the well (reducing yield and increasing well losses) and may increase the stress on the pump, leading to more severe wear and tear and, ultimately, failure.

In certain types of groundwater, corrosion of metal components in pumps and pipework may be severe. Recognition of such conditions is vital when specifying equipment – for example, maximizing the use of plastic pipework can reduce the problems, leaving only the submersible pumps made from metal.

The nature of these problems and appropriate avoidance and mitigation measures are discussed in Section 23.5.

19.6.1.3 Borehole Electro-Submersible Pumps – Operational Problems

There are three common causes of motor failure:

a) Wear of the seal between the motor and the impellers. In time, wear would allow water to enter the sealed motor, thus causing failure of the motor windings.
b) Uneven wear of the bottom thrust bearing. This will lead to uneven wear of the upper seal also – result as (a).
c) Overheating of the motor. This will damage the windings of the motor.

Failure due to seal wear usually results from pumping of water containing significant concentrations of fine particles. The well design, installation and development should aim to ensure that the water pumped is clean. However, if the grading of the filter media around the well screen is too coarse to retain the soil particles (or the well has not been adequately developed; see Section 16.7), the pumped water may contain fine particles in suspension. In time, this will cause wear of the impellers and stator, leading to reduced output. Also, the seal between the impellers and the motor will be worn, eventually causing failure of the motor windings. The rewinding of a pump motor is both costly and time consuming.

It is important that the pumpset hangs vertically. Otherwise, the thrust on the bottom bearing pad will be uneven, leading to uneven wear of the seal between the motor and the impellers. Eventually, the bearing will fail, allowing water to penetrate the windings of the electric motor, causing it to fail and stopping the pump.

In operation, the pump casing is immersed in the groundwater, and also, the motor windings are surrounded by a jacket filled with oil or water-based emulsion coolant fluid. The groundwater around the motor plays a key role in dissipating heat generated by the motor. Provided that the inflow to the well is significant compared with the volume of water surrounding the pump motor casing, overheating of the motor windings is unlikely. However, if the rate of pumping is low (e.g. in a low-permeability soil/rock), the pump motor casing may be in 'dead water'. The heat generated by the motor will gradually raise the temperature of the dead water in the very bottom of the well above the generally cool temperature of the inflowing groundwater. Eventually, the motor windings will overheat and burn out.

The risk of encountering dead water is more prevalent where the very bottom of a well is formed in the top of an underlying very low-permeability layer (e.g. clay or unfractured bedrock). Often, it is necessary to design the base of the well to penetrate into the very low-permeability stratum to achieve maximum drawdown by positioning the pump intake beneath the level of the base of the aquifer. This is particularly relevant where the proposed formation level is close to an aquifer/aquitard interface, and the wetted screen length per well is limited.

Often, the cause of motor overheating is simply that the capacity of the submersible pump is too large for the yield of the well. This is a problem that may occur in some hydrogeological conditions, when well yields may decrease significantly after long periods of dewatering pumping. Ideally, in those circumstances, after the initial period of drawdown, the submersible pumps installed at the start of the project would be replaced with units of lower capacity. The labour cost of the pump changeout operation will be recovered in reduced energy usage over future years. This approach is described in more detail by Rea and Monaghan (2009).

Apart from changing to a smaller-output pump – technically the most satisfactory course of action – there are two expedients that can be used where there is risk of overheating of the motor windings. Fitting a shroud over the pump intake and motor casing will ensure that water from the cool groundwater source flows upwards over the motor casing before reaching the intake and ensures that the motor is cooled (Figure 19.10a). An alternative is to fit a small bore (3 mm or 6 mm) bypass pipe tapped into the riser pipe above the pump outlet but below the non-return valve to divert some of the cool intake flow into the very bottom of the well around and below the motor casing (see Figure 19.10b). This small and continuous flow of intake water will maintain cool water around the pump motor casing.

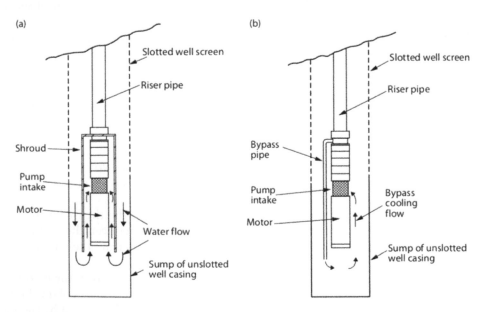

Figure 19.10 Methods to prevent overheating of borehole electro-submersible pump motors. (a) Motor shroud. The shroud is a tube, open only at its base, enclosing the electro-submersible pump. Water can only reach the pump intake by flowing over the motor, helping to keep it cool. (b) Bypass pipe. The small flow diverted from the riser pipe via a bypass pipe is discharged near the base of the motor to provide a cooling flow.

There is an insidious variation on motor overheating – though the discharge water may appear clear to the naked eye, this is no guarantee that there are no fines in suspension. When operating a low-yield well, there is a risk that the water in the very base of the well is dead. Small fine particles, not visible to the naked eye, may settle in the dead water and gradually build up around the outside of the motor casing with the result that there will be no mobile water to cool the motor casing. In time, the windings will fail due to overheating. The fitment of either a shroud or a small bore bypass pipe, as described earlier, should prevent sediment build-up around the motor.

19.6.2 Vertical Lineshaft Turbine Pumps

Lineshaft pumps (also known as spindle pumps) are a type of submersible turbine centrifugal pump (Figure 19.11). The wet-end unit of a lineshaft pumpset (which consists of one or more impellers in a bowl assembly), like that of the submersible pump, is located near the bottom of the well, submerged below water. However, unlike the electro-submersible pumpset, the prime mover unit is located on top of, or immediately adjacent to, the well head. The prime mover can be either petrol-driven (small output pumps only) or diesel-driven, with a separate energy source for each well. However, the prime movers are often electrically driven from a common power source in a similar manner to electro-submersible borehole pumps. This type of pump is particularly suitable for high volume outputs at low heads and for high horsepower duties and is commonly used in developing countries, probably due to its greater simplicity in operation.

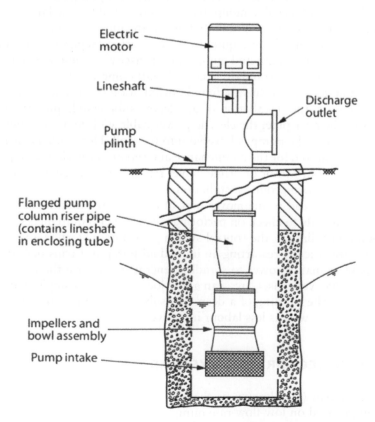

Figure 19.11 Vertical lineshaft turbine pump. The pump is powered by an electric motor mounted above the pump column.

The impellers of the pump unit are powered via a vertical lineshaft drive (with bearing assemblies) inside the pump column riser pipe, which is connected to the prime mover unit located at ground level. The verticality and straightness of the well are as important to the trouble-free operation of the lineshaft pump as for the submersible pump. The connection from the prime mover to the drive shaft depends on the type of prime mover and may be a direct coupling, a belt drive or right angle gear. As with a borehole electro-submersible pump, it is prudent to incorporate a non-return valve at the pump outlet.

19.6.2.1 Lineshaft Pumps – Operational Problems

If the pumped water contains fine particles in suspension (either because the well was not developed adequately or because the grading of the filter media is too coarse), in time, the impeller stages and casing will wear, and the pump output will decline. Otherwise, the lineshaft unit is generally trouble-free and reliable in use provided that the pump unit with its lineshaft assembly has been properly installed in the well. All the foregoing assumes that the pump unit sizing is appropriate to the actual well yield.

19.6.2.2 Comparison of Merits of Lineshaft Pumps vs. Electro-Submersible Pumps

The mechanical efficiency of lineshaft pumps is often slightly greater than that of electro-submersible pumps. The initial cost of small output lineshaft pumps tends to be greater than for a comparably sized submersible pump, but this is reversed for the large output units.

The installation procedures for lineshaft pumps are more onerous. The well must be plumb. Also, skilled personnel are required for installation. Separate connections have to be made to each drive shaft length and pump column riser pipe joint assemblies as these are being installed in the well. The standard length of shaft components may not be the same as those of the riser pipes; this can cause tedium in installation of the riser pipe and drive shaft in the well. In contrast, the installation of an electro-submersible pump unit entails only the connections of the riser pipe; the electric power cable will be in one continuous length and is simply paid out as the pump is lowered into the well. However, a considerably lower standard of skills is required for the maintenance and repair of lineshaft pumps as compared with the submersible unit. Motor or prime mover repairs are particularly easy for lineshaft pumps because of the above-ground installation.

Each individual well head prime mover unit of a lineshaft pump installation should be visited at regular intervals to check on performance and fuel requirements. However, for a submersible pump installation, the control switch gear, pump starters and associated process timers (for automatically restarting an individual pump if it trips out) and the change-over switchgear from mains power to standby generators (in case there is a failure of the mains supply, for whatever cause), etc., can all be located inside a single switch house. This makes supervision of the running of a deep well submersible pump installation easier and more straightforward as well as less labour intensive.

19.7 PUMPS FOR EJECTOR WELL SYSTEMS

The ejector well system (also known as the eductor or jet-eductor system) is a relatively specialist technique used on low-flow-rate pumping systems. It is often used on pore water pressure reduction projects in low-permeability soils such as very silty sands, silts or clays with permeable fabric (see Section 17.2). The system works by recirculating high-pressure water (from a tank and supply pumps at ground level) down riser pipes to ejectors (a type of jet pump) in each well. There are two pumping elements necessary for an ejector system:

  i. High-pressure supply pumps
  ii. Ejectors (jet pumps)

19.7.1 Ejector Supply Pumps

Ejector supply pumps are typically located at ground level and connected to a recirculation system which links a reservoir of water (the recirculation tank – typically of a few cubic metres' capacity) to the array of ejector wells, each containing a jet pump (Figure 17.1). The supply pumps are typically high-speed single or multi-stage centrifugal pumps and are operated with a flooded suction, fed by the recirculation tank. The pump and motor may be either horizontally or vertically coupled (Figure 17.4). Pumps are typically electrically powered, but units with diesel prime movers can be used in remote locations or for emergency projects.

The pumps must be sized to feed sufficient supply water at adequate pressure (typically 750 to 1500 kPa measured at ground level) to the relevant number of ejectors. If the system consists of relatively few ejector wells, one pump may be used to supply the whole system. However, large systems (more than 15 to 25 ejector wells) could be supplied by one large duty pump (Figure 17.4a) or a bank of several smaller duty pumps connected in parallel (Figure 17.4b). The use of several smaller pumps can allow a more flexible approach. with additional pumping capacity being easy to add as needed, and allows more economical use of standby pumps.

19.7.2 Ejectors

An ejector is a form of jet pump. This is a hydraulic device, which despite having no moving parts, acts as a pump of the velocity type. Jet pumps were first proposed in the mid-nineteenth century for the removal of water from water-wheel sumps (Thomson, 1852). Outside civil engineering, jet pumps are used in a wide range of other applications.

An ejector well system comprises multiple wells. An ejector is suspended in each well on the riser pipe and receives water at high pressure from the supply pumps. Within the ejector, water is forced through a narrow nozzle, followed by the gradually expanding cross section of a venturi, before returning to the surface and being recirculated back to the supply pumps. The constriction of flow through the nozzle converts pressure head to velocity head (the water passes through the nozzle at high velocity), thereby creating a zone of low pressure and generating a vacuum of up to 9.5 m of water at the level of the ejector. The vacuum draws groundwater into the well through the well screen, where it joins the water passing through the nozzle and is carried along with it. The expansion of the flow within the venturi converts velocity head back to pressure to lift the water out of the well and back to the supply pumps. Detailed analysis of ejectors of the type used on groundwater control projects can be found in work by Miller (1988) and Powrie and Preene (1994b).

Two different ejector types are commonly used:

  (i) Single (or concentric) pipe (Figure 17.2), where the supply and return risers are arranged concentrically with the return riser inside the supply riser. The supply flow passes down the annulus between the pipes and through the ejector and then returns up the central pipe.
  (ii) Twin (or dual) pipe (Figure 17.3), where the supply and return risers are separate, typically being installed parallel to each other.

In both cases, a non-return valve is present within the ejector to prevent any injected water flowing into the well screen and thence into the ground.

Ejectors can pump both air and water; as a result, when installed in a sealed well in low-permeability soil, if the ejector capacity exceeds the water inflow, a vacuum will be developed in the well. Drawdowns of 20–30 m below pump level can be achieved with commonly available equipment, and drawdowns in excess of 50 m have been achieved with systems capable of operating at higher supply pressures.

Ejector systems are designed to pump 'clean' water containing very low levels of suspended solids. Due to the high water velocities through the nozzle, even slightly dirty water can lead to rapid nozzle wear (erosion and enlargement of the nozzle openings – see Figure 17.6), leading to loss of performance. Ejectors can also be affected by clogging or blockage due to chemical encrustation and biofouling. These factors are discussed in Section 17.2.5.

19.7.3 Ejector Performance Curves

Ejectors work on the velocity principle, and like any pump of that type, their output will vary significantly with discharge head. An additional complication is that in order to function, the ejector must be supplied with sufficient supply water at adequate pressure (typically 750 to 1500 kPa measured at ground level). Each design of ejector will have different operational characteristics, so performance curves will be needed for the model being used (Figure 19.12). It is important that the performance curves are representative of the conditions in a well, including the depth of submergence of the ejector (see Miller, 1988 and Powrie and Preene, 1994b for further details). At a given supply pressure and ejector setting depth, the net pumping capacity (the induced flow rate) can be estimated from the performance curves. If this is sufficient to deal with the anticipated well inflow, then the required supply flow per ejector (at the specified pressure) and the number of ejectors to be fed by each pump can be used to select the supply pumps, which should be sized to supply the required flow at the necessary pressure, taking into account friction losses in the system.

If the ejector induced flow rate is not large enough to deal with the predicted well inflow, it may be possible to increase the capacity of the ejector by increasing the supply pressure. However, Figure 19.12 shows that as the supply pressure is increased, the induced flow rate of the ejector tends to plateau. This phenomenon occurs when the pressure at the exit of the ejector nozzle generates a vacuum of around 9.5 m, and the pressure reduces to the vapour pressure of water. This causes cavitation in the ejector, and additional increases in supply pressure beyond that point do not give a corresponding increase in ejector capacity. With some ejector designs, it is possible to increase capacity by fitting a larger-diameter nozzle and venturi. This gives an increased induced flow rate at a given supply pressure at the expense of an increase in the required supply flow per ejector.

Compared with other pumping systems, the ejector method has a low mechanical (or energy) efficiency. In low- to moderate-permeability soils, where flow rates are small, this may not be a major issue, but in higher-permeability soils, the power consumption and energy costs may be huge in comparison to other methods.

19.8 OTHER PUMPING METHODS

19.8.1 Airlift Pumping

Airlift pumping is widely used for well development (see Section 16.7) but is rarely utilized for continuous dewatering pumping. It is based on the buoyancy principle of pumping, where compressed air is bubbled through water, thereby reducing the density of the affected volume of water and causing it to rise through the remaining body of fluid.

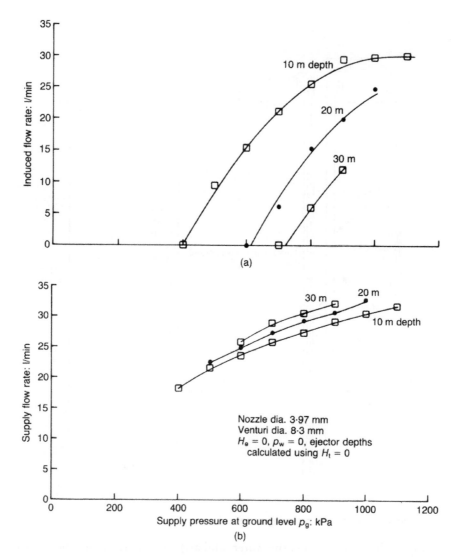

Figure 19.12 Example ejector performance curves. (a)The upper graph is used to determine the supply pressure necessary to achieve the desired induced flow rate per ejector at the specified ejector setting depth (10, 20 and 30 m shown). (b) The lower graph is then used to determine the supply flow rate per ejector necessary to maintain the specified supply pressure. (From Powrie, W and Preene, M, *Proceedings of the Institution of Civil Engineers, Geotechnical Engineering*, 107, 143–154, 1994. With permission.)

When used to pump from a well, airlifting involves lowering an eductor tube and compressed air line below groundwater level. Various different configurations can be used, including airline inside the eductor tube (Figure 16.7a) or airline outside the eductor tube, with a special nozzle or cross-over fitting used to allow the air to be discharged inside the eductor tube (Figure 19.13). In smaller-diameter wells, it is possible to airlift without an eductor pipe by simply hanging the airline within the well – in effect, the well liner is used as the eductor tube. The airline can be a flexible hose, or a sectional steel or plastic pipe, of suitable pressure rating. In all cases, the method works by the introduction of air from a compressor located at ground level. This reduces the effective density of the water, causing it to rise within the eductor tube. If a pipe elbow and a short length of discharge pipe are

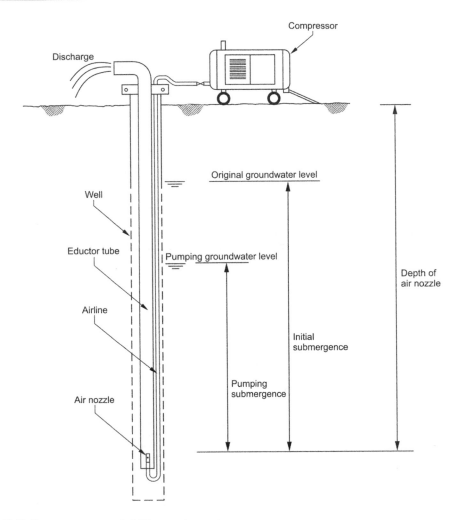

Figure 19.13 Key parameters of airlift pumping system.

fitted to the top of the eductor tube, the water (and air) will emerge from the end of the pipe (Figure 19.14).

In high-yielding wells with groundwater levels close to the surface, the flow from an airlifting system can be quite 'steady' with only modest fluctuations in flow rate as the air and water separate in the pipe between the elbow and the discharge point. However, in low-yielding or shallow wells, airlifting can produce very irregular flow, where the discharge may alternate between discrete slugs of air and water. This behaviour can be explained in terms of 'airlift submergence'.

Figure 19.13 shows the geometry of a well being airlifted. Ideally, the pumping submergence should be around 60 per cent or greater of the depth of the air nozzle. If the submergence reduces significantly below 60 per cent, the pumping rate will reduce notably, and the flow can become very irregular, with alternating slugs of air and water produced (further details of airlift pumping can be found in Sterrett (2008)). The compressor used for airlifting must meet two requirements:

- Sufficient air pressure to overcome the water pressure corresponding to the initial submergence at the air nozzle near the base of the eductor tube (i.e. approximately 10 kPa for every metre of the initial submergence of the air nozzle).

Figure 19.14 Discharge from high-capacity airlift pumping during well development.

- Sufficient air volume to lift significant volumes of water, provided that the pumping submergence is adequate. Sterrett (2008) indicates that, as a very crude guide, around 0.06 m³/s (12 cubic feet per minute [cfm]) of air is required for each litre per second of water pumping rate.

Although airlift pumping is widely used for well development, it is an inefficient method and is rarely applied in continuous pumping for groundwater control purposes. One of the characteristics of the method is that it can pump water containing significant concentrations of suspended solids without damage to equipment. On a very few occasions, airlifting has been used for dewatering pumping in cases where the pumped water would be very aggressive to conventional pumps because of water chemistry or suspended solids.

19.8.2 Bailing of Deep Shafts

One of the earliest forms of pumping was the direct lift method – such as hauling a bucket from a well using a rope. This crude method has the advantage that it can work to great depths, limited only by the capacity of the hauling rope or cable and the power (and time) available to haul the bucket to the surface. In the nineteenth century, when the Industrial Revolution resulted in deeper and deeper coal mines in the United Kingdom, many underground mines were drained using horse-powered gins (winches) hauling large steel buckets (known as 'kibbles') to the surface. Younger (2004) reports that the limiting flow rate using a 2.7 m³ kibble in a 48 m deep mine was around 4.5 l/s and that for deeper collieries, the maximum bailing rate was around 2 to 3 l/s. For inflows greater than this, mines will have been either pumped or else abandoned.

Surprisingly, bailing by kibble is still used in the twenty-first century when sinking very deep mine shafts, provided that rates of groundwater inflow are small. In this method, a sump pump in the shaft bottom is used to remove water and pump it to an empty kibble sitting at that level. Periodically, the kibble is swapped for an empty one, and the full kibble is hoisted to the surface to dispose of the water. This avoids the cost and complication of using very high-head pumps and long lengths of high-pressure pipework to lift water to the surface. Bailing rates are largely controlled by kibble size and hoisting rates, and modern rates are not so different from historical values. Mining industry experience indicates that bailing can be practicable up to around 2 l/s, and in the short term, to deal with localized water problems, inflows of around 4 l/s can be handled. In both cases, rates of shaft sinking will be slowed due to the raising of the water kibble reducing the hoist time available for personnel, materials, spoil, etc.

19.9 PUMP CONTROL SYSTEMS

Control systems are an essential requirement to allow the operation of pumps. Controls can vary from the very simple (e.g. a basic on/off switch for a pump running at constant speed) to more complex systems that allow pumps to run at variable speeds in response to changing water levels or other parameters. Control systems can monitor and record energy usage and transmit data via mobile data connections (energy usage is discussed in Section 19.11.4). Alarm systems and automation are also discussed in Section 23.4.

19.9.1 Electrical Control Systems

Most electrical pumps are driven by three-phase alternating current (AC) induction motors, which are simple, rugged and reliable. In normal operation, these motors operate at a constant speed, determined by the number of poles in the motor and the frequency of the electrical supply. The simplest and cheapest control system is the direct-on-line (DOL) starter, where the pump is started on full voltage. However, this approach requires a large starting current (several times the running current) to overcome inertia and spin the motor (and pump) up to speed; high-capacity generators or mains electrical supplies are needed to handle the starting current. DOL starters are typically used on relatively small motors and are rarely used on units greater than 15 kW. Larger motors can be started by the star-delta system, which requires a smaller starting current than the DOL system. The star-delta system starts the pump at a lower voltage (when the pump is connected to the supply by a 'star' connection), and then, when the motor reaches around 80 per cent of full speed, the control unit switches the motor to a 'delta' connection to run at full speed. In both these cases, the motors (and hence the pump wet end connected to it) runs at constant speed, and the flow rate vs. head relationship varies along a fixed pump curve (Figure 19.2).

More sophisticated control units can be used to vary the frequency of the electrical supply reaching the pump, hence varying the motor speed. These control units (known by various names, including variable speed drives, variable frequency drives and inverter drives) are connected between the pump and the power source (generator or mains supply) and effectively provide an electrical supply to the motor that allows the speed to be varied between maximum and minimum values. As the speed varies, the pump output will vary (at a given total head). The effect of varying the pump speed can usefully be visualized as changing the shape of the pump curve to control the output. On a fixed speed pump, the same effect requires the total head to be changed, for example by using a valve to throttle the discharge side of the pump.

The maximum speed possible with a variable speed drive may be significantly higher than the rated speed at the initial fixed frequency. The minimum pump speed is typically around 30 to 50 per cent of the rated speed (so the pump cannot be 'turned down' to zero – there is a minimum pumping rate associated with variable speed operation). Due to the cost and complexity of variable speed drives, they are rarely applied on groundwater pumping operations. They are more commonly applied on pumping systems for sewer bypass and overpumping schemes.

19.9.2 Intermittent Operation of Pumps

In the vast majority of groundwater control applications, the pumps operate continuously, 24 hours per day. One of the few cases when intermittent operation is a viable option is when wells or deep sumps are operated in low-permeability soil or rock. Each well or sump will have a low yield and will be quickly emptied of water, even if a small pump is used. If the pump is left to operate continuously, it will run 'dry', resulting in cavitation and over-heating. This may cause the pump to burn out or be stopped by protective circuits in the control unit.

In very low-yielding wells and sumps, one option is to control the pumps so that they operate intermittently, using sensors in the well or sump to trigger the pump to start when the water level is high and stop when the water level is too low. Three types of sensor are commonly used:

- Float switch: This device is commonly used to control sump pumps. A small buoyant plastic float (around 100 mm long) is attached to the top of the pump by a short cable or tether. When the water level in the sump is high, the float will rise, and when it reaches the end of the tether, it will float vertically. In the vertical position, a simple switch (often a small ball dropping downwards to complete a circuit) inside the float will activate and start the pump. When the sump is emptied by the pump, the float will sink to a low level, and the change in position will break the circuit and stop the pump.
- High- and low-level electrodes: This approach is suitable for use in deep wells and takes advantage of the electrical conductivity of water. A pair of electrodes are placed in the well, typically secured to the riser pipe. Each electrode is part of a low-voltage circuit (analogous to a dipmeter used to measure water levels in a well – see Section 22.6.2). When each electrode is submerged, the water completes the circuit. This causes the pump to start when the water level rises above the upper electrode and stop when the water level falls below the lower electrode.
- Pressure sensor: This involves a pressure transducer of the vibrating wire piezometer (VWP) type (similar to those used for groundwater level monitoring – see Section 22.6.3). When used to control deep wells and sumps, the transducer is placed in the well. The depth of the sensor is noted, and the water pressure recorded can be converted to a water level. A programmable logic controller (PLC) in the pump control unit can be set to start and stop the pump when observed water pressure corresponds to the specified high and low water levels, respectively.

Typically, the low-level trigger is set to switch the pump motor off just before the water level is drawn down sufficiently to cause the pump to 'draw air'. The high level trigger is set to switch the pump on when the water level in the well or sump has recovered to a pre-determined level (which must obviously be below excavation formation level).

The average level of lowering achievable by this mode of pumping will be approximately the mean level between the two trigger levels. The distance between the upper and lower levels should be the minimum practicable to allow the maximum amount of water lowering by intermittent pumping. The disadvantage of this is that it will cause more frequent starts of the pump motor, increasing the risk of motor failure. Most sump pumps and borehole pumps operated in this arrangement will be electrically driven and controlled using DOL starters. With this arrangement, the high starting current can generate significant heat on start-up. Thus, if the motor is frequently switched on and off in response to trigger level settings, with time the motor windings may tend to become overheated, eventually resulting in failure of the motor. There is a risk of this happening if there is on/off switching more than about six times per hour. In fact, even six starts per hour is a lot, and the number of starts should be minimized if at all possible.

19.10 DISCHARGE PIPEWORK AND EQUIPMENT

Discharge pipework is the system of pipes, valves, fittings and tanks used to carry water from the pumps to the ultimate disposal point. It is a simple but vital part of any dewatering system and should not be neglected.

Due to the temporary nature of most dewatering systems, discharge systems most commonly comprise sectional pipework, delivered to site in short lengths (typically less than 6 m, but occasionally longer). Pipework is typically available in standard diameters, often based on old imperial sizes. Common sizes for discharge pipework are nominal 6 inch (150 mm) and 8 inch (200 mm) diameter, although nominal 4 inch and 12 inch diameter (100 mm and 300 mm, respectively) are also widely available. A wide range of fittings – bends, tees, valves and reducers (to allow pipes of different sizes to be joined) – are available and allow pipework to be laid to follow irregular routes or pass around obstacles (Figure 19.15).

Pipework can be made from a range of materials, although the most common materials used are steel and high-density polyethylene (HDPE). In most short-term applications, pipework is joined by flanged and bolted connections or proprietary 'quick action' couplings fitted to the end of the pipes. On long-term or permanent projects, HDPE may be joined by butt welding or electro-fusion welding.

(a)   (b)

Figure 19.15 Pipework used for dewatering discharges. (a) Sectional pipework used for dewatering discharge systems. (b) Pipe bridge used to carry dewatering discharge over access road. (Courtesy of Theo van Velzen BV, Heiloo, The Netherlands.)

Discharge pipework must also incorporate flow measurement equipment such as flow meters or weir tanks (see Section 22.7) and, if necessary, settlement tanks or other means to remove suspended solids in the water (see Section 21.9).

19.11 DESIGN OF PUMPING SYSTEMS

The design of the pumping elements of a groundwater control scheme involves the selection of a suitable combination of appropriate pumps, pipework and control systems, based on the anticipated flow rates and drawdowns from a dewatering design (Chapter 13).

19.11.1 Assessing the Pump Duty

The design process should have produced:

1. A preferred technology (sump pumps, wellpoints, deep wells, ejector wells, etc.)
2. Estimates of pumped flow rate (total and, for some technologies, a flow rate per well)

The first element is an essential output of design, because, as outlined in this chapter, different pump types are required for different technologies.

Based on the selected technology, the required pump 'duty' should be determined, based on the flow rate per pump and the total head against which the pump must operate.

For deep well systems, the key question is the capacity of the pump in each well. For sump pumping and wellpoint systems, there is an additional step, before sizing the pumps, to decide how many pumps will be used (this may depend on the size of the project and the equipment available locally). Ejector well systems require a hybrid approach, where the number of supply pumps must be assessed (and then sized) as well as selecting a suitable ejector capacity to deal with the flow rate per well.

In addition to the flow rate, the other element of the pump duty is the total head against which the water must be pumped. As described in Section 19.2.3, the total head comprises the sum of the height the water is lifted and the friction losses (Equations 19.4 and 19.5). In most cases, the dominant element of the total head is the static lift of the water (the physical height the water is lifted), with friction losses being a secondary element. In many cases, assessing the static lift is simple, but for more complex pumping arrangements, some thought may be needed to determine what static lift applies to the pump in question (Figure 19.16).

19.11.2 Estimating Friction Losses

'Friction losses' is the term used to describe the energy lost by turbulence and friction where water passes along pipes and through the constructions and deviations of valves, bends, etc., which comprise part of the total head against which pumps must discharge. For dewatering systems using standard pump and pipework sizes, friction losses are often small in comparison with the other elements of total head and are rarely a major issue in the design. Cases where friction losses may be significant and may reduce the output from the pumps include:

i. High-flow-rate systems (greater than around 50 l/s) where the water has to be discharged considerable distances (greater than 100 m)
ii. In-tunnel pumping systems where space limitations mean that small-diameter pipework is used, and where the discharge length from the pumping location to the disposal point (such as a shaft or portal) is significant

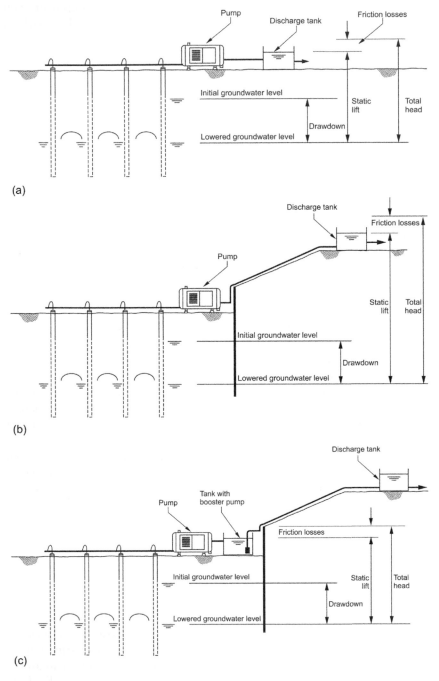

Figure 19.16 Total head for pumps in dewatering systems. (a) Simple system with no variation in ground level. In this case, the pump discharge is at essentially the same level as the pump, and the total head is the static suction lift plus friction losses. (b) Complex system with change in ground level. In this case, the arrangement of the excavation means that pump discharge is above the level of the pump. The total head is the static suction lift plus the static discharge head plus friction losses. In this case, the pump output will be reduced compared with (a). (c) Booster pumps used to reduce static lift for dewatering pumps. In this case, the system is arranged so that the dewatering pump discharges to a booster pump. This reduces the total head against which the dewatering pump is working and will allow an increased pumping rate compared with (b).

Friction losses can be estimated from published charts and tables such as Figure 19.17 (from Preene *et al.*, 2016), which allow losses to be estimated for commonly used pipe sizes and fittings. Figure 19.17 shows that such generic tables typically report the friction loss as metres head per 1000 m of pipework (for a given flow rate in a pipe of specified diameter).

Friction loss f can also be estimated using the empirical Hazen–Williams formula, which (using SI units) is

$$f = l_p \frac{10.67 Q^{1.85}}{\left(C_{hw}\right)^{1.85} \left(d\right)^{4.87}} \tag{19.6}$$

where:

l_p is the effective length of pipework (in metres) through which water is pumped, including an additional length of pipework to allow for the presence of fittings (see Figure 19.17)

d is the internal diameter (in metres) of the pipework

Q is the flow rate (in m³/s)

C_{hw} is the Hazen–Williams roughness coefficient (dimensionless)

For steel pipework, C_{hw} is usually assumed to be between 100 and 120, and for plastic pipework, 140 can be used.

If friction losses are perceived to be large enough to detrimentally affect pump outputs, then a number of remedial measures are possible:

(i)   Replace the pumpsets with units rated at higher heads.
(ii) Provide additional pumpsets connected in parallel, so that a greater number of pumps producing a lower output can collectively achieve the required total flow rate.
(iii) Modify the discharge pipework by:
  a. Planning the pipework layout carefully to avoid unnecessary bends, junctions or constrictions.
  b. Using discharge pipework of larger diameter, if available.
  c. Using additional lines of discharge pipework laid in parallel, reducing the flow rate taken by an individual pipe.
(iv) Reduce the distance that the discharge water must be pumped (e.g. by locating an alternative discharge point).
(v)   Provide 'booster pumps' between the dewatering pumps and the discharge point. This will reduce the total head on the dewatering pumps, increasing their output (Figure 19.16c).

19.11.3 Sizing of Pumps

The pumps must be selected to be an appropriate match for the pump duty (flow rate and total head). It is rare for a dewatering designer to have the luxury of being able to specify and purchase a pump to exactly match the pump duty for a specific project. Most commonly, pumpsets must be chosen from standard or 'off the shelf' sizes either bought new, hired-in or re-used from a previous project. Thus, dewatering pumps are inevitably an imperfect match to the required duty. This problem is further compounded by the fact that the pump duty will change during the project. Typically, the required pumping rate will be higher during the initial drawdown period, and the longer-term steady-state pumping rate will be lower, with a higher static lift.

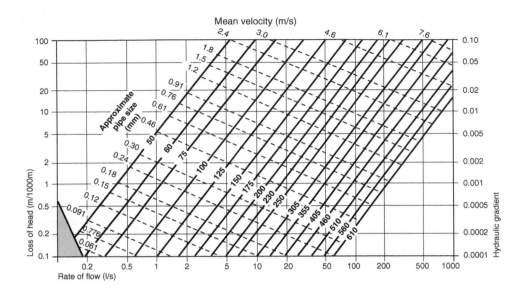

Friction losses in valves and fittings as an equivalent length of straight pipe in metres

Type of Fitting		Nominal pipe diameter (mm)								
		150	200	250	300	450	600	750	1065	1200
Gate valve	Open	1.1	1.4	1.7	2.0	2.8	4.3	5.2	6.4	7.6
	¼ closed	6.1	7.9	10.1	12.2	18.3	24.4	30.5	41.2	48.8
	½ closed	30.5	39.6	51.8	59.5	91.5	122.0	152.0	213.0	244.0
	¾ closed	122.0	159.0	213.0	244.0	366.0	488.0	610.0	854.0	976.0
Standard Tee	Flow in line	2.9	4.3	5.0	5.9	9.1	11.9	15.1	22.0	24.7
	Flow to/from branch	9.8	12.8	16.8	19.8	30.5	39.6	50.3	73.2	82.3
Medium Sweep 90° elbow		4.3	5.5	6.7	7.9	12.2	15.9	21.3	28.0	32.0
Long Sweep 90° elbow		3.2	4.3	5.3	6.1	9.1	12.2	15.2	21.3	24.4
Square 90° elbow		9.8	12.8	16.8	19.8	30.5	39.6	50.3	73.2	82.3
45° elbow		2.3	3.1	3.7	4.6	6.4	8.5	10.7	15.2	18.3
Sudden enlargement	d/D = ¼	4.9	6.4	8.4	9.9	15.2	19.8	25.2	36.6	41.2
	d/D = ½	3.2	4.3	5.3	6.1	9.1	12.2	15.2	21.3	24.4
	d/D = ½	2.9	3.7	4.9	5.6	8.4	11.0	13.7	19.8	22.9
Sudden contraction	d/D = ¼	2.3	3.1	3.7	4.6	6.4	8.5	10.7	15.2	18.3
	d/D = ½	1.7	2.3	2.9	3.4	4.9	6.4	8.2	11.3	12.8
	d/D = ½	1.1	1.4	1.7	2.0	2.8	4.3	5.2	6.4	7.6

Figure 19.17 Friction losses in pipework. (Reproduced from Preene, M, *et al.*, *Groundwater Control – Design and Practice*, 2nd edition. Construction Industry Research and Information Association, CIRIA Report C750, London, 2016. With permission from CIRIA: www.ciria.org)

Section 19.2.3 introduced the concept of a 'pump curve' – a relationship unique to each model of pumpset, which describes how the pumped flow rate varies with the total head against which the pump must act. Using the required flow rate from the dewatering design and an assessment of the total head against which a pumpset must work, the manufacturer's performance charts allow estimation of the pump output under field conditions. If the predicted output of the chosen pump is lower than required, larger pumpsets could be used, or (if the system design permits) multiple units could be employed.

As has been discussed elsewhere, the natural variability of the ground and its permeability can result in considerable uncertainty in pumping rates. The whole philosophy of this book is intended to ensure that designers and practitioners develop and implement groundwater control systems that are robust and flexible enough to deal with credible variations in conditions. Among non-specialists, groundwater control is widely perceived to be about 'dealing with water' and keeping the excavation 'dry' – in reality, as described in Chapters 6, 7 and 8, the real focus should be on avoiding groundwater-induced instability. Consequently, inexperienced designers often fear high flow rates – 'more water than the pumps can handle' – and their natural inclination is to increase the capacity of individual pumps significantly above the design values to give a margin of safety.

It is true that in some hydrogeological settings – such as highly permeable gravels or intensely fractured rock, especially if there is a close source of recharge – there may be a risk of flows much higher than design values, which can cause systems to be overwhelmed. However, Figure 13.2 shows that suitably designed systems can often cope with significantly greater flows with only relatively modest changes. Unfortunately, providing excessively oversized pumps brings its own problems – the sizing of dewatering pumps is one case where bigger is not always better. Many types of pump experience operational problems (including overheating and cavitation damage) if operated for long periods at very small fractions of their maximum rated flow. In any case, pumps operated in this way will have a very low efficiency. A much better approach is to use pumps more closely matched to the design flow rates but provide flexibility to change the pump unit, or provide additional pumps, if higher flow rates are encountered.

19.11.4 Energy Used in Pumping

By definition, a pump requires energy to lift water. The energy consumption E of a unit pumping flow rate Q against a total head H for a run time t with a combined efficiency of the pump and motor of η is

$$E = \frac{Q \times H \times \rho_w \times g \times t}{\eta} \tag{19.7}$$

On a small project, where the duration of pumping is short, energy costs may not be a large proportion of overall groundwater control costs. However, if high flow rates are pumped for long periods, the energy costs will be proportionately larger; for permanent dewatering systems (Chapter 20), energy costs may be very significant. It is good engineering practice to reduce energy consumption, as this will result in lower overall costs as well as lower CO_2 emissions from diesel fuel or electricity use. This may be part of a wider strategy to optimize dewatering systems (see Chapter 26).

Equation 19.7 shows that the energy required for dewatering pumping can be reduced by a range of measures (which can be used in combination):

a) Reducing the flow rate Q; this can be achieved by:
   • Use of low permeability cut-off walls or groundwater barriers to seal off significant permeable zones, thereby reducing total dewatering pumping rate.

- Effective monitoring of groundwater levels during dewatering to allow pumping to be controlled to avoid excessive lowering of groundwater levels below the level assumed at design stage.

b) Reducing total head H by good design of pumping pipework to reduce friction losses and minimize the required total head. This can be achieved by minimizing the lengths of suction and discharge pipework, using larger pipe diameters and lower friction materials, and avoiding unnecessary bends and restrictions in fittings.

c) Reducing the hours run, t, for example by the use of effective control systems to allow pumps to switch on/off in response to groundwater levels (see Section 19.9.2).

d) Increasing the efficiency η of the pumpsets. Possible measures include:

- Selection of appropriate pumps for the duty, so that they run at an efficient point in their performance curve. One example is in low-yielding well systems – with yields in the range of 0.3 to 1.0 l/s per well. These yields could be handled by either ejectors or borehole submersible pumps. In this case, the submersible pumps have much higher efficiency and would consume less energy than an ejector system.
- Use of suitable control systems (e.g. variable speed drives) to allow pumps to operate at high efficiency in a range of duty conditions (see Section 19.9.1).
- Swapping of pumps during the project if the duty changes. One option is that, after the initial period of drawdown, the pumps installed at the start of the project would be replaced with units of lower capacity to deal with the steady-state pumping rate.

Chapter 20

Permanent Groundwater Control Systems

Water is the driving force of all nature.

Leonardo da Vinci

20.1 INTRODUCTION

The vast majority of groundwater control applications are temporary. Dewatering or groundwater exclusion is carried out for a defined period, typically for the construction period of a structure or the production period of a mine or underground facility. The period for which groundwater control is required may be an extensive period, perhaps several months, even several years, but most systems are definitively temporary.

This chapter describes the issues associated with long-term groundwater control systems, which are to be in operation for so long that they can be considered 'permanent'. There is no hard and fast definition of when a groundwater control system can be considered permanent, but for the purposes of this chapter, a system might be considered permanent if it is intended to operate for more than 5 to 10 years without major changes and is not part of an active construction or mining project. Permanent systems are typically an element of completed structures, facilities or engineering infrastructure.

This chapter outlines the types of systems that can be used, describes the objectives of long-term or permanent groundwater control systems, and discusses practical problems.

Case histories 27.11 and 27.12 describe, respectively, long-term groundwater control systems used as part of a groundwater remediation scheme and to stabilize a subway tunnel.

20.2 TYPES OF PERMANENT GROUNDWATER CONTROL SYSTEMS

Like temporary systems, permanent groundwater control systems can be characterized as those based on the principles of groundwater pumping or exclusion (see Section 9.3).

20.2.1 Permanent Groundwater Pumping Systems

Systems based on pumping (Figure 20.1) work on the principle that groundwater is pumped (i.e. abstracted or removed from the ground) in order to lower groundwater levels or groundwater pressures. Systems may be based on conventional wells pumped by, for example, wellpoint or deep well systems (Figure 20.1a).

On some occasions, 'passive' drainage systems may be used, where the drainage elements such as relief wells (see Section 17.8) or drainage blankets are not pumped directly but have

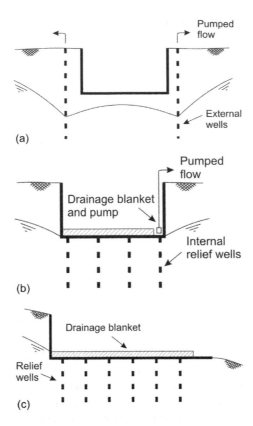

Figure 20.1 Permanent groundwater pumping systems. (a) Pumping system. The wells are pumped directly to lower groundwater levels. (b) Passive pumping system. The wells are not pumped directly but are allowed to overflow to a drainage blanket, from where water is pumped away by a sump pump. (c) Fully passive pumping system. The topography of the ground is used to allow water to flow away from the wells or drainage features without the need for pumping.

outlets at a sufficiently low level, relative to original groundwater level, that they flow without the need for pumping. The water is then typically collected in sumps or drainage systems and pumped away (Figure 20.1b).

Where the terrain or geometry allows, such as in slope stabilization applications, it may be possible to allow the water to flow away by gravity without the need for pumping (Figure 20.1c).

20.2.2 Permanent Groundwater Barrier Systems

Barrier systems are used to exclude groundwater in order to reduce, and in some cases to effectively eliminate, seepage of groundwater into structures or other infrastructure that is below ground level. Examples of technologies used to form very low-permeability groundwater cut-off barriers are described in Chapter 18, together with background on the advantages and disadvantages of this approach.

The most common application of barrier systems involves the use of artificial barriers (typically low-permeability vertical cut-off walls) in combination with natural 'geological barriers' such as very low-permeability layers of clay or unfractured rock (Figure 20.2a). In combination, the artificial and natural barriers act to form a complete exclusion system around the structure.

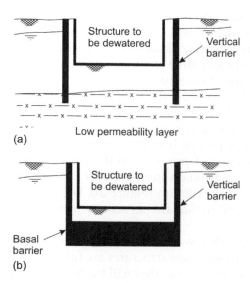

Figure 20.2 Permanent groundwater barrier systems. (a) Artificial barrier in combination with geological barrier. (b) Artificial barrier with basal plug.

Where there is no suitable geological barrier, it is necessary to install an artificial barrier around the entire structure. This normally involves forming a very low-permeability seal or 'basal seal' beneath the structure, most commonly by some form of grouting method (see Section 18.3.1). Forming such a basal seal can be complex and expensive.

No barrier system is completely impermeable. Some residual seepage through the barrier should be anticipated, either through joints or imperfections in the barrier system or through the mass of the barrier. Some modest pumping capacity to drain any residual seepage is often associated with permanent barrier systems.

20.3 OBJECTIVES OF PERMANENT GROUNDWATER CONTROL SYSTEMS

Permanent groundwater control systems can be used to meet a range of objectives, including:

i. To reduce groundwater uplift loads on structures
ii. To control anticipated rises in groundwater levels
iii. To reduce leakage into below-ground structures such as tunnels
iv. To stabilize slopes and other earth structures
v. To form part of a system to remediate groundwater contamination

Each type of system is described in the following sections.

20.3.1 Reduction of Groundwater Uplift Loads on Structures

Any sealed structure constructed below groundwater level will be subject to groundwater pressures acting on its external faces. Where a structure is constructed to enclose a large void (such as an underground tank, water treatment plant or metro station) in permeable strata, the structure will experience a net uplift force and will be effectively buoyant (see Section 7.5).

If suitable engineering measures are not taken to counteract or reduce the net uplift force, there is a risk that the structure will become overstressed, may experience distortion or damage, or may, in extreme cases, physically rise up out the ground until the buoyancy forces match the deadweight and other resisting forces. Any of these events would have serious consequences for the use of the structure.

One possible approach is to resist the uplift forces by the addition of deadweight to the structure (for example by increasing the thickness of concrete base slabs or walls) or by use of structural elements such as tension piles or ground anchors to 'hold down' the structure. An alternative approach is to use a permanent groundwater pumping system to lower groundwater levels and pressures so that there is no excess uplift pressure on the structure. Reduction in groundwater pressures is achieved by either pumped wells, located inside or outside the structure (Figure 20.3a), or by relief wells flowing into a drainage blanket or gallery within the structure, from which water is pumped away (Figure 20.1b).

Dry docks and certain other water-filled or fluid-filled structures present particular problems. In normal operations, these structures are full of water and are not buoyant, but at some point in their operational cycle, they will be emptied of fluid and will become buoyant. One solution is to have a permanent dewatering system installed around the structure. The system is normally left 'on standby' (i.e. fully or partly commissioned but not pumping); it is pumped only when the structure is to be emptied and is used to lower groundwater levels for the period until the structure is re-filled. An alternative approach, sometimes used in dry docks, is to equip the base of the structure with an array of relief wells, which will flow to pumped sumps when the structure is emptied. Care should be taken in the

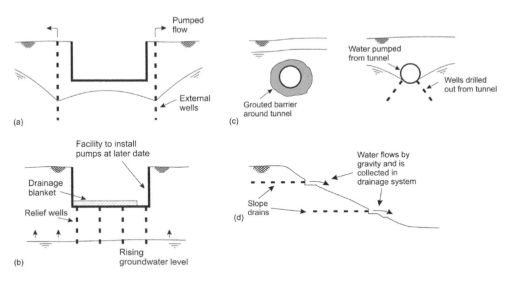

Figure 20.3 Objectives of permanent groundwater control systems. (a) Reduction of groundwater uplift loads on structures. Lowering of groundwater levels will result in a reduction of groundwater uplift forces on the structure. (b) Control of anticipated rises in groundwater levels. Wells and other drainage infrastructure are installed in anticipation of future rises in groundwater level. The system is not pumped following installation but has the capability to commence pumping at a later date if groundwater levels rise to problematic levels. (c) Control of leakage into below-ground structures. Leakage can be reduced by forming a barrier around the structure or by pumping from within or around the structure. (d) Stabilization of slopes and earth structures. Slope drains are used to lower pore water pressures and thereby to increase effective stress levels and hence improve stability.

design of such systems to provide suitable seals or non-return devices on top of the relief wells to ensure that when the structure is filled, the fluid inside does not leak out into the groundwater.

An example of the application of a permanent dewatering system to address groundwater uplift loads is the Stratford Station box in London, described by Whitaker (2004) and Casey *et al.* (2015). The project involved the construction of a large underground structure to house a new station on the High Speed 1 railway line (formerly known as the Channel Tunnel Rail Link [CTRL]). A value engineering exercise identified that a permanent dewatering solution would be a cheaper way to mitigate the effect of groundwater uplift pressures than structural alternatives such as tension piles. Subsequently, a system of 22 pumped wells was installed and commissioned to maintain groundwater levels at a pre-determined level below the base of the station box structure.

20.3.2 Control of Anticipated Rises in Groundwater Levels

In many parts of the world, groundwater levels have been influenced by human activities and are subject to long-term change. The phenomenon of rising groundwater levels (see Section 3.5.1) is recognized in several major cities around the world (Wilkinson and Brassington, 1991). This phenomenon typically occurs when water levels below cities have been historically lowered due to high levels of groundwater abstraction as the cities grow and develop. Rising groundwater occurs when abstraction quantities are reduced, either as a result of changing water use in the city or as a result of water quality and availability issues resulting from the lowering of groundwater levels. The reduced levels of abstraction pumping allow groundwater levels to return slowly, probably over years or decades, to a higher level in equilibrium with the reduced abstraction regime. A classic example of this is the lower aquifer beneath London, as described by Simpson *et al.* (1989).

Rising groundwater levels can cause a range of problems for structures that are above groundwater level when first constructed but are below the levels to which groundwater levels are expected to recover in the long term. Increased groundwater levels during the life of a structure can result in increased seepage into basements, tunnels and subterranean structures; increased buoyancy forces; and changes in effective stresses (and hence ground strength) around the structure.

Where a structure is anticipated to be affected by rises in groundwater level, it may be necessary to include drains or relief wells in new structures (or retrofit them to existing facilities) to limit future groundwater rises to acceptable levels. Options include pumped wells located outside the structure or relief wells located inside the structure (Figure 20.3b). Nicholson and Harris (1993) give an example of relief wells installed in the basement of a new structure in London. Williams (2008) describes groundwater drainage measures installed in anticipation of rises in groundwater levels associated with impounding of water behind a tidal barrage constructed across Cardiff Bay, United Kingdom (Crompton and Heathcote, 1993; Heathcote *et al.*, 1997).

20.3.3 Reduction in Leakage into Below-Ground Structures

It is not unusual for structures (such as basements or tunnels) that extend below groundwater level to 'leak' and suffer from inflow of groundwater through imperfections in the structure. Structures may leak throughout their life (as a result of poor or inappropriate design and construction). More commonly, leakage begins to occur partway through a facility's life as a result of cracking or other deterioration in structural condition, or in response to a change in external groundwater conditions (see Section 20.3.2).

It is relatively rare for groundwater leakage to completely inundate an existing structure; rates of groundwater leakage are often perceived to be merely a 'nuisance' causing minor wet patches and puddles. However, there are many cases where apparently minor seepages have caused significant problems, for example where sensitive goods are stored in basements, or where electrical systems are used in railway tunnels.

Options for controlling the leakage into underground structures include exclusion (for example by using grouting to form low-permeability barriers around the structure) or by pumping (using wells inside or outside the structure). These methods are shown in Figure 20.3c.

There have been several cases where underground railway tunnels have suffered from problems of groundwater ingress, causing problems for electrical traction systems. Gallagher and Brassington (1993) describe groundwater problems on the Merseyrail tunnel system in the United Kingdom. Case History 27.12 describes a permanent dewatering system installed to control leakage into a tunnel on the Glasgow underground railway.

20.3.4 Stabilization of Slopes and Earth Structures

Reduction in pore water pressures by some form of drainage is widely used to improve the stability of slopes and earth structures, either as a planned design measure or as part of a remedial scheme following slope failure or ground movements.

The drainage effect reduces pore water pressures, thereby increasing effective stress and, in essence, improving soil/rock strength. This means that interruptions in drainage flows can, in some cases, result in instability of the earth structure. Therefore, in many cases, passive drainage systems (where water flows by gravity) are used in preference to pumped systems, where interruption in pumping could cause instability issues (Figure 20.3d). Relevant techniques for slope drainage are described in Sections 17.10 and 17.11.

20.3.5 Remediation and Containment of Contaminated Groundwater

Long-term control of groundwater, using pumping and/or exclusion technologies, can form part of schemes to remediate or contain contaminated groundwater (Figure 17.36). Further details of groundwater control technologies used on contaminated sites are given in Section 17.14. Case History 27.11 describes the use of groundwater cut-off walls and pumping systems as part of the remediation system at Derby Pride Park, United Kingdom.

20.4 DESIGN ISSUES FOR PERMANENT GROUNDWATER CONTROL SYSTEMS

In principle, the design issues for permanent or long-term dewatering systems are no different from those for temporary systems with a relatively short-term life. Long-term systems still need to be designed to meet the project objectives and must be matched to the hydrogeological conditions at the site (see Chapter 13 for an overview of design issues).

However, the design process for permanent dewatering systems must address some additional issues, which are not typically major concerns for short-term systems. Five of the most important issues are longevity, reliability, efficiency, environmental impacts and legacy issues.

20.4.1 Longevity of Permanent Groundwater Control Systems

An obvious issue with permanent dewatering systems is that they must be designed and installed to a standard that will provide the system with a long life. Permanent systems must have operational lives measured in decades, which contrasts with most construction dewatering systems, where the operational life is measured in months.

With a little foresight, it is not difficult to design a dewatering system for a long life. This has long been the norm in the water supply industry, where public water supply wells and associated pumping systems are often designed for, and routinely achieve, operational lives in excess of 50 years. Longevity can be achieved by the application of basic engineering principles: selection of long-lasting, corrosion-resistant materials; design to reduce the severity of chemical and bacterial clogging (see Section 23.5) and to allow access for cleaning and rehabilitation of wells and pipework; provision of redundancy in design where possible; and design to allow easy replacement and upgrading of key elements of the system.

20.4.2 Reliability of Permanent Groundwater Control Systems

A dewatering system is a complex interconnected engineering system with *inter alia* mechanical, electrical, electronic and human elements. If any of these elements fail to work correctly together, the dewatering system will effectively 'fail'. It is a significant challenge to keep a normal dewatering system operating reliably for a few months on a construction project. It is a much bigger challenge to ensure adequate reliability of a permanent dewatering system.

A key issue is to understand the system and identify any 'bottlenecks' or key elements of the system. These are elements where the system would stop or would cease to be effective due to damage or failure of a single element or component, and where that element or component cannot be quickly or easily replaced or bypassed. It is sometimes not straightforward to identify the key bottlenecks. Power supply failure is an obvious key risk to the reliability of a dewatering system but is relatively easy to address by the provision of standby generators with automatic re-start systems (see Section 23.4). In practice, some less obvious reliability issues can be harder to mitigate.

Reliability of water transmission and disposal is sometimes overlooked. For example, permanent dewatering systems often have their main water collection pipework buried so that it does not take up surface space and restrict access. This means that the pipework is out of sight, and so there is a risk that during the life of the system, perhaps the connection to the ultimate water disposal point (e.g. a municipal sewer) may become blocked, or a pipeline might be accidently damaged, say by a contractor digging a new utility trench. The dewatering designer should consider this risk, and it may be appropriate to double or duplicate key sections of pipework to allow rapid diversion of water flows in the event of damage or blockage to pipework.

Measures to ensure reliability can only be rationally developed if the implications and timing of 'system failure' are fully understood. For example, how long can a system be inoperative before remedial action is needed? It makes more sense to invest in substantial back-up systems in cases where groundwater levels recover rapidly (as might occur in a confined aquifer) compared with systems where water levels might take days or weeks to rise significantly following an interruption in pumping. This assessment of failure issues should be carried out at design stage and might be supplemented by data gathered during commissioning and initial operation of the system. In some cases, a programme of monitoring and testing is carried out during the commissioning period, including a 'switch off test', wherein pumping is stopped in a controlled manner and the speed of recovery of water level is observed.

20.4.3 Efficiency of Permanent Groundwater Control Systems

For most temporary construction dewatering systems, the capital cost of equipment and the cost of installation will be much greater than the operating costs (e.g. power costs, spare parts and other consumables). On most temporary dewatering projects, little effort is made to maximize efficiency of operation, the focus often being on speed of installation and use of standardized equipment.

For permanent dewatering systems, the situation may be reversed. Total operating costs during the life of the system may have a much bigger impact on overall costs than capital costs. The designer may need to consider bespoke equipment and sophisticated control systems to get the desired performance at minimal operating costs.

One obvious way to reduce costs is to power pumps with electricity from the municipal grid system, since in urban areas, unit power costs tend to be cheaper than if generators are used for supply. Furthermore, it may be that the electricity tariff varies during a 24 hour period, being cheaper at night when demand is lowest. If pumps do not have to run continuously, it may be possible to reduce operating costs by running pumps preferentially at night and maximizing daytime idle periods when pumping is not needed.

Another strategy is to change the pumps for smaller units during the early part of the system life. If hydrogeological analysis indicates that the pumped flow rates will initially be high but will reduce after several weeks or months of pumping as aquifer storage depletion occurs (see Section 13.8.6), it may be appropriate to plan to use larger pumps for initial pumping and then swap them for smaller pumps. The labour cost of the pump changeout operation will be recovered in reduced energy usage over future years. This approach is described in more detail by Rea and Monaghan (2009).

Optimization of groundwater control systems is discussed in wider terms in Chapter 26.

20.4.4 Environmental Impacts of Permanent Groundwater Control Systems

Because permanent dewatering systems will be in operation for a long time, environmental impacts should be a potential concern. What may be an inconsequential impact from a temporary dewatering system in operation for 6 months may be a real issue of concern for a system operating for 20 years. The different types of possible impacts and methods of assessment are discussed in Chapter 21.

It is not unusual for numerical modelling to be used to predict the long-term impacts on the surrounding groundwater regime that may be caused by a permanent dewatering system. This is discussed further in Section 21.10.

20.4.5 Legacy Issues for Permanent Groundwater Control Systems

For a typical temporary dewatering scheme for a construction project, throughout the operational life, there will always be someone involved in the project who is familiar with the dewatering system and knows some of its history and features. In contrast, there is a real risk that in the later years of the life of a permanent dewatering system, it will become 'forgotten about' and the owners or occupiers of the site will have little detailed knowledge of the system – staff will have left or retired, or the site may have been sold to new owners. Indeed, a reliable and smoothly operating permanent dewatering system may become a victim of its own success. If little maintenance or external intervention is required, detailed knowledge of the system will gradually be lost. This loss of knowledge can be a serious issue in the event of a future problem with the system, when rapid intervention and repairs may be required. It can also cause problems if forgotten parts of the system are damaged or dug up, say during building works on site.

The role of the dewatering designer is to try to ensure good communication with future site owners and occupiers (some of whom may not take possession of the site until decades hence). The communication can be via conventional means, such as an operation and maintenance (O&M) manual containing simple and clear plans, diagrams and instructions for the site owners. The dewatering designer can also communicate in a literally more concrete way by deliberately making pipework, well heads, etc. visible where possible. Where equipment must be buried or covered over, substantial signs and markers can be set in walls and surfaces to show the locations of important equipment.

20.5 PRACTICAL ISSUES FOR PERMANENT GROUNDWATER CONTROL SYSTEMS

In addition to the design issues described in the previous sections, permanent or long-term dewatering systems also face a range of practical challenges, which must be overcome in order to ensure an effective and successful system.

20.5.1 Location of Dewatering System

The location of wells and equipment is a fundamental issue for permanent dewatering systems. If the locations are not picked carefully, a range of problems can arise. The most obvious is that maintenance may be difficult, for example if wells are located inside a structure where low headroom would restrict the use of cranes or lifting equipment to remove pumps. Another issue is simply that the elements of the dewatering system will take up valuable space that the site owner would rather use for business purposes. For that reason, dewatering wells and equipment are often placed in otherwise unproductive corners of the site, or they may be installed in sub-surface chambers below areas such as car parks, which can be temporarily cordoned off when access is needed for maintenance.

20.5.2 Maintainability

A fundamental aspect of permanent dewatering systems is that they will be in operation long enough that every mechanical, electrical or other active part of the system will require significant maintenance, and probably replacement, during the system's life. The dewatering designer should have designed an efficient and reliable system to minimize maintenance needs, but practical issues should also be addressed. For example:

- How will the owner or occupier know that maintenance or replacement of equipment is needed? Instead of having a responsive maintenance system where equipment is replaced when it fails, it may be better to have a monitoring system (see Chapter 22) that measures key parameters such as flow rates and water levels and will issue an alarm when key parameters reach critical values, possibly suggesting that equipment is approaching the end of its life.
- Can failed equipment be replaced quickly and easily? It is important to ensure that there is adequate access for lifting equipment and to physically transport large pieces of equipment (such as borehole pumps and rising main sections) to and from the work locations.
- Can periodic well cleaning and rehabilitation (see Section 23.5) be carried out effectively? A key issue is: can the dirty or discoloured water typically generated by well

rehabilitation be disposed of in an acceptable manner? The ultimate water disposal point should be assessed for its ability to accept dirty water. There may be a requirement to provide an alternative water disposal point during well rehabilitation, or to provide room to use a mobile water treatment system (see Section 21.9) to reduce sediment loading in water pumped during rehabilitation.

20.5.3 Decommissioning

Nothing is truly permanent, so even very long-term dewatering systems will have a finite life, or will ultimately no longer be needed if the structure or facility they are associated with reaches the end of its useful life. It is important that some consideration be given to the requirements of decommissioning at the end of service life. Further details on decommissioning are given in Chapter 24.

20.6 OPPORTUNITIES ASSOCIATED WITH PERMANENT GROUNDWATER CONTROL SYSTEMS

Permanent dewatering systems that involve pumping are effectively a commitment to continued pumping of water and to the cost associated with it. Once the requirement for continued pumping is accepted, it may be appropriate to consider the potential opportunities – the potential to put the water to positive use. For example:

- The pumped water can be supplied (subject to water quality constraints and the requirements of the abstraction licensing system; see Section 25.4) as raw water to be used for a range of purposes. The water may be suitable for use in irrigation systems or industrial processes. Subject to treatment by disinfection, the water may be passed into local and regional drinking water supply networks.
- Brandl (2006) describes how wells used for groundwater lowering can also be an energy source by harvesting heat energy using the open loop principle (Figure 3.26a), described by Banks (2012), to provide heating and cooling for buildings and industrial processes.
- In areas where rising groundwater is a potential concern, permanent dewatering systems can be incorporated into strategies to control regional groundwater levels. An example is the permanent dewatering system for the High Speed 1 (formerly Channel Tunnel Rail Link [CTRL]) Stratford Station Box in London, which is intended to contribute to the strategy to control groundwater levels in the deep aquifer beneath London (Whitaker, 2004).

Chapter 21

Environmental Impacts from Groundwater Control

We never know the worth of water till the well is dry.

Thomas Fuller, *Gnomologia*, 1732

21.1 INTRODUCTION

Groundwater control operations, whether using pumping or exclusion methods, have the potential to cause adverse impacts on the groundwater environment. These impacts may in some circumstances extend for considerable distances beyond the construction site itself. In many cases, this will not cause problems, but there are situations when the potential impacts may be significant enough to merit careful monitoring and if necessary, mitigation. This chapter outlines some potential environmental impacts that can result from groundwater control, the conditions in which they may occur and possible mitigation measures.

The potential impacts discussed in this chapter are:

(i) Impacts that may result from abstraction or pumping of groundwater. These include ground settlement, impacts on groundwater-dependent features such as rivers and wetlands, derogation or depletion of groundwater sources, and changes in groundwater quality, including movement of contamination plumes and saline intrusion.

(ii) Impacts that may result from the recovery of groundwater levels after dewatering pumping ceases.

(iii) Impacts that may result from the creation of artificial groundwater pathways, such as poorly sealed boreholes.

(iv) Impacts that may result when low-permeability groundwater barriers (such as cut-off walls) are created.

(v) The impact of discharges (such as artificial recharge or pollution leaks) on the groundwater environment.

(vi) The impact of discharge flows on the surface water environment.

This chapter addresses the direct impacts from manipulation of the groundwater regime when constructing below-ground engineering works. There will be additional indirect impacts (such as noise, dust, emissions from plant, etc.) associated with the physical construction activities, such as well drilling or pumping. These indirect impacts are not discussed here.

21.2 WHY ARE IMPACTS FROM GROUNDWATER CONTROL OF CONCERN?

This book is primarily aimed at engineers and construction professionals who need to gain an understanding of how groundwater might affect below-ground engineering works and how it can be suitably controlled by means of pumping and exclusion. In this context, most engineers would view groundwater as a *problem*. The presence of groundwater in water-bearing strata will bring complications and additional cost relative to a comparable but 'dry' site. In effect, most people working in construction would, consciously or unconsciously, view groundwater as a 'bad thing'.

In stark contrast, those professionals (such as hydrogeologists) working in the field of water resources or environmental protection view groundwater quite differently. To them, groundwater is a 'good thing'. It is a *resource*, worth protecting and managing. In many countries around the world, groundwater is an essential source of water for direct human consumption, agricultural use, industrial processes and a myriad of other purposes. Taking the United Kingdom as an example, in England, around one-third of the public water supply is obtained from groundwater. However, these figures can hide considerable local variations. Even in regions with low overall groundwater usage, there may be communities dependent on groundwater for most or all of their water supply. In addition to groundwater abstractions for public supply, groundwater is relied upon by many rural families and communities for domestic water supplies from private springs or boreholes.

In addition to its value as a resource for water supply, groundwater has a strong interaction with many surface water features such as rivers and wetlands. As a consequence, changes in groundwater levels or quality can have detrimental environmental impacts. In the United Kingdom, as in many other countries, groundwater protection policies, such as Environment Agency (2018), have been applied by governmental regulatory bodies to prevent:

i. Over-abstraction of aquifers.
ii. Derogation of individual sources.
iii. Damage to environmental features dependent on groundwater levels (e.g. river baseflows).
iv. Unacceptable risk of pollution of groundwater from point and diffuse sources. This includes the delineation of Source Protection Zones (SPZs) around individual groundwater abstraction sources, within which various activities (including some techniques used for groundwater control) are either strictly controlled or prohibited.

Environmental regulation of groundwater control is discussed in Section 25.4.

21.3 POTENTIAL ENVIRONMENTAL IMPACTS FROM GROUNDWATER CONTROL

Activities either directly or indirectly associated with groundwater control can cause a range of environmental impacts. These impacts can be grouped into five main categories (after Preene and Brassington, 2001):

1. Pumping or abstraction from aquifers
2. Physical disturbance of aquifers creating pathways for groundwater flow
3. Physical disturbance of aquifers creating barriers to groundwater flow
4. Discharges to groundwaters
5. Discharges to surface waters

The categories of impacts are summarized in Table 21.1 and described in more detail in the following sections.

21.4 IMPACTS FROM GROUNDWATER ABSTRACTION

Abstraction or pumping of groundwater is a natural part of most groundwater control systems. Groundwater is typically pumped on a temporary basis. Occasionally, dewatering systems operate for long enough that they can be considered 'permanent' (see Chapter 20).

A number of groundwater impacts may result from groundwater abstraction for dewatering purposes. These include:

i.   Ground settlement
ii. Depletion of groundwater-dependent features
iii. Effects on water levels and water quality in the aquifer as a whole
iv. Derogation of individual borehole or spring sources
v.   Other, less common effects

21.4.1 Settlement Due to Groundwater Lowering

Ground settlements are an inevitable consequence of every groundwater lowering exercise. In the great majority of cases, the settlements are so small that no distortion or damage is apparent in nearby buildings. However, occasionally settlements may be large enough to cause damaging distortion or distress of structures, which can range from minor cracking of architectural finishes to major structural damage. In extreme cases, these effects have extended several hundred metres from the construction site itself and have affected large numbers of structures.

If there is any concern that groundwater lowering (or any other construction operation) may result in ground settlements beneath existing structures, it is essential that a pre-construction building condition survey be carried out. This exercise (sometimes known as a dilapidation survey) involves recording the current state of any structures that may be affected by settlement. This should provide a detailed record of any pre-existing defects, so that if damage is alleged, there is a basis for judging the veracity of the claims. Unfortunately, for groundwater lowering projects, where the influence of the works may extend for several hundred metres from the site, the building condition survey area selected is often too small and does not cover all the structures that may be significantly affected. Ideally, the extent of the survey should be finalized following a risk assessment exercise.

Settlements caused by groundwater lowering may be generated by a number of different mechanisms, some easily avoidable, some less so:

1. Settlement resulting from the instability of excavations when groundwater is not adequately controlled
2. Settlement caused by loss of fines
3. Settlement induced by increases in effective stress

21.4.2 Settlement Due to Poorly Controlled Groundwater

This book describes the philosophy and methods whereby groundwater can be controlled to provide stable excavations for construction works. However, sometimes groundwater is not adequately controlled, leading to instability of excavation, uncontrolled seepages and

Table 21.1 Potential Environmental Impacts from Groundwater Control Activities

	Category	Potential impacts	Duration	Relevant activities
1	Abstraction	Ground settlement Derogation of individual sources Effect on aquifer – groundwater levels Effect on aquifer – groundwater quality Depletion of groundwater-dependent features	Temporary	Dewatering of excavations and tunnels using wells, wellpoints and sumps Drainage of shallow excavations or waterlogged land by gravity flow
			Permanent	Permanent drainage of basements, tunnels, road and rail cuttings, both from pumping and from gravity flow
2	Pathways for groundwater flow	Risk of pollution from near-surface activities Change in groundwater levels and quality	Temporary	Vertical pathways created by site investigation and dewatering boreholes, open excavations, trench drains, etc. Horizontal pathways created by trenches, tunnels and excavations
			Permanent	Vertical pathways created by inadequate backfilling and sealing of site investigation and dewatering boreholes and excavations and by permanent foundations, piles and ground improvement processes Horizontal pathways created by trenches, tunnels and excavations
3	Barriers to groundwater flow	Change in groundwater levels and quality	Temporary	Barriers created by temporary or removable physical cut-off walls such as sheet-piles or artificial ground freezing
			Permanent	Barriers created by permanent physical cut-off walls or groups of piles forming part of the foundation or structure or by linear constructions such as tunnels and pipelines Barriers created by reduction in aquifer hydraulic conductivity (e.g. by grouting or compaction)
4	Discharge to groundwaters	Discharge of polluting substances from construction activities	Temporary	Leakage and run-off from construction activities (e.g. fuelling of plant) Artificial recharge (if used as part of the dewatering works)
			Permanent	Leakage and run-off from permanent structures Discharge via drainage soakaways
5	Discharge to surface waters	Effect on surface waters due to discharge water chemistry, temperature or sediment load	Temporary	Discharge from dewatering systems
			Permanent	Discharge from permanent drainage systems

After Preene, M and Brassington, F C, Water and Environmental Management Journal, 17, 59–64, 2003.

perhaps base failure (sometimes called a groundwater 'blow'; see Section 7.5). These problems may result from several causes, such as failure to appreciate the need for groundwater control; a mis-directed desire to reduce costs by scaling down or deleting groundwater control from the temporary works; inadequate standby or back-up facilities to prevent interruption in pumping; and ground or groundwater conditions not anticipated by the site investigation or design and not identified by construction monitoring.

Ideally, with adequate investigation and planning, most of these issues can normally be avoided, especially the last one. Sadly, these problems still occur, leading to the failure of excavations and both significant additional costs and delays to the construction project (see Bauer *et al.*, 1980 and Greenwood, 1984 for case histories). An example is also described in Case History 27.6. If there is a sudden 'blow' or failure of an excavation, material (from the soil or rock) will be washed into the excavation. This can create large and unpredictable settlements around the excavation, much larger than the effective stress settlements associated with groundwater control. Any buildings in the area where the uncontrolled settlements occur are likely to be severely damaged.

The aftermath of an extreme case of settlement caused by poorly controlled groundwater is shown in Figure 21.1. An excavation was being dug into a stratum of fine sand within the sheet-piled cofferdam visible in the left of the figure. Dewatering comprised wellpoints located within the dig, and this allowed excavation to the target depth, several metres below original groundwater level. However, problems occurred when a power failure caused all pumps to stop. Within minutes, the base of the excavation began to 'boil', causing water and sand to flow upwards, and the excavation had to be evacuated. Subsequently, a large sinkhole opened up outside the excavation. The probable mechanism is base instability by fluidization due to upward seepage gradients (Section 7.5.2). Large quantities of sand were washed into the excavation from below. This created voids or loosened zones around the toes of sheet-piles, which led to the sinkhole and distortion of the cofferdam.

Figure 21.1 Large sinkhole outside a sheet-piled excavation following the failure of a groundwater control scheme.

21.4.3 Settlement Due to Loss of Fines

Settlement can also occur if a groundwater lowering system continually pumps 'fines' (silt and sand sized particles) in the discharge water – a problem known as 'loss of fines'. Most dewatering systems will pump fines in the initial stages of pumping as a more permeable zone is developed around the well or sump (see well development, Section 16.7). However, if the pumping of fines continues for extended periods, the removal of particles will loosen the soil and may create sub-surface erosion channels (sometimes known as 'pipes'; see Section 7.5.4). Compaction of the loosened soil or collapse of such erosion channels may lead to ground movements and settlement.

Continuous pumping of fines is not normally a problem with pre-drainage methods such as wellpoints, deep wells or ejector wells provided that adequate filter packs have been installed, developed and monitored for fines in their discharge (see Section 10.7.1). Occasionally, a sand pumping well may be encountered, perhaps caused by a cracked screen or poor installation techniques. Such wells should be taken out of service immediately.

The approach that most commonly causes loss of fines is open pumping in the form of sump pumping (see Chapter 14). This is because the installation of adequate filters around sumps is often neglected, allowing fine particles in the soil to become mobile as groundwater is drawn towards the pump. Examples of such problems are given in Section 14.8 and Case History 27.7. Powers (1985) lists soil types where the use of sump pumping is fraught with risk. These include:

- Uniform fine sands
- Soft non-cohesive silts and soft clays
- Soft and weathered rocks where fractures can erode and enlarge due to high water velocities
- Rocks where fractures are filled with silt, sand or soft clay, which may be eroded
- Sandstone with uncemented layers that may be washed out

In these soil types, even the best-engineered sump pumping systems may encounter problems. Serious consideration should be given to carrying out groundwater lowering by a pre-drainage method using wells (wellpoints, deep wells or ejector wells) with correctly designed and installed filters.

21.4.4 Settlement Due to Increases in Effective Stress

Lowering of groundwater levels will naturally reduce pore water pressures and hence increase effective stress (see Section 6.5). This will cause the layer of soil or rock to compress, leading to ground settlements. In practice, however, for the great majority of cases, the effective stress settlements are so small that no damage to nearby structures results.

The magnitude of effective stress settlements will depend on a number of factors:

(i) The presence and thickness of highly compressible strata below the groundwater level, which will be affected by the pore water pressure reduction. Examples include soft alluvial silts and clays or peat deposits. The softer a stratum (and the thicker it is), the greater the potential settlement.

(ii) The amount of drawdown. The greater the drawdown of the groundwater level, the greater the resulting settlement.

(iii) The period of pumping. In general, at a given site, the longer pumping is continued, the greater the settlement.

Powers *et al.* (2007) state that the most significant of these factors is the presence of a layer of highly compressible soil. It is certainly a truism that damaging settlements are unlikely to result from groundwater lowering on sites where highly compressible soils are absent. The corollary of this is that the potential for damaging settlements should be investigated carefully on any sites where there is a significant thickness of highly compressible soils.

It is important to realize that consolidation settlements occur in both soil and rock, but because rocks are generally much stiffer, the corresponding ground settlements are much smaller. For most projects where dewatering pumping is carried out in rock, settlements due to compression of the rock are often so small that they are not a concern, although consolidation of any soil strata in hydraulic connection with the rock can be more problematic. However, there are occasions where groundwater lowering in rock can cause settlements that require further assessment, for example where very sensitive structures are located in areas where groundwater drawdowns are expected to be large.

Effective stress settlements can be calculated using basic soil mechanics theory. The final (or ultimate) compression ρ_{ult} of a soil layer of thickness D is:

$$\rho_{ult} = \frac{\Delta u D}{E'_o} \tag{21.1}$$

where

Δu is the reduction in pore water pressure

E'_o is the stiffness of the soil in one-dimensional compression (which is equal to $1/m_v$, where m_v is the coefficient of volume compressibility)

E'_o can be estimated using several techniques; values used in calculations should be selected with care (see Preene *et al.*, 2016).

Equation 21.1 is based on the assumption that total stress remains constant as the groundwater level is lowered. This is a reasonable simplifying assumption, since the difference between the unit weight of most soils in saturated and unsaturated conditions is generally small and, given the uncertainties in other parameters, can be neglected without significant error.

To be useful in practice, Equation 21.1 needs to be written in terms of drawdown s. This will be different for aquicludes, aquifers and aquitards (see Section 3.4):

1. Aquicludes. This type of stratum is of very low permeability. During the period of pumping from an adjacent aquifer, no significant pore water pressure reduction (and hence compression) will be generated.
2. Aquifers. This type of stratum is normally of significant permeability and is pumped directly by the wells or wellpoints. Pore water pressure reductions will occur effectively at the same time as the drawdown. In practice, this means that compression and settlement of aquifers occur instantaneously once drawdown occurs. The rate at which drawdown occurs, and the resulting pattern of drawdown around a groundwater lowering system, can be estimated using the methods given in Section 13.11.
3. Aquitards. These strata are at least one to two orders of magnitude less permeable than aquifers and are not generally pumped directly by wells or wellpoints. Aquitards will tend to drain vertically into the aquifer(s) being dewatered at a rate controlled by the vertical permeability of the aquitard. In practice, this means that even if the drawdown and compression of the aquifer has stabilized, water may still be draining slowly out of the aquitard and will continue to do so until the pore water pressures equilibrate with those in the aquifer. The pore water pressure reductions and compressions of aquitard

layers will tend to lag behind the aquifer and may take weeks, months or even years to reach their ultimate value. Of course, if the groundwater lowering system operates for only a short period of time, the ultimate compression will not develop fully.

Figure 21.2 shows that where aquifers and aquitards are present, the distribution of pore water pressure reduction with depth will depend on the nature of the strata at the site and will change with time. This should be taken into account when estimating ground settlements.

Settlements caused by groundwater lowering will generally increase with time and will be greatest at the end of the period of pumping. There are two separate time-dependent effects at work:

(i) Increasing aquifer drawdown with time. When groundwater is pumped from an aquifer by a series of wells, a zone of drawdown propagates away from the system at a rate controlled by the pumping rate and aquifer properties. This means that at a given

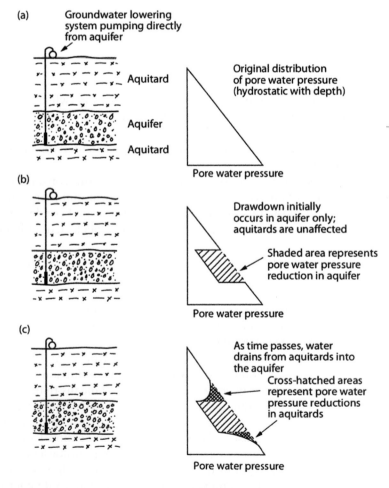

Figure 21.2 Pore water pressure reductions in response to groundwater lowering. (a) Prior to pumping. Hydrostatic conditions prevail in the aquifer and aquitards. (b) After short-term pumping. Pore water pressure reductions have not yet occurred in the aquitards. (c) After long-term pumping. Drainage from aquitards is occurring – as pumping continues, the pore water pressure reduction will propagate further from the aquifer into the aquitards.

point, some distance from the system, drawdown in the aquifer will increase with time, as will compression of the aquifer. For all practical purposes, compression of an aquifer occurs contemporaneously with drawdown.

(ii) Slow drainage from aquitards. When drawdown occurs in an aquifer, any adjacent aquitards will begin to drain vertically into the aquifer. This drainage may occur quite slowly. It follows that the compression of aquitards will lag behind the drawdown in the aquifer.

For a simple case (Figure 21.3) with wells fully penetrating a confined or semi-confined aquifer, groundwater flow will be horizontal, Δu will be constant with depth and $\Delta u = \gamma_w s$, where γ_w is the unit weight of water and s is the drawdown of the piezometric level, giving the ultimate compression of a soil layer as:

$$\rho_{ult} = \frac{\gamma_w s D}{E_o'} \tag{21.2}$$

At the location under consideration, drawdown s can be estimated *approximately* using standard solutions for the shape of drawdown curves, such as those given in Section 13.11.

Equation 21.2 shows the ultimate compression. However, because compression may lag behind aquifer drawdown, settlement calculations should concentrate on the effective compression ρ_t at a time t after pumping commenced. For an aquifer, compression occurs rapidly, so the effective compression is the ultimate compression:

$$(\rho_t)_{aquifer} = \rho_{ult} \tag{21.3}$$

However, the effective compression of an aquitard layer may be lower than the ultimate compression:

$$(\rho_t)_{aquitard} = R.\rho_{ult} \tag{21.4}$$

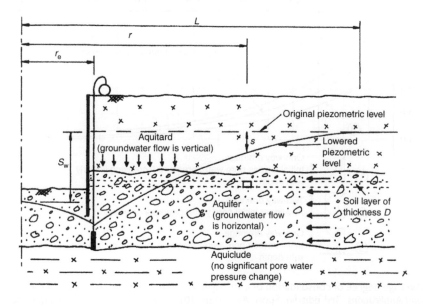

Figure 21.3 Groundwater flow to fully penetrating wells.

where R is the average degree of consolidation (with a value of between zero and one) of the aquitard layer, determined in terms of a non-dimensional time factor T_v. Powrie (2013) gives approximate solutions for R that can be applied to vertical consolidation of aquitard layers:

For $T_v < 1/12$

$$R = \left(2/\sqrt{3}\right) \times \sqrt{T_v} \qquad (21.5)$$

For $T_v > 1/12$

$$R = 1 - \left(2/3\right) \times \exp\left(0.25 - 3T_v\right) \qquad (21.6)$$

where
  T_v is time factor $= (c_v \times t)/(h^2)$
  t is elapsed time since drainage began
  h is the drainage path length (which for the 'double drainage' case, where the aquitard layer can drain both upwards and downwards, is half the layer thickness)
  c_v is the coefficient of consolidation

Some common applications of average degree of consolidation are plotted on Figure 21.4. The ultimate compression calculated in Equation 21.2 is based on pore water pressure reductions consistent with a confined aquifer, where the piezometric head is not drawn down below the top of the aquifer. However, for greater drawdowns in confined aquifers, for unconfined aquifers and for aquitards, the pore water pressure distribution will be different. Therefore, the effective compression estimated from Equations 21.2 through 21.6 needs to be converted to corrected compression ρ_{corr}:

$$\rho_{corr} = C_d \cdot \rho_t \qquad (21.7)$$

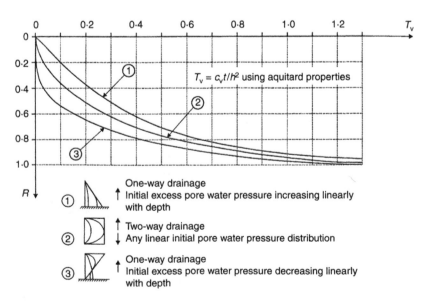

Figure 21.4 Average degree of consolidation of a soil layer vs. time. (After Powrie, W, *Soil Mechanics: Concepts and Applications*, 3rd edition, Spon, Abingdon, 2013.) Note: $T_v = c_v t/h^2$ using aquitard parameters, where c_v = coefficient of consolidation, t = time since pumping began, h = length of drainage path.

where C_d is a correction factor for effective stress (with values of between 0.5 and 1.0) shown in Figure 21.5. Note that this correction will tend to reduce the compression compared with the uncorrected values.

The preceding calculations are for the compression of individual soil or rock layers. The result of interest to the designer is the resulting ground settlement ρ_{total}. This is determined by summing the corrected compressions of all aquifer and aquitard layers affected by the groundwater lowering:

$$\rho_{total} = \left(\rho_{corr} \right)_{all\ aquifers} + \left(\rho_{corr} \right)_{all\ aquitards} \tag{21.8}$$

21.4.5 Settlement Damage to Structures

From a practical point of view, it is necessary to consider the magnitude of settlement that will result in varying degrees of damage to structures and other infrastructure such as services and utilities – predicted settlements will be of less concern if they will not damage nearby structures. Considerable work has considered the damage to structures due to self-weight settlements (Burland and Wroth, 1974) or due to tunnelling-induced settlements (Burland et al., 2001). There is a paucity of data on damage that may result from groundwater-lowering-induced settlements. This may be because pre-construction building condition surveys are often not carried out over a wide enough area around a groundwater lowering system – damaging settlements may occur up to several hundred metres away. If damage occurs (or is alleged by property owners), the actual damage caused by groundwater lowering can be difficult to assess, since the original condition of the property is not known. This is in contrast to settlements from tunnelling or deep excavation, which rarely extend more than a few times the excavation depth; building condition surveys normally cover almost all the structures at risk.

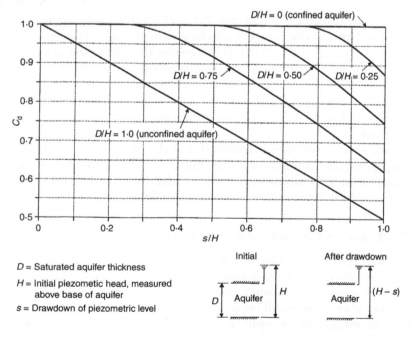

Figure 21.5 Correction factor for effective stress. (From Preene, M, *Proceedings of the Institution of Civil Engineers, Geotechnical Engineering*, 143, 177–190, 2000. With permission.)

Structures are not generally damaged by settlements *per se* but rather, by differential settlement or distortion across the structure (Figure 21.6). In uniform soil conditions, the typical convex upward drawdown curve will create a settlement profile that will distort structures in hogging (Figure 21.6b). The drawdown curve will propagate away from the groundwater lowering system with time; at a given location, the slope of the drawdown curve (and hence the differential settlements and distortion) will increase while pumping continues.

In fact, most groundwater lowering operations in relatively homogeneous soils do not cause damaging settlements (unless the soils are very compressible). This is because the distortions and ground slopes resulting purely from the drawdown curve are generally slight. Variations in soil conditions (Figure 21.6c) or foundation type (Figure 21.6d) can allow more severe differential settlements or distortions to occur.

Damage risk assessment exercises are often carried out for tunnel projects by assessing the maximum settlement and tilt that a structure will experience (Burland *et al.*, 2001). A similar approach can be applied to groundwater lowering projects (Preene, 2000). Table 21.2 shows tentative values to be used in initial damage risk assessments for settlements caused by groundwater lowering.

21.4.6 Risk Assessment of Settlement Damage from Groundwater Lowering

Once the settlement at various distances from the dewatering system has been estimated (using the equations outlined earlier and drawdown curves from numerical modelling or the methods of Section 13.11.1), the values shown in Table 21.2 can be used to delineate risk zones. These are defined as areas where structures may experience particular levels of settlement and hence degrees of damage. The simplest form of risk zones assumes that soil conditions do not vary with distance. For radial flow, the risk zones will be a series of concentric circles centred on the groundwater lowering system (Figure 21.7a); for planar flow to pipeline trenches, the risk zones will be parallel lines on either side of the trench. The risk zones can be refined by including any known changes in ground conditions with distance. If settlements are significant, it is likely that compressible alluvial or post-glacial soils are present. Geological mapping (from published data or from site investigation) can help determine the extent of these soils; damaging settlements are much less likely where these soils are absent. The application of geological mapping data will tend to produce non-circular risk zones for radial flow (Figure 21.7b) and may reduce the extent of the zones and the number of structures at risk.

When the risk zones have been determined and the number and type of structures at risk identified, appropriate action is required. Initial actions are summarized in Table 21.3. Additional, more detailed assessments may be needed for structures in the moderate and severe risk zones and for sensitive structures in the slight risk zone.

21.4.7 Mitigation and Avoidance of Settlement

Depending on the type of project and the number and nature of structures at risk, some (or perhaps all) of the anticipated damage may be deemed unacceptable. Table 21.4 lists settlement mitigation or avoidance measures, including the use of artificial recharge (see Section 17.13). Mitigation of environmental impacts is discussed in more detail in Section 21.10.

In addition to mitigation or avoidance, there is a third option, rarely considered explicitly – that is the acceptance of settlement. Powers *et al.* (2007) have suggested that if damage risk is no more than slight, it may be more economical to accept third party claims

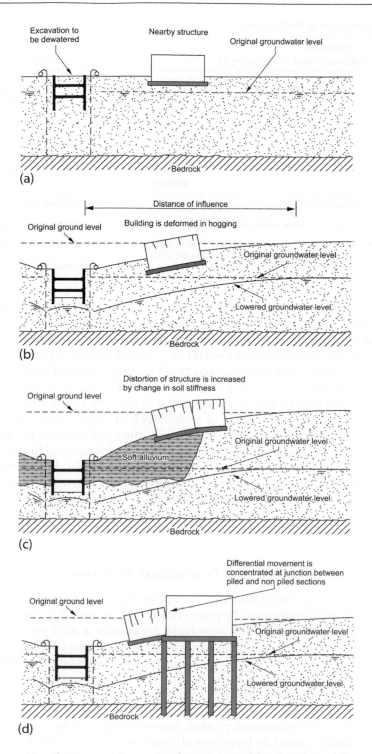

Figure 21.6 Deformation of structure due to settlement caused by drawdown of groundwater levels. (a) Before groundwater levels are lowered. (b) Uniform ground conditions (vertical ground movements exaggerated). (c) Change in ground conditions (vertical ground movements exaggerated). (d) Change in foundation type (vertical ground movements exaggerated).

Table 21.2 Tentative Limits of Building Settlement and Tilt for Damage Risk Assessment

Risk category	Maximum settlement (mm)	Building tilt	Anticipated effects
Negligible	<10	<1/500	Superficial damage unlikely
Slight	10–50	1/500–1/200	Possible superficial damage, unlikely to have structural significance
Moderate	50–75	1/200–1/50	Expected superficial damage and possible structural damage to buildings; possible damage to rigid pipelines
Severe	75	>1/50	Expected structural damage to buildings and expected damage to rigid pipelines or possible damage to other pipelines

From Preene, M, *Proceedings of the Institution of Civil Engineers, Geotechnical Engineering*, 143,–190, 2000. With permission.

Notes: 1 Maximum settlement is based on the nearest edge of the structure to the groundwater control system.

   2 Tilt is based on rigid body rotation, assuming that all of the maximum settlement occurs as differential settlement across the width of the structure or across an element of the structure.
   3 The risk category is to be based on the more severe of the settlement or tilt criteria.

rather than deploying large-scale mitigation measures. This might be quite a controversial approach on many projects, and would present a public relations challenge, but could be appropriate where relatively few structures are classified as being slightly at risk. A pre-construction building condition survey would be essential for this approach.

Powers (1985) suggests that a pre-construction building condition survey should include a photographic and narrative report on the interior and external condition of buildings. Particular attention should be paid to the condition of concrete foundations, structural connections, brickwork, and the condition of plasterwork or other architectural finishes that are particularly susceptible to cracking. Additionally, the condition of other structures must also be documented – examples include bridges, utility enclosures and historic monuments. Paved surfaces (roads, pavements and hardstandings) should also be examined and their condition recorded. Ideally, the survey should be carried out by an independent organization (to avoid later charges of bias in the event of claims).

The legal position under United Kingdom law related to settlement damage resulting from the abstraction of groundwater is outlined in Section 25.4.4.

21.4.8 Impact on Groundwater-Dependent Features

Groundwater can play an important role in supporting many surface water features. In many rivers, the flow is 'supported' by groundwater (see Section 3.6). In other words, the rivers receive a contribution to their flow from groundwater (this groundwater-derived contribution is termed 'baseflow'). Similarly, wetlands and ponds (which can be important ecological habitats) often exist because of the presence of groundwater flows or springs. Rivers, wetlands and other phenomena whose stability and existence are partly controlled by the availability of groundwater are collectively termed 'groundwater-dependent features'.

The degradation of groundwater-dependent features (such as reduction in river flows and drying up of wetlands) caused by long-term abstraction for water supply is an issue that is widely recognized in water resource planning (Cunningham, 2001). However, Acreman *et al.* (2000) noted that in some cases, degradation of the aquatic environment believed to be linked to long-term groundwater abstraction may be due, at least in part, to other factors such as changes in land drainage, river channelization and climate change.

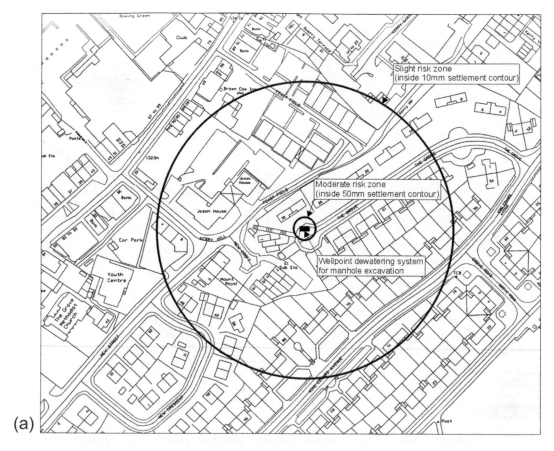

(a)

Figure 21.7 Settlement risk zones. (a) Idealized settlement risk zones for radial flow to a groundwater lowering system. (b) Settlement risk zones for radial flow to a groundwater lowering system based on variation in soil type from geological mapping. (From Preene, M, *Proceedings of the Institution of Civil Engineers, Geotechnical Engineering*, 143, 177–190, 2000. With permission.) Panel (a): (1) Predicted settlement at excavation is less than 75 mm, so no severe risk zone is generated. (2) No buildings lie in the moderate risk zone. (3) Numerous buildings lie within the slight risk zone. Building condition surveys should be considered for this zone. Any sensitive structures should be identified. (4) Settlement assessment assumes thickness of Alluvium is constant. Thickness based on borehole at manhole location.

    Wetlands (areas of marsh, fen or peatland, or areas covered with shallow water or poorly drained areas subject to intermittent flooding) can be particularly vulnerable to impacts from longer-term groundwater pumping. To assess each case, the interaction between the surface water and groundwater will need to be quantified. Some wetlands are directly supported by groundwater seepages, while others (if the soil is of low permeability) may receive little contribution from groundwater. Wetlands tend to vary with the seasons (and also from year to year), so the additional influence of groundwater pumping may or may not be significant in comparison. Merritt (1994) gives a thorough background to the creation and management of wetlands.

    While the impacts on groundwater-dependent features are commonly assessed during the development of water resource abstraction schemes, they are rarely considered during the development of groundwater control schemes. In practice, this probably reflects the fact that

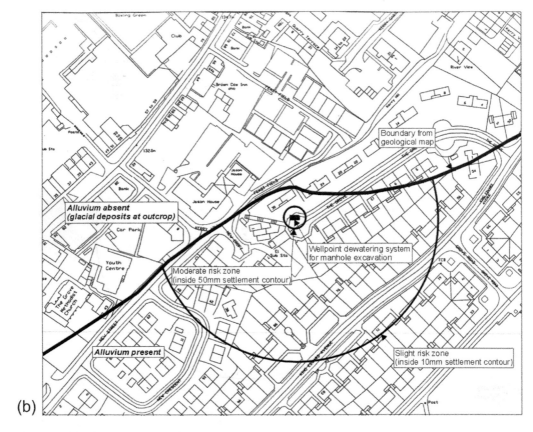

(b)

Figure 21.7 (Continued) Panel (b): (1) Geological boundary taken from published geological mapping, confirmed by nearby boreholes. (2) Predicted settlement in area where Alluvium is absent is less than 10 mm. (3) Settlement assessment assumes thickness of Alluvium is constant when present. Thickness based on borehole at manhole location.

Table 21.3 Suggested Actions for Settlement Risk Categories

Risk category	Description of likely damage	Actions required
Negligible	Superficial damage unlikely	None, except for any buildings identified as being sensitive, for which a detailed assessment should be made.
Slight	Possible superficial damage, unlikely to have structural significance	Building condition survey to identify any pre-existing cracks or distortions. Identify any buildings or pipelines which may be sensitive, and carry out detailed assessment. Determine whether mitigation or avoidance measures are required locally.
Moderate	Expected superficial damage and possible structural damage to buildings; possible damage to rigid pipelines	Building condition survey and structural assessment. Assess buried pipelines and services.
Severe	Expected structural damage to buildings and expected damage to rigid pipelines or possible damage to other pipelines	Determine whether anticipated damage is acceptable or whether mitigation or avoidance measures are required.

From Preene, M, *Proceedings of the Institution of Civil Engineers, Geotechnical Engineering*, 143,–190, 2000. With permission.

Table 21.4 Measures to Mitigate or Avoid Groundwater-Lowering-Induced Settlement Damage

Mitigation of settlement	Possible measures
Protect individual structures	Prior to the works, underpin the foundations of some or all of the structures at risk
Reduce the number of structures at risk	Reduce drawdowns by reducing the depth of excavation below groundwater level
	Reduce the extent of the risk zones by minimizing the period of pumping
	Use cut-off walls to reduce external drawdowns
	Use an artificial recharge system to minimize external drawdowns
Avoidance of settlement	Relocate excavation away from vulnerable structures
	Re-design project to avoid excavation below groundwater level, or excavate underwater
	Carry out excavation within a notionally impermeable cut-off structure[a]

From Preene, M, *Proceedings of the Institution of Civil Engineers, Geotechnical Engineering*, 143,–190, 2000. With permission.

[a] If a cut-off structure is used to avoid external drawdowns, it is essential that a groundwater monitoring regime is in place to allow any leaks in the cut-off to be identified before significant settlements can occur.

most dewatering schemes are of short duration with relatively modest pumped flow rates, and the potential for such impacts will be minimal on most sites. Nevertheless, there will be circumstances when these impacts should be assessed, particularly for long-term or 'permanent' groundwater control systems.

Pumping of groundwater for dewatering schemes can impact on groundwater-dependent features in one of two principal ways. First, the pumping may draw water from the feature, for example by lowering groundwater levels so that water losses increase from the base of a pond or wetland. Second, pumping may intercept water that would otherwise have reached the feature. An example is where groundwater pumping reduces baseflow reaching a river, thereby reducing flow in the river. Assessment of water losses from groundwater-dependent features is not straightforward and can be complicated by uncertainties in the properties of semi-permeable sediments in the base of rivers and ponds. Analytical methods for the assessment of impacts on groundwater-dependent features are discussed in Kirk and Herbert (2002).

The degradation of groundwater-dependent features is most likely to be an issue for projects where:

a) The proposed excavation is located close to the groundwater-dependent feature (Figure 21.8a), and groundwater is likely to be drawn from the feature even in the very short term.
b) The proposed dewatering system is likely to operate for a very long time (as is often the case with dewatering systems for open pit mines). These impacts should be considered for 'permanent' dewatering systems (see Chapter 20).
c) The proposed dewatering system is operating at very high pumped flow rates, which have the potential to lower groundwater levels over a very wide area, and therefore may affect a larger number of sensitive sites.

When assessing the potential impact on groundwater-dependent features, specialist ecological and hydrological advice is likely to be required. It may be possible to reduce impacts by constructing a groundwater cut-off barrier between the dewatering system and the feature to be protected (Figure 21.8b). An alternative approach would be to use artificial recharge (see Section 17.13) of groundwater or surface water or to pipe a portion of the discharge

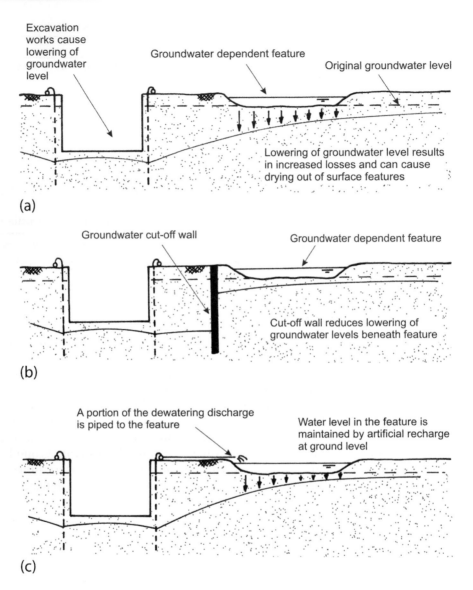

Figure 21.8 Depletion of groundwater-dependent features. (a) Depletion due to lowering of groundwater levels. (b) Cut-off wall used to mitigate impact. (c) Artificial recharge used to mitigate impact. (Redrawn from Preene, M and Brassington, F C, *Water and Environmental Management Journal*, 17, 59–64, 2003.)

water directly to the feature (Figure 21.8c). If the latter approach is adopted, the temperature, chemistry and sediment content of the discharge water must be assessed to ensure that there will not be adverse reactions between the water and ecology of the feature; the risk of additional erosion must also be considered. Impacts from the discharge of groundwater are discussed in Section 21.9.

It is worth noting that in many countries, including the United Kingdom, many wetlands and other sensitive groundwater-dependent features are 'designated sites' under environmental regulations and as such are protected by law. Negotiations will have to be opened with the appropriate regulatory authorities when planning work near such sites. Similar protected environments exist in other parts of the world. This is an example where a desk study will be of immense value at the site investigation stage (see Section 11.6). In many countries, government bodies and environmental regulators maintain registers of designated sites. Identification of the presence of such a site during the desk study stage can allow further investigations to be made to prepare for future monitoring and mitigation requirements.

21.4.9 Effect on Groundwater Quality

Groundwater quality (i.e. the chemical composition of the water) varies naturally from place to place and from aquifer to aquifer (see Section 3.7). In some cases, groundwater is almost pure enough to be potable with only minimal treatment in the form of chlorination to destroy any harmful bacteria. In other locations, the groundwater may contain considerable impurities, which could be naturally occurring or man-made. It is important to realize that pumping from groundwater lowering systems will change natural groundwater flow in aquifers and may cause existing contamination plumes or zones to migrate. Two of the most important cases to be considered are contaminated groundwater, often left over from industrial land use, and intrusion of saline water in coastal areas.

21.4.10 Movement of Contaminated Groundwater

The study of groundwater contamination is a major field in itself, and the reader is commended to texts such as the one by Fetter *et al.* (2018) to obtain the full background on the subject. Contaminants interacting with groundwater flow exist in one of three forms (or phases):

1. Dissolved (or aqueous) phase. A wide range of substances are soluble in water and so become part of the water itself, travelling with it.
2. Non-aqueous phase. This describes liquids that are immiscible with water. They may have densities lower than that of water (light non-aqueous phase liquids or LNAPLs) and will float on top of the water table (examples are petrol and diesel compounds). Dense non-aqueous phase liquids (DNAPLs) also exist, which are denser than water and tend to sink below the water table until they meet a low-permeability layer (examples include chlorinated hydrocarbons such as trichloroethylene). The non-aqueous phase is sometimes described as 'free product', meaning that it is a form of contamination existing separately (in a different phase) from the water.
3. Vapour phase. Volatile compounds in the contaminant can move in gaseous form in the unsaturated zone above the water table.

Some contaminants (e.g. hydrocarbons such as petroleum products) may create all three phases when they reach groundwater.

Contamination may be caused by a variety of land uses:

- Industrial processes – mainly from spillages or leakages from stored materials. Sites of concern include not only manufacturing and chemical production sites but also vehicle storage areas, airports and other locations where fuel and detergents are used.
- Landfilling and waste disposal – from both official and unofficial disposals.
- Agricultural practices – such as fertilizer or pesticide use.
- Urban use – including leaking sewers, fuel spills, etc.

The absence of active use of a site does not provide assurance that the site is uncontaminated. A legacy of contamination may exist in the ground and groundwater for years or decades after the pollution is stopped. A guide to the types of pollution that can be expected from former industrial sites can be found in CIRIA reports on contaminated land (Harris *et al.*, 1995). A desk study (see Section 11.6) should investigate former uses of a site to determine the risk of contaminants being present at problematic levels.

As described in Chapter 3, groundwater is constantly in motion, and where contamination exists, that will tend to move too, gradually forming a plume stretching away from the original source of contamination. The rate and direction of movement of contamination depend on many factors, including hydraulic gradients, the geological structure of the aquifer, the nature of the contaminant and any chemical changes in the contaminant with time. Detailed consideration of these factors is beyond the scope of this book, and the references cited earlier are recommended for further study. However, it is vital that anyone designing or carrying out a groundwater lowering understands that pumping may change considerably the existing groundwater gradients and velocities, affecting both the magnitude (generally increasing flow velocities) and the direction. This means that groundwater lowering can cause the extent of a contamination plume to change, perhaps much more rapidly than previously. If movement of contamination is of real concern, a thorough site investigation followed by development of a groundwater flow and contaminant transport model is essential, and specialist advice should be obtained at an early stage.

When planning groundwater lowering on or near a contaminated site, there are two important issues to be addressed in addition to the dewatering design itself:

1. How can the influence of groundwater lowering on the contamination plume be minimized or controlled? The use of physical cut-off barriers (Chapter 18) to hydraulically separate the groundwater lowering system from adjacent contaminated sites is a method often used (see Section 17.14).
2. How can the discharge be disposed of? The water pumped from the wells may contain problematic levels of contamination, preventing direct discharge to watercourses or sewers. On occasion, it has been necessary to establish a temporary water treatment plant on site to clean up the discharge water quality (see Section 17.14).

Methods based on groundwater lowering technology can be used to help clean up sites. Pumping of groundwater and treatment of discharge prior to disposal, with the aim of reducing contamination levels, is known as the 'pump and treat' method. This is a specialist method, and its effectiveness should be compared with other competing clean-up techniques (see Holden *et al.*, 1998). Further details are given in Section 17.14. A project example is discussed in Case History 27.11.

21.4.11 Saline Intrusion

Saline intrusion describes the way more mineralized water is drawn into freshwater aquifers under the influence of groundwater pumping. This is a particular problem where large volumes of groundwater are abstracted for potable supply, because if saline water reaches the well, it may have to be abandoned. Saline intrusion principally affects coastal aquifers, but saline water can sometimes be found in inland aquifers where the water has become highly mineralized at depth.

Saline intrusion is a complex process affected by aquifer permeability, rate of recharge, natural groundwater gradients and the effect of any existing pumping wells. Any significant groundwater lowering operations will affect the boundary between fresh and saline water. If saline water is drawn to the groundwater lowering system, that may not be a problem in itself (provided that the water can be disposed of), but any saline water drawn towards nearby supply wells is of much greater concern. The risk of saline intrusion may need to be investigated using numerical modelling to assess the effect on local and regional water resources; the reader is referred to hydrogeological texts such as Younger (2007) for further details.

Although not strictly a case of saline intrusion, in arid countries, there is a risk of affecting the salinity of wells used by local communities. In some hydrogeological settings, fresh or brackish water lenses may exist above a generally saline water table and are often exploited by shallow wells to provide water for irrigation and livestock. Preene and Fisher (2015) highlight that in those circumstances, even small changes in groundwater levels caused by dewatering pumping could cause significant changes in water quality in the shallow wells. It is possible that the salinity of the water in the well may increase dramatically, rendering the well unusable.

21.4.12 Effect on Groundwater Borehole or Spring Supplies

This book mainly deals with groundwater as a problem needing to be controlled to allow construction excavations to proceed, but, as was highlighted at the start of this chapter, groundwater is also a resource used by many. Groundwater is obtained from wells and springs as part of public potable water supplies and for private supplies for domestic dwellings and industrial users such as breweries, paper mills, etc. and by farmers for irrigation and watering of livestock. If temporary works groundwater lowering is carried out in the vicinity of existing well or spring abstractions, there is a risk that the abstractions will be 'derogated' – in other words, it will be harder for the user to abstract water, and in extreme cases, the source may even dry up completely (Figure 21.9). The interaction between groundwater supplies and civil engineering works is discussed further by Brassington (1986).

Occasionally, water quality from a spring or well source may deteriorate as a result of changes in the groundwater flow direction. This can occur by various mechanisms:

- Changing the mix of water pumped from the water source (well or spring) This can occur when the water source taps into multiple aquifer zones – e.g. a shallow and a deeper water bearing zone – but where the dewatering works affect only one zone. This will affect the chemical make-up of the water from the source. As noted earlier, in arid countries where wells tap into both fresh and brackish water zones, there is a risk that nearby dewatering activities may reduce the freshwater flow, increasing the salinity and rendering the source unusable.
- Drawing in of different water quality towards the water source so that it eventually yields water of inferior chemistry. Examples include by drawing in contaminated groundwater (Section 21.4.10), saline water from coastal waters (Section 21.4.11) or poorer-quality water from abandoned mine workings (Neymeyer *et al.*, 2007).

The effects are often temporary, and may cease soon after the end of groundwater control pumping, but can cause considerable inconvenience and cost to groundwater users. The legal issues must also be considered, since in England and Wales, licensed groundwater abstractors have a legal right to continue to obtain water (see Section 25.4).

When groundwater lowering is carried out for a construction project, the primary effect on nearby abstractions is likely to be a general lowering of water levels, which will affect

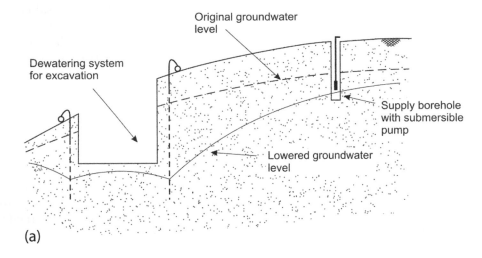

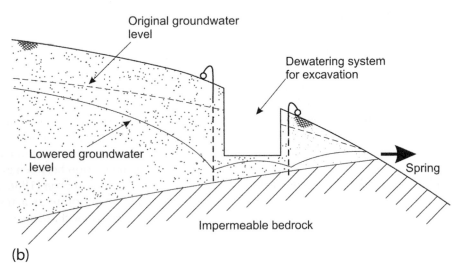

Figure 21.9 Derogation of groundwater sources. (a) Impact on borehole. Dewatering system lowers groundwater level at water supply borehole. (b) Impact on spring. Dewatering system reduces flow from spring. (After Preene, M and Brassington, F C, *Water and Environmental Management Journal*, 17, 59–64, 2003.)

operating water levels in existing wells, with a corresponding reduction in output. The magnitude of the reduction in output will depend on factors including:

1. Aquifer characteristics, including permeability and storage coefficient
2. Distance between groundwater lowering wells and supply wells, and their location in relation to any existing hydraulic gradients in the aquifer
3. The dewatering pumping rate and period of pumping (low-flow-rate and short-duration pumping systems will have less of an effect on supply wells)
4. The depth, design and condition of the supply wells and associated pumping system

Any rational assessment of the effect of groundwater lowering on supply wells will require some form of conceptual groundwater model to be developed. This could then be used as the basis for a numerical model, or an initial assessment could be made using the methods of Chapter 13, treating the groundwater lowering system as an equivalent well. The assessment of potential impacts is discussed in Section 21.10.

Permanent dewatering systems (see Chapter 20) can also have an impact on nearby groundwater sources. Impacts may be caused not only by systems that are actively pumped by wells but also by projects where linear engineered features (such as road or rail cuttings) are drained by gravity (Figure 21.10).

If the estimated effects on the supply wells are small, this may be deemed acceptable with no further mitigation measures. However, if the effects are more severe, Powers (1985) suggests the following mitigation measures:

1. If only a few low-volume users are affected, and the dewatering period is short, the lost supply might be replaced by a temporary tanker supply.
2. If the supply well is deep but with the pump set at a fairly high level, it may be possible to install higher-head pumps at greater depth in the supply well. This would allow abstraction to continue even with the additional drawdown generated by groundwater lowering.
3. If the supply wells are shallow, it may be necessary to deepen the wells, perhaps into another aquifer. Alternatively, for small-diameter shallow wells, it may be more economical to simply drill a new, deeper well.
4. A portion of the dewatering discharge may be piped to the affected user by temporary pipeline. Depending on the water quality, point of use treatment may need to be provided to ensure that the water is suitable for use.
5. Public water mains may be extended into the affected area, giving a permanent benefit for the money spent.

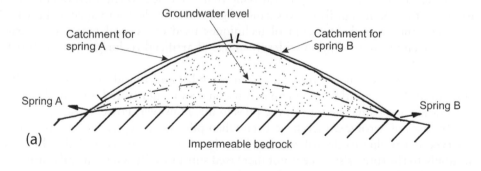

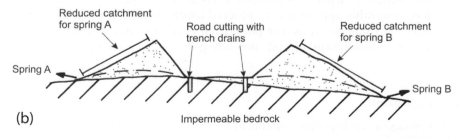

Figure 21.10 Groundwater abstraction from linear construction projects. (a) Flow to springs prior to construction (b) Reduced flow to springs following construction. (Redrawn from Preene, M and Brassington, F C, *Water and Environmental Management Journal*, 17, 59–64, 2003.)

Some of these measures have huge cost, time and public relations implications and clearly need to be compared with the alternative of constructing the project without groundwater lowering or even relocating the project away from the supply wells.

A desk study (see Section 11.6) will be valuable in identifying the presence of any vulnerable water abstractions (well or spring sources) that may be impacted by a proposed groundwater control scheme. National and regional governmental bodies and environmental regulators often hold records of the location of water sources. Larger sources may have SPZs delineated around them, within which engineering works require special permission from the regulators.

If the effects on nearby groundwater abstractions are of real concern, it is essential that they are addressed early in the planning of a project, because it is unrealistic to expect the contractor to bear all the costs and risk of some of these measures. The project client will have to face up to the potential need for some of these measures and perhaps allow for them when negotiating with landowners for wayleaves, etc. The mitigation measures might be included at the very start of site works as part of the enabling works. Alternatively, the supply wells may be monitored during the works, with a contingency in place that the mitigation measures will be applied if the well is affected beyond a certain pre-defined level.

21.4.13 Other Effects

Occasionally, other, less common side effects may be of concern.

21.4.13.1 Damage to Timber Piles

It is widely recognized that timber piles supporting older structures may be detrimentally affected by drawdown of water levels. This is a particular issue in Scandinavia, where buildings founded on timber piles are commonplace (Peek and Willeitner, 1981). In cities such as Copenhagen, this is such an important issue that it has led to the introduction of local regulations that prohibit significant lowering of groundwater levels in specified areas. This has influenced the groundwater control techniques used on several major infrastructure projects in Copenhagen, leading to widespread use of artificial recharge systems (Bock and Markussen, 2007).

Powers (1985) states that the damage to timber piles and foundations may result from fungi present in the timber thriving in an aerobic environment created if groundwater levels are drawn down, exposing the tops of the piles to air. However, Powers also states that the most severe cases of aerobic attack have been when piles were exposed in excavations, and that observed decay due to drawdown has been less severe. This is probably because the oxygen supply to the timber surface is not increased substantially when the piles are in dense or fine-grained soils, even when groundwater levels are lowered.

Nevertheless, a sensible approach is to proceed cautiously when working in areas when older structures are founded on timber piles. Even if aerobic attack does not compromise pile stiffness, soil consolidation and pile downdrag due to negative skin friction should be considered.

21.4.13.2 Vegetation

It is rare for groundwater lowering systems to have a noticeable effect on vegetation. This is mainly due to the short-term nature of pumping and the fact that plants generally draw their water from immediately below the surface, above the water table. This zone is much more likely to be affected by changes in precipitation and infiltration than by deeper pumping. For longer-term pumping (for certain types of quarries or open pit mines), this issue may need to be considered further, and the services of an experienced ecologist can be very useful in this regard.

21.4.13.3 Impact on Archaeological Remains

The continued in situ preservation of archaeological remains may also be dependent on stable groundwater levels, and there have been cases of degradation associated with large-scale dewatering works (French and Taylor, 1985). Numerical modelling has been used to assess groundwater lowering beneath areas of archaeological interest (Garrick et al., 2010).

21.5 IMPACTS FROM RECOVERY OF GROUNDWATER LEVELS

Much of the discussion in this chapter relates to the potential impacts when groundwater levels are lowered on a temporary basis to allow construction below groundwater level. There are, however, some circumstances when problems can be caused by the recovery of water levels after groundwater pumping is stopped.

One example is where a supposedly watertight structure has been constructed within a dewatered excavation, but where defects are revealed when pumping is stopped and the groundwater levels recover. Such problems have occurred with concrete tanks for wastewater treatment plants and for basement structures for underground car parks and the like. Problems with structures can be apparent as water ingress (if the cause is simply failure of waterproofing measures) or can include cracking and structural distress if the structure is unable to resist the groundwater pressures after water levels recover. These problems can be tricky to resolve. Possible measures include retrofitting waterproofing and/or structural improvements (often in combination with temporary dewatering) or the use of permanent dewatering systems (Chapter 20).

Another case is where rapid recovery of water levels causes the migration of fine particles from the ground into coarse-grained backfill below structures and around utility pipelines. Rowe (1986) describes a case where a pipeline was constructed in a dewatered trench through silty fine sand. Very coarse gravel (particle size 100–150 mm), compacted by ramming, was used to provide a base onto which the pipes were laid. Rowe states that

> a temporary rise in groundwater level occurred due to a pump breakdown. The pipes then sank by amounts up to 150 mm. Re-excavation proved the sunken stone beds contained silty sand in their voids over a greater depth than to be expected from the ramming action. The reason was that as the water level rose it carried fines into the stone and gravel surround since the filter rules had not been applied. Once the fault was diagnosed the pipe line was completed satisfactorily using a bed of graded sandy gravel only 300 m thick.

In another case, a sewer was constructed by open cut methods through fine sand, using deep well dewatering, with the sewer pipes surrounded by a coarse gravel bedding. The sewer was successfully constructed and the dewatering stopped as planned by turning off all the pumps simultaneously. Shortly afterwards, it was noticed that the several sections of the pipeline had settled irregularly by more than 100 mm in places. The pipes were dug out, revealing that sand had migrated into the gravel bedding around the pipes. The sewer was reconstructed with geotextile filter mesh placed between the gravel and the natural ground. Additionally, when dewatering was no longer needed, the pumps were switched off sequentially, over a few days, in an attempt to slow the rate of groundwater level recovery. On this occasion, the pipes did not settle, and the sewer was successfully completed.

21.6 IMPACTS FROM GROUNDWATER PATHWAYS

One potential impact that is sometimes overlooked is the potential for groundwater to flow along permeable pathways created by wells, piles, excavations or even a structure itself. Flow along these pathways can affect groundwater quality or the quantity of water available to groundwater sources, or it may increase leakage rates into existing below-ground structures.

Some of the groundwater pathways may be temporary (such as investigation and dewatering boreholes) and can be sealed on completion. Other pathways are formed by parts of the structure or works and may exist in perpetuity. Examples of permanent pathways include the granular bedding of pipelines (which may allow horizontal flow) and some types of piling or ground improvement processes (which can form vertical pathways). Open excavations such as road or rail cuttings may themselves form vertical pathways.

It is now recognized that the vulnerability of aquifers to pollution resulting from surface sources (e.g. surface run-off, fuel spills, etc.) will depend on the nature of the aquifer and any overlying strata. For example, a high-permeability unconfined aquifer will be much more vulnerable to contamination than a deep confined aquifer overlain by a thick clay layer, which can act as a barrier to pollution. In the United Kingdom, aquifer vulnerability maps have been produced for many areas (Palmer and Lewis, 1998).

21.6.1 Vertical Pathways via Boreholes and Wells

The installation of wells and boreholes may puncture low-permeability layers, increasing the risk of surface pollution finding its way down into the aquifer. This is of particular concern if the near-surface strata have been contaminated by historical or ongoing polluting industries.

Changes in groundwater quality may result if pathways are formed between different aquifer units. For example, poorly sealed investigation boreholes could allow mixing of fresh and more saline water in aquifers where groundwater quality is stratified, or polluted groundwater at shallow depth may be able to flow into deeper aquifers (Figure 21.11).

If changes in aquifer vulnerability are of concern, a number of mitigation measures should be considered:

1. The well design should include appropriate grout seals to prevent vertical seepage around the outside of the well casing (see Section 13.10.4).
2. The well casing should stand sufficiently proud of ground level to prevent surface waters being able to pass into the well casing (and thence into the aquifer) in the event of localized flooding around the well.
3. The top of the well casing should be capped when not in use and sealed around the pumping equipment when in use. This will reduce the chances of noxious substances being dropped or poured down the well, either maliciously or by accident.
4. When groundwater control is no longer required, the wells should be adequately decommissioned by capping or sealing (see Chapter 24).
5. Where flowing artesian conditions (see Section 3.4.2) exist in the aquifer, there is a risk that a well may overflow naturally at ground level. Such 'uncontrolled' artesian discharge should be avoided. The design, installation, operation and decommissioning of works in flowing artesian conditions should be planned and executed by experienced personnel. Wells (for both dewatering and monitoring purposes) should include suitable surface casings, grout seals and headworks to allow the control of artesian pressures and flows.

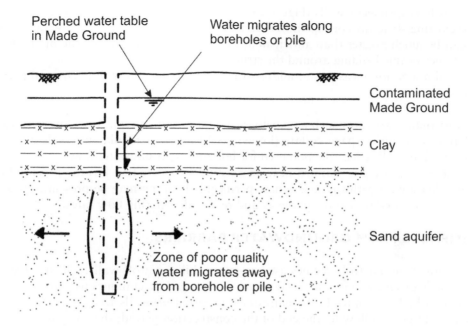

Figure 21.11 Vertical pathways for groundwater. (Redrawn from Preene, M and Brassington, F C, *Water and Environmental Management Journal*, 17, 59–64, 2003.) Borehole or pile punctures clay aquiclude and may allow water from contaminated near-surface layers to seep downwards and contaminate lower aquifer.

21.6.2 Vertical Pathways via Excavations and Structures

In addition to the drilling of wells and boreholes, excavations or structures also have the potential to create vertical pathways. It is rare that sufficient groundwater level monitoring is carried out to identify changes in groundwater levels due to installation of piles or substructures. Ervin and Morgan (2001) recorded a temporary reduction in piezometric level of up to 1.5 m (over a period of 3 months) in a low-permeability aquitard underlain by a more permeable gravel aquifer. They attributed this change to the installation of bored piles into the gravel through the aquitard. Piezometric levels recovered following the end of the piling works, implying that the impact was associated with the pile construction and not the long-term presence of the piles.

Where aquifer conditions are sensitive (for example in the inner catchments around public water supply wells), deep structures such as shafts or basements should be designed to limit the potential for creation of vertical flow paths. This can be done by using raft foundations in preference to piles, which would puncture low-permeability aquitard layers.

If piling or ground improvement methods have to be used, methods should minimize the formation of vertical flow paths. Guidance is given in Westcott *et al.* (2001).

21.6.3 Horizontal Pathways

Linear horizontal underground structures such as pipelines may act as horizontal conduits for groundwater flow. A classic example is where a sewer pipeline is laid on granular bedding material. The bedding is likely to be of high permeability and will allow preferential flow of groundwater along its length. This can divert flow away from nearby groundwater sources such as springs or wells (Figure 21.12). Problems can also occur when dewatering near to existing structures or pipelines that act as pathways. Dewatering flow rates in these areas can be much greater than anticipated, as water is drawn towards the pumping wells along the permeable bedding around the structure or pipeline.

Potential mitigation measures to address the issues associated with horizontal pathways include:

i. Horizontal structures such as pipelines should be constructed with low-permeability barriers or anti-seepage collars (known as 'stanks') at regular intervals along their route. This will reduce the potential for horizontal groundwater flow.

ii. Where dewatering is planned near an existing structure that has the potential to act as a horizontal pathway (e.g. where permeable bedding may be present), grouting is sometimes used to seal the pathway and reduce dewatering flow rates.

21.7 IMPACTS FROM GROUNDWATER BARRIERS

Many groundwater control schemes use low-permeability cut-off walls, which act as barriers to exclude groundwater from excavations. The range of techniques used to form cut-off walls is described in Chapter 18. In many cases, these cut-off walls are effectively permanent and remain in place following the end of the construction period; they may interrupt horizontal groundwater flow, causing a damming effect (Figure 21.13a). These effects may not be significant unless large structures fully penetrate significant aquifer horizons. It is rare that sufficient groundwater monitoring is carried out to allow these effects to be quantified; Barton (1995) recorded groundwater level rises of 0.2–0.8 m upstream of a structure that fully penetrated a valley gravel aquifer.

In addition to cut-off walls installed with the objective of blocking groundwater flow, groundwater barriers may inadvertently be formed where extensive heavy-duty foundations are installed into aquifers that are shallow or of limited thickness (Figure 21.13b).

In reality, the impact of groundwater barriers will be modest for most engineering structures. The exception is where very long linear structures (such as metro stations or cuttings for roads or railways) are contained within low-permeability walls. If such structures are located across the direction of natural groundwater flow, then groundwater flow will be diverted around the sides of the structure. This can reduce the supply to nearby groundwater sources or cause flooding of adjacent basements upstream of the structure.

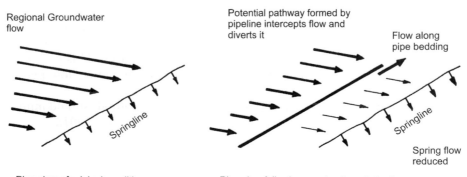

Figure 21.12 Horizontal pathways for groundwater flow. (Redrawn from Preene, M and Brassington, F C, *Water and Environmental Management Journal,* 17, 59–64, 2003.) Groundwater flow is diverted along the permeable bedding of the pipeline, reducing the flow to the springs.

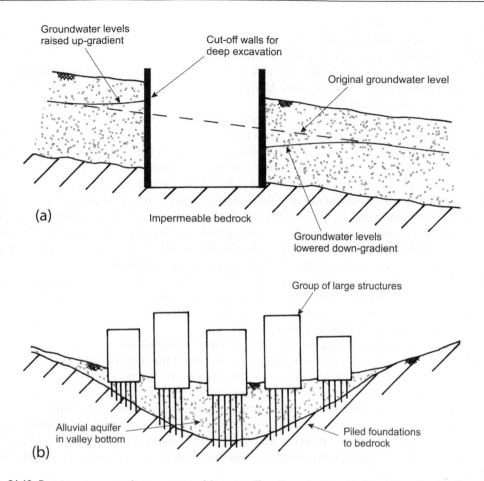

Figure 21.13 Barriers to groundwater created by cut-off walls and piles. (a) Cut-off walls. (b) Groups of piles. (After Preene, M and Brassington, F C, *Water and Environmental Management Journal*, 17, 59–64, 2003.)

Where impacts are a potential concern, it may be appropriate to use numerical groundwater modelling to assess changes in groundwater level. If impacts are considered to be significant, then consideration should be given to modifying the cut-off wall or piled foundation design to limit the depth of piles or cut-off walls, or using cut-off walls that are temporary in nature (such as artificial ground freezing or steel sheet-piles removed at the end of construction). Section 24.5 notes that, very occasionally, groundwater cut-off walls are deliberately breached at the end of construction to avoid the risk of a permanent barrier effect.

An additional impact occasionally associated with groundwater barriers is groundwater contamination derived from the materials used in the barrier. This is a potential issue for grout barriers formed from certain types of grout (see Section 18.10), where the groundwater chemistry may have the potential to leach polluting chemicals from grout into the groundwater.

21.8 IMPACTS FROM DISCHARGE FLOWS TO THE GROUNDWATER ENVIRONMENT

The construction activities associated with civil engineering excavations can create the potential for discharges to groundwaters with the consequent risk of pollution and degradation of groundwater quality. The main sources of potentially polluting discharges are leakages and spills of fuels and lubricants from plant and vehicles; run-off from operations such as concrete placement or grouting; and run-off of turbid surface water as a result of topsoil removal and excavation. Normally, the risk of polluting discharges can be reduced by the adoption of good practice (for example Murnane *et al.*, 2006) and guidance from the environmental regulators for the site locality.

The risk of pollution is increased if pathways for groundwater flow (see Section 21.6) are associated with the works (Figure 21.14). Often, open excavations form a ready pathway for inadvertent discharges to groundwater. Good site practice should include prohibition of refuelling of plant (and storage of fuels) in or near excavations. Surface water drainage (see Section 9.2) should be arranged to reduce the risk of spills or run-off entering the excavation.

Structures with deep basements or below-ground spaces may also provide potential for discharges to groundwater in the longer term. If the structures are not watertight and penetrate confining beds over aquifers, leaks, spillages or surface water flooding may be able to percolate more freely into groundwater. Individually, such leakages may be small, but their combined effect may lead to significant groundwater contamination.

The use of artificial recharge (see Section 17.13) also creates the risk of groundwater pollution if the re-injected water were to become contaminated as a result of its pumping and transfer through the recharge system. Great care must be taken to reduce the risk of recharge water becoming contaminated.

There have been rare occasions when the release of chemicals from processes used in ground engineering has also been a focus of concern. This may be relevant when works are carried out in aquifers that are used for drinking water supply or are very sensitive for ecological reasons. This was a particular concern during the 2000s on the Copenhagen Metro project, which involved extensive tunnel construction. Raben-Levetzau *et al.* (2004) describe how chemical grouts were prohibited and numerical groundwater modelling studies were carried out to assess whether migration of products used in tunnel construction (soil conditioning products, lubrication and sealing projects, dispersants, etc) might affect surrounding water quality.

On another project, a tunnel was being driven by tunnel boring machine (TBM) through a fractured carbonate rock aquifer relatively close to a high-capacity public drinking water supply borehole. There was a concern that the cutting action of the TBM might create

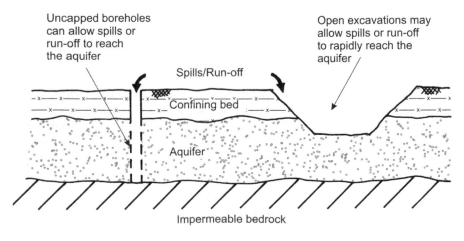

Figure 21.14 Potentially polluting discharges to groundwater. (Redrawn from Preene, M and Brassington, F C, *Water and Environmental Management Journal*, 17, 59–64, 2003.)

loose particles, which could be drawn to the pumping borehole along enlarged fractures believed to be feeding the borehole. This could create turbidity in the pumped water, making it unsuitable for drinking water use. The strategy adopted was to take the water supply borehole out of service while the TBM traversed the area. When the borehole was re-started after completion or tunnelling, it was pumped to waste for an extended period and closely monitored for turbidity to ensure that any turbid water did not find its way into the supply.

21.9 IMPACTS FROM DISCHARGE FLOWS TO THE SURFACE WATER ENVIRONMENT

Any system that lowers groundwater by pumping will produce a discharge flow of water that must be disposed of. The most commonly used routes for the disposal of discharge water from dewatering systems include:

- Discharge to surface waters (e.g. river, watercourse, lake, sea)
- Artificial recharge (see Section 17.13) to groundwater (e.g. via recharge wells or recharge trenches)
- Discharge to an existing sewerage network

Legal permissions necessary for the discharge of groundwater are outlined in Section 25.4.

Poorly managed discharges may have adverse impacts on the environment. The following sections describe good practice for discharges to the surface water environment.

21.9.1 Erosion Caused by Discharge Flows

Poorly located discharges can cause erosion of riverbanks or watercourses if the flow is concentrated in one location – particularly if the flow rate is large. A scour hollow will form under the discharge point, possibly undermining the bank. Problems may be created downstream as the scoured material is redeposited, blocking or changing the flow in the watercourse.

This problem can be avoided by designing the discharge system to reduce the potential for scour. Materials such as a blanket of coarse gravel (Figure 21.15), geotextile mattresses,

Figure 21.15 Blanket of coarse gravel used to prevent erosion at where discharge from dewatering system enters watercourse.

gabions or even straw bales can be placed at the discharge point to dissipate the energy of the water before it passes into the watercourse proper.

21.9.2 Suspended Solids

Suspended solids, in the form of clay, silt and occasionally sand-sized particles, are a common problem resulting from dewatering discharges. Silt discharges are a highly visible aesthetic problem (Figure 21.16), but silt also harms aquatic plant, fish and insect life and can build up in watercourses, blocking flow.

The suspended solids in sediment-laden discharges can be difficult to deal with economically, so the best approach is to tackle the problem at source and avoid silt being drawn into the discharge water. This requires appropriately designed and installed filters to be included in the system design. This is the norm for pre-drainage methods (wellpoint, deep well and ejector wells), and such systems do not commonly produce discharges with high sediment loads except for a short period during initial well development.

Most problems with suspended solids in discharge water arise when open pumping is carried out by sump pumping (Section 14.9). In many cases, adequate filters are not installed around the sumps, and fine particles can be drawn out of the ground and entrained in the discharge water. Where sump pumping is carried out in soil/rock containing a significant proportion of fine particles, the discharge will need to be treated to reduce any suspended solids to acceptable levels prior to discharge. If treatment is not possible, a change to a pre-drainage dewatering method (e.g. by using wellpoints with adequate filters) should be considered.

The most common form of treatment for suspended solids is by passing through settlement tanks or lagoons. In essence, settlement tanks provide an environment (a tank or lagoon) of relatively still water through which the discharge flow passes very slowly and hence has a long 'retention time'. This allows solid particles to settle out of the water and be deposited in the base. The water, now with a reduced load of suspended solids, is discharged from the outlet side of the tank, typically via a weir or high-level outlet pipe. The solids will be retained in the base of the tank or lagoon. Periodically, the tank may need to be drained and the solids removed.

The settling rate of a solid particle is controlled by Stokes' law – the smaller the particle, the slower its rate of settling. Hence, the larger the tank, the longer the retention time,

Figure 21.16 Silt plume in a watercourse resulting from the discharge of water from poorly controlled sump pumping. (Courtesy of T O L Roberts.)

allowing more time for smaller particles to settle out. In order to settle out silt and clay size particles by conventional means, very large tanks or lagoons may be needed.

The efficiency of settlement can be improved by chemical treatment, whereby coagulants and flocculants are added to encourage groups of particles to coalesce together and form larger 'flocs' of particles, which will settle much more quickly (Binnie and Kimber, 2009). Chemical treatment of water is a complex process, and specialist advice should be obtained if this approach is being considered.

The most common form of settlement tanks used on dewatering projects are small portable steel tanks (typically 3 m by 1.5 m in plan and up to 1.5 m deep). Purpose-built tanks are available (Figure 21.17a), or improvised tanks can be made from waste skips available on site (Figure 21.17b). Such tanks can be effective in removing sand-size particles, but silt and clay-size particles settle slowly and will pass through these small tanks.

In recent years, some innovative approaches have been developed to allow more effective settlement of solids via small tanks. One example is the Siltbuster® unit (Figure 21.17c) based on the principle of lamella plate clarification, a process widely used in the treatment of drinking water. In this approach, the dirty water is passed upwards between inclined parallel plates, which are separated by a small gap. Solid particles settle onto the inclined plates and fall to the base of the tank. The lamella plate approach allows a small tank, comparable in plan size to a conventional steel settlement tank, to have the settling capacity of a lagoon several times its size.

Large lagoons can be used to provide sufficient retention time for silt and clay size particles to settle out. A typical lagoon is shown in Figure 21.17d and consists of an earthwork bund or pit with some form of waterproof lining on the base and sides; edge protection is necessary to reduce hazards to personnel. It may be necessary to operate two or more lagoons in parallel. When one lagoon is full, the other receives the discharge, allowing more time for settlement in the first. Water from each lagoon is decanted and disposed of in turn; sediment needs to be removed periodically from the lagoons.

21.9.3 Discharge Water Chemistry and Temperature

Even where the groundwater discharged from the dewatering system is uncontaminated and free from suspended solids, there can still be impacts where the water is discharged to surface water or groundwater:

- Groundwater chemistry: If the pH and salinity of the discharge water are significantly different from those of the receiving water body, then there is the potential for adverse impacts. Problems also occasionally occur when natural dissolved substances (such as carbonates or iron-related compounds) in the pumped water precipitate out in the discharge flow. For discharges to surface watercourses, this can result in unsightly deposits and possible adverse impacts on surface water quality and aquatic ecology. For discharges to groundwater, this can result in clogging of recharge wells and trenches (see Section 17.13.5).
- Groundwater temperature. The temperature of the discharge water will be similar to that of groundwater. As discussed in Section 3.8, the temperature of groundwater varies little year round, typically being close to the mean air temperature at a site. Conversely, the temperature of a surface watercourse receiving the discharge will be more affected by seasonal temperature variations. In winter, the discharge from a dewatering system will tend to be warmer than surface water, and in summer, the discharge will tend to be cooler than surface water. If the receiving surface water environment contains ecological habitats (e.g. plants or aquatic wildlife) that are sensitive to temperature, there is a risk of adverse impacts on local habitats. In such cases, specialist ecological and hydrological advice should be obtained.

Figure 21.17 Treatment of discharges by settlement. (a) Modular settlement tanks. (b) Small settlement tanks improvised from waste skips. (c) Specialist lamella settlement tank. (d) Large settlement lagoons. (Panel (a) courtesy of Hölscher Wasserbau GmbH, Haren, Germany; panel (c) courtesy of Siltbuster Limited.)

21.9.4 Oil and Petroleum Products

Oil and petroleum-based products may find their way into dewatering discharges as a result of spills or leaks from plant or fuel storage areas. This is a particular risk with sump pumping, because any spills or leaks in the base of the excavation will be carried to the pumps by surface water. Petroleum products may also occur in discharges if contaminated groundwater is pumped.

Petroleum products are generally of lower density than water and do not mix well with it. They are known as light non-aqueous phase liquids (LNAPLs) and will tend to appear as floating films or layers on top of ponds or tanks of water. The petroleum products must be separated from the discharge before disposal. This can be accomplished by passing the discharge through proprietary 'petrol interceptors' or 'phase separators'. The water is then discharged as normal, and the product (which collects in the interceptor) is disposed of (e.g. to a waste oil company) at appropriate intervals. Very thin oil layers may be removed by the use of floating skimmer pumps or sorbent booms or pillows placed in discharge tanks or lagoons.

21.9.5 Contaminated Groundwater

If groundwater is pumped from a contaminated site, or from adjacent to such a site, the resulting dewatering discharge may itself be contaminated. Unless the flow is discharged to a sewage treatment works capable of dealing with the contaminants, some form of on-site treatment will be required prior to discharge.

A wide range of treatment methods is available. Often, a given contaminant could be treated by several quite different methods; the choice of method will depend on the concentration of contaminants, the discharge flow rate, the duration of pumping, and the availability of treatment equipment and technologies. Some of the available technologies are described in Section 17.14.

The scale of treatments used in practice varies greatly. Occasionally, on very heavily contaminated sites with low pumped flow rates, the discharge may be pumped directly to special road tankers for off-site disposal at a licensed facility. In other cases, on-site treatment may be feasible. This ranges from simple dosing with caustic soda for pH adjustment of acidic groundwater through to the construction of modular treatment plants on heavily contaminated sites (see Section 17.14 and Case History 27.11).

A programme of chemical testing will be needed to monitor discharge water quality, and the environmental authorities (see Section 25.4) should be kept fully informed. Often, the targets for treatment of the discharge will be set by the environmental regulators, based on a consideration of the risk of environmental impact from the discharge.

21.10 ASSESSMENT OF POTENTIAL ENVIRONMENTAL IMPACTS

For any groundwater control project of any substantial size, it will be necessary to assess the potential environmental impacts. It may also be necessary to consider potential impacts for smaller groundwater control systems in potentially sensitive locations (for example on construction sites near water supply wells or groundwater-dependent features).

There are four main stages to the assessment and management of potential environmental impacts from groundwater control activities. These are identification; prediction; mitigation; and monitoring.

21.10.1 Identification of Potential Environmental Impacts

The first stage is to assess, in principle, the potential for significant impacts to occur. It would be extremely useful to practising engineers if generic 'key indicators' of potential impacts could be identified. This could allow early screening of projects to determine whether groundwater control works have the potential to cause significant impact and therefore, whether special monitoring and mitigation measures may be required. Unfortunately, in practice, potential impacts will be individual to each project site; it is difficult to provide such simple indicators of the risk of potential impacts.

The essential starting point when assessing potential impacts is the development of an appropriate conceptual model (see Sections 5.4 and 13.4). The conceptual model must include factors related to the ground and groundwater conditions, to the groundwater control technique to be used, and to the presence and location of environmental features that may be impacted by the range of impacts listed in Section 21.3.

A desk study (see Section 11.6) can be very useful at this stage to help identify the existence of environmental features that have the potential to be impacted. Where features such as wetlands, private water supply wells, etc. are identified, it may be appropriate to carry out a site inspection visit of the feature to gather further information.

At this stage, it is also worth investigating the existence of sources of background or baseline monitoring data (see Section 21.10.4) to determine existing groundwater levels, etc.

This initial stage of assessment should have an additional objective of assessing the level of uncertainty in the information relating to environmental impacts. The identified uncertainties should be used to help plan future monitoring and/or investigations to fill any data gaps.

21.10.2 Prediction of Potential Environmental Impacts

Based on the previously identified potential for impacts, it is necessary to make some kind of assessment of the magnitude of the impacts to determine whether any mitigation may be necessary.

The choice of methods used to quantify the impacts will depend on the conceptual model and on the scale and type of impact anticipated. For complex projects, numerical groundwater modelling may be used, while for many projects it may be appropriate to use conventional closed-form analytical design equations to assess drawdown impacts. It is important to realize that often, the objective of calculations or modelling is not necessarily to make precise predictions of the magnitude of impacts. Instead, the objective of the predictions is to aid decision making by determining whether the impacts are likely to be significant enough to require mitigation.

21.10.3 Mitigation of Potential Environmental Impacts

The selection of any mitigation measures should be based on the conceptual model and the predicted impacts.

Three principal types of mitigation can be applied to pumped dewatering schemes:

i. *Return of pumped water to the environment.* Many of the potential impacts derive directly from the removal of water from the aquifer system – for example ground settlement, impacts on groundwater-dependent features and loss of yield of water supply wells. These impacts can be mitigated by returning the water back to the area of impact. This could include artificial recharge to the aquifer (see Section 17.13) or discharge of water to affected wetlands or rivers.

ii. *Prevention of removal of groundwater from sensitive areas.* This can involve the use of groundwater cut-off barriers (Chapter 18) to prevent groundwater flow in unfavourable directions, to prevent seepage from rivers and wetlands, and so on. Additionally, appropriate design of dewatering wells and monitoring wells with screens and grout seals at appropriate levels can help control vertical flow of groundwater, for example to prevent excessive drawdowns in shallow strata when dewatering is carried out in deeper layers.

iii. *Avoidance or reduction of groundwater lowering.* The ultimate mitigation is to consider changing the scope of the overall project to reduce or avoid the need for groundwater control. Most commonly, this would involve redesign of the project to reduce the depth of excavation, or relocation of the project to increase the distance to any identified environmentally sensitive features. Obviously, such a mitigation measure is only feasible in the very early planning stages of a project, before location and design are finalized, but this approach is sometimes used during the route planning stage for tunnel projects or linear infrastructure projects such as new roads or railways. On tunnel projects, the shafts (which may require dewatering) may be deliberately located in areas distant from potentially sensitive groundwater-dependent features.

Mitigation measures for specific categories of impacts are described earlier in this chapter alongside description of the relevant impacts.

A distinction should be drawn between those mitigation measures that are intended to reduce impacts over a wide area and those that are targeted to protect specific locations, structure or features. For example, compare Case Histories 27.8 and 27.9, which both included artificial recharge of groundwater. On a project in Cairo, Case History 27.8 describes recharge targeted to avoid damaging settlements at specific nearby structures. In contrast, Case History 27.9 describes an application of artificial recharge intended to minimize drawdown impacts across a large area. When assessing possible mitigation measures, it can be useful to consider these two different approaches. If there are only a small number of isolated features that are vulnerable to impacts, it may be more cost-effective to prevent the problem directly at the feature, for example by underpinning the foundations of a sensitive structure or by replacing a residential water supply well with a piped supply where lowering of groundwater levels has reduced the yield.

21.10.4 Monitoring of Potential Environmental Impacts

Monitoring of the groundwater regime around a site is an essential part of identifying and mitigating potential impacts. In addition to the direct monitoring of the groundwater control system for routine parameters such as groundwater levels in the excavation, pumped flow rates, and temperature/chemistry/suspended solids of the discharge water, additional monitoring is often required to assess external impacts.

Typical parameters monitored include:

i. Ground levels at defined locations (to allow ground settlements to be estimated).
ii. Condition of existing structures or services that may be affected by lowering of groundwater levels (to allow any movement, distortion and damage to be estimated).
iii. Groundwater levels in wells and boreholes. Ideally, piezometers and wells should have response zones in the strata where the drawdown of piezometric levels is anticipated.
iv. Surface water levels in wetlands, streams, etc.
v. Flow rate from springs and in associated watercourses.

vi. Flow rate in rivers where groundwater baseflow is anticipated to be a key component of flow.

vii. Water quality parameters at watercourses, springs or boreholes, including the use of geophysical fluid logging in boreholes with stratified water quality.

Monitoring of rainfall and barometric pressure can also be useful to help identify any natural changes in groundwater conditions that may occur during the monitoring period. Monitoring methods for groundwater control projects are discussed in detail in Chapter 22.

The location of monitoring points should be controlled by the conceptual model of the anticipated impacts. It is clear that monitoring points must be located in areas and aquifer units where impacts are expected. However, it is also prudent to carry out monitoring in aquifer units where no impacts are expected (e.g. horizons that are hydraulically isolated from the works by very low-permeability strata).

It is also essential that the type of monitoring installation or device is appropriate to the conceptual model. In particular, for the case where the impact of concern is ground settlement resulting from pore water pressure reductions in low-permeability alluvial soils, conventional standpipe piezometers may not provide realistic monitoring results. Standpipe piezometers require a relatively large flow of water into or out of the soil in order to register a change in groundwater level. In low-permeability soils such as silts or clays, these instruments will respond only very slowly to changes in pore water pressures. Where it is necessary to accurately monitor pore water pressures in low-permeability compressible soils, consideration should be given to the use of 'rapid response' instruments such as vibrating wire piezometers (VWP), which use electronic pressure transducers (see Sections 11.7.6.3 and 22.6.3).

In the past, monitoring has been done manually by visiting each location. This has restricted the frequency at which data can be obtained economically. The availability of simple, cheap and reliable datalogging systems (see Section 22.4) now provides the option of obtaining almost continuous records of parameters.

Figure 21.18 shows an example of the use of datalogging systems to monitor a private water supply borehole assessed to be potentially impacted by a nearby dewatering system. In normal operation, the groundwater level in the supply borehole fluctuates during the day in response to the pump being turned on and off to meet the user's water demand. This means that when manual readings of water level are taken on a daily basis (a commonly used frequency of manual readings), there is a lot of scatter in the water levels depending on whether the pump was operating or not at the time of the reading. Such manual readings are shown in Figure 21.18a. This shows that while a general reduction in groundwater level is indicated, the scatter in the data makes it difficult to estimate precisely the drawdown impact at the supply borehole. In contrast, Figure 21.18b shows data for the same period but with readings taken at hourly intervals by a datalogger. This dataset clearly shows the drawdown impact on the supply borehole and allows a much better estimation of the magnitude of the impact.

Where groundwater impacts are a potential concern, it is essential that adequate 'background' or baseline data are obtained to provide reference levels against which any impacts can be measured. For major groundwater control projects, an extensive programme of data collection often forms part of the design and planning stage of a scheme.

Streetly (1998) has pointed out the difficulties of establishing the true baseline conditions against which to assess impacts such as changes in groundwater level. Typically, even in the absence of dewatering activities, groundwater levels will vary in the short term (due to barometric changes, rainfall, abstraction, etc.). In the longer term, groundwater levels may fluctuate due to variations in recharge and, ultimately, climate change. Without some

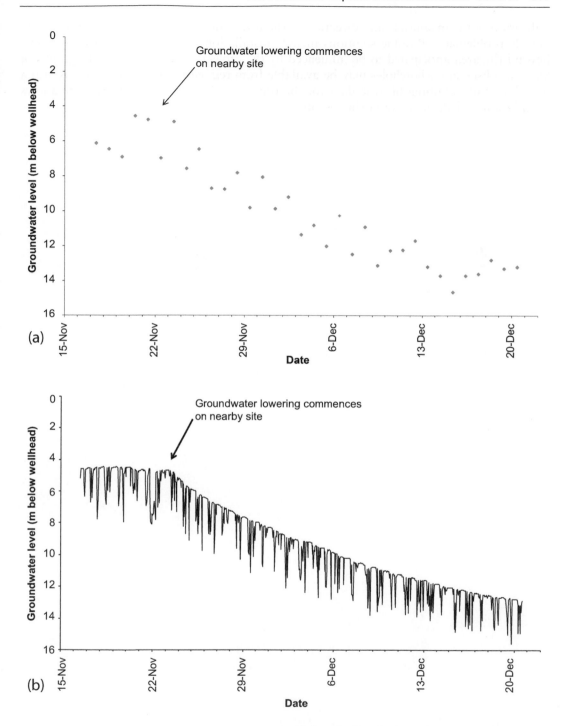

Figure 21.18 Monitoring of impacts on private water supply borehole. (a) Monitoring manual readings of water level taken once per day. The scatter in the data results from variations in water level in the borehole caused by the pump switching on and off in response to the user's water demand. (b) Monitoring using datalogger equipment taking water level readings every hour. The more frequent readings show more clearly the variations in water level in the borehole caused by the pump switching on and off. These data allow the drawdown impact to be better identified.

indication of the magnitude and direction of these variations, interpretation of monitoring data is problematic. Possible solutions include installation of 'control' monitoring points beyond the area anticipated to be influenced by the project. Alternatively, hydrographs of 'distant' observation boreholes may be available from regulatory authorities; this is also a useful way of obtaining historic data for the time period before access to the site allows installation of dedicated monitoring points.

Chapter 22

Monitoring of Groundwater Control Systems

You see, but you do not observe. The distinction is clear.

Sherlock Holmes in *A Scandal in Bohemia, The Adventures of Sherlock Holmes*, Sir Arthur Conan Doyle, 1892

22.1 INTRODUCTION

Any groundwater control system will need to be monitored, to some degree, to provide feedback on the operation, which can be used to help identify the need for any modifications or maintenance. This chapter describes the methods commonly used for monitoring of parameters relevant to groundwater control systems, including groundwater levels and pumped flow rates, as well as other performance indicators. This chapter discusses methods of manual data recording, automatic monitoring using dataloggers and the means to disseminate data. Maintenance of groundwater control systems is discussed in Chapter 23.

22.2 THE NEED FOR MONITORING

Once in operation, a groundwater control system is the end result of a lot of effort by a lot of people. It is a complex system dependent on a diverse range of hydrogeological, hydraulic, chemical, mechanical and human factors, but it will have a clear aim – to lower groundwater levels sufficiently to allow the construction works to proceed. It would be silly, having invested so much time and money in a system, not to monitor it to check that initially, and on a continuing basis, these aims are achieved. Yet many systems are not adequately monitored, leading to poor performance and loss of time and money, not to mention a stressful time for those concerned.

Groundwater control systems should have specific targets – maximum allowable groundwater levels at particular locations. They achieve these targets by pumping groundwater, sometimes in combination with groundwater cut-off walls or barriers. The two most important parameters to be monitored are groundwater levels (Section 22.6) and pumped flow rate or discharge (Section 22.7). Other parameters may also need to be monitored (Sections 22.8 through 22.11).

The requirement for monitoring will change during the life cycle of a groundwater control system (see Section 22.3); timely and accurate monitoring can be vital in identifying and remedying potential problems. In some cases, particularly on large projects or where site investigation information is limited, monitoring of system performance can be used as part of the observational method of design (Section 13.3.2). The observational method involves developing a design based on initial design assumptions about ground conditions plus contingency plans to allow modification of the system in light of conditions actually encountered. Adoption of the observational method can allow fine-tuning of the system in difficult ground conditions. There may be a temptation to install only the bare minimum pumping capacity or number of wells to achieve the drawdown. This temptation must be resisted and allowance made for standby equipment (Section 23.4) and long-term deterioration of the system (Section 23.5). Optimization of groundwater control systems is described in a wider context in Chapter 26.

22.2.1 Manual Monitoring or Automated Monitoring?

A key choice is between manual and automated monitoring:

- Manual monitoring involves technicians moving around the site, manually dipping water levels, reading flowmeters, etc. and recording the results on paper (in notebooks or on record sheets) or on hand-held electronic devices such as tablets.
- Automated monitoring uses electronic datalogging equipment, connected to appropriate sensing devices (such as pressure transducers and flowmeters), to record the required data. The data is either stored in a datalogger (for later download) or transmitted to a server.

On many projects, particularly smaller ones, the primary methods for collecting monitoring data are manual, and this can be a perfectly acceptable approach. Automated monitoring systems are more commonly used on larger or more complex projects but should not be discounted for smaller or short-term schemes. As each year passes, available datalogging equipment is becoming cheaper, smaller and more versatile. Equipment that, only a few years ago, might have been suitable for large-scale projects only is now cost-effective on even simple schemes. The datalogging systems used as part of automated monitoring are described in Section 22.4.

22.3 PHASES OF MONITORING

Monitoring is important throughout the operational life of a system but especially soon after the start of pumping. There have been many cases where carefully designed systems have not initially achieved the target water levels. This potentially embarrassing eventuality may result from some very simple problems with the pumping equipment or the way it is operated. Occasionally, problems arise if ground and groundwater conditions differ significantly from those indicated by the site investigation – so-called 'unforeseen ground conditions'. Monitoring during the early stages of pumping is vital to allow potential problems to be identified and solved.

There are five principal phases of monitoring of a dewatering system:

1. Installation: During this period, monitoring commences, and more and more monitoring locations become available as new wells are installed.
2. Commissioning and drawdown phase: This is the period immediately before and after pumping starts (sometimes known as the 'pumping down' period), when groundwater levels are changing rapidly, and monitoring readings are taken relatively frequently.
3. Maintenance phase: This is the period once the target drawdown has been achieved, and groundwater levels are changing only slowly. Monitoring readings are taken less frequently.

4. Recovery phase: After pumping is stopped, groundwater levels will recover, and monitoring should be appropriate to the rate of change of groundwater levels. More frequent monitoring will be required compared with the maintenance phase.
5. Decommissioning: There is sometimes a requirement for additional monitoring during the decommissioning phase to help identify any long-term impacts.

22.3.1 Using the Data

Whatever monitoring is carried out, merely taking and storing the readings is not enough. The data need to be communicated to, and reviewed by, personnel who can interpret them appropriately. Increasingly, data are stored electronically on web-based data management platforms (data taken manually are typically 'keyed into' a spreadsheet to allow electronic storage). Monitoring records can then be accessed on various electronic devices, including laptop computers, tablets and smartphones – dissemination of data is discussed in Section 22.5.

However, it is vital that the data are actually used to ensure that the works are carried out safely and efficiently. CIRIA Report C750 (Preene *et al.*, 2016) states:

> Monitoring should not be undertaken as a matter of course or because it seems the 'right' thing to do. The monitoring should be an integral part of the safety and quality management system on site. Merely taking the readings and filing them away is not sufficient; the results should be plotted in a way that highlights the performance of the system and be displayed for engineering and management staff. In addition, they should be regularly reviewed by a nominated member of the site management team, and any observed changes or trends in the data investigated.

One of the simplest and easiest ways of using the data is to plot long-term trends of groundwater levels or pumped flow rates to aid the identification of potential problems or anomalies. On many projects, this need not be done by a groundwater control specialist. The main contractor's site engineer could review them – provided they have been briefed by the system designer or installer as to what factors are important and whether there are any particular targets that must be achieved.

Whoever reviews the data, specialist advice should be sought if there is any uncertainty about the effectiveness of system performance.

A common practice that has developed on tunnelling projects is for key geotechnical data, including dewatering parameters, to be reviewed daily by a meeting of designers' representatives and construction managers – collectively known as the 'Shift Review Group (SRG)' – named after the shift work pattern used routinely on tunnelling projects. Where geotechnical data are formally reviewed by an SRG meeting or equivalent, a widely used approach for geotechnical data is the trigger level method, applied in a 'traffic light' system of green, amber and red trigger levels (Devriendt, 2012). A trigger level is a pre-defined value of a measured parameter, such as the groundwater level in a specific monitoring well or the flow rate from a specific sub-section of a dewatering system. If an observed reading crosses the trigger value, then a pre-defined action is carried out.

1. Green condition: Monitoring data are within pre-defined acceptable limits, and work can proceed as normal. The boundary between the green condition and the amber condition is known as the 'amber trigger'.
2. Amber condition: Some monitoring data have passed the pre-defined amber trigger and are at values that are not yet problematic but are closer to troubling values. Examples include rises in groundwater levels within excavations or a reduction in pumped flow

rate (potentially indicating clogging of wells or pump failure). In the amber condition, additional and more frequent monitoring is carried out, and the data are reviewed more regularly, sometimes with specialist expertise called in. Decisions may be made to implement contingency measures or remedial works. The boundary between the amber condition and the red condition is known as the 'red trigger'.

3. Red condition: Some monitoring data have passed the pre-defined red trigger and are at problematic values. Examples include excessively high groundwater levels in the excavation or evidence of uncontrolled seepage associated with hydraulic failure of the base of an excavation. Actions in the red condition (beyond those in the amber condition) include stopping excavation work at the first safe hold point, obtaining specialist advice, and rapidly identifying and implementing the relevant remedial measures.

22.4 DATALOGGING SYSTEMS

A datalogger is an electronic device that receives, stores and sometimes transmits data from one or more sensors and is used as part of automated monitoring systems (see Section 22.2.1). In the context of groundwater control systems, many monitoring sensors produce electrical outputs that can be linked to datalogging systems. The most common types of sensors are vibrating wire (VWP) pressure transducers used to monitor groundwater levels and flowmeters used to record pumping rates. Many other instruments can also be datalogged, including water quality sensors, rain gauges, etc.

Datalogging systems vary greatly in capability, complexity and cost, but there are some attributes common to all systems:

- Datalogging systems comprise electronic devices and require power from mains electricity, batteries, photovoltaic cells or a combination of these. This can affect how the systems can be used, particularly if external power is not available. However, even when external power is not available, many systems now have a very long battery life (months or even years) and can be used even in remote locations.
- Datalogging systems comprise two main elements – the sensors that record the parameter in question and the 'logger unit' that typically includes the power source and carries out the data storage and communication role. Some dataloggers have multiple channels, serving several sensors, linked by data cables or wireless connections. Some dataloggers are single channel (sometimes known as 'stand-alone' loggers) and serve only one sensor. Often, in these cases, the sensor and logger elements are combined into a single unit.
- All datalogging systems include a clock to record the date and time each sensor reading was taken. It is important that the logger clock is set up with the correct time, including any daylight saving time changes at the project location.
- While dataloggers are considered part of 'automated' monitoring, the systems only act in response to instructions from the user. Therefore, all such systems are 'programmable'; the user sets the parameters of the monitoring programme, such as which sensors are to be monitored, at what frequency, etc. Programming is normally done via interface software on a laptop computer or tablet, which may be linked to the datalogger wirelessly or by a cabled connection.
- The parameters recorded from each sensor are typically stored on the logger unit and then transmitted (a process usually referred to as 'downloading') either to an external device such as a laptop computer or to an external server. Downloading can be

done by personnel on site, via a cabled or wireless connection, or can be done entirely remotely via mobile phone data networks.

- A final point about dataloggers and the associated sensors is that they are relatively delicate devices that must operate in the harsh environment of a construction site. These systems are vulnerable to damage, vandalism and the effects of weather (extreme hot and cold temperatures, flooding by surface water ponding, etc). The equipment should be selected to be robust for the working environment and should be installed and located to reduce the risk of accidental damage and vandalism.

The use of dataloggers for automated monitoring can reduce the number of personnel needed for monitoring, especially if regular monitoring during day and night shifts is required. By downloading the data directly, there is no need for someone to laboriously key in manually recorded readings, and the time and effort required to go from data gathering to plotting and analysis of data can be minimized. However, automated monitoring will not completely remove the requirement for manual readings. Data collected manually are needed when first setting up the monitoring system and for periodic spot checks to validate the output of the automated system.

The use of datalogging systems has the potential to generate huge volumes of data, which can be disseminated very rapidly via email and web-based data management systems to various parties. This is not necessarily a good thing! There is a real danger that 'data overload' will occur and that the use of dataloggers will actually be counterproductive and will make identification of key factors and potential problems more difficult. It is essential that adequate procedures are put in place for the management and review of field data. Holmes *et al.* (2005) give a good example of data collation and management from a large project in the United Kingdom.

As well as for operational monitoring, datalogging systems can be useful for gathering data during site investigation pumping tests (see Section 12.8.5 and Appendix 4).

22.5 DATA MANAGEMENT AND DISSEMINATION

The advent of automated monitoring and the use of datalogger systems have vastly increased the volume of data that can be processed, stored and disseminated. On simple projects, it is still possible to manage the data on spreadsheets and to use these to provide the necessary plots of data, key values, etc. However, on larger or more complex projects, this can be cumbersome and inefficient. Such projects now commonly use web-based data management tools to manage and disseminate geotechnical monitoring data, including those from groundwater control systems.

A wide range of web-based systems are available from various geotechnical software providers, and they vary in nature and capability, but features typically include:

- The monitoring data from site are stored on off-site servers, solving many of the security, storage and back-up issues that occur when the data are managed on site. Users are provided with password protected accounts to allow them to access the data from anywhere with a web connection. Many data management systems can be accessed on hand-held devices via 'apps' for tablets and smartphones.
- Data from sensors are automatically processed into the relevant engineering units and uploaded to the system and can be viewed in real time – the most recent reading on a given instrument can be viewed soon after it is recorded.
- The user interfaces within the data management system are often linked to a geographical information system (GIS), which is a digital representation of the locations, depths, dimensions and functional characteristics of each monitoring point. This can be a great aid to interpreting and understanding the data.

On groundwater control projects, the key advantage of these systems is the potential to make the analysis of data more efficient by reducing repetition of work updating spreadsheets and plots. Standard plots of groups of instruments, automatically updated each time the system is accessed, can be created by a user and shared with interested parties. These plots can be very useful to inform regular reviews by SRG meetings or their equivalent. Alerts can be set up to contact key personnel by text message or email if trigger levels (See Section 22.3.1) are breached.

22.6 MONITORING OF GROUNDWATER LEVELS

There are three commonly used methods for measuring groundwater levels on dewatering projects:

   i. Manual monitoring of groundwater levels
  ii. Monitoring groundwater levels using pressure transducers
 iii. Monitoring of artesian groundwater levels

On a given project, these methods may be used in combination, including the use of manual readings to validate data from pressure transducers.

22.6.1 Datums Used for Groundwater Level Monitoring

In the context of this chapter, the objective of groundwater level monitoring is to determine the groundwater level in relation to the current or proposed excavation level as an aid to the control and planning of excavation and dewatering works. It is therefore necessary to ensure that groundwater levels can be related to the depth of excavation using a common 'datum' or reference system.

On very small projects on sites where ground level is essentially horizontal, it can be acceptable to record groundwater levels relative to ground level (as metres below ground level, mbgl). However, on larger and more complex sites, this is unsatisfactory, as ground level may vary across the site and may change during the project due to excavation and/or construction. A better approach is to process the data gathered in the field so that groundwater levels are reported relative to a standard datum for the project. Most commonly, groundwater levels are reported in relation to mean sea level for the locality (generically known as metres above sea level, masl). In the United Kingdom, this is reported as metres above Ordnance Datum (mAOD); under this system, negative numbers indicate levels below mean sea level. Similar nationally defined datums exist in many other countries. Other datums are sometimes used:

   • Chart datum. This is commonly the lowest astronomical tide level at a given location, such as a port, and is a few metres below mean sea level. Referred to as metres Chart Datum (mCD) these datums are sometimes used for construction works at tidal ports or waterways.
   • Tunnel datum. On tunnel projects in areas where ground level is close to mean sea level, all the works may be below sea level and hence have negative levels. This can sometimes lead to confusion, so some projects report levels as metres above Tunnel Datum (mATD), where tunnel datum is a defined level, typically 100 m below mean sea level. As an example, if a project was sinking a 20 m deep shaft from a ground level of 5 masl, the base of the shaft would be –15 masl or 85 mATD (where tunnel datum is –100 masl).

- Site datum. Where there is no immediate need to refer groundwater levels to external datums, an arbitrary datum level (often a representative ground level) can be selected for a particular project or site.

The key point is that all groundwater level data should be reported relative to a common datum, and the level of proposed excavation should be known in the same system. Confusion and problems will result if this is not done.

22.6.2 Manual Monitoring of Groundwater Levels

Manual measurement of groundwater levels in a monitoring well or dewatering well is most commonly done using a 'dipmeter' or 'dipper' or 'water level indicator' (Figure 22.1); the process is known as 'taking dip readings'. The dipmeter consists of a reel of graduated cable or tape with a probe at its tip. The probe is lowered down the well until it touches the water surface; the water completes a low-voltage circuit (powered by a battery housed inside the dipmeter reel), and a buzzer or light is activated on the dipmeter reel. Commonly available dipmeters can be used in bores of 19 mm or greater internal diameter.

There is an art to obtaining reliable dip readings. It is best to just gently lower the probe to the water surface (when the signal will activate), then raise the probe a little (the signal should cease) and lower the probe back into the water to confirm the reading. The graduated tape is used to determine the depth to water below a defined level, which for convenience is usually taken as the top of the well casing. If the top of the well casing has been surveyed, its level will be known, so the depth to water recorded in the field can be converted to one of the datum levels described in Section 22.6.1.

The most useful records of groundwater levels are those taken from unpumped wells or monitoring wells, not from pumped wells. This is because a pumped well may show a significantly lower water level than in the surrounding aquifer due to the effect of well losses (see Section 5.8.4). Monitoring wells constructed as standpipe piezometers respond to water levels in one stratum only (see Section 11.7.6) and tend to give more representative groundwater levels than unpumped dewatering wells, which may exhibit a hybrid or average water level influenced by more than one stratum. If taking dip readings in a pumped well, it is best to install a plastic dip tube of 19–50 mm diameter, down which the dipmeter probe is lowered. This avoids the dip tape getting stuck around the pump, riser pipe or power cable.

While the measurement of groundwater levels by dipping is a very simple process, it is possible that misleading readings may be generated. Possible causes for rogue readings include:

a) Clogged monitoring wells. If the well has a clogged screen, the water level inside it will not be representative of the groundwater around it. Once a few days' readings are available, clogged wells are often easy to identify, because readings will tend to be constant with time and will not reflect changes in drawdown shown in other, unclogged wells. Wells can sometimes be rehabilitated by flushing out with clean water or compressed air.

b) Cascading of water into a well. If a well has a large depth of screen, and the water level inside the well is drawn down, water may cascade into the well from the upper parts of the screen above the true water level. When the dipmeter probe is lowered down the well, it may signal when it touches the cascading water, indicating a water level higher than the true level. Dipmeters with 'shrouded' probes are less prone to this problem and should be used if at all possible.

c) Saline water. The dipmeter relies on the groundwater acting as an electrolyte to conduct the low-voltage current and trigger the signal. Saline or heavily mineralized

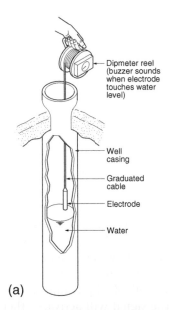

(b)

Figure 22.1 Dipmeter for measuring depth to water in a well or piezometer. (a) Schematic view of dipmeter in use. (b) Dipmeter used to record water level in a monitoring well. (Panel (a) reproduced from Preene, M. *et al.*, *Groundwater Control – Design and Practice*, 2nd edition, Construction Industry Research and Information Association, CIRIA Report C750, London, 2016. With permission from CIRIA: www.ciria.org; panel (b) courtesy of WJ Groundwater Limited, Kings Langley, UK.)

waters are more conductive and can cause problems. The dipmeter may be triggered erroneously by small moisture droplets of the side of the well casing or may signal continuously once water gets onto the probe. Obtaining reliable readings in these conditions depends on the skill (and patience!) of the operator. It helps to dry the probe on a cloth between readings and to repeat the reading at each well several times to be sure that a 'true' reading has been obtained. Some dipmeters have an internal sensitivity adjustment. If this is turned to a minimum setting, the dipmeter may be more reliable in saline wells. Occasionally, poorly conducting water (with low total dissolved solids

[TDS]) may be encountered, where contact with the water does not complete the circuit and activate the dipmeter. This has been solved by the simple expedient of adding a packet of salt to the monitoring well!

d) Errors in recording readings. The person taking the readings may have made an error. The graduated tape of a well-worn dipmeter can be difficult to read when used in the field. It is not uncommon for the centimetre part of the reading to be correct but the metre part to be in error by 1 metre. If the operator is recording all the readings on a sheet or notebook, this error may be repeated for subsequent readings. This is because the operator may expect each new reading to be similar to the last and will copy the metre portion of the reading from one record to the next. When the data are checked, any sudden 1 metre changes in level may indicate this type of error.

e) Malfunctioning dipmeters. Dipmeters sometimes malfunction. A dipmeter that does not work (perhaps because the battery is flat) is a practical problem but does not generate false readings. A more subtle problem is when the dipmeter tape is damaged and the conductors are broken and exposed above the probe. The dipmeter will not signal when the probe touches water but may do so when the exposed conductor reaches the water, indicating a lower water level than is actually present.

f) Modified dipmeters. Dipmeters are sometimes repaired by shortening the graduated tape and re-jointing the probe. If the operator is not aware of this, the recorded water levels will be deeper than actual levels, with the difference being equivalent to the shortening of the tape. It is not good practice to shorten dipmeter tapes; they should be replaced with a new tape of the correct length.

22.6.3 Monitoring Groundwater Levels Using Pressure Transducers

Automated monitoring systems using datalogging equipment typically monitor groundwater levels via electronic pressure transducers. This requires a transducer to be installed below the water level in the well (Figure 22.2). Such devices typically operate on the vibrating wire principle (and are commonly known as VWP instruments), where the transducer contains

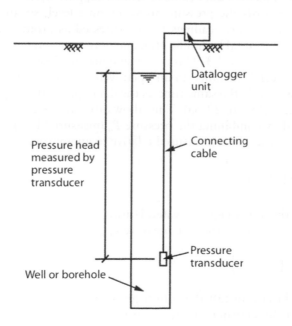

Figure 22.2 Schematic view of water level datalogger.

a diaphragm in hydraulic connection, on one side, with the groundwater. Changes in water pressure deflect the diaphragm and an associated tensioned steel wire (the vibrating wire). The deflection changes the frequency response of the vibrating wire when excited electrically (via a connecting data cable) by a readout unit or datalogger. The transducer records the pressure above its own level, so an on-site calibration is required to allow pressure readings to be converted to water levels.

There are two common configurations for pressure transducers used to monitor groundwater levels in wells:

a) Combined fully submersible transducer and datalogger for a single well. This is housed in a compact waterproof cylindrical casing (around 30 mm in diameter) designed to be placed below water level with no surface equipment. The unit is suspended in the well by a support wire.

b) Multi-channel datalogger (located at surface) with pressure transducers installed in several wells. Each sensor is linked to the surface by data cable and thence either to the datalogger or to a wireless transmitter connecting to the datalogger.

Pressure transducers are used to obtain groundwater level data as follows:

1. For systems with separate transducers and dataloggers, the system must be programmed with the transducer characteristics. The relationship between water pressure and electrical output is unique to each instrument. The manufacturer provides a calibration sheet for each instrument, which gives specific 'factors' used to convert the transducer signal to a pressure in the requisite units (typically kilopascals in metric units). When the system is set up, these factors must be programmed into the relevant channel on the datalogger. The manufacturer's guidelines should be followed during installation.

2. The pressure transducer is installed inside the well at a significant depth below the lowest anticipated drawdown level. Typically, for shallow installations, the transducer is suspended from its own data cable or from a support wire.

3. The transducer records the pressure above its own level, so an on-site conversion is required to allow pressure readings to be expressed as groundwater levels, using the relationships between total hydraulic head and pore water pressures discussed in Section 3.3.2 and shown in Figure 3.3. This conversion may be done in one of two ways:

    a. If the transducer is installed in an open well, the water level should be manually measured at exactly the same time as the datalogger takes a pressure reading from the transducer (Figure 22.3a). This allows the effective transducer depth z_{trans} to be estimated by combining the pressure P_o (measured by the transducer) with the observed water level H_o (measured relative to an appropriate datum, e.g. masl)

$$z_{\text{trans}} = H_o + \left(\frac{P_o}{\gamma_w} \right) \tag{22.1}$$

    where γ_w is the unit weight of water. During monitoring, the groundwater level H_w can then be estimated from the pressure P_{trans} recorded at the transducer by

$$H_w = z_{\text{trans}} - \left(\frac{P_{\text{trans}}}{\gamma_w} \right) \tag{22.2}$$

    b. If the transducer is installed in a fully grouted installation (see Section 11.7.6.3 and Figure 11.7), direct water level readings will not be available. In this case, z_{trans} must

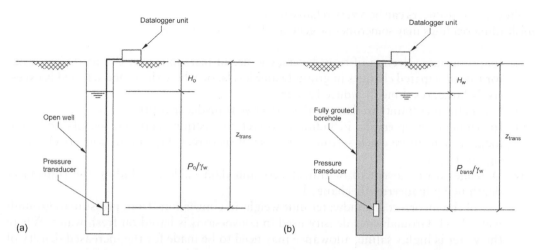

Figure 22.3 Effective depth of pressure transducer used for groundwater level monitoring. (a) Effective pressure transducer depth determined at time of set-up from manual water level observation and pressure sensor reading. (b) Effective pressure transducer depth determined from transducer depth reported at time of installation.

be taken as the transducer level reported at the time of installation (Figure 22.3b). Groundwater level is then estimated from Equation 22.2. This approach is not as accurate as method (a) due to errors and inaccuracies in the measurement of the transducer installation level, especially for deeper installations.

The most commonly used type of pressure transducer is the unvented (or absolute pressure) type. These respond to the total pressure (atmospheric pressure plus water pressure due to submergence), and water pressures are reported relative to a standard atmospheric pressure. Changes in barometric pressure during the monitoring period will affect the recorded water pressure. This is commonly addressed by including a 'control' instrument as part of the system, sitting in a small reservoir of water at ground level in the vicinity of the monitoring locations. This instrument will record the background fluctuations in observed pressure due to barometric effects. The trace from the control instrument is then used to correct the data from the instruments in the wells in order to obtain the true groundwater pressure changes. Price (2009) discusses the issues related to barometric pressure changes that affect the water pressures determined by pressure transducers.

Pressure transducers are also available in a vented design (where one side of the sensor diaphragm is connected to a thin venting tube connected to atmosphere), avoiding the need for barometric corrections. However, vented instruments are rarely used as part of monitoring for groundwater control systems.

One of the most important considerations when selecting pressure transducers is accuracy – defined in this context as the difference between the measured value, reported by the transducer, and the actual pressure. Pressure transducers are manufactured with a pre-set pressure range and a finite accuracy. The instrument for a particular application should be selected with care; manufacturer's specification sheets provide the necessary detail. The accuracy of a VWP pressure transducer is the same (typically 0.05 to 0.1 per cent) for each pressure range; therefore, sensors with a smaller range have better accuracy. For example, a transducer with a 10 m water level range and 0.05 per cent accuracy is accurate to ±5 mm, but the same transducer type with a 100 m range will only be accurate to ±50 mm.

Pressure transducers can be a very reliable means of monitoring groundwater levels; however, misleading readings may sometimes be generated. Possible causes for unusual readings include:

i. Inappropriate selection of transducers, e.g. use of instruments with insufficient range for the anticipated changes in groundwater levels, or where the range selected is excessively large, resulting in reduced accuracy.

ii. Errors or uncertainty in determining effective transducer depth.

iii. Instrument set-up errors, including incorrect connections between sensors and data-loggers; inaccurate clock settings; and use of incorrect 'factors' to convert electrical signals to pressure.

iv. Damage to or movement of transducers and data cables, including the transducer depth being inadvertently changed.

v. Use of the incorrect groundwater unit weight in conversions from pressure to groundwater level. Groundwater density used in conversions is based on fresh water. Where the water is highly saline, allowance may need to be made for the increased density of the groundwater.

vi. Instrument drift. Pressure transducers are generally reliable, but occasionally defects arise, with the most common being instrument drift. This can be a problem for long-term monitoring. Transducers installed in open wells can be replaced, but transducers installed by the fully grouted method cannot.

22.6.4 Monitoring of Artesian Groundwater Levels

Where flowing artesian conditions exist (see Section 3.4.2), a well left open will overflow at ground level. This complicates obtaining groundwater level data. The pre-requisite of obtaining representative data in these conditions is to prevent overflow from the well (which would affect the observed groundwater levels) – even minor leakage can influence groundwater levels in the well. Provided that there is a sufficient grout seal around the well casing (to prevent upward flow outside the casing), this can be achieved by two methods:

i. By extending the well casing upwards until the casing is higher than the groundwater head, at which stage the overflowing will stop. The situation is then the same as a conventional well, and water levels can be determined either by dipmeter or by pressure transducer, subject to scaffolding or mobile work platforms being used to provide safe access to the top of the well. In practice, this approach is limited to groundwater heads of 2 to 3 m above ground level.

ii. By sealing the top of the well casing using a watertight flange, blanking plate and gaskets. If the blanking flange is fitted with threaded outlets of 25 to 50 mm diameter, these can allow some forms of monitoring. Groundwater levels can be monitored by:

    a. Attaching a small-diameter vertical pipe to one of the threaded outlets and extending it vertically upwards until it no longer overflows. Groundwater levels can then be determined conventionally. This method has similar height restrictions to the method where the well casing is extended upwards.

    b. Attaching a Bourdon type pressure gauge to one of the threaded outlets. This will record the groundwater pressure above the level of the gauge. This method has the disadvantage that if the well becomes non-artesian (e.g. if water levels are lowered due to dewatering pumping from other wells), the gauge will simply record zero pressure and cannot measure water level in the well.

    c. Lowering a pressure transducer into the well via one of the threaded outlets and fitting a watertight seal (known as a 'gland') to allow the data cable from the transducer to pass out of the well, where it is connected to a datalogger. This has the advantage relative to a pressure gauge that it may still be possible to measure groundwater levels if the well becomes non-artesian.

22.6.5 Validating Groundwater Level Data

Whatever the source of the data (by manual or automated readings), groundwater level data should never be taken at face value. Some degree of validation checking should be done. Oftentimes, only a brief 'sanity check' is needed to try to identify gross errors. Such errors can include:

- Random errors caused by mistakes. These errors are randomly distributed around the 'true' values and are often easy to identify because they stand out of the general data trends. These errors are much more prevalent in manual monitoring data; for example, misreading of dip tapes or incorrect transcribing of data from field notebooks to electronic formats.
- Systematic errors that are based on the methods and equipment and how they are applied. These errors affect the whole dataset (or large parts of it) and can be difficult to identify, as they can affect the overall trend of data. These errors can affect both manual and automated readings. Examples include use of incorrect datums or conversion factors; incorrect transducer depths; and readings from damaged or repaired dipmeters where the tape length is inaccurate.

It is not always straightforward to spot unrepresentative readings, but it becomes easier with experience. If in doubt, treat the readings with caution and investigate how they were taken. A lot of information can be gained by speaking directly to the person who took the manual readings or who installed and set up the datalogger system. If the long-term trends of water levels are plotted and compared with any changes in pumping, rogue readings are often obvious.

In the field of hydrogeology for water supply, water levels are often measured at weekly or even monthly intervals (see Brassington, 2017), which is sufficient to identify long-term trends. For temporary works groundwater lowering applications, water level monitoring is normally carried out much more frequently. This is because short-term effects (such as the rate of drawdown during the pumping-down phase or fluctuation in drawdowns resulting from pump failures) are of greater interest. Typically, groundwater levels in all monitoring wells or piezometers should be recorded at least once per day, on every day of operation. On sites where the groundwater lowering system is critical to the stability of the works, groundwater levels may be recorded more frequently, perhaps several times per shift. If very frequent monitoring is necessary (perhaps on sites where water levels vary due to tidal effects; see Section 3.6.3), consideration should be given to the use of automatic datalogging equipment (see Section 22.4 and Figure 21.18).

22.7 MONITORING OF DISCHARGE FLOW RATE

The discharge flow rate pumped by a system is another vital measure of performance. The most commonly used monitoring methods are:

i. Proprietary flowmeters
ii. Weir tanks
iii. Timed volumetric methods

22.7.1 Monitoring Total Flow Rate or Flow from Individual Wells

The way in which flow rate is typically measured depends on the characteristics of the pumping method in use:

1. Wellpoint systems and ejector well systems typically use a small number of pumps to draw on many wells simultaneously. Flow rate is typically measured at the common discharge point(s) from the pumps rather than at individual wells.
2. Deep well systems have multiple pumps (one per well), and flow rate may be measured at the common discharge point(s) or measured for individual wells and then added together.
3. Sump pumping systems often produce water with some suspended sediment. This can cause problems for some flow measurement equipment, and flow rate is typically measured at the common discharge point(s) after the water has been treated to remove suspended solids.

22.7.2 Proprietary Flowmeters

Proprietary flowmeters, installed into pipework downstream of pumps, are widely used. Flowmeters may be of the totalizing type (which record total volumes of flow – flow rate is calculated from two readings at known time intervals) or the transient type (which measure flow rate directly). Some meters can perform both functions. Commonly used flowmeters are based on one of the following principles:

- Propeller or turbine meters, where the velocity of water flow rotates blades inside the meter.
- Magnetic flowmeters. These are based on the phenomenon whereby a voltage is induced in an electrical conductor moving through a magnetic field. In this case, the conductor is water, which induces a voltage proportional to the flow rate.
- Ultrasonic flowmeters. These operate on the principle that the transit time of an acoustic signal is altered by fluid velocity. Signal generators and receivers are located on the pipe and send ultrasonic signals through the flowing water.

Most modern flowmeters can be set up to produce a suitable electronic output and thereby allow automatic monitoring by datalogger.

All meters should be installed in the pipework in accordance with the manufacturer's recommendations, and should be selected so that the operational range is appropriate for the expected flow rates. Flowmeters should generally be located away from valves and with adequate lengths of straight pipe on either side (typical requirements are a straight length of 10 pipe diameters upstream and five downstream).

Flowmeters may require periodic maintenance or recalibration. In particular, some designs are not tolerant of suspended solids, biofouling detritus and other debris and can be easily affected sufficiently to give spurious readings.

22.7.3 Weir Tanks

Thin plate weirs are a well-established method to measure flow in open channels. The weir is a plate installed perpendicular to the direction of flow to cause a restriction by reducing the size of the flow channel. This increases the water level on the upstream side, and calibration charts are then used to convert the upstream water level to a flow rate over the weir.

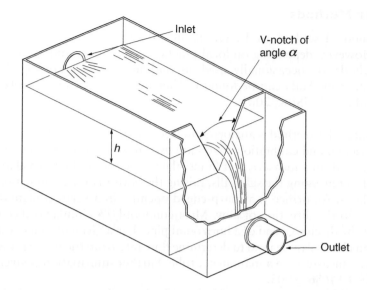

Figure 22.4 V-notch weir for measurement of discharge flow rate. The depth of water *h* over the weir is measured above the base of the V-notch. The position of the measurement should be upstream from the weir plate by a distance of approximately 0.1 to 0.7 m but not near a baffle or in the corner of a tank. Baffles may be required to smooth out any surges in the flow. (From Preene, M. *et al.*, *Groundwater Control – Design and Practice*, 2nd edition, Construction Industry Research and Information Association, CIRIA Report C750, London, 2016. With permission from CIRIA: www.ciria.org.)

In their original application, to measure flow in watercourses, accurate set-up requires the section of channel immediately upstream of the weir (the 'approach channel') to be uniform and linear. This is necessary to ensure smooth flow over the weir. When applied on groundwater control systems, a suitable approach channel is ensured by using a weir plate placed near the downstream end of a steel tank a few metres long – a 'weir tank'.

The weir tank method is rugged and reliable in the field and can deal with a wide range of flow rates. The most common configurations use a V-notch (Figure 22.4) or rectangular weir with measuring devices or sensors used to record the depth of water over the weir. Calibration charts are available to allow this measurement to be converted to flow rate. Calibration charts for V-notch weirs are given in Appendix 6. The method is suitable for routine estimation of flow rate for monitoring purposes, but it can be difficult to achieve high precision with weirs installed in small tanks. Details of the method are given in ISO 1438: 2017 and Appendix C of BS ISO 14686: 2003. It is important that the weir plate is undamaged and clean. Even small defects in the edge of the weir, or build-up of encrustations, can cause large errors in the assessed flow rate.

22.7.4 Timed Volumetric Methods

This simple method estimates flow rate by recording the time taken to fill a container of known volume (the method is sometimes called the container or bucket method). It is important that a sufficiently large container is used, to avoid too much water being lost by splashing, and to ensure that the container fills slowly enough for accurate timing. For low flows (generally less than 5 l/s, but perhaps up to 10 l/s), this technique gives reasonable accuracy using equipment that is cheap, rugged and easily portable. Ideally, timing should be by stopwatch, and a relatively large container (40 to 200 litres) should be used. The measurement should be repeated three times and the average flow recorded.

22.7.5 Other Methods

The three methods described are by far the most commonly used for groundwater control systems. However, depending on local practices, equipment and availability of expertise, other methods are occasionally used, as outlined in the following. Further details on these and other methods can be found in United States Bureau of Reclamation *Water Measurement Manual* (USBR, 2001).

- Flume tanks. This method is analogous to weir tanks in that the flow of water in a channel is restricted, and calibration charts are used to determine flow rate from the water heads observed. In a measurement flume, the restriction of flow in a channel or tank is achieved by converging the sidewalls, raising the bottom or a combination of both.
- Orifice plates. An orifice is a sharp-edged opening in a vertical plate that is set perpendicular to the direction of flow. Most commonly, a circular orifice is used and is located at the discharge end of a horizontal pipe. For a given orifice size and geometry, calibration charts can be used to determine flow rate from the water pressure upstream of the plate, measured in a manometer tube. Further information is given in Appendix C of BS ISO 14686: 2003.
- Pipe velocity/trajectory methods. These methods involve measuring the horizontal and vertical coordinates of a point in the jet of water flowing from the end of a horizontal (or vertical) pipe. The challenge with these methods is the difficulty of obtaining accurate and repeatable measurements of the geometry of the stream of flowing water.

22.7.6 Validating Flow Rate Measurement Data

Flow rate data should be validated in a similar fashion to groundwater level data (Section 22.6.5), essentially as a 'sanity check' to look for gross errors. The data can be subject to random errors and systematic errors:

- Random errors related to flow rate measurements commonly include errors in manual readings (from flowmeters, weir tank measurements or the timing of volumetric measurements) or transposition of figures and other errors when manual readings from field notebooks are keyed into spreadsheets.
- Systematic errors can affect flow rate measurements in several ways, including inappropriate set up of flowmeter, weir tanks and other devices; reporting of flow rates in the wrong units; and use of incorrect calibration charts or correlation factors for flowmeters and weirs.

In some ways, flow rate measurements are harder to validate than groundwater level data. For example, a depth to water level is easy to visualize – a reported water level of 25 mbgl around a 5 m deep excavation dewatered by wellpoints would immediately seem incongruous and would be investigated. Flow rates are harder to visualize and sense check – a small wellpoint system dewatering an excavation in a stratum of fine sand might pump 5 m³/h, but only an experienced eye would spot the error if the flow rate was reported as 5 m³/d (1/24 of the actual flow rate). The authors are aware of projects where the pumping rates were misreported by a factor of 1000 due to confusion between units of litres and cubic metres. In this case, the success of the dewatering was not affected, as it was the monitoring data that were in error, not the scale of pumping. The problem that emerged after several months was that the discharge permission obtained from the environmental regulator to dispose of the pumped water was based on a rate 1000 times smaller than actual flow rates, which caused considerable administrative headaches.

When validating flow rate data, the following questions are pertinent:

1. Is the equipment set up correctly? This includes a check that the devices are installed in accordance with manufacturer's guidance and the relevant standards. Things to address include: is there adequate straight pipe upstream of flowmeters, or are weir tanks set so the weir plate is level and vertical? Equipment can be very sensitive to set-up conditions. Many years ago, a technician was working on a system that included several high-capacity deep wells, each with its own propeller flowmeter. He decided that his job would be easier if the flowmeters were not at ground level, connected to the tangle of pipework across the site, but instead were connected directly to the 90° bend on each well head at a convenient height to take readings. When the system was off for maintenance, he laboriously moved several flowmeters and bolted them to the well heads. When the pumps were re-started, he was amazed to see that the indicator dials on the flowmeters were rotating three times as fast as when installed in straight pipe, despite the flow rate being the same! The explanation was that the water at the outer radius of the 90° bend was travelling much faster than in straight pipe, and this water fed straight into the flowmeter, completely invalidating its readings. This newly gained knowledge did not soften the blow of having to reinstate the flowmeters to their original conditions. This salutary tale highlights that some equipment is very sensitive to how it is installed and used.

2. Is the flow rate consistent with pump capacities? A straightforward check is whether the recorded flow rates seem plausible when compared with the pump output predicted from pump curves (Section 19.2). If the flow rate is significantly greater than the predicted pump capacity, this indicates a problem, perhaps caused by inappropriate equipment set-up or the flow rate being reported in the wrong units.

3. Is the flow rate being measured in the right places? Larger and complex systems will have multiple pumps, and a check should be made that the flow from all pumps is measured; if some pumps are not monitored, then the flow rate may be under-estimated. It is also possible for flow rates to be over-estimated if there are multiple flowmeters in the system, if the pipework arrangement means that some water passes through more than one flowmeter. If this is not recognized and accounted for when summing flows from all flowmeters, then some water may be 'double counted'.

4. Are changes in flow rates natural, or are they due to changes in operation? Where a dewatering pumping system is operating continuously, it is unusual for there to be large sudden changes in flow rates, unless something has changed. In general, any rapid changes in flow rate should be investigated. Possible causes of sudden changes include pumps stopping or starting; blockage or failure of flow rate monitoring equipment; or significant changes in hydrogeological conditions.

5. Is the equipment damaged, blocked or worn out? It is unavoidable that equipment will be vulnerable to damage or will gradually degrade in accuracy. For work in developing countries or in remote locations, the weir tank or volumetric methods are to be preferred for their simplicity. A jammed or broken flowmeter could be an embarrassment when located remote from the nearest service agent or replacement meter.

22.8 MONITORING OF DISCHARGE WATER QUALITY

Often, on sites where there are potential concerns about changes in groundwater quality caused by groundwater pumping, perhaps due to saline intrusion or migration of groundwater contamination (see Section 21.4), it is necessary to monitor certain chemical parameters of the pumped water directly in the field. In practice, the chemistry of the water discharged from site is relevant on all projects, as basic knowledge of water quality is needed by environmental regulatory bodies when assessing permissions for water discharge (Section 25.4). Methods of groundwater sampling during site investigations are discussed in Section 3.7 and Appendix 4.

On simple projects, there may be little water quality monitoring carried out during dewatering pumping. However, on larger projects, or where contamination may be encountered, groundwater quality may be monitored routinely. Water samples may be taken from the discharge flow rate and sent for testing in an off-site laboratory. The results of these tests may not be available for days or weeks after sampling, so periodic sampling and testing are often supplemented by field monitoring.

Field monitoring of water quality can be carried out manually by diverting a portion of the flow to a sampling container and then using hand-held devices. Monitoring can also be done via sensors installed into tappings in the discharge pipework (known as in-line monitoring), linked to dataloggers to allow very frequent readings to be taken. Two common aspects of water quality monitoring are:

  i. Monitoring of well head water chemistry. Section 3.7.2 describes the method of determining water quality at the well head. Well head chemistry parameters include specific conductivity; water temperature; pH; redox potential; and dissolved oxygen.
  ii. Suspended solids and turbidity. The concentration of suspended solids can be determined from water samples taken from the discharge stream using an Imhoff cone or by manual readings from a Rossum sand sampler connected to the discharge pipe (these methods are described in Section 16.7). In-line monitoring of suspended solids is harder to achieve. An approach sometimes used is to deploy an in-line turbidity meter linked to a datalogger. Turbidity is an optical measure of water clarity (measured in Nephelometric Turbidity Units [NTU]) and is obviously related to the concentration of suspended solids. However, because turbidity values are affected by the size, shape and reflectivity of the solid matter in the water, there is no generic relationship between the concentration of suspended solids and turbidity. The value of turbidity records from in-line meters is that they can provide data on trends of suspended solids. If the turbidity increases notably, this may indicate an increase in suspended solids, and further testing can be carried out by manual methods.

22.9 MONITORING OF SURFACE SETTLEMENTS

It is well known that groundwater control works can cause ground settlements due to increases in effective stress (Section 21.4.4) or loss of fines (Section 21.4.3). On the vast majority of projects, these settlements are so small that no effort is made to monitor them. However, on large or complex projects, or if the designers consider that significant impacts are possible, a settlement monitoring programme may be implemented. Settlement monitoring is based on the principle of recording the level of defined points on the ground relative to a stable datum. The level is recorded at various times, and the difference in level between readings is reported as settlement (if the monitoring point has moved down) or heave (if movement is upwards). Two main methods are used for such monitoring:

• Conventional surveying techniques
• Remote sensing techniques

Whatever method is used, it is important to obtain reliable baseline information on ground levels before dewatering starts. A commonly used approach is to carry out three baseline level surveys over a few weeks and average the results to give baseline values.

22.9.1 Conventional Surveying Techniques

Subject to the accuracy that can be achieved, conventional surveying techniques (optical or Global Positioning System [GPS]) can be used to record ground levels at defined locations. A suitable benchmark must be used as the level datum for the settlement monitoring. This benchmark is a point or feature considered unlikely to be affected by any dewatering-related settlement, against which the level of the monitoring points is surveyed. Because the zone of influence of a groundwater control system can extend for several hundred metres from the pumping locations, this datum may need to be distant from the dewatering location.

Settlement monitoring locations can be defined points on existing surface features such as kerbs, pavements, etc. However, such locations may experience vertical movements due to temperature effects or traffic loadings, which can make dewatering-related settlements harder to discern. An alternative approach is to install a vertical steel rod (reinforcing bar is suitable), 2 to 3 metres in length, set below ground in a borehole, with the top of the bar protruding at ground level. The lower section of the bar is surrounded by concrete or grout, and the upper metre or so is surrounded by sand or other weak backfill (to decouple it from surface layers). The top of the bar is the levelling point and is more representative of movements below surface layers.

Whatever type of ground level monitoring point is used, a visual inspection should be undertaken during routine monitoring of the survey network. If this inspection indicates that a ground level monitoring point is damaged or otherwise altered, this should be reported to avoid false indications of ground movements.

22.9.2 Remote Sensing Techniques

Remote sensing techniques are now sufficiently advanced that ground levels can be determined by satellite surveillance with sufficient accuracy to be useful to assess the relatively small-scale settlements associated with groundwater lowering. The principle is the same as conventional surveying, whereby ground level is recorded at various times at specific locations, and the calculated difference in level between readings is reported as settlement or heave. The advantage of remote sensing over conventional methods is that because no physical monitoring points need to be installed, the method can determine the level of a large number of densely distributed points around a construction project.

An example of a remote sensing technique used for this purpose is interferometric synthetic aperture radar (InSAR), as described by Giardina *et al.* (2018). This uses microwave frequency signals from overflying satellites to derive ground levels and hence surface displacements. The technique can gather data from a location by day or night irrespective of any obscuring cloud cover.

22.10 MONITORING OF SURFACE WATER BODIES

Occasionally, there may be concerns regarding the impact of dewatering pumping or groundwater barriers on nearby surface water features such as watercourses, ponds, lakes and wetlands. In these cases, it may be necessary to measure surface water levels or flow

Table 22.1 Observable Parameters for Dewatering Systems

Parameter	Commonly used methods	Comment
Soil/rock stratification	Well borehole log Observation of jetting water returns	Should always be checked against site investigation
Groundwater level in monitoring well or unpumped well	Dipmeter monitoring or pressure transducer datalogging of monitoring wells (standpipe or standpipe piezometer) Dipmeter monitoring or pressure transducer datalogging of unpumped well	Generally essential, should be monitored daily
Water level in pumped well	Dipmeter monitoring of dip tube in well	Useful check on well performance
Flow rate, system total	V-notch weir Flowmeter Volumetric measurement	Generally important, should be monitored daily
Flow rate, individual well	V-notch weir Flowmeter Volumetric measurement	Important for well and ejector systems only, monitoring intervals 1–6 months
Damage and obstruction to system	Visual inspection	Important to avoid unnecessary interruptions in system operation
Mechanical performance	Vacuum (wellpoints) Supply pressure (ejectors) Discharge back pressure Engine speed and fuel consumption (diesel-driven pumps) Voltage, current and energy usage (electrically driven pumps) Power supply alarms	Important for maintenance and/or monitoring continuous running, but data required only if flow rate or drawdown unsatisfactory
Water quality	On-site testing (pH, specific conductivity, etc.) Off-site testing (laboratory testing)	Necessary to assess clogging potential and environmental impact and for compliance with discharge permissions
Suspended solids	Condition of settlement tank or lagoon, turbidity or suspended solids content of discharge water	Always recommended to check for fines removal
Settlement	Pre-construction building condition survey Level monitoring of selected locations Crack monitoring of selected locations	Sometimes necessary if risk of damaging settlements is significant
Tidal effects	Regular groundwater level monitoring (at 15–60 minute intervals) of drawdown for a minimum period of 24 hours Data from public domain tidal gauges	Provides a useful check on data, detailed analysis of significance is complex
Rainfall	Rain-gauge on site Data from regional weather stations	Can be important for pumping test, or for assessing the impact of dewatering pumping on regional groundwater levels
Barometric pressure	Barometer Barometric pressure sensor Data from regional weather stations	Can be relevant for pumping test or for assessing the impact of dewatering pumping on regional groundwater levels

Adapted from Roberts, T O L and Preene, M, *Géotechnique*, 44, 727–734, 1994.

rates at specific locations. Brassington (2017) describes various monitoring methods that can be used in these circumstances.

If groundwater control works are carried out where there is the potential for tidal influence (see Section 3.6.3), it can be useful to have data on water levels in nearby tidal bodies of water. If there is safe access to the water's edge, a pressure transducer, linked to a datalogger, can be used to record the water level. To protect the transducer and reduce short-term variations due to waves, the transducer should be set in a stilling tube clamped against a jetty or sea wall. If access is not available to install a transducer, it is often possible to source tidal data from public domain sources or the local port authorities.

22.11 OTHER PARAMETERS THAT MAY BE MONITORED

In addition to groundwater levels and flow rate, more complex projects may require other parameters to be recorded. Table 22.1 is a comprehensive list of parameters that can be observed; it is unlikely that the full list would be monitored on a single project. Increasingly in the future, there may be a focus on energy consumption (e.g. diesel fuel usage and electricity consumption) and the associated CO_2 emissions, but at the time of writing, this is often not addressed directly by designers.

An often neglected element of monitoring is a simple visual inspection of the condition of the groundwater control system. Examples of conditions that should be reported during a site inspection include:

- Wells and pumps that are out of commission or are operating unusually (e.g. a diesel-powered pump with a very smoky exhaust plume).
- Minor leaks from pipework, joints, etc.
- Sediment or washout from slopes that has built up around pipes and other equipment (see Figure 7.12).
- Construction waste and detritus on or near pipework and equipment – examples include discarded sections of falsework and formwork.
- Visible damage or leaks in exposed sections of groundwater cut-off walls.

Observations of these conditions can be very useful in proactive maintenance programmes to try to address minor problems before they become serious issues.

Chapter 23

Maintenance of Groundwater Lowering Systems

Don't spoil the ship for a ha'p'orth of tar.

British proverb

23.1 INTRODUCTION

Any groundwater control system will need some maintenance measures to ensure continued effective operation and to reduce the risk of sudden interruptions in pumping due to equipment failure. This chapter describes common maintenance activities. The problems of corrosion, encrustation and biofouling, which sometimes result in the gradual deterioration of system performance, are also described. Decommissioning measures, to be applied when the groundwater control system is no longer needed, are discussed in Chapter 24.

23.2 THE NEED FOR MAINTENANCE

As has been said elsewhere in this book, any groundwater control scheme is a complex system dependent on a diverse range of hydrogeological, hydraulic, chemical, mechanical and human factors. It is clear that any system operating for any but the shortest period of time will require maintenance to allow it to continue to function. However, maintenance is often considered an unglamorous activity, and there have been occasions when it has been neglected, and operational problems have followed.

There are two dictionary definitions of 'maintenance' that are relevant to groundwater control systems:

1. The first definition is what most people would think of in this context:

   Maintenance – The work of keeping a machine or system in proper condition.

   This maintenance work includes the upkeep of pumps and other equipment and can be divided into preventative maintenance, carried out to avoid something breaking down, and corrective maintenance, to bring something back to working order.
2. The second definition is slightly more subtle:

   Maintenance – the process of preserving a condition or making sure that something continues in the same way or at the same level.

   This is relevant because groundwater control systems create a condition where groundwater levels are lowered and the excavation is stable and workably dry. The important point is that these conditions must be 'maintained'; otherwise, groundwater levels will rise, and the excavation will become unstable – this could occur if pumping is interrupted for an extended period or if a groundwater cut-off wall is breached inadvertently.

An effective maintenance programme must address the practicalities of Objective 1 – to keep equipment running – but the overall aim must be Objective 2 – to maintain the beneficial stabilizing effect of the groundwater control system.

23.3 PRINCIPAL MAINTENANCE REQUIREMENTS

An effective maintenance programme should include the elements necessary to ensure the continued effective and efficient operation of the system. The precise nature of the maintenance programme will vary from project to project but should be focused on preventative measures and may include some or all of the following elements. These are listed in the approximate order of how widely they are applied to groundwater control systems, with the most common listed first.

i. Pump maintenance
ii. Upkeep of pipework
iii. Maintenance of monitoring equipment
iv. Rehabilitation of poorly performing wells
v. Protection against accidental damage
vi. Upkeep of water disposal routes
vii. Upkeep of groundwater barriers
viii. Weather protection

The maintenance regime required on a given project should be developed on a risk basis – i.e. what are the consequences of failures of one or more elements of the system, and what maintenance activities can reduce the probability of events that could have serious consequences? This same risk-based approach should be applied to decisions on the prudent level of standby pump provision (Section 23.4).

23.3.1 Pump Maintenance

Every groundwater control system includes pumping equipment of some sort, and it is essential that mechanical and electrical plant in the system (including pumps and controls) is appropriately maintained as a preventative measure to reduce the risk of sudden failure of pumps. In most cases, groundwater control designs rely on continuous pumping, so any maintenance measures must minimize periods when pumps are stopped for oil changes, etc. Where standby pumps are provided, continuous pumping can be achieved by running the standby pumps while the duty pumps are off for maintenance.

Diesel-powered pumps or generators will require regular fuelling and checking/replenishment of lubricant and coolant levels. Electrically powered equipment typically requires less maintenance but should be tested regularly (see BS 7671:2018). Some pumps and control systems can provide data remotely via mobile data networks, analogous to datalogger systems used for monitoring. Data from such systems can be used to help plan for regular maintenance.

23.3.2 Upkeep of Pipework

Pipework plays a key role in moving water around – suction pipework delivers water to pumps, and discharge pipework carries water away to the disposal point. If set up correctly at the outset, pipework should require little maintenance. The principal requirement is to carry out regular visual inspections in order to:

- Identify and fix any minor leaks before they become major problems
- Remove any sediment or washout from slopes that has built up around pipes and other equipment before it causes excessive stress on the pipework
- Remove any construction waste and detritus on or near pipework and equipment before it restricts access or damages the pipework

23.3.3 Maintenance of Monitoring Equipment

Monitoring equipment (such as monitoring wells, flowmeters, etc) can be very important to the successful operation of groundwater control systems (Chapter 22). Such equipment often does not require much maintenance other than checking that it is functioning and is not at risk of damage or obstruction.

23.3.4 Rehabilitation of Poorly Performing Wells

Dewatering wells of all kinds (wellpoints, deep wells and ejector wells) may suffer from a gradual reduction in performance. This may be caused by clogging due to encrustation (precipitation of chemical compounds) or biofouling (bacterial growth and associated processes). Wells may need to be periodically rehabilitated by physical or chemical cleaning processes (see Section 23.5).

23.3.5 Protection against Accidental Damage

Regular visual inspection of the various elements of the system can provide a very useful indicator of potential problems where equipment may be damaged, buried or obstructed by construction activities. Good communication with site supervisors and operatives can be very useful. If they can be impressed with the value of the dewatering system in providing good working conditions, then they are more likely to work carefully around the equipment, reducing the risk of accidental damage.

23.3.6 Upkeep of Water Disposal Routes

Occasionally, where water from groundwater control systems is discharged into small watercourses (e.g. ditches and streams), some work may be needed to maintain the channel. Examples include clearing excessive growths of vegetation or removing debris in the channel placed there by vandals.

23.3.7 Upkeep of Groundwater Barriers

With the exception of artificial ground freezing, the techniques used to form groundwater barriers described in Chapter 18 do not require maintenance in the physical sense. However, there is a need to prevent any cut-off walls being advertently breached by excavation, piling or drilling. These risks can be reduced by clearly marking the location of the barriers; good communication with site teams; and regular visual inspections of the relevant areas of the site.

23.3.8 Weather Protection

In some climates, the weather may impact the operation of groundwater control systems:

- In very hot conditions, electrical switchgear may overheat and trigger safety cut-outs. To avoid this problem, it may be necessary to provide additional cooling and forced ventilation in control cabins.
- In very cold climates, where temperatures are below freezing for long periods, it may be necessary to 'winterize' pumping and pipework systems. This can include providing thermal insulation and heating (either electrical or steam) around vulnerable equipment such as wellpoint pumps, ejector pumps, deep well headworks and all pipework containing water. Any pipework and equipment that is not operating (and therefore contains static water) will be at risk of freezing and should be drained down as soon as pumping stops.

Even in temperate climates, problems may occur during severe storms, such as lightning strikes (which can cause interruptions in electrical supplies) or heavy rainfall (which can lead to flooding in watercourses that receive dewatering discharge water). It may be prudent to prepare for such issues if bad weather is forecast.

23.4 STANDBY PUMPS AND AUTOMATION

23.4.1 Standby Pumps

In general, groundwater pumping systems should operate continuously 24 hours per day, 7 days per week. Hence, it is imperative that the installed system incorporates facilities to ensure that pumping is indeed continuous apart from very short interruptions in pumping for maintenance purposes. The provision of standby pumps (and/or standby power supplies) should be considered for all groundwater control systems where relatively short interruptions in pumping will cause instability or flooding of the excavation.

The provision of standby pumps will have a cost associated with it, and in the commercial world, there can be pressure to operate systems without standby pumps as a cost-saving measure. In practice, in a small number of cases, it can be acceptable to operate without standby pumps, but in most cases, it is a prudent measure to provide suitable standby pumping capacity. The relevant question before commencing site work is – what are the likely effects of a short-term interruption of pumping? Short-term can mean the time to make an on-site fix for a simple problem (perhaps a few hours) or the time to get new pumps delivered to site and swapped out with the failed pumps (perhaps 24 hours on an urban site, but possibly much longer on a remote site). The decision on standby pumps should be based on the following two criteria for the likely period of 'down time' while pumps are fixed:

(i) If interruption of pumping (in one or more pumps) will cause instability and flooding of the work, then the provision of standby pumping and/or power facilities is essential.
(ii) If cessation of pumping will create only a relatively minor mess that can be cleaned up afterwards at an acceptable cost in terms of both inconvenience and delay, and provided there is no safety risk, a judgement is sometimes made to accept the cost savings by not providing standby pumps while recognizing that a risk is being taken.

Where standby pumps are required, the degree of standby provision typically varies with the pumping technique:

1. Sump pumping systems are often used on shallow excavations for relatively simple projects, and in many cases, standby pumps are not provided. Because sump pumps are simple in design and widely available, contractors sometimes take a view that it will be a quick and easy process to either repair or replace a failed pump.
2. Wellpoint systems typically use a small number of pumps to draw on many wellpoints simultaneously. Wellpoint systems often have standby pumps provided – one standby for every two duty pumps is typical, often arranged in sets of three pumps (two duty and one standby) connected into the header main, ready for immediate start-up (Figure 15.4b).
3. Deep well systems comprise multiple pumps (one per well) and often do not have standby pumps provided. This is acceptable provided that the system design has allowed for a few extra wells over and above the minimum number required. In this way, the failure of one or two submersible pumps will not cause a problem while the pumps are replaced. On remote sites, a few spare borehole electro-submersible pumps are often stored on site to reduce the down time while an individual pump is replaced.
4. Ejector well systems typically use a small number of high-pressure supply pumps to feed water to many wells simultaneously. It is normal to provide standby pumps connected into the supply header main ready for immediate start-up (Figure 17.4).

Where standby pumps are configured for immediate start-up, they should be equipped with suitably placed valves for a swift changeover of operation from the running pump to the standby unit as and when required. This might be necessary in the event of an individual running pump failure or maintenance stoppage to check oil levels, etc. Automated switch over to standby pumps is described in Section 23.4.3. Standby pumps should be tested regularly (at least weekly) by running on load. If possible, it is good practice to use equipment in rotation, whereby one week a unit is run as the duty unit, next week it acts as standby, and so on.

23.4.2 Standby Electrical Power Supplies

Most groundwater control pumping methods – sump pumping, wellpoints, deep wells and ejector wells – can be based on electrically driven pumps. Electrical power can be provided by mains supply (from the local utility provider) or by on-site generators (typically diesel powered). In most developed countries, mains power is very reliable, and the same can be said for a well-maintained modern generator. Interruptions in power supply are rare. However, the consequences of power failure are significant. Typically, all pumps will stop, leading to more severe risk of excavation instability than when a single pump fails. For this reason, standby power supplies are provided for the vast majority of electrically powered systems:

- Systems operated by mains power can be provided with a standby generator.
- Where duty generators provide the power to pumps, separate standby generators can be provided.

The duty/standby power source can be linked to the system via a changeover switch, which can be operated manually or automatically (Section 23.4.3). Standby generators should be tested regularly (at least weekly) by running on load. Where both duty and standby generators are present, it is good practice to use them in rotation, whereby one week a unit is run as the duty unit, next week it acts as standby, and so on.

23.4.3 Alarm Systems and Automation

Electronic equipment is available to allow systems to be alarmed and, to some degree, automated. Sensors, similar to those used in datalogging, can be used to monitor groundwater levels, pump vacuum, ejector supply pressure, problems with the power supply or failure of individual pumps. The output from these sensors can be linked to monitoring equipment, which can send an alarm signal (which could be via a siren, flashing beacon, radio, text message or email) to warn staff of problems and alert them to the need for remedial action. The remedial action might typically be to switch over to standby pumps or standby generators and re-start the system.

While this could be done manually by a technician called out to site, the next logical step is to automate this process and allow an automated system to activate the standby plant in response to the alarm signal. This is the basis of automatic mains failure (AMF) systems. Such systems are used on only a minority of groundwater control systems but have been established practice on some of the larger water supply wells since the 1980s. Electronics and computer technology are advancing constantly; it is probable that in the near future, automation and remote monitoring will be applied more widely to groundwater control systems than at present.

23.5 ENCRUSTATION, BIOFOULING AND CORROSION

Wells pumping groundwater for extended periods of time may suffer a reduction in performance (in the form of loss of yield or increase in drawdown in the well) gradually during operation. The loss of performance may result from several processes, including clogging due to encrustation (precipitation of chemical compounds) or biofouling (bacterial growth and associated processes). Additionally, if the groundwater is saline or brackish, corrosion of metal components (such as pumps or pipework) may be of concern.

These effects have long been recognized in water supply wells, which have working lives of several decades. Such wells (and associated pumps and pipework) are typically rehabilitated at periodic intervals to ensure that performance does not reduce to unacceptable levels (see Howsam *et al.*, 1995; Houben, 2001). Gradual reduction in performance is less of a problem for temporary works groundwater lowering systems, largely because of the shorter pumping periods involved. Nevertheless, there have been a number of instances where systems operated for periods between several months and a few years have been affected. An understanding of the factors involved is useful when planning for the maintenance of systems, particularly for permanent or long-term systems (Chapter 20).

Three principal clogging processes occur in and around wells whereby material is rearranged or deposited to plug flow paths and restrict water flow:

- Physical clogging, where particulate matter is rearranged
- Chemical clogging, where mineral compounds derived from dissolved material in the water are deposited
- Bacterial clogging (commonly known as biofouling), where bacterial colonies grow in the well, feeding from dissolved material in the water and excreting a biomass

The focus of the discussion in this chapter is on chemical and bacterial problems. These processes, including iron oxide encrustations, iron bacteria and calcium carbonate, are the most common causes of well performance problems in dewatering systems. Clogging effects can be a problem in wells of all types – wellpoints, deep wells, and ejector wells – but difficulties can also occur due to the encrustation or biofouling of pump internal components, pipework and flowmeters. Some of the observed symptoms of well clogging and encrustation are shown in Table 23.1.

23.5.1 Chemical Encrustation

Groundwater contains, to varying degrees, chemical compounds in solution (see Section 3.7). These compounds may precipitate in and around the well, being deposited as insoluble compounds to form deposits of scale on well screens and pumps. As groundwater enters a well, it experiences a drop in pressure and may be aerated by cascading; these are ideal conditions for precipitation.

The principal indicators of the encrustation of potential of groundwater (from Wilkinson, 1986) are:

  (i) pH greater than 8.0
 (ii) Total hardness greater than 330 mg/l
(iii) Total alkalinity greater than 300 mg/l
 (iv) Iron content greater than 2 mg/l

The most commonly reported chemical deposits are iron oxyhydroxides (sometimes associated with manganese deposits), iron sulphides and calcium carbonates. Carbonate clogging is a different process from iron-related clogging. The natural carbon dioxide dissolved in solution is released, resulting in an increase in water pH. As the pH increases in waters with high levels of calcium carbonates, rapid precipitation of white or pale grey calcareous deposits occur in the well and pump (Figure 23.1).

For temporary works applications, chemical encrustation on its own is rarely severe enough to affect operation – problems normally occur when encrustation is enhanced by bacterial action, as described in subsequent sections.

Table 23.1 Possible Observed Symptoms of Well Clogging

Problem	Observed symptoms
Iron oxide, iron oxyhydroxide and iron bacteria	Red-brown slime inside pipes Reduced specific capacity Cloudy rusty water at pump start-up Slimy deposits blocking main lines and laterals. Smelly and poor quality water
Manganese oxide	Blackish-brown deposits blocking pipes Reduced specific capacity Cloudy water at pump start up. Smelly and poor quality water
Calcium carbonate	Deposits of distinct white scale more layered than iron oxide deposits Reduced specific capacity Iron oxide could also be present indicated by a red-brown coloured calcification. Smelly water if iron-related bacteria or biofouling also present

From Deed, M E R and Preene, M, Managing the clogging of water wells, in *Proceedings of the XVI ECSMFGE, Geotechnical Engineering for Infrastructure and Development* (Winter, M C, Smith, D M, Eldred, P J L and Toll, D G, eds), ICE Publishing, London, pp. 2787–2792, 2015. With permission.

Figure 23.1 Carbonate deposits on well pumping equipment. (From Deed, M E R and Preene, M, Managing the clogging of water wells, in *Proceedings of the XVI ECSMFGE, Geotechnical Engineering for Infrastructure and Development* (Winter M C, Smith D M, Eldred P J L and Toll D G, eds). ICE Publishing, London, pp2787–2792, 2015. With permission.)

When problems are first suspected, visual inspection, either of equipment removed from the well or (via closed-circuit television survey) of the well itself, can help identify the nature of the problem. The colour of the encrustation deposits can give some indication of the type of deposit (see Table 23.1). Such visual inspections should be supported by chemical testing of the encrustation deposits and of the groundwater.

If groundwater analyses suggest that it may be a problem, the wells should be designed to have as low a screen entrance velocity (see Section 13.9) as possible. If encrustation becomes a problem once the system is in operation, acidization or chemical treatment of the wells, followed by airlifting or clearance pumping, may help loosen and remove deposits (see Howsam *et al.*, 1995).

23.5.2 Bacterial Growth and Biofouling

In many groundwater systems, bacteria are naturally present. Even if bacteria are not present, they are invariably introduced into the well via the water added during drilling or by unavoidable contamination on the surfaces of drilling tools and pumping equipment. Problems occur because the wells and pipework forming a groundwater lowering system may offer an environment in which these bacteria can thrive – the water in the well is often aerated, and the water flowing to the well provides the nutrients and chemical ingredients to sustain the growth processes.

The most common, and hence most problematic, bacteria are the iron-related species such as *Gallionella* or *Crenothrix*. They derive the energy they need by oxidizing the soluble ferrous iron (Fe^{2+}) present in the groundwater to an insoluble ferric form (Fe^{3+}). According to Howsam and Tyrrel (1990), these are sessile bacteria; this means that they attach themselves to surfaces. The action of the bacteria will cause a biofilm to develop on a surface. A biofilm consists of not only bacterial cells but also proportionately large volumes of extra-cellular slime. This can trap particulate matter and detritus from the water flowing by and provides an environment for the precipitation of iron and manganese oxides and oxyhydroxides. The most common form of biofilm (also known as biomass) is a thick red-brown gelatinous slime or paste that builds up within the well filters, the well screen and the well bore, and inside pumps and pipework.

The practical result of this is that if conditions are favourable for bacterial growth, a system may become biofouled. The biofilm can be surprisingly tenacious and if not removed or controlled in some way may clog a system, dramatically reducing performance (Figure 23.2). As biofouling deposits build up, the discharge flow rate will decrease due to the restriction in flow and reduction in hydraulic efficiency that result. If no action is taken, the groundwater levels within the excavation will rise, and instability or flooding may result. Regular monitoring of flow rate and water levels is essential for diagnosis of these problems. A programme of periodic rehabilitation of the system may be necessary to ensure continued satisfactory operation.

Howsam and Tyrrel (1990) state that the following conditions may give an increased risk of biofouling:

- The use of iron or mild steel in the system (increasing iron availability)
- Intermittent pumping, large well drawdowns and cascading of water into the well (all increasing oxygenation)
- High flow velocities, such as through well screen slots or at valves (increasing nutrient uptake)

Figure 23.2 Biofouling of submersible pump due to iron-related bacteria. The submersible pump and riser pipe shown have been removed from a dewatering well after several weeks of operation. The upper part of the pump is coated with a thick red-brown biofilm slime.

Unfortunately, many of these conditions are almost unavoidable in groundwater lowering systems! Therefore, biofouling should be considered as an operational risk for most systems.

Consider the conditions needed for the growth of iron-related bacteria (Howsam and Tyrrel, 1990):

1. Nutrients. Bacteria in general need carbon, nitrogen, phosphorus and sulphur. Many natural groundwaters contain sufficient levels of these nutrients to sustain significant biofilm growth.
2. The presence of an aerobic/anaerobic interface. This is typically formed by a well, where anaerobic aquifer water can come into contact with oxygen. This interface is the point at which aerobic bacterial activity and chemical iron oxidation are initiated.
3. The presence of iron in the groundwater. The presence of dissolved iron is necessary because the oxidation of the iron provides energy for the bacteria's metabolism. This action precipitates the insoluble iron compounds, giving the characteristic red-brown colour to the biofilm.
4. Water flow. The flow of water transports nutrients to the biofilm. The faster the flow of water, the more food is available for bacterial growth, and the faster the biofilm will grow. This is a key point, because it means that high-flow wells may biofoul more quickly than low-flow wells.

Conditions 1 and 2 will be present for most systems, so the likelihood of biofouling should be assessed from conditions 3 and 4, dissolved iron and rate of flow. Based on a number of case histories, Powrie *et al.* (1990) produced some guidance in terms of those two factors (Table 23.2).

Table 23.2 Tentative Trigger Levels for Susceptibility to *Gallionella* Biofouling

Pumping technique	Susceptibility to biofouling	Concentration of iron in groundwater (mg/l)	Frequency of cleaning
Wellpoints	Low	<10	Biofouling unlikely to present difficulties under normal operating conditions and times of less than 12 months
		>10	Biofouling may be a problem for long-term systems
Submersible pumps	Moderate	<5	6–12 months
		5–10	0.5–1 month
		>10	Weekly (system may not be viable)
Ejector wells (low flow rate; <10 l/min)	Moderate	<5	6–12 months
		5–10	Monthly
		10–15	Weekly (system may not be viable)
Ejector wells (high flow rate; >20 l/min)	High	<2	6–12 months
		2–5	Monthly
		5–10	Weekly (system may not be viable)
Recharge wells	Very high	Recharge wells are extremely prone to biofouling, which is likely to occur even if iron concentrations are below 0.5 mg/l	
		To minimize biomass growth and encrustation, extreme care should be taken to avoid aerating the recharge water	
		It is not uncommon for recharge wells to require cleaning on a weekly or monthly basis	
		Recharge wells may not be viable at high iron concentrations	

Modified after Powrie, W, et al., Biofouling of site dewatering systems, in *Microbiology in Civil Engineering* (Howsam, P, ed.), Spon, London, pp. 341–352, 1990.

Table 23.2 shows that the risk of biofouling problems is also dependent on the type of system. Wellpoint systems (where much of the pipework is under vacuum) do not provide a good environment for the growth of aerobic bacteria and are not especially prone to biofouling. Deep well systems using submersible pumps are prone to biomass building up on the outside of the pump and riser pipe, inside the pump chambers and inside the riser pipe. Because the riser pipe is normally of relatively large diameter, clogging of the riser pipe is only a problem in severe cases; biofouling inside the pump is more of a problem, often leading to pump damage if not removed by cleaning. Ejector well systems are particularly susceptible to clogging from biofouling; this is because the smaller-diameter pipework generally used is easily blocked by biomass. Artificial recharge systems (Section 17.13) are extremely vulnerable to biofouling – an example is described in Case History 27.9.

23.5.3 Maintenance Strategies to Manage Clogging and Encrustation

If clogging or encrustation occurs, and significant loss of performance has been identified by regular monitoring of groundwater levels and discharge flow rates, the system may need to be rehabilitated. Possible methods include:

(i) Cleaning of pumps and pipework. Submersible pumps, ejectors and risers can be removed from wells and cleaned at ground level by jet washing or scrubbing. Pumps and ejectors may need to be dis-assembled to allow cleaning or replacement of internal components. Riser, header and discharge pipes may be cleaned by jetting or the use of pipe cleaning moles.

(ii) Physical treatment of wells by agitation, scrubbing or flushing. Biomass may be removed from the wells by development techniques (see Section 16.7) that surge and agitate the well, thereby loosening the clogging material. Airlift surging and pumping, water jetting (using fresh water or a chlorine solution) and scrubbing (using a tight-fitting wire brush hauled up and down inside the well) have all proved effective (see Case History 27.9).

(iii) Chemical treatment of wells and pipework. In recent years, the use of chemical treatments to remove encrustation and to reduce the rate of subsequent re-growth has become widespread. Typically, this involves dosing the wells and rest of the system (pipework, etc) with specialist chemical solutions to kill the bacteria, and then using method (i) or (ii) to remove the residue of biomass. Traditionally, many of the chemical treatments were chlorine based. However, well cleaning products (typically based on organic acids) are now on the market that are highly effective yet are biodegradable and relatively environmentally friendly (Deed, 2009; Deed and Preene, 2015). Some of these products are approved for use in drinking water wells. Chemical treatment is a specialist technique and should be carried out with care and due consideration for the handling and disposal of chemicals and effluents to reduce the health risks to the workforce and to minimize any environmental impacts. Some available chemical treatments are described in Table 23.3.

Systems operating for long periods may need rehabilitation several times during their working lives. There is some evidence that regular cleaning on a 'preventative' basis, i.e. before the clogging and encrustation become very severe, is more effective than cleaning regimes that wait until the clogging reaches problematic levels before cleaning is carried out.

Table 23.3 Chemical Treatments Used in Well Rehabilitation, Adapted from US Army Corps of Engineers 2000

Chemical	Advantages	Disadvantages
Hydrochloric Acid (also known as Muriatic Acid) (HCl)	Effective against a range of mineral deposits and highly effective at removing scale. Widely used in groundwater well rehabilitation.	Corrosive to most metals, particularly stainless steel because of chloride content. Not effective against iron biofouling. Produces toxic fumes, requires careful handling, purity levels needed be defined before handling, lowers pH levels.
Sulfamic Acid (H_2NSO_3H)	Strong acid, which reacts very quickly against carbonate scales. Powder form should be dissolved in water before adding. Safer to handle than muriatic acid.	Not effective against iron or manganese deposits. More effective as a combination chemical treatment against biofouling or metal oxides.
Phosphoric Acid (H_3PO_4)	Less corrosive than hydrochloric acid but slower acting. Effective against iron and manganese deposits.	Requires careful handling. Leaves phosphates behind, which can provide nutrients for microbial growth.
Sodium hypochlorite (NaOCl)	Liquid product. Good disinfectant capabilities. Effective at oxidizing and killing bacteria.	Not effective against mineral deposits. Short shelf life. Can increase the redox potential of the aquifer.
Acetic Acid (CH_3COOOH)	Effective biocide and biofilm dispersing acid. Relatively safe to handle.	Glacial acetic is very corrosive to the skin and produces a pungent vapour that can cause mild to severe lung damage.
Oxalic Acid $(COOH)_2$	Strong, reducing acid and is excellent against iron and manganese oxide. Biodegradable. As a combination chemical, works with even greater power.	Salts of the acid are poisonous, but during a treatment converts to inert elements, with any residues easily removed from water body.

From Deed, M E R and Preene, M, Managing the clogging of water wells, in *Proceedings of the XVI ECSMFGE, Geotechnical Engineering for Infrastructure and Development* (Winter, M C, Smith, D M, Eldred, P J L and Toll, D G, eds), ICE Publishing, London, pp. 2787–2792, 2015. With permission.

23.5.4 Corrosion

Metal components in the system (such as pumps, pipework, flowmeters, steel well casings, etc.) may be subject to corrosion. Corrosion is a relatively minor problem in water supply wells and is rarely severe in the short term, as the abstracted groundwater is usually relatively fresh (i.e. low in chlorides). Groundwater lowering systems, on the other hand, may sometimes pump water that is significantly brackish or saline (see Section 3.7) – this is likely in coastal areas or aquifers where saline intrusion has occurred, or in very arid climates (such as the Middle East) where groundwater can be highly mineralized with a very high salinity.

Systems pumping brackish or saline groundwater may be subject to very severe corrosion during even a few months of pumping. Figure 23.3 shows corrosion of submersible pumps (made from grade 304 stainless steel) after between 6 and 12 months' operation at a site with highly saline groundwater. This is an extreme example, but it does show that corrosion may occasionally be a problem.

The indicators of the corrosion potential of groundwater (from Wilkinson, 1986) are:

(i)   pH less than 7.0
(ii) Dissolved oxygen (which will accelerate corrosion even in slightly alkaline waters)
(iii) The presence of hydrogen sulphide
(iv) Carbon dioxide exceeding 50 mg/l

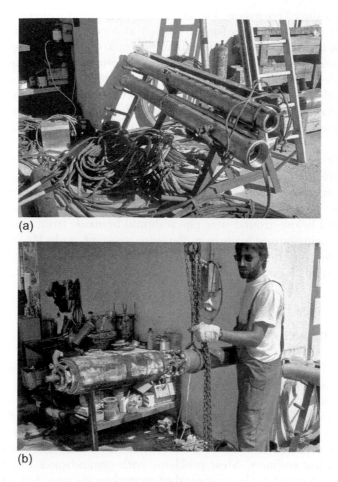

(a)

(b)

Figure 23.3 Extreme corrosion of submersible pumps in highly saline groundwater. (a) Pumps in pristine condition prior to use. (b) Heavily corroded pumps. Note that the previously shiny stainless steel pumps are now blackened and badly pitted.

   (v) Total dissolved solids (TDS) exceeding 1000 mg/l
  (vi) Chlorides exceeding 300 mg/l
 (vii) High temperatures

It should be recognized that these indicators are a guide only. Corrosion of groundwater abstraction systems is complex and may be influenced by other factors, including:

1. Electrochemical corrosion. This is when corrosion is facilitated by the flow of an electric current. Two conditions are necessary: water containing enough dissolved solids to act as a conducting fluid (electrolyte); and a difference in electrical potential on metal surfaces. Differences in potential can occur on the same metal surface in heat- or stress-affected areas where the metal has been worked (such as threads or bolt holes) or at breaks in surface coatings such as paint. Such conditions allow both a cathode and an anode to develop; metal is removed from the anode. Bimetallic corrosion occurs when two different metals are in contact and immersed in an electrolyte. This can affect pumps made from different metals; the more susceptible metal will corrode preferentially and may be severely affected.

2. Microbially induced corrosion. The growth of a biofilm (formed by iron-related aerobic bacteria) on the surface of metal pumps and pipework can allow an anaerobic environment to form below the film. If sulphates are present in the groundwater, the anaerobic condition can allow the growth of sulphate-reducing bacteria. These produce sulphides, which are very corrosive to cast iron and stainless steel. Even though stainless steels are normally very resistant to corrosion, they become susceptible to attack by chloride ions under such conditions. The passive oxide layer, which normally prevents corrosion, cannot re-form in the anaerobic zone beneath the biofilm.

If corrosive conditions are anticipated, the materials used in the system should be chosen with care. Mild steel (as is commonly used for pipework) is very susceptible to corrosion, stainless steel (from which many submersible pumps are made) is less so, and plastics are inert to corrosion. To reduce corrosion risk, as much pipework as possible (above-ground discharge pipes and below-ground riser pipes) should be made from plastic. Consideration should be given to using pumps made from the more resistant grade 316 stainless steel rather than the more common grade 304. If microbially induced corrosion is suspected, regular cleaning of the system to remove the biofilm will remove the environment for this form of corrosion.

23.6 FAULT FINDING AND PROBLEM SOLVING

As was discussed in Section 23.2, the objective of maintenance should be to ensure the continuation of the benefits of groundwater lowering in order to give stable and workably dry conditions within the excavation. However, in the real world, there will be times when a groundwater lowering scheme does not adequately achieve its desired aims when switched on, or if it does work effectively at first, its performance may deteriorate with time. Identifying system problems and figuring out how to put them right is a vital skill for any practical groundwater engineer. Most problems with groundwater lowering systems will probably have a mechanical, hydrogeological or geotechnical cause, but it is worth thinking about all the factors that can have an effect:

 (i) The ground. The nature of the soil or rock beneath the site will clearly affect the way a groundwater lowering system performs.
 (ii) The groundwater. The water chemistry can lead to corrosion, clogging and encrustation of wells and pipework, causing a reduction in the effectiveness of a groundwater lowering system.
 (iii) The pumps. Inappropriate or poorly performing pumps may cause problems.
 (iv) The power supply. Every pump needs a power supply, generally diesel or electric. Unreliable or underpowered units may be the cause of difficulties.
 (v) Control systems. There have been cases where malfunctioning control systems have caused otherwise serviceable pumps to stop or perform significantly below their rated capacity.
 (vi) The pipework. If the water cannot get to or from the pump, a system will perform poorly. This can occur when valves are jammed in the wrong setting (either open or closed, depending on conditions).
 (vii) The environment. External factors, such as changes in groundwater levels, extreme weather, etc., may have an influence.
 (viii) Human factors. Systems do not generally run themselves; the way they are operated will affect their performance.

As with any daunting problem, the secret is to identify the possible causes and eliminate the irrelevant ones until you are left with those directly linked to your woes. It requires a logical approach and might even be enjoyable if it weren't invariably carried out under pressure to solve the problem as soon as possible and allow the excavation to proceed.

Typical operational problems requiring correction include:

During initial period of running:

High-flow problem. The flow rate is greater than the pump capacity, and the target drawdown is not achieved.

Low-flow problem. The flow rate is lower than the pump capacity, and the target drawdown is not achieved.

Lack of dry conditions in excavation, when monitoring of groundwater levels indicates the target drawdown has been achieved.

After extended running:

Sudden loss of performance. The system operates satisfactorily, but after some time, the flow rate or drawdown changes suddenly.

Gradual loss of performance. The system operates satisfactorily at first, but performance deteriorates gradually with time.

To diagnose any significant problems, the raw material is monitoring data of the sort described in Chapter 22. If a system has not been adequately monitored, the first stage of any problem-solving process will involve gathering data about the system performance. This will take time, probably taxing the patience of all concerned – neglecting regular monitoring of groundwater lowering systems is a false economy in terms of both time and money.

Chapter 24

Decommissioning of Groundwater Control Systems

When you drink the water, remember the spring.

Chinese Proverb

24.1 INTRODUCTION

Groundwater control systems are designed with the specific intent of manipulating or influencing groundwater flow in order to allow construction or engineering work below groundwater level. However, once construction is complete, it is possible that wells or cut-off walls may continue to affect groundwater conditions. Furthermore, such wells or groundwater barriers, if left in place, could act as hazards or impediments to future construction projects. Therefore, decommissioning measures may be required at the end of a project. This chapter describes the background to decommissioning and describes commonly used methods.

24.2 THE NEED FOR DECOMMISSIONING

If left in place without the appropriate decommissioning measures, the elements of a groundwater control system have the potential to cause long-term impacts or disturbance of the groundwater regime in the area as well as presenting a physical hazard. Decommissioning should be designed and implemented on a risk-based basis: there may be cases where only very limited works are needed, but on other, more sensitive sites, extensive decommissioning may be required.

The pre-requisite for decommissioning is that groundwater control is no longer required. Commonly, at this stage, pumping is stopped and groundwater levels allowed to recover. Usually, the stage at which pumping can be discontinued is selected not by the dewatering designer but by the designer of the structure under construction. Dewatering pumping should only be stopped when:

i. The structure is completed sufficiently to have the necessary structural integrity and deadweight (or other resistance to uplift) to adequately balance the buoyancy forces when groundwater levels recover to original levels.
ii. The structural waterproofing measures should be sufficiently complete that the structure does not experience problematic leakage or inflow when groundwater levels recover.

In many hydrogeological settings, groundwater levels will recover only slowly after pumping is stopped. Occasionally, the rapid recovery of groundwater levels may cause problems (see Section 21.5). In some circumstances, dewatering pumping is discontinued gradually by turning off wells or pumps in stages, spread over several days or weeks, to control the rate of groundwater level recovery.

24.2.1 Hazards from Dewatering Wells

At the end of the groundwater control phase, it is normal for all the above-ground equipment (surface pumps, header pipe, etc.) and below-ground equipment (submersible pumps, riser pipe, etc.) to be removed. After the equipment is removed, the well itself can potentially present a range of hazards, including:

   (i) The risk of having an open borehole, which is a safety hazard.
  (ii) The well acting as a conduit for surface water (which may be contaminated) to reach the aquifer (this is especially important for wells which penetrate aquifers used for public supply; see Section 21.6).
 (iii) Uncontrolled flow of groundwater between strata penetrated by the well, which can lead to the mixing of groundwaters of variable chemistry from different aquifers. This is a particular concern if the near-surface zones may be contaminated, where there is a risk that pollutants may migrate downwards via the pathway provided by the well (see Section 21.6).
  (iv) Uncontrolled artesian overflow at ground level where a well penetrates a flowing artesian aquifer (see Section 3.4.2) or where the post-construction well head is located within an excavation or structure, below original groundwater level.
   (v) Hazards to future excavation works at the site. These include the possibility of increased groundwater inflows to future excavations on site if unsealed wells provide an artificial pathway for upward vertical seepage (see Section 8.6.2). This is a particular concern if confined aquifers may be present below the excavation.
  (vi) Hazards to future tunnelling works at the site. These include:
      a. The presence of well liners and well screen (of steel or plastic) causing problems for excavation by tunnel boring machines (TBM).
      b. Unsealed wells in the vicinity of closed face TBM or compressed air tunnel drives that can act as vertical conduits, allowing the ejection of fluid (slurry/paste/compressed air) at ground level and a resultant loss of face pressure. An example of loss of compressed air via a piezometer borehole is given in Section 10.4.1.
      c. Unsealed wells in the vicinity of open face TBM tunnels driven in free air acting as artificial pathways for vertical groundwater flow, locally increasing inflow to the tunnel.

Therefore, on completion of the groundwater lowering works, it is often necessary to carry out decommissioning works, in addition to removal of pumping equipment, to mitigate these risks.

24.2.2 Hazards from Cut-Off Walls

Very low-permeability cut-off walls or zones of ground treatment are widely used as part of groundwater exclusion strategies (cut-off methods are described in Chapter 18). With two principal exceptions (steel sheet-piles and artificial ground freezing), the cut-off methods provide a permanent barrier to groundwater flow, which will remain in place after construction is completed. The hazards associated with cut-off walls include:

(i) Cut-off walls can act as a barrier to natural groundwater flow across the site, causing groundwater levels to increase and decrease on the upgradient and downgradient side, respectively.

(ii) Where a site is completely enclosed within cut-off walls that toe into very low-permeability strata below the site, there is a risk that groundwater levels on site will rise due to infiltration from surface precipitation being unable to dissipate into the wider groundwater regime (this is discussed in Case Histories 27.10 and 27.11).

(iii) Cut-off walls can act as an obstruction to future excavation or tunnelling works at the site.

On some occasions, it will be necessary to carry out decommissioning measures on groundwater cut-off walls.

24.3 DECOMMISSIONING OF WELLS EXTERNAL TO STRUCTURES

Where wells and sumps are located outside the structure being constructed, the principal decommissioning measure is backfilling and sealing to fill any voids and minimize the potential for groundwater pathways. In some cases, wells can be capped off and used as monitoring wells. If the wells pass through an engineered structure, additional decommissioning measures may be required, as discussed in Section 24.4.

24.3.1 Capping of Dewatering Wells

The simplest form of decommissioning is to fit a secure, vandal-proof cap or headworks to the well (Figure 24.1a). The objectives of the cap can include:

a) Prevent the open well from being a physical hazard.

b) Preserve the well to allow it to be used in the future as a dewatering or monitoring well.

c) Prevent surface water from flowing down the well.

d) Prevent any contamination or debris being dropped (either accidentally or maliciously) down the well.

Capping of a well without backfilling is only an acceptable option when the well design and sequence of strata mean that the potential impacts in Section 24.2.1 are not assessed to be problematic.

24.3.2 Conversion of Dewatering Well to Monitoring Well

Although a capped dewatering well can be used as a monitoring well, such wells tend to have a relatively large diameter, which means that they may respond only slowly to changes in external groundwater level (the response time of piezometers is discussed in Section 11.7.6.2). To give a better response, a smaller-diameter well screen (typically 50 mm diameter) can be installed inside the original well screen, surrounded by filter gravel over the relevant sections (Figure 24.1b). The well is then fitted with a secure headworks.

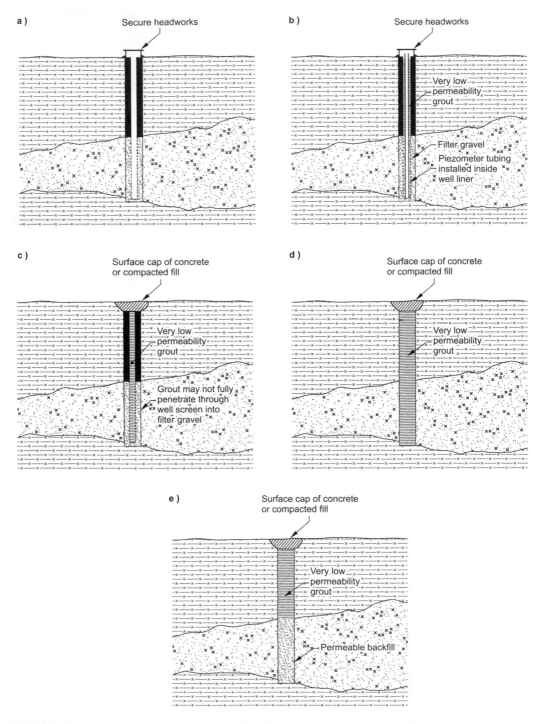

Figure 24.1 Decommissioning of dewatering wells. (a) Capping of dewatering well with secure vandal-proof headworks. (b) Conversion of dewatering well to monitoring well. (c) Borehole backfilled with very low-permeability grout, screen and casing left in place. (d) Borehole backfilled with very low-permeability grout, screen and casing removed. (e) Borehole backfilled with materials to mimic geology.

24.3.3 Backfilling and Sealing of Wells

If a dewatering well (or a relief well that has been installed with a well screen and casing) is not required for monitoring purposes, it is common practice to seal it by backfilling with appropriate materials. The well casing and screen are normally left in place but are cut off just below surface reinstatement level. A key principle of well decommissioning is that the backfill materials must be clean, inert and non-polluting. Suitable materials include clean sands and gravels, bentonite pellets or cement-bentonite grout.

It may be necessary to obtain agreement from the environmental regulatory authorities that the sealing methods proposed are adequate. In England, the Environment Agency (2012) has produced guidelines on best practice that include various options depending on site conditions. Typical well sealing arrangements include:

i. Borehole backfilled with very low-permeability grout, screen and casing left in place (Figure 24.1c). Typically, cement-bentonite grouts are used. The addition of a small percentage of bentonite reduces potential shrinkage and improves the sealing properties of the grout.

ii. Borehole backfilled with very low-permeability grout, screen and casing removed (Figure 24.1d). This approach is sometimes used when very high standards of sealing are required (e.g. on very contaminated sites) and there is concern that the backfill grout may not fully penetrate into the filter gravel around the well screen. In this method, a drill rig is used to mill out the well screen and casing and flush out as much of the filter gravel as possible. This can be slow and difficult work – steel well casing is very strong and can be drilled out only very slowly. Conversely, plastic casings are sufficiently flexible that they are also difficult to break up and remove with the drill bit.

iii. Borehole backfilled with materials to mimic geology (Figure 24.1e). Typically, this involves the well screen sections being backfilled with permeable material (such as sand and gravel) and the sections of well casing being backfilled with very low-permeability material such as bentonite pellets or cement-bentonite grout.

If wells experience flowing artesian conditions (where water from a confined aquifer overflows naturally at the well head), well decommissioning is more difficult and should be planned and executed by experienced personnel. The methods of decommissioning that can be applied will be controlled by the well design, including the type and depth of near-surface casing. It is important to realize that a well in the base of an excavation or structure can experience the same conditions (water overflowing from the well head) as a 'truly' artesian well, and will present the same difficulties in decommissioning – some of these issues are discussed in Section 24.4.

24.3.4 Sealing of Shallow Wells or Wellpoints

In some cases, particularly shallow wellpoint systems penetrating one aquifer only on uncontaminated sites, shallow wells may be left unsealed with the wellpoint riser simply cut off below surface reinstatement level. In contrast, because of their greater depth, almost all deep wells and many ejector wells will need to be sealed.

24.3.5 Sealing of Sumps

Sumps are generally relatively shallow and are commonly decommissioned by backfilling with clean, inert and non-polluting backfill materials. Where sumps have been constructed below the proposed structure, the backfill materials and degree of compaction should be of a standard suitable for founding of the structure.

24.4 DECOMMISSIONING OF WELLS AND SUMPS LOCATED WITHIN STRUCTURES

Dewatering wells, relief wells and sumps are sometimes located within the footprint of the structure being constructed. In many cases, the wells or sumps will need to remain in operation until construction has advanced sufficiently far that the structure can resist buoyancy forces when groundwater levels recover to original levels. Typically, this requires that the wells or sumps pass through the reinforced concrete base slab of the structure to allow pumping to continue even after the slab is cast. The detail where the well/sump penetrates the slab should allow subsequent decommissioning and should not create a vertical pathway for seepage through the slab or otherwise compromise the water-resisting properties of the structure.

A commonly used arrangement is shown in Figure 24.2. Vertical steel ducts are set inside the reinforcement cage of the slab with waterproofing measures fitted to the outside of the ducts to reduce the risk of creating a pathway through the slab. The ducts are then set in

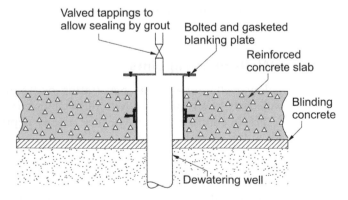

Figure 24.2 Example detail for sealing a well through the base slab of a structure. Similar details can be used where sumps pass through slabs. (a) During pumping. (b) When sealed after pumping stops.

concrete. An example of this type of construction is shown in Figure 17.20, where four relief wells inside a shaft are shown passing through a base slab under construction.

The arrangement shown in Figure 24.2 comprises several elements, including:

  i. The existing well liner or sump structure
  ii. A steel duct (large enough to fit over existing well liner or sump pipe) with a flange at the upper end
  iii. A proprietary waterproofing system (sometimes called a puddle flange) attached to the outside of the steel duct
  iv. Blanking plate, gasket and bolts

The arrangement in Figure 24.2 leaves the flanged top of the duct standing proud of the slab. If the slab is sufficiently thick, it may be possible to set the duct in a box out below top of slab level (see Figure 17.19c).

A common detail for the waterproofing around the steel duct is a proprietary puddle flange, comprising a flexible seal of EPDM (ethylene propylene diene monomer) rubber, clamped to and extending out from the conductor pipe, to act as a flexible water stop in the event of concrete shrinking from the conductor pipe. The sealing system should be rated for the full upward groundwater pressure that will exist when groundwater levels recover fully. In deep structures where high groundwater pressures act on the slab, it may be necessary to have shear key or shear bar extensions (outwards from the steel duct into the concrete) to reduce the risk of the conductor pipe moving upwards under the effect of groundwater pressures.

The ducting of the wells of sumps through the slab allows them to continue to operate until the structural designer indicates that the concrete structures have sufficient strength and integrity to allow pumping to be stopped. At that point, the wells can be capped off. The flange of the steel duct is below the original groundwater level, so as water levels recover, the duct will begin to overflow. In relatively low-permeability materials and in shallow structures, this can be done by quickly removing the well or sump equipment and rapidly capping the top of the conductor pipe with the blanking flange, bolts and gaskets. Where high-permeability aquifers are present and/or the potential groundwater heads above flange level are high, capping of the ducts may be difficult due to the rapid onset of overflowing conditions soon after the pumps are stopped. One option is, during the pumping period, to attach a large-diameter valve to the top of the duct and pass the pumping equipment through the valve. When pumping is stopped, the equipment can be removed rapidly by a crane or hoist, while the well overflows, with temporary pumps at slab level used to remove the excess water. The valve can then be closed to seal off the flow. If the wells or sumps are to be decommissioned, this will require pressure grouting through the duct to inject grout backfill against the groundwater pressure.

24.5 BREACHING OF CUT-OFF WALLS

Where very low-permeability groundwater cut-off walls are used, the deliberate intent is to block groundwater flow around the excavation as part of an exclusion strategy. Cut-off walls can comprise physical structures, such as diaphragm walls, or zones of treated ground formed by techniques such as grouting or artificial ground freezing. In general, the barrier effect of the cut-off wall will remain after construction is complete. As discussed in Section 21.7, this can lead to long-term changes in groundwater levels and flows.

In practice, the changes in groundwater levels and flow due to the barrier effect of groundwater cut-off walls are often very small, and such walls are rarely decommissioned in any formal manner. Occasionally, cut-off walls or grout curtains are decommissioned. One approach is to physically breach the barrier over part of its length by digging out with an excavator and replacing with permeable backfill. To be effective, the wall needs to be breached down to a level below typical seasonal low groundwater levels. An example is described in Case History 27.10.

Two groundwater exclusion methods that can be applied in a temporary fashion are steel sheet-piles (Section 18.5) and artificial ground freezing (Section 18.12).

- Steel sheet-piles can be extracted at the end of construction. This is sometimes done for the whole length of the wall (often with the intention of re-using the piles on another project) or just over a limited section of wall (to reduce any long-term barrier effect).
- Artificial ground freezing. The freezewall will gradually dissipate in the weeks after circulation of refrigerant is discontinued. No permanent barrier effect should remain.

Chapter 25

Safety, Contracts and Environmental Regulation

> Quality is never an accident; it is always the result of intelligent effort.
>
> John Ruskin

25.1 INTRODUCTION

Groundwater control is the means to an end – the excavation and construction of below-ground works in stable and workable conditions. It will be one of many diverse activities carried out on a construction site during the life of a project. As such, dewatering must:

  (i) Be carried out safely and in such a way so as to minimize (as far as is reasonably practicable) the risk of harm to the workers or others who might be affected by the works
 (ii) Have an appropriate contractual arrangement, so that adequate information is available to allow design and construction, and to ensure that all parties are aware of their rights and responsibilities
(iii) Be executed with a view to minimizing or controlling any adverse environmental impacts and conforming to any specific legal restrictions imposed by environmental regulators.

This chapter presents brief details of the health and safety legislation in force in the United Kingdom and outlines key issues relevant to dewatering works. Obviously, legislation will vary from country to country, but the regulatory framework in the United Kingdom covers safety matters that are relevant to construction projects internationally. This chapter also outlines typical contractual arrangements used to procure and manage groundwater control works, including the traditional contractor and sub-contractor relationships, as well as alternative forms of contract. Finally, the environmental law in relation to dewatering is discussed, and the legal implications of dewatering works are outlined (potential environmental impacts from groundwater control, are discussed in Chapter 21).

25.2 HEALTH AND SAFETY

In the United Kingdom, it is a legal requirement that all company and site management personnel ensure that the construction operations carried out by their company conform to safe working conditions and noise abatement regulations.

Since the beginning of the twentieth century, there has been concern over accidents and working conditions on building works. In 1904 a Public Enquiry was held, but it was not till 1926 that the first Building Regulations were passed into law. From time to time since then, building and construction regulations have been augmented and strengthened by various specific laws. Currently, the major legislation covering occupational health and safety of construction works includes:

a) *Health and Safety at Work Act etc. 1974*
b) *Management of Health and Safety at Work Regulations 1999*
c) *Construction (Design and Management) Regulations 2015*

These pieces of legislation, which cover the whole construction process, including design, are supported by more detailed legislation covering specific types of activities or risks.

Legislation will change with time, as technology and working practices evolve. This chapter provides only a brief overview of some significant issues. It is essential that the reader checks to see what legislation is current, consults the original regulations and obtains specialist advice. This legislation applies to England and Wales, with Scotland and Northern Ireland having similar legislation. Naturally, this legislation is not applicable in other countries, but the legal duties of the various parties defined in the *Health and Safety at Work Act etc. 1974* are reflected in the legal systems of many other countries.

This section will deal specifically with health and safety issues particular to groundwater control works. Generic guidance on construction health and safety can be found in various publications (see, for example, Health and Safety Executive, 2006).

The remainder of this section on health and safety will describe issues associated with:

*   CDM regulations – which set out a philosophy of eliminating or reducing risk throughout the construction process
*   Working with hazardous substances
*   Use of electricity on site
*   Lifting operations
*   Working at height
*   Work equipment and drilling rigs
*   Safety of excavations
*   Noise
*   Working with wells and boreholes
*   Interaction with third-party tunnels and infrastructure
*   The risk from unexploded ordnance (UXO)

25.2.1 CDM Regulations

The *Construction (Design and Management) Regulations 2015* (known as the CDM regulations) define the legal duties of the various parties to the construction process, including clients, their agents, designers and contractors. In simple terms, the regulations require health and safety to be considered at every stage of a project, from the pre-construction phase (which includes site investigation, design, planning and feasibility studies) through to the construction phase, subsequent maintenance, and ultimate demolition or decommissioning. This is a change in focus from the traditional approach to health and safety, which concentrated on the construction phase and the role of the contractor.

The CDM approach requires designers to consider whether it is reasonably practicable to exclude or reduce potential hazards. The risk to health and safety arising from a particular aspect of the design has to be weighed against the cost of excluding that feature by:

*   Designing to eliminate or reduce foreseeable risks to health and safety
*   Tackling the causes of risk at source

- Reducing and controlling risks by means that will protect everyone (rather than just the individual) and so yield the greatest benefit

The cost is based not just on financial considerations but also on fitness for purpose, environmental impact and 'buildability'. The purpose of the regulations is not only to reduce possible hazards and risks but also to identify and record those which are unavoidable as a warning to subsequent parties in the investigation/design/construction/maintenance/demolition process.

Detailed guidance on the application of the CDM regulations is given in the *Guidance on Regulations* by the Health and Safety Executive (2015) and in CIRIA Report C755 (Ove Arup & Partners, 2015). The application of CDM to the design of groundwater lowering systems is discussed in Preene *et al.* (2016).

In essence, the CDM regulations formalize what a good dewatering designer should be doing as a matter of course. The designer should first consider the full range of options for a design and then, as far as is practicable, design systems that will be effective and can be installed without unnecessary risks to the workers or others – in other words, consider the 'buildability' of a system. The final element is ensuring that the relevant information is made available to others in the current project team and to those parties that will take over or occupy the site in the future.

When considering options, the designer must look at detail – for example, if a line of wellpoints requires the jetting crew to work next to an open deep cofferdam, could the line of wellpoints be relocated to avoid the risk? The designer must also, however, look at the bigger picture. If a sheet-pile cofferdam is dewatered but relies on continued dewatering for stability, is that too risky an option? What if the power supply fails? What if the standby generator fails too? Such risks will need to be assessed against the factors discussed earlier. In this case, the options might be to avoid pumped dewatering altogether and use a system based on groundwater exclusion. A multiple redundant pumping and power supply system may increase the cost but may produce a significant reduction in the risk of catastrophic failure of the excavation. These are the kind of questions that should be part of the design process.

25.2.2 Working with Hazardous Substances

The *Control of Substances Hazardous to Health Regulations 2002* (known as the COSHH regulations) lay down the requirements for the assessment of risks from hazardous substances and for the protection of people exposed to them.

The regulations cover almost all potentially harmful substances (although some substances, such as lead and asbestos, are covered in separate legislation). Suppliers of any chemical or hazardous substance must provide a data sheet to the purchaser, which must contain sufficient information to allow an assessment to be made of the risk to health from the use of the substance. Based on these assessments, employers have to ensure that as far as is reasonably practicable, the exposure of employees to hazardous substances is prevented, or if this is not possible, then controlled. This may involve the use of personal protective equipment (gloves, eye protection, respirators, etc.) appropriate to the hazard.

25.2.3 Use of Electricity on Site

The use of electrical equipment on site is controlled by:

- *Electricity at Work Regulations 1989*
- *Electrical Equipment (Safety) Regulations 2016*

Portable equipment used on construction sites is generally limited to 110 V supply, arranged so that no part of the installation is at more than 55 V (single phase) or 65 V (three phase) to earth. These voltage limits are intended to reduce the severity of injuries if electric shock occurs. Many dewatering systems are electrically powered (particularly deep well and ejector systems) and use 415 V three-phase system in which each phase is at 240 V to earth. In the event of electric shock, injuries will be much more severe than with conventional equipment; unfortunately, fatal incidents do still occur in the construction industry.

Due to the high voltages involved, most dewatering electrical systems should be treated as permanent installations and should comply with the *Institution of Electrical Engineers Wiring Regulations* (which have been issued as BS 7671:2018). Electrical switchgear must be installed, commissioned and regularly tested by an 'authorized person' defined under the regulations.

Many injuries caused by electricity arise from equipment being worked on by inexperienced or unauthorized personnel, or by poor communication between those working on the system. This can lead to personnel working inside control panels or switchgear which they believe to be safe but which are, in fact, live. A significant number of accidents occur in this way. It is imperative that this situation is avoided by good communication, training and management.

25.2.4 Lifting Operations

Lifting operations, in one form or another, are an integral part of most dewatering or groundwater control projects. Some forms of lifting operation may be very obvious, such as the use of a crane to lower pumps or plant down into an excavation, or where an excavator is used to lift a placing tube during wellpoint jetting. But there are many other operations involving lifting where sometimes insufficient attention is given to safe equipment and methods of working. Examples of less high-profile lifting operations include the use of lorry loaders to unload equipment from flatbed trucks or the use of hoists on drilling rigs to lift casings or drill tools. It is essential to realize that all lifting operations are potentially hazardous and must be carried out safely.

Lifting operations are governed by the *Lifting Operations and Lifting Equipment Regulations 1998* (known as the LOLER Regulations). The regulations cover all lifting equipment that might be used on site for lifting or lowering loads, including attachments used for anchoring, fixing or supporting the equipment. Lifting equipment may include cranes, excavators, drilling rigs, forklift trucks and telehandlers, hoists and mobile elevating work platforms, as well as lifting accessories such as chains, slings, eyebolts, etc.

The LOLER Regulations require that:

- Lifting operations are carried out safely and are planned, organized and performed by competent people.
- Lifting equipment used is strong and stable enough for the particular use and is marked to indicate safe working loads.
- Lifting equipment is positioned and installed to minimize any risks.
- Lifting equipment is subject to ongoing thorough examination and, where appropriate, inspection by competent people.

Further information is given in Health and Safety Executive (2014b).

25.2.5 Working at Height

Falls from height are one of the most common causes of serious injuries on construction sites. Working at height is defined as work in any place where a person could be injured by falling from it. In relation to dewatering activities, working at height can include working next to open excavations, standing on the load bed of a delivery truck to unload equipment, or activities where operatives need to adjust or repair pieces of equipment above ground level, such as on the mast of a drilling rig or on top of a groundwater treatment plant.

These activities are governed by the *Work at Height Regulations 2005*. The regulations establish the hierarchy for safe working at height as:

- Avoid work at height where practicable.
- Use work equipment or other measures to prevent falls where working at height cannot be avoided.
- Where the risk of a fall cannot be eliminated, use work equipment or other measures to minimize the distance and consequences of a fall should one occur.

Specialist equipment that may be relevant to working at height includes means of access for work at height, collective fall prevention (e.g. guardrails and working platforms), collective fall arrest (e.g. nets, airbags, etc.) and personal fall protection (e.g. work restraints, fall arrest and rope access).

25.2.6 Work Equipment and Drilling Rigs

Dewatering activities and the associated construction projects involve a wide of work equipment, from large piling and drilling rigs, trucks and plant down to hand tools, ladders and knives. All these types of equipment present potential health and safety hazards and must be used safely.

Work equipment is subject to the *Provision and Use of Work Equipment Regulations 1998* (known as the PUWER Regulations). The regulations require that equipment is:

- Suitable for the intended use
- Safe for use, maintained in a safe condition and, in certain circumstances, inspected to ensure that this remains the case
- Used only by people who have received adequate information, instruction and training
- Accompanied by suitable safety measures, e.g. protective devices and markings

While the PUWER Regulations apply to a wide range of equipment used to install and operate a dewatering system, there are particular issues associated with drilling rigs. Drilling rigs use a lot of kinetic energy in the form of rotating or falling drill tools to break up soil or rock to form boreholes. If the human body interacts in the wrong way with the moving parts of a drilling rig, then serious injury or death can result. There have been cases where drilling crew members have lost fingers as a result of crush injuries caused by heavy drill tools, or where operators have been killed or seriously injured by being entangled in the rotating parts of drilling rigs.

Since the late 1990s, there has been a focus, in line with the risk avoidance and risk reduction ethos of the CDM Regulations (see Section 25.2.1), on modifying the design of drilling equipment to improve safety. Most commonly, this involves the provision of mesh guards or other protective devices around rotating or other dangerous parts of rigs. The guards must fully enclose the rotating parts in any areas where they could come into contact with personnel and must be interlocked with the rig controls to prevent operation of the equipment while the guards are open to allow, for example, addition of further rods to the drill string.

Further information is given in Health and Safety Executive (2014a) in relation to the PUWER Regulations and in British Drilling Association (2015) in relation to safety issues associated with drilling equipment.

25.2.7 Safety of Excavations

A significant number of accidents occur in excavation works, particularly as a result of falls or resulting from collapses of poorly supported excavations. Legal requirements for safety in excavations are contained in the CDM Regulations 2015. A key requirement is that a competent person is required to inspect the excavation at the start of each shift while persons are working in the excavation. Additional inspections are required after any event that may have affected the strength and stability of the excavation. Further details of requirements for safe working are given in Health and Safety Executive (1999). Practical guidance on methods of work in trenches up to 6 m in depth is given in Irvine and Smith (2001).

25.2.8 Noise

Pumps and dewatering plant are normally operational continuously day and night, 7 days a week. During installation, considerable noise will probably be generated by drilling rigs, jetting pumps and so on. Any noise from the plant will be a potential health hazard to those working on site and an annoyance to the public around the site.

The exposure of employees to noise is governed by the *Noise at Work Regulations 2005*, which sets various action levels in relation to daily exposure to peak noise levels; noise levels need to be assessed in the working area. Noise should be reduced at source, but if this is not reasonably practicable, then appropriate hearing protection must be provided and worn.

The effect of noise on the public outside the site is covered by the *Control of Pollution Act 1974*, which gives local authorities statutory powers to set noise limits and allowable working hours.

The selection of dewatering pumping plant has the greatest impact on ambient noise levels:

a) The quietest systems are those electrically driven from mains power. Deep well systems thus powered will be almost silent (since the pumps are below ground), and well-point or ejector systems will only generate a low 'hum' from the rotation of the pump and motor.

b) If mains power is not available, electrical plant should be run from super-silenced diesel-powered generators – modern equipment is very quiet.

c) For wellpoint systems, diesel-powered pumps are sometimes the only practicable option. Pumps with silenced or 'hushed' prime movers should be used whenever possible, not only to reduce ambient noise but also to reduce exposure to operatives. If pumps with unsilenced prime movers are used, noise levels should be reduced by reducing the engine revs as far as possible.

d) Acoustic screens can also be constructed around pumps, but these can cause problems with ventilation and overheating of pumps, which have resulted, in extreme cases, in pumps catching fire. It is far better to start with suitably silenced equipment in the first place.

e) Noise from all items of plant can often be reduced if regular maintenance and servicing of plant are undertaken to try to ensure that it is running smoothly and efficiently.

25.2.9 Working with Wells and Boreholes

Many groundwater control works will involve the construction, commissioning, maintenance and ultimately, decommissioning of wells and boreholes used for dewatering and monitoring purposes. The hazards associated with wells and boreholes can broadly be subdivided into three types:

(i)   Hazards during drilling, relating to the drilling and construction works themselves. These include crushing; penetration or impact injuries caused by equipment breakages or inappropriate handling of casing, drilling tools, etc; and health issues arising from ingestion or dermal contact with soil, groundwater or gases on potentially contaminated sites.

(ii) Hazards during operation. During operation, pumping equipment will be installed in most wells, reducing the risk of falls into all but the largest-diameter wells. However, if any relatively large-diameter wells are left without pumps during part of the construction period, they should be temporarily covered or capped off.

(iii) Hazards following decommissioning. These only arise if the wells are not adequately decommissioned by backfilling or capping off at the end of the project as outlined in Chapter 24. Larger-diameter wells, if left uncovered, clearly present a hazard. Even the smaller-diameter wells (typically of a few hundred millimetres open bore) common on dewatering projects may present a hazard to small children if they can gain access to the site. It is essential that wells be decommissioned responsibly.

Guidance on safety during drilling works is given in publications by the British Drilling Association (2008, 2015). Particular guidance on working with large-diameter wells is given in *Safety in Wells and Boreholes* (Institution of Civil Engineers, 1972).

25.2.10 Interaction with Third-Party Services, Tunnels and Infrastructure

By definition, groundwater control works will involve some activities below ground level, such as the drilling of wells and the installation of cut-off walls. If works are not carefully planned, there is a risk that drilling rigs and other equipment may 'hit' or penetrate any below-ground third-party infrastructure present under the site. Examples include shallow services and utilities (water/sewage/electricity/gas/telecoms/data) or deeper tunnels (road/rail/subway) and mine workings. If such infrastructure is struck, there is a direct safety risk to the drilling crews involved (e.g. if a live electricity cable is hit) and potentially to the wider public (e.g. if a gas main was struck and an explosion resulted). Even if no direct safety hazard results, the damaged infrastructure will probably be put out of use and require repair, possibly inconveniencing many people and resulting in significant costs as well as considerable embarrassment for all involved.

The authors are aware of several examples where shallow utilities have been damaged by the drilling of dewatering wells. This includes one case where a dewatering well was installed by cable percussion drilling rig and passed completely through a foul sewer a few metres below ground level. The drilling crew later reported that they had not noticed anything unusual while drilling down, but when the temporary casing was withdrawn (after the well screen and filter gravel had been installed), the foul smell and floating 'matter' inside the well revealed what had happened!

In this case, the consequences of the incident were not life threatening but resulted in considerable cost and disruption while the sewer was repaired. But the potential consequences of holes drilled below construction sites can be much greater. Although not directly related to groundwater control, a report by the United Kingdom Rail Accident Investigation Branch (2014) reports three serious incidents affecting railway tunnels beneath London between 1986 and 2013.

1. In 1986, a 200 mm diameter investigation borehole penetrated the crown of a London Underground Tube tunnel near Wanstead. The drilling tool protruded into the top of the tunnel itself and was hit by two trains, causing damage to both trains and minor injuries to the driver of the first train.
2. In 2007, a London Underground Tube tunnel at Kennington 14 m below ground was hit, but not penetrated, by a ground investigation borehole. Drilling was stopped and the incident reported. No delay or injuries resulted from the incident.
3. In 2013, a piling rig penetrated a railway tunnel 13 m below ground north of Old Street station. Two sections of auger (each 350 mm in diameter and 2 m long) detached from the drill string and lay on the line, along with considerable debris. A train driver reported seeing water flowing into the upper part of the tunnel. An out-of-service passenger train was then driven into the tunnel at low speed to check for damage. The driver was able to stop the train short of the blockage and report the problem. No injuries were caused to staff or passengers, and there was no damage to trains.

These were very serious incidents, and it is purely by chance that serious injuries were avoided. The Rail Accident Investigation Branch (2014) identifies that possible causal factors in these incidents include the construction site teams not being aware of the presence of the tunnel (due to not using all relevant information sources when checking for services and buried infrastructure); and, in at least one case, the presence of the tunnel being known but errors being made in positioning the drilling location at ground level.

The risk of hitting buried services or infrastructure should be reduced by a combination of the following methods:

• Carrying out a thorough review of the available information on buried utilities, services, tunnels and other infrastructure. In many cases, the published plans are only approximate, and some utility companies offer a free service to come to construction sites and mark actual locations on the ground surface. It is important to note that deep tunnels are not always marked on maps of surface features. If the locality is expected to have railway, subway or other tunnels, or deep-level sewers, then the organizations responsible for such infrastructure should be contacted directly. It should be borne in mind that privately owned assets, or infrastructure installed by the military or security services, may not be included on databases and maps maintained by the utility companies or public sector organizations.
• Using non-intrusive methods – such as ground-penetrating radar (GPR) – from ground surface to detect shallow utilities.
• Careful positional control when setting out drilling locations to reduce the risk of errors.
• Starting each hole with an inspection pit dug by hand or by vacuum excavation methods.
• Briefing the drilling crews to stop and ask for instruction if they encounter unexpected obstructions or voids.

The greatest likelihood of encountering services and tunnels is obviously in cities and urban areas. However, even in rural areas, major buried infrastructure may be present, including long-distance gas transmission lines and buried power cables. No site should be assumed to be free of buried infrastructure.

25.2.11 Risk from Unexploded Ordnance (UXO)

A particular problem that affects some areas of the United Kingdom as well as northern and central Europe is the legacy of UXO from the Second World War. During Allied and German air raids, some bombs failed to explode on impact and may have penetrated to several metres below ground level and remain there to this day. An example of dewatering used to aid the defusing and recovery of a bomb during the Second World War is shown in Figure 2.9.

Even more than 75 years after the end of the war, several times each year, UXO is encountered on construction sites. Most incidents are dealt with safely, but there have been injuries and even deaths when UXO has been encountered unexpectedly.

For groundwater control schemes, the UXO risk is principally related to hitting a UXO such as a buried bomb when drilling for dewatering wells, monitoring wells or ground treatment boreholes or when installing cut-off walls. A drilling rig could repeatedly hit a buried bomb with considerable force before the crew realizes an artificial obstruction is present. The action of the drilling rig could detonate the bomb immediately or in some cases could activate a delayed timer, causing detonation several hours later.

For sites in known heavily bombed areas or those situated near wartime installations (e.g. airfields and industrial complexes), specialist advice should be obtained. Possible risk mitigation measures include:

- Desk-based review (including archive records of air raids and bomb damage) of UXO risk for the site.
- Non-intrusive surveys, including traversing magnetometer sensors across the site to detect the metallic casings of shallow buried bombs.
- Intrusive surveys. These can include the use of cone penetration testing (CPT) equipped with magnetometer sensors to detect deeper unexploded bombs. On high-risk sites, it may be appropriate to specify a magnetometer CPT probe hole at every well or borehole location to prove the absence of UXO in advance of drilling.
- Monitoring of magnetic flux during the construction of boreholes and wells. Typically. the measurements are made at 1 m intervals using a down-hole gradiometer until the likely maximum depth of UXO penetration is reached.

Further guidance is given in CIRIA Report C681 (Stone *et al.*, 2009) and CIRIA Report C785 (Bowman *et al.*, 2019).

25.3 CONTRACTS FOR GROUNDWATER CONTROL WORKS

Groundwater control is a specialist process, which when measured in cost or manpower terms, is often only a tiny part of the overall construction project – a study by Roberts and Deed (1994) showed that the direct cost of construction dewatering systems was typically less than 1 per cent of total costs on a large civil engineering project. Despite this, groundwater (and its inadequate or inappropriate control) has historically been the cause of many construction disputes. Suitable contractual arrangements are an important part of managing dewatering works for positive results. This section discusses the background to contractual issues and outlines some arrangements used in practice.

25.3.1 The Need for Contracts

As has been described throughout this book, groundwater control is highly dependent on the ground conditions, over which the designer has no control. It follows that groundwater control must run a greater risk of poor performance than, say, reinforced concrete construction (where the designer can specify and control the materials used). The dewatering designer must gain his design information from the results of the site investigation (see Chapter 11). Sadly, it is still the case that not all investigations attain the standards that designers would aspire to – see *Inadequate Site Investigation* (Institution of Civil Engineers, 1991) and *Without Investigation Ground is a Hazard* (Site Investigation Steering Group, 1993).

Because groundwater control works are often carried out at the very start of a construction project (e.g. for construction of foundations), any problems or delays at that stage can have serious knock-on effects for the rest of the project. A review of cost overruns on groundwater lowering projects is given in Roberts and Deed (1994).

The specialized and perceived 'risky' nature of groundwater lowering works has led many clients and main contractors to view dewatering as a 'black art' best left to the cognoscenti. Apart from on the very largest projects, many contractors prefer to sub-contract dewatering works to specialist organizations that provide the expertise, experience and equipment to carry out such work. It is important that an appropriate contract exists between the various parties, and that the rights and responsibilities of each are clearly identified.

25.3.2 Traditional Contract Arrangements

In the United Kingdom, the traditional form of construction contract involved the client appointing a client's representative (called the Engineer under some forms of contract) to administer and supervise the works. Traditionally, the client's representative would also design the permanent works but not the temporary works, which were the remit of the contractor. The client's representative would, via a bidding or negotiation process of some sort, arrange for a contractor to undertake the works. The contractor would be employed directly by the client under a form of contract such as one of the editions of the Institution of Civil Engineers conditions of contract (known as the ICE conditions).

Groundwater control works are almost always classed as temporary works and are the main contractor's responsibility. These works are commonly sub-contracted, and the contractor would employ a dewatering sub-contractor. The sub-contract between the dewatering company and the main contractor would typically be 'back to back', meaning that the rights and responsibilities of each party apply to the other. In essence, this means that the dewatering sub-contractor is effectively acting on behalf of the contractor and takes over their responsibilities relevant to the dewatering. It also means that in the event of any changes or problems, the dewatering sub-contractor has the same rights as the contractor to apply for additional time or money via the clauses in the contract.

One of the most common types of serious disputes in traditional dewatering contracts occurs when the groundwater lowering system does not achieve the target lowering of groundwater levels, and it is believed that the ground conditions may not be as represented in the site investigation data provided at tender stage. The ICE conditions contain clause 12 (commonly known as the 'unforeseen ground conditions' clause), which allows the sub-contractor and contractor to apply to the client's representative for additional time or payment. The 'claims' process is often protracted, whereby the contractors have to demonstrate that the physical conditions or artificial obstructions encountered could not reasonably have been foreseen by an experienced contractor. This process can sometimes distract from the real problem of trying to finish the project and to deal with the conditions actually present in the ground. Some claims cannot be quickly resolved, may go to court and are only finally resolved (one way or the other) several years after the end of construction.

25.3.3 Alternative Forms of Contract

From the 1990s onwards, in the United Kingdom, there were moves away from traditional contracts, which were viewed as being too adversarial and leading to many costly and time-consuming 'claims'. When construction problems occur, prompt and open sharing of information can be vital in developing solutions; this did not always happen under traditional contracts. Sometimes, in the past, these contracts were applied in such a way that rather more effort was spent on trying to apportion blame than on solving the problem.

At that time, reports by Latham (1994) and Egan (DETR, 1998) reviewed the then performance of the UK construction industry and recommended specific improvements to planning and execution. They promoted increased efficiency and integration between the different parties involved in projects. With hindsight, the concept of having more integrated planning on projects has often improved the way that groundwater control has been carried out.

Under the old, often adversarial, contractual system, the need to control groundwater was often left as a last-minute temporary works fix for the contractor (after many other aspects of the project had been finalized) – described as a 'distress purchase' by Preene (2016) – and was procured on a lowest-cost basis. If the integrated approach of Latham and Egan is followed, it is more likely that key constraints, such as the need to control groundwater, will be identified as risks early during construction planning. This can allow rational assessment, and open discussion between the various parties to construction, of the potential risks and the way they could be managed. This opens up a wide range of options to control groundwater, including, for example, redesign of the permanent works to reduce (or avoid completely) the need for groundwater control. The High Speed 1 rail line (HS1, formerly known as the Channel Tunnel Rail Link), constructed in the United Kingdom from the mid-1990s onwards is a good example of how geotechnical engineering requirements, including the need to control groundwater, were among the key factors considered throughout the design process when assessing options for structures below ground level (O'Riordan, 2003).

In recent years, the nature of contracts has also changed as a result of the increased use of so-called 'design and build' contracts. These involve the contractor designing the permanent works as well as the temporary works, changing the nature of the relationships between client, client's representative and contractor.

Various non-adversarial forms of contract have been developed. A number of different schemes are possible, including:

a) 'Partnering' – which implies the development of longer-term relationships between the various parties, including client, permanent works designer, contractor and specialist sub-contractors.
b) 'Open book' contracts – where information is shared, and all parties are kept informed of what is going on and are able to have input into relevant decisions.

These forms of contract can allow the 'risks' of unforeseen ground conditions (or other factors) to be shared between the various parties in an open and transparent way. Dispute resolution procedures exist within such contracts to allow problems to be quickly highlighted and examined without the need for claims or other confrontational procedures.

When these forms of contract are used, the aim should be to control risks to the project. Any sub-contractors should be selected on the basis not merely of cost but also in terms of quality, health and safety, environmental management and ability to meet the programme timescale. It is also important that by involving all the parties, expertise and experience can be pooled to solve problems as quickly as possible. The need to overcome problems in a timely manner cannot be over-emphasized – on modern construction projects, many cost overruns result mainly from time delays rather than changes in methods. The control of geotechnical risks in construction is discussed by Clayton (2001).

25.3.4 Dewatering Costs

Because of the varied nature of groundwater lowering works and the wide variety of ground and groundwater conditions, the development of generic costs is not easy. It is not possible to estimate dewatering costs, even on an approximate basis, from the quantity of water pumped, the volume of soil dewatered or the depth of drawdown. Dewatering costs are more commonly broken down on an 'activity schedule' basis. Some of the activities will be costed on a unit basis (e.g. per well, per metre of header pipe), while the costs during the pumping period – pump hire, fuel, supervision, etc. – will be time-related charges (e.g. per day or per week).

25.4 ENVIRONMENTAL REGULATION OF GROUNDWATER CONTROL

As described in Section 21.2, in relation to construction works, groundwater is viewed primarily as a problem; hence the need for groundwater control. In other contexts, groundwater is a resource, used for public and private drinking water and for industrial use. Furthermore, in many locations, natural groundwater flows play an important role in supporting rivers, lakes, wetlands and other ecosystems. In many countries, environmental regulations or laws exist to help safeguard groundwater resources (Charalambous, 2011). This section describes the environmental regulatory regime applicable in England and the implications for the planning of groundwater control works.

Similar regulations apply elsewhere in the United Kingdom and in other locations internationally. In addition to national regulations, there may be additional legal requirements at state/province level and at city/municipal level. An example of local regulation is given in Section 25.4.6.

Because the pumping and discharge or disposal of groundwater are regulated by law, the person or organization responsible for the groundwater control has certain legal obligations to ensure that consents and permissions are obtained from the regulators. Under the normal forms of contract, the party responsible is either the contractor or the client. The dewatering sub-contractor or pump hirer is not normally responsible for the consents, but they should satisfy themselves, before work commences, that the necessary consents and permissions have been obtained.

There are two main facets to the legal requirements. The first deals with pumping of groundwater (termed *abstraction*) and is intended to make sure that the regulators can control groundwater abstraction to ensure that groundwater lowering systems do not cause nearby groundwater users to lose their supplies. The second deals with disposal of groundwater (termed *discharge*) and is intended to ensure that the pumped water does not itself cause pollution or other impacts.

25.4.1 Groundwater Protection

Groundwater protection is the collective term for the policy and practice of safeguarding the groundwater environment. This covers protection of both groundwater resources – the quantity of groundwater available for use (to ensure that the aquifers do not 'dry up') – and groundwater quality (to ensure that groundwater is not contaminated by human activities). In England, the regulator for groundwater is the Environment Agency, and the regime for groundwater protection is described in *The Environment Agency's Approach to Groundwater Protection* (Environment Agency, 2018).

One way to protect groundwater resources is to require each significant abstraction to be licensed via an application system. The licensing process allows the regulators (typically a governmental or quasi-governmental body) to scrutinize the applications and set limits on flow rates, or perhaps even prohibit abstraction in certain areas. Licensing of abstractions is a relatively modern concept, dating from the second half of the twentieth century. The history of the Chalk aquifer beneath London was described in Section 3.5.1. Here, unregulated abstraction in the late nineteenth and early twentieth centuries resulted in large and widespread lowering of the piezometric level, followed by a gradual rise in groundwater levels when abstraction rates reduced (Figure 25.1). Such a situation would have been unlikely

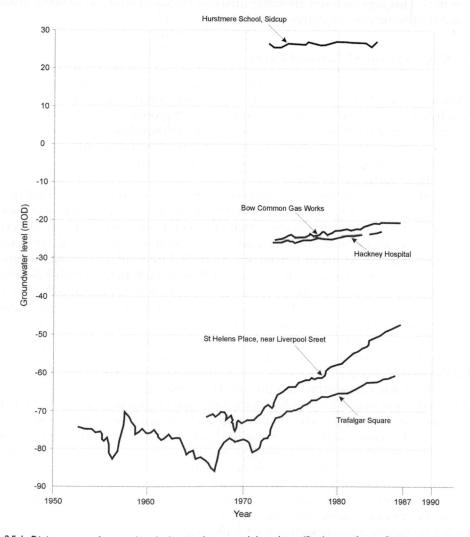

Figure 25.1 Rising groundwater levels beneath central London. (Redrawn from Simpson, B, *et al.*, *The Engineering Implications of Rising Groundwater in the Deep Aquifer below London*, Construction Industry Research and Information Association, CIRIA Special Publication 69, London, 1989. With permission from CIRIA: www.ciria.org)

to occur had a system of regulation been in place at the time – modern regulators are, of course, left with the legacy of managing rising groundwater levels beneath London.

Over-abstraction of groundwater resources has occurred in several other British cities and in other countries. In Thailand, for example, literally thousands of wells were drilled beneath Bangkok between the 1950s and the 1990s. Abstraction lowered groundwater levels from almost ground level down to 50 m depth, depleting groundwater resources as well as causing considerable ground settlement (Eddleston, 1996). This resulted in the Thai government passing laws, for the first time, to regulate groundwater abstractions. Similar regulations exist in many other countries.

The protection of groundwater quality is important, because if groundwater becomes polluted, it can be very difficult to rehabilitate. It is therefore better to avoid or reduce the risk of contamination than to deal with its consequences. Groundwater quality is protected by the imposition of controls on discharges to the water environment (groundwater and surface water). This approach sets allowable limits on the concentrations of many dissolved or suspended substances in discharge water.

25.4.2 Abstraction of Groundwater

The abstraction of groundwater in England is restricted by law. The relevant legislation is the *Water Resources Act 1991*, which replaced earlier legislation dating from 1945 and 1963. This establishes that an abstraction licence is required if groundwater is to be pumped for subsequent use, for example by a farmer or by a factory (although very low-volume abstractions do not require a licence). The licence will set limits on the quantity to be pumped, and the requirement to obtain a licence ensures that the regulator can keep records of total abstractions from particular aquifers.

The regulator responsible for the protection of groundwater resources in England, the Environment Agency, considers the merits of the application in relation to groundwater resources in the area affected. Applications that raise significant environmental concerns or are in areas where groundwater resources are over-utilized may be rejected. Abstractions in Wales, Scotland and Northern Ireland are controlled by similar licensing systems but managed by different regulatory bodies – Natural Resources Wales, Scottish Environment Protection Agency and the Environment Agency Northern Ireland, respectively. The latest information on the policies and regulations of each regulator can be found on their websites.

The water abstraction licensing system set out in the *Water Resources Act 1991* provides a small number of exemptions, which can allow groundwater to be abstracted without the need for a licence or consent. These exemptions apply if the total abstracted volume is very low or if the water is being abstracted for one of the purposes exempt under the Act.

Historically, dewatering abstractions (for construction, mining or quarry activities) were exempt from the licensing process, although the Environment Agency did have indirect powers to set limits on rates of abstraction and/or to require environmental mitigation measures.

This historical exemption from licensing likely developed because in practice, for the great majority of construction dewatering systems, any impacts outside the immediate area of the works were minimal. This is probably because the rate of abstraction from dewatering pumping was a tiny fraction of the recharge available to the aquifer, or the area of aquifer dammed by a cut-off wall was small in relation to the extent of the aquifer. However, in a small number of cases, often involving large-scale construction dewatering abstractions from highly permeable aquifers such as the Chalk, significant lowering of groundwater levels has occurred over a very wide area, perhaps 1 km or more from the site being dewatered.

However, following the adoption by the United Kingdom of the *EU Water Framework Directive* (WFD), any significant construction dewatering systems were included in the abstraction licensing system. The aim of the WFD is to promote long-term protection of groundwater quality (by preventing and remediating groundwater pollution) and groundwater quantity (by controlling abstraction volumes to prevent over-exploitation of aquifers).

Cumulatively, it is possible that on a local level, abstractions for dewatering may form a significant proportion of total groundwater abstractions. The management of water resources in line with the WFD is much easier to regulate now that dewatering abstractions are licensed in a similar way to other groundwater abstractions.

In England, dewatering abstractions are licenced subject to a tiered system, whereby very small abstractions do not require licensing, but systems with greater potential to cause impacts (due to higher flow rates and/or longer durations of pumping) are subject to more stringent requirements. Such a tiered approach is common to many licensing systems for dewatering in other countries. Under the system in England, the requirements of the licensing system (as at December 2019) for dewatering abstractions where the water is not put to a beneficial use are:

- Where the total pumping rate from the dewatering system is less than 20 m^3/d (0.23 l/s), no permission is required.
- Where the total pumping rate from the dewatering system is more than 20 m^3/d (0.23 l/s), and the duration of pumping is 27 days or less, a Temporary Licence is required.
- Where the total pumping rate from the dewatering system is more than 20 m^3/d (0.23 l/s), and the duration of pumping is 28 days or greater, a Transfer Licence is required.

As part of the process to obtain a licence from the regulatory bodies, there will be a likely requirement to assess the groundwater impacts (see Section 21.10). This process of impact assessment is sometimes known as a hydrogeological impact appraisal or HIA (Boak et al., 2007).

It is not unusual for the complete regulatory process to take several months from initial applications to when the necessary permissions are received. There is often insufficient time for the contractor to apply for permission from scratch once the site works are about to begin. Ideally, the project client or their design team should commence initial discussions (sometimes termed 'early consultation') with the regulatory authorities at the planning stage and carry out any initial technical studies at that stage. This allows the regulators additional time and can allow draft consent information to be incorporated into the tender documents, so that potential contractors are aware of any likely constraints. The successful contractor should then finalize the consent with the regulators.

25.4.3 Discharge of Groundwater

Once abstracted from the ground, the pumped water must be disposed of (volumes generated are normally too great to be stored on site). The abstracted water from a groundwater control system is legally classified as *trade effluent*. In urban areas, it may be possible to dispose of water into sewers or surface water drains – permission must be obtained from the water utility or its agents before this can be done. Such permissions normally take the form of trade effluent licences, which may set limits on the quantity and quality of water that can be discharged. Charges (based on a cost per cubic metre) are generally levied for disposal of water in this way. For long-duration discharge of substantial flow rates, these charges may add up to considerable sums.

However, in many cases, the abstracted water must be disposed of to surface water (including rivers, lakes and the sea) or to groundwater (via recharge wells or trenches).

In England, all surface water and groundwater are legally classified as *controlled waters* under the *Water Resources Act 1991*. In these cases, if the duration of discharge is more than 3 months, an Environmental Permit is required from the Environment Agency. It is a criminal offence to discharge poisonous, noxious or polluting material into any controlled waters either deliberately or accidentally. Polluting materials include silt, cement, concrete, oil, petroleum spirit, sewage, or other debris and waste materials. Measures to prevent the discharge of polluting materials are described in Section 21.9.

Even if the discharge flow is to be disposed of by artificial recharge (see Section 17.13) back into the same aquifer, an environmental permit may still be required. This means that the recharge water quality will have to be monitored to ensure that it is within the quality limits set on the environmental permit.

25.4.4 Settlement Resulting from the Abstraction of Groundwater

The lowering of groundwater levels resulting from the abstraction of groundwater will cause an increase in vertical effective stress in the strata around the excavation. This will inevitably lead to some ground settlement, possibly over a wide area. Occasionally, the settlements may be large enough to cause distortion or damage to nearby structures or services. Engineering mitigation techniques are available to minimize the extent and effects of these settlements. Section 21.4 describes these problems and their potential solutions.

There has been considerable litigation over many years regarding settlement damage that has been alleged to result from groundwater abstractions for water supply or dewatering purposes. The law related to ground settlements arising from groundwater abstraction is found in common law rights, which are based on the precedents of judgements laid down in previous cases in the civil courts.

According to a number of reviews of the legal issues (such as Akroyd, 1986; Powrie, 1990), it is well established in English common law that a landowner has no right of support from the water percolating in undefined channels beneath his land. Groundwater in almost all aquifers is effectively flowing in undefined channels and is therefore considered by common law to be a reservoir or source of supply, which is no one's property, but from which everyone has the right to abstract as much as they wish in so far as it is physically possible to do so.

The courts have found that if groundwater abstraction causes settlement damage to a neighbour's property, there is no right of action against the abstractor. The courts have also found that since there is no duty of care to avoid damage resulting from groundwater abstraction, there is also no cause of action under the law of nuisance or negligence.

Possible conditions where groundwater would not be considered to be in undefined channels include well-established subterranean solution channels known to exist in karst aquifers (see Sections 4.6.2 and 8.3.5) or water flowing in abandoned mine workings. From a legal point of view, if water is believed to be flowing in a defined underground channel, the presence of the channel must be known. It is not necessary for it to be revealed by excavation or exposure, but its presence must be a reasonable inference from the available information.

This means that if dewatering abstractions pump clear groundwater (with no suspended solids), from most aquifers, there is no legal liability if settlement damage occurs, even if mitigation measures could have been employed to prevent damage. This position is quite different from settlement damage arising from excavation or tunnelling, where the primary mechanism is the removal of the support from the soil, and a legal liability exists.

However, for dewatering schemes, if the water pumped is not perfectly clear and if it contains suspended silt or sand particles, the support of the ground is being removed. A landowner does have the right to the support of the ground beneath his land. If the pumping of silt or sand led to settlement damage, there would be a legal right of action against the abstractor. A consequence of this is that all wells and sumps must be equipped with effective filters to prevent the flow of groundwater removing fine particles from the soil. In practice, this is most likely to be a problem with sump pumping systems, and these should be employed with caution if structures are present near the excavation (see Case History 27.7).

Despite there being no legal liability for settlement damage resulting from the abstraction of groundwater, in the great majority of projects, construction methods are designed to be sensitive to the risks of consequential damage. If there is believed to be a significant risk of settlement damage, the project client or designer will generally require the use of settlement mitigation or avoidance measures (such as artificial recharge – see Section 17.13).

25.4.5 Environmental Regulation of Major Projects

Very large projects are, in principle, subject to the same environmental constraints as smaller projects. However, the sheer size and high-profile nature of the projects may result in a slightly different approach being taken. This can be illustrated by the case history of the Medway Tunnel, constructed in southern England in the 1990s (Lunniss, 2000; Thorn, 1996).

The tunnel was of immersed tube design and required significant dewatering to allow the construction of a casting basin and cut-and-cover tunnel sections on either side of the river. At planning stage, the dewatering was anticipated to require extensive pumping from the Chalk aquifer, which provided public water supply from a number of nearby wells. There was concern that the dewatering would derogate these sources and that saline intrusion might be promoted by river dredging during construction.

As with many large infrastructure projects, construction of the tunnel was covered by a special Act of Parliament, in this case the *Medway Tunnel Act 1990*. Such acts require concerned bodies to petition for their 'rights' in the drawing up of the bill and request special protective provisions. The environmental regulator at the time – National Rivers Authority, the predecessor body to the Environment Agency – obtained, via the Act, protective measures as follows:

> The wardens (of the scheme) … shall so design and construct the tunnel as to ensure by all reasonably practicable means that saline or other contaminating intrusion into water resources in underground strata does not occur by reason of such construction.

The construction contract gave the contractor the responsibility of developing a dewatering and monitoring scheme acceptable to the National Rivers Authority. Detailed liaison followed, and a dewatering and monitoring strategy was developed; this strategy was robust enough to cope with an increase in anticipated pumping rates from around 250 l/s up to 400 l/s. The monitoring exercise was intensive. The data gathered in the initial stages of the project allowed a numerical model of the aquifer to be developed by the contractor to support their proposals for the later stages of dewatering.

While the extent of monitoring and discussion with the National Rivers Authority was considerably more extensive than is routine, this project highlights that any major dewatering works are likely to require significant ongoing liaison with the environmental regulators. It is not unusual for the planned dewatering works to be varied in light of the regulator's requirements.

25.4.6 Local Regulations

In addition to regulations at national and state/province level, occasionally, local regulations exist to address specific environmental issues. One example is the protection of sensitive or historic structures, which could be damaged by dewatering-related settlements. This can be illustrated by the example of St Paul's Cathedral in central London.

The cathedral, designed by Sir Christopher Wren, was built between 1675 and 1710 on the site of an earlier, even larger cathedral destroyed in the Great Fire of London in 1666. According to Thomas and Fisher (1974), it is founded on a layer of clay underlain by the extensive sand and gravel stratum of the River Terrace Deposits associated with the flood plain of the River Thames. Since the mid-nineteenth century, drainage works associated with the growth of the metropolis have lowered groundwater levels in the sands and gravels, leading to consolidation of the strata below the cathedral. This resulted in settlement and damage to the fabric of the cathedral. In the 1920s and 1930s, a major programme of restoration included a geotechnical investigation led by the eminent engineers Sir Alexander Gibb and Mr (later Sir) Ralph Freeman. Thomas and Fisher (1974) report that Gibb and Freeman concluded that the cathedral was stable, but that

> any interference whatsoever with either the subsoil or the water levels in the vicinity might 'upset the equilibrium' and cause further settlement.

Based on the recommendations of Gibb and Freeman, *The City of London (St Paul's Cathedral Preservation) Act 1935* was established, which gives the Dean and Chapter of St Paul's control over subterranean works below prescribed depths within a defined area around the cathedral. In relation to dewatering, the Act states:

> No person shall pump or otherwise abstract water from the subsoil of the prescribed area at or below the prescribed depth and above the level of the London clay except with the approval of the Dean and Chapter and in accordance with any conditions attached to such approval.

Similar regulations exist in other cities that contain potentially vulnerable historic structures.

Chapter 26

Optimization of Groundwater Control Systems

Engineering problems are under-defined, there are many solutions, good, bad and indifferent. The art is to arrive at a good solution. This is a creative activity, involving imagination, intuition and deliberate choice.

Sir Ove Arup

26.1 INTRODUCTION

It is axiomatic that every groundwater control scheme is different. The almost infinite combinations of ground conditions and project requirements will never be repeated. However, to make progress, a pragmatic approach is required, and usually a groundwater control project can be matched to a particular pumping or exclusion technology (or a combination of the two types) and a site-specific solution developed. Typically, there are multiple options that can effectively control groundwater. The preferred solution (in terms of number and location of wells, cut-off walls, pump capacity and other system parameters) might be selected based on a combination of factors including cost, time, and local availability of expertise and materials. Because a single 'perfect' option rarely exists, there is often a desire to optimize a groundwater control system – either at design stage or during implementation in the field. This chapter describes some possible optimization strategies and discusses advantages and limitations of these approaches

26.2 WHY OPTIMIZE?

Groundwater control is required at the very start of most below-ground engineering projects. It is therefore one of first geotechnical processes required on site and is often the first that must be proved in order for work to proceed. Any problems or delays caused by groundwater at this early stage can delay the whole project. Roberts and Deed (1994) highlight that the cost of resultant delays can be many times greater than the direct cost of the groundwater control works themselves.

There are often multiple possible groundwater control techniques that could, in theory, be applied at a given site. Furthermore, dewatering design is not 'codified' in detail by geotechnical design codes. Such dewatering guidance documents as do exist are typically in the form of 'toolkits' of design methods and construction techniques. Examples of this non-prescriptive approach include, in the United Kingdom, CIRIA Report C750 *Groundwater Control Design and Practice* (Preene *et al.*, 2016); in the United States, *Dewatering and Groundwater Control* (Unified Facilities Criteria, 2004); and in the

Middle East, in Abu Dhabi, *Dewatering Guidelines. Geotechnical, Geophysical, and Hydrogeological Investigation Project (GGHIP)* (Abu Dhabi City Municipality 2014) and in Qatar, *Management of Construction Dewatering – Construction Dewatering Guidelines* (ASHGHAL, 2014). At the start of a project, the design team can be faced with a bewildering choice of design and implementation options, and a rational optimization approach can look attractive.

26.3 DIFFERENT OPTIMIZATION METHODS

Geotechnical design teams typically have access to significant computing power in terms of both proprietary numerical modelling packages and spreadsheet-based analytical models. It is a straightforward, and not uncommon, practice to analyse multiple groundwater flow scenarios (using numerical groundwater models) and apply the results to the design of groundwater pumping or exclusion systems. It is a logical step to use the results of multiple analyses as part of a process to develop an 'optimal' groundwater control design, typically based on optimizing the number of wells, the pumped flow rate or the depth of cut-off walls. This approach of using numerical solutions to optimize dewatering design first dates from at least as far back as the 1970s (Aguado *et al.*, 1974) and has subsequently developed along with emerging numerical decision-making tools of their time, such as expert systems (Davey-Wilson, 1994) and multi-attribute decision analysis (Golestanifar and Ahangari, 2012), among others.

Many of the published studies have taken a fundamentally mathematical approach to optimization, perhaps in an attempt to provide better reliability or consistency in dewatering design, in part by reducing the role of 'expert judgement'. The authors consider that this is a rather narrow approach and suggest that some of the non-numerical optimization strategies discussed by Preene and Loots (2015) are worthy of consideration.

Four possible optimization methods were proposed by Preene and Loots (2015); these are discussed in the following sections. Each of these is a based on a method of analysis or modelling described elsewhere in this book.

1. Empirical methods: Design based largely on experience, local knowledge and 'rules of thumb'
2. Numerical and analytical methods: Use of numerical groundwater flow models or hydrogeological design equations
3. Observational approaches: Use of construction observations to design and refine the dewatering system
4. Troubleshooting methods: Application of immediate and urgent changes when problems occur on site

26.3.1 Empirical Optimization

Empirical models (see Section 5.5.1) are not derived from theory or analysis of the physical laws of a problem; rather, they are based on observation and previous experience. There is a long history of the use of empirical methods on groundwater control projects, and optimization by empirical methods has been successfully used on many simple projects. A simple project can be defined as one where the hydrogeological conditions have been adequately characterized and are relatively straightforward; where the excavation is relatively small and shallow; and where environmental impacts are not a key concern. Examples might include shallow basements, pipeline projects, sewers, etc., especially where other dewatering projects have been executed successfully nearby.

This approach uses experience of groundwater control methods and their performance on previous projects to optimize another project where the conditions are comparable.

When geotechnical engineers become involved in the design of groundwater control systems, empirical methods are sometimes viewed as being less rigorous compared with numerical or analytical methods. However, there is a huge track record of empirical methods providing successful outcomes. One of the reasons why this is the case is that provided a suitable groundwater control method is selected, a given dewatering technology can often successfully deal with modest variations in ground conditions. This is illustrated for groundwater exclusion techniques by Figure 9.10 and for pumped groundwater control methods by Figure 9.11, which show that individual techniques are appropriate for a relatively wide range of permeability and drawdown. Conversely, this highlights the need for designers to understand the performance limits of each method. It is essential to select an appropriate groundwater control technology for a project.

The empirical method requires sufficient site investigation data to allow the hydrogeological conditions to be identified, and relevant experience from comparable projects. Adequate site investigation data are essential to characterize site conditions; otherwise, it cannot be known whether the previous sites, from which experience is drawn, are comparable. In practice, when problems occur with the empirical method, this is often because empirical rules were applied between sites where underlying conditions are different.

26.3.2 Numerical Modelling or Analytical Optimization

Numerical modelling (see Section 5.5.3) uses computer software to solve multiple iterations of the governing groundwater flow equations over a model 'domain' that is discretized into numerous elements or cells. The necessary investments in software, hardware and training for these methods have reduced dramatically in recent decades, and this approach is increasingly used in the design of groundwater control systems. Numerical modelling offers the flexibility to take into account known or inferred variations in the geological or hydrogeological conditions within the range of influence. This might include assessing the effects of a nearby river, another dewatering project, or a natural barrier in the aquifer.

The analytical approach uses closed-form hydrogeological equations (see Section 5.5.2) to estimate pumped flow rates, drawdowns, seepage beneath cut-off walls, etc. This approach is often applied in relatively simple hydrogeological conditions with simple boundary conditions, where the complexity of a numerical model is not required. Each type of analytical equations is only applicable to a relatively narrow range of hydrogeological conditions, and gross errors can result if it is used in the wrong conditions.

Both numerical and analytical approaches need to be applied based on a conceptual model (see Section 5.4), which captures the important features of the groundwater system at the site and its environs. The conceptual model will normally be developed directly from the site investigation data, including a hydrogeological desk study. If the conceptual model is inaccurate or incomplete, the results of any subsequent analysis are likely to be erroneous.

26.3.3 Observational Optimization

An obvious example of optimization is to use observations from field monitoring – this is termed the observational method (see Section 13.3.2). In this approach, construction observations (such as groundwater levels and pumped flow rates) are used to guide optimization of the system as part of a deliberate process of design, construction control, monitoring and review (Nicholson *et al.*, 1999). Observational optimization can be combined with 'inverse numerical modelling', where a series of numerical modelling scenarios are prepared in advance for a range of possible hydrogeological conditions and then compared with the field data (Section 12.6.3).

The observational approach can be useful to deal with local variations in hydrogeological conditions. On larger projects, it may be the best solution to address these variations locally (using the flexibility of the observational method) instead of engineering the overall system based on the worst-case conditions, as might be necessary if the dewatering system was conservatively designed at the start with little flexibility.

26.3.4 Optimization in the Field (Troubleshooting)

Ground and groundwater conditions can vary across a site, and in any case, most ground investigations can test or sample only a tiny volume of the soil or rock affected by dewatering. Therefore, it is no surprise that occasionally, groundwater control schemes are not fully effective when initially installed, and 'troubleshooting' measures may be needed. This approach takes place during construction and so has access to field data (e.g. dewatering well logs, pumped flow rates, drawdown water levels, as well as anecdotal reports from drillers and other construction staff) that were not available to the original designer. The aim is to identify whether the lack of performance is related to various causes, including:

- 'Unexpected ground conditions' (i.e. ground conditions different from the original conceptual model)
- Operational problems with the current system (e.g. existing pumps and wells not delivering their design capacity, insufficient depth of cut-off walls, etc)
- The wrong dewatering technology or approach being used

The objective of troubleshooting is to develop a plan of action in order to provide an effective dewatering system at the site, suitable for current conditions.

26.4 PROBLEMS WITH OPTIMIZATION

The principle of optimization seems straightforward, but there are a number of challenges to obtaining a successful outcome. Potential problems are outlined in the following sub-sections.

26.4.1 Lack of Clarity in Objectives of Optimization

The benefits of optimization will be limited unless the objective of the process is clearly defined and relevant. Sometimes, those planning the optimization of groundwater control schemes fail to recognize that optimizing in one aspect may require compromises in other matters.

Often, the focus is on optimizing dewatering pumping rates (i.e. to avoid pumping water unnecessarily) while still achieving the required lowering of groundwater levels. This has the advantage that it will likely also minimize operational costs and energy consumption. However, if pursued single-mindedly, this approach could result in a dewatering system with little spare pumping capacity to deal with modest changes in ground conditions that may require higher flow rates. Also, such a system might be designed without consideration of environmental impacts on the groundwater regime; increasingly, the minimization of impacts is a necessary design consideration.

26.4.2 Insufficient Data Quality and Quantity

Data are the lifeblood of geotechnical design. Without information from site investigation (and previous projects) there can be little confidence in the conceptual model and all subsequent calculations, modelling and design. If the relevant data are inadequate in quality or quantity then problems will likely follow. No modelling effort can correct false or poorly determined parameters. Potential problems include:

- Data quality: This can be a very subjective matter and relates to how reliable the data are perceived to be. There can be issues with the source of the data (e.g. by whom the work was carried out and how it is reported) or questions over internal consistency of the data – e.g. if borehole logs describe a sandy gravel (which should be highly permeable), but permeability test results include some very low values.
- Data quantity: Relevant questions include 'Is there enough data?' and 'Are the relevant information requirements and uncertainties addressed?' There should be sufficient data to develop some understanding of the likely variations in ground conditions. A geological desk study (Section 11.6) can be of great value in this regard. The relevance of the data relates to whether the necessary information is provided. For example, a common issue is where the site investigation boreholes are not deep enough to identify the presence of any confined aquifers beneath the base of the excavation that could cause an uplift hydraulic failure of the base (see Section 7.5.1 and Section 13.7).

One outcome of optimization could be to recommend additional ground investigation to plug any identified data gaps and/or to recommend that the dewatering system be implemented by the observational method (Section 13.3.2) to provide flexibility against variations in ground conditions.

26.4.3 Errors in Conceptual Model

Conceptual models that are realistic representations of the key elements of the groundwater system are fundamental to the design of groundwater control systems (this is discussed in Sections 5.4 and 13.4). Conceptual models play a similarly important role in the optimization process. Experience shows that the causes of many poorly performing groundwater control schemes can ultimately be traced back to an inappropriate conceptual model that either leads the designer down the wrong design avenue or causes the designer to ignore a design condition that is, in fact, important. Examples include:

- Failure to identify a confined aquifer beneath an excavation, which could cause hydraulic failure of the base of the excavation
- Failure to identify that the permeability range of the soil/rock potentially includes lower-permeability zones that may limit the flow rates yielded by pumped wells
- Failure to identify pre-existing groundwater contamination in the vicinity of a dewatering system that may be mobilized by pumping

If conditions such as these are not identified, then subsequent modelling or analysis cannot address the relevant questions or will be based on unrealistic parameters.

26.4.4 Inappropriate Groundwater Control Technique

As discussed earlier, and shown in Figure 9.10 for groundwater exclusion techniques and Figure 9.11 for pumped groundwater control methods, each technique is applicable to a finite range of ground conditions. If an unsuitable technique is selected at the outset of design (e.g. if ejector wells are used in a high- permeability soil, or a grout curtain is attempted with an inappropriate grout mix), then even extensive and detailed optimization measures are likely to be futile.

Designers and analysts should have an understanding of the limits of performance of the chosen groundwater control method and select the technique based on these limits. For example, if the chosen technique is appropriate not just for the 'design value' of permeability but also for the 'highest credible' and 'lowest credible' values, then the design is likely to be robust. However, if relatively small changes in soil/rock permeability may require a change in technique, this can cause major delays and cost overruns to a project.

26.5 POSSIBLE PRIORITIES FOR OPTIMIZATION

Groundwater control projects are carried out in the commercial world of construction and engineering projects, and can cost significant sums of money to design, install and operate, and it is natural for a project team to try to obtain 'value for money'. However, a drive for lowest cost can be counterproductive. Preene (2016) stated:

> There is sometimes a perception that construction dewatering is a distress purchase – in other words 'a good or service purchased only because there are no alternatives, or the alternatives are all far inferior, rather than having any desire to actually purchase this good or service'. Also, if it cannot be avoided altogether, there is a temptation to focus purely on minimising the cost of dewatering.

In fact, the direct costs of installation and operation of groundwater control schemes can be a small proportion of the total construction costs – a study by Roberts and Deed (1994) showed that the direct cost of construction dewatering systems was typically less than 1 per cent of total costs on a large civil engineering project. On most large projects, if there are project delays caused by inadequate or ineffective control of groundwater, the delay costs will dwarf the costs of doing it right in the first place. There have been examples where, in construction disputes arising from the alleged poor performance of groundwater control systems, delay costs of more than £10 million have been claimed on sites where the direct cost of effective groundwater control would have been significantly less than £100,000 (assuming the dewatering was done to high standards).

The most effective and relevant optimization measures look beyond pure cost measures. Optimization has been applied to meet various criteria, including:

- Regulatory requirements, such as when a limit has been set on the maximum permitted discharge rate of pumped water
- Practical constraints, such as sites where available water disposal routes can only handle a finite flow rate of pumped water
- Defined limits on external environmental impacts (such as ground settlement from effective stress increases caused by lowering of groundwater levels, or increases in groundwater levels caused by the barrier effect of cut-off walls)

Table 26.1 summarizes several different ways that systems can be optimized.

Table 26.1 Possible Aspects of Groundwater Control for Optimization

Optimization priority	Comments
Lowest pumping rate	There is a risk that systems optimized in this way will have insufficient spare pumping capacity to handle modest increases in flow rate above design values.
Capped pumping rate	This applies where there is a specified maximum limit on the flow rate of water that can be pumped. The limit may be based on regulatory limits or on physical constraints of the water disposal route. This case often applies to very high-flow-rate systems, where the available water disposal routes cannot easily handle the theoretical maximum pumping rates estimated at design stage.
Lowest energy usage	Will tend to favour lowest pumping rate solutions with the same risks. May involve swapping pumps for smaller units for steady-state pumping once initial drawdown has been achieved.
	May favour groundwater exclusion solutions that use low-permeability cut-off walls to avoid or minimize pumping.
Lowest CO_2 emissions	Needs to account for the CO_2 emissions both during installation (e.g. wells, cut-off walls or ground treatment) and during operation (e.g. pumping of water or other ongoing running equipment such as freezeplants for artificial ground freezing).
	For systems where groundwater control is required for short periods, the emissions from installation will dominate. As the required dewatering period increases, operational CO_2 emissions will become increasingly important. Where electrically driven pumps are used, the carbon intensity of the electricity supply is a key factor. Assessment of CO_2 emissions associated with dewatering systems is discussed in Casey *et al.* (2015).
Minimal external impacts	May favour groundwater exclusion solutions that use low-permeability cut-off walls to avoid or minimize pumping.
Minimal capital cost	Will tend to favour lowest pumping rate solutions with the same risks.
Minimal operating cost	Will tend to favour lowest pumping rate solutions with the same risks. May be a significant overlap with lowest energy usage systems.
Shortest dewatering period	Shorter dewatering periods can be achieved by higher initial pumping rates to rapidly achieve drawdown of groundwater levels. May be appropriate for emergency groundwater control systems to recover a project after a failure or inundation, or for projects where the direct costs of dewatering are small relative to project weekly on-costs.
Maximum certainty of outcome	May be appropriate for projects where achieving deadlines within the construction programme with certainty is vital and the dewatering must be fully effective without time-consuming modifications.

Adapted from Preene, M and Loots, E, Optimisation of dewatering systems, in *Proceedings of the XVI ECSMFGE, Geotechnical Engineering for Infrastructure and Development* (Winter, M C, Smith, D M, Eldred, P J L and Toll, D G, eds), ICE Publishing, London, pp. 2841–2846, 2015.

A common factor with many pumped groundwater control systems is that a higher flow rate is needed during the early period (the first days or weeks) of pumping while drawdown is being achieved (see Section 13.8.6). It is not currently widespread practice to modify a dewatering system to reduce pumping capacity, for example by removing some pumps or reducing pump size, after the initial drawdown period (this approach is described in more detail by Rea and Monaghan, 2009 for a mine dewatering project). It is possible to use field measurements, inverse numerical modelling and risk assessments to estimate the reduction in pumping capacity that can be achieved while still being capable of handling the worst credible hydrogeological conditions. There are significant potential savings of energy usage (and CO_2 emissions) by optimizing long-term pumping capacity in this way, which may use a hybrid of optimization methods. Energy usage of pumping systems is discussed in Section 19.11.4.

There is no perfect optimization method that will address all the possible priorities for a groundwater control system. In reality, different aspects of optimization may conflict, and trade-offs will be needed between different priorities of design. For example, a dewatering system designed for minimum installation cost may not offer the lowest environmental impact or the shortest period to achieve initial drawdown. The message is that meaningful optimization of groundwater control systems is more complicated than might appear on first impression – but in many cases, there are real benefits to accepting the challenge.

Chapter 27

Case Histories

Every Civil Engineer, however academically endowed, has to start his practical career in a state of ignorance and then gradually accumulate experience. The lessons learned from the past are not always out of date so an intelligent interest in the antics of his predecessors can speed up the process.

Sir Harold Harding

27.1 INTRODUCTION

Each application of groundwater control will be different due to variations in ground conditions, the different depths and layouts of excavations, and other project constraints. Nevertheless, review of previous projects can provide some insight into the groundwater problems that may be encountered, and the solutions deployed, on future projects. This chapter presents 11 case histories from the authors' experience, with the hope that sharing these experiences will help the reader.

The following case histories are presented:

27.2 Sump pumping of large excavation, Aberdeen, UK
27.3 Wellpointing at Derwent outlet channel, Northumberland, UK
27.4 Deep wells and relief wells at Tees Barrage, Stockton-on-Tees, UK
27.5 Ejector wells for shaft dewatering, with high-permeability feature present, UK
27.6 Hydraulic failure of base of shaft, UK
27.7 Groundwater control by sump pumping and deep wells for shaft, UK
27.8 Artificial recharge to control settlement: MISR Bank, Cairo, Egypt
27.9 Artificial recharge to protect drinking water source, UK
27.10 Low-permeability cut-off wall at Sizewell B Power Station, UK
27.11 Contamination remediation using cut-off walls and groundwater abstraction at Derby Pride Park, UK
27.12 Permanent groundwater control system at Govan Underground Tunnel, Glasgow, UK

Each case history is relevant to different groundwater control techniques and different types of groundwater problems. The key issues are highlighted at the start of each case history. Where project-specific references exist, they are listed for each case.

27.2 CASE HISTORY: SUMP PUMPING OF LARGE EXCAVATION, ABERDEEN, UK

This Case History is relevant to:

- Dewatering methods (Section 9.5)
- Sump pumping (Chapter 14)

27.2.1 Summary

This Case History describes a project where groundwater was adequately controlled by sump pumping methods in relatively high-permeability soils – alluvial gravels with some sand and having little or no 'fines'.

27.2.2 Case History

Morrison Construction Limited formed an open reservoir upstream of the city of Aberdeen beside the River Dee, to the instructions of Mott MacDonald, the engineer appointed by the client, Grampian Regional Council Water Services. The water surface area of the reservoir is of the order of 8 hectares with a storage capacity of 240,000 m³. The reservoir was created by forming a horseshoe-shaped embankment to marry with the existing flood embankment of the river Dee and all to similar height – the length of the reservoir embankment constructed by Morrison Construction Limited was 1150 m. The flood plain soils beneath the reservoir were alluvial sandy gravels of very high permeability – one pumping test indicated a permeability of the order of 2.5×10^{-2} m/s. The levels of the groundwater were much influenced by the river level. This is not surprising in view of the very high permeability of the flood plain deposits!

The engineer required the formation level of the floor of the reservoir to be some 1.5 m below normal river level. The floor and the sides of the reservoir were required to be lined with an impermeable membrane of low-density polyethylene (LDPE) sheeting laid on a 50 mm thick sand bedding. Thus, in order to lay the impermeable membrane and satisfactorily joint adjacent sheets, it was necessary to lower the groundwater level to some 2 m below normal river level.

A large central pumping sump (Figure 27.1) was excavated to a depth of about 4 m below the level of the underside of the membrane. Three drainage trenches (approximately 80 to 120 m long) were excavated to about 2 m depth below the level of the underside of the impermeable membrane, sited to radiate out from the central sump to the riverside perimeter of the reservoir (Figure 27.2).

Into these drainage trenches were placed 150 mm diameter perforated uPVC drainage pipes, and the trenches were backfilled with 75 mm cobble size material. This facilitated the general drawdown of the groundwater beneath the area of the reservoir by pumping from the one central sump, though there was some additional pumping towards the end of membrane laying from an auxiliary sump sited close to the inside toe of the embankment.

The discharge from two 200 mm vacuum-assisted self-priming centrifugal pumpsets plus three 150 mm pumpsets – approximately 1000 m³/h (around 280 l/s) – was discharged to the River Dee. This sump pumping installation enabled Morrison Construction Limited to lay the specified sand bedding and membrane satisfactorily.

Figure 27.1 Dee Reservoir, central pumping sump. The sump pumps comprise one electrically powered unit, supplied from a portable generator, and two diesel-powered units. (Courtesy of Grampian Regional Council, Aberdeen, UK.)

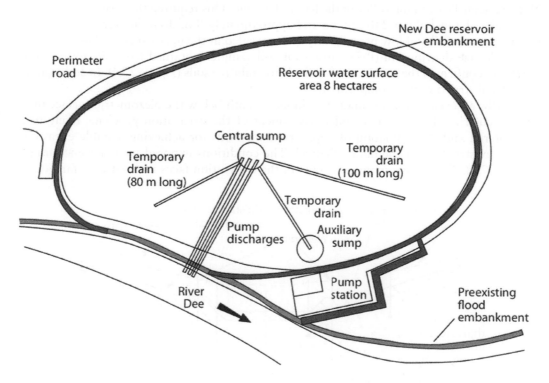

Figure 27.2 Dee Reservoir, plan showing positions of sump and temporary drainage trenches. (Courtesy of Grampian Regional Council, Aberdeen, UK.)

27.3 CASE HISTORY: WELLPOINTING AT DERWENT OUTLET CHANNEL, NORTHUMBERLAND, UK

This Case History is relevant to:

- Dewatering methods (Section 9.5)
- Pore water pressure control systems in fine-grained soils (Section 9.5.4)
- Wellpoint systems (Chapter 15)
- Wellpoints with long risers and lowered header mains (Section 15.10.1)
- Drainage measures within the slope (Section 17.10.3)

27.3.1 Summary

This Case History describes a sealed wellpoint system installed to a high standard in low-permeability laminated Glaciolacustrine deposits. The wellpoints lowered pore water pressures in the fine-grained soils and dramatically improved the stability and working conditions of the excavation.

27.3.2 Case History

In the early 1960s, the Sunderland & South Shields Water Co. awarded to John Mowlem & Co. Ltd. a contract for the construction of an earth dam about 15 miles to the south west of Newcastle on Tyne, in the valley of the River Derwent. This required the formation of a spill-way outlet channel about 240 m long and approximately 8 m deep on average. Laminated Glaciolacustrine silts of low permeability were revealed at the upstream end (Figure 27.3).

Particle size distribution (PSD) curves for soil samples obtained from site investigation boreholes confirmed the existence of difficult-to-stabilize soils (Figure 27.4). In addition, the varved soil structure was complicated.

Initially, an excavation was made to a modest depth below the piezometric surface in the Glaciolacustrine deposits. This aided assessment of the excavation problems likely to be encountered and determination of appropriate methods for achieving a stable excavation for the construction of the outlet channel. The conditions exposed were not encouraging (Figure 27.5). There were many washouts in the excavation faces and outwash fans due to

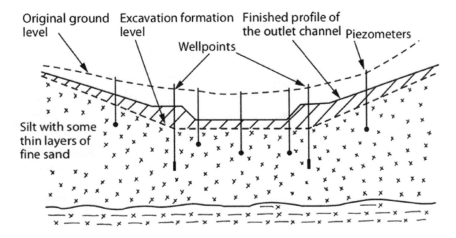

Figure 27.3 Typical cross section of Derwent outlet channel. (Courtesy of Northumbrian Water Limited, Durham, UK.)

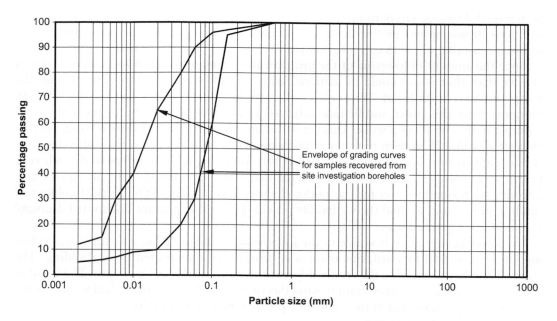

Figure 27.4   Envelope of gradings of glacial lake soils, the Derwent outlet channel. (Courtesy of Northumbrian Water Limited, Durham, UK.)

Figure 27.5   Initial excavation (showing unstable conditions) for Derwent outlet channel. (Courtesy of Northumbrian Water Limited, Durham, UK.)

seepages that continuously transported fines. Also, a series of piezometer tubes installed in the excavation had to be extended to some height above the excavated level because the pore water pressure at depth was artesian relative to the excavation floor. The excavation surface was so unstable that duckboards were needed in order to be able to take the measurements in these piezometers.

Careful undisturbed sampling of these glacial soils revealed, within the layered soil structure of these deposits, some very thin layers of more permeable material that could act as preferential drainage layers to influence the reduction of pore water pressures of the total soil mass. It was agreed by the client, his specialist advisors and the contractor that a trial wellpoint installation, with careful attention to sanding-in, should be undertaken.

Since the amount of lowering of the groundwater required to reach formation level was greater than that which could be achieved with a single-stage wellpoint installation, the wellpoints were installed at 1·8 m centres on extra-length riser pipes (see Section 15.10.1). Pumping from this initial installation achieved about 3 m of lowering of the water level within a period of about a week. This created a surface of sufficient firmness to support an excavator, and 2·75 m deep trenches were then opened beside the two lines of wellpoints. A duplicate suction header main was laid at this lower level, and progressively, each riser length was shortened and connected into the lower active pumping main. By this means, lowering to formation level for the outlet channel was achieved. The vertical column of sanding-in material around the wellpoints provided the all-important downward drainage for multi-layer perched water tables.

The lower header main is visible in the foreground of Figure 27.6. As excavation to formation level moved forward, this main was progressively supported on scaffolding. The channel had been shaped to the required formation and side slopes (see Figure 27.3), and as shaping proceeded, a blanket drain of sand (Section 17.10.3) was quickly laid to permit the drainage of pore water but at the same time prevent the removal of fines.

The principles observed were:

- Shape the excavation to just below formation level and form the required side slopes.
- As the side slopes are formed, immediately and progressively blanket them to allow relief of water seepages, thus preventing any build-up of pore pressures.
- Do not allow continuous transportation of fines; this can only lead to slope instability.
- Remove all seepage waters by controlled (i.e. filtered) sump pumping.

The message is – study and understand the soil structure, and observe the basic installation guidelines. Thereby, the almost impossible can become quite possible.

Figure 27.6 Workable conditions, following wellpoint pumping, under which the channel was actually formed. (Courtesy of Northumbrian Water Limited, Durham, UK.)

27.3.3 Sources and References

BUCHANAN, N. (1970). Derwent Dam – construction. *Proceedings of the Institution of Civil Engineers*, 45, March, pp401–422.

CASHMAN, P M. (1971). Discussion on Derwent Dam. *Proceedings of the Institution of Civil Engineers*, 48, March, pp487–488.

ROWE, P W. (1968). Failure of foundations and slopes in layered deposits in relation to site investigation practice. *Proceedings of the Institution of Civil Engineers*, Supplement, pp73–131.

27.4 CASE HISTORY: DEEP WELLS AND RELIEF WELLS AT TEES BARRAGE, STOCKTON-ON-TEES, UK

This Case History is relevant to:

- Base instability by buoyancy uplift (Section 7.5.1)
- Dewatering methods (Section 9.5)
- Observational method (Section 13.3.2)
- Deep well systems (Chapter 16)
- Relief wells (Section 17.8)

27.4.1 Summary

This Case History describes an application of deep wells and relief wells to depressurize a deep confined aquifer in order to prevent a buoyancy uplift failure of the base of the excavation, which is founded in very low-permeability Glacial Till.

27.4.2 Case History

In the early 1990s, a barrage was constructed across the River Tees at a site between Stockton-on-Tees and Middlesbrough. The barrage was constructed by Tarmac Construction Limited on behalf of the Teeside Development Corporation. The barrage substructure (a reinforced concrete slab 70 m wide, 35 m long and 5 m thick) was formed in a large construction basin (Figure 27.7), which provided dry working conditions, while the river flow was diverted to one side during the works.

The construction basin was approximately 70 m by 150 m in plan dimensions and was contained within bunds containing sheet-pile cut-off walls; the formation level of the basin was at −8 mOD. The bunds were formed 'wet' across the river, and it was intended that the water that remained trapped in the basin when the bunds were closed would be removed by sump pumping. Ground investigations indicated the soil in the floor of the basin to be very stiff clay of Glacial Till, an ideal material on which to found the barrage. However, a few metres below the floor of the basin, a confined aquifer of Fluvio-glacial Sand and gravel was present (see Figure 27.8) with a mean piezometric level of +1 to 2 mOD (up to 10 m above the floor of the excavation). The high piezometric level in this aquifer (and the possible presence of gravel lenses in the Glacial Clay) meant that when the basin was pumped dry, if the piezometric pressure was not lowered significantly, there would be a risk of heave caused by a buoyancy uplift failure of the base of the excavation (Section 7.5.1).

The solution adopted was to install a system of deep wells around the perimeter of the bund to lower the piezometric level down to formation level and ensure that factors of safety against heave were acceptably high. Because there was limited time available for additional ground investigation when the problem was identified, the system design was finalized using the observational method (see Section 13.3.2).

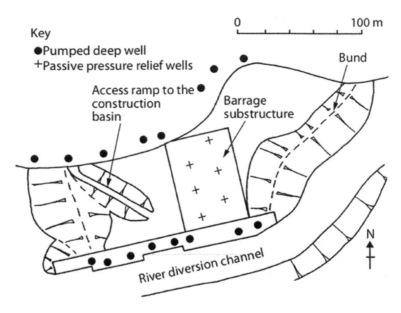

Figure 27.7 Schematic plan of construction basin for Tees Barrage. Deep wells are located to the north and south of the basin. Pressure relief wells are located in the deepest part of the basin. (After Leiper, Q J and Capps, C T F, Temporary works bund design and construction for the Tees Barrage, in *Engineered Fills* (Clarke, B G, Jones, C J F P and Moffat, A I B, eds), Thomas Telford, London, pp. 482–491, 1993.)

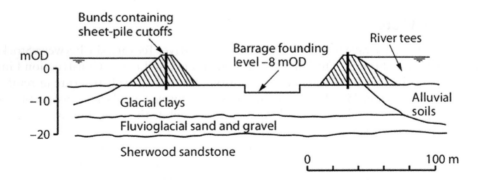

Figure 27.8 Section through construction basin for Tees Barrage showing ground conditions. (After Leiper, Q J and Capps, C T F, Temporary works bund design and construction for the Tees Barrage, in *Engineered Fills* (Clarke, B G, Jones, C J F P and Moffat, A I B, eds), Thomas Telford, London, pp. 482–491, 1993.)

Two wells were installed and test pumped (by constant rate pumping and recovery tests) in turn, with the other well being used as an monitoring well. Although these wells were intended as trial or test wells, they were located carefully so that if the trial was successful, the wells could be incorporated into the final deep well system. The trial allowed flow rate and drawdown data to be gathered, and the system was designed using the cumulative drawdown method (see Section 13.8.2). A particular feature of the test data was that when pumping was interrupted, piezometric pressures recovered rapidly to close to their original levels. This is a characteristic feature of confined aquifers and means that any design should strive to ensure that any breakdowns or interruptions in pumping are minimized.

The final pumped system employed had the following key elements, several of which were specified based on the expectation of the rapid recovery of piezometric levels:

(i) A system of 16 deep wells was installed to abstract from the sand and gravel aquifer. This included an additional two wells over the minimum number required to allow for maintenance or individual pump failure. The wells were drilled by cable percussion methods at 300 mm boring diameter to allow installation of 200 mm nominal diameter well screen and casing, which in turn allowed installation of electro-submersible pumps of 10 l/s nominal capacity.

(ii) The power supply was split into two separate systems, with each part feeding half the pumps. This was to reduce the risk of a power system failure rendering the whole system inoperable.

(iii) Standby generators were permanently connected into the system, ready for immediate start-up, and were connected to alarm systems.

(iv) Twenty-four-hour supervision was provided by the dewatering sub-contractor.

(v) Passive relief wells were installed in the base of the excavation to provide additional pressure relief capacity in the event of a total system failure. In such an event, the wells would have overflowed and flooded the excavation in a controlled manner but without the risk of heave at formation level.

(vi) All piezometers, deep wells and pressure relief wells were monitored regularly to ensure that the system was operating satisfactorily. A positive management system was established for the monitoring, with a proper inspection record kept by an approved and suitably experienced and qualified individual.

The system produced a total yield of 95 l/s in steady-state pumping (compared with a nominal installed pumping capacity of 16 wells at 10 l/s each) and operated for the 11 month life of the basin. On completion, the pumps were removed, and the deep wells and relief wells were decommissioned by backfilling with gravel, bentonite and concrete.

27.4.3 Sources and References

FRANKLIN, J B and CAPPS, C T F. (1995). Tees Barrage – construction. *Proceedings of the Institution of Civil Engineers, Municipal Engineering*, 109, September, pp196–211.

LEIPER, Q J and CAPPS, C T F. (1993). Temporary works bund design and construction for the Tees Barrage. *Engineered Fills* (Clarke, B G, Jones, C J F P and Moffat, A I B, eds), Thomas Telford, London, pp482–491.

27.5 CASE HISTORY: EJECTOR WELLS FOR SHAFT DEWATERING, WITH HIGH-PERMEABILITY FEATURE PRESENT, UK

This Case History is relevant to:

- Localized groundwater problems (Section 7.6)
- Dewatering methods (Section 9.5)
- Groundwater control methods for shafts (Section 10.5)
- Ejector well systems (Section 17.2)

27.5.1 Summary

This Case History describes an ejector well system used to lower groundwater levels for an 8 m deep shaft sunk through an aquifer of Fluvio-glacial Sand. During shaft sinking, persistent water seepages were encountered on one side of the excavation. Investigation drilling revealed that this was due to the presence of a shoestring of highly permeable gravel within the sand.

27.5.2 Case History

Section 7.6.3 described how if a relatively permeable aquifer (such as sand) contains lenses or zones of highly permeable water-bearing material (such as clean gravel), then dewatering becomes more difficult. The highly permeable zone can form a source of copious recharge, requiring additional pumping capacity to be installed to abstract groundwater directly from the more permeable zone.

A small shaft of approximately 4 m diameter was constructed by underpinning methods in northern England through an extensive stratum of fine to coarse sand of fluvio-glacial origin. The final shaft depth was around 8 m, penetrating to 4 m below groundwater level, and groundwater levels were controlled by a pumped well system. The excavation was too deep to be dewatered by a single stage of wellpoints, and the chosen method was a ring of five ejector wells installed externally around the shaft (Figure 27.9).

The initial ejector well system had a nominal capacity of approximately 2.5 l/s and yielded around 1 l/s in steady-state pumping, with the objective of lowering groundwater levels by around 4 m. Observations in a monitoring well adjacent to one side of the shaft indicated that the groundwater level had been lowered by the required amount. However, during excavation, persistent troublesome seepage was encountered on one side of the shaft just below the original groundwater level. This seepage destabilized the sand, and shaft sinking was forced to stop. Several additional internal ejector wells were jetted into the base of the shaft (Figure 27.10). This improved conditions slightly, but the localized seepage and instability persisted, preventing further shaft-sinking progress.

A site investigation rig was brought to site and used to drill investigation boreholes outside the shaft. The drilling programme was concentrated outside the 'wet' side of the shaft. Logs of the investigation boreholes eventually revealed the presence of a 'shoestring' of permeable gravel present within the main stratum of Fluvio-glacial Sand (Figure 27.11). 'Shoestring' is a term for a long, thin gravel lens that may result from paths of former streams or flow

Figure 27.9 External ejector wells around the shaft.

Figure 27.10 Additional ejector wells installed inside the shaft. The problematic seepage is visible in the right hand side of the excavation.

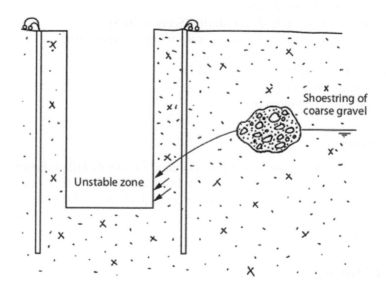

Shoestring of coarse gravel

Unstable zone

Figure 27.11 Effect of close source of recharge on effectiveness of groundwater lowering for shaft excavation. A permeable gravel shoestring acts as a close source of recharge and concentrates seepage on one side of the shaft, leading to local instability. (From Preene, M and Powrie, W, *Proceedings of the Institution of Civil Engineers, Geotechnical Engineering,* 10717–26, 1994. With permission.)

channels within the geological deposits. Like modern-day watercourses, these gravel lenses may follow an irregular, difficult-to-predict path. These may easily pass undetected during site investigation – this seems to have been the case here.

Additional ejector wells were installed into the shoestring to intercept the water before it reached the excavation, the seepage was eliminated and the shaft was successfully completed to final depth. This sounds very straightforward, but this was far from the full story.

Because the location of the shoestring was not known precisely, the dewatering contractor had to keep drilling wells until a few of them directly intercepted the permeable zone. Up until that time, even when many additional wells had been installed and pumped, the seepage persisted. The total number of wells increased from five to 22! In reality, only the handful pumping directly from the shoestring was truly effective. The total pumped flow rate increased from around 1 l/s to almost 4 l/s when the shaft was completed. Most of this flow was coming from a handful of ejector wells that directly penetrated the shoestring.

27.5.3 Sources and References

PREENE, M and POWRIE, W. (1994). Construction dewatering in low permeability soils: some problems and solutions. *Proceedings of the Institution of Civil Engineers, Geotechnical Engineering*, 107, January, pp17–26.

27.6 CASE HISTORY: HYDRAULIC FAILURE OF BASE OF SHAFT, UK

This Case History is relevant to:

- Base instability by buoyancy uplift (Section 7.5.1)
- Dewatering methods (Section 9.5)
- Groundwater control methods for shafts (Section 10.5)
- Groundwater control methods for recovery works (Section 10.8)
- Determination of ground profile (Section 11.9)
- Assessment of base stability (Section 13.7)
- Sump pumping (Chapter 14)
- Deep well systems (Chapter 16)

27.6.1 Summary

This Case History describes a project where the base of a shaft failed during construction, resulting in flooding of the shaft. Post-failure investigations revealed that the cause was unrelieved water pressures in a confined aquifer beneath the base of the shaft, which had not been identified prior to construction. Recovery works involved a pumped deep well system to reduce groundwater pressures at depth.

27.6.2 Case History

A 19 m deep shaft, approximately 6 m in diameter, was constructed at a site in northern England. Very low-permeability Glacial Till was indicated to be present for the full depth of the shaft, and the shaft-sinking method was by underpinning with a segmental concrete lining. The only groundwater control measure planned was sump pumping to deal with any localized seepages from any sand lenses that might be present within the generally clay-dominated Till deposit.

The shaft was successfully dug to full depth (19 m) with only very limited sump pumping required. The shaft was lined incrementally with concrete segments, and when final depth was reached at the end of a working week, a thin layer of blinding concrete was placed in the base of the shaft. When work finished on that Friday evening, the shaft was dry, and the crews planned to start work after the weekend, placing the reinforcement steel for the concrete base slab.

However, when workers returned to site on the following Monday, they were surprised to discover that the shaft was full of water to just below ground level! The 'failure' was not observed, but it was clear that there had been a significant problem – more than 500 m³ of water had entered the shaft over the weekend, despite the Till being of very low permeability.

Immediately after the failure, the original ground investigation information was reviewed, and this revealed that the deepest pre-construction investigation borehole went to 23 m, only 4 m below the base of the shaft (Figure 27.12a). This is much shallower than good practice, which recommends that boreholes in the area of the excavation should penetrate to a depth of

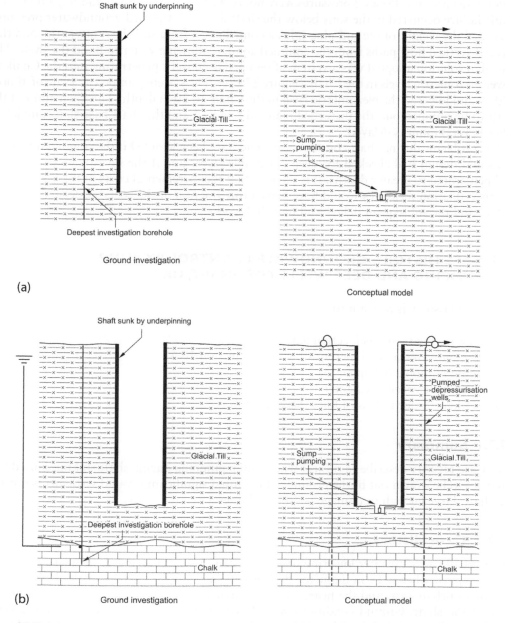

Figure 27.12 Information on ground conditions at the shaft location. (a) Pre-construction ground investigation and conceptual model. (b) Post-failure ground investigation and conceptual model.

1.5 to 2 times the depth of the excavation. A post-failure desk study revealed that the bedrock beneath the site was Chalk (a fractured limestone), which formed a confined aquifer in the region. The depth to the Chalk was not known when the shaft was designed and constructed. Deeper piezometer boreholes drilled post-failure showed that the top of the Chalk was 6 m below the base of the shaft, with a piezometric level just below ground level (Figure 27.12b).

The cause of the problem was that the original investigation did not recognize the presence of a confined aquifer below the base of the shaft, and their conceptual model implicitly assumed that low-permeability strata existed for a significant depth below excavation (Figure 27.12a). This led to a sump pumping solution with no consideration of pressure relief of deep strata. Because pressures were not reduced in the confined aquifer, a buoyancy uplift failure occurred in the soils below the shaft dig level. Upward groundwater pressures exceeded the weight of the soil plug (plus a contribution from soil shear strength), and this created vertical flow paths from the confined aquifer, allowing water to flood the shaft. The remedial solution adopted was a system of pumped deep wells penetrating into the Chalk to lower piezometric levels in the Chalk (Figure 27.12b). The system pumped at a rate of almost 100 l/s and lowered the piezometric level in the Chalk to 20 m depth. This then allowed the shaft to be pumped out, the disturbed material removed and the shaft base slab completed, albeit several months delayed.

The problems during shaft construction could have been avoided if the original investigations and design had checked for deep permeable layers and if the investigation boreholes had penetrated to greater depths. It is obvious that the incremental costs of these additional site investigation measures would have been a tiny fraction of the costs associated with the failure and the subsequent delays and remedial works.

27.7 CASE HISTORY: GROUNDWATER CONTROL BY SUMP PUMPING AND DEEP WELLS FOR SHAFT, UK

This Case History is relevant to:

- Dewatering methods (Section 9.5)
- Groundwater control methods for shafts (Section 10.5)
- Sump pumping (Chapter 14)
- Deep well systems (Chapter 16)
- Settlement due to loss of fines (Section 21.4.3)

27.7.1 Summary

This Case History describes a project where sump pumping during shaft sinking caused fine particles to be drawn out of the ground. This caused significant ground settlements outside the shaft, large enough to damage nearby buildings.

27.7.2 Case History

Occasionally, groundwater lowering operations can cause damaging settlements. One such case occurred when a 15 m diameter shaft was being sunk by caisson methods. The shaft was in an urban area, and the nearest existing structure was located around 5 m from the edge of the shaft. Ground conditions were Glaciolacustrine Deposits that were predominantly very low-permeability clay with some laminations or thin layers/lenses of sand were expected. The Glaciolacustrine Deposits were underlain by fractured bedrock of higher permeability, which formed a confined aquifer (Figure 27.13a).

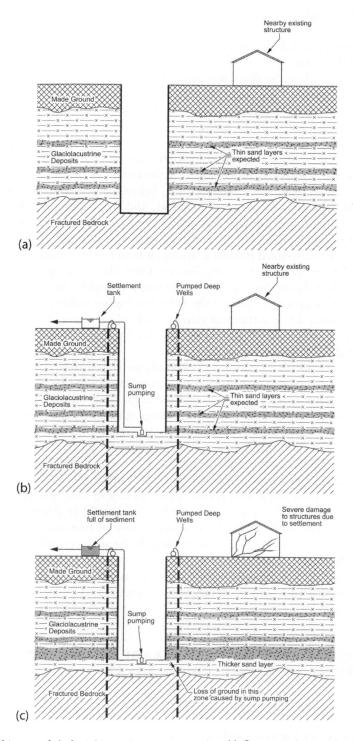

Figure 27.13 Case history of shaft sinking using sump pumping. (a) Conceptual ground conditions, showing shaft completed to full depth. (b) Anticipated ground conditions when shaft sinking through the Glaciolacustrine Deposits. (c) Ground conditions encountered when shaft sinking through the Glaciolacustrine Deposits.

The proposed groundwater control measures are shown in Figure 27.13b and included:

i. A system of pumped deep wells to lower piezometric levels in the bedrock to stabilize the base of the shaft.
ii. Sump pumping while the shaft was sunk through the Glaciolacustrine Deposits. The investigation boreholes had recorded very little water inflow from sandy zones with the clay, and it was planned that the shaft would be sunk as a 'dry caisson' (Figure 10.2c) with excavating plant working in the base of the shaft.

The deep well system was established, and shaft sinking progressed. Inflows to the shaft were generally small. However, inflows were reported to be rather greater when sinking through a more sandy zone, a few metres thick, below 10 m depth. As the shaft was dug further, it passed back into more clayey material, and sump pumping flow rates reduced. Although the sump pumping allowed the shaft to be sunk, the site team observed that the settlement tanks through which the discharged water was pumped had become filled with sediments of silt and sand-sized particles. So much material had built up in the tanks that the sediment had to be dug out by operatives on more than one occasion. As is often the case, no detailed observations were made during sump pumping, but it is likely that most of the sediment had been generated during the period of higher sump pumping.

Over the subsequent days, settlement damage was observed in the nearest structure (a residential house of masonry construction). Eventually, the cracking and damage to the structure rendered the building uninhabitable.

Review of the ground investigation information indicated that all the strata were relatively stiff, and it was unlikely that damaging settlements could have resulted purely from the increases in effective stress caused by lowering of groundwater levels. The likely cause of the ground settlement was 'loss of fines' when digging through the thicker sand zone. The mechanism is that fine-grained (silt and sand-sized) particles are removed from the ground by flowing groundwater during sump pumping (Figure 27.13c).

The loss of fines mechanism resulted in a void or loosened zone immediately outside the shaft at the level of the main sandy zone below 10 m. The reports of the sediment being dug out of the settlement tank imply that several cubic metres of sand/silt were removed during sump pumping. Even though the material was lost from a 10 m deep zone, surface settlement still occurred. The probable mechanism is collapse or relaxation of the original void/loosened zone, which then migrates upwards towards the surface. With this mechanism, the void may reach the surface, and a 'crown hole' may open up. In other cases, a crown hole may not form, but a surface depression may develop. In either case, the foundation support to any buildings in this zone will be significantly reduced, and building damage is likely, as occurred on this project.

27.8 CASE HISTORY: ARTIFICIAL RECHARGE TO CONTROL SETTLEMENT: MISR BANK, CAIRO, EGYPT

This Case History is relevant to:

- Base instability by buoyancy uplift (Section 7.5.1)
- Dewatering methods (Section 9.5)
- Wellpoint systems (Chapter 15)
- Artificial recharge systems (Section 17.13)
- Settlement due to groundwater lowering (Section 21.4.1)

- Mitigation of potential environmental impacts (Section 21.10.3)
- Encrustation, biofouling and corrosion (Section 23.5)

27.8.1 Summary

This Case History describes an example of an artificial recharge system used to reduce the magnitude of ground settlements on a congested urban site.

27.8.2 Case History

An interesting application of recharge was carried out in the centre of Cairo, Egypt. A two-storey basement was to be constructed for the new Bank Misr headquarters immediately adjacent to the existing bank building. The basement was to be formed inside a sheet-piled cofferdam to a depth of 7.5 m below ground level and approximately 4.5 m below original groundwater level.

Although the sides of the excavation were supported by sheet-piles, and a stratum of fissured clay was expected at final formation level, the presence of permeable sand layers below the excavation required reduction of pore water pressures at depth to prevent base heave. In planning the groundwater lowering operations, two issues were of concern:

(i) To limit drawdown of groundwater levels beneath adjacent structures (which were founded on compressible alluvial soils) in order to prevent damaging settlements
(ii) To dispose of the discharge water, which could not be accepted by the local sewer network, which was heavily overloaded

A system of recharge wells, used in combination with wellpoints pumping from within the excavation, was used to solve the problem. Without such a system, it is unlikely that the Cairo authorities would have allowed the work to proceed.

The sequence of strata revealed in the site investigation is shown in Figure 27.14. Falling head tests in the silty sand below formation level gave results of the order of 10^{-5} m/s, but

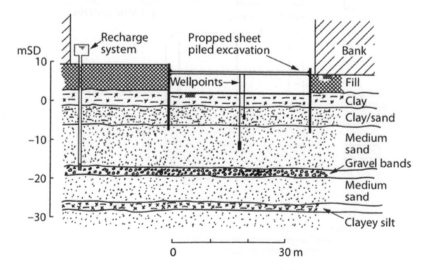

Figure 27.14 Bank Misr, Cairo: section through dewatering and recharge system. (After Troughton, V M, Groundwater control by pressure relief and recharge, *Groundwater Effects in Geotechnical Engineering* (Hanrahan, E T, Orr, T L L and Widdis, T F, eds), Balkema, Rotterdam, pp. 259–264, 1987.)

pumping tests at the site indicated that the permeability was more probably of the order of 10^{-4} m/s, consistent with Hazen's method applied to particle size tests on the sand. Due to the complexity of the interaction between pumping and recharge, a finite element numerical groundwater model was set up for the site. Based on the soil fabric in the alluvial soils, the horizontal permeability was assumed to be significantly greater than the vertical permeability.

The numerical model was used to predict the pore water pressure reduction needed to safeguard the base of the excavation and to model the artificial recharge system required to limit the drawdown beneath adjacent structures to less than 1 m. Calculations indicated that a 1 m drawdown would result in 12 mm settlement of the fill and alluvial soils, with differential movements within acceptable limits. Figure 27.15 shows a schematic plan view of the system, with an abstraction wellpoint system inside the cofferdam and recharge wells between the excavation and critical buildings.

Abstraction wellpoints were installed to pump from strata at two levels at 13.5 and 21.5 m below ground level. The recharge wells were installed in three lines with wells at 3 m spacing. The recharge wells (which consisted of 125 mm diameter screen and casing installed in a 250 mm diameter borehole) were designed to make use of the soil stratification at the site by injecting water into the most permeable horizon present, a layer of gravel at 28 m below ground level. Water could be relatively easily injected into this permeable layer, from where it would feed to the other soil layers – analogous to the underdrainage dewatering effect (see Section 13.6.2) in reverse. The discharge water was fed to the recharge wells from a header tank 2 m above ground level; total flow rate was in the range of 25–40 l/s. An array of pneumatic piezometers (a type of rapid-response piezometer) was installed to monitor pore water pressure reductions in the compressible alluvial clays. Conventional standpipe piezometers were used in the sands.

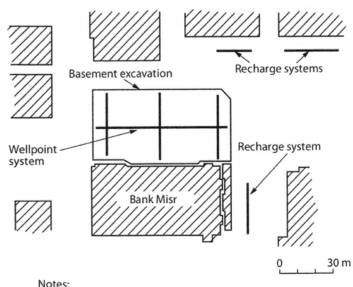

Notes:
Recharge system consisted of recharge wells at 3 m centers
Pumping system consisted of wellpoints at 3 m centers

Figure 27.15 Bank Misr, Cairo: plan layout of dewatering and recharge system. (After Troughton, V M, Groundwater control by pressure relief and recharge, *Groundwater Effects in Geotechnical Engineering* (Hanrahan, E T, Orr, T L L and Widdis, T F, eds), Balkema, Rotterdam, pp. 259–264, 1987.)

In operation, the maximum drawdowns observed below structures were limited to 0.7 m, which resulted in settlements of 3 mm, around half the value predicted by calculation. As is often the case, the efficiency of the recharge wells reduced with time, and they had to be redeveloped by airlifting at monthly intervals to maintain performance. Two of the lines of recharge wells were located in a car park used during daytime business hours – these had to be re-developed at night.

As mentioned earlier, one of the aims for the recharge system was to dispose of the discharge water without overloading the antiquated sewer system. However, as reported by Cashman (1987), the dewatering field supervisor did not have a lot of faith in recharge and tapped into the sewer system with a hidden discharge pipe. So, in fact, much of the discharge water was actually going down the sewer. Unfortunately, over the Christmas period, one of the main Cairo pumping stations broke down, and everything flooded back. The chairman of the main contractor received a telephone call from the Mayor of Cairo Municipality demanding his personal presence on site immediately. He was told that if this ever happened again, he, the chairman, would be immediately put in jail. This was a pretty convincing argument to ensure that things were put right, and the recharge system stayed in operation.

27.8.3 Sources and References

CASHMAN, P M. (1987). Discussion. *Groundwater Effects in Geotechnical Engineering* (Hanrahan, E T, Orr, T L L and Widdis, T F, eds), Balkema, Rotterdam, p1015.

TROUGHTON, V M. (1987). Groundwater control by pressure relief and recharge. *Groundwater Effects in Geotechnical Engineering* (Hanrahan, E T, Orr, T L L and Widdis, T F, eds), Balkema, Rotterdam, pp259–264.

27.9 CASE HISTORY: ARTIFICIAL RECHARGE TO PROTECT DRINKING WATER SOURCE, UK

This Case History is relevant to:

- Dewatering methods (Section 9.5)
- Deep well systems (Chapter 16)
- Artificial recharge systems (Section 17.13)
- Effect on groundwater quality (Section 21.4.9)
- Effect on groundwater borehole or spring supplies (Section 21.4.12)
- Vertical pathways via excavations and structures (Section 21.6.2)
- Mitigation of potential environmental impacts (Section 21.10.3)
- Encrustation, biofouling and corrosion (Section 23.5)

27.9.1 Summary

This Case History describes an example of artificial recharge used to reduce impacts on nearby public water supply boreholes.

27.9.2 Case History

Some of the challenges of artificial recharge systems were highlighted during a project in northern England in the 1990s. A large excavation was to be constructed to form the portal of a proposed road tunnel beneath a river. This involved dewatering of, and excavation into, a confined aquifer, which was extensively exploited for public drinking water supply via high-capacity wells. In particular, one major public water supply well was located within a few kilometres of the site.

During planning of the dewatering, it was identified that the project posed two potentially significant risks to the aquifer and the public water supply well. First, in the location of the tunnel, the aquifer was confined by a thick sequence of Quaternary deposits of very low vertical permeability. This layer would be punctured by the excavation, meaning that a site pollution incident (such as a fuel spill) could result in pollution of the public supply aquifer. Second, the pumping of groundwater during dewatering would lower groundwater levels in the aquifer over a wide area and could potentially affect the public supply well, reducing its yield. As a result, the public water supply in the region could have been detrimentally affected.

The first risk was dealt with by good site management practices to minimize the likelihood of a site pollution incident in the areas where the aquifer was temporarily exposed. Additionally, the permanent works were designed to ensure that they formed a seal through the Quaternary deposits to replicate the natural very low vertical permeability.

The second risk was mitigated in consultation with the environmental regulator, resulting in the use of artificial recharge. Unlike Case History 27.8, where the objective was to minimize settlement around defined structures, in this case, the objective was to reduce net abstraction of groundwater from the aquifer. Therefore, there was no requirement to cluster the recharge wells around specific buildings or areas; instead, the aim was to return the water to the aquifer over a relatively wide area. By agreement with the environmental regulator, a defined flow rate (significantly lower than the maximum dewatering pumping rate) could be discharged to the river, with the remainder returned to the aquifer via recharge wells.

This type of application of an artificial recharge system allows some flexibility in the location of recharge wells. However, consideration must be given to the interaction between the dewatering and the recharge system. The dewatering flow rate was substantial (in excess of 100 l/s). Numerical groundwater modelling showed that if recharge wells were located within approximately 750 m of the dewatering wells, then a significant proportion of the recharged water would be drawn to the dewatering system, decreasing drawdown at the excavation and increasing pumped flow rates in a feedback loop.

Based on the numerical modelling, it was determined that the recharge wells should be located more than 750 m from the dewatering system around the tunnel portal. This raised the problem that it would require recharge wells in land not under the control of the construction project. This was addressed by making the best use of the land available to the project to develop four arrays of recharge wells (Figure 27.16a). The route of the proposed road was used as much as possible as a pipeline route to transmit the water from the dewatering system to the recharge wellfields. Some recharge wells were located in the verge of the partially constructed road (Figure 27.16b). Other recharge wellfields were located in land leased from local landowners.

The distance between the dewatering system and the recharge wellfields meant that it was not practicable or cost effective to use the borehole submersible pumps to drive the water to the recharge wells. An array of booster pumps (Figure 27.17) was used to boost the water along pipelines to the recharge wells. In an attempt to minimize clogging by iron-related compounds, the system was operated as a 'sealed system' to reduce contact between the pumped water and air. The use of a sealed system also had the advantage that it reduced the risk of external contamination affecting the recharge water. To reduce the risk of contamination of the aquifer and the public water supply wells, it was essential that the water recharged into the aquifer was uncontaminated, and this was checked by a programme of regular chemical testing.

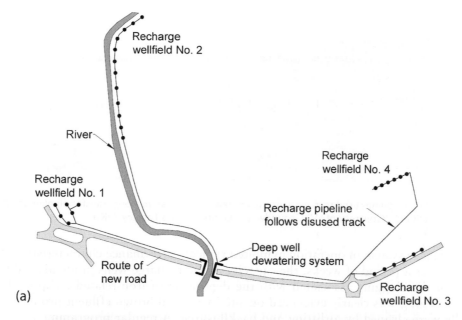

(a)

(b)

Figure 27.16 Location of recharge wells on road construction project. (a) Location of recharge wells. (b) Recharge wells located in verge of partially constructed road. (Panel (b) courtesy of WJ Groundwater Limited, Kings Langley, UK.)

Figure 27.17 Booster pumps used to drive water from dewatering system along transfer pipeline to recharge wellfield. (Courtesy of WJ Groundwater Limited, Kings Langley, UK.)

The recharge system in operation exhibited typical performance characteristics of such systems. Over a period of weeks, the capacity of the recharge wells gradually reduced as a result of the clogging associated with the deposition of iron-related compounds. The type of clogging was clearly evidenced on site by the red-brown effluent produced when the wells were cleaned by airlifting and backflushing. A regular programme of rehabilitating wells by these methods was adopted, which typically restored each recharge well back to close to its original capacity for a few weeks. However, after a further few weeks of operation, the reduction in recharge capacity again became significant, and the well was rehabilitated again. In each recharge wellfield, the process of well rehabilitation was almost continuous, with a crew scheduled to move sequentially from well to well in a regular cycle.

27.10 CASE HISTORY: LOW-PERMEABILITY CUT-OFF WALL AT SIZEWELL B POWER STATION, UK

This Case History is relevant to:

- Exclusion methods (Section 9.4)
- Dewatering methods (Section 9.5)
- Exclusion methods to reduce external groundwater impacts (Section 9.6.2)
- Deep well systems (Chapter 16)
- Concrete diaphragm walls (Section 18.8)
- Verification of groundwater exclusion methods (Section 18.14)
- Settlement due to groundwater lowering (Section 21.4.1)
- Impact on groundwater-dependent features (Section 21.4.8)
- Breaching of cut-off walls (Section 24.5)

27.10.1 Summary

This Case History describes a project where a deep and extensive very low-permeability cut-off wall was used to enclose a construction site with the objective of reducing impacts on neighbouring sites.

27.10.2 Case History

In the late 1980s, Sizewell B Power Station was constructed in Suffolk, United Kingdom. The new infrastructure, termed the 'B Station', was being constructed immediately adjacent to the existing 'A Station'. The A Station had been constructed in the 1960s, and its foundations had been dewatered using a relatively small number of wells pumping from a 40 to 50 m thick stratum of fine to medium sand of Quaternary age, known as the 'Crag Sand', where horizontal permeability was of the order of 10^{-4} m/s.

Haws (1970) reports that during the early stages of the B Station design in the 1960s, dewatering of a large open cut excavation was modelled by an analogue method using Teledeltos paper. The results indicated that for a required drawdown of around 10 m, a system of 25 wells pumping an initial rate of 400 l/s could achieve drawdown in around 22 days. Long-term pumping rates were predicted to be around 100 l/s. These predicted dewatering flow rates are large but manageable. However, a drawdown of around 6 m was predicted beneath the A Station. This implied that effective stress settlements due to groundwater lowering should be of concern given the sensitive operating requirements of the neighbouring power station. During the 1970s, a series of well pumping tests (Figure 27.18) were carried out, typically pumping at 36 l/s for 5 days' duration.

Later stages of design in the 1980s modelled the dewatering system using another type of analogue model – a resistance-capacitance network. During planning for construction of the B Station, modelling and geotechnical calculations indicated that the effective stress increase in the Crag Sands caused by lowering of groundwater levels by dewatering for the new foundations would result in unacceptable ground settlements beneath the A Station. There were also concerns that the drawdown of groundwater levels might have a detrimental effect on nearby marshland and farm irrigation systems.

The solution adopted was to enclose the entire perimeter of the B Station construction site within a 1259 m long, 0.8 m wide concrete diaphragm wall cut-off with a target permeability of 1×10^{-8} m/s. The diaphragm wall extended to 56 m depth to key into the very low-permeability London Clay Formation beneath the Crag Sands (Figure 27.19). A system of nine dewatering wells, each with a maximum pumping capacity of 28 l/s, was operated within the area enclosed by the cut-off and lowered the water level to 16 m below sea level.

Figure 27.18 Well pumping test at Sizewell, 1976. (Courtesy of David Hartwell.)

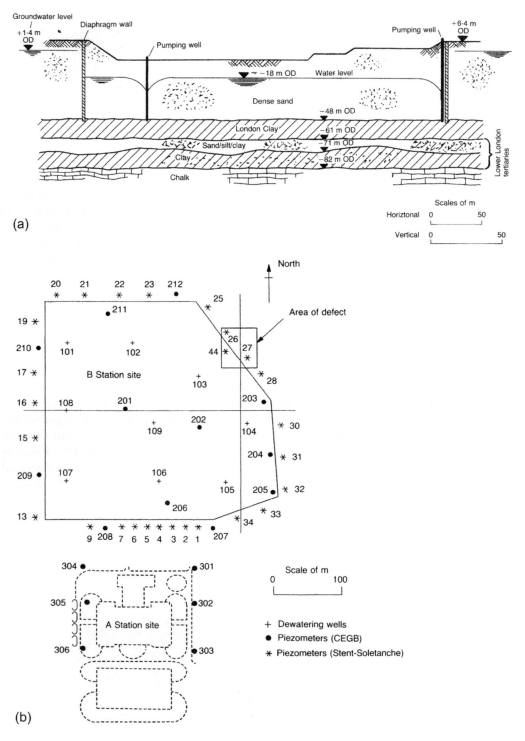

Figure 27.19 Use of diaphragm wall at Sizewell B Power Station. (a) Typical geological section showing diaphragm wall and dewatering wells. (b) Position of dewatering wells and monitoring piezometers. (From Howden, C and Crawley, J D, *Proceedings of the Institution of Civil Engineers, Civil Engineering, Sizewell B Power Station*, 108, 48–62, 1995. With permission.)

The total dewatering pumping rate peaked at approximately 250 l/s during the pumping-down phase and then reduced rapidly, ultimately stabilizing at approximately 30 l/s during long term pumping. Dewatering pumping continued for just over 4 years, from late 1987 to early 1992 (Figure 27.20).

During the pumping-down phase, the groundwater levels in a small number of external piezometers began to fall, and subsequent investigations indicated that there was a permeable defect in the north east section of the cut-off wall. The mitigation measures deployed included tube à manchette grouting in the vicinity of the suspected area of cut-off wall. Knight *et al.* (1996) state that back analysis indicates that over the life of the dewatering, the leak contributed around 15 l/s of the long-term dewatering flow rate. Once the wall defect was sealed, groundwater levels outside the dewatered excavation remained very close to their natural level of around 1 m above sea level. Construction was completed with no significant dewatering-related external impacts.

During design of the works, there was a concern that after construction, rainfall infiltration would cause water levels to rise to excessively high levels beneath the B Station within the area enclosed by the cut-off wall. At the end of the dewatering period, the cut-off wall was deliberately breached by excavating out over a 350 m length down to −0.8 mAOD (an area of wall of 525 m²). The level of the breach was designed to be below background groundwater level so that in the long term, groundwater beneath the B Station would be in connection with the external groundwater regime.

Figure 27.20 View of excavation work by the diaphragm wall on the construction site of Sizewell B Nuclear Power Station. (©Historic England Archive, John Laing Photographic Collection.)

27.10.3 Sources and References

BARRATT, G O and VOWLES-SHERIDAN, M J. (1995). Construction of the cooling water tunnels using immersed tube techniques. *Proceedings of the Institution of Civil Engineers, Civil Engineering, Sizewell B Power Station*, 108, February, pp63–72.

GAMMON, K M and PEDGRIFT, G F. (1962). The selection and investigation of potential nuclear power station sites in Suffolk. *Proceedings of the Institution of Civil Engineers*, 21, 1, January, pp139–160.

HAWS, E T. (1970). Discussion. *Proceedings of the Conference on Ground Engineering*. Institution of Civil Engineers, London, pp68–70.

HOWDEN, C and CRAWLEY, J D. (1995). Design and construction of the diaphragm wall. *Proceedings of the Institution of Civil Engineers, Civil Engineering, Sizewell B Power Station*, 108, February, pp48–62.

KNIGHT, D J, SMITH, G L and SUTTON, J S. (1996). Sizewell B foundation dewatering – system design, construction and performance monitoring. *Géotechnique*, 46, No. 3, pp473–490.

27.11 CASE HISTORY: CONTAMINATION REMEDIATION USING CUT-OFF WALLS AND GROUNDWATER ABSTRACTION AT DERBY PRIDE PARK, UK

This Case History is relevant to:

- Exclusion methods (Section 9.4)
- Dewatering methods (Section 9.5)
- Deep well systems (Chapter 16)
- Dewatering and groundwater control technologies used for the control or remediation of contaminated groundwater (Section 17.14)
- Slurry trench walls (Section 18.7)

27.11.1 Summary

This Case History describes a case where a slurry trench cut-off wall was used to isolate a contaminated zone of a shallow gravel aquifer as part of a groundwater remediation scheme that included a groundwater abstraction and treatment system.

27.11.2 Case History

During the 1990s, extensive contamination remediation works were carried out on a 96 hectare site close to the centre of Derby, United Kingdom. The largely derelict site had formerly been used for domestic/industrial landfill, coke, gas and heavy engineering works, and gravel extraction. It was heavily contaminated with pollutants, including oils, tars, heavy metals, phenols, ammonium and boron. The site was remediated on behalf of Derby City Council, with Arup acting as designers. The remediated site was to be known as Derby Pride Park.

It was intended to reclaim the site for mixed commercial, leisure and residential use. The remediation strategy included an extensive groundwater cut-off wall and associated groundwater abstraction system.

A comprehensive programme of ground investigation indicated that ground conditions comprised fill and Alluvium over highly permeable gravels of the River Terrace Deposits, which were in turn underlain by lower-permeability bedrock of the Mercia Mudstone Group. Extensive areas of soil contamination existed on the site, and groundwater contamination was detected in the gravels of the River Terrace Deposits. There was concern that existing contamination could migrate from the site and enter an adjacent river.

As part of the wider remediation strategy, a groundwater cut-off wall was constructed around the most contaminated area of the site. The solution finally adopted was a slurry trench wall, approximately 3 km long and a maximum of 10 m deep, around the perimeter of the identified area to create an enclosing low-permeability barrier, cutting off the fill, Alluvium and River Terrace Deposits and keying into the Mercia Mudstone bedrock by 1 m (Figure 27.21).

The wall was constructed as a 600 mm thick cement-bentonite slurry trench wall, dug by backactor. The slurry mix design comprised cement, bentonite and ground granulated blast furnace slag (GGBS) with a target permeability of 1×10^{-8} m/s at 28 days. Additionally, an HDPE membrane with a design permeability of 1×10^{-9} m/s was placed within the wall. The membrane was installed in panels, using a special handling frame, each panel being connected to its neighbours using pre-fitted interlock joints. The wall alignment crossed 36 pre-existing underground services (pipes, sewers and ducts), which had to be incorporated through the wall (via welded seals in the HDPE membrane) without compromising design permeability. Once the wall was at least 7 days old, the top 0.5 m was dug out and replaced with a clay cap.

Numerical groundwater modelling at design stage predicted that the presence of the cut-off walls would have a modest impact on external groundwater levels. The barrier effect of the wall was predicted to increase external groundwater levels by approximately 0.7 m on the upgradient side of the site.

Groundwater modelling also predicted that without control measures, groundwater levels within the area enclosed by the wall would rise, partly due to infiltration from surface precipitation and partly due to upward seepage of groundwater from the underlying Mercia Mudstone bedrock. The solution adopted was to install an array of groundwater abstraction wells to maintain groundwater levels inside the wall slightly (approximately 0.25 m) lower than external groundwater levels. As well as preventing the problems associated with groundwater level rises within the site, this solution had two other advantages. First, it provided an element of 'hydraulic containment', so that if there was a leak or breach in the wall, water would flow into the site rather than contaminated water flowing out. Second, pumping of the contaminated water from within the site would, over time, remove much of the dissolved contamination from the site, resulting in a gradual clean-up of contamination levels.

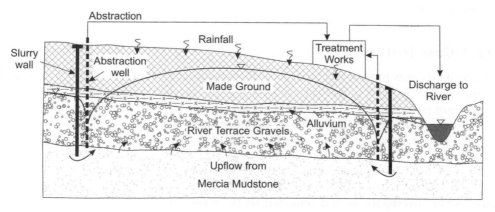

Figure 27.21 Schematic section through Derby Pride Park remediation project. (After Braithwaite, P, *et al.*, *Arup Journal*, 31, 13–15, 1996.)

The groundwater abstraction wells were connected via a ring main to a water treatment plant located in one corner of the site. The water was treated to a standard determined by the regulatory permissions for water discharge at the site, and the treated water was then disposed of to the river. The groundwater abstraction and treatment system was intended to operate for 15 years. Due to the relatively long design life, the treatment plant was not of the temporary modular tank type commonly used on groundwater control projects but was instead constructed to similar standards as permanent wastewater treatment plants.

27.11.3 Sources and References

BARKER, P, ESNAULT, A and BRAITHWAITE, P. (1998). Containment at Pride Park, Derby, England. *Land Contamination and Reclamation*, 6, 1, pp51–58.
BRAITHWAITE, P, WADE, S and WEBB, G. (1996). Pride Park, Derby. *Arup Journal*, 31(1), pp13–15.

27.12   CASE HISTORY: PERMANENT GROUNDWATER CONTROL SYSTEM AT GOVAN UNDERGROUND TUNNEL, GLASGOW, UK

This Case History is relevant to:

- Dewatering methods (Section 9.5)
- Groundwater control methods for tunnels (Section 10.4)
- Drilling out from existing tunnels (Section 10.6.1.2)
- Wellpoint systems (Chapter 15)
- Deep well systems (Chapter 16)
- Collector wells (Section 17.6)
- Permanent groundwater control systems (Chapter 20)
- Encrustation, biofouling and corrosion (Section 23.5)

27.12.1   Summary

This Case History describes a permanent groundwater control system using collector wells to ensure the long-term stability and operation of a railway tunnel that forms part of the Glasgow Subway.

27.12.2   Case History

The City of Glasgow in Scotland has the third oldest underground railway in the world, dating from the 1890s. Now operated by Strathclyde Partnership for Transport (SPT), it forms an important part of the public transport network in the city; any prolonged interruption in its operation would be very disruptive to the citizens of Glasgow.

In the 1980s, part of the tunnel near Govan Station, where the layout was a twin tunnel with a crown and invert of concrete and brick sides, experienced inflows of water and fine sand through minor defects in the tunnel walls. The loss of material from beneath the tunnel caused settlement of around 75 mm, and that part of the railway had to be closed for several months while investigations and remedial works were carried out. The original remedial works included the installation of a permanent dewatering system, comprising vertical wellpoints through the track bed (Figure 27.22 a and b), pumped by wellpoint pumps located

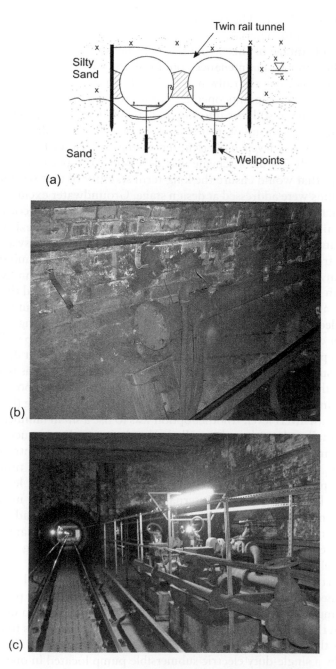

Figure 27.22 Internal wellpoint system within tunnel. (a) Cross section through internal wellpoint system. A line of wellpoints is installed through the trackbed of each of the tunnels. The wellpoints are connected to header mains mounted within the tunnels. (b) Wellpoint header main mounted on internal face of railway running tunnel. The flexible connecting pipe to one of the wellpoints is visible in the foreground. (c) Wellpoint pumps within the tunnel. Electrically driven duty and standby wellpoint pumps are situated in an existing space at one end of the tunnels. The railway tracks and one of the tunnels are visible in the left of the picture. (Courtesy of Strathclyde Partnership for Transport, Glasgow, UK.)

at track level (Figure 27.22c) The wellpoint system was effective in lowering piezometric levels in the silty fine sand materials around the tunnel, with pumped flow rates of the order of 20 l/s. One disadvantage of this layout was that because the wellpoints and pump were located within the tunnels, maintenance could only be carried out during the short overnight track possessions when the railway was shut down.

The system of vertical wellpoints operated for more than 20 years but experienced some pump failures, resulting in soil loss into the tunnels. In later years, problems with clogging of well screens and the build-up of iron-related deposits in the wellpoints had reduced efficiency, and pumped water levels had risen, increasing the risk of further sand ingress into the tunnel. In the early 2000s, a scheme was developed to replace the existing wellpoints with a new system that would ensure the long-term stability of that section of tunnel.

A range of options were addressed at design stage. Groundwater exclusion solutions based on grouting, either from the surface or from inside the tunnel, were considered unsuitable due to the uncertainty of the condition of the tunnel and because of the disruption to railway operation associated with grouting from within the tunnel. Pumping systems based on new wells drilled out from inside the tunnel were deemed unsuitable for similar reasons.

The solutions investigated in most detail were based on lowering groundwater levels using a dewatering system located outside the tunnel. The sand deposits around the tunnel were relatively permeable, and the depth to tunnel invert was approximately 8 m below ground level. A conventional deep well system would have been effective in lowering groundwater levels. However, while it was possible to find surface space and access for deep wells on a temporary basis, a system of permanent deep wells was not acceptable to the client due to the disruption to access to their surface operations. It was necessary to consider options with a much smaller surface footprint than conventional deep wells. Two options were proposed by the tendering contractors. One involved wells drilled by horizontal directional drilling (HDD) along the line of the tunnel beneath invert level. Another option was to use a small number of collector wells.

The solution finally adopted, and installed on a design and construction basis by Keller Ground Engineering (with WJ Groundwater Limited as their sub-contractor), included the installation of an array of four collector wells around the line of the tunnel. The collector wells were to be pumped to hold piezometric levels down along the length of the tunnel. Four segmentally lined shafts of 4 m internal diameter and up to 20 m deep were sunk to form the wells themselves, and a total of 32 lateral wells were drilled radially outwards from the shafts to extend beneath the tunnel (Figure 27.23). When the water level in the shaft of the collector well is lowered by pumping, groundwater flows along the lateral wells and enter the main well shaft, from where it can be removed.

An elegant part of the system was that drilling rig for the lateral wells was also used to drill interconnecting pipes between the shafts. As a result, it was possible to pump the entire system using a single set of duty and standby pumps located in one shaft. Lowering the water level in the pumped shaft allows water from the other shafts to flow to the pumps by gravity along the interconnecting drains. The design flow rate for the system was 20 to 25 l/s, handled by a single-duty electric submersible pump located in one shaft.

The drilling of the lateral wells presented practical challenges. Rotary drilling was carried out at 127 mm diameter with a maximum drilled length of 35 m (Figure 27.24). The original groundwater level was of the order of 10 m above the level of drilling out from the shaft. Considerable thought was given to the method of drilling the lateral wells out from the shafts. If drilling was carried out against the original external groundwater head, considerable volumes of water would enter the shaft and have to be pumped away, even if a stuffing box was used to try to seal around the rotating drill rods, This meant there was a risk that fine sand could be washed into the tunnel, possibly causing problematic ground settlements. The solution adopted was to use a temporary dewatering system of conventional vertical deep wells to

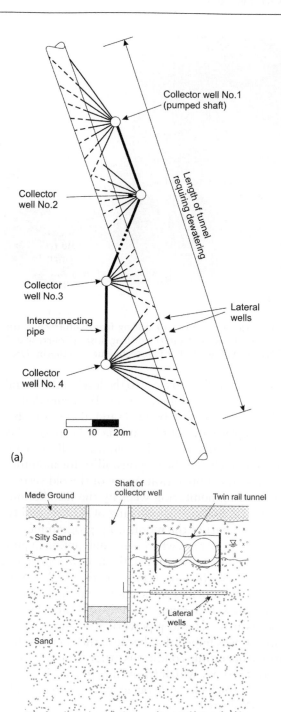

(a)

(b)

Figure 27.23 Collector wells installed around tunnel. (a) Plan view. The tunnel length is dewatered by lateral wells drilled out from four collector well shafts. (b) Cross section. Lateral wells are drilled radially outward from the collector well shafts. (Courtesy of Strathclyde Partnership for Transport, Glasgow, UK and Keller Geotechnique, Wetherby, UK.)

Figure 27.24 Installation of lateral wells by rotary drilling from within collector well shafts. Sump pumps are used to control the water from the drilling flush. (Courtesy of Strathclyde Partnership for Transport, Glasgow, UK and Keller Geotechnique, Wetherby, UK.)

lower piezometric levels during drilling to close to the level of the lateral wells. The lateral wells were completed with 60 mm diameter slotted screens. To ensure that an effective filter could be installed, the screens had a pre-fitted resin bonded sand filter of suitable grading (Figure 16.3c). This avoided problems with placement of filter media in horizontal boreholes. The temporary deep well system was decommissioned once the collector wells were in operation.

The construction contract included the responsibility for monitoring and maintenance of the system for the first 20 years of operation. Some of the old vertical wellpoints in the tunnel trackbed were converted to monitoring wells by the installation of vibrating wire pressure transducers linked to dataloggers. Monitoring of tunnel condition and ground levels to check for settlement was also part of the monitoring programme.

Unlike the previous wellpoint system, where any interruption in pumping resulted in almost immediate sand and water inflows to the tunnel, the new system holds groundwater levels sufficiently low that if duty and standby pumps fail, a period of 2 to 3 days elapses before water levels rise to tunnel invert. This provides sufficient time to make arrangements for repair.

Based on the experience of well clogging from the old tunnel wellpoint system, the build-up of iron-related deposits in wells and pipework was anticipated and allowed for in the maintenance programme. It was hoped that because the lateral wells were kept permanently submerged under normal operating conditions, they would suffer less from iron-related clogging than the previous system, which from time to time had drawn in air. However, in the early years of operation, some clogging occurred, and an annual programme of flushing of lateral wells had to be instigated.

27.12.3 Sources and References

KIRKLAND, C J, ROWDEN, I J and SMYTH-OSBOURNE, K R. (1987). A permanent dewatering system installed in an underground railway. Groundwater Effects in Geotechnical Engineering (Hanrahan, E T, Orr, T L L and Widdis, T F, eds), Balkema, Rotterdam, pp179–182.
SCHÜNMANN, D. (2005). Well sorted. *Ground Engineering*, October, pp34–35.

Chapter 28

The Future

Toby Roberts

> If you can look into the seeds of time and say which grain will grow and which will not, speak then unto me.
>
> William Shakespeare, *Macbeth*, Act I, Scene 3

28.1 INTRODUCTION

This chapter explores the drivers for change and the future of groundwater control works. Demand for underground space, notably for transport infrastructure and services, continues to expand around the world, and so too will the need to control groundwater where construction works to build these facilities reach below the groundwater level. Lowering groundwater levels has the potential to deliver significant project savings in terms of ground support, programme and cost. However, the application of dewatering processes can be constrained by the prevailing hydrogeological conditions, access constraints, regulatory compliance, available discharge capacity and risk of adverse impact on third-party assets. Where these constraints apply, physical exclusion using cut-offs to reduce or eliminate groundwater inflow may be required.

Can there be any fundamental improvement in groundwater lowering systems, given that their hydro-mechanical performance is governed by the basic laws of physics? In practice, we find that advances in materials, equipment design and technology lead to continual, although increasingly marginal, improvements in plant efficiency and reliability. Developments and improvements in safe working practices remain a key driver for change in the construction industry. Change is also being driven by society's demands to protect the environment and by the search for sustainability, which are increasingly underpinned by evolving regulation and legislation. Alongside all this, there is the unfolding technology revolution in communication, information and the Internet of Things (IoT).

My experience in the construction industry is consistent with Amara's Law, which states:

> We tend to overestimate the effect of a technology in the short run and underestimate the effect in the long run.
>
> (Roy Amara)

When I first started in the construction dewatering industry nearly 40 years ago,

- All pumps, pipework and well screens were made of steel, with virtually no plastic components or flexible pipes.

- Most dewatering projects involved steel wellpoints installed manually; deep wells with submersible borehole pumps were cutting edge, and ejector wells were found only in textbooks.
- Steel sheet-piling was the dominant option for temporary retaining walls and cut-offs. Concrete diaphragm walls and secant piles, which can act as a retaining wall and be integrated into the permanent works, were gaining ground. Slurry walls were in occasional use as a long-term hydraulic barrier. Permeation grouting was very much a 'black art' and jet grouting a novel technique with little traction.
- There was minimal constraint on abstraction or discharge of groundwater and no visible enforcement of regulations.
- Monitoring of flow rates and drawdowns was an expensive luxury.
- Contracts and project management were openly confrontational, with scant regard for health, safety or the environment.
- Construction activities were for the most part sequential, and so considered in isolation rather than as part of a highly integrated process.
- The fax had just replaced the telex as the most rapid form of written communication.
- If you wanted to speak to someone on site, you had to wait until they came back to the site office – if the site office had a phone at all. Staff on overseas sites in remote areas were effectively on their own, with the best form of communication being air mail!
- Computers had entered the workplace, but only for word processing by secretarial staff; it was more than 10 years before most engineers had a personal computer on their desk.

The earlier chapters have shown that much has changed. This chapter explores recent innovations, regulations and technology advances, and assesses the potential impact of these on the future development of groundwater control techniques.

28.2 REGULATION AND LEGISLATION

Workplace safety has been a key driver for change in the construction industry, with environmental regulation and legislation becoming increasingly important. Although Europe is at the forefront of effective regulation and enforcement, regulation of groundwater control activities is evident in all the jurisdictions that I have worked in over the last 20 years and almost certainly now applies anywhere in the world. The following requirements are ubiquitous:

- Abstraction consent: in areas where groundwater is not considered a resource, this may be included with the discharge consent or excavation planning consent.
- Discharge consent: this generally includes controls on flow rate and increasingly stringent controls on quality (suspended solids and contaminants). Monitoring and reporting of flow and quality are generally obligatory.
- Consent for artificial recharge back to the ground (including shallow soakaway pits), which is generally considered a form of discharge. Recharge is often encouraged to preserve water resources, although stringent water quality requirements may apply.
- Dewatering system designers and installers, particularly drillers, need to be suitably qualified. This is primarily to ensure that staff understand the requirement to protect aquifers by applying appropriate drilling methods, installing annular seals, and sealing wells and piezometers when no longer required.
- General construction health and safety regulations to protect site workers and the public.

Consent for dewatering may be subject to further local legislation and controls, for example:

- Cities that are built on weak soils, especially where there are many older, heritage buildings at risk of settlement, may have a general moratorium on groundwater lowering. Examples include Copenhagen, Amsterdam, Hong Kong and the area around St Paul's Cathedral in the City of London (see Section 25.4.6).
- Where groundwater is an important resource, local controls and legislation to restrict abstraction and promote recharge are common. These controls recognize that groundwater quality and availability are at risk from over-abstraction, contamination and, potentially, climate change.
- The available options for discharge can be an important constraint, particularly where these are of restricted capacity, distant or prohibitively costly. Sewer capacity is being eroded by increases in population, increases in water usage and increased storm intensity due to climate change. The capacity of streams and rivers may also be limited due to flood risk arising from a combination of development and climate change.

These types of regulation and control, which are often a response to local conditions and experience, are likely to become more common in the future. It is worth noting that the demand for underground space is such that these regulatory constraints have not curbed development but rather, have driven innovation in groundwater control strategies, including cut-off wall technology, dewatering techniques, and monitoring and control systems. Construction arrangements that use cut-offs and groundwater lowering systems in combination can be the most cost-effective approach to meeting regulatory requirements.

One implication of the evolving regulation is that a good understanding of the groundwater regime is needed to develop the strategies and schemes to meet the imposed constraints. This can only be achieved with high-quality geotechnical investigations, including site-specific pumping tests combined with relevant experience. Often, a sophisticated conceptual model of the ground, together with numerical groundwater modelling, will also be needed to obtain the permission and consents from the regulatory authorities required to proceed with the works. For major projects, at least, this can go some way towards addressing one of the geotechnical community's oldest laments: that site investigations are too often inadequate; see Institution of Civil Engineers (1991).

It might be thought that environmental legislation could represent a future threat to the use of dewatering systems, but compared with alternative processes, the environmental impact of groundwater lowering systems is relatively benign. The energy consumption for dewatering systems is modest, the impact on groundwater flows should be temporary, and only the well screens are left behind on completion (even these can be removed with a bit of ingenuity; see Section 24.3.3). Well screens and piezometers are routinely sealed to avoid the risk of cross-links between aquifers. In comparison, artificial ground freezing has a heavy energy draw and uses refrigerants with questionable environmental credentials. Grout curtains, concrete diaphragm walls and secant pile cut-offs leave permanent barriers to groundwater flow paths and may restrict aquifer storage capacity. Cement and steel also have a high embedded carbon footprint, with Casey et al. (2015) arguing that long-term dewatering can have a significantly lower capital cost and carbon footprint than the steel and concrete for piles and anchors needed to resist uplift pressures for a large station box. Tools are being developed for assessing the carbon footprint and wider sustainability of construction activities and materials; see, for example, Hughes et al. (2011). Some project clients are now taking a lead in demanding an assessment of the embedded carbon in construction works, and in due course, legislation may encourage or enforce the wider adoption of these assessments.

Environmental regulation also focuses on other issues, such as poor air quality in cities, which is recognized as a significant health hazard. Control measures have focused mainly on road-vehicle emissions, but construction equipment is also a concern. Emission standards for new diesel-powered construction plant are in force in many countries, and major projects may only accept equipment that meets the latest standards, effectively outlawing the use of older, more polluting equipment. Historically, older plant might have been sold or relocated overseas to a country with lower emissions controls, but with standards rising around the world, this option is limited. The logical result of this will be a switch to electric-powered machinery; this is already common for pumping equipment but is likely to be a future requirement for installation plant as well. Also relevant is the European Union's Ecodesign Directive 2009/125/EC, which establishes a framework for energy-using and energy-related products. This has spawned regulations for design requirements for electric motors (regulations 640/2009 and 4/2014) and for water pumps (regulation 547/2012), which aim to further reduce energy consumption.

28.3 TECHNOLOGY AND THE INTERNET OF THINGS

The penetration of computer technology into groundwater control activities in the UK started with word processing in about 1980 and progressed to cost estimating and accounts, before reaching the desks of design engineers around 1990. Dataloggers (initially desktop computers with appropriate software and sensors) were used on site for monitoring occasional high-specification pumping tests as early as the mid-1980s but were not in regular use until the mid-1990s. By 2000, datalogger systems were being widely used to provide cost-effective remote automated monitoring and alarms for dewatering systems. The rapid advance of mobile data networks continues to impact all areas of our working and private lives, with high-definition video phones, video conferencing and remote site-viewing technology widely available. The use of these systems, including direct access to real-time information and performance-monitoring data, is now routine for many dewatering systems and construction sites.

Alongside this has been the development of smart devices, the Internet of Things (IoT) and machine learning algorithms. The application of IoT devices for mine water management is considered by Losavio et al. (2019). IoT devices typically have a network interface, an external control system that can communicate with the device via the network interface, actuators that operate the device, and a microprocessor with software that can monitor the device operation, communicate with the external control system and respond to instruction so as to serve as a functional remote device. Construction plant, including monitoring equipment, pumps and generators, plus rigs for drilling, piling and diaphragm walls, is routinely available with at least some IoT functionality. Such devices are in widespread use for remote performance monitoring of plant, with automated remote operation and control increasingly an option. This presents both an opportunity and a risk. The opportunity is in applications such as automated control of pumping systems, which can respond almost instantly to changes in the groundwater regime to track a target level. This has the potential to improve efficiency and reduce risk, but, as Losavio et al. (2019) point out, there is also increased risk if these systems are not programmed correctly or respond in an unintended way to an unforeseen event or circumstance. A simple example might be the removal of a transducer from a piezometer, which the control system will interpret as a sudden drop in water level, triggering pumps to stop or start inappropriately, with potentially serious safety-critical or environmental consequences. In addition to unintended operational risk, there are also security concerns, with the potential for malicious hacking of systems. These new hazards and risks will require appropriate management and control measures. It is an understatement to say that the potential liability, contractual and legal implications of these technological developments for the companies and staff involved are complex.

Advances in communication and technology have had a profound impact on my working life, and widespread adoption of the technologies currently available, let alone new ones, will bring substantial changes in the future. The pace of change is moderated only by a lack of common standards, including effective methods for visualizing large, complex datasets, and by our own ability to absorb and respond to information. The opportunities offered by IoT and machine learning software offer a way forwards but also raise new hazards and risks.

28.4 DESIGN AND NUMERICAL MODELLING

Current groundwater numerical models are immensely sophisticated and can account for both unsaturated flow and negative pore pressures in two dimensions, three dimensions and four dimensions (i.e. with time-dependent effects). These programmes are moving beyond mere modelling towards 'virtual reality'. Clever interfaces and element building tools allow models to be put together quickly. Numerical modelling is widely used to investigate the regional impact of groundwater control schemes and is increasingly demanded by regulatory authorities. Numerical modelling is also used when the geometry of a scheme is not amenable to analysis using closed-form solutions for planar or radial flow, such as in the case of flow under a partially penetrating cut-off (see Section 13.12) and for large projects where time for drawdown is a factor; see, for example, Roberts *et al.* (2009). Despite these advances, it remains the case today that the majority of small and medium-sized dewatering schemes, with a straightforward geological setting, are designed on the basis of experience, local knowledge and empirical 'rules of thumb'. Two-dimensional or radial steady-state closed-form analyses may be undertaken to justify the design. Three-dimensional time-dependent modelling remains too time consuming for most medium and smaller-sized projects and in any case, is often not justified.

A further shortcoming of two- and three-dimensional models is the inability to realistically model conditions in the immediate vicinity of dewatering wells, which commonly have a significant seepage face. This is generally overcome by making simplifying assumptions, such as combining several wells into one sink and ignoring any seepage face at the wells. For the purposes of assessing total seepage flows to an excavation, these simplifications are reasonable, and the resulting errors in flow estimates are generally small. However, such simplifications effectively prevent direct assessment of the number, size and depth of the dewatering wells required. The datasets obtained for individual projects, together with back analysis, could be used to increase our understanding of the flow regime in the vicinity of well arrays, which could lead to an advancement in modelling by improving the representation of individual dewatering wells.

A numerical model is much less costly than a fully instrumented field test, and modelling software has been exploited to some extent as a way of developing accessible design tables or charts for particular geometries or conditions. Good examples are the design charts given by Powrie and Preene (1992), which bridge the gap between radial analysis and two-dimensional plan analysis. A fruitful area of further study would be to establish the scale of the errors resulting from applying two-dimensional or radial analysis to problems that are strictly three-dimensional. This would put the relatively simple two-dimensional and radial analysis on a firmer footing and make it clearer where a three-dimensional model is needed.

Numerical modelling techniques are having an important impact on the design of groundwater control systems, but they remain a long way from challenging the central role of judgement based on experience, local knowledge and empirical methods in the design process. In the meantime, numerical modelling will continue to be used to support design assumptions (principally for parametric analysis), to examine environmental and settlement risks, and as a tool to back-analyse, visualize and interpolate monitoring data obtained during the works.

28.5 APPLICATIONS AND TECHNIQUES

To some extent, dewatering systems and cut-offs are competing strategies for groundwater control. However, complete cut-offs can be difficult or costly to achieve in some geological settings, and a more cost-effective approach may be the integration of a dewatering system and partial cut-off. Changes to the principal techniques used are incremental, generally aimed at improving safety, efficiency and reliability with a corresponding reduction in cost. Other advancements may be focused on increasing the installation depth of cut-offs, which often requires a combination of more powerful equipment and improved tolerance to ensure the interlock of adjacent cut-off elements. One of the most promising developments is rota-sonic drilling, which uses high-frequency vibration on the drill string to enhance penetration in both soils and rock. This has been a niche technique used for coring in site investigation over the last 20 years but is finding wider application as a method of rapid drilling, especially in difficult formations such as glacial outwash deposits.

A key area of challenge for dewatering systems is in the control of pore water pressure in fine-grained low-permeability soils. These soils also present a challenge to other techniques such as grouting. The wider use of ejector systems and vacuum deep wells in the UK over the last 30 years has extended the application of dewatering systems to fine-grained soils; see, for example, Roberts *et al.* (2007). There is some evidence from detailed monitoring of dewatering systems in fine-grained soils that there can be a significant variation in the yield from individual wells, installed in a similar manner, in apparently similar ground conditions. It is not clear whether this is due to some installation effect or to undetected variations in the soil fabric local to the wells, or both. There would be significant cost and performance benefits if a system could be devised to ensure that all the wells in a system were as efficient as the best-performing well. Any solution will probably involve modification to some or all of the current methods of well drilling, well installation, screen specification, filter-pack specification/placement and well development.

Siphon wells are described in Section 17.11. This is a rare example of a previously unused method of pumping being applied to groundwater control. The attributes of siphons make then particularly suited to slope stability problems, which are often caused by excess pore water pressures. This is clearly a niche technique but offers the potential for considerable cost saving over a 'hard' retaining wall or piled solution.

Changes in project requirements and other construction processes and methods can also offer opportunities:

- Access constraints. Construction of new and improved infrastructure in cities must be carried out while minimizing disruption to existing services, transport routes and infrastructure. This requirement often introduces significant access constraints; see, for example, McNamara *et al.* (2008). The use of cut-offs in such conditions may be challenging and costly. Sometimes, constraints can be overcome by dewatering using inclined or horizontal wells or wells installed directly out of tunnels; see, for example, Roberts *et al.* (2015).
- Horizontally directionally drilled (HDD) wells. Directional drilling techniques were developed to allow pipelines and services to be installed without trench works, but

these techniques have also been used to install horizontal wells (see Section 17.4). The majority of applications have been for water supply and ground remediation, but long-term dewatering schemes using horizontal wells drilled from collector shafts are in use to prevent groundwater and sand ingress at Govan Station on the Glasgow subway (Case History 27.12) and to mitigate against rises in groundwater levels as a result of the Cardiff Bay Barrage; see Williams (2008). Costs are significant, and installation of filter packs and development is challenging.

- Compressed air working in tunnels. Health and safety regulations mean that there are significant costs and health risks associated with working in air pressures above 1 bar. In cases where pressures above 1 bar would normally be required, it has sometimes proved cost effective to use a dewatering system to lower pore pressures so that working air pressures can be reduced to below 1 bar.
- Pressure relief on retaining walls. Controlling pore water pressures can reduce the need for temporary propping or anchoring of retaining walls. Controlling pore water pressures on the external 'active' side will reduce loading. Lowering groundwater levels on the internal 'passive' side to significantly below excavation level can improve the soil properties and increase the passive resistance.
- Soil remediation. An effective method of removing volatile contaminants from soils is by soil vapour extraction. A high water table prevents movement of vapour and curtails the effectiveness of this technique. Dewatering systems can be used to reduce the water level in order to improve the performance and reach of the remediation process.
- Sprayed concrete lining (SCL) technology is an efficient method for sinking large-diameter shafts and for forming irregularly shaped caverns underground, for example for underground stations during the construction of a metro system. When combined with the sequential excavation method (SEM), the technique relies on short advances of excavated soil or weak rock to be self-supporting for a few hours while the sprayed concrete is applied and sets. This is not a problem in weak rock and stiff clays but is an issue in water-bearing granular soils. Pore water pressure control using wells from the surface or installed from the heading can be used to extend the application of SCL techniques into otherwise marginal soils.

These are some examples of industry demands and innovations that can widen the scope and application of groundwater control systems, and the ingenuity of engineers will doubtless find more opportunities in the future.

28.6 WHERE DO WE GO FROM HERE?

In this chapter, I have drawn attention to the areas of innovation and regulation that I anticipate will be the main drivers for future changes. Amara's law highlights the difficulty of assessing the short-term versus the long-term impact of changes in technology. Innovation in construction generally moves forwards at a stately pace, and as a result, I anticipate that changes in technology and the IoT, together with environmental regulation, will have the most significant impact on the application of groundwater control techniques over the next 10 years.

28.7 THE NEXT GENERATION OF DEWATERING PRACTITIONERS

As this book shows, the successful application of groundwater control strategies requires a sound understanding of engineering, construction methods and hydrogeology, together with a wealth of experience. The value of this book is that it distils more than a lifetime's experience of two engineers whom I greatly respect. However, we sometimes fail to recognize and appreciate the vital contribution made by the site managers, site supervisors, rig operators and site staff who implement these schemes and can rightly claim more raw practical experience than any design engineer. There are few short cuts to gaining experience, and I worry about the loss to our industry each time one of our site staff moves on or retires. There are some counter-trends: for example, it was recently pointed out to me that younger rig operatives are sometimes more comfortable using the new generation of rigs, which may be remote-controlled with reliance on data interfaces. Nevertheless, we need to identify and develop the next generation of dewatering specialists and equip them with the necessary skills, both theoretical and practical, to allow them to take groundwater technology and practice forwards over the coming decades.

Appendix 1: Estimation of Permeability by Correlations with Particle Size Distribution (PSD) of Soil

This appendix provides background to methods of estimating permeability of soil by correlation with particle size distribution (PSD) analyses. Methods of obtaining soil samples and the subsequent laboratory testing processes are presented briefly, and some potential limitations with the correlation methods are discussed. Detailed background is then given on the Hazen method and Kozeny–Carman method. Particle size correlations are discussed in the main text in Section 12.6.1.

A1.1 NOTES ON OBTAINING SAMPLES USED FOR PSD ANALYSES

The most common source of samples used for PSD analysis is ground investigation boreholes. It is important to recognize that the drilling and sampling process may affect the material obtained and therefore influence the permeability values subsequently estimated from particle size correlations.

(i) Bulk or disturbed samples are typically recovered from below the water level in a borehole. Samples obtained in this way can be subject to 'loss of fines', a phenomenon where finer particles are washed out of the sample as it is removed from the borehole. Because the finer soil particles have a disproportionate effect on permeability, samples affected in this way will tend to give over-estimates of permeability. Loss of fines is particularly prevalent in samples taken from the drilling tools during light cable percussion boring. This can be minimized by placing the whole contents (water and soil) of the tool into a tank or tray and allowing the fines to settle before decanting clean water and recovering the remaining soil. Unfortunately, in practice, this is rarely done. Loss of fines is usually less severe for tube samples; these methods may give more representative samples in fine sands.

(ii) Core samples are commonly recovered from rotary drilled boreholes. Typically, the borehole is drilled with the aid of drilling fluids. While the main body of the cylindrical core sample will be relatively isolated from the drilling fluid, there is a risk that fine particles from the drilling fluid (either from drilling additives or from silty material suspended in water in the borehole) may contaminate the external faces of core samples in coarse-grained soil. If these fine particles are included in the particle size analysis, the results of correlations will tend to under-estimate permeability. It may be possible to avoid these problems if there is sufficient material in the sample to allow the material from the edges to be discarded. Samples from rota-sonic drilling are less prone to this problem provided that the borehole can be drilled without the addition of water or drilling fluids.

AI.2 NOTES ON LABORATORY PROCESSES TO DETERMINE PSD ANALYSES

The correlations between permeability and PSD are based on data from the PSD curve (sometimes known as a grading curve) of a soil sample (Figure A1.1). The PSD curve represents the cumulative percentage (by weight) of soil mass retained on each sieve size and allows estimation of particle size parameters such as D_{10}, D_{15}, D_{50}, D_{60} (and so on), where D_n is the particle sieve aperture through which n per cent of a soil sample will pass. D_n is determined by the intercept between the PSD curve and the horizontal line of n per cent passing (Figure A1.1 shows the D_{10} intercept as an example).

A PSD curve is obtained by laboratory analysis of a soil sample. Particle size analysis (PSD analysis) is typically carried out on relatively small samples (100 to 500 g mass). Laboratory methods are described in detail in Head (2006). PSD analysis comprises three phases:

1. Pre-treatment of sample
2. Mechanical analysis of sample
3. Calculation and presentation of results

AI.2.I Pre-Treatment of Sample

During pre-treatment, the sample will be disaggregated (and, if necessary, washed) and then oven-dried before mechanical analysis. If soils contain more than a few per cent of fine particles (defined as smaller than 0.063 mm in British practice), washing is used to separate the fine and coarse fractions to allow them to be subjected to different forms of mechanical analysis. The net effect of pre-treatment is to destroy any soil structure, fabric or cementation of particles present in the in situ soil even if it has survived the sampling process. Therefore, PSD analysis is carried out on an effectively homogenized sample, and

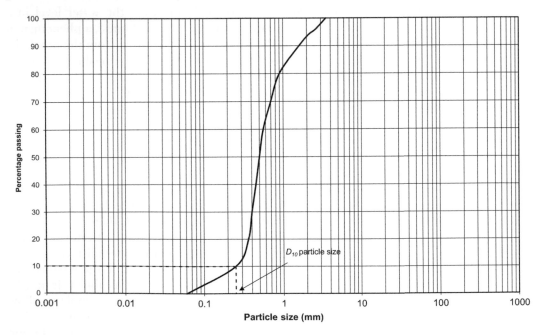

Figure AI.I Particle size distribution (PSD) curve for a soil sample. D_{10} particle size is shown.

the results may not represent the material as present in the ground. Figure A1.2a shows a hypothetical clean sand deposit containing some laminations of silty clay; the horizontal permeability will be dominated by the properties of the sand and vertical permeability by the silty clay. Even if high-quality core samples are obtained, during pre-treatment, these two materials will become mixed, and the resulting PSD curve will indicate a clayey or silty sand (Figure A1.2b). For the anisotropic soil shown in Figure A1.2, the permeability estimate derived from PSD correlations will probably more closely represent the vertical permeability k_v than the horizontal permeability k_h.

A1.2.2 Mechanical Analysis of Sample

Mechanical analysis is the term given to the process of determining the relative fractions of different particle size within a soil sample expressed as a percentage of the total dry weight. Two separate procedures are used to process the wide range of particle sizes that can be present (finer and coarse particles having been separated by washing during pre-treatment):

- Sieve analysis is used for coarse particles (gravel and sand-size) and involves shaking the oven-dried soil through a series of sieves of standard sizes that have progressively smaller openings.
- Sedimentation analysis is used for the finer particles (silt and clay), which are too small to be separated by dry sieving. Measurements are made on a suspension of the fine particles in water, either by sampling using a special pipette or by determining the density of the suspension using a special hydrometer.

A1.2.3 Calculation and Presentation of Results

In geotechnical practice, the results of particle size analysis are most commonly presented graphically in the form of a PSD curve such as that shown in Figure A1.1 The test results are graphed to show percentages by weight finer than a given size, plotted against particle size on a logarithmic scale.

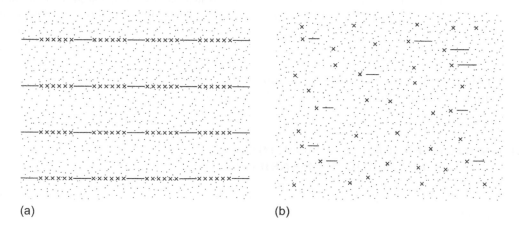

(a)   (b)

Figure A1.2 Limitations of particle size analysis of samples in soil containing structure or fabric. (a) In situ soil: clean sand containing fabric of discrete silt and clay layers. (b) Sample subjected to particle size analysis after sample has been disaggregated.

AI.3 POTENTIAL ERRORS WHEN ESTIMATING PERMEABILITY FROM PARTICLE SIZE DISTRIBUTION (PSD) DATA

Given the common limitations of sampling and PSD analysis methods described earlier, any method that determines permeability from correlations with particle size analyses of soils has several potential limitations:

1. The particle size distribution of the samples subjected to PSD analysis may be unrepresentative of the in situ soil. This can be caused by a variety of factors, including:
   - Heterogeneity of the in situ soil, which can mean that the sample was obtained from a zone that is not representative of the wider soil mass (see Figure 12.4).
   - Sample contamination or losses. This can include contamination from drilling fluids or dirty water in the borehole, or loss of fines from samples.
   - Loss of soil structure/fabric and mixing of different soil fractions during pre-treatment or mechanical analysis (see Figure A1.2).
   - A combination of more than one of the above.
2. The results of a PSD analysis give no direct information on the stress state and hence the porosity of the in situ soil. Some correlations require parameters that represent stress state or porosity. However, the necessary input values for such terms are often uncertain (for example, porosity used in the Kozeny–Carman method is rarely measured in the field). There are methods to estimate porosity and related parameters from other data, but this introduces further potential errors in the assessed permeability.
3. The permeability values provided by this method should be considered to be very small-scale (see Figure 12.1). PSD analysis is carried out on only relatively small samples, typically with the maximum physical dimension of the sampled zone being some few hundred millimetres or less. If the soil is heterogeneous or has significant fabric, estimates from PSD correlations may not be representative of the mass permeability of the strata. Large-scale tests (such as pumping tests) may give more representative results for use in dewatering design.
4. The correlations discussed in this Appendix are all empirical. This means that they are not derived from theory or analysis of the physical laws of a problem; rather, they are based on observation and previous experience. Any empirical model is based on a finite underlying dataset and/or series of assumptions and can only be used with confidence within the limits of the observations upon which it is based. Furthermore, many empirical correlations are specific to certain combinations of units, and applying input parameters in the wrong units can result in further errors. A wider discussion of empirical models in general is given in Section 5.5.1.

AI.4 SOME EMPIRICAL RELATIONSHIPS BETWEEN PSD AND PERMEABILITY

Bricker and Bloomfield (2014) summarize a generic form of correlation between PSD and permeability as:

$$k = \left(\frac{\gamma_w}{\mu_w} \right) C f(n) D_e^2 \tag{A1.1}$$

where

γ_w is the unit weight of water
μ_w is the dynamic viscosity of water (which will vary with temperature)
C is a sorting coefficient
n is porosity (which is the ratio of voids to the total volume of soil – see Section 4.3)
$f(n)$ is a porosity function
D_e is the effective particle size

As discussed in Section 4.3, the finer fraction of particles has a disproportionate effect on the permeability of a soil mass, and in these correlations, the effective particle size is often taken as D_{10} or D_{15}.

Two commonly used correlations are Hazen's method and the Kozeny–Carman method.

AI.4.1 Hazen's Method

In geotechnical practice, Hazen's method is perhaps the best-known empirical correlation used to estimate permeability from PSD analyses. This is named after Allen Hazen, an American water works and sanitary engineer from New England, who was one of the first to propose an empirical correlation for the permeability of a sand from its PSD curve (Hazen, 1892, 1900, 1911).

Hazen's formula is often applied in its simplest form, as shown in Equation A1.3, which does not include the effects of temperature. However, several reviews of particle size correlations (such as Feng *et al.*, 2019) quote the formulation used to estimate permeability k, which includes a temperature correction:

$$k = (0.7 + 0.03T)C(D_{10})^2 \tag{A1.2}$$

where

T is the temperature in degrees Celsius
D_{10} is the 10 per cent particle size taken from the particle size distribution curve
C is a calibration factor

As is often the case with empirical relations, the calibration factor is not dimensionless and will vary for different combinations of input units. The temperature correction, which gives an approximate 30 per cent increase in permeability between 10 °C and 20 °C, is consistent with the corresponding decrease in the dynamic viscosity of water, based on the generic form of correlations shown in Equation A1.1.

The temperature of shallow groundwater in the United Kingdom varies little between about 5 °C and 15 °C, so in geotechnical practice, the temperature correction is often neglected, and Hazen's formula to estimate permeability is simplified to

$$k = C(D_{10})^2 \tag{A1.3}$$

Hazen stated in his work that his rule was applicable over the range of D_{10} particle size from 0.1 to 3.0 mm and for soils having a uniformity coefficient U lower than 5, where

$$U = \frac{D_{60}}{D_{10}} \tag{A1.4}$$

and D_{10} and D_{60} are sieve apertures through which 10 per cent and 60 per cent, respectively, of a soil sample will pass.

Hazen stated that (when k is in metres per second and D_{10} is in millimetres) his calibration factor C could vary between about 0.007 and 0.014. In this formulation, C has the units of (m/mm²s). In practice, presumably for reasons of simplicity, and in the stated set of units, C is normally taken to be 0.01, so Hazen's formula becomes

$$k = 0.01(D_{10})^2 \tag{A1.5}$$

where k is in metres per second and D_{10} is in millimetres. This equation is generally taken to apply at a temperature of 10 °C, which is similar to shallow groundwater temperatures in the United Kingdom.

AI.4.2 Kozeny–Carman Method

The Kozeny–Carman method is sometimes favoured in hydrogeological investigations, perhaps because its formula includes a greater number of parameters compared with Hazen's rule and therefore accounts for the wider PSD curve, not just D_{10}. This does, however, mean that site-specific values may not be available for all necessary parameters, which will need to be estimated by other means. This can introduce further uncertainty in estimated permeability.

The Kozeny–Carman method (Kozeny, 1927; Carman, 1937, 1956) relates permeability to porosity and various indicators of particle size and shape. Several different formulations are available in the published literature, with the differences explained by the use of different particle size indicators and different units. Interestingly, unlike the Hazen method and the Prugh method (see Section 12.6.1.4), which were originally derived for granular filter media and soils, respectively, the Kozeny–Carman formula was originally developed for a solid medium containing closely spaced pipe conduits rather than for a granular medium.

A generic form of the Kozeny–Carman equation is given in Carrier (2003) to determine permeability k:

$$k = \left(\frac{\gamma_w}{\mu_w}\right)\left(\frac{1}{C_{KC}}\right)\left(\frac{1}{S_o^2}\right)\left(\frac{e^3}{(1+e)}\right) \tag{A1.6}$$

where
  γ_w is the unit weight of water
  μ_w is the dynamic viscosity of water
  S_o is the specific surface of particles (surface area per unit volume of soil particles)
  e is the void ratio (defined as the ratio of voids to solids in a soil, which is related to porosity)
  C_{KC} is the Kozeny–Carman coefficient, which accounts for the tortuosity and shape of the particles

C_{KC} is usually taken as 5. Carrier (2003) provides a version of this equation, based on the properties of water at 20 °C, which is presented here, modified to give permeability in metres per second when S_o is in units of 1/cm:

$$k = 200\left(\frac{1}{S_o^2}\right)\left(\frac{e^3}{(1+e)}\right) \tag{A1.7}$$

Applying the temperature correction of the same form as used in Equation A1.2 to account for the increase in the viscosity of water at lower temperatures, the version of the Kozeny–Carman equation applicable to groundwater temperatures of 10 °C is

$$k = 150 \left(\frac{1}{S_o^2} \right) \left(\frac{e^3}{(1+e)} \right) \tag{A1.8}$$

where permeability k is in metres per second when S_o is in units of 1/cm. Note that to avoid implying a false impression of precision, the overall coefficient is given to 2 significant figures (SF) only rather than the 3 SF used in Carrier (2003).

In practical terms, it is inconvenient that these equations do not include any of the particle size parameters D_n that are obtained during conventional PSD analyses. However, if it is assumed that the soil particles are spheres of effective diameter D_e, then specific surface S_o can be derived from basic geometry:

$$S_o = \frac{\text{Area}}{\text{Volume}} = \frac{\pi D_e^2}{\left(\pi D_e^3 / 6 \right)} = \frac{6}{D_e} \tag{A1.9}$$

Of course, very few real soils have particles that are even approximately spherical. Loudon (1952) discusses that the actual specific surface S_o of a soil is influenced by the angularity factor f (also known as the coefficient of rugosity). The specific surface of non-spherical particles is equivalent to the specific surface of spheres (from Equation A1.9) multiplied by the angularity factor f:

$$S_o = \frac{6f}{D_e} \tag{A1.10}$$

For a given particle size, the more angular the shape of the particles, the greater the specific surface will be. According to Loudon, a visual estimate (using a hand lens) of the angularity factor seems to be quite good enough for the standard of accuracy generally needed. Loudon's estimates of f values are given in Table A1.1. This shows that in practice, the range in angularity factor of natural soils is not great, being from about 1.1 to 1.4 – a full range variation for naturally occurring soils of about 25 per cent. Loudon further indicated that very angular materials such as crushed marble, crushed quartzite or crushed basalt may have angularity factors of 1.5 to 1.7.

Combining Equations A1.8 and A1.10, the Kozeny–Carman equation (for water at 10 °C), with k in metres per second and D_e in centimetres, becomes

$$k = 4.2 \left(\frac{D_e^2}{f^2} \right) \left(\frac{e^3}{(1+e)} \right) \tag{A1.11}$$

Table A1.1 Angularity Factor of Soil Particles

Material type	Angularity factor f
Spherical glass beads	1.0
Rounded sand	1.1
Sand of medium angularity	1.25
Angular sand	1.4

Based on Loudon, A G, *Géotechnique*, 3, 165–183, 1952.

If D_e is in millimetres, as commonly recorded in laboratory tests, this becomes

$$k = 4.2 \times 10^{-2} \left(\frac{D_e^2}{f^2} \right) \left(\frac{e^3}{(1+e)} \right) \tag{A1.12}$$

Using the relationship between void ratio and porosity n given in Section 4.3,

$$n = \frac{e}{(1+e)} \tag{A1.13}$$

the Kozeny–Carman equation can be presented in terms of porosity, a parameter that many groundwater specialists will find more familiar:

$$k = 4.2 \times 10^{-2} \left(\frac{D_e^2}{f^2} \right) \left(\frac{n^3}{(1-n)^2} \right) \tag{A1.14}$$

This formulation is for water at 10 °C with k in metres per second and D_e in millimeters. n should be input as a fraction, not a percentage.

D_e can be assessed from the complete particle size curve using the method given by Carrier (2003), where the fractions of the sample retained between adjacent sieve sizes are used to assess an effective particle size.

$$D_e = \frac{1}{\left[x_1 / D_1 + x_2 / D_2 + \dots + x_i / D_i + \dots + x_n / D_n \right]} \tag{A1.15}$$

where
   $x_1 \dots x_n$ are the fractions (not percentages) of the total mass (i.e. $x_1 + x_2 + \dots + x_i + \dots + x_n = 1$) retained on each sieve
   $D_1, D_2, \dots D_i, \dots D_n$ are the mean particle sizes between the respective adjacent sieves

The mean particle size D_i between the nominal sizes D_x and D_y of adjacent sieves is taken as the geometric mean of the two sizes:

$$D_i = \sqrt{(D_x D_y)} \tag{A1.16}$$

Where raw data of a PSD curve are available electronically, it is a simple matter to evaluate D_e by means of a spreadsheet calculation. Carrier (2003) notes that Equation A1.16 can be improved slightly by calculating the mean particle size D_i between adjacent sieves by assuming a log-linear distribution, so that

$$D_i = (D_x)^{0.404} \times (D_y)^{0.595} \tag{A1.17}$$

where D_x and D_y are the larger and smaller of the adjacent sieve sizes, respectively. In practice, given the other uncertainties in the method (such as values of porosity and particle shape), the additional complexity is rarely justified, and Equation A1.16 is used widely.

One of the problems with applying this formulation is that porosity n is difficult to determine, and site-specific values are rarely available for sand-dominated soils, which are often of interest to the designers of groundwater control systems. While the porosity of a sample can be determined in the laboratory, at that point a sample of a sandy soil will have been significantly disturbed, and its porosity is likely to be very different from in situ conditions. In practice, porosity is sometimes estimated from published values (such as from Table A1.2) appropriate to the soil description, or from empirical correlations such as that given by Bricker and Bloomfield (2014), which quotes a formula from Vukovic and Soro (1992) whereby porosity can be approximately related to the particle size distribution using the uniformity coefficient U (from Equation A1.4):

$$n = 0.255\left(1 + 0.83^{U}\right) \tag{A1.18}$$

Permeability estimates can then be obtained by combining the porosity values with D_e values into the Kozeny–Carman formula in Equation A1.14.

This requirement to use porosity values that may themselves be estimates or have been derived empirically is a limitation on the usefulness of Kozeny–Carman and other similar formulations and is an explanation for the somewhat erratic results that they sometimes give.

Carrier (2003) summarized some specific limits on the applicability of the Kozeny–Carman method:

1. The method is not applicable in soils that behave as clays. This is because the formula assumes that there are no electrochemical bonds between the soil particles and the water. Carrier indicates that the method can be used in non-plastic silts.
2. The method assumes the Darcian condition (laminar flow). This is appropriate for silts, sands and even gravelly sands, but in very coarse soils, the method is invalid. An upper limit on the effective particle size D_e is around 3 mm.
3. The method assumes that the soil particles are not excessively elongated and is not appropriate for use in soils containing platy particles such as mica. This is another reason why the method is not appropriate in clayey soils.
4. The method is not appropriate if the PSD has a long, flat tail in the fine fraction with a significant non-zero percentage shown at the smallest recorded size. For D_e to be calculated accurately, the smallest particle size present, D_o, must be measured or assumed.

Table AI.2 In Situ Porosity and Void Ratio of Typical Soils in Natural State

Description	Porosity n	Void ratio e
Uniform sand, loose	0·46	0·85
Uniform sand, dense	0·34	0·51
Mixed-grained sand, loose	0·40	0·67
Mixed-grained sand, dense	0·30	0·43

Based on Terzaghi, K, et al., *Soil Mechanics in Engineering Practice*, 3rd edition, Wiley, New York, 1996.

Appendix 2: Execution and Analysis of Variable Head Permeability Tests in Boreholes

A2.1 GENERAL COMMENTS ON PERFORMING PERMEABILITY TESTS

Many factors may influence the results obtained from in situ permeability tests. Great care must be exercised to ensure that tests are made under controlled conditions; some tests are reliable only when performed according to specific conditions, and these should be strictly adhered to. Some general guidance is given in the following:

a) When making tests in boreholes, prior to the test, the borehole should be carefully cleaned out to remove loose or disturbed material, the presence of which can introduce significant errors in the results obtained. Any sediment-laden water in the borehole should be flushed out and replaced with clean water.

b) When carrying out tests where water is added to the borehole, the water used should be clean, and attention should be given to the cleanliness of tanks, buckets, pumps, etc., so that the water does not become contaminated by dirt in the equipment, since very small amounts of silt may cause serious errors in the results.

c) When monitoring by manual methods, water level should be measured with a dipmeter relative to a stable datum such as the top of the borehole casing or a firmly positioned rod across the top of the borehole. If monitoring of water levels is by a pressure transducer linked to a stand-alone datalogger (see Section 22.6), the transducer should be installed at a suitable depth below the lowest expected water level, arranged so that its position will not change during the test.

d) In the early stages of rising and falling head tests, it is desirable to take readings of the water level at frequent intervals of about 30 seconds. To obtain reasonable accuracy in these circumstances using manual readings, it is necessary to have two persons taking the readings: one to adjust the dipmeter to keep pace with the change in water level and to take the depth readings, and the other to call out the time intervals from a stopwatch and record the water levels. Alternatively, a pressure transducer linked to a stand-alone datalogger (see Section 22.6) may be installed in the borehole and programmed to take pressure readings every few seconds.

e) The equations for calculating permeability are all based on the assumption that flow through the tested material is laminar (and therefore Darcy's law is valid). To be sure of laminar flow, velocity should not exceed 0.03 m/s. Therefore, the tests are only applicable when the flow rate (into or out of the borehole) divided by the area of the test section is less than 0.03 m/s. The open area is taken as the peripheral area of the filter section around a well, or the intake area of the sides and base in the case of an unsupported hole.

f) Tests should, if possible, be made during dry weather. If a test has to be made while it is raining, measures must be taken to prevent surface run-off entering the test borehole. If the borehole casing is close to ground level, this may involve using sandbags to build a small dam around the top of the borehole.

Where the test is carried out in an unconfined aquifer, and surface conditions are relatively pervious, rainfall during the test may produce a change in the groundwater level, which could affect the results obtained. In these circumstances, careful note should be made of any heavy showers immediately preceding or during a test. The groundwater level should be measured before and after the test.

A2.2 VARIABLE HEAD (FALLING HEAD) TEST – TEST PROCEDURE

Each test is carried out at a specific depth interval known as the 'test section' (Figure A2.1). The whole test section must be below the groundwater level. The initial groundwater level should be measured before making the test. It is important that the 'pre-test' groundwater level is representative of the natural, background groundwater level. If the water level has been influenced by drilling or other activities, it will be necessary to wait until water levels have recovered to their original levels.

Prior to the test itself, the borehole is advanced to the base of the test section. Where casing is not necessary to support the borehole, it must nevertheless be inserted to the top of the test section so that only the test section is exposed to the strata. The casing must be tight against the side of the borehole; otherwise, the test water will leak around the casing. This test is suitable for testing at various depths during the progress of boring where the test section will stand unsupported.

If casing is required to support the sides of the hole, it must be taken down initially to the base of the test section. Fine gravel backfill is then placed in the borehole while the casing is withdrawn to expose the required length of unlined hole for the test; subsequently, the

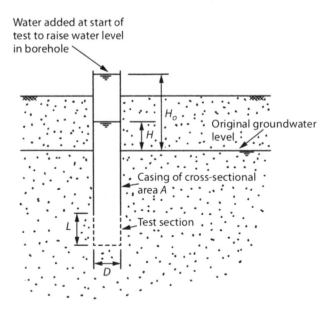

Figure A2.1 General arrangement of variable head tests in boreholes (falling head test shown).

gravel fill should be brought up with it and always kept just above the bottom of the casing as this is withdrawn. It is good practice, prior to placing the gravel, to replace any water standing in the borehole with clean water; this helps avoid suspended silt becoming trapped in the test section, which would reduce permeability. This type of test can also be used at various depths during boring, but the gravel backfill will need to be drilled out following each test – this will slow down drilling progress.

It is best if these tests are carried out at the start of the day's work, as the initial ground-water level will be less affected by drilling activities. If they are carried out during the day's drilling, the pre-test water level recorded may not be a representative background water level. If necessary, the start of the test should be delayed until readings show that the pre-test groundwater level has stabilized.

To start the test, clean water is rapidly added to the borehole to raise the water level as high as possible in the casing; no more water is added during the test. It is essential that the water is added as rapidly as possible – the analysis methods assume that the initial change in head in the borehole is effectively instantaneous. Black (2010) suggests that test results will be compromised if the test initiation phase (when water levels are changed) exceeds 10% of the duration of the subsequent equilibration period.

Once the water has ceased to be added, readings of the water level inside the borehole are taken at frequent time intervals, making a note of the time corresponding to each reading. Time $t = 0$ is taken as the time of the first reading. No fixed time intervals can be specified, as the frequency of readings will depend on the rate of fall of the water level. As a guide, an attempt should be made to take the readings so that each level differs from the previous reading by about equal increments. Since the water level will fall most rapidly at the start of the test and then at a decreasing rate, readings will become less frequent as the test progresses. If the initial excess head of water in the borehole relative to background ground-water level is, say, 2 m, take readings so that the difference in successive readings is about 100 mm; if the head is smaller, the difference in level readings should be smaller, say 25 to 50 mm for initial head of 0.5 to 1 m. Ideally, readings should be continued until the head of water in the borehole above groundwater level is less than one-fifth of the initial head above initial groundwater level.

The following information should be recorded:

  (i)   Diameter of unlined borehole being tested and diameter of casing
 (ii)   Depth to the base of the borehole (bottom of test section)
(iii)   Depth to bottom of casing (top of test section)
 (iv)   Depth to top of gravel backfill, if used
  (v)   Date and time when water level readings are started
 (vi)   Water level readings and time of each reading
(vii)   Depth to the initial groundwater level

All depths and water level readings should be measured relative to a clearly defined datum, ideally ground level. If ground level is uneven or flooded, the top of the casing can be used as a datum, provided that the casing is secured so that it will not slip or sink during the test.

A2.3 VARIABLE HEAD (RISING HEAD) TEST – TEST PROCEDURE

This test should be used only in lined boreholes or in monitoring wells, provided that baling or pumping out is practicable. Each test is carried out at a specific depth interval known as the 'test section'. The whole test section must be below the groundwater level; ideally, the borehole should penetrate below groundwater level by at least 10 times its diameter. The initial groundwater level should be measured before making the test.

Preparation of the borehole and test section is similar to that for a falling test. At the start of the test, water is rapidly bailed or pumped out of the borehole to just above the bottom of the casing. No more water is removed during the test. It is essential that the water is removed as rapidly as possible – the analysis methods assume that the initial change in head in the borehole is effectively instantaneous. Black (2010) suggests that test results will be compromised if the test initiation phase (when water levels are changed) exceeds 10 per cent of the duration of the subsequent equilibration period.

Once the water has ceased to be removed, readings of the water level inside the borehole are taken at frequent time intervals, making a note of the time corresponding to each reading. Time $t = 0$ is taken as the time of the first reading. Readings should be taken at the same general intervals as a falling head test and the same data recorded. Readings should be continued until the difference between the water level in the borehole and initial groundwater level is less than one-fifth of the initial difference between these levels.

It is best if these tests are carried out at the start of the day's work, as the initial groundwater level will be less affected by drilling activities. If carried out during the day's drilling, the initial water level recorded may not be a representative background water level. If necessary, the start of the test should be delayed until readings show that the pre-test groundwater level has stabilized.

A2.4 VARIABLE HEAD (FALLING AND RISING HEAD) TESTS – CALCULATION OF PERMEABILITY VALUES

The permeability k may be determined for a variable head test using the following formula:

$$k = \frac{A}{FT} \tag{A2.1}$$

where
  A is the cross-sectional area of the borehole casing (at the water levels during the test)
  T is the basic time lag
  F is a shape factor dependent on the geometry of the test section

A2.5 DETERMINATION OF SHAPE FACTOR F

The geometry of the test should be compared with the analytical solutions of Hvorslev given in Figure 12.10 and the appropriate shape factor calculated.

A2.6 DETERMINATION OF BASIC TIME LAG T

1. Where the initial groundwater level is considered representative of background conditions, plot values of H/H_o on a logarithmic scale against corresponding values of elapsed time t on an arithmetic scale (Figure A2.2). H_o is defined as the excess head of water (measured relative to the background water level) at the start of the test (the initial reading at $t = 0$). H is the excess head of water at time t during the test. Draw the best-fitting straight line through the experimental points. In some cases, the

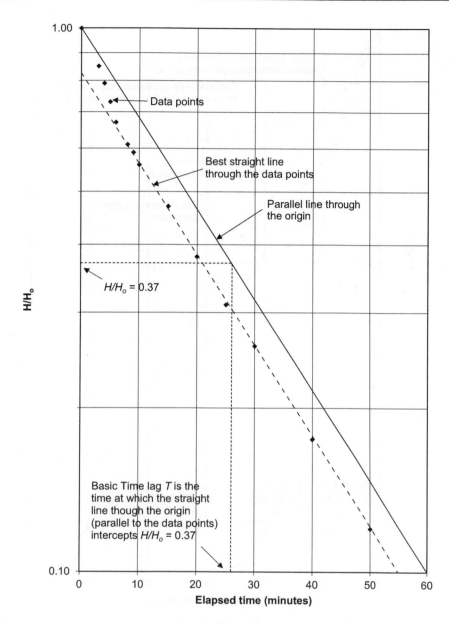

Figure A2.2 Estimation of basic time lag (original water level known).

experimental points for values of H/H_o near 1.0 may follow a curve; these should be ignored and the straight line drawn through the remaining points; then, draw a parallel straight line through the origin. The basic time lag T is obtained by reading off the value of t when $H/H_o = 0.37$ using the straight line through the origin.

2. If the background groundwater level is only approximately known, calculate values of excess head H from an assumed groundwater level (making as accurate an estimate as is possible). Plot the resulting values of H/H_o on a logarithmic scale against corresponding values of t on an arithmetic scale (Figure A2.3). If the slope of the line through these points decreases with increasing t (curve A), the assumed background groundwater level was too low. If the slope increases with increasing t, the assumed

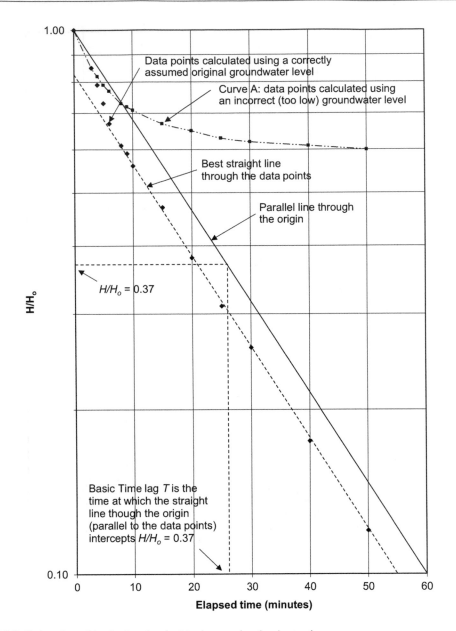

Figure A2.3 Estimation of basic time lag (original water level unknown).

groundwater level was too high. By trial and error, a groundwater level can be determined from which values of H/H_o plot as a straight line against t, at least for the higher values of t. Experimental points for small values of t may follow a short curve due to the flow settling down into equilibrium with the soil. Basic time lag T can then be determined in the same way as for Step 1.

The values of A (in square metres), F (in metres) and T (in seconds) are applied into Equation A2.1 to produce the calculated permeability k in metres per second.

Appendix 3: Execution and Analysis of Packer Permeability Tests in Boreholes in Rock

A3.1 GENERAL COMMENTS ON PERFORMING PACKER PERMEABILITY TESTS

The test results and derived permeability values from packer permeability tests in rock are affected by many factors – the host rock, the borehole and any associated zone of disturbance, water quality (injected water and water in the host rock), the packers or isolation system, and the head/flow rate measurement system – packer permeability tests are described in detail by Preene (2019). Some general guidance is given here:

1. The direct data gathered during packer tests are water injection rates and pressure responses for a borehole test section of finite length. Test response is controlled by both the properties of the rock (permeability k) and the test geometry (length L and diameter D of the test section). Commonly used test analyses calculate transmissivity T, which means that the resulting permeability is effectively $k_{average}$, an average permeability for the test section ($k_{average} = T/L$). It is not always straightforward (without reference to the geological model and other data, such as borehole geophysics) to determine whether outflow is via a single fracture, multiple fractures, or more general percolation through the rock mass.

2. The packer test response can be influenced by equipment issues, including leakage of water past packers intended to isolate the test section. When high injection flow rates are observed, the test data should be reviewed for any evidence to indicate whether packer leakage is a possible factor.

3. The application of high injection flow rates or high excess heads may affect the rock properties by erosion, plugging or jacking of fractures. This can cause the apparent permeability to change during the test.

4. Injected water quality can affect the test response. A particular problem occurs if the water used for injection is even slightly dirty or turbid – the suspended fine-grained particulate matter in the water can clog or plug fractures and intergranular flow paths on the walls of the test section. A lesser problem is geochemical clogging, where the injected water may react with the host water or rock.

5. Flow of water into the rock is driven by the applied 'excess head' above the background groundwater level in the test section. Because tests are carried out at different depths in a borehole, it is important that the water level used is representative of the particular test section. The pre-test groundwater level should be measured after the packers have been inflated to isolate the test section and ideally after sufficient time has passed to allow the borehole to equilibrate after the influence of drilling or other activities.

6. Packer tests typically involve multiple injection phases (five in the case of an A1–B1–C–B2–A2 test), and analysis produces multiple values of permeability. For some test responses, it is clear that the test has modified the permeability around the borehole. The most 'representative' permeability value should be selected carefully.

A3.2 PACKER PERMEABILITY TEST – TEST PROCEDURE

An uncased section of a borehole in rock is isolated by inflatable packers to form a 'test section', and water is injected to induce flow of water out of the test section, with the flow rate and pressure head in the test section monitored. Between tests, the packers can be deflated and relocated to different depths. This allows packer tests to be carried out at different depths (and with different test lengths) in the same borehole, either during drilling or after completion of the borehole. The test section can be isolated above and below by packers (double packer test, Figure A3.1a) or between a packer and the base of the borehole (single packer test, Figure A3.1b).

A common format of packer permeability tests is five consecutive injection phases (each phase of the same duration – typically 10 or 15 minutes; longer durations are sometimes used). Injection pressures are varied rapidly at the end of each phase, giving a stepwise transition of pressure between phases in three ascending and descending test pressures, A, B and C, in the sequence A1–B1–C–B2–A2. Example test sequences are shown in Table A3.1. Suitable values of maximum injection pressure P_{max} are discussed in Section A3.4.

A3.3 FIELD MEASUREMENTS

In addition to recording the depth and dimensions of the test section, field data recorded during the test are:

- Background groundwater level (prior to the test, after the packers have been inflated and water levels have been given time to equilibrate)
- Time duration of each test phase
- Pressure achieved for each test phase
- Injection flow rate in each test phase

The field data during each test phase can be collected by manual recording (typically at 30 second or 1 minute intervals) or can be recorded by datalogging systems, providing an effectively continuous data record.

Flow rates are typically recorded by flowmeters in the above-ground injection pipework, although timed level changes in holding tanks of known dimensions can also be used. Depending on the test equipment, water pressures in the test section can be measured via a pressure gauge in above-ground injection pipework (Figure A3.2a) or via a pressure transducer within the test section (Figure A3.2b). In both cases, the pressure measured in the field must be corrected to determine the excess head H in the test section (excess head is the head above background groundwater level – this is the head driving the flow into the strata).

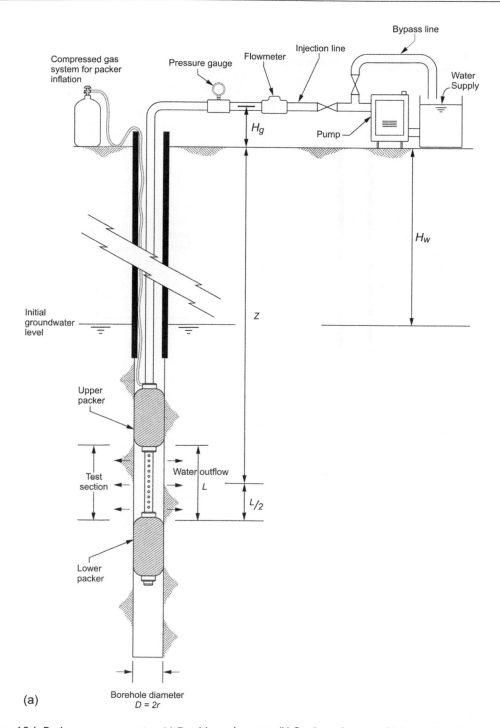

Figure A3.1 Packer test geometries. (a) Double packer test. (b) Single packer test. (Adapted from Preene, M, *Quarterly Journal of Engineering Geology and Hydrogeology*, 52, 182–200, 2019.) Note: above-ground pressure measurement system shown.

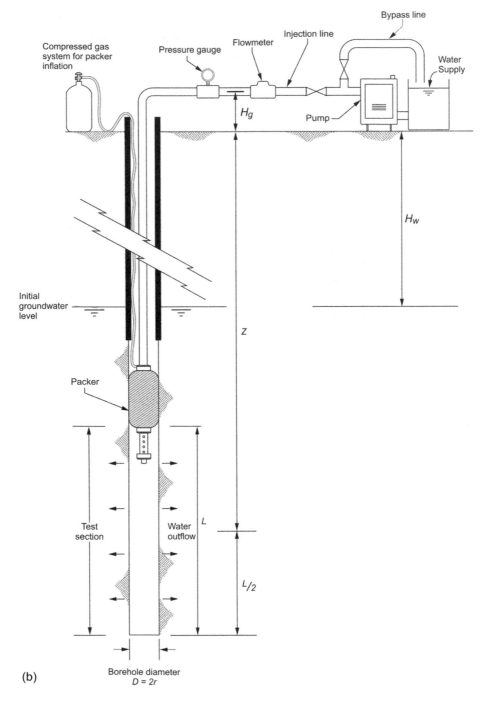

(b)

Borehole diameter
$D = 2r$

Figure A3.1 (Continued)

Table A3.1 Typical Sequence for a Packer Permeability Test

Test phase	Phase type	Typical ratios of specified excess pressure at midpoint of test section		
A1	Ascending	$0.25 \, P_{max}$	$0.33 \, P_{max}$	$0.5 \, P_{max}$
B1	Ascending	$0.50 \, P_{max}$	$0.67 \, P_{max}$	$0.75 \, P_{max}$
C	Peak	P_{max}	P_{max}	P_{max}
B2	Descending	$0.50 \, P_{max}$	$0.67 \, P_{max}$	$0.75 \, P_{max}$
A2	Descending	$0.25 \, P_{max}$	$0.33 P_{max}$	$0.5 \, P_{max}$

Note: P_{max} is the maximum excess pressure applied to the test section.

From Preene, M, *Quarterly Journal of Engineering Geology and Hydrogeology*, 52, 182–200, 2019. With permission.

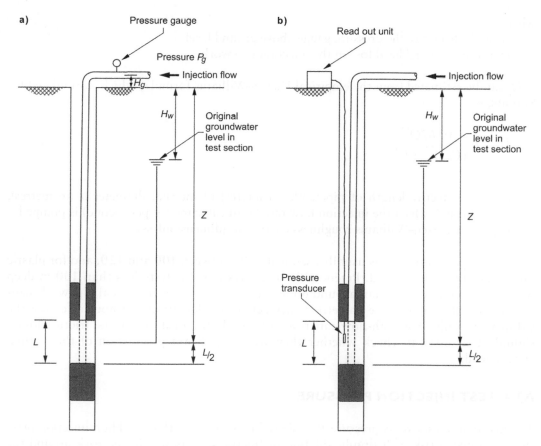

Figure A3.2 Measurement of applied excess head in packer test. (a) Pressure gauge in the above-ground injection pipework (double packer test shown). (b) Pressure transducer within the test section (double packer test shown).

Excess head H is determined as follows. For the arrangement shown in Figure A3.2b, with a pressure sensor located at the midpoint of test section, where the pressure measured is P_t:

$$H = \left(\frac{P_t}{\gamma_w}\right) - \left(Z - H_w\right) \tag{A3.1}$$

where

    H_w is the initial depth to groundwater level
    Z is the depth to the midpoint of the test section

When an above-ground pressure gauge is used, as shown in Figure A3.2a, the excess head H is related to the gauge pressure P_g by

$$H = \left(\frac{P_g}{\gamma_w}\right) + H_g + H_w - H_f \tag{A3.2}$$

where

    H_g is the height of the pressure gauge above ground level
    H_f is the frictional head loss in the injection pipework

H_f can be estimated from the empirical Hazen–Williams formula, which, formulated for SI units, is

$$H_f = l_p \frac{10.67 Q^{1.85}}{\left(C_{hw}\right)^{1.85} \left(d\right)^{4.87}} \tag{A3.3}$$

where

    l_p is the effective length of pipework (in metres) of internal diameter d (in metres),
        through which the injection flow rate Q (in cubic metres per second) is pumped
    C_{hw} is the Hazen–Williams roughness coefficient (dimensionless)

For steel pipework, C_{hw} is usually assumed to be between 100 and 120, and for plastic pipework, 140 can be used. Preene (2019) reports that for tests less than 100 m deep with injection flows less than around 60 l/min, the head losses in typical pipework sizes are small (<0.5 m) and are sometimes neglected in calculations without affecting the validity of results. For high-flow-rate tests ($Q > 60$ l/min) or deeper tests, friction losses should be assessed to check whether they are a significant proportion of the applied excess head.

A3.4 TEST INJECTION PRESSURE

The maximum test excess pressure P_{max} should be selected with care. High injection pressures can give a risk of hydraulic dilation of fractures/joints within the rock around the borehole. Bjerrum *et al.* (1972) recommended that in rock subject to isotropic stress conditions, the maximum excess test pressure P_{max} should not exceed the vertical effective stress σ'_v; this is also consistent with the recommendations of Walthall (1990):

$$P_{max} < \sigma'_v \tag{A3.4}$$

The test pressure estimated from Equation A3.4 is a theoretical maximum based on a criterion to reduce the risk of hydraulic dilation of fractures/joints. In many cases, such high pressures are not needed to determine representative permeability values. Preene (2019) indicates that for geotechnical investigations, tests with applied excess heads in the range 5 to 25 m (50 kPa < P_t < 250 kPa) can give good results.

A3.5 PRESENTATION OF TEST RESULTS

The visualization of test results can be aided by graphing injection flow rate Q vs. applied excess head H (Q–H plots), as shown in Figure A3.3; it is essential that the origin (0,0) is included. The Darcian permeability (i.e. the permeability assuming laminar flow; see Section 3.3.3) of any (Q, H) point is proportional to the gradient of a line joining the point to the origin. Such plots show whether data points for later test phases have a steeper gradient to the origin; this indicates that permeability is increasing during the test. If data from later phases show a shallower gradient to the origin, permeability is indicated to be decreasing in later phases.

The packer test response is controlled by the whole test system – the host rock, the borehole and any associated zone of disturbance, the packers or isolation system, and the head/flow rate measurement system. For example, if a test shows an apparent increase in permeability between phases (i.e. greater Q/H values), this could be due to the opening/erosion of rock fractures (changes in hydraulic conditions in the rock) or could be due to leakage past packers (changes in equipment performance).

The Darcian permeability of any (Q, H) point is proportional to the slope of a line from the point to the origin. For a given test geometry, point iii has a higher indicated permeability than point ii, which in turn has a higher indicated permeability than point i.

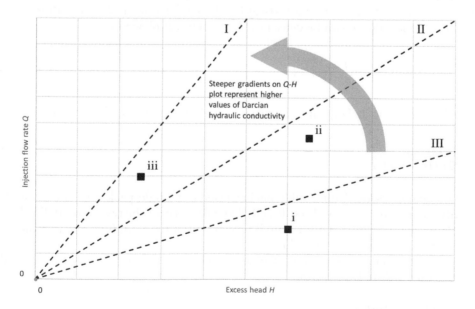

Figure A3.3 Example plot of injection flow rate Q vs. applied excess head H. (From Preene, M, *Quarterly Journal of Engineering Geology and Hydrogeology*, 52, 182–200, 2019. With permission.)

A3.6 PACKER PERMEABILITY TESTS – CALCULATION OF PERMEABILITY VALUES

A packer test of the A1–B1–C–B2–A2 pattern is typically analysed by treating each phase as a steady-state constant head injection test with each phase independent of the others; the test will generate five pairs of data points for Q and H. The following equations can be used to determine a permeability value for each point. For packer permeability tests in rock, where flow is predominantly via fine fracture networks, Darcian flow conditions should prevail, and the Q vs. H plot of the five data points should be approximately linear. However, where more open fractures are present, allowing higher flow rates, non-Darcian (turbulent) flow may occur; the flow rate will increase under-proportionally with excess head as energy is lost to turbulence. The Q vs. H plot will be non-linear for at least part of the test (example plots are shown in Figure A3.4).

Packer permeability tests can be analysed by similar methods to variable head tests in boreholes (Appendix 2) using Hvorslev's formula (Hvorslev, 1951) to determine permeability k from excess head H (measured relative to original groundwater level), the injection flow rate Q and shape factor F (a function of the geometry of the test zone):

$$k = \frac{Q}{FH} \tag{A3.5}$$

H can be related to the excess pressure P_t in the test section by $H = P_t/\gamma_w$, where γ_w is the unit weight of water.

Shape factors relevant to the geometry (length L and diameter D) of the test section are based on the work of Hvorslev (1951). The following equations below can be used with any combination of SI units. However, to obtain permeability values in metres per second, normal practice is to use L, D and H in metres and Q in cubic metres per second (not in litres per minute, which is the unit commonly recorded in the field).

For a test section of length L in a vertical borehole of diameter D in a uniform isotropic aquifer, the permeability k can be determined for each test phase by

$$k = \frac{Q}{2\pi HL} \ln\left[\frac{L}{D} + \left(1 + \frac{L^2}{D^2}\right)^{0.5}\right] \tag{A3.6}$$

Most packer tests have L/D greater than 4, which allows Equation A3.6 to be simplified to

$$k = \frac{Q}{2\pi HL} \ln\left(\frac{2L}{D}\right) \tag{A3.7}$$

The same equation is used for double packer tests and single packer tests (where there is potentially some 'end effect' outflow directly from the base of the test section). For $L/D > 4$, the flow from the end effect is a very small proportion of the total flow and does not have a major effect on calculated permeability.

If permeability conditions are anisotropic, the horizontal permeability k_h can be determined (for tests with $L/D > 4$) from

$$k_h = \frac{Q}{2\pi HL} \ln\left(\frac{2mL}{D}\right) \tag{A3.8}$$

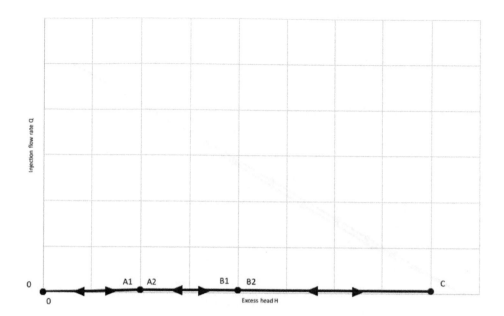

Figure A3.4 Q–H plots for conventional packer test responses. (a) Type 1: Zero water take. Effectively zero water injection rate at all excess heads. (b) Type 2: Linear Q–H relationship with small hysteresis loop. Injection flow rate and excess head have an approximate linear relationship (including the lIne to the origin), with only a small hysteresis between the ascending and descending test phases. (c) Type 3: Non-linear (under-proportional) Q–H relationship with small hysteresis loop. Injection flow rates are not linearly related to excess head, and the apparent permeability reduces as excess head increases, with only a small hysteresis between the ascending and descending test phases. (d) Type 4: Non-linear (over-proportional) Q–H relationship with small hysteresis loop. Injection flow rates are not linearly related to excess head, and the apparent permeability increases as excess head increases, with only a small hysteresis between the ascending and descending test phases. (e) Type 5: Non-linear (over-proportional) Q–H relationship with large hysteresis loop. Apparent permeability increases for each phase, including descending heads; this gives a significant hysteresis loop, where permeability is greater in the descending A2, B2 phases compared with the ascending A1, B1 phases. (f) Type 6: Non-linear (under-proportional) Q–H relationship with large hysteresis loop. Apparent permeability decreases for each phase, including descending heads; this gives a significant hysteresis loop, where permeability is lower in the descending A2, B2 phases compared with the ascending A1, B1 phases. (g) Type 7: Water take limited by equipment pumping rates with low excess head achieved. Injection rate quickly reaches close to the maximum injection flow rate Q_{max} for the test equipment, which is not able to establish an excess head in the test section. (From Preene, M, *Quarterly Journal of Engineering Geology and Hydrogeology*, 52, 182–200, 2019. With permission.)

where

$$m = \left(\frac{k_h}{k_v} \right)^{0.5} \tag{A3.9}$$

and k_v is the vertical permeability.

These equations calculate the average permeability $k_{average}$, assuming that the injected water leaves the test section uniformly across the cylindrical boundary. If the test section includes different beds with two to three orders of magnitude difference in permeability between zones of high- and low-permeability strata, then the water flowing into the

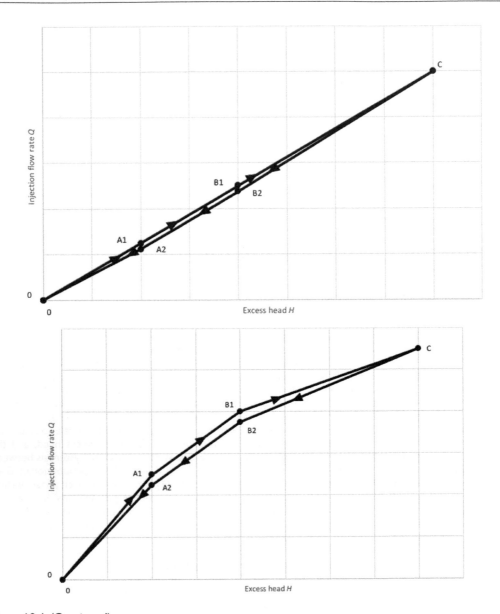

Figure A3.4 (Continued)

low-permeability stratum can be neglected, and the approximate equivalent permeability k' of the permeable zones can be estimated as:

$$k' = k_{average} \times \left(\frac{L}{L'}\right) \qquad (11)$$

where L' is the assessed total thickness of the permeable stratum within the test section.

A3.7 CATEGORIES OF PACKER TEST RESPONSE

A plot of flow rate against excess head (a Q–H plot) can be prepared for test phases of the A1–B1–C–B2–A2 pattern. The data points represent the Q and H for each phase, and the

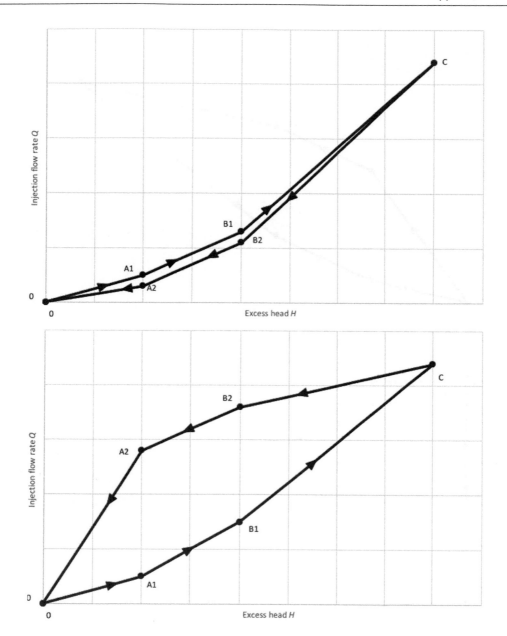

Figure A3.4 (Continued)

points are linked in sequence by straight lines. The plot must also include origin ($Q = 0$, $H = 0$) at both the start and the end of the test, when the applied excess head is zero. This can allow further interpretation of test responses and categorization against standard responses.

Preene (2019) shows seven conventional test responses types, as summarized in Figure A3.4 and discussed in the following.

Type 1: Zero Water Take
Categorized by effectively zero water injection rate at all excess heads. Indicative of a test carried out in rock of very low permeability with good packer seals achieved.

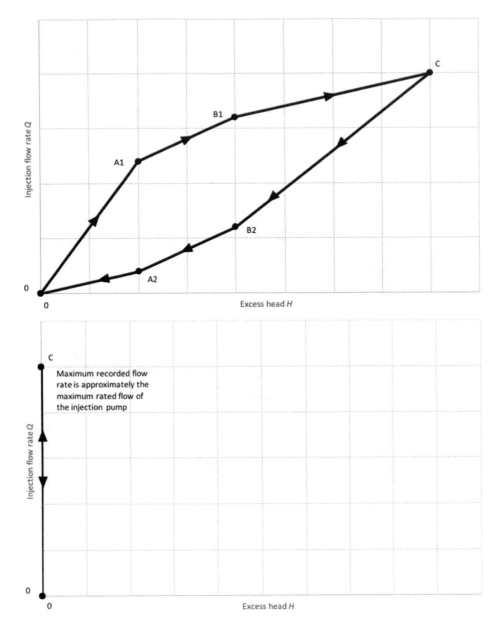

Figure A3.4 (Continued)

Type 2: Linear Q–H Relationship with Small Hysteresis Loop

Categorized by injection flow rate and excess head having an approximate linear relationship (including the line to the origin), with only a small hysteresis between the ascending and descending test phases. Permeability is essentially independent of excess head. A possible interpretation is Darcian flow out from the test section.

Type 3: Non-Linear (Under-Proportional) Q–H Relationship with Small Hysteresis Loop

Categorized by injection flow rates and excess head having a non-linear relationship, with the apparent permeability reducing as excess head increases, with only a small hysteresis between the ascending and descending test phases, indicating that permeability is not permanently reduced during the test. A possible interpretation is turbulent (non-Darcian) flow at higher injection rates.

Type 4: Non-Linear (Over-Proportional) Q–H Relationship with Small Hysteresis Loop
Categorized by injection flow rates and excess head having a non-linear relationship, with the apparent permeability increasing as excess head increases, with only a small hysteresis between the ascending and descending test phases, indicating that permeability is not permanently increased during the test. Possible interpretations include that existing bedding planes or other discontinuities in the rock are opened up by the applied pressure and close when pressure is removed, and/or packer leakage or movement causes the test section to lose water at higher heads but closes with reduced excess head.

Type 5: Non-Linear (Over-Proportional) Q–H Relationship with Large Hysteresis Loop
Categorized by the apparent permeability increasing for each phase, including descending heads, this gives a significant hysteresis loop, where permeability is greater in the descending A2, B2 phases compared with the ascending A1, B1 phases. Possible interpretations include an increase in permeability of the rock caused by the test due to movement/erosion of infill in fractures in such a way that they do not block flow paths, or permanent rock movements caused by the testing; and/or leakage past the packers that disturbs or erodes the rock, so that leakage paths do not close with reduced excess head.

Type 6: Non-Linear (Under-Proportional) Q–H Relationship with Large Hysteresis Loop
Categorized by the apparent permeability decreasing for each phase, including descending heads, this gives a significant hysteresis loop, where permeability is lower in the descending A2, B2 phases compared with the ascending A1, B1 phases. Possible interpretations include a decrease in permeability of the rock caused by the test, with possible mechanisms including a) water filling and pressurizing of voids or discontinuities not linked to a wider network, b) movement or swelling of infill in fractures in such a way that they become trapped and block flow paths, or c) clogging of rock fractures due to use of dirty water for injection.

Type 7: Water Take Limited by Equipment Pumping Rates with Low Excess Head Achieved
Categorized by the injection rate quickly reaching close to the maximum injection flow rate Q_{max} for the test equipment, which is not able to establish an excess head in the test section. Possible interpretations include the test section intersecting highly permeable fractures or discontinuities, and/or excessive water leakage past the packers, and/or poor selection of test equipment with an under-rated injection pump.

Preene (2019) also identifies three non-conventional packer test responses (Figure A3.5):

Type 8: Sudden Excess Head Drop to Zero
Categorized by injection flow rate showing a conventional $Q–H$ relationship up to a given head, after which the flow rate increases rapidly, and the excess head drops to almost zero. Possible interpretations include packer failure or a major fracture opening suddenly as the applied excess head clears a blockage.

Type 9: Flow Rate Initiates at Significant Non-Zero Excess Head
Categorized by injection flow rates that are effectively zero at lower excess heads and then increase suddenly under the higher heads of later phases. Possible interpretations include hydrojacking or fracture dilation in a 'tight' rock of very low permeability, and/or clearing of a blockage in a sediment-filled fracture, and/or packer leakage or movement, and/or the rest water level assumed in analysis being significantly higher than assumed (by an amount greater than the excess head in phases A1/A2).

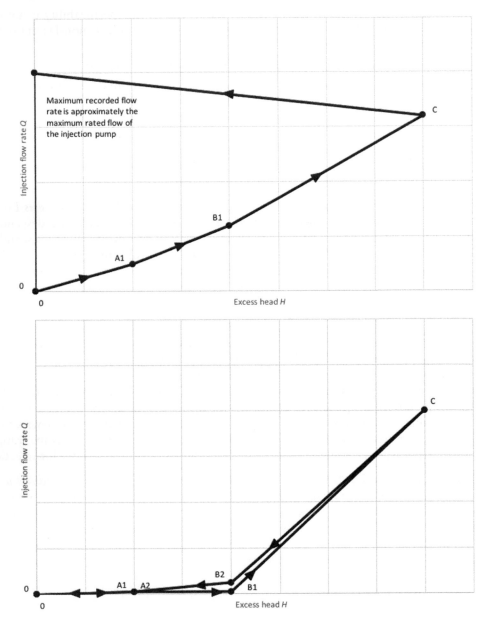

Figure A3.5 Q–H plots for non-conventional packer test responses. (a) Type 8: Sudden excess head drop to zero. Injection flow rate shows a conventional Q–H relationship up to a given head, at which point the flow rate increases rapidly, and the excess head drops to almost zero. (b) Type 9: Flow rate initiates at significant non-zero excess head. Injection flow rate is effectively zero at lower excess heads and increases suddenly under the higher heads of later phases. (c) Type 10: Excess head does not build until significant non-zero flow rate. Excess head is effectively zero at lower injection rates and increases suddenly under the higher flow rates of later phases. (From Preene, M, *Quarterly Journal of Engineering Geology and Hydrogeology*, 52, 182–200, 2019. With permission.)

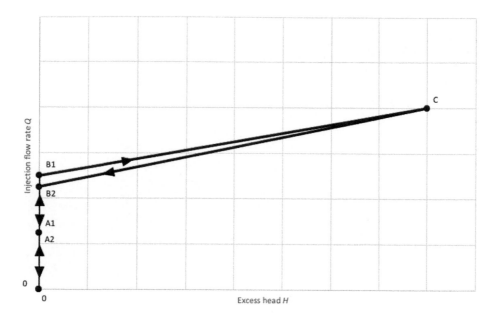

Figure A3.5 (Continued)

Type 10: Excess Head Does Not Build until Significant Non-Zero Flow Rate
Categorized by excess heads of effectively zero at lower injection rates, then increasing suddenly under the higher flow rates of later phases. A possible interpretation is that the rest water level assumed in analysis is significantly lower than assumed (by an amount greater than the excess head in phases A1/A2).

Appendix 4: Execution of Well Pumping Tests

A4.1 PLANNING OF PUMPING TESTS

The simplest form of pumping test involves controlled pumping from a single well – the 'test well' – and monitoring the discharge flow rate from the well and the drawdown in monitoring wells at varying radial distances away (Figure A4.1). A pumping test is one of the more expensive in situ tests and requires careful planning to ensure that time, effort and money are not wasted.

A pumping test consists of a number of phases:

1. Pre-pumping monitoring: This involves monitoring natural groundwater levels in the monitoring wells and test well for a period from a few days to several weeks prior to commencing pumping. The aim is to determine any natural or artificial variations in groundwater level that may affect the drawdowns observed during pumping.
2. Equipment test: This is a short period of pumping (typically 15–120 minutes) to allow the correct operation of the pumps, flowmeters and dataloggers to be ascertained and to check for leaks in the discharge pipework. It is normal to take some crude readings of flow rate and drawdown in the well to allow selection of flow rate for subsequent phases.
3. Step-drawdown test: This is a period of continuous pumping, typically lasting 4–8 hours, during which the flow rate is increased in a series of steps. Each step is of equal duration (normally 60 or 100 minutes), and an appropriately designed test should have four or five steps at roughly equal intervals of flow rate. Water levels are normally allowed to recover for at least 12 hours before the constant rate pumping phase can be started.
4. Constant rate pumping phase: This is often considered to be the main part of the test and involves pumping the well at a constant flow rate (chosen following the equipment test [or step-drawdown test, if carried out]). The constant rate phase normally lasts between 1 and 7 days, although test durations of up to several weeks are not unknown (see Table 12.2).
5. Recovery phase: Water levels in the monitoring well and test well are monitored as they recover following cessation of pumping. This phase often lasts between 1 and 3 days.

Guidance on the execution of pumping tests is given in BS ISO 14686: 2003, BS EN ISO 22282-4:2012 and Kruseman and De Ridder (1990).

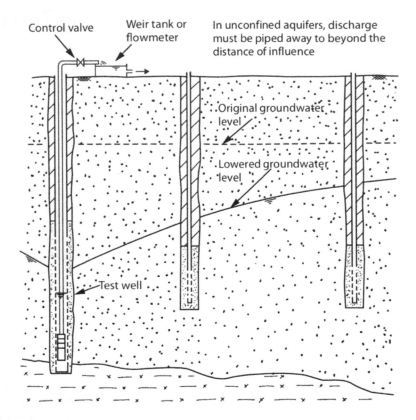

Figure A4.1 Components of a pumping test

Pumping tests as part of groundwater lowering investigations need to be designed to meet certain criteria consistent with providing high-quality data to the dewatering designer.

a) The aim should be to generate significant drawdowns in monitoring wells. Ideally, at the end of the constant rate phase, the drawdown in nearby monitoring wells should be at least 10 per cent of that in the proposed dewatering system.

b) If the site is subject to natural groundwater level variations (such as tidal effects), the test should aim to achieve drawdowns significantly greater than the background variations.

c) It is essential that pumping is absolutely continuous during the initial hours of pumping. If pumping is interrupted due to whatever cause, such as a pump or power failure, it is essential that the test is suspended until groundwater levels return to equilibrium or those levels recorded prior to the commencement of any pumping. Only then should pumping recommence. Once the test pumping has been going on for 24 hours or more, occasional short interruptions (e.g. for daily checks of generator oil levels) are permissible. These interruptions should be as short as possible (ideally no longer than a few minutes); any longer interruptions may require the test to be suspended and re-started later. Ideally, pumping should continue until steady-state drawdown conditions are established in monitoring wells near the test well.

d) Most of the simpler methods used to analyse pumping test data assume that the well fully penetrates the aquifer. Ideally, the test well should be fully penetrating. However, provided the screened length is of the order of 70 per cent to 80 per cent of the aquifer thickness, the accuracy of the assessment of permeability is generally acceptable. Where the amount of penetration is lower, the flow will not be predominantly horizontal, and more special corrections may need to be applied to the drawdown data.

e) Sufficient monitoring wells should be installed, in suitable locations, to allow groundwater flow patterns to be identified. Ideally, there should be least three lines of monitoring wells radiating out from the test well, with each line spaced 120 degrees apart. These need not penetrate the full depth of the aquifer but must be deeper than the expected drawdown (a monitoring well that becomes dry is not acceptable!). Each line should consist of three or more monitoring wells. Depending upon the permeability of the aquifer, the distance from the test well to the nearest monitoring well should be in the range of 2 to 3 m (in a low- to moderate-permeability aquifer) or 5 to 10 m (in a high-permeability aquifer) from the test well. More distant monitoring wells should be located within the anticipated distance of influence of the test well.

A4.2 GENERAL COMMENTS ON PERFORMING PUMPING TESTS

The test well should be designed, installed and developed according to the guidelines of Chapter 16, and monitoring wells should ideally be consistent with the good practice outlined in Section 11.7.6. The most common form of test well is a deep well pumped by a borehole electro-submersible pump. Suction pumps can be used, but only in highly permeable soils with a high water table; otherwise, as drawdown occurs, the pump may become starved of water, giving a varying flow rate. Whatever the method of pumping, it is desirable to measure the water inside the test well during pumping. This will require the installation of a dip tube in the well so that the dipmeter tape does not become entwined around the pump riser pipe due to the swirl of water near the pump intake.

The principal parameters to be measured during a test are water levels in the test well and monitoring wells, and the discharge flow rate from the well. All readings need to be referenced to the time elapsed since the start of pumping (or recovery) for that test phase. The frequency of monitoring must be matched to the rate of change of the measured parameter. During a test, the flow rate should remain approximately constant, but the water levels will fall rapidly immediately following the start of pumping and will fall more slowly as pumping continues. Measurements of water levels must be taken very frequently at the start of the test and become less frequent as time passes. A commonly used schedule of timings for readings is given in BS ISO 14686: 2003. This is summarized as follows:

- Readings to be taken immediately before the discharge flow rate is started, changed or stopped
- Readings to be taken every minute for the first 10 minutes of the test phase (if practicable, readings should be taken every 30 seconds during this period)
- Readings to be taken every 2 minutes thereafter up to an elapsed time of 20 minutes
- Readings to be taken every 5 minutes thereafter up to an elapsed time of 60 minutes
- Readings to be taken every 10 minutes thereafter up to an elapsed time of 100 minutes
- Readings to be taken every 20 minutes thereafter up to an elapsed time of 300 minutes
- Readings to be taken every 50 minutes thereafter up to an elapsed time of 1000 minutes
- Readings to be taken every 100 minutes thereafter up to an elapsed time of 3000 minutes
- Readings to be taken every 200 minutes thereafter unless other influences warrant more frequent measurement

These timings are for guidance only and should be applied with judgement. The key point is to ensure that sufficient readings are taken to allow the rate of change of water levels to be clearly identified. If water levels are only changing slowly, it may be possible to increase the intervals between readings without reducing the quality of the data. This would avoid generating unnecessary readings and can reduce the work involved in data analysis.

Water levels can be measured by manual dipping with a dipmeter (see Section 22.6.2). However, at the start of the test, readings need to be taken very frequently. If more than a few monitoring wells are to be monitored manually, several people may be needed to take the frequent readings at the start of the test, creating potential problems of co-ordination between personnel. There is also the problem of actually gathering enough capable people together in the first place. This can lead to poor recording of the precise timing of measurements taken during the first few minutes of pumping, leading to difficulties in subsequent analysis. If taking manual water level readings with limited numbers of observers, it is best to concentrate initial monitoring on points nearest the test well. As time passes, and the time period between readings increases, more monitoring wells can be included in the monitoring.

The use of electronic datalogging equipment, linked to pressure transducers in the monitoring wells (see Sections 22.5 and 22.6.3), may help to overcome such staffing problems. Once programmed to take readings at the appropriate intervals, such equipment not only reduces the personnel requirements but also will produce the test measurements in electronic form, allowing rapid analysis using spreadsheet programs.

If a test is carried out in an aquifer with significant tidal response (Section 3.6.3), the monitoring intervals later in the test (after several days of pumping) must be selected with care. Standard tables of monitoring intervals (such as in BS ISO 14686: 2003) allow monitoring frequencies of one reading every few hours. If such large time gaps between readings are permitted, data will be very difficult to interpret, because the tidal responses will not be accurately recorded. For tidal pumping tests, the monitoring intervals for the first few hours of the test should be as per published guidelines, but the remainder of the test should be monitored at 15–30 minute intervals to ensure that the tidal fluctuations are fully resolved. This requirement means that a datalogging system is almost essential for pumping tests on tidal sites.

The pump discharge can be measured by means of a tank or gauge box fitted with a V-notch or rectangular weir (see Section 22.7) or by installing an integrating flowmeter into the discharge pipeline. Even if flowmeters are used in preference to weir tanks, it is good practice to include a settlement tank in the discharge line so that it can be visually checked for suspended solids in the discharge water. In general, once adjusted at the start of the test phase, the flow rate should not vary significantly. Immediately following the commencement of pumping, flow rate does not need to be monitored as frequently as water levels for the first hour or so of pumping. The outlet of the discharge should be well away from the test area to reduce the risk of return seepage affecting the drawdown levels.

It is normal to take water samples from the pumped discharge during the test. Obtaining a sample is relatively straightforward if it is possible to fill a sample bottle directly at the discharge tank. However, when taking groundwater samples from the discharge flow, the following factors should be considered.

a) Try to minimize the exposure of the sample to the atmosphere. Try to obtain it directly at the point where the pump discharges into the tank. Totally fill the bottle, and try to avoid leaving any air inside when it is sealed. If the pump discharge is 'cascading' before the sampling point, the water will become aerated, and oxidation may occur. The discharge arrangements should be arranged so that the sample can be obtained before aeration occurs.

b) Samples may degrade between sampling and testing. The samples should be tested as soon as possible after they are taken and ideally, should be refrigerated in the meantime. The bottles used for sampling should be clean with a good seal. However, the sample may degrade while in the bottle (for example by trace metals oxidizing and precipitating out of solution). Specialists may be able to advise on the addition of suitable preservatives to prevent this from occurring. The choice of sample bottle (glass or plastic) should also be discussed with the laboratory, since some test results can be influenced by the material of the sample bottle.

The wellhead chemistry (see Section 3.7.2) can be determined using probes or sensors immersed in the flowing discharge water. These can be portable water quality meters, recorded manually, or may be linked to datalogging systems.

A4.3 EXECUTION OF PUMPING TESTS

Before any pumping is carried out, water levels in monitoring wells must be recorded regularly over a period of days. This is to try to determine whether any natural (or artificially induced) variations in water level are occurring. There are no generic guidelines on the duration and frequency of background monitoring. On sites where it is anticipated that variations will be small (e.g. remote inland sites), monitoring three times daily for 3 to 5 days is the minimum acceptable. On coastal sites with a significant tidal response, it has been necessary to use dataloggers to record water levels every 15 minutes, 24 hours per day for up to 30 days. If in doubt, it is best to do more than the minimum monitoring. The time to find out about background variations is prior to test pumping, not during the test itself.

The equipment test phase should be used to determine the most suitable pump discharge rate for subsequent phases of the test, so that at the end of pumping, the water in the borehole is not drawn down to the pump intake. It should also be used to ensure that flow measuring devices function and that discharge pipework is not blocked and does not leak – to check, in general, whether the test is ready to proceed. Ideally, these preparations should be tried out at least a day before test pumping is commenced so that the test well and monitoring wells can be left to ensure that the groundwater level is realistically re-established.

The following general information should be recorded prior to testing:

 (i) Elevation of ground surface at the test well and at each monitoring well.
 (ii) Elevation of reference datum for water levels at each well (the top of the well casing is often used as a datum). Datums for groundwater level measurement are discussed in Section 22.6.1.
 (iii) Depth of the well screen in the test well and the depth of response zones in all monitoring wells.
 (iv) Distances from the centre of the test well to all monitoring wells.

A step-drawdown test (if carried out) has the aim of investigating the performance of the well at increasing flow rates (see Clark, 1977). The well is pumped in a number of steps (ideally four, or an absolute minimum of three); the flow rate in each step is constant, with the rate for each step greater than the last. Increments for flow rate should be roughly equal. For example, a four-step test might be designed as:

Step 1: one-quarter of maximum well yield
Step 2: one-half of maximum well yield

Step 3: three-quarters of maximum well yield

Step 4: maximum well yield (or maximum pump output), estimated from the equipment test

Step-drawdown pumping test results can be analysed to provide information about well efficiency. This is a function of friction head losses through the filter pack and the well screen. It can also be affected by the techniques used to bore and develop the test well.

When starting any pumping (step-drawdown or constant rate) or recovery phase, it is important that the monitoring team has been adequately briefed and is ready. For manual monitoring, checks include ensuring that all the dipmeters work and that all observers have a pen or pencil and paper or an electronic tablet to record manual readings. These checks may seem obvious, but it can be very embarrassing if they are overlooked! If datalogging equipment is being used, it should be checked in advance for battery function and for accuracy of clock setting. Above all, one person should be responsible for deciding when to start pumping, and he or she should resist being pressurized to start before everyone is ready. It is not necessary to start a test on the hour of local time, but it makes recording and analysis easier if the test starts on, say, a multiple of 10 minutes past the hour.

The time when the pump is started should be recorded. The control valve of the test pump should be adjusted to achieve the desired flow rate as quickly as possible after the start of pumping (or the step in a step-drawdown test). Once the flow rate has been set, the valve should not be further adjusted, as this will affect the drawdown and complicate analysis of results. At the start of the test phase, when readings are being taken frequently, it is important that readings in the test well and the monitoring wells are taken at as nearly the same instant as possible. If all the wells are close together, one of the observers may be able to make a visible or audible signal to the others that it is time for a reading. Otherwise, each observer will need to have a clock of some sort, and the clocks will all need to be synchronized. If there are few observers and many wells, it is sometimes best to abandon a rigid schedule of monitoring and just take readings as rapidly as possible but record the precise time each reading was taken.

Monitoring during the test phases normally comprises:

(i) Water levels in the test well and monitoring well, with the time of each reading recorded. Readings are normally taken at specified intervals (very frequently at the start of the test and less frequently as pumping proceeds). Guidance is given in BS ISO 14686: 2003.
(ii) Pump discharge flow rate, recorded at the same time as the water levels (during the first hour or so of pumping, it is acceptable to record the flow rate less frequently than water levels, especially if the number of observers is limited).
(iii) Clarity of discharge water (as a check on any sand or silt in the discharge water).
(iv) Discharge water temperature and chemistry (by taking samples for testing or using portable water quality meters; see Sections 3.7 and 22.8). It is normal to take a water sample during the constant rate phase immediately after pumping commences, again after a few hours, and just before pumping ceases.

Upon cessation of pumping, monitoring of water levels in piezometers should be continued until full recovery of water level is approached. During the initial recovery period, readings should be taken at similar intervals to the start of pumping. As the rate of water rise slows down, the time intervals between readings may be extended progressively.

A4.4 PRESENTATION OF TEST RESULTS

All water level and flow rate data gathered during the test should be plotted in graphical form while the test is proceeding. Even if this is done on site in rough form (on a graph pad or on a spreadsheet on a laptop computer), it will help to identify any anomalies or inconsistencies. These may be due to occasional human error. If readings are plotted on site as the test is in progress and an anomaly shows up, a further check reading can be made immediately. If the second reading confirms the earlier reading, it is possible that some fault has developed with the dipmeter, datalogger transducer or monitoring well, and so immediate remedial measures can be implemented.

Plotting of the test results also allows the aquifer drawdown response to be observed almost in 'real time'. This can be a useful guide when deciding whether a pumping phase needs to be extended or whether pumping can be stopped early.

On groundwater control projects, the most useful method of plotting data is to use the Cooper–Jacob straight line method described in Section 12.8.5.1. Drawdown is plotted on the vertical axis (linear scale) against elapsed time on the horizontal logarithmic scale (Figure A4.2). The first few readings curve upwards and, in theory, then form a straight line. Obviously, there will be times when the data will not conform precisely to the theory, but this form of presenting the data is nevertheless very useful for viewing trends in the drawdown data.

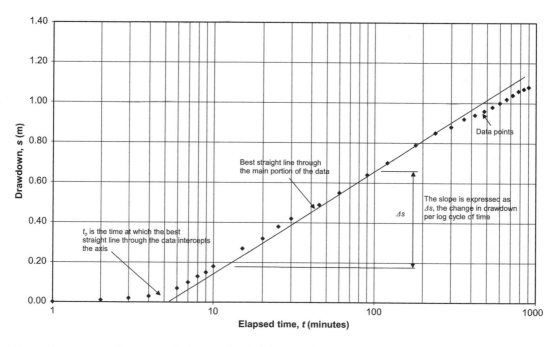

Figure A4.2 Cooper–Jacob straight line method of plotting data

A.4 PRESENTATION OF TEST RESULTS

Appendix 5: Design Examples

Four design examples based on the methods of Chapter 13 are presented here. For ease of reference, the relevant equation and figure numbers from the main text are also given.

A5.1 DESIGN EXAMPLE 1

This example is an application of the equivalent well method to analyse a system of fully penetrating deep wells used to lower the piezometric level in a confined aquifer beneath a rectangular excavation. A sensitivity analysis is carried out to assess the impact on the calculated flow rate of various possible values of permeability.

A5.1.1 Conceptual Model

A rectangular excavation is to be made to a depth of 9 m. The details of the conceptual model can be summarized as shown in Figure A5.1:

- Excavation dimensions are: 35 by 15 m in plan, 9 m depth to the deepest part of excavation. The excavation is to have vertical sides supported by sheet-piles. Dewatering is required for 6 months.
- A confined aquifer, consisting of a medium sand, extends from 10 m depth to 19 m depth. The confining layer above the aquifer is a stiff clay. Maximum piezometric level in the aquifer is 1 m below ground level.
- No pumping test was carried out during site investigation, but particle size distributions (PSD) and falling head test data can be used to estimate permeability.
- No recharge boundaries are believed to exist. Flow is likely to be radial to the array of wells and can be idealized as a distant circular source.
- No compressible strata are believed to exist. Groundwater-lowering-related settlement is not anticipated to be a problem.

A5.1.2 Selection of Method

The presence of a confined aquifer at shallow depth beneath the excavation would result in a risk of base heave if the piezometric level is not lowered (see Section 13.7.1). The conservative case is to lower the piezometric level to 9.5 m below ground level (i.e. 0.5 m below formation level). This requires a drawdown of 8.5 m below original groundwater level.

Without assessing the available permeability data in detail, typical values of permeability given in Table 4.1 suggest that the permeability k of a medium sand would be in the range of 1×10^{-4} to 5×10^{-4} m/s.

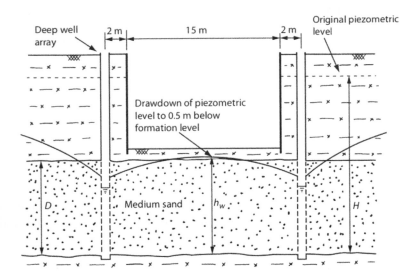

Figure A5.1 Conceptual model for Design Example 1.

Inspection of Figure 9.11 suggests that either deep wells (Chapter 16) or two-stage well-points (Section 15.10.2) would be suitable for this combination of drawdown and permeability. In this case, the deep well method will be used, because the contractor wishes to excavate to full depth in one operation, and two-stage wellpointing requires a pause in excavation while the second stage is installed.

A system of relief wells (Section 17.8) might have been considered. However, the anticipated permeability suggests that the flow rate from the relief wells would have been too great to be handled by sump pumping without interfering with construction operations.

A5.1.3 Estimation of Steady-State Discharge Flow Rate

The wells are to be installed in a regular pattern around the excavation. This geometry is amenable to solution as an equivalent well. If it is assumed that the wells will be fully penetrating, the total flow rate Q can be estimated from Equation 13.3.

$$Q = \frac{2\pi k D (H - h_w)}{\ln\left[\frac{R_o}{r_e}\right]}$$

(13.3)

where
k is the aquifer permeability. As described earlier, Table 4.1 suggests a likely permeability in the range 1×10^{-4} to 5×10^{-4} m/s. The site investigation data include PSD data showing a 10 per cent particle size (D_{10}) of 0.1 to 0.3 mm. The sand is relatively uniform, and Hazen's formula (Equation 12.4) can be used to estimate permeability as 1×10^{-4} to 9×10^{-4} m/s. These permeability estimates are broadly consistent with Table 4.1. The higher end of the range is rather greater than might be expected in a medium sand; the sample may have been affected by 'loss of fines' or may represent a coarser layer within the aquifer. Falling head tests in boreholes gave permeabilities in the range 1×10^{-8} to 1×10^{-6} m/s. Comparison with the soil description suggests that these results are unrepresentatively low and are likely to have been affected by silting up during the tests. The falling head test results are

not used in subsequent assessments of permeability. It is probably reasonable to assume that design values of permeability are between 1×10^{-4} and 9×10^{-4} m/s and to carry out sensitivity analyses when estimating flow rate.

D is the aquifer thickness: $D = 19 - 10 = 9$ m.

$(H - h_w)$ is the drawdown: $(H - h_w) = 9.5 - 1 = 8.5$ m.

r_e is the equivalent radius of the array of wells: assuming that the wells are located 2 m outside the edge of the sheet-piles, the overall dimensions of the system will be 39 by 19 m. r_e can be estimated from either Equation 13.1

$$r_e = \frac{(a+b)}{\pi} = \frac{(39+19)}{\pi} = 18.5 \text{ m} \tag{13.1}$$

or Equation 13.2

$$r_e = \sqrt{\frac{ab}{\pi}} = \sqrt{\frac{(39 \times 17)}{\pi}} = 15.4 \text{ m} \tag{13.2}$$

Because r_e appears in a log term in Equation 13.3, these two values will produce very similar estimates of flow rate, so a value of $r_e = 18.5$ m will be used in subsequent calculations.

R_o is the radius of influence. In the absence of pumping test data, R_o can be estimated for equivalent wells from Equation 13.15 (shown here for R_o, r_e and $(H - h_w)$ in metres and k in metres per second):

$$R_o = r_e + 3000(H - h_w)\sqrt{k} \tag{13.15}$$

The total flow rate Q can then be estimated from Equation 13.3 by a sensitivity analysis within the selected range of permeability. Using the units quoted earlier, Q will be calculated in cubic metres per second; when commenting on results, Q is normally quoted in litres per second (1 m³/s = 1000 l/s) to make the numbers easier to read and interpret.

k (m/s)	r_e (m)	R_o (m)	Q (m³/s)	Q (l/s)
1×10^{-4}	18.5	274	0.018	18
2×10^{-4}	18.5	379	0.032	32
5×10^{-4}	18.5	589	0.069	69
9×10^{-4}	18.5	784	0.12	120

A5.1.4 Estimation of Number of Wells

For soils of permeability greater than 1×10^{-4}, q, the maximum yield of a well, can be estimated from Equations 13.35 and 13.36, combined into the following equation (or alternatively taken from Figure 13.19):

$$q = \frac{2\pi r l_w \sqrt{k}}{15}$$

where

l_w is the wetted screen length of wells

r is the radius of the well borehole

For each case of the sensitivity analysis, the total wetted length (all wells in combination) can be estimated.

The number of wells and the corresponding well yield can then be estimated once certain assumptions have been made about the dimensions of the well. In this case, the diameter of the well borehole (not the diameter of the well screen) is taken to be 0.305 m. The wetted depth per well must also be assumed. The wells fully penetrate the aquifer and have a total screen length of 9 m. However, drawdown within the excavation is to 9.5 m below ground level compared with the top of the aquifer at 10 m depth. The drawdown in the wells will be greater than the drawdown in the general excavation area, so the wetted length per well will be less than the aquifer thickness of 9 m. In this case, the wetted length per well will be assumed to be 6 m (two-thirds of the aquifer thickness).

The number of wells is the total wetted screen length divided by the wetted screen length per well, with the answer rounded up to the next whole number. The nominal well spacing (assuming that the wells are evenly spaced) is then determined from the plan dimensions of the well system (39 by 19 m). The calculations are shown in the following:

k (m/s)	Q (m³/s)	Total l_w for all wells (m)	Number of wells and yield	Nominal well spacing (m)
1×10^{-4}	0.018	28	5 no. at 3.6 l/s	23
2×10^{-4}	0.032	36	6 no. at 5.3 l/s	19
5×10^{-4}	0.069	48	8 no. at 8.6 l/s	15
9×10^{-4}	0.12	63	11 no. at 10.9 l/s	11

A5.1.5 Empirical Checks

It is vital that the basic design is checked against experience or the 'normal' range of dewatering systems.

In Section 13.9.3, it is stated that most deep well systems have a well spacing of between 10 and 60 m. The design fits within these limits, although if the permeability is at the higher end of the analysed range, it can be seen that the well spacing is at the edge of the normal range – this implies that if permeability is actually rather greater than the analysed range, deep wells may not be the most appropriate method.

Another rule of thumb is that where the aquifer depth allows, dewatering wells should generally penetrate to at least one and a half to two times the depth of the excavation (Section 13.9.2). In this case, the wells are 19 m deep compared with an excavation depth of 9 m, so this condition is satisfied.

In Table 16.1, the minimum diameters of well bore for a given yield are listed. This table indicates that wells drilled at 305 mm diameter can accommodate pumps of capacity up to 15–20 l/s, within the range of the anticipated yield.

These empirical checks confirm, in the gross sense, the validity of the concept of the design.

A5.1.6 Final Design

Design calculations have produced a range of flow rates and well yields and spacings. Ultimately, a decision has to be made based on engineering judgement. Any records of local experience may help in forming an opinion. In this case, the nominal system of six or eight wells would seem to be the most appropriate option. If the construction programme is tight and cannot cope with delays, the larger system would be prudent. However, in this case, it is assumed that there is time in the programme to install a few additional wells (if crude testing of the first wells shows that flow rates are higher than the design value). Accordingly, the nominal six well system is appropriate.

If the aquifer was unconfined, the calculated steady-state flow rates would need to be increased to allow for the additional water from storage release in the early stages of pumping. However, the aquifer is initially confined, and after drawdown, it will only become unconfined locally to the wells. Confined aquifers have very small storage coefficients, so the water released from storage will not be significant and is not normally allowed for in design.

It is normally prudent to increase the number of wells by around 20 per cent to provide some allowance for individual wells being temporarily out of service. In this case, this increases the number of wells from six to eight. At the design yield of 5.3 l/s, this gives a total design capacity of around 42 l/s. In practice, the system capacity will be slightly larger, because submersible pumps are manufactured with discrete capacities. The pump chosen will normally have a slightly greater capacity than the design yield. These factors should all help provide a robust design capable of coping with modest changes in the predicted flow rate.

Because the excavation is fairly narrow, it is likely that a well system of adequate capacity will achieve drawdown everywhere within the excavation area. In many cases, no formal calculation is made of the drawdown distribution within the excavation. If necessary, the methods given in Section 13.11 can be used to obtain *approximate* estimates of drawdown within the excavation.

A5.2 DESIGN EXAMPLE IA

This example shows an alternative method for the estimation of steady-state flow rate from a system of deep wells in a confined aquifer. The well array is still modelled as an equivalent well, but the flow rate is determined using published 'shape factors'. The case analysed is exactly the same as in Design Example 1.

A5.2.I Conceptual Model

As Design Example 1.

A5.2.2 Selection of Method

As Design Example 1.

A5.2.3 Estimation of Steady-State Discharge Flow Rate

Flow to rectangular arrays of wells in confined aquifers can be determined by the shape factor method of Equation 13.13:

$$Q = kD(H - h_w)G \tag{13.13}$$

where

k is the aquifer permeability. See Design Example 1 for discussion of selected values of permeability between 1×10^{-4} and 9×10^{-4} m/s.

D is the aquifer thickness: $D = 19 - 10 = 9$ m.

$(H - h_w)$ is the drawdown: $(H - h_w) = 9.5 - 1 = 8.5$ m.

G is a geometry shape factor, obtained from Figure 13.11. To determine the appropriate value of G for a given case, it is necessary to evaluate the following parameters:

Array aspect ratio a/b: if the array of wells is of plan dimensions a by b, the aspect ratio a/b will have an effect on the flow rate. In this case, $a/b = 39/19 = 2.1$.

L_o/a: The ratio of distance of influence L_o to the long dimension of the well array a. In the absence of pumping test data, L_o can be estimated from either Equation 13.14 or Equation 13.16. In this case, the excavation is not very long, and narrow and radial flow is likely to be the dominant flow regime. Therefore, Equation 13.14 is most appropriate:

$$L_o = 3000(H - h_w)\sqrt{k} \tag{13.14}$$

Note that Equation 13.15 (which includes the radius r_e of the equivalent well) should not be used with this method, because Figure 13.11 is based on distances of influence from the edge of the well array, not the centre. If the well array had been long and narrow (i.e. a had been much greater than b), planar flow would predominate, and L_o should have been estimated from Equation 13.16.

The flow rate is then calculated from Equation 13.13 using values of G from Figure 13.11. The permeability sensitivity analysis of Design Example 1 has been repeated here.

k (m/s)	L_o (m)	L_o/a	a/b	G	Q (m³/s)	Q (l/s)
1×10^{-4}	255	6.5	2.1	2.4	0.018	18
2×10^{-4}	361	9.2	2.1	2.1	0.032	32
5×10^{-4}	570	14.6	2.1	1.9	0.073	73
9×10^{-4}	765	19.6	2.1	1.7	0.12	120

It is unrealistic to expect the flow rates calculated by two different methods to be precisely the same, but in this case, there is good agreement between the methods used here and in Design Example 1. When rounded to two significant figures (as above), the greatest difference between methods is only 3 l/s.

The remainder of the design process is carried out in the same way as for Design Example 1.

A5.3 DESIGN EXAMPLE 2

This example describes the design of a partially penetrating wellpoint system for trench works in an unconfined aquifer. The line of wellpoints is analysed as an equivalent slot under planar flow conditions, and the contribution from radial flow to the ends of the slot is assessed. A sensitivity analysis is carried out to assess the effect of varying the depth of the aquifer on the calculated flow rate.

A5.3.1 Conceptual Model

A narrow trench excavation is required to allow the laying of a shallow pipeline that extends for several hundred metres. The formation level for the trench is 3 m below ground level. The details of the conceptual model can be summarized as shown in Figure A5.2:

- The excavation is 1 m wide and 3 m in depth. The excavation has near-vertical sides, with the pipelaying operatives working in the trench, protected by a 'drag box' – a temporary trench support system. The trench is several hundred metres long, but it is anticipated that no more than 20–30 m will be open at any one time.

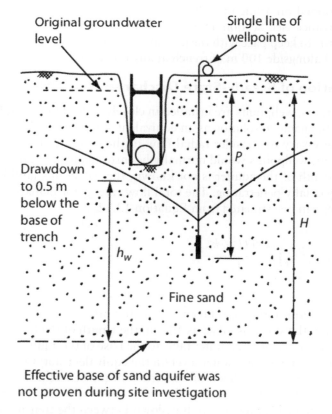

Figure A5.2   Conceptual model for Design Example 2.

- An unconfined aquifer is present beneath the site. The aquifer comprises a uniform fine sand, and groundwater level is generally 0.5 m below ground level.
- Back-analysis of previous groundwater lowering nearby indicates that the sand permeability is approximately 1×10^{-4} m/s. Boreholes at the current site indicate that ground conditions are very similar to the nearby site, and analysis of PSD results from the current site (using Hazen's formula, Equation 12.4) confirm this permeability.
- The base of the aquifer was not determined during the investigation. The deepest borehole penetrated to 12 m depth and did not encounter any underlying impermeable stratum. Published geological maps of the area indicate that the stratum of fine sand may extend to 20 m depth or more.
- No recharge boundaries are believed to exist. Flow is likely to be planar to the line of wellpoints and can be idealized as a distant source.
- No compressible strata are believed to exist. Groundwater-lowering-related settlement is not anticipated to be a problem.

A5.3.2 Selection of Method

The required drawdown (to 0.5 m below formation level) is 3 m (from 0.5 to 3.5 m depth), and the anticipated permeability is 1×10^{-4} m/s. Figure 9.11 suggests single-stage wellpointing, which is the method commonly used for dewatering of pipeline trenches (Section 15.8).

Although the trench is very long, only 20–30 m will be open or being worked on at any time; it will take around a week to excavate, lay and backfill this length of trench. In such circumstances, it is not common practice to dewater the whole length of trench. Typically, pumping is maintained on a line of wellpoints of length equal to three or four times the weekly rate of advance (see Section 15.8.3). The wellpoints and pumping equipment are progressed forwards to keep pace with the pipelaying. In this design, it is assumed that wellpoints are pumped alongside 100 m of trench at any one time.

A5.3.3 Estimation of Steady-State Discharge Flow Rate

For trench excavations, wellpoints are installed in closely spaced lines parallel to the trench. This geometry is amenable to solution as an equivalent slot – although, as will be described later, it may be necessary to consider radial flow to the end of a line of wellpoints. For a trench depth of 3 m, wellpoints would normally be installed to the standard depth of 6 m. Even though the depth of the aquifer is not known, the wellpoints (and hence the equivalent slot) will be partially penetrating. The total flow rate Q for planar flow to a slot in an unconfined aquifer can be estimated from Equation 13.12:

$$Q = \left[0.73 + 0.23 \left(P/H \right) \right] \frac{kx \left(H^2 - h_w^2 \right)}{L_o} \tag{13.12}$$

where
  k is the aquifer permeability: k is taken as 1×10^{-4} m/s.
  x is the length of the slot: x is taken as 100 m, the length of the line of wellpoints.
  H is the depth from the original water table to the base of the aquifer, and h_w is the depth from the lowered groundwater level (in the equivalent slot) to the base of the aquifer. Note that the drawdown of concern is in (or beneath) the trench itself rather than at the wellpoints. In reality, the wellpoints will be so close to the trench that there will be little difference in drawdown between the trench and the wellpoints. In calculations, h_w is taken as the head beneath the trench (see Figure A5.2).
  P is the penetration of the slot below the original water table: $P = (6 - 0.5) = 5.5$ m.
  L_o is the distance of influence. In the absence of pumping test data, L_o can be estimated for equivalent slots from Equation 13.16 (shown here for L_o, H and h_w in metres and k in metres per second).

$$L_o = 1750 \left(H - h_w \right) \sqrt{k} \tag{13.16}$$

For $k = 1 \times 10^{-4}$ m/s and $(H - h_w) = 3$ m (drawdown from 0.5 to 3.5 m), L_o is estimated to be 53 m.

If the depth of the aquifer was known, the total flow rate Q could then be estimated from Equation 13.12. The problem of the unknown aquifer depth can be overcome by carrying out a sensitivity analysis – and of course, it is known that the base of the aquifer is more than 12 m below ground level. Using the units quoted earlier, Q will be calculated in cubic metres per second; when commenting on results, Q is normally quoted in litres per second (1 m³/s = 1000 l/s) to make the numbers easier to read and interpret.

Depth to base of aquifer (m)	H (m)	h_w (m)	Q (m³/s)	Q (l/s)
12	11.5	8.5	9.6×10^{-3}	9.6
14	13.5	10.5	1.1×10^{-2}	11
16	15.5	12.5	1.3×10^{-2}	13
18	17.5	14.5	1.4×10^{-2}	14
20	19.5	16.5	1.6×10^{-2}	16

This shows that in this case, assuming a deeper base to the aquifer increases the flow by almost 70 per cent, but because the initial flow rate was modest, the actual increase in predicted flow is only 6.5 l/s. This small increase in flow will not significantly affect the design of the system. The effect of a deeper base to the aquifer would have been more problematical if the aquifer permeability had been higher, because the initial flow rate would have been larger, and a 70 per cent increase would have resulted in a much greater increase in flow.

The preceding calculations assume planar flow to the sides of the slot, but because the line of wellpoints is of finite length (100 m), there will be some contribution from radial flow to the ends (see Figure 13.4b). The total flow rate to the ends of the slot is the same as the flow rate to a well of radius r_e equal to half the width of the slot. For a partially penetrating well in an unconfined aquifer, the flow rate from such a well can be estimated from Equations 13.5 and 13.8, combined as:

$$Q = \frac{B\pi k \left(H^2 - h_w\right)}{ln\left[\frac{L_0}{r_e}\right]}$$

where all the terms are as defined previously, apart from B, which is a partial penetration factor for radial flow to wells, determined from Figure 13.10b. This equation has been used to calculate the contribution from radial flow to the ends of the slots – the flow rate is theoretically split 50–50 at either end of the slot, but it is the total flow that is relevant now. In calculations, it has been assumed that the slot is a single line of wellpoints of width 0.2 m (a typical width of the jetted hole formed by a placing tube), so r_e was taken as 0.1 m. In this case, the contribution from radial flow to the ends of the slot is small (between 8 and 15 per cent of the flow to the sides). Obviously, if the slot had not been so long, the percentage contribution from the ends would have been greater.

Depth to base of aquifer (m)	Q Planar flow to both sides only (l/s)	Q Radial flow to ends (l/s)	Q Total flow: planar plus radial (l/s)
12	9.6	1.4	11
14	11	1.4	12
16	13	1.4	14
18	14	1.4	15
20	16	1.4	17

If it was anticipated that the wellpoint system would have been installed as a double-sided system, consisting of two parallel lines of wellpoints, the flow to the ends would have been greater. If the two lines of wellpoints were, say, 5 m apart, the radial flow calculation would have used r_e = 2.5 m, and the total flow to both ends of the slot would have been between 25 and 45 per cent of the flow to the sides. This highlights that it is more important to consider the flow to the ends of a double-sided wellpoint system than it is for a single-sided system.

A5.3.4 Determination of Wellpoint Spacing and Pump Capacity

It is likely that a single-sided wellpoint system (consisting of a line of wellpoints alongside one side of the trench only) will be appropriate for this excavation. This is because this case satisfies the conditions favourable for single-sided wellpointing set out in Section 15.8.1:

a) A narrow trench
b) Effectively homogeneous, isotropic permeable soil conditions that persist to an adequate depth below formation level
c) Trench formation level not more than about 5 m below standing groundwater level

Table 15.1 gives typical wellpoint spacing in sands as 1 to 2 m. The length of the proposed wellpoint system is 100 m. Assuming an initial spacing of 2 m, 50 wellpoints will be pumped at any one time. If wellpoints of standard screen length 0.7 m are installed by placing tube, Figure 13.20 suggests that for a permeability of 1×10^{-4} m/s, each wellpoint would have a capacity of 0.26 l/s. This gives a maximum yield of 13 l/s for a 50-wellpoint system. This is lower than the predicted flow rate if the aquifer is 20 m deep, so it would be prudent to install the wellpoints at 1.5 m centres, giving a system of 67 wellpoints with a maximum yield of 17 l/s, which is acceptable for the maximum predicted flow rate. On routine projects, wellpoint spacings tend to use increments of 0.5 m. If a spacing of 1.5 m was not satisfactory, the next case to be tried would be 1.0 m spacings. It is rare to consider spacings such as 1.4 m, 1.3 m and so on.

If the yield from each wellpoint is acceptable for the steady-state case, it is almost certainly adequate to deal with the additional flow rate from storage release during the initial drawdown period. This is because until the steady-state drawdown develops, the wetted screen length of the wellpoints will be much greater, allowing them to yield more water.

The wellpoint pump(s) must also be selected. The choice of pump may be influenced by the equipment available locally, and sometimes much larger pumps than strictly necessary are provided purely because they are readily to hand. Whatever pumps are used, the pump capacity must be adequate for the predicted flow rate, not just the steady-state discharge but also the additional water released from storage during initial pumping.

A simplistic calculation using Equation 13.34 (and assuming that the sand has a storage coefficient S of 0.2) estimates that the design steady-state distance of influence of 53 m will take around half a day to 1 day to develop. This gives a very crude estimate of the time during which water released from storage will be significant and also roughly correlates with the time to achieve drawdown close to the line of wellpoints. A drawdown period of a day or so may seem fairly quick, but pipelaying is a progressive operation, moving forwards the whole time. Each wellpoint may only be pumped for a week or so until it is turned off after the trench has passed. New wellpoints are continually being installed and commissioned ahead of pipelaying. A rapid drawdown period is essential to avoid the pipelaying operation moving too fast and advancing ahead of the dewatered area into 'wet' ground, where the target drawdown has not yet been achieved. This event should be avoided, as it can waste a lot of time and money. This problem does occur from time to time, and when it does, the pipelaying operatives are often forthright (to say the least!) in their criticism of the groundwater lowering operation. For 'static' or non-progressive excavations, there is often a little less pressure to achieve very rapid drawdowns.

In this case, it is assumed that double-acting piston pumps are available. In Section 19.5, a 100 mm unit is quoted as having a capacity of up to 18 l/s. This could handle the predicted steady-state flow rate but may not be able to cope with the higher flows to establish drawdown. A larger 125 mm unit has a capacity of 26 l/s, around 50 per cent greater than the maximum predicted steady-state discharge, giving useful spare capacity to deal with water released from storage. If 125 mm pumpsets were not available, two 100 mm units could be provided as an alternative. Both units would be pumped during the initial drawdown period, but later, when the flow rate has reduced, it may be possible to maintain drawdown using one pump only. The other pump would remain connected into the system as a standby.

Even if large-capacity pumps are available, the need for standby pumps must be considered; conditions when standby pumps are required are outlined in Section 15.7.

A5.4 DESIGN EXAMPLE 3

This example is an application of cumulative drawdown analysis (theoretical method) to analyse a system of fully penetrating deep wells used to lower the piezometric level in a confined aquifer beneath a rectangular excavation. Aquifer parameters determined from a pumping test are used to allow estimation of drawdown at specified locations around the excavation area.

A5.4.1 Conceptual Model

A rectangular excavation is to be made to a depth of 12 m. The details of the conceptual model can be summarized as shown in Figure A5.3:

- Excavation dimensions are 50 by 50 m in plan at formation level, with the deepest part of the excavation at 12 m depth. The sides of the excavation are battered back at 1 in 1.5, giving overall dimensions at ground level of 86 by 86 m. To try to keep any dewatering wells slightly closer to the deepest part of the excavation, they will be installed on a bench in the excavation batters at a level of 2 m below ground level. The plan dimensions of the well array will be 80 by 80 m. The excavation is to be dewatered for a period of 5 months, and the construction programme requires that drawdown be achieved within 2 weeks of commencing pumping.
- A confined aquifer, consisting of a sandy gravel, extends from 10 m depth to 26 m depth. The confining layer above the aquifer is a stiff clay. Maximum piezometric level in the aquifer is 5 m below ground level.
- A well pumping test was carried out, pumping at a rate of 15 l/s for 7 days. The discharge during the test was limited by the capacity of the pump; if a larger pump had been available, a greater flow rate would have been possible. Analysis of the pumping test data gave an aquifer permeability k of 6×10^{-4} m/s and a storage coefficient S of 0.001.
- No compressible strata are believed to exist. Groundwater-lowering-related settlement is not anticipated to be a problem.

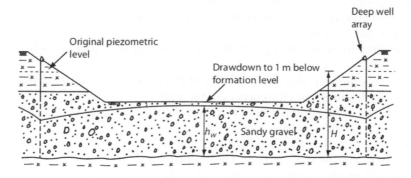

Figure A5.3 Conceptual model for Design Example 3.

A5.4.2 Selection of Method

The excavation extends through the stiff clay aquiclude and into the upper few metres of the confined aquifer. The piezometric level in the confined aquifer will need to be lowered prior to excavation to prevent base heave during excavation through the clay and then to provide a workable excavation when the excavation penetrates into the top of the aquifer.

The target drawdown is to lower the groundwater level to 1 m below formation level. This is 13 m depth, or a drawdown of 8 m below the original piezometric level.

For a drawdown of 8 m and the design permeability of 6×10^{-4} m/s, inspection of Figure 9.11 suggests that either deep wells or two-stage wellpoints would be suitable for this combination of drawdown and permeability. In this case, the deep well method will be used, because the contractor wishes to excavate rapidly to full depth.

A5.4.3 Estimation of Steady-State Discharge Flow Rate and Estimation of Number of Wells

The cumulative drawdown method (using the Cooper–Jacob simplification) can be used in confined aquifers. It has also been successfully applied in unconfined aquifers where the final drawdown is less than around 20 per cent of the initial saturated aquifer thickness. In this case, because drawdown is required to 13 m depth compared with the top of the aquifer at 10 m depth, the initially confined aquifer will become unconfined. The aquifer thickness will be reduced by 3 m out of 16 m, or 19 per cent. Therefore, this problem will be analysed assuming confined behaviour throughout.

The cumulative drawdown is calculated using Equation 13.20:

$$(H-h) = \sum_{i=1}^{n} \frac{2.3 q_i}{4 \pi k D} \log \left[\frac{2.25 k D t}{r_i^2 S} \right] \tag{13.20}$$

where

$(H - h)$ is the cumulative drawdown (at the point under consideration) resulting from n wells each pumped at constant flow rate q_i.

k is the aquifer permeability: k is taken as 6×10^{-4} m/s.

S is the aquifer storage coefficient: S is taken as 0.001.

D is the original aquifer saturated thickness: $D = 26 - 10 = 16$ m.

t is the time since pumping began. In this case, the target drawdown is required within 14 days. It is always prudent to design to obtain the drawdown a little more quickly than planned – this allows for minor problems during commissioning. In design, we will aim to achieve the target drawdown within 10 days. $t = 86{,}400$ seconds will be used in calculations.

r_i is the distance from each pumped well to the point where drawdown is being estimated.

Equation 13.20 is valid provided that $u = (r^2 S)/(4kDt)$ is lower than 0.05. In this case, taking r to be 113 m (the distance from the corners of the well array to the centre of the excavation), u is lower than 0.05 after around 2 hours – therefore, this method can be used for all values of t greater than 2 hours.

The method requires that the plan layout of the well array be sketched and the x–y co-ordinates of each well be determined. The co-ordinates then allow the radial distances r_i (from each well to the point where drawdown is being checked) to be calculated. An initial guess is made of the number of wells and well spacing and the resulting x–y co-ordinates determined. In this case, the initial guess was 16 wells evenly spaced at 20 m centres (Figure A5.4).

A spreadsheet program is then used to evaluate Equation 13.20 for the cumulative draw-down at selected locations within the excavation. For circular or rectangular excavations with evenly spaced wells, it is normally sufficient to determine the drawdown in the centre of the excavation, because drawdown everywhere else will be greater. This is the method used here. If the well array is irregular in shape (or if the depth of excavation is not constant), it will be necessary to determine the drawdown in a number of locations to ensure that the target drawdown is achieved at all critical locations.

The results from a spreadsheet calculating the drawdown in the centre of the excavation for a 16-well system are shown in the following. The radial distance r_i from each well (at location x_i, y_i) to the location (x_c, y_c) where the drawdown is being determined is calculated from:

$$r_i = \sqrt{\left(\left[x_i - x_c\right]^2 - \left[y_i - y_c\right]^2\right)}$$

For simplicity, the flow rate q_i from each well has been assumed to be the same, but if it was intended to use pumps of different sizes in certain wells, this can easily be incorporated in the calculation. In the spreadsheet, different values of q_i were tried until the target draw-down of 8 m was just achieved in the centre of the excavation. The total flow rate is simply the sum of all the well flow rates.

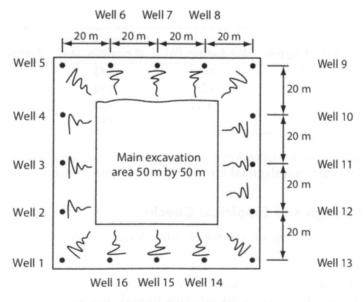

Figure A5.4 Schematic plan of 16-well system.

Well	x co-ordinate (m)	y co-ordinate (m)	Well flow rate q_i (m^3/s)	Radial distance r_i (m)	Drawdown $(H - h)_i$ (m)
1	0	0	7.0×10^{-3}	56.5	0.50
2	0	20	7.0×10^{-3}	44.7	0.53
3	0	40	7.0×10^{-3}	40.0	0.54
4	0	60	7.0×10^{-3}	44.7	0.53
5	0	80	7.0×10^{-3}	56.6	0.50
6	20	80	7.0×10^{-3}	44.7	0.53
7	40	80	7.0×10^{-3}	40.0	0.54
8	60	80	7.0×10^{-3}	44.7	0.53
9	80	80	7.0×10^{-3}	56.6	0.50
10	80	60	7.0×10^{-3}	44.7	0.53
11	80	40	7.0×10^{-3}	40.0	0.54
12	80	20	7.0×10^{-3}	44.7	0.53
13	80	0	7.0×10^{-3}	56.6	0.50
14	60	0	7.0×10^{-3}	44.7	0.53
15	40	0	7.0×10^{-3}	40.4	0.54
16	20	0	7.0×10^{-3}	44.7	0.53
		Total flow rate	0.112 m^3/s (112 l/s)	Total drawdown	8.43 m

This calculation indicates that a system of 16 wells, each discharging 7 l/s (total flow rate 112 l/s), will achieve the target drawdown in the centre of the excavation after 10 days. During the pumping test, the test well produced 15 l/s and could have yielded more if a larger pump has been used. It therefore makes sense to repeat the above calculations assuming fewer wells of greater discharge rate. The results of these calculations are summarized in the following:

No. of wells	Well spacing (m)	Well discharge (l/s)	Drawdown in centre of excavation (m)	Total flow rate (l/s)
16	20	7.0	8.43	112
13	25	8.5	8.33	110.5
10	32	11.0	8.28	110
8	40	13.5	8.07	108

It is apparent that the target drawdown can be achieved by various combinations of well numbers and yields, but that the total flow rate remains approximately constant.

A5.4.4 Final Design and Empirical Checks

The number and yield or wells chosen for the final design will depend on a number of factors, including:

(i) The need for redundancy in a well system. Any system relying on relatively few wells is vulnerable to one or two wells suffering from damage or pump failure, leading to loss of drawdown and flooding or instability of the excavation. A system consisting of a greater number of wells will lose proportionately less drawdown if one or two wells are lost.

(ii) Each well must be able to yield the discharge flow rate q_i assumed in design. Because the pumping test well produced 15 l/s (and could have yielded more), and since all the current designs use q_i of less than 15 l/s, it is likely that all the current designs are feasible. In any case, the theoretical maximum well yield can be estimated from Equations 13.35 and 13.36 (or from Figure 13.19). Assuming a well borehole diameter of 0.305 mm and a wetted screen length of 10 to 12 m (i.e. a drawdown at the wells of 1 to 3 m below the excavation target drawdown level), the maximum well yield is estimated as 16 to 19 l/s. However, in practice, problems can occur if the dewatering wells are not designed, installed and developed in exactly the same way as the test well – this may cause the production wells to have lower yields than the test wells. Also, in fractured rock aquifers, well yields may vary significantly. Some wells can have high yields, and yet others, poorly connected into fractures, may be almost 'dry'. In fractured aquifers, the cumulative drawdown method needs to be applied with care.

In this case, it is assumed that due to the availability of pumps of suitable capacity, the nominal system of 10 wells, each discharging 11 l/s, will be adopted. It is normal practice to apply an empirical superposition factor J; the system capacity is increased by a factor of $1/J$ (see Equation 13.21). This empirical factor allows for interference between wells and also provides some allowance for additional drawdown around the wells and water released from storage when the aquifer becomes unconfined. Where aquifers become unconfined, and drawdowns are small (less than 20 per cent of the initial saturated aquifer thickness), the empirical superposition factor J is normally taken as 0.8 to 0.95. In this case, because the drawdown will reduce the thickness of the aquifer by almost 20 per cent, the greatest superposition factor of 0.8 will be applied, so the system capacity (and hence the number of wells) will need to be increased by $1/0.8 = 1.25$.

The final system design is, therefore, for 13 wells, each of 11 l/s capacity. Total system capacity is 143 l/s. Table 16.1 indicates that to accommodate a pump of suitable capacity, a minimum well bore diameter of 300 mm is required. The corresponding well screen and liner diameter is 165 mm.

The design is then verified with some simple empirical checks. The well spacing for a 13-well system is approximately 25 m; this is within the 'normal' 10 to 60 m range quoted in Section 13.9.3. The wells are intended to fully penetrate the aquifer, and so are 26 m deep, just over twice the depth of the excavation. This is also consistent with guidelines given in Section 13.9.2. These empirical checks confirm, in a gross sense, the validity of the concept of the design.

Appendix 6: Estimation of Flow Rate Using V-Notch Weirs

V-notch weirs are a common method for estimation of discharge flow rate in the field. They comprise a thin plate weir where the area of flow is a notch cut in the shape of a 'V' with an internal angle of α (see Figure A6.1), normally installed in a tank. The flow rate over the weir is a function of the head over the weir, the size and shape of the discharge area, and an experimentally determined discharge coefficient.

BS 3680:1981 provides formulae and discharge coefficients for V-notch weirs (and the less commonly used rectangular notch weirs). These formulae have been used to produce calibration charts for various standard V-notch angles; these charts are presented at the end of this appendix.

Knowing the angle of the V-notch, a measurement of the head of water above the base of the notch allows the flow rate to be estimated from the appropriate chart. The position of measurement should be upstream from the weir plate by a distance of approximately 1.1 to 0.7 m but not near a baffle or in the corner of the tank. The tank should be positioned on a firm, stable base with timber packing used to ensure that the V-notch plate is vertical and that the top of the notch is horizontal.

The calibration charts given here are based on generic formulae and discharge coefficients and are intended for use only to provide field estimates of flow rate from groundwater lowering systems. BS 3680:1981 gives guidance on the use of thin plate weirs for more accurate measurements. Calibration charts are provided for the following notch angles:

Name	V-notch angle
90° V-notch	90°
½ 90° V-notch	53° 8′
¼ 90° V-notch	28° 4′
60° V-notch	60°
45° V-notch	45°
30° V-notch	30

The angles of ½ 90° and ¼ 90° V-notch weirs are not 45° and 22.5°, as might be expected. These weirs are so-called because they can pass one-half and one-quarter as much flow as a 90° V-notch weir (Figure A6.2 through Figure A6.7).

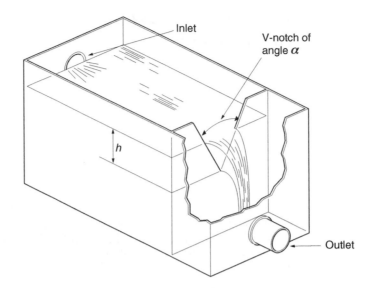

Figure A6.1 V-notch weir for measurement of discharge flow rate. (Reproduced from Preene, M, *et al.*, *Groundwater Control – Design and Practice*, 2nd edition, Construction Industry Research and Information Association, CIRIA Report C750, London, 2016. With permission from CIRIA: www.ciria.org)

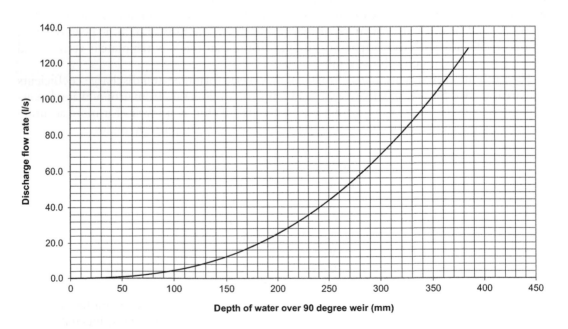

Figure A6.2 Calibration chart for 90 degree V-notch.

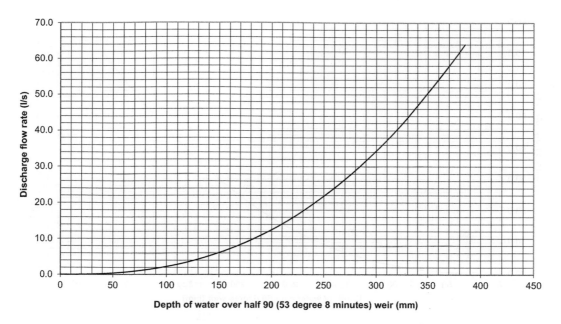

Figure A6.3 Calibration chart for half-90 degree V-notch.

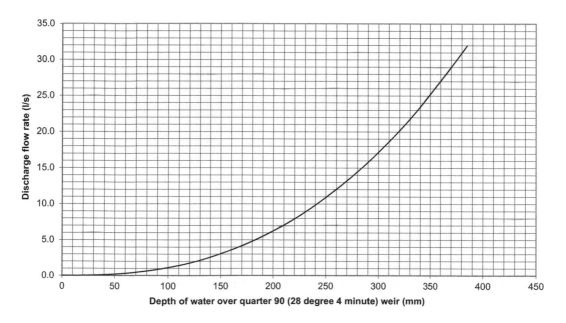

Figure A6.4 Calibration chart for quarter-90 degree V-notch.

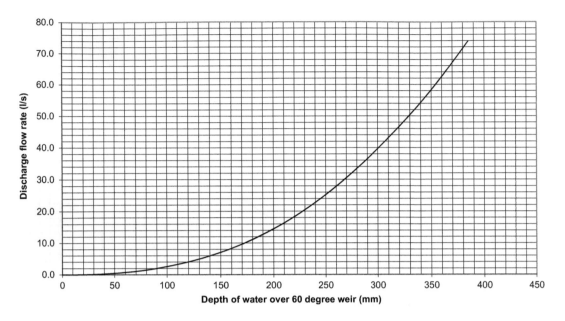

Figure A6.5 Calibration chart for 60 degree V-notch.

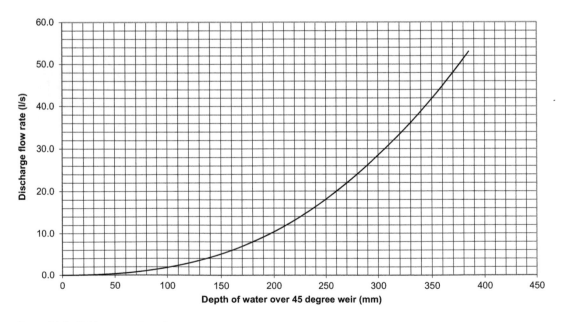

Figure A6.6 Calibration chart for 45 degree V-notch.

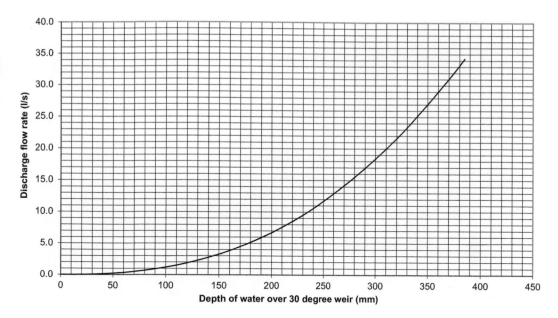

Figure A6.7 Calibration chart for 30 degree V-notch.

Notation

A	Area; cross-sectional area of borehole casing
A_p	Plan area enclosed within cut-off walls
a	Calibration coefficient; length of well array; length of excavation
B	Calibration coefficient; partial penetration factor for radial flow
b	Calibration coefficient; fracture aperture; width of well array; width of excavation
C	Calibration coefficient; sorting coefficient
C_d	Correction factor for effective stress
C_{hw}	Hazen–Williams roughness coefficient
C_{kc}	Kozeny–Carman coefficient
c_h	Coefficient of consolidation of soil for vertical compression of soil under horizontal drainage
c_v	Coefficient of consolidation of soil
D	Aquifer thickness; depth of excavation; diameter of borehole test section; depth of soil plug below excavation; diameter of tunnel; thickness of soil layer
D_1, D_i, D_n	Mean particle size between adjacent sieves in particle size analysis
D_e	Effective particle size of a soil sample
D_n	Sieve aperture through which n per cent of a soil sample will pass
D_o	Smallest particle size in a soil sample
D_x, D_y	Nominal size of sieve in particle size analysis
D_{10}	Sieve aperture through which 10 per cent of a soil sample will pass
D_{15}	Sieve aperture through which 15 per cent of a soil sample will pass
D_{40}	Sieve aperture through which 40 per cent of a soil sample will pass
D_{50}	Sieve aperture through which 50 per cent of a soil sample will pass
D_{60}	Sieve aperture through which 60 per cent of a soil sample will pass
D_{70}	Sieve aperture through which 70 per cent of a soil sample will pass
D_{85}	Sieve aperture through which 85 per cent of a soil sample will pass
D_{95}	Sieve aperture through which 95 per cent of a soil sample will pass
d	Diameter of the section of tubing within which the water level rises and falls during a variable head test; depth of excavation; internal diameter of pipework; penetration of cut-off wall below lowered groundwater level in cofferdam
d_1	Drawdown of groundwater level in cofferdam
E	Energy consumption of a pumpset
ΔE	Electrical potential difference
E'_o	Stiffness of soil in one-dimensional compression

EC	Specific conductivity
e	Void ratio
F	Factor of safety; Hvorslev's shape factor for permeability test in borehole or monitoring well
$F_{destabilizing}$	Partial factor for vertical destabilizing forces
$F_{stabilizing}$	Partial factor for vertical stabilizing forces
f	Friction loss; angularity factor of soil particles
f_d	Friction loss in discharge pipework
f_s	Friction loss in suction pipework
f_{sub}	Friction loss in submerged pipework
$f(n)$	Porosity function
G	Geometry shape factor for flow to rectangular equivalent wells in confined aquifers; geometry factor for flow to tunnels
g	Acceleration due to gravity
H	Excess head in test section; initial groundwater head; initial saturated thickness of an aquifer; head of groundwater above tunnel axis; total head against which a pump works (also known as the total dynamic head)
H_c	Constant excess head during constant head permeability test
H_f	Frictional head loss in injection pipework
H_g	Height of the pressure gauge above ground level in a packer permeability test
H_o	Excess head in test section at $t = 0$; depth to groundwater level at the time of pressure transducer depth calibration
H_w	Depth to groundwater level; depth to initial groundwater level before a packer permeability test
h	Total hydraulic head; groundwater head; maximum drainage path length for vertical drainage; height of water over V-notch weir
h_d	Static discharge head
h_s	Static suction lift
h_w	Groundwater head in pumped well or slot; groundwater head in excavation
dh	Difference in hydraulic head
Δh	Difference in hydraulic head
$(H - h)$	Drawdown
$(H - h_w)$	Drawdown in a pumped well or slot
i	Hydraulic gradient
i_{crit}	Critical hydraulic gradient
i_{max}	Maximum hydraulic gradient at entry to a well; maximum hydraulic exit gradient into an excavation
i_{maxr}	Maximum hydraulic exit gradient into an excavation when a lower-permeability layer is present
J	Empirical superposition factor
K	Coefficient of earth pressure
k	Permeability (also known as coefficient of permeability or hydraulic conductivity)
$k_{average}$	Average permeability calculated assuming that the flow of injected water is distributed uniformly across the cylindrical boundary of test section
k_e	Calibration factor between total dissolved solids and specific conductivity; electro-osmotic permeability (coefficient of electro-osmotic permeability)
k_f	Fracture permeability
k_h	Permeability in the horizontal direction
k_i	Intrinsic permeability
k_r	Permeability of lower-permeability layer

k_v	Permeability in the vertical direction
k_x	Permeability in the direction of the x axis
k_y	Permeability in the direction of the y axis
k_z	Permeability in the direction of the z axis
k'	Equivalent permeability of the permeable zones within a test section
L	Length of borehole test section
L'	Assessed total thickness of the permeable stratum within the test section of a packer permeability test
L_o	Distance of influence for planar flow; distance to line source
Lu	Lugeon coefficient
l	Length of flow path
l_p	Effective length of pipework
l_w	Wetted screen length of a well
dl	Length of flow path
m	permeability transformation ratio
m_v	Coefficient of volume compressibility of soil
N	Standard penetration test (SPT) blow count
NPSHa	Available net positive suction head
NPSHr	Required net positive suction head
n	Porosity; number of pumped wells
P	Depth of penetration into aquifer of partially penetrating well or slot; power requirement to pump a water flow rate
P_g	Water pressure recorded in surface pipework during packer permeability test
P_{max}	Maximum excess water pressure (above ambient groundwater level) at the mid-point of the test section
P_o	Pressure measured by pressure transducer at time of depth calibration
P_t	Water pressure in test section
P_{trans}	Pressure measured by pressure transducer
P_{tunnel}	Air pressure in tunnel
Q	Flow rate; discharge flow rate from an equivalent well or slot; injection flow rate to the test section during a packer permeability test
Q_{fp}	Discharge flow rate from a fully penetrating well
Q_{max}	Maximum injection flow rate possible with a given set of equipment during a packer permeability test
Q_{pp}	Discharge flow rate from a partially penetrating well
Q_{1000}	Water take of borehole (in litres per minute) at an excess pressure of 1000 kPa during a Lugeon test
q	Flow rate; discharge flow rate from a well; discharge flow rate per unit length of excavation perimeter
q_c	Calculated steady-state dewatering flow rate
q_r	Steady-state dewatering flow rate recorded in the field; discharge flow rate per unit length of excavation when a lower-permeability layer is present
R	Average degree of consolidation of a soil layer; degree of dissipation of pore water pressure
R_o	Radius of influence for radial flow
R_{uplift}	Factored additional resistance to uplift
r	Distance from the test well to a monitoring well; radial distance from a well; radius of a well borehole
r_e	Equivalent radius of a well array
r_o	Equivalent radius of a penetrometer

S	Storage coefficient of aquifer
S_o	Specific surface of soil particles
S_s	Specific storage of aquifer
S_y	Specific yield of unconfined aquifer
s	Drawdown; distance from base of cut-off wall to deeper impermeable layer
s_w	Drawdown at a pumped well or slot
Δs	Change in drawdown; change in drawdown per log cycle
T	Temperature; transmissivity of aquifer; basic time lag for permeability test in borehole or monitoring well
T_f	Fracture transmissivity
T_r	Time factor for radial groundwater flow
T_v	Time factor for vertical drainage
T_{50}	Time factor for 50 per cent dissipation of pore water pressure
TDS	Total dissolved solids
t	Time; time since pumping began; time since permeability test started; pump run time; thickness of lower-permeability layer
t_o	Time at which a straight line through the monitoring well data intercepts the zero drawdown line in a Cooper–Jacob time–drawdown plot
t_{50}	Time for 50 per cent dissipation of pore water pressure
U	Uniformity coefficient
u	Pore water pressure; argument of Theis well function
u_a	Air entry pressure of a soil
u_i	Pore water pressure at the start of a permeability test
u_o	Equilibrium pore water pressure
u_t	Pore water pressure at time t
Δu	Change in pore water pressure
V	Volume of water trapped in soil/rock within the area enclosed within cut-off walls
$V_{destabilizing}$	Factored vertical destabilizing forces
$V_{stabilizing}$	Factored vertical stabilizing forces
v	Darcy velocity (Darcy flux)
v_d	Water velocity in discharge pipework
v_s	Water velocity in suction pipework
v_x	Average linear groundwater velocity
W	Weight
$W(u)$	Theis well function
x	Linear distance from a slot; length of a pumped slot; length of tunnel; fraction of the total mass of a soil sample
x_1, x_i, x_n	Fraction of soil mass retained on sieve in particle size analysis
Z	Depth below ground level; depth to midpoint of test section during a packer permeability test; depth to tunnel axis
z	Depth below the water table in an unconfined aquifer
z_{trans}	Effective transducer depth
α	Internal angle of V-notch in thin plate weir
ϕ'	Angle of shearing resistance of soil
ε	Skin factor
γ_s	Unit weight of soil
γ_w	Unit weight of water
λ	Partial penetration factor for confined slots
μ	Dynamic viscosity of a fluid; flow reduction factor for seepage into a cofferdam

μ_w	Dynamic viscosity of water
ρ	Density of fluid
ρ_{corr}	Corrected compression of a soil layer
ρ_t	Effective compression of a soil layer at time t
ρ_{total}	Ground settlement
ρ_{ult}	Ultimate compression of a soil layer
ρ_w	Density of water
η	Combined efficiency of pump and motor
σ	Total stress
σ_h	Horizontal total stress
σ_t	Internal air or fluid pressure in tunnel
σ_v	Vertical total stress
σ'	Effective stress
σ'_h	Horizontal effective stress
σ'_v	Vertical effective stress
τ	Shear stress
τ_f	Shear stress at failure

Conversion factors

LENGTH

In the metric system: 1 km = 1000 m; 1 m = 100 cm; 1 m = 1000 mm.

Conversion from metric to imperial			Conversion from imperial to metric		
From	To	Multiply by	From	To	Multiply by
km	mile	0.622	Mile	km	1.609
m	yd	1.094	Yd	m	0.914
m	ft	3.281	Ft	m	0.305
cm	in	0.394	In	cm	2.54
mm	in	0.0394	In	mm	25.4

VOLUME

In the metric system: $1 \text{ m}^3 = 1000$ litres; $1 \text{ m}^3 = 10^6 \text{ cm}^3$; $1 \text{ cm}^3 = 1$ ml.

Conversion from metric to imperial			Conversion from imperial to metric		
From	To	Multiply by	From	To	Multiply by
m^3	cu yd	1.309	cu yd	m^3	0.764
m^3	cu ft	35.32	cu ft	m^3	0.028
m^3	gal (UK)	220	gal (UK)	m^3	0.0045
m^3	gal (US)	264	gal (US)	m^3	0.0038
litre	gal (UK)	0.220	gal (UK)	litre	4.546
litre	gal (US)	0.264	gal (US)	litre	3.785
litre	pint (UK)	1.76	pint (UK)	litre	0.568
litre	pint (US)	2.11	pint (US)	litre	0.473

FLOW RATE

In the metric system: 1 m³/s = 1000 l/s; 1 m³/h = 3.6 l/s; 1 m³/d = 0.0115 l/s

Conversion from metric to imperial			Conversion from imperial to metric		
From	To	Multiply by	From	To	Multiply by
m³/d	gal/d (UK)	220	gal/d (UK)	m³/d	0.0045
m³/d	gal/d (US)	264	gal/d (US)	m³/d	0.0038
m³/h	gal/h (UK)	220	gal/h (UK)	m³/h	0.0045
m³/h	gal/h (US)	264	gal/h (US)	m³/h	0.0038
l/s	gal/min (UK)	13.2	gal/min (UK)	l/s	0.076
l/s	gal/min (US)	15.85	gal/min (US)	l/s	0.063

PERMEABILITY (OR HYDRAULIC CONDUCTIVITY OR COEFFICIENT OF PERMEABILITY)

In the metric system: 1 m/s = 8.64×10^4 m/d; 1 m/s = 3.16×10^7 m/year

Conversion from metric to imperial			Conversion from imperial to metric		
From	To	Multiply by	From	To	Multiply by
m/s	ft/d	2.83×10^5	ft/d	m/s	3.53×10^{-6}
m/d	ft/d	3.281	ft/d	m/d	0.305
m/year	ft/year	3.281	ft/year	m/year	0.305

PRESSURE AND HEAD

In the metric system: 1 m H_2O = 9.789 kPa; 1 bar = 100 kPa = 102 m H_2O; 1 bar = 750 mm Hg

Conversion from metric to imperial			Conversion from imperial to metric		
From	To	Multiply by	From	To	Multiply by
mm Hg	in Hg	0.0394	in Hg	mm Hg	25.4
m H_2O	in Hg	2.896	in Hg	m H_2O	0.345
m H_2O	ft H_2O	3.281	ft H_2O	m H_2O	0.3048
m H_2O	psi	1.42	psi	m H_2O	0.703
kPa	psi	0.1451	psi	kPa	6.884
bar	in Hg	29.551	in Hg	bar	0.0339
bar	ft H_2O	33.48	ft H_2O	bar	0.0299
bar	psi	14.51	psi	bar	0.0689

Abbreviations

AC	Alternating current
AMF	Automatic mains failure
ASR	Aquifer storage and recovery
BEP	Best efficiency point
BMR	Borehole magnetic resonance
BOP	Blow out preventer
CDM	Construction Design and Management
CIRIA	Construction Industry Research and Information Association
CPT	Cone penetration testing
CSM	Cutter soil mixing
DC	Direct current
DNAPL	Dense non-aqueous phase liquid
DO	Dissolved oxygen
DOL	Direct-on-line
EBS	Excavation before support
EC	Electrical conductivity (specific conductivity)
EoC	End of casing
EPB	Earth pressure balance
EPDM	Ethylene propylene diene monomer
GAC	Granular activated carbon
GGBS	Ground granulated blast furnace slag
GIN	Grouting intensity number
GIS	Geographical information system
GPS	Global Positioning System
GSI	Geological strength index
HDD	Horizontal directional drilling
HDPE	High-density polyethylene
HIA	Hydrogeological impact appraisal
ICE	Institution of Civil Engineers
InSAR	Interferometric synthetic aperture radar
IoT	Internet of Things
LDPE	Low-density polyethylene
LN	Liquid nitrogen
LNAPL	Light non-aqueous phase liquid
MDPE	Medium-density polyethylene
MWD	Measurement while drilling
NATM	New Austrian tunnelling method
NPSH	Net positive suction head

NTU	Nephelometric Turbidity Units
OPC	Ordinary Portland cement
PFA	Pulverised fuel ash
PLC	Programmable logic controller
PRB	Permeable reactive barrier
PSD	Particle size distribution
PVD	Pre-fabricated vertical drain
RMR	Rock mass rating
RQD	Rock quality designation
SBE	Support before excavation
SCL	Sprayed concrete lining
SEM	Sequential excavation method
SF	Significant figures
SGI	Spheroidal graphite iron
SPT	Standard penetration test
SPZ	Source protection zone
SRG	Shift review group
TAM	Tube à manchette
TBM	Tunnel boring machine
TDS	Total dissolved solids
uPVC	Unplasticized polyvinyl chloride
UXO	Unexploded ordnance
VCW	Vertical circulation well
VSM	Vertical shaft boring machine
VWP	Vibrating wire piezometer
WFD	Water Framework Directive

km	kilometre
m	metre
cm	centimetre
mm	millimetre
l	litre
ml	millilitre
d	day
h	hour
min	minute
s	second
J	Joule
W	Watt
kW	kilowatt
kWh	kilowatt hour
K	Kelvin
°C	degrees celsius
Kg	kilogram
mg	milligram
µg	microgram
MPa	megapascal
kPa	kilopascal
mm Hg	millimetres of mercury
m H_2O	metres head of water

V	Volt
S	Siemen
μS	microsiemen
ft	foot
in	inch
yd	yard
gal	gallon
cu ft	cubic feet
cu yd	cubic yard
in Hg	inches of mercury
ft H_2O	feet head of water
psi	pounds per square inch
cfm	cubic feet per minute
mbgl	metres below ground level
mAOD	metres above Ordnance Datum
mATD	metres above Tunnel Datum
masl	metres above sea level
mCD	metres Chart datum

Glossary

Abstraction: The pumping or removal of groundwater (such as from a *well* or *sump*) or surface water (such as from a lake or river).

Airlift: A means of pumping water using compressed air to aerate the water so that the air–water mixture rises to the surface. The method is capable of pumping silt and sand with the water and is often used as a method of *well development*.

Anisotropic: Having one or more physical properties that vary with the direction of measurement – the converse of *isotropic*.

Aquiclude: Soil or rock forming a stratum, group of strata or part of stratum of very low permeability, which acts as an effective barrier to groundwater flow.

Aquifer: Soil or rock forming a stratum, group of strata or part of stratum from which water in usable quantities can be *abstracted*. By definition, an aquifer will be water-bearing (i.e. *saturated* and permeable).

Aquitard: Soil or rock forming a stratum, group of strata or part of stratum of intermediate to low permeability, which yields only very small groundwater flows or allows groundwater to pass through very slowly.

Artificial ground freezing: A type of *ground treatment* that reduces the temperature of the ground sufficiently to form a very low permeability freezewall of frozen soil. The freezewall can be used as a *cut-off wall* and support structure for below-ground works.

Artificial recharge: The deliberate re-injection of water (via pits, trenches or wells) into aquifers or aquitards. Sometimes used as a means of reducing drawdowns in the aquifer outside an excavation to minimize *consolidation settlements*.

Barrier boundary: An aquifer boundary that is not a source of water.

Base instability: Uplift, heave or instability of the floor of an excavation as a result of unrelieved or uncontrolled *pore water pressures* at depth.

Bentonite slurry: A suspension of bentonite (a clay mineral consisting of sodium montmorillonite) in water. The slurry is commonly used to support the sides of temporary trench excavations used to construct *diaphragm walls* and *slurry trench walls*.

Biofouling: The clogging or deterioration of performance of wells, pumps and pipework as a result of bacterial growth.

Boil: The turbulent jet of soil and water that rises to the surface around a placing tube when a wellpoint or well is installed by *jetting*. Also used to describe an uncontrolled upward seepage into an excavation that may wash *fines* out of the soil, leading to instability.

Borehole: Strictly, a borehole is a hole drilled into the ground for any purpose, including site investigation boreholes. In groundwater terminology, a borehole is often taken to mean a relatively small-diameter *well*, particularly one used for water supply.

Buoyancy uplift: A form of *base instability* that can occur when a permeable stratum exists at relatively shallow depth beneath an excavation. If the upward force from the groundwater pressures exceeds the weight of the 'plug' of soil or rock beneath the excavation, there is a risk that buoyancy forces may cause heave or instability of the base of the excavation.

Caisson: A form of structure used in shaft sinking, where the perimeter walls of the shaft are constructed (as a reinforced concrete monolith) or assembled (from pre-cast segments) at ground level, and then sunk into the ground in a controlled manner.

Capillary saturated zone: The zone immediately above the *water table* in an unconfined aquifer where the soil remains *saturated* at negative (i.e. less than atmospheric) *pore water pressures*. The water in this zone is continuous with the pore water below the water table. The height of the capillary saturated zone is greater in finer-grained soils than in coarse-grained soils. The zone rises or falls with any variations in the water table.

Cement-based grout: A type of *suspension grout* consisting of a mixture of cement and water, sometimes with additives such as bentonite or PFA.

Cement-bentonite wall: A type of *slurry trench wall* where the trench is supported during excavation by *bentonite slurry*, which is then replaced with a self-hardening cement-bentonite slurry to form the final *cut-off wall*.

Chemical grout: Specialist *grouts* that are not based on cement suspensions and typically gel (solidify) on setting. These materials are most commonly silicate compounds, resins or colloidal suspensions of silica particles.

Coefficient of permeability: See *permeability*.

Cofferdam: A temporary retaining wall structure used to exclude groundwater and surface water from an excavation.

Collector well: A pumping technique used for water supply and *groundwater lowering* where a shaft is excavated below the groundwater levels. Lateral wells are drilled horizontally out from the shaft. Pumping water from the shaft draws water into the shaft from the lateral wells.

Conceptual model: A theory-based description of groundwater and hydrogeological conditions that describes the relevant conditions that will affect the design and implementation of a *groundwater lowering* scheme. The elements of the conceptual model include geometry, boundary conditions, groundwater levels and permeability values.

Cone of depression: A depression in the *water table* or *piezometric level* that, in theoretically idealized conditions, has the shape of an inverted cone and develops radially around a well from which water is being abstracted. It defines the *radius of influence* of a well.

Confined aquifer: An *aquifer* in which the groundwater is isolated from the atmosphere by an overlying (or confining) *aquiclude* or *aquitard*, and where the *piezometric level* is above the top of the aquifer. A confined aquifer is *saturated* throughout. Also known as an artesian or sub-artesian aquifer.

Consolidation settlements: Ground settlements resulting from increases in vertical *effective stress* when groundwater levels are lowered or *pore water pressures* are reduced.

Constant head test: A permeability test, typically carried out in a *borehole* or *monitoring well*, where water is added or removed to the well at a constant flow rate. Observation of the change in water level in response to the flow rate can be used to estimate the *permeability* of the surrounding ground.

Construction dewatering: See *dewatering*.

Contiguous pile wall: A type of bored pile wall used as a retaining structure where successive concrete piles are bored in close proximity. In this application, the piles are not intended to intersect. Typically, a small gap remains between piles, and the wall is not watertight unless sealed by *grouting* or other means.

Cut-off wall: A very low-permeability artificial barrier constructed to prevent groundwater from flowing into an excavation as part of a *groundwater exclusion* strategy.

Darcian flow: A flow regime where flow is relatively smooth without any turbulence and where *Darcy's law* describes the relationship between driving head and flow rate. It is characteristic of most groundwater flow through porous media and finely fractured rock formations – the converse of turbulent flow or *non-Darcian flow*.

Darcian velocity: The mean flow velocity across a unit area of a porous medium as defined by *Darcy's law*. When it is divided by the *porosity*, the mean groundwater flow rate through the pores may be calculated.

Darcy's law: The expression, developed by Henri Darcy in 1856, relating the flow rate through a porous medium to the *hydraulic gradient*, the cross-sectional area of flow and the *permeability* of the medium.

Deep well: A groundwater abstraction well for the purpose of *groundwater lowering* (as opposed to a well for the purpose of water supply, which is termed a *borehole*).

Depressurization: See *groundwater depressurization*.

Derogation: The reduction in yield or other adverse effect on a water supply *borehole* or *spring* as a result of groundwater pumping by others.

Desk study: The review of all available information (such as maps, aerial photographs and historical records) as part of the *site investigation* process.

Development: See *well development*.

Dewatering: A colloquialism for *groundwater lowering* that originated in America and is now widely used in Europe. Also known as construction dewatering.

Diaphragm wall: A technique used to form *cut-off walls* or retaining structures where cast in situ concrete walls are formed within a trench supported by a slurry (most commonly, but not always, *bentonite slurry*) during excavation. Diaphragm walls are commonly known as D-walls.

Dipmeter: A portable device for measuring the depth to water in a *well*, *borehole*, *standpipe* or *piezometer*. Also known as a dipper or water level indicator.

Discharge: The flow of water from a pump or a groundwater lowering system.

Drawdown: The lowering of the *water table* (in *unconfined aquifers*) or the lowering of the *piezometric level* (in *confined aquifers*) as a result of the *abstraction* of groundwater. It is measured as the vertical distance between the original water table (or piezometric level) and the current level during pumping.

Drift: Geologically recent superficial strata of sand, gravel silt or clay (see also *soil*). Excavations below the groundwater level in drift deposits may be unstable without adequate *groundwater control*.

Eductor: See *ejector*.

Effective stress: The difference between the total stress (due to self-weight and any external loading) and the *pore water pressure* at a point in a soil mass. The effective stress is a vital controlling factor in the strength and compressibility behaviour of soils.

Ejector: A water jet pump that pumps water (and air) via a nozzle and venturi arrangement. When it is used in low-yielding wells, a vacuum can be developed if the top of the well casing is sealed.

Ejector well: A small-diameter *groundwater lowering* well pumped by an *ejector*.

Electro-osmosis: A rarely used method of *pore water pressure control* applicable to very low-permeability soils such as silts and clays. An electric potential is applied across the area of soil to induce groundwater flow from the anode to the cathode.

Equivalent well: A conceptual large well, used in design to represent a *groundwater lowering* system that consists of a ring of many wells. By considering the real system as an equivalent well, simpler and more accessible methods of analysis can be used.

Falling head test: A type of *variable head test*, typically carried out in a *borehole* or *monitoring well*, used to estimate *permeability*. At the start of the test, the water level in the well is raised rapidly, and the rate of fall of the water level in the well is measured.

Filter pack: Sand or gravel placed around a *well screen* to act as a filter. This allows water to freely enter the well while preventing movement of *fines* towards the well.

Fines: The smaller particles in a soil stratum or analysis. Generally taken to refer to clay, silt and fine sand-size particles. See *loss of fines*.

Flow, steady-state: The flow regime when the magnitude and direction of the groundwater flow rate are constant.

Flow, transient-state: The flow regime when the magnitude or direction of the groundwater flow rate is changing with time.

Flowing artesian conditions: A special case of *confined aquifers*, where the *piezometric level* of the aquifer is above ground level. A well drilled through the confining bed into the aquifer will overflow at ground level without the need for pumping.

Fluidization due to upward seepage gradients: A form of *base instability* that can occur when upward *hydraulic gradients* into an excavation cause a significant reduction in *effective stress* in the soil immediately beneath an excavation. If effective stress approaches zero, the soil can fluidize and lose all strength.

Formation level: The final dig level of an excavation, after all digging and filling, but before any concreting or construction.

Fractures: Natural cracks, fissures, joints or discontinuities caused by mechanical stresses in *rock*. Fractures can allow groundwater to pass and affect the strength and stability of most rocks.

Fully grouted piezometer: A method sometimes used to install *rapid response piezometers* (such as *vibrating wire piezometers*), where the instrument or sensor is installed in the *borehole* surrounded by cement-bentonite grout with no sand response zone.

Fully penetrating: The case where a well penetrates the full thickness of an *aquifer* – the converse of *partially penetrating*.

Gap-graded: Granular materials, such as gravels, in which a specific range of particle sizes is entirely or substantially absent.

Geotechnical investigation: See *site investigation*.

Ground investigation: The physical investigation of a site as part of the *site investigation* process. Ground investigation includes drilling and probing, in-site tests and laboratory testing.

Ground treatment: Geotechnical processes used to reduce the *permeability* and/or increase the strength of in situ soil and rock. Techniques that can be used as part of *groundwater exclusion* strategies include *artificial ground freezing* and *grouting*.

Groundwater: Water contained in and flowing through the pores and fractures of soil and rock in the *saturated zone*. In this zone, the water body is essentially continuous except for an occasional bubble of air. In legal documents, groundwater is sometimes described as water in underground strata.

Groundwater control: Methods and techniques to control groundwater to allow stable excavations to be formed for construction purposes. The two main approaches are *groundwater exclusion* and *groundwater lowering*.

Groundwater depressurization: Groundwater *abstraction* carried out with the objective of lowering groundwater pressures in *soil* and *rock* with the primary objective of improving stability rather than reducing inflows to an excavation or tunnel. This approach is often relevant for projects in strata of relatively low *permeability*. Also known as *pore water pressure control*.

Groundwater exclusion: The construction of artificial barriers to prevent groundwater from flowing into an excavation. The barriers can be formed from physical *cut-off walls* or by reducing the permeability of the in situ soil and rock by *ground treatment*. See also *cofferdam*.

Groundwater lowering: The temporary reduction (or *drawdown*) of groundwater levels and *pore water pressures* around and below an excavation by some form of pumping sufficiently to enable excavation and foundation construction to be carried out safely and expeditiously in water-bearing ground.

Groundwater treatment: Processes applied to groundwater to remove contamination or otherwise improve water quality prior to disposal of the water.

Grout: A pumpable material injected into the ground as part of the *grouting* process. Common types of grout are *cement-based grouts* and *chemical grouts*.

Grouting: A type of *ground treatment* based on the controlled injection of material (termed *grout*), usually in a temporary fluid phase, into soil or rock, where it stiffens to improve the physical characteristics of the ground for geotechnical engineering reasons. The fluids injected are most commonly *cement-based grouts* or *chemical grouts*. Grouting is used to reduce the permeability of the ground to form a low-permeability barrier as part of *groundwater exclusion* methods.

Heterogeneous: Non-uniform in structure or composition or composed of diverse elements – the converse of *homogeneous*.

Homogeneous: Uniform in structure or composition – the converse of *heterogeneous*.

Horizontal directional drilling (HDD): A form of directional boring, using steerable drilling techniques, to drill and install wells in a shallow arc to allow dewatering beneath inaccessible areas and without the need for vertical wells.

Hydraulic conductivity: The hydrogeological term for *permeability*. In general, engineering literature uses permeability, while hydrogeological references use hydraulic conductivity.

Hydraulic failure: *Base instability* caused by groundwater.

Hydraulic gradient: The change in *total hydraulic head* between two points divided by the length of flow path between the points.

Hydraulic head: See *total hydraulic head*.

Hydrogeology: The study of groundwater or underground waters. Also known as groundwater hydrology.

Hydrological cycle: The interlinked processes by which water is circulated from the oceans to the atmosphere to the ground surface and thence returned to the oceans as surface water and groundwater flow.

Hydrostatic conditions: Groundwater conditions when there are no vertical hydraulic gradients and hence no vertical flow of water. The *total hydraulic head* is constant with depth, so that *monitoring wells* installed at different depths show the same groundwater level.

Internal erosion: Movement of soil particles in granular soils under the influence of flowing groundwater. Where this occurs in or around an excavation, it can lead to *base instability* or *hydraulic failure*.

Intrinsic permeability: The expression of permeability of a porous medium that is independent of the permeating fluid. *Hydraulic conductivity* and *permeability* are special cases, where the permeating fluid is water.

Isotropic: Having the same physical properties in all directions – the converse of *anisotropic*.

Jet grouting: A form of *grouting* where the action of water/grout jets is used to disrupt existing soil structure and effect some in situ mixing, and replacement by, injected *cement-based grout*. Jet grouting is a *ground treatment* method that can be used to form low-permeability barriers as part of *groundwater exclusion* methods.

Jetting: A method used to install *wellpoints* and, less commonly, *deep wells* or *ejector wells*. A jet of high-pressure water is used to allow the penetration of a steel placing tube into the ground, within which the wellpoint or well is installed, following which the placing tube is removed. The jetting water returns to the surface as a *boil*, bringing with it the displaced soil.

Laminar flow: see *Darcian flow*.

Leaky aquifer: An *aquifer* confined by a low-permeability *aquitard*. When pore water pressures in the aquifer are lowered by pumping, water will flow (or leak) from the aquitard and recharge the aquifer. Also known as a semi-confined aquifer.

Loss of fines: The uncontrolled movement of *fines* carried by groundwater flow towards a well, sump or soil face where filters are inadequate or absent. Also used to describe the washing of fines from samples recovered from water-filled *boreholes* during site investigation drilling.

Mix-in-place walls: A method of *ground treatment* used to form *cut-off walls* where low-permeability barriers are produced in situ by *soil mixing*, where the mechanical action of an auger or cutter, in combination with *grout* injection into the mixing zone, creates a zone or block of treated material. See also *soil mixing*.

Monitoring well: An instrument (a *standpipe*, *piezometer* or unpumped well) installed into the ground in a selected location for the purpose of measuring the level of the *water table*, *piezometric level* or *pore water pressure*. Also known as an observation well.

Non-Darcian flow: A flow regime where the flow lines are turbulent and heterogeneously mixed, and the relationship between driving head and flow rate does not follow *Darcy's law*. It is typical of flow in rocks dominated by discrete fractures and of some surface water bodies – the converse of laminar flow or *Darcian flow*.

Numerical model: A method of analysing a groundwater *conceptual model* using computer software. The elements of the conceptual model are expressed as the geometry, boundary conditions and numerical values of the problem. The resulting equations are solved numerically (often by iteration) using the software, generally run on a personal computer.

Observation well: See *monitoring well*.

Open area: The proportion of the surface area of a *well screen* that is perforated or slotted and allows water to pass. Open area is normally expressed as a percentage.

Open pumping: The methods that lower groundwater levels by allowing groundwater to enter an excavation, from where it is then removed by pumping. The most commonly used method of this type is *sump pumping*. The converse of open pumping is *pre-drainage*, where groundwater levels are lowered in advance of excavation.

Overbleed: A commonly used colloquial term to describe residual groundwater seepages into an excavation, as might occur when a *perched water table* is penetrated or the groundwater level is drawn down below a low-permeability layer. Overbleed normally describes low rates of seepage that are a nuisance rather than causing significant flooding. Nevertheless, even at small rates of seepage, if *fines* are being continuously transported by the seepage, this should be counteracted immediately. The guiding principle is: do not try to stop the flow of seepage water, as this will cause a build-up of pore pressures, but do prevent continuous *loss of fines*, perhaps by placing a suitable filter.

Packer permeability test: A *permeability* test, typically carried out in a *borehole* in rock, where a test section of the borehole is isolated by packers, and tests are carried out on the isolated zone. In the most common configuration, *constant head tests* are carried out by injecting water into the test section.

Partially penetrating: The case where a well does not penetrate the full thickness of an *aquifer* – the converse of *fully penetrating*.

Particle size distribution: The relative percentages by dry weight of the different particle sizes of a soil sample, determined in the laboratory by mechanical analysis (e.g. by sieving). Also known as mechanical analysis, PSD, sieve analysis and soil grading.

Perched water table: Water trapped (or 'perched') in an isolated saturated zone above the *water table*. Perched water tables may exist naturally or may be caused when groundwater levels are lowered by dewatering; very low-permeability silt or clay layers will tend to inhibit downward seepage of groundwater, trapping water above them. When an excavation is dug through a perched water table, water will enter the excavation as residual seepage or *overbleed*.

Permeability: A measure of the ease with which water can flow through soil or rock. Also known as coefficient of permeability or *hydraulic conductivity*.

Permeation grouting: A form of *grouting* where *grout* is injected to fill or partially fill the permeable pores within a soil without disturbing the structure of the soil. Permeation grouting can be used as *ground treatment* to form low-permeability barriers as part of *groundwater exclusion* methods.

Phreatic level: The level at which pore water pressure is zero (i.e. equal to atmospheric pressure) and thus is at the base of the *capillary saturated zone*. See also *water table*.

Piezometer: An instrument installed into the ground to act as a *monitoring well* to measure the *pore water pressure* or *piezometric level* in a specific stratum, layer or elevation. This is achieved by having only the relevant section of the *borehole* exposed to the ground, the remainder of the borehole being sealed with grout or other very low-permeability material. See also *fully grouted piezometer*.

Piezometric level: The level representing the *total hydraulic head* of groundwater in a *confined aquifer*. Also known as potentiometric surface.

Piping: The washing out of fine particles from soils or weak rocks under the action of flowing groundwater to create preferential flow paths. This can lead to the creation of discrete open voids, termed 'pipes', within which groundwater flow is concentrated. Where this occurs in or around an excavation, it can lead to *base instability* or *hydraulic failure*. See also *internal erosion*.

Planar flow: A two-dimensional flow regime in which flow occurs in a series of parallel planes. This type of flow occurs toward a pumped slot or long line of closely spaced *wellpoints* when the direction of flow is perpendicular to the slot or line of wellpoints.

Pore water pressure: The pressure of groundwater in the pores of a soil, measured relative to atmospheric pressure. Positive and negative pore water pressures are higher and lower than atmospheric, respectively.

Pore water pressure control: The application of *groundwater lowering* to low-permeability soils. The fine-grained nature of such soils means that relatively little water drains from the pores, and desaturation does not occur. Nevertheless, pumping from vacuum-assisted wells can reduce *pore water pressures*, thereby controlling effective stress levels and preventing instability. See also *groundwater depressurization*.

Porosity: The ratio of the volume of voids (or pore space) to the total volume of a soil sample or mass.

Pre-drainage: The methods that lower groundwater levels in advance of excavation. Pre-drainage methods include *wellpoints*, *deep wells* and *ejectors*. The converse of pre-drainage is *open pumping*, where water is allowed to enter the excavation, from whence it is removed.

Pumping test: A method of in situ *permeability* testing involving controlled pumping from a *well* while the *discharge* flow rate and *drawdown* in the aquifer are recorded.

Radial flow: A two-dimensional flow regime in which the flow occurs in planes, which converge on an axis of radial symmetry. This type of flow occurs towards an individual well or sump when flow lines converge as they approach the well.

Radius of influence: The radial distance outwards from the centre of the pumping well to the point where there is no lowering of the *water table* or reduction of the *piezometric level* – the edge of the *cone of depression*. Also known as the distance of influence.

Rapid response piezometer: A *piezometer* that uses specialist instruments (such as *vibrating wire piezometers*) where only a very small volume of water must flow into or out of the sensor in order to record a change in pressure. This type of piezometer is suitable to observe *pore water pressure* changes in silts, clays, laminated soils and low-permeability rocks. Rapid response piezometers may be installed as *fully grouted piezometers*.

Recharge: Water that flows into an aquifer to increase or maintain the quantity of groundwater stored therein. Recharge may be from several sources: infiltration from surface waters; precipitation; seepage from other *aquifers* or *aquitards*; or *artificial recharge*.

Relief well: A well installed within an excavation to act as a preferential pathway for upward flow of water to relieve any *pore water pressures* trapped beneath the excavation. As the excavation is deepened, the well will overflow, relieving pressures. Also known as a bleed well.

Rheology: The study of the flow of materials, including fluids used in ground engineering, such as *grouts* and *bentonite slurry*.

Rising head test: A type of *variable head test*, typically carried out in a *borehole* or *monitoring well*, used to estimate *permeability*. At the start of the test, the water level in the well is lowered rapidly, and the rate of rise of the water level in the well is measured.

Rock: A geological deposit formed from mineral grains or crystals cemented together – this is distinct from uncemented *soil* or *drift*. Groundwater flow in rock is often dominated by flow through *fractures*.

Rock grouting: A form of *grouting* where *grout* is injected to fill or partially fill *fractures*, fissures or joints in a *rock* mass without creating new fractures or opening existing fractures. Rock grouting can be used as a *ground treatment* method to form low-permeability barriers as part of *groundwater exclusion* methods.

Running sand: A colloquial term for the conditions when saturated granular soils are so unstable that they are unable to support a cut face or slope and become an almost liquid slurry. Running sand is not a type or property of the material; it is a condition in which a granular soil can exist under unfavourable seepage conditions. Effective *groundwater lowering* can change running sand into a stable and workable material.

Saline intrusion: The movement of saline water into a fresh water aquifer as a result of *hydraulic gradients*. Problems of saline intrusion can be created or exacerbated as a result of pumping from wells.

Saturated: The condition when all the pores and fractures of a soil or rock are completely filled with water – the converse of *unsaturated*.

Saturated zone: The zone of *saturated* soil or rock, all parts of which are in hydraulic connection with each other. In a *confined aquifer*, the saturated zone comprises the entire aquifer. In an *unconfined aquifer*, the saturated zone is the area below the *water table* plus the *capillary saturated zone* above the *water table*.

Secant pile wall: A type of bored pile wall used to form *cut-off walls* or retaining structures where successive concrete piles are bored in close proximity. In this application, alternate piles are drilled at spacings of less than one diameter, so that they intersect and form a continuous wall.

Settlement lagoon: A large lagoon or pit through which the pumped discharge is passed to settle out sand and silt-sized particles prior to disposal.

Settlement tank: A small self-contained tank (often containing baffles) through which the pumped discharge is passed to settle out sand-sized particles prior to disposal. Such relatively small tanks are ineffective at settling out silt-sized particles due to the short retention time.

Sheet-piling: A technique used to form *cut-off walls* or retaining structures where a series of interlocked steel sections (typically of a 'Z' or 'U' profile) are driven or pushed into the ground to form a continuous barrier.

Siphon drain: A specialist pumping technique, occasionally used for *groundwater lowering*, which allows water to be pumped by gravity (without the need for external energy input) via permanently primed siphon pipes.

Site investigation: The overall process of obtaining all relevant information about a site to allow the design and construction of a civil engineering or building project. Elements of site investigation include the *desk study* and the *ground investigation*.

Slug test: A *variable head test*, typically carried out in a *borehole* or *monitoring well*, used to estimate *permeability*. At the start of a slug test, the water in the well is subjected to a rapid change in water level. This is achieved by inserting a heavy rod into the well or removing it from the well. When the rod passes the water level, it will displace water and hence rapidly change the water level. Observation of the subsequent rate of change of water level in the borehole as the head change dissipates can be used to estimate the *permeability* of the surrounding ground. See also *falling head test*; *rising head test*.

Slurry trench wall: A type of *cut-off wall* formed by the excavation of a trench supported by a slurry (most commonly, but not always, *bentonite slurry*) during excavation.

Soil: The term used in civil engineering to describe uncemented deposits of mineral (and occasionally organic) particles such as gravel, sand, silt and clay – this is distinct from cemented *rock*. Groundwater flow in soils is typically intergranular – i.e. through the interconnected pore spaces between the soil particles. Excavations below the groundwater level in soils may be unstable without adequate *groundwater control*. See also *drift*.

Soil-bentonite wall: A type of *slurry trench wall* where the trench is supported during excavation by a *bentonite slurry*, which is then replaced with a mixture of soil and bentonite to form the final *cut-off wall*.

Soil mixing: A *ground treatment* technique used to form *cut-off walls* by the *mix-in-place* method, where the existing soil structure is disrupted by mechanical tools (e.g. augers or cutters) and then the soil is mixed in situ with injected grout.

Solution grout: *Grout* that does not contain solid particles and generally gels (solidifies) on setting. The most common form of solution grouts is *chemical grouts*.

Specific yield: The *storage coefficient* in an *unconfined aquifer*.

Spring: A natural outflow of groundwater at the ground surface. In rural areas, springs may be used as water supplies.

Standpipe: An instrument, typically consisting of a perforated casing or pipe, installed into the ground to act as a *monitoring well*. A standpipe is open to water inflows from all strata that it penetrates and may give a water level reading that is a hybrid of levels in different aquifers. A *piezometer* should be used to determine water levels or pore water pressures in a specific stratum.

Storage coefficient: The volume of water released by gravity drainage from a volume of *aquifer*. Normally expressed as a dimensionless ratio or percentage. Also known as storativity.

Storage release: The release of water from storage in an aquifer. During the early stages of pumping (the first few hours or days), storage release can significantly increase the discharge flow rate above the steady-state value. Storage release is most significant in *unconfined aquifers* of high permeability.

Stratum: A geological term for a layer of any deposited *soil* or *rock*. The plural of stratum is strata.

Sub-drain: A perforated or open-jointed pipe sometimes found immediately beneath sewers or pipelines that were laid in trenches. During construction, the sub-drain is laid in the trench just ahead of the main pipe to carry water away from the working area.

Submersible pump: A pump designed to operate wholly or partly submerged in water. The most common forms of submersible pumps are electrically powered. Special slimline submersible pumps are suitable for use in *wells* or *boreholes*.

Sump: A pit, usually located within an excavation, in which water collects prior to being pumped away. See also *sump pumping*.

Sump pump: A robust pump, capable of handling solids-laden water, used to pump groundwater and surface water from *sumps*.

Sump pumping: A method of *groundwater lowering* by *open pumping* that allows water to enter the excavation, from whence it is pumped away. In some circumstances, the flow of water into the excavation may have a destabilizing effect, and the use of *pre-drainage* methods (in preference to sump pumping) may reduce the risk of instability.

Surface water: Water contained in rivers, streams, lakes and ponds.

Suspension grout: *Grout* that comprises solids suspended in water without being dissolved. The most common form of suspension grouts is *cement-based grouts*.

Total hydraulic head: The potential energy of water due to its height above a given level. The total hydraulic head controls the height to which water will rise in a piezometer. The total hydraulic head at a given point is the sum of the elevation head (i.e. the height of the point above an arbitrary datum) and the pressure head (i.e. the height of water above the point recorded in a standpipe piezometer). Also known as hydraulic head, total head or total hydraulic potential.

Transmissivity: A term used in *hydrogeology* to describe the ease with which water can flow through the saturated thickness of the *aquifer*. Transmissivity is equal to the product of *permeability* and saturated aquifer thickness.

Tube à manchette (TAM): A technique used as part of *permeation grouting* where special grouting tubes are installed into *boreholes*. The TAM pipes comprise plastic or metal pipes with grout ports drilled in them at specific intervals. Rubber sleeves cover the grout ports, and the TAM pipes are sealed into the boreholes with weak sleeve grout. *Grout* can be injected into the TAM pipes via special packers, which allow grout to be injected into the ground from specific levels in the boreholes.

Turbulent flow: See *non-Darcian flow*.

Unconfined aquifer: An *aquifer* whose upper surface is not confined by an overlying impermeable stratum and whose upper surface is directly exposed to atmospheric pressure. Also known as a water table aquifer.

Underdrainage: A method of *groundwater lowering* that uses soil layering to advantage by pumping directly from a deeper, more permeable stratum in preference to the overlying lower-permeability layers. Pumping from the deeper layer promotes vertical downward drainage from the lower-permeability layer into the stratum being pumped.

Uniformly graded: A *soil* containing a narrow range of particle sizes, so that most of the particles are of similar size – the converse of *well-graded*. Also known as poorly graded.

Unsaturated: The condition when the pores and fractures of a soil or rock are not completely filled with water and may contain some air or other gases – the converse of *saturated*.

Unsaturated zone: The zone of *unsaturated* soil or rock. In an *unconfined aquifer*, the saturated zone is the area above the *capillary saturated zone*. Within the unsaturated zone, the forces of surface tension render the water pressure negative (i.e. lower than atmospheric). Also known as the vadose zone.

Vadose zone: See *unsaturated zone*.

Variable head test: A *permeability* test, typically carried out in a *borehole* or *monitoring well*, where at the start of the test the water level in the well is subjected to a rapid (ideally instantaneous) change in water level. Observation of the subsequent rate of change of water level in the borehole as the head change dissipates can be used to estimate the *permeability* of the surrounding ground. See also *falling head test*; *rising head test*; *slug test*.

Vertical drain: A permeable vertical feature installed to act as a preferential pathway for downward flow of water to allow shallow strata to drain more rapidly.

Vibrating wire piezometers (VWPs): *Piezometers* where vibrating wire pressure transducers are used to measure pore water pressures. Vibrating wire instruments contain a metal diaphragm in hydraulic connection with the groundwater. A taut wire is stretched between the diaphragm and a stable datum. When the instrument is read, it is 'plucked' by passing a controlled-frequency electrical pulse along it. The taut wire resonates at a frequency related to its tension, which can be related to the deflection of the diaphragm and hence the water pressure on the diaphragm. See also *rapid response piezometer*.

V-notch weir: A thin plate *weir* where a V-shaped notch of a specified angle is cut. Calibration charts allow the flow rate to be estimated from measurements of the height of water flowing over the weir.

Void ratio: The ratio of the volume of voids to the volume of solids in a soil mass.

Water table: The level in an *unconfined aquifer* at which the pore water pressure is zero (i.e. atmospheric). See also *phreatic level*.

Weir: A structure used to control the flow of water, often so that it can be measured. See also *V-notch weir*.

Well: A hole sunk into the ground for the purposes of abstracting water. Wells for groundwater lowering purposes are generally categorized by their method of pumping as *deep wells*, *ejector wells* or *wellpoints*. In water supply terminology, a well is often taken to mean a large-diameter shaft, such as may be dug by hand in developing countries. A smaller-diameter well, constructed by a drilling rig, is termed a *borehole*.

Well casing: The unperforated section of the *well liner*, installed at depths where any groundwater present is to be excluded from the well (e.g. if several aquifers exist, but drawdown is not required in all of them). Also known as plain casing or solid casing.

Well development: Increasing or maximizing the yield of a well by removing drilling residue and fine particles from the well and the aquifer immediately outside the well. Normally carried out following well construction but prior to the installation of pumping equipment.

Well-graded: A *soil* containing a wide range of particle sizes – the converse of *uniformly graded*. A soil formed from a mixture of various proportions of gravel, sand, silt and clay-sized particles would be described as well-graded.

Well liner: A generic term for *well casing* and *well screen*.

Well loss: The head loss (or additional drawdown inside the well) that occurs when water flows from the aquifer, through the *well screen* and *filter pack*, and into the well itself.

Wellpoint: A small-diameter shallow *well* for groundwater lowering purposes. Normally installed by jetting and pumped by a suction pump of some kind.

Well screen: The perforated or slotted section of the well liner installed to allow water to enter a well where it penetrates an *aquifer*. Also known as perforated casing or slotted casing.

Yield: The discharge flow rate of a well. The yield is generally taken to be controlled by the well rather than the pump – a high-yielding well would produce a low flow rate if an under-sized pump were installed. Also known as well yield.

References

Abu Dhabi City Municipality. (2014). *Dewatering Guidelines. Geotechnical, Geophysical, and Hydrogeological Investigation Project (GGHIP)*. Report 13-5015, Draft B, Abu Dhabi.

Acreman, M C, Adams, B, Birchall, P and Connorton, B. (2000). Does groundwater abstraction cause degradation of rivers and wetlands? *Journal of the Chartered Institution of Water and Environmental Management*, 14, June, pp. 200–206.

AGS. (2006). *A Client's Guide to Desk Studies*. Association of Geotechnical & Geoenvironmental Specialists, Beckenham.

Aguado, E, Remsin, I, Pikul, M F and Thomas, W A. (1974). Optimal pumping for aquifer dewatering. *Journal of the Hydraulics Division, Proceedings of the American Society of Civil Engineers*, 100, HY7, pp. 869–877.

Akbarimehr, D and Aflaki, E. (2019). Site investigation and use of artificial neural networks to predict rock permeability at the Siazakh Dam, Iran. *Quarterly Journal of Engineering Geology and Hydrogeology*, 52, pp. 230–239.

Akroyd, D S. (1986). The law relating to groundwater in the United Kingdom. *Groundwater: Occurrence, Development and Protection* (Brandon, T W, ed). Institution of Water Engineers and Scientists, Water Practice Manual No. 5, London, pp. 591–607.

Alder, D, Jones, C J F P, Lamont-Black, J, White, C, Glendinning, S and Huntley, D. (2015). Design principles and construction insights regarding the use of electrokinetic techniques for slope stabilisation. *Proceedings of the XVI ECSMFGE, Geotechnical Engineering for Infrastructure and Development* (Winter, M C, Smith, D M, Eldred, P J L and Toll, D G, eds). ICE Publishing, London, pp. 1531–1536.

Allenby, D and Kilburn, D. (2015). Overview of underpinning and caisson shaft-sinking techniques. *Proceedings of the Institution of Civil Engineers, Geotechnical Engineering*, 168, February, pp. 3–15.

Anderson, M P, Woessner, R H and Hunt, R. (2015). *Applied Groundwater Modelling, Simulation of Flow and Advective Transport*, 2nd edition. Academic Press, New York.

Anon. (1976). Dewatering for Andoc's Hunterson platform site. *Ground Engineering*, 9, 7, October, pp. 42–43.

Anon. (1998). Vacuum packed: A new rapid and cost effective method of pre-consolidation is being used to combat the rapid settlement of soft, compressible soils in the Netherlands. *Ground Engineering*, 31, 2, February, pp. 18–19.

Asadi, A, Huat, B B K, Nahazanan, H and Keykhah, H A. (2013). Theory of electroosmosis in soil. *International Journal of Electrochemical Science*, 8, pp. 1016–1025.

ASHGHAL. (2014). *Management of Construction Dewatering – Construction Dewatering Guidelines*. Public Works Authority, Quality and Safety Department, Doha, Qatar.

Atkinson, J. (2008). Rules of thumb in geotechnical engineering. *Soil-Structure Interaction: From Rules to Reality, 18th New Zealand Geotechnical Society 2008 Symposium* (Chin, C Y, ed). The Institution of Professional Engineers New Zealand, Wellington, pp. 9–21.

Auld, F A, Belton, J and Allenby, D. (2015). Application of artificial ground freezing. *Proceedings of the XVI ECSMFGE, Geotechnical Engineering for Infrastructure and Development* (Winter, M C, Smith, D M, Eldred, P J L and Toll, D G, eds). ICE Publishing, London, pp. 901–906.

Banks, D. (2012). *An Introduction to Thermogeology: Ground Source Heating and Cooling*, 2nd edition. Wiley, New York.

Banks, D, Cosgrove, T, Harker, D, Howsam, P J and Thatcher, J P. (1993). Acidisation: Borehole development and rehabilitation. *Quarterly Journal of Engineering Geology*, 26, pp. 109–125.

Baptiste, N and Chapuis, R P. (2015). What maximum permeability can be measured with a monitoring well? *Engineering Geology*, 184, pp. 111–118.

Bar, N and Baczynski, N. (2018). Ground characterization and design of a drainage gallery tunnel network for a large open pit mine in a high rainfall environment. *10th Asian Rock Mechanics Symposium, ARMS 2018, Singapore*, International Society for Rock Mechanics (ISRM).

Barker, P, Esnault, A and Braithwaite, P. (1998). Containment at pride park, Derby, England. *Land Contamination and Reclamation*, 6, 1, pp. 51–58.

Barker, R D. (1986). Surface geophysical techniques. *Groundwater, Occurrence, Development and Protection* (Brandon, T W, ed). Institution of Water Engineers and Scientists, Water Practice Manual No. 5, London, pp. 271–314.

Barratt, G O and Vowles-Sheridan, M J. (1995). Construction of the cooling water tunnels using immersed tube techniques. *Proceedings of the Institution of Civil Engineers, Civil Engineering, Sizewell B Power Station*, 108, February, pp. 63–72.

Barton, M E. (1995). The Bargate Centre, Southampton: Engineering geological and geohydrological aspects of the excavation for basement construction. *Engineering Geology of Construction* (Eddleston, M, Walthall, S, Cripps, J C and Culshaw, M G, eds). Geological Society Engineering Geology Special Publication 10, London, pp. 67–77.

Barton, N, Lien, R and Lunde, J. (1974). Engineering classification of rock masses for the design of tunnel support. *Rock Mechanics and Rock Engineering*, 6, 4, pp. 189–236.

Bauer, G E, Scott, J D, Shields, D H and Wilson, N E. (1980). The hydraulic failure of a cofferdam. *Canadian Geotechnical Journal*, 17, pp. 574–583.

Beale, G and Read, J. (2013). *Guidelines for Evaluating Water in Pit Slope Stability*. CRC Press/ Balkema, Leiden, The Netherlands.

Beesley, K. (1986). Downhole geophysics. *Groundwater, Occurrence, Development and Protection* (Brandon, T W, ed). Institution of Water Engineers and Scientists, Water Practice Manual No. 5, London, pp. 315–352.

Behroozmand, A A, Keating, K and Auken, E. (2014). A review of the principles and applications of the NMR Technique for near-surface characterization. *Surveys in Geophysics*, 36, 1, pp. 27–85.

Bell, A L. (2012). Geotechnical grouting and soil mixing. *ICE Manual of Geotechnical Engineering, Volume II Geotechnical Design, Construction and Verification* (Burland, J, Chapman, T, Skinner, H and Brown, M, eds). ICE Publishing, London, pp. 1323–1342.

Bell, F G and Cashman, P M. (1986). Groundwater control by groundwater lowering. *Groundwater in Engineering Geology* (Cripps, J C, Bell, F G and Culshaw, M G, eds). Geological Society Engineering Geology Special Publication No. 3, London, pp. 471–486.

Bennett Associates. (1996). *Case Study: Shaft Excavator*. Bennett Associates, Rotherham, UK.

Bevan, M A, Powrie, W and Roberts, T O L. (2010). Influence of large-scale inhomogeneities on a construction dewatering system in chalk. *Géotechnique*, 60, 8, pp. 635–649.

Bieniawski, Z T. (1973). Engineering classification of jointed rock masses. *Transactions of the South African Institution of Civil Engineers*, 15, 12, pp. 335–344.

Biggart, A R and Sternath, R. (1996). Storebaelt Eastern Railway Tunnel: Construction. *Proceedings of the Institution of Civil Engineers*, 114, Special Issue 1, pp. 20–39.

Binnie, C and Kimber, M. (2009). *Basic Water Treatment*, 4th edition. Thomas Telford, London.

Binns, A. (1998). Rotary coring in soils and soft rocks for geotechnical engineering. *Proceedings of the Institution of Civil Engineers, Geotechnical Engineering*, 131, April, pp. 63–74.

Bjerrum, L, Nash, J K T L, Kennard, R M and Gibson, R E. (1972). Hydraulic Fracturing in field permeability testing. *Géotechnique*, 22, pp. 319–332.

Black, J H. (2010). The practical reasons why slug tests (including falling and rising head tests) often yield the wrong value of hydraulic conductivity. *Quarterly Journal of Engineering Geology and Hydrogeology*, 43, 3, pp. 345–358.

Blackwell, J. (1994). A case history of soil stabilisation using the mix-in-place technique for the construction of deep manhole shafts at Rochdale. *Grouting in the Ground* (Bell, A L, ed). Thomas Telford, London, pp. 497–509.

Bliss, J and Rushton, K. (1984). The reliability of packer tests for estimating the hydraulic conductivity of aquifers. *Quarterly Journal of Engineering Geology*, 17, pp. 81–91.

Blyth, F G H and De Freitas, M H. (1984). *A Geology for Engineers*, 7th edition. Edward Arnold, London.

Boak, R, Bellis, L, Low, R, Mitchell, R, Hayes, P, Mckelvey, P and Neale, S. (2007). *Hydrogeological Impact Appraisal for Dewatering Abstractions*. Science Report SC040020/SR1, Environment Agency, Bristol.

Bock, M and Markussen, L M. (2007). Dewatering control in central Copenhagen. *Proceedings of the 14th European Conference on Soil Mechanics and Geotechnical Engineering*, Madrid, Vol. 2, pp. 715–720.

Bomont, S, Fort, D S and Holliday, J. (2005). Two applications of deep drainage using siphon and electro-pneumatic drains: Slope works for Castlehaven Coast Protection Scheme, Isle of Wight (UK) and slope stabilisation for the Railways Agency, France: Objectives, design, construction and post construction monitoring. *Proceedings of the International Conference on Landslide Risk Management*, 18th Annual Vancouver Geotechnical Society Symposium, Vancouver.

Borthwick, M A. (1836). Memoir on the use of cast iron in piling, particularly at Brunswick Wharf, Blackwall. *Transactions of the Instruction of Civil Engineers*, 1, 1836, pp. 195–205.

Boulton, N S. (1951). The flow pattern near a gravity well in a uniform water bearing medium. *Journal of the Institution of Civil Engineers*, 36, 10, pp. 534–550.

Bourne, J C. (1839). *Drawings of the London and Birmingham Railway*. Collection of the Library of the Institution of Civil Engineers, London.

Bowman, R, Davies, P and Baptie, P. (2019). *Unexploded Ordnance (UXO) Risk Management Guide for Land-Based Projects*. Construction Industry Research and Information Association, CIRIA Report C785, London.

Box, G E P and Draper, N R. (1987). *Empirical Model-Building and Response Surfaces*. Wiley, New York.

Boyd–Dawkins, W. (1898). On the relation of geology to engineering, James Forrest lecture. *Minutes of the Proceedings of the Institution of Civil Engineers*, 134, Part 4, pp. 2–26.

Braithwaite, P, Wade, S and Webb, G. (1996). Pride Park, Derby. *Arup Journal*, 31, 1, pp. 13–15.

Brand, E W and Premchitt, J. (1982). Response characteristics of cylindrical piezometers. *Géotechnique*, 32, 3, pp. 203–216.

Brandl, H. (2006). Energy foundations and other thermo-active ground structures. *Géotechnique*, 56, 2, pp. 81–122.

Brassington, F C. (1986). The inter-relationship between changes in groundwater conditions and civil engineering construction. *Groundwater in Engineering Geology* (Cripps, J C, Bell, F G and Culshaw, M G, eds). Geological Society Engineering Geology Special Publication No. 3, London, pp. 47–50.

Brassington, F C and Preene, M. (2003). The design, construction and testing of a horizontal wellpoint in a dune sands aquifer as a water source. *Quarterly Journal of Engineering Geology and Hydrogeology*, 36, Part 4, pp. 355–366.

Brassington, F C and Walthall, S. (1985). Field techniques using borehole packers in hydrogeological investigations. *Quarterly Journal of Engineering Geology*, 18, pp. 181–193.

Brassington, F C and Younger, P L. (2010). A proposed framework for hydrogeological conceptual modelling. *Water and Environment Journal*, 24, pp. 261–273.

Brassington, R. (1995). *Finding Water: A Guide to the Construction and Maintenance of Private Water Supplies*, 2nd edition. Wiley, Chichester.

Brassington, R. (2017). *Field Hydrogeology*, 4th edition. Wiley, New York.

Brawner, C O. (1982). Control of groundwater in surface mining. *International Journal of Mine Water*, 1, 1, pp. 1–16.

Brice, G J and Woodward, J C. (1984). Arab Potash solar evaporation system: Design and development of a novel membrane cut-off wall. *Proceedings of the Institution of Civil Engineers*, 76, February, pp. 185–205.

Bricker, S H and Bloomfield, J P. (2014). Controls on the basin-scale distribution of hydraulic conductivity of superficial deposits: A case study from the Thames Basin, UK. *Quarterly Journal of Engineering Geology and Hydrogeology*, 47, pp. 223–236.

British Drilling Association. (2008). *Guidance for Safe Intrusive Activities on Contaminated or Potentially Contaminated Land*. British Drilling Association, Northampton.

British Drilling Association. (2015). *Health and Safety Manual for Land Drilling 2015: A Code of Safe Drilling Practice*. British Drilling Association, Northampton.

British Tunnelling Society. (2012). *A Guide to the Work in Compressed Air Regulations 1996*. British Tunnelling Society, London.

Bromehead, C N. (1956). Mining and quarrying in the seventeenth century. *A History of Technology* (Singer, C, Holmyard, E J, Hall, A R and Williams, T L, eds). Oxford University Press, Oxford, Vol. 2, pp. 1–40.

Brown, D A. (2004). Hull wastewater flow transfer tunnel: Recovery of tunnel collapse by ground freezing. *Proceedings of the Institution of Civil Engineers, Geotechnical Engineering*, 157, April, pp. 77–83.

Bruce, D A. (2005). Pregrouting in rock: Contemporary basic principles. *Proceedings of 17th Rapid Excavation and Tunneling Conference (RETC)*, Seattle.

BS 3680. (1981). *Measurement of Liquid Flow in Open Channels: Part 4A Method Using Thin Plate Weirs*. British Standards Institution, London.

BS 5930. (2015). *Code of Practice for Ground Investigations*. British Standards Institution, London.

BS 7022. (1988). *Geophysical Logging of Boreholes for Hydrogeological Purposes*. British Standards Institution, London.

BS 7671. (2018). *Requirements for Electrical Installations: IEE Wiring Regulations*, 18th edition. British Standards Institution, London.

BS EN 1536. (2010). *Execution of Special Geotechnical Works – Bored Piles*. British Standards Institution, London.

BS EN 1538. (2010). *Execution of Special Geotechnical Works – Diaphragm Walls*. British Standards Institution, London.

BS EN 1997-1. (2004). *Eurocode 7: Geotechnical Design – Part 1: General Rules*. British Standards Institution, London.

BS EN 1997-1:2004 + A1. (2013). *UK National Annex to Eurocode 7. Geotechnical Design: General Rules*. British Standards Institution, London.

BS EN 1997-2. (2007). *Eurocode 7: Geotechnical Design – Part 2: Ground Investigation and Testing*. British Standards Institution, London.

BS EN 12063. (1999). *Execution of Special Geotechnical Works – Sheet Pile Walls*. British Standards Institution, London.

BS EN 12715. (2010). *Execution of Special Geotechnical Works – Grouting*. British Standards Institution, London.

BS EN 12716. (2001). *Execution of Special Geotechnical Works – Jet Grouting*. British Standards Institution, London.

BS EN 14679. (2005). *Execution of Special Geotechnical Works – Deep Mixing*. British Standards Institution, London.

BS EN 14688-1. (2002). *Geotechnical Investigation and Testing – Identification and Classification of Soil. Part 1 – Identification and Description*. British Standards Institution, London.

BS EN 14689-1. (2003). *Geotechnical Investigation and Testing – Identification and Classification of Rock. Part 1 – Identification and Description*. British Standards Institution, London.

BS EN ISO 22282-1. (2012). *Geotechnical Investigation and Testing – Geohydraulic Testing, Part 1: General Rules*. British Standards Institution, London.

BS EN ISO 22282-2. (2012). *Geotechnical Investigation and Testing – Geohydraulic Testing, Part 2: Water Permeability Tests in a Borehole Using Open Systems*. British Standards Institution, London.

BS EN ISO 22282-3. (2012). *Geotechnical Investigation and Testing – Geohydraulic Testing, Part 3: Water Pressure Tests in Rock*. British Standards Institution, London.

BS EN ISO 22282-4. (2012). *Geotechnical Investigation and Testing. Geohydraulic Testing, Part 4: Pumping Tests*. British Standards Institution, London.

BS ISO 14686. (2003). *Hydrometric Determinations — Pumping Tests for Water Wells — Considerations and Guidelines for Design, Performance and Use.* British Standards Institution, London.

Buchanan, N. (1970). Derwent Dam – Construction. *Proceedings of the Institution of Civil Engineers*, 45, March, pp. 401–422.

Burland, J B. (2008a). The founders of Géotechnique. *Géotechnique*, 58, 5, pp. 327–341.

Burland, J B. (2008b). Reflections on Victor de Mello, friend, engineer and philosopher. *Soils and Rocks*, 31, 3, Brazilian Association for Soil Mechanics and Geotechnical Engineering, pp. 111–123.

Burland, J B, Standing, J R and Jardine, F M. (2001). *Building Response to Tunnelling: Case Studies from Construction of the Jubilee Line Extension, Volume 1 Projects and Methods.* Construction Industry Research and Information Association, CIRIA Special Publication, London.

Burland, J B and Wroth, C P. (1974). Settlement of buildings and associated damage. *Proceedings of the Conference on Settlement of Structures, Cambridge,* British Geotechnical Society, Pentech Press, pp. 611–654.

Card, G B and Roche, D P. (1989). The use of continuous dynamic probing in ground investigation. *Penetration Testing in the UK,* Thomas Telford, London, pp. 119–122.

Carder, D R, Watts, G R A, Campton, L and Motley, S. (2008). *Drainage of Earthworks Slopes.* TRL Project Report PPR341, Transport Research Laboratory, Wokingham.

Carey, M A, Fretwell, B A, Mosley, N G and Smith, J W N. (2002). *Guidance on the Design, Construction, Operation and Monitoring of Permeable Reactive Barriers.* National Groundwater and Contaminated Land Centre Report NC/01/51, Environment Agency, Bristol.

Carman, P C. (1937). Fluid flow through granular beds. *Transactions of the Institution of Chemical Engineers*, 15, pp. 155–166.

Carman, P C. (1956). *Flow of Gases through Porous Media.* Butterworths Scientific Publications, London.

Carrier, W D. (2003). Goodbye, Hazen; Hello, Kozeny–Carman. *Journal of Geotechnical and Geoenvironmental Engineering*, 129, pp. 1054–1056.

Casagrande, L. (1952). Electro-osmotic stabilisation of soils. *Journal of the Boston Society of Civil Engineers*, 39, pp. 51–83.

Casagrande, L, Wade, N, Wakely, M and Loughney, R. (1981). Electro-osmosis projects, British Columbia, Canada. *Proceedings of the 10th International Conference on Soil Mechanics and Foundation Engineering,* Stockholm, Sweden, pp. 607–610.

Cashman, P M. (1971). Discussion on Derwent Dam. *Proceedings of the Institution of Civil Engineers*, 48, March, pp. 487–488.

Cashman, P M. (1973). Groundwater control. *Ground Engineering*, 6, September, pp. 26–29.

Cashman, P M. (1987). Discussion. *Groundwater Effects in Geotechnical Engineering* (Hanrahan, E T, Orr, T L L and Widdis, T F, eds). Balkema, Rotterdam, p. 1015.

Cashman, P M. (1994). Discussion of Roberts and Preene (1994a). *Groundwater Problems in Urban Areas* (Wilkinson, W B, ed). Thomas Telford, London, pp. 446–450.

Casey, G, Pantelidou, H, Whitaker, D, O'Riordan, N, Soga, K and Guthrie, P. (2015). Capital & operational carbon – an assessment of the permanent dewatering solution at Stratford International Station. *Proceedings of the XVI ECSMFGE, Geotechnical Engineering for Infrastructure and Development* (Winter, M C, Smith, D M, Eldred, P J L and Toll, D G, eds). ICE Publishing, London, pp. 2511–2516.

Cedergren, H R. (1989). *Seepage, Drainage and Flow Nets,* 3rd edition. Wiley, New York.

Chandler, R J, Leroueil, S and Trenter, N A. (1990). Measurements of the permeability of London Clay using a self-boring permeameter. *Géotechnique*, 40, 1, pp. 113–124.

Chapman, D N, Metje, N and Stärk, A. (2017). *Introduction to Tunnel Construction,* 2nd edition. CRC Press, Boca Raton, Florida.

Chapman, T G. (1957). Two-dimensional ground-water flow through a bank with vertical faces. *Géotechnique*, 7, 1, March, pp. 35–40.

Chapman, T G. (1959). Groundwater flow to trenches and wellpoints. *Journal of the Institution of Engineers, Australia*, October–November, pp. 275–280.

Chapman, T G. (1960). Capillary effects in a two-dimensional ground-water flow system. *Géotechnique*, 10, 2, June, pp. 55–61.

Chapuis, R P. (2013). Permeability scale effects in sandy aquifers: A few case studies. *Proceedings of the 18th International Conference on Soil Mechanics and Foundation Engineering*, Paris, France, pp. 505–510.

Charalambous, A N. (2011). Groundwater and the law. *Quarterly Journal of Engineering Geology and Hydrogeology*, 44, pp. 147–158.

Clark, L. (1977). The analysis and planning of step drawdown tests. *Quarterly Journal of Engineering Geology*, 10, pp. 125–143.

Clarke, R P J and Mackenzie, C N P. (1994). Overcoming ground difficulties at Tooting Bec. *Proceedings of the Institution of Civil Engineers, Civil Engineering, Thames Water Ring Main*, pp. 60–75.

Clay, R B and Takacs, A P. (1997). Anticipating the unexpected – Flood, fire overbreak, inrush, collapse. *Proceedings of the International Conference on Tunnelling Under Difficult Ground and Rock Mass Conditions*, Basel, Switzerland, pp. 223–242.

Clayton, C R I. (1995). *The Standard Penetration Test (SPT): Methods and Use*. Construction Industry Research and Information Association, CIRIA Report 143, London.

Clayton, C R I. (2001). *Managing Geotechnical Risk: Improving Productivity in UK Building and Construction*. Thomas Telford, London.

Clayton, C R I, Matthews, M C and Simons, N E. (1995). *Site Investigation*, 2nd edition. Blackwell, London. http://www.geotechnique.info

Clayton, C R I and Smith, D M. (2013). *Site Investigation Steering Group. Effective Site Investigation*. ICE Publishing, London.

Cliff, M I and Smart, P C. (1998). The use of recharge trenches to maintain groundwater levels. *Quarterly Journal of Engineering Geology*, 31, pp. 137–145.

Coe, R H, Foreman, W, Harrison, N. (1989). Construction of shafts for the Greater Cairo Wastewater project. *Proceedings of the Conference on Shaft Engineering*, Institution of Mining & Metallurgy, pp. 109–128.

Cole, M. (1996). At the cutting edge. *Construction News*, 11 July 1996, Harrogate, England.

Cole, R G, Carter, I C and Schofield, R J. (1994). Staged construction at Benutan Dam assisted by vacuum eductor wells. *Proceedings of the 18th International Conference on Large Dams*, Durban, South Africa, pp. 625–640.

Contreras, I A, Grosser, A T and Verstrate, R H. (2007). The use of the fully-grouted method for piezometer installation. *Proceedings of the Seventh International Symposium on Field Measurements in Geomechanics*, FMGM 2007, Boston, MA. ASCE Geotechnical Special Publication 175.

Cooper, H H and Jacob, C E. (1946). A generalised graphical method for evaluating formation constants and summarising well field history. *Transactions of the American Geophysical Union*, 27, pp. 526–534.

Cox, S and Powrie, W. (2001). Horizontal wells for leachate control in landfill. *Ground Engineering*, 34, 1, pp. 18–19.

Crippa, C and Manassero, V. (2006). Artificial ground freezing at Sophiaspoortunnel (The Netherlands) — freezing parameters: Data acquisition and processing. *GeoCongress 2006: Geotechnical Engineering in the Information Technology Age*, American Society of Civil Engineers, Reston, VA, pp. 254–259.

Cripps, J C, Bell, F G and Culshaw, M G, eds. (1986). *Groundwater in Engineering Geology*. Geological Society Engineering Geology Special Publication No. 3, London.

Croce, P, Flora, A and Modoni, G. (2014). *Jet Grouting: Technology, Design and Control*. CRC Press, Boca Raton, Florida.

Crompton, D M and Heathcote, J. (1993). Cardiff Bay barrage and groundwater rise. *Groundwater Problems in Urban Areas* (Wilkinson, W B, ed). Thomas Telford, London, pp. 172–196.

Cruse, P K. (1986). Drilling and construction methods. *Groundwater, Occurrence, Development and Protection* (Brandon, T W, ed). Institution of Water Engineers and Scientists, Water Practice Manual No. 5, London, pp. 437–484.

Cunningham, R. (2001). The groundwater-ecosystem link: The case for wetlands. *Protecting Groundwater*. Environment Agency, National Groundwater and Contaminated Land Centre Project NC/00/10, Solihull, supplementary papers.

Darcy, H. (1856). *Les Fontaines Publique de la Ville de Dijon*. Dalmont, Paris.

Davis, G M and Horswill, P. (2002). Groundwater control and stability in an excavation in Magnesian Limestone near Sunderland, NE England. *Engineering Geology*, 66, pp. 1–18.

Davey-Wilson, I E G. (1994). A knowledge-based system for selecting excavation groundwater control methods. *Ground Engineering*, 27, pp. 42–46.

Daw, G P. (1984). Application of aquifer testing to deep shaft investigations. *Quarterly Journal of Engineering Geology*, 17, pp. 367–379.

Daw, G P and Pollard, C A. (1986). Grouting for ground water control in underground mining. *International Journal of Mine Water*, 5, 4, pp. 1–40.

Deed, M E R. (2009). The Italian job. *Ground Engineering*, 42, July, p. 26.

Deed, M E R and Preene, M. (2015). Managing the clogging of water wells. *Proceedings of the XVI ECSMFGE, Geotechnical Engineering for Infrastructure and Development* (Winter, M C, Smith, D M, Eldred, P J L and Toll, D G, eds). ICE Publishing, London, pp. 2787–2792.

Deere, D, Hendron, A, Patton, F and Cording, E. (1966). Design of surface and near-surface construction in rock. *Proceedings of the 8th U.S. Symposium on Rock Mechanics – Failure and Breakage of Rock*, American Institute of Mining, Metallurgical and Petroleum Engineers, Inc, New York. pp. 237–302.

Denies, N and Huybrechts, N. (2018). *Handbook – Soil Mix Walls: Design and Execution*. CRC Press, Boca Raton, Florida.

Department of the Environment, Transport and the Regions: London (DETR). (1998). *Rethinking Construction*. The Report of the Construction Task Force. HMSO, London.

Devriendt, M. (2012). Trigger levels for displacement monitoring. *Geotechnical Instrumentation News*, March, pp. 23–25.

Ding, L Y, Chen, E J, Luo, H B, Zhou, C and Guo, J. (2015). Artificial ground freezing in cross passage construction for the Yangtze River Tunnel of Wuhan Metro Line 2. *Proceedings of the XVI ECSMFGE, Geotechnical Engineering for Infrastructure and Development* (Winter, M C, Smith, D M, Eldred, P J L and Toll, D G, eds). ICE Publishing, London, pp. 431–436.

Doran, S R, Hartwell, D J, Kofoed, N and Warren, S. (1995). Storebælt Railway tunnel – Denmark: Design of cross passage ground treatment. *Proceedings of the 11th European Conference on Soil Mechanics and Foundation Engineering*, Copenhagen, Denmark.

Doran, S R, Hartwell, D J, Roberti, P, Kofoed, N and Warren, S. (1995). Storebælt Railway tunnel – Denmark: Implementation of cross passage ground treatment. *Proceedings of the 11th European Conference on Soil Mechanics and Foundation Engineering*, Copenhagen, Denmark.

Drinkwater, J S. (1967). The Ranney method of abstracting water from aquifers. *Water and Water Engineering*, July, pp. 267–274.

Dumbleton, M G and West, G. (1976). *Preliminary Sources of Information for Site Investigations in Britain*. Transport and Road Research Laboratory, LR403, Crowthorne.

Duncan, J M, Wright, S G and Brandon, T L. (2014). *Soil Strength and Slope Stability*, 2nd edition. Wiley, Hoboken, New Jersey.

Dupuit, J. (1863). *Etudes Théoretiques et Practiques sur les Mouvement des Eaux dans les Canaux Decouverts et a Travers les Terrains Permeable*. Dunod, Paris.

Eddleston, M. (1996). Structural damage associated with land subsidence caused by deep well pumping in Bangkok, Thailand. *Quarterly Journal of Engineering Geology*, 29, pp. 1–4.

Edwards, R J G. (1997). A review of the hydrogeological studies for the Cardiff Bay Barrage. *Quarterly Journal of Engineering Geology*, 30, 1, pp. 49–61.

Environment Agency. (2012). *Good Practice for Decommissioning Redundant Boreholes and Wells*. Reference LIT 6478 / 657_12, Environment Agency, Bristol.

Environment Agency. (2018). *The Environment Agency's Approach to Groundwater Protection, Version 1.2*. Environment Agency, Bristol.

Ervin, M C and Morgan, J R. (2001). Groundwater control around a large basement. *Canadian Geotechnical Journal*, 38, pp. 732–740.

Essler, R D. (1995). Applications of jet grouting in civil engineering. *Engineering Geology of Construction* (Eddleston, M, Walthall, S, Cripps, J C and Culshaw, M G, eds). Geological Society Engineering Geology Special Publication No. 10, London, pp. 85–93.

Essler, R D. (2012). Design principles for ground improvement. *ICE Manual of Geotechnical Engineering, Volume II Geotechnical Design, Construction and Verification* (Burland, J, Chapman, T, Skinner, H and Brown, M, eds). ICE Publishing, London, pp. 911–938.

Fannin, R J and Moffat, R. (2006). Observations on internal stability of cohesionless soils. *Géotechnique*, 56, 7, pp. 497–500.

Farmer, I W. (1975). Electro-osmosis and electrochemical stabilisation. *Methods of Treatment of Unstable Ground* (Bell, F G, ed). Newnes-Butterworth, London, pp. 26–36.

Faustin, N E, Elshafie, M Z E B and Mair, R J. (2018). Case studies of circular shaft construction in London. *Proceedings of the Institution of Civil Engineers, Geotechnical Engineering*, 171, 5, pp. 391–404.

Faustin, N E, Mair, R J, Elshafie, M Z E B, Menkiti, C and Black, M. (2017). Field measurements of ground movements associated with circular shaft construction. *Proceedings of the 9th International Symposium on Geotechnical Aspects of Underground Construction in Soft Ground*, Sao Paulo.

Feng, S, Vardanega, P J, Ibraim, E, Widyatmoko, I and Ojum, C. (2019). Permeability assessment of some granular mixtures. *Géotechnique*, 69, 7, pp. 646–654.

Fernie, R, Puller, D and Courts, A. (2012). Embedded walls. *ICE Manual of Geotechnical Engineering, Volume II Geotechnical Design, Construction and Verification* (Burland, J, Chapman, T, Skinner, H and Brown, M, eds). ICE Publishing, London, pp. 1271–1288.

Fetter, C W. (2000). *Applied Hydrogeology*, 4th edition. MacMillan, New York.

Fetter, C W, Boving, T and Kreamer, D, eds. (2018). *Contaminant Hydrogeology*, 3rd edition. Waveland Press, Inc., Long Grove, Illinois.

Fetzer, C A. (1967). Electro-osmotic stabilization of West Branch Dam. *Journal of the Soil Mechanics and Foundations Division, Proceedings of the American Society of Civil Engineers*, 93, SM4, pp. 85–106.

Fitzpatrick, L, Kulhawy, F H and O'Rourke, T D. (1981). Flow patterns around tunnels and their use in evaluating problems. *Soft Ground Tunnelling* (Resendiz, D and Romo, M P, eds). Balkema, Rotterdam, pp. 95–103.

Forchheimer, P. (1886). Uber die ergibigkeit von brunnenanlagen und sickerschitzen. *Der Architektenund IngenieurVerein*, 32, 7.

Franklin, J B and Capps, C T F. (1995). Tees Barrage – construction. *Proceedings of the Institution of Civil Engineers, Municipal Engineering*, 109, September, pp. 196–211.

Freeze, R A. (1994). Henri Darcy and the Fountains of Dijon. *Groundwater*, 32, 1, January–February, pp. 23–30.

Freeze, R A and Cherry, J A. (1979). *Groundwater*. Prentice Hall Inc, Englewood Cliffs, New Jersey.

Fredlund, D G, Xing, A and Huan, S. (1994). Predicting the permeability function for unsaturated soils using the soil-water characteristic curve. *Canadian Geotechnical Journal*, 31, 4, pp. 533–546.

French, C and Taylor, M. (1985). Desiccation and destruction: The immediate effects of de-watering at Etton, Cambridgeshire. *Oxford Journal of Archaeology*, 4, 2, pp. 139–155.

Gallagher, N J and Brassington, F C. (1993). Merseyside groundwater – regional overview and specific example. *Groundwater Problems in Urban Areas* (Wilkinson, W B, ed). Thomas Telford, London, pp. 310–319.

Gammon, K M and Pedgrift, G F. (1962). The selection and investigation of potential nuclear power station sites in Suffolk. *Proceedings of the Institution of Civil Engineers*, 21, 1, January, pp. 139–160.

Garrick, H, Davison, R and Digges La Touche, G. (2010). Assessing the impact of quarry dewatering on archaeological assets. *Proceedings of the 15th Extractive Industry Geology Conference* (Scott, P W and Walton, G, eds), EIG Conferences Limited, pp. 169–164.

Garshol, K F. (2003). *Pre-Excavation Grouting in Rock Tunnelling*. MBT International Underground Construction Group.

Giardina, G, Milillo, P, Dejong, M, Perissin, D and Milillo, G. (2018). Evaluation of InSAR monitoring data for post-tunnelling settlement damage assessment. *Structural Control and Health Monitoring*, 26, 2, pp. 1–19.

Glossop, R. (1950). Classification of geotechnical processes. *Géotechnique*, 2, 1, pp. 3–12.

Glossop, R. (1960). The invention and development of injection processes part I: 1902–1850. *Géotechnique*, 10, 3, September, pp. 91–100.

Glossop, R. (1961). The invention and development of injection processes part II: 1850–1960. *Géotechnique*, 11, 4, December, pp. 255–279.

Glossop, R. (1968). The rise of geotechnology and its influence on engineering practice. *Géotechnique*, 18, 2, pp. 105–150.

Glossop, R. (1976). The invention and early use of compressed air to exclude water from shafts and tunnels during construction. *Géotechnique*, 26, 2, June, pp. 253–280.

Glossop, R and Collingridge, V H. (1948). Notes on groundwater lowering by means of filter wells. *Proceedings of the 2nd International Conference on Soil Mechanics and Foundation Engineering*, Rotterdam, Vol. 2, pp. 320–322.

Glossop, R and Skempton, A W. (1945). Particlesize in silts and sands. *Journal of the Institution of Civil Engineers*, 25, pp. 81–105.

Golder, H Q and Seychuk, J L. (1967). Soil problems in subway construction. *3rd Pan-American Conference on Soil Mechanics and Foundation Engineering*, Caracas, Venezuela, pp. 203–240.

Golestanifar, M and Ahangari, K. (2012). Choosing an optimal groundwater lowering technique for open pit mines. *Mine Water and the Environment*, 31, pp. 192–198.

Goodman, R E, Moye, D G, Van Schalkwyk, A and Javandel, I. (1965). Groundwater inflows during tunnel driving. *Engineering Geology*, 2, 1, January, pp. 39–56.

Gray, I. (2017). Effective stress in rock. *Deep Mining 2017: Eighth International Conference on Deep and High Stress Mining* (Wesseloo, J, ed). Australian Centre for Geomechanics, Perth.

Greenwood, D A. (1984). Re-levelling a gas holder at Rhyl. *Quarterly Journal of Engineering Geology*, 17, pp. 319–326.

Greenwood, D A. (1989). Sub-structure techniques for excavation support. *Economic Construction Techniques*, Thomas Telford, London, pp. 17–40.

Greenwood, D A. (1994). Engineering solutions to groundwater problems in urban areas. *Groundwater Problems in Urban Areas* (Wilkinson, W B, ed). Thomas Telford, London, pp. 369–387.

Griffiths, J S and Martin, C J, eds. (2017). *Engineering Geology and Geomorphology of Glaciated and Periglaciated Terrains – Engineering Group Working Party Report*. Engineering Geology Special Publications No. 28, London.

Guatteri, G, Koshima, A, Lopes, R, Ravagalia, A and Pieroni, M R. (2009). Historical cases and use of horizontal jet grouting solutions with 360° distribution and frontal septum to consolidate very weak and saturated soils. *Geotechnical Aspects of Underground Construction in Soft Ground* (Ng, C W W, Huang, N W and Liu, G B, eds). Taylor & Francis Group, London.

Hamilton, A, Riches, J, Realey, G and Thomas, H. (2008). 'Elred': New water for London from old assets. *Proceedings of the Institution of Civil Engineers, Civil Engineering*, 161, February, pp. 26–34.

Hanrahan, E T, Orr, T L L and Widdis, T F, eds. (1987). *Groundwater Effects in Geotechnical Engineering*. Balkema, Rotterdam.

Hantush, M S. (1962). On the validity of the Dupuit-Forchheimer well-discharge formula. *Journal of Geophysical Research*, 67, 6, pp. 2087–2594.

Hantush, M S. (1964). Hydraulics of wells. *Advanced Hydroscience*, I, pp. 281–431.

Harding, H J B. (1935). Groundwater-lowering and chemical consolidation of foundations. *Concrete & Constructional Engineering*, 30, 1, pp. 3–12.

Harding, H J B. (1938). Correspondence on Southampton docks extension. *Journal of the Institution of Civil Engineers*, 9, pp. 562–564.

Harding, H J B. (1946). *The Principles and Practice of Groundwater Lowering*. Institution of Civil Engineers, Southern Association.

Harding, H J B. (1947). The choice of expedients in civil engineering construction. *Journal of the Institution of Civil Engineers*, Works Construction Paper No. 6.

Harding, H J B. (1981). *Tunnelling History and My Own Involvement*. Golder Associates, Toronto.

Harding, H J B and Davey, A. (2015). *It's Warmer Down Below: The Autobiography of Sir Harold Harding, 1900–1986*. Tilia Publishing UK, Sussex.

Hardwick, N. (2010). Pick 'n' mix. *Ground Engineering*, 43,December, pp. 18–20.

Hardy, S, Nicholson, D, Ingram, P, Gaba, A, Chen, Y and Biscontin, G. (2018). New observational method framework for embedded walls. *Geotechnical Research*, 5, 3, pp. 122–129.

Harris, C S, Hart, M B, Varley, P M and Warren, C, eds. (1996). *Engineering Geology of the Channel Tunnel*. Thomas Telford, London.

Harris, J S. (1995). *Ground Freezing in Practice*. Thomas Telford, London.

Harris, M R, Herbert, S M and Smith, M A. (1995). *Remedial Treatment for Contaminated Land*. Construction Industry Research and Information Association, CIRIA Special Publications 101–112 (12 Volumes), London.

Hartwell, D J. (2001). Getting rid of the water. *Tunnels & Tunnelling International*, January, pp. 40–42.

Hartwell, D J. (2015). Permeability testing problems in rock. *Proceedings of the XVI ECSMFGE, Geotechnical Engineering for Infrastructure and Development* (Winter, M C, Smith, D M, Eldred, P J L and Toll, D, eds). ICE Publishing, London, pp. 3657–3662.

Hartwell, D J, Kofoed, M and Unterberger, W. (1994). Great Belt Tunnel: Cross passage construction. *Proceedings of the 3rd Symposium on Strait Crossing* (Krokeborg, J, ed). Balkema, Rotterdam, The Netherlands, pp. 327–335.

Hartwell, D J and Nisbet, R M. (1987). Groundwater problems associated with the construction of large pumping stations. *Groundwater Effects in Geotechnical Engineering* (Hanrahan, E T, Orr, T L L and Widdis, T F, eds). Balkema, Rotterdam, pp. 691–694.

Haws, E T. (1970). Discussion. *Proceedings of the Conference on Ground Engineering*, Institution of Civil Engineers, London, pp. 68–70.

Haydon, R E V and Hobbs, N B. (1977). The effect of uplift pressures on the performance of a heavy foundation on layered rock. *Proceedings of Conference on Rock Engineering*, British Geotechnical Society, pp. 57–472, Newcastle-on-Tyne, England.

Hayward, D. (2000). Big price tag for small subway. *European Foundations*, Winter 2000, pp. 13–14.

Hazen, A. (1892). Some physical properties of sands and gravels with special reference to their use in filtration. *24th Annual Report. Massachusetts State Board of Health*, Wright & Potter Printing, Boston, MA, pp. 539–534.

Hazen, A. (1900). *The Filtration of Public Water Supplies*. Wiley, New York.

Hazen, A. (1911). Discussion of 'Dam on Sand Foundation' by A. C. Koenig. *Transactions of the American Society of Civil Engineers*, 73, pp. 199–203.

Head, K H. (2006). *Manual of Soil Laboratory Testing. Volume I: Soil Classification and Compaction Tests*, 3rd edition. Whittles Publishing, Caithness, UK.

Head, K H and Epps, R J. (2011a). *Manual of Soil Laboratory Testing. Volume II: Permeability, Shear Strength and Compressibility Tests*, 3rd edition. Whittles Publishing, Caithness, UK.

Head, K H and Epps, R J. (2011b). *Manual of Soil Laboratory Testing. Volume III: Effective Stress Tests*, 3rd edition. Whittles Publishing, Caithness, UK.

Health and Safety Executive. (1999). *Health and Safety in Excavations: Be Safe and Shore*. Document HSG185, HMSO, London.

Health and Safety Executive. (2006). *Health and Safety in Construction*. Document HSG150, HMSO, London.

Health and Safety Executive. (2014a). *Safe Use of Work Equipment: Provision and Use of Work Equipment Regulations 1998, Approved Code of Practice and Guidance*, 4th edition. Document L22, HMSO, London.

Health and Safety Executive. (2014b). *Safe Use of Lifting Equipment: Lifting Operations and Lifting Equipment Regulations 1998, Approved Code of Practice and Guidance*, 2nd edition. Document L113, HMSO, London.

Health and Safety Executive. (2015). *Managing Health and Safety in Construction: Construction (Design and Management Regulations) 2015, Guidance on Regulations*. Document L153, HMSO, London.

Heathcote, J A, Lewis, R T, Russell, D I and Soley, R W N. (1997). Cardiff Bay Barrage: Investigating groundwater control in a tidal aquifer. *Quarterly Journal of Engineering Geology*, 30, 1, pp. 63–77.

Hencher, S R. (2015). *Practical Rock Mechanics*. CRC Press, Boca Raton, Florida.

Hencher, S R and McNicholl, D P. (1995). Engineering in weathered rock. *Quarterly Journal of Engineering Geology and Hydrogeology*, 28, pp. 253–266.

Herbert, R and Rushton, K R. (1966). Groundwater flow studies by resistance networks. *Géotechnique*, 16, 1, pp. 53–75.

Herrera, E and Garfias, J. (2013). Characterizing a fractured aquifer in Mexico using geological attributes related to open-pit groundwater. *Hydrogeology Journal*, 21, pp. 1323–1338.

Heuer, R E. (1976). Catastrophic ground loss in soft ground tunnels. *Proceedings of the Rapid Excavation and Tunnelling Conference (RETC)*, Las Vegas, pp. 278–295.

Heuer, R E. (1995). Estimating rock tunnel water inflow. *Proceedings of the Rapid Excavation and Tunnelling Conference* (Williamson, G and Gowring, I M, eds). Society for Mining Metallurgy and Exploration, pp. 41–60, San Francisco, CA.

Heuer, R E. (2005). Estimating rock tunnel water inflow – II. *Proceedings of the Rapid Excavation and Tunnelling Conference* (Hutton, J D and Rogstad, W D, eds). Society for Mining Metallurgy and Exploration, pp. 394–407, Seattle, WA.

Hoek, E. (1994). Strength of rock masses. *ISRM News Journal*, 2, 2, pp. 4–16.

Holden, J M W, Jones, M A, Fernado, M-W and White, C. (1998). *Hydraulic Measures for Control and Treatment of Groundwater Pollution*. Construction Industry Research and Information Association, CIRIA Report 186, London.

Holmes, G, Roscoe, H and Chodorowski, A. (2005). Construction monitoring of cut and cover tunnels. *Proceedings of the Institution of Civil Engineers, Geotechnical Engineering*, 158, October, pp. 187–196.

Holzbecher, E, Yulan, J and Ebneth, S. (2011). Borehole pump & inject: An environmentally sound new method for groundwater Lowering. *International Journal of Environmental Protection*, 1, pp. 53–58.

Hong, Y, Ng, C W W and Wang, L Z. (2015). Initiation and failure mechanism of base instability of excavations in clay triggered by hydraulic uplift. *Canadian Geotechnical Journal*, 52, 5, pp. 599–608.

Hoover, H G and Hoover, L H. (1950). *De Re Metalica* by Georgius Agricola. Translated from the first Latin edition of 1556. Dover Publications, New York.

Houlsby, A C. (1990). *Construction and Design of Cement Grouting: A Guide to Grouting in Rock Foundations*. Wiley Interscience, New York.

Houben, G J. (2001). Well ageing and its implications for well and piezometer performance. *Impact of Human Activity on Groundwater Dynamics*, International Association of Hydrological Sciences (IAHS) Publication 269, IAHS, Wallingford, pp. 297–300.

Houben, G J. (2015). Review: Hydraulics of water wells—flow laws and influence of geometry. *Hydrogeology Journal*, 23, 8, pp. 1633–1657.

Howden, C and Crawley, J D. (1995). Design and construction of the diaphragm wall. *Proceedings of the Institution of Civil Engineers, Civil Engineering, Sizewell B Power Station*, 108, February, pp. 48–62.

Howell, R. (2013). Airlift testing in exploration coreholes. *International Mine Water Association Conference – Reliable Mine Water Technology*, Denver, Colorado, August 2013, Vol. I, pp. 145–150.

Howsam, P, Misstear, B and Jones, C. (1995). *Monitoring, Maintenance and Rehabilitation of Water Supply Boreholes*. Construction Industry Research and Information Association, CIRIA Report 137, London.

Howsam, P and Tyrrel, S F. (1990). Iron biofouling in groundwater abstraction systems: Why and how? *Microbiology in Civil Engineering* (Howsam, P, ed). Spon, London, pp. 192–197.

Hsu, S-M, Lo, H-C, Chi, S-Y and Ku, C-Y. (2011). Rock Mass Hydraulic conductivity estimated by two empirical models. *Developments in Hydraulic Conductivity Research* (Dikinya, O, ed). Intech Europe, Rijeka, Croatia.

Hughes, A K, Bruggemann, D A, Kettle, C and Justino, C. (2018). *Grouting for Reservoir Dams – A Guide to Good Practice*. Construction Industry Research and Information Association, CIRIA Report C774, London.

Hughes, L, Phear, A, Nicholson, D, Pantelidou, H, Soga, K, Guthrie, P, Kidd, A and Fraser, N. (2011). Carbon dioxide from earthworks: A bottom-up approach. *Proceedings of the Institution of Civil Engineers, Civil Engineering*, 164, May, pp. 66–72.

Humpheson, C, Fitzpatrick, A J and Anderson, J M D. (1986). The basements and substructure for the new headquarters of the Hongkong and Shanghai Banking Corporation, Hong Kong. *Proceedings of the Institution of Civil Engineers, Part 1*, 80, August, pp. 851–883.

Hunt, H. (2002). American experience in installing horizontal collector wells. *Riverbank Filtration: Improving Source Water Quality* (Chittaranjan, R, Melin, G and Linsky, R B, eds). Kluwer Academic Publishers, Dordrecht, pp. 29–34.

Hunt, H, Schubert, J and Chittaranjan, R. (2002). Conceptual design of riverbank filtration systems. *Riverbank Filtration: Improving Source Water Quality* (Chittaranjan, R, Melin, G and Linsky, R B, eds). Kluwer Academic Publishers, Dordrecht, pp. 19–27.

Hutchinson, J N. (1977). Assessment of the effectiveness of corrective measures in relation to geological conditions and types of slope movement. *Bulletin of the International Association of Engineering Geology*, 16, pp. 131–155.

Hvorslev, M J. (1951). *Time Lag and Soil Permeability in Groundwater Observations*. Waterways Experimental Station, Corps of Engineers, Bulletin No. 36, Vicksburg, Mississippi.

ICOLD. (2016). *Internal Erosion of Existing Dams, Levees and Dikes, and their Foundations*. International Commission on Large Dams, Bulletin 164, Paris.

Ineson, J. (1956). Darcy's law and the evaluation of permeability. *International Association of Hydrological Sciences, Symposia Darcy, Dijon*. Publication 41, pp. 165–172.

Ineson, J. (1959). The relation between the yield of a discharging well at equilibrium and its diameter, with particular reference to a chalk well. *Proceedings of the Institution of Civil Engineers*, 13, pp. 299–316.

Institution of Civil Engineers. (1972). *Safety in Boreholes and Wells*. Institution of Civil Engineers, London.

Institution of Civil Engineers. (1982). *Vertical Drains: Géotechnique Symposium in Print*. Thomas Telford, London.

Institution of Civil Engineers. (1991). *Inadequate Site Investigation*. Thomas Telford, London.

Institution of Civil Engineers. (1999). *Specification for Construction of Slurry Trench Cut-off Walls as Barriers to Pollution Migration*. Thomas Telford, London.

Irvine, D J and Smith, R J H. (2001). *Trenching Practice*, revised edition. Construction Industry Research and Information Association, CIRIA Report 97, London.

ISO 1438. (2017). *Hydrometry — Open Channel Flow Measurement Using Thin-plate Weirs*. International Standards Organisation, Geneva.

Jackson, C R, Bloomfield, J P and Mackay, J D. (2015). Evidence for changes in historic and future groundwater levels in the UK. *Progress in Physical Geography*, 39, 1, pp. 49–67.

Jacob, C E. (1946). Drawdown test to determine effective radius of artesian well. *Proceedings of the American Society of Civil Engineers*, 72, pp. 629–646.

Jefferis, S A. (1993). In-ground barriers. *Contaminated Land – Problems and Solutions* (Cairney, T, ed). Blackie, London, pp. 111–140.

Jones, C J F P, Lamont-Black, J, Glendinning, S and Pugh, R C. (2005). New applications for smart geosynthetics. *Proceedings of the Sessions of the Geo-Frontiers 2005 Congress, Waste Containment and Remediation*, ASCE Conference Proceedings. doi: 10.1061/40789(168)33.

Karplus, W J. (1958). *Analog Simulation*. McGrawHill, London.

Kavvadas, M, Giolas, A and Papacharambous, G. (1992). Drainage of supported excavations. *Geotechnical and Geological Engineering*, 10, 2, pp. 141–157.

Kinnear, J K, Davis, J A, Wiegand, F and Smith, R. (2018). Cross passage ground treatment for the Crossrail C310 Thames tunnel. *Engineering in Chalk: Proceedings of the Chalk 2018 Conference* (Lawrence, J A, Preene, M, Lawrence, U L and Buckley, R, eds). ICE Publishing, London, pp. 155–160.

Kinnear, J K, Nicholls, B G, Davis, J A and Wiegand, F. (2018). Cross passage construction on the Crossrail C310 Thames Tunnel. *Engineering in Chalk: Proceedings of the Chalk 2018 Conference* (Lawrence, J A, Preene, M, Lawrence, U L and Buckley, R, eds). ICE Publishing, London, pp. 161–167.

Kirk, S and Herbert, A W. (2002). Assessing the impact of groundwater abstractions on river flows. *Sustainable Groundwater Development* (Hiscock, K M, Rivett, M O and Davison, R M, eds). Geological Society Special Publication 193, London, pp. 211–233.

Kirkland, C J, Rowden, I J and Smyth-Osbourne, K R. (1987). A permanent dewatering system installed in an underground railway. *Groundwater Effects in Geotechnical Engineering* (Hanrahan, E T, Orr, T L L and Widdis, T F, eds). Balkema, Rotterdam, pp. 179–182.

Klimczak, C, Schultz, R A, Parashar, R and Reeves, D M. (2010). Cubic law with aperture-length correlation: Implications for network scale fluid flow. *Hydrogeology Journal* 18, 4, pp. 851–862.

Knight, D J, Smith, G L and Sutton, J S. (1996). Sizewell B foundation dewatering – system design, construction and performance monitoring. *Géotechnique*, 46, 3, pp. 473–490.

Kofoed, N and Doran, S R. (1996). Storebaelt Eastern Railway Tunnel, Denmark – Ground freezing for tunnels and cross passages. *Proceedings of the Symposium on Geotechnical Aspects of Underground Construction in Soft Ground* (Mair, R J and Taylor, R N, eds). Balkema, Rotterdam, pp. 391–398.

Kolpakov, V B and Zhdanov, S V. (2019). Implementation of slope drainage system for optimal slope design at Anfisa open pit, Khabarovsk Territory, Russian Federation. *International Mine Water Association Conference –Mine Water: Technological and Ecological Challenges*, Perm, Russia, pp. 570–576.

Kozeny, J. (1927). Über kapillare Leitung des Wassers im Boden (Aufstieg, Versickerung und Anwendung auf die Bewässerung). *Sitzungsberichte der Kaiserlichen Akademie der Wissenschaften in Wien*, 136a, pp. 271–306 (in German).

Kruseman, G P and De Ridder, N A. (1990). *Analysis and Evaluation of Pumping Test Data*, 2nd edition. International Institute for Land Reclamation and Improvement, Publication 47, Wageningen, The Netherlands.

Kummerer, C. (2015). HDD drillings for special applications in infrastructure works and environmental geotechnics. *Proceedings of the XVI ECSMFGE, Geotechnical Engineering for Infrastructure and Development* (Winter, M C, Smith, D M, Eldred, P J L and Toll, D G, eds). ICE Publishing, London, pp. 619–624.

Latham, M. (1994). *Constructing the Team*. Final Report of the Government/Industry Review of Procurement and Contractual Arrangements in the UK Construction Industry. HMSO, London.

Lazarus, D. (2013). Water, water everywhere: Ingress in underground structures. *Proceedings of the Institution of Civil Engineers. Forensic Engineering*, 166, 4, November, pp. 189–197.

Leech, S and McGann, M. (2008). Open pit slope depressurization using horizontal drains – a case study. *Mine Water and the Environment* (Rapantova, N and Hrkal, Z, eds). Proceedings of the 10th International Mine Water Association Congress, Karlsbad, Czech Republic.

Leiper, Q J and Capps, C T F. (1993). Temporary works bund design and construction for the Tees Barrage. *Engineered Fills* (Clarke, B G, Jones, C J F P and Moffat, A I B, eds). Thomas Telford, London, pp. 482–491.

Leiper, Q J, Roberts, T O L and Russell, D. (2000). Geotechnical engineering for the Medway Tunnel and approaches. *Proceedings of the Institution of Civil Engineers, Transport*, 141, 1, pp. 35–42.

Linde-Arias, E, Harris, D and Davis, A. (2015). Depressurisation for the excavation of Stepney Green cavern. *Proceedings of the Institution of Civil Engineers, Geotechnical Engineering*, 168, 3, pp. 215–226.

Linney, L F and Withers, A D. (1998). Dewatering the Thanet beds in SE London: Three case histories. *Quarterly Journal of Engineering Geology*, 31, pp. 115–122.

Lloyd, J W and Heathcote, J A. (1985). *Natural Inorganic Hydrochemistry in Relation to Groundwater: An Introduction*. Clarendon Press, Oxford.

Lombardi, G. (2003). Grouting of rock masses. *Grouting and Ground Treatment, Proceedings of the Third International Conference* (Johnsen, L F, Bruce, D A and Byle, M J, eds). Geotechnical Special Publication No. 120, American Society of Civil Engineers, pp. 164–197.

Lomizé, G M, Netushil, A V and Rzhanitzin, B A. (1957). Electro-osmotic processes in clayey soils and dewatering during excavations. *Proceedings of the 4th International Conference on Soil Mechanics and Foundation Engineering*, London, Vol. 1, pp. 62–67.

Losavio, M, Lauf, A and Elmaghraby, A. (2019). The Internet of things and issues of mine water management. *International Mine Water Association Conference: Mine Water – Technological and Ecological Challenges*, Perm State University, Russia, pp. 678–683.

Loudon, A G. (1952). The computation of permeability from simple soil tests. *Géotechnique*, 3, pp. 165–183.

Lugeon, M. (1933). *Barrages et Geologie*. Dunod, Paris.

Lunardi, P. (1997). Ground improvement by means of jet grouting. *Ground Improvement*, 1, pp. 65–85.

Lunne, T, Robertson, P K and Powell, J J M. (1997). *Cone Penetration Testing in Geotechnical Practice*. Blackie, London.

Lunniss, R C. (2000). Medway Tunnel – planning and contract administration. *Proceedings of the Institution of Civil Engineers, Transport*, 141, February, pp. 1–8.

Macdonald, A M, Brewerton, L J and Allen, D J. (1998). Evidence for rapid groundwater flow and karst-type behaviour in the Chalk of southern England. *Groundwater Pollution, Aquifer Recharge and Vulnerability* (Robins, N S, ed). Geological Society Special Publication No. 130, pp. 95–106.

Macdonald, A M, Burleigh, J and Burgess, W G. (1999). Estimating transmissivity from surface resistivity soundings: An example from the Thames Gravels. *Quarterly Journal of Engineering Geology*, 32, pp. 199–205.

Machon, A and Stevens, S. (2004). Wastewater flow transfer tunnel: Design and construction. *Proceedings of the Institution of Civil Engineers, Geotechnical Engineering*, 157, 3, pp. 101–106.

Madabhushi, S P G. (2007). Ground improvement methods for liquefaction remediation. *Proceedings of the Institution of Civil Engineers, Ground Improvement*, 11, 4, pp. 195–206.

Mair, R J, Rankin, W J and Essler, R D. (1995). Compensation grouting. *Proceedings of the Institution of Civil Engineers, Geotechnical Engineering*, 113, January, pp. 55–67.

Mansel, H, Drebenstedt, C, Jolas, P and Blankenburg, R. (2012). Dewatering of opencast mines using horizontal wells. *International Mine Water Association Symposium – Mine Water and the Environment*, Bunbury, Australia, pp. 574A–574H.

Mansur, C I and Kaufman, R I. (1962). Dewatering. *Foundation Engineering* (Leonards, G A, ed). McGraw-Hill, New York, pp. 241–350.

Marefat, V, Duhaime, F, Chapuis, R and Le Borgne, V. (2019). Performance of fully grouted piezometers under transient flow conditions: Field study and numerical results. *Geotechnical Testing Journal*, 42, 2, pp. 433–456.

McCann, D M, Eddleston, M, Fenning, P J and Reeves, G M, eds. (1997). *Modern Geophysics in Engineering Geology*. Geological Society Engineering Geology Special Publication No. 12, London.

Mchaffie, M G J. (1938). Southampton docks extension. *Journal of the Institution of Civil Engineers*, 9, pp. 184–219.

Mckenna, G T. (1995). Grouted-in installation of piezometers in boreholes. *Canadian Geotechnical, Journal*, 32, pp. 355–363.

McNamara, A M, Roberts, T O L, Morrison, P R J and Holmes, G. (2008). Construction of a deep shaft for Crossrail. *Proceedings of the Institution of Civil Engineers, Geotechnical Engineering*, 161, October, pp. 299–309.

Mcwhirter, G. (2000). Vibwall installation – Cardiff Bay Barrage. *Proceedings of the 25th Annual Members' Conference and Eighth International Conference and Exposition*, Deep Foundations Institute, New York, NY.

Meigh, A C. (1987). *Cone Penetration Testing: Methods and Interpretation*. CIRIA Ground Engineering Report, Butterworths, London.

Meinzer, O E, ed. (1942). *Hydrology*. McGrawHill, New York.

Menkiti, C O, Davis, J A, Semertzidou, K, Abbireddy, C O R, Hight, D W, Williams, J D and Black, M. (2015). The geology and geotechnical properties of the Thanet Sand Formation – an update from the Crossrail Project. *Proceedings of the XVI ECSMFGE, Geotechnical Engineering for Infrastructure and Development* (Winter, M C, Smith, D M, Eldred, P J L and Toll, D G, eds). ICE Publishing, London, pp. 855–860.

Merritt, A. (1994). *Wetlands, Industry & Wildlife: A Manual of Principles and Practices*. Wildfowl & Wetlands Trust, Slimbridge, Gloucester.

Miller, E. (1988). The eductor dewatering system. *Ground Engineering*, 21, 1, September, pp. 29–34.

Michalski, A. (1989). Application of temperature and electrical-conductivity logging in ground water monitoring. *Groundwater Monitoring and Remediation*, 9, 3, pp. 112–118.

Misstear, B, Banks, D and Clark, L J. (2006). *Water Wells and Boreholes*. Wiley, Chichester.

Moon, J and Fernandez, G. (2010). Effect of excavation-induced groundwater level drawdown on tunnel inflow in a jointed rock mass. *Engineering Geology*, 110, pp. 33–42.

Moore, J F A and Longworth, T I. (1979). Hydraulic uplift of the base of a deep excavation in Oxford Clay. *Géotechnique*, 29, 1, pp. 35–46.

Mortimore, R N and Pomerol, B. (1996). Chalk Marl: Geoframeworks and engineering appraisal. *Engineering Geology of the Channel Tunnel* (Harris, C S, Hart, M B, Varley, P M and Warren, C, eds). Thomas Telford, London, pp. 455–466.

Mrvik, O, Bomont, S and Carastonian, D. (2010). Experience with innovative methods of ground-water lowering in construction. *From Research to Design in European Practice, Proceedings of 14th Danube-European Conference on Geotechnical Engineering*, Bratislava, Slovakia.

Munks, S J, Chamley, P and Eddie, C. (2004). Ground freezing and spray concrete lining in the recon-struction of a collapsed tunnel. *North American Tunneling 2004: Proceedings of the North American Tunneling Conference 2004*, 17–22 April 2004, Balkema, Leiden, The Netherlands, pp. 277–284.

Murnane, E, Heap, A and Swain, A. (2006). *Control of Water Pollution from Linear Construction Projects – Technical Guidance*. Construction Industry Research and Information Association, CIRIA Report C648, London.

Muskat, M. (1935). The seepage of water through dams with vertical faces. *Physics*, 6, p. 402.

Muskat, M. (1937). *The Flow of Homogeneous Fluids Through Porous Media*. McGrawHill, New York.

New Civil Engineer. (1998). Bulkhead location blamed for DLR blast. February 1998, pp. 3–4.

New Civil Engineer. (2004). Docklands tunnel blowout down to "elementary error", says Judge. January 2004, pp. 8–9.

Newman, R L, Essler, R D and Covil, C S. (1994). Jet grouting to enable basement construction in difficult ground conditions. *Grouting in the Ground* (Bell, A L, ed). Thomas Telford, London, pp. 385–401.

Newman, T. (2009). The impact of adverse geological conditions on the design and construction of the Thames Water Ring Main in Greater London, UK. *Quarterly Journal of Engineering Geology and Hydrogeology*, 42, pp. 5–20.

Neymeyer, A, Williams, R T and Younger, P L. (2007). Migration of polluted mine water in a public supply aquifer. *Quarterly Journal of Engineering Geology and Hydrogeology*, 40, pp. 75–84.

Nicholson, D P and Harris, S J. (1993). The use of gravity relief wells below basements in London Clay to control the rise in groundwater level. *Groundwater Problems in Urban Areas* (Wilkinson, W B, ed). Thomas Telford, London, pp. 424–442.

Nicholson, D P, Tse, C-M and Penny, C. (1999). *The Observational Method in Ground Engineering*. Construction Industry Research and Information Association, CIRIA Report C185, London.

Norbury, D R. (2015). *Soil and Rock Description in Engineering Practice*, 2nd edition. Whittles Publishing, Caithness, UK.

Norton, P J. (1987). Geotechnical benefits of advance dewatering in mining excavations and backfill. *Groundwater Effects in Geotechnical Engineering* (Hanrahan, E T, Orr, T L L and Widdis, T F, eds). Balkema, Rotterdam, pp. 217–220.

Nyer, E K. (2009). *Groundwater Treatment Technology*, 3rd edition. Wiley, New York.

Oakes, D B. (1986). Theory of groundwater flow. *Groundwater, Occurrence, Development and Protection* (Brandon, T W, ed). Institution of Water Engineers and Scientists, Water Practice Manual No. 5, London, pp. 109–134.

O'Riordan, N. (2003). Channel Tunnel Rail Link section 1: Ground engineering. *Proceedings of the Institution of Civil Engineers, Civil Engineering, Channel Tunnel Rail Link Section 1*, 156, November, pp. 28–31.

Ove Arup & Partners. (2015). *CDM2015 – Construction Work Sector Guidance for Designers*, 4th edition. Construction Industry Research and Information Association, CIRIA Report C755, London.

Packer, M, Newman, R, Prangley, C and Heath, I. (2018). Permeation grouting and excavation at Victoria station, London. *Proceedings of the Institution of Civil Engineers, Geotechnical Engineering*, 171, 3, pp. 267–281.

Paisley, A C, Wullenwaber, J and Bruce, D.A. (2017). Practical aspects of water pressure testing for rock grouting. *5th International Grouting, Deep Mixing, and Diaphragm Walls Conference*, International Conference Organization for Grouting (ICOG), Honolulu, Oahu, Hawaii.

Pallett, P F and Filip, R. (2018). *Temporary Works: Principles of Design and Construction*, 2nd edition. ICE Publishing, London.

Palmer, R C and Lewis, M A. (1998). Assessment of groundwater vulnerability in England and Wales. *Groundwater Pollution, Aquifer Recharge and Vulnerability* (Robins, N S, ed). Geological Society Special Publication 130, London, pp. 191–198.

Palmström, A and Stille, H. (2015). *Rock Engineering*, 2nd edition. ICE Publishing, London.

Parker, A H, West, L J and Odling, N E. (2019). Well flow and dilution measurements for characterization of vertical hydraulic conductivity structure of a carbonate aquifer. *Quarterly Journal of Engineering Geology and Hydrogeology*, 52, pp. 74–82.

Parker, A H, West, L J, Odling, N E and Bown, R T. (2010). A forward modeling approach for interpreting impeller flow logs. *Groundwater*, 48, pp. 79–91.

Parsons, S B. (1994). A re-evaluation of well design procedures. *Quarterly Journal of Engineering Geology*, 27, pp. S31–S40.

Peck, R B. (1969). Advantages and limitations of the observational method in applied soil mechanics. *Géotechnique*, 19, 2, pp. 17–187.

Peek, R-D and Willeitner, H. (1981). Behaviour of wooden pilings in long time service. *Proceedings of the 10th International Conference on Soil Mechanics and Foundation Engineering*, Stockholm, Sweden, pp. 147–152.

Peter Cowsill Limited. (2001). Lateral thinking. *Tunnels & Tunnelling International*, 33, November, pp. 51–53.

Pike, D, Banks, D, Waters, A and Robinson, V K. (2013). Regional distribution of temperature in the Chalk of the western London Basin syncline. *Quarterly Journal of Engineering Geology and Hydrogeology*, 46, pp. 117–125.

Placzek, D. (2009). The main principles of tunneling under compressed air. *Proceedings of the 17th International Conference on Soil Mechanics and Geotechnical Engineering* (Hamza, M, Shahien, M and El-Mossallamy, Y, eds). IOS Press, Amsterdam, The Netherlands, pp. 2465–2468.

Popielak, R, Moreno, J and Striegl, A. (2013). Design of a dewatering system for a geothermally influenced underground gold mine. *Mine Water Solutions in Extreme Environments, Chapter 5: Extremely Complex Hydrogeology*, April 2013, Lima, Peru, pp. 538–548.

Powers, J P. (1985). *Dewatering – Avoiding Its Unwanted Side Effects*. American Society of Civil Engineers, New York.

Powers, J P, Corwin, A B, Schmall, P C and Kaeck, W E. (2007). *Construction Dewatering and Groundwater Control: New Methods and Applications*, 3rd edition. Wiley, New York.

Powrie, W. (1990). *Legal Aspects of Construction Site Dewatering for Temporary Works*. Unpublished MSc Dissertation, Kings College, London.

Powrie, W. (2013). *Soil Mechanics: Concepts and Applications*, 3rd edition. Spon, Abingdon.

Powrie, W and Preene, M. (1992). Equivalent well analysis of construction dewatering systems. *Géotechnique*, 42, 4, pp. 635–639.

Powrie, W and Preene, M. (1994a). Time-drawdown behaviour of construction dewatering systems in fine soils. *Géotechnique*, 44, 1, pp. 83–100.

Powrie, W and Preene, M. (1994b). Performance of ejectors in construction dewatering systems. *Proceedings of the Institution of Civil Engineers, Geotechnical Engineering*, 107, July, pp. 143–154.

Powrie, W and Roberts, T O L. (1990). Field trial of an ejector well dewatering system at Conwy, North Wales. *Quarterly Journal of Engineering Geology*, 23, pp. 169–185.

Powrie, W and Roberts, T O L. (1995). Case history of a dewatering and recharge system in chalk. *Géotechnique*, 45, 4, pp. 599–609.

Powrie, W, Roberts, T O L and Jefferis, S A. (1990). Biofouling of site dewatering systems. *Microbiology in Civil Engineering* (Howsam, P, ed). Spon, London, pp. 341–352.

Preene, M. (1996). Ejector feat. *Ground Engineering*, 29, 4, May, p. 16.

Preene, M. (2000). Assessment of settlements caused by groundwater control. *Proceedings of the Institution of Civil Engineers, Geotechnical Engineering*, 143, October, pp. 177–190.

Preene, M. (2004). Robert Stephenson (1803–59) – the first groundwater engineer. *200 Years of British Hydrogeology* (Mather, J D, ed). Geological Society Special Publication 225, London, pp. 107–119.

Preene, M. (2015). Techniques and developments in quarry and surface mine dewatering. *Proceedings of the 18th Extractive Industry Geology Conference 2014 and Technical Meeting 2015* (Hunger, E and Brown, T J, eds). EIG Conferences Ltd, London, pp. 194–206.

Preene, M. (2016). *Groundwater Control – A Distress Purchase that Is Worth Getting Right.* Construction Industry Research and Information Association, CIRIA Briefing, Ref 02-02-16, London.

Preene, M. (2019). Design and interpretation of packer permeability tests for geotechnical purposes. *Quarterly Journal of Engineering Geology and Hydrogeology*, 52, pp. 182–200.

Preene, M and Brassington, F C. (2001). The inter-relationship between civil engineering works and groundwater protection. *Protecting Groundwater.* Environment Agency, National Groundwater and Contaminated Land Centre Project NC/00/10, Solihull, pp. 313–320.

Preene, M and Brassington, F C. (2003). Potential groundwater impacts from civil engineering works. *Water and Environmental Management Journal*, 17, 1, March, pp. 59–64.

Preene, M and Fisher, S. (2015). Impacts from groundwater control in urban areas. *Proceedings of the XVI ECSMFGE, Geotechnical Engineering for Infrastructure and Development* (Winter, M C, Smith, D M, Eldred, P J L and Toll, D G, eds). ICE Publishing, London, pp. 2846–2852.

Preene, M and Loots, E. (2015). Optimisation of dewatering systems. *Proceedings of the XVI ECSMFGE, Geotechnical Engineering for Infrastructure and Development* (Winter, M C, Smith, D M, Eldred, P J L and Toll, D G, eds). ICE Publishing, London, pp. 2841–2846.

Preene, M and Powrie, W. (1993). Steady-state performance of construction dewatering systems in fine soils. *Géotechnique*, 43, 2, pp. 191–206.

Preene, M and Powrie, W. (1994). Construction dewatering in low permeability soils: Some problems and solutions. *Proceedings of the Institution of Civil Engineers, Geotechnical Engineering*, 107, January, pp. 17–26.

Preene, M and Powrie, W. (2009). Ground energy systems – delivering the potential. *Proceedings of the Institution of Civil Engineers, Energy*, 162, May, pp. 77–84.

Preene, M and Powrie, W. (2019). Assessment of permeability for design of groundwater control systems. *Proceedings of the XVII ECSMGE, Geotechnical Engineering: Foundation of the Future*, Reykjavik, September 2019.

Preene, M and Roberts, T O L. (1994). The application of pumping tests to the design of construction dewatering systems. *Groundwater Problems in Urban Areas* (Wilkinson, W B, ed). Thomas Telford, London, pp. 121–133.

Preene, M and Roberts, T O L. (2002). Groundwater control for construction in the Lambeth Group. *Proceedings of the Institution of Civil Engineers, Geotechnical Engineering*, 155, October, pp. 221–227.

Preene, M and Roberts, T O L. (2017). Construction dewatering in Chalk. *Proceedings of the Institution of Civil Engineers, Geotechnical Engineering*, 170, 4, August, pp. 367–390.

Preene, M, Roberts, T O L and Hartwell, D J. (2018). Pumping tests for construction dewatering in chalk. *Engineering in Chalk: Proceedings of the Chalk 2018 Conference* (Lawrence, J A, Preene, M, Lawrence, U L and Buckley, R, eds). ICE Publishing, London, pp. 631–636.

Preene, M, Roberts, T O L and Powrie, W. (2016). *Groundwater Control – Design and Practice*, 2nd edition. Construction Industry Research and Information Association, CIRIA Report C750, London.

Preene, M, Roberts, T O L, Powrie, W and Dyer, M R. (2000). *Groundwater Control – Design and Practice*. Construction Industry Research and Information Association, CIRIA Report C515, London.

Price, D G. (1995). Weathering and weathering processes. *Quarterly Journal of Engineering Geology and Hydrogeology*, 28, pp. 243–252.

Price, M. (1994). A method for assessing the extent of fissuring in double porosity aquifers, using data from packer tests. *Future Groundwater Resources at Risk, Proceedings of the Helsinki Conference*, 13–16 June 1994, International Association of Hydrological Sciences (IAHS) Publication No. 222, Helsinki, pp. 271–278.

Price, M. (1996). *Introducing Groundwater*, 2nd edition. Chapman & Hall, London.

Price, M. (2009). Barometric water-level fluctuations and their measurement using vented and non-vented pressure transducers. *Quarterly Journal of Engineering Geology and Hydrogeology*, 42, pp. 245–250.

Price, M and Williams, A T. (1993). A pumped double-packer system for use in aquifer evaluation and groundwater sampling. *Proceedings of the Institution of Civil Engineers, Water, Maritime and Energy*, 101, April, pp. 85–92.

Privett, K D, Matthews, S C and Hodges, R A. (1996). *Barriers, Liners and Cover Systems for Containment and Control of Land Contamination*. Construction Industry Research and Information Association, CIRIA Special Publication 124, London, pp. 59–60.

Prugh, B J. (1960). New tools and techniques for dewatering. *Journal of the Construction Division, Proceedings of the American Society of Civil Engineers*, 86, CO1, pp. 11–25.

Puller, M. (2003). *Deep Excavations: A Practical Manual*, 2nd edition. Thomas Telford, London.

Pyne, R D G. (1994). *Groundwater Recharge and Wells: A Guide to Aquifer Storage Recovery*. Lewis Publishers, Boca Raton.

Raben-Levetzau, J, Markussen, L M, Bitsch, K and Nicolaisen, L L. (2004). Copenhagen Metro: Groundwater control in sensitive urban areas. *Proceedings of World Tunnel Congress and 13th ITA Assembly*, Singapore, Paper C04.

Rail Accident Investigation Branch. (2014). *Penetration and Obstruction of a Tunnel between Old Street and Essex Road Stations, London, March 2013*. Report 03/2014, Department for Transport, London.

Ramelli, A. (1588). *Le Diverse et Artificiose Machine del Capitano Agostino Ramelli*. Paris, France. https://digital.sciencehistory.org/works/4b29b614k

Rao, D B. (1973). Construction dewatering by vacuum wells. *Indian Geotechnical Journal*, 3, 3, pp. 217–224.

Rawlings, C G, Hellawell, E E and Kilkenny, W M. (2000). *Grouting for Ground Engineering*. Construction Industry Research and Information Association, CIRIA Report C514, London.

Rea, I and Monaghan, D. (2009). Dewatering bore pumps – reducing costs and emissions by maximising pump efficiency over time. *Mining Technology*, 118, pp. 220–224.

Reiffsteck, P, Haza-Rozier, E, Dorbani, B and Fry, J-J. (2010). A new Hydraulic profiling tool including CPT measurements. *2nd International Symposium on Cone Penetration Testing*, Huntington Beach, California.

Richards, D J, Wiggan, C A and Powrie, W. (2016). Seepage and pore pressures around contiguous pile retaining walls. *Géotechnique*, 66, 7, pp. 523–532.

Riley, E, Bellhouse, M, Mortimore, R and Condron, A. (2018). Use of rota-sonic drilling to investigate deep London Basin geology: A case study for an urban East London environment, including comparison to a conventionally drilled borehole. *Engineering in Chalk: Proceedings of the Chalk 2018 Conference* (Lawrence, J A, Preene, M, Lawrence, U L and Buckley, R, eds). ICE Publishing, London, pp. 431–436.

Roberts, T O L, Botha, C P and Welch, A. (2009). Design and operation of a large dewatering system in Dubai. *Proceedings of the 17th International Conference on Soil Mechanics and Geotechnical Engineering* (Hamza, M, Shahien, M and El-Mossallamy, Y, eds). IOS Press, Amsterdam, The Netherlands, pp. 2556–2559.

Roberts, T O L and Deed, M E R. (1994). Cost overruns in construction dewatering. *Risk and Reliability in Ground Engineering* (Skipp, B O, ed). Thomas Telford, London, pp. 254–265.

Roberts, T O L and Hartwell, D J. (2018). Chalk permeability. *Engineering in Chalk: Proceedings of the Chalk 2018 Conference* (Lawrence, J A, Preene, M, Lawrence, U L and Buckley, R, eds). ICE Publishing, London, pp. 425–430.

Roberts, T O L and Holmes, G. (2011). Case study of a dewatering and recharge system in weak Chalk rock. *Proceedings of the XV ECSMFGE, Geotechnics of Hard Soils – Weak Rocks* (Anagnostopoulos, A Pachakis, M and Tsatsanifos, C H, eds). IOS Press, Amsterdam, The Netherlands.

Roberts, T O L, Linde, E and Sutton, M. (2015). In-tunnel dewatering for a cross passage in London. *Proceedings of the third Arabian Tunnelling Conference, Dubai, United Arab Emirates*, 23–25 November 2015.

Roberts, T O L and Preene, M. (1994a). Range of application of construction dewatering systems. *Groundwater Problems in Urban Areas* (Wilkinson, W B, ed). Thomas Telford, London, pp. 415–423.

Roberts, T O L and Preene, M. (1994b). The design of groundwater control systems using the observational method. *Géotechnique*, 44, 4, pp. 727–734.

Roberts, T O L, Roscoe, H, Powrie, W and Butcher, J E. (2007). Controlling clay pore pressures for cut-and-cover tunnelling. *Proceedings of the Institution of Civil Engineers, Geotechnical Engineering*, 160, 4, pp. 227–236.

Roberts, T O L, Smith, R, Stärk, A and Zeisig, W. (2015). Sub-surface dewatering for an inclined SCL tunnel. *Proceedings of the XVI ECSMFGE, Geotechnical Engineering for Infrastructure and Development* (Winter, M C, Smith, D M, Eldred, P J L and Toll, D G, eds). ICE Publishing, London, pp. 2853–2858.

Robinson, C and Bell, A. (2018a). Diaphragm walls. *Temporary Works: Principles of Design and Construction*, 2nd edition (Filip, R and Pallett, P F, eds). ICE Publishing, London, pp. 175–192.

Robinson, C and Bell, A. (2018b). Contiguous and secant piled walls. *Temporary Works: Principles of Design and Construction*, 2nd edition (Filip, R and Pallett, P F, eds). ICE Publishing, London, pp. 193–205.

Roscoe, H and Twine, D. (2001). Design collaboration speeds Ashford Tunnels. *World Tunnelling*, 14, 5, pp. 237–241.

Roscoe Moss Company. (1990). *Handbook of Ground Water Development*. Wiley, New York.

Rose, E P F. (2012). Groundwater as a military resource: Pioneering British military well boring and hydrogeology in World War I. *Military Aspects of Hydrogeology* (Rose, E P F and Mather, J D, eds). Geological Society Special Publication No. 362, London, pp. 49–72.

Rossum, J R. (1954). Control of sand in water systems. *Journal American Water Works Association*, 46, 2, pp. 123–132.

Rowe, P W. (1968). Failure of foundations and slopes in layered deposits in relation to site investigation practice. *Proceedings of the Institution of Civil Engineers*, Supplement, pp. 73–131.

Rowe, P W. (1972). The relevance of soil fabric to site investigation practice. *Géotechnique*, 22, 2, pp. 195–300.

Rowe, P W. (1986). The potentially latent dominance of groundwater in ground engineering. *Groundwater in Engineering Geology* (Cripps, J C, Bell, F G and Culshaw, M G, eds). Geological Society Engineering Geology Special Publication No. 3, London, pp. 27–42.

Rowles, R. (1995). *Drilling for Water: A Practical Manual*, 2nd edition. Avebury, Aldershot.

Rushton, K R. (2003). *Groundwater Hydrology: Conceptual and Computational Models*. Wiley, Chichester.

Rushton, K R and Redshaw, S C. (1979). *Seepage and Groundwater Flow: Numerical Analysis by Analog and Digital Methods*. Wiley, Chichester.

Schöpf, M. (2004). CSM for cut-off and retaining walls in deep foundations. *GeoDrilling International*, December, pp. 18–19.

Schmäh, P. (2007). Vertical shaft machines. State of the art and vision. *Acta Montanistica Slovaca*, 12, 1, pp. 208–216.

Schmall, P and Dawson, A. (2017). Ground-freezing experience on the east side access Northern Boulevard crossing, New York. *Proceedings of the Institution of Civil Engineers, Ground Improvement*, 170, 3, pp. 159–172.

Schünmann, D. (2004). Seal of approval. *Ground Engineering*, 37, July, pp. 24–25.

Schünmann, D. (2005). Well sorted. *Ground Engineering*, 38, October, pp. 34–35.

Shirlaw, N. (2012). Setting operating pressures for TBM tunnelling. *Geotechnical Aspects of Tunnelling for Infrastructure Development*. Proceedings of the 32nd Annual Seminar Geotechnical Division, The Hong Kong Institution of Engineers, pp. 7–28.

Siddle, H J. (1986). Groundwater control by drainage gallery at Aberfan, South Wales. *Groundwater in Engineering Geology* (Cripps, J C, Bell, F G and Culshaw, M G, eds). Geological Society Engineering Geology Special Publication No. 3, London, pp. 533–540.

Simpson, B, Blower, T, Craig, R N and Wilkinson, W B. (1989). *The Engineering Implications of Rising Groundwater in the Deep Aquifer Below London*. Construction Industry Research and Information Association, CIRIA Special Publication 69, London.

Site Investigation Steering Group. (1993). *Site Investigation in Construction, Volume 1: Without Investigation Ground Is a Hazard*. Thomas Telford, London.

Skempton, A W. (1964). Long-term stability of clay slopes. *Géotechnique*, 14, 2, pp. 77–102.

Skempton, A W and Brogan, J M. (1994). Experiments on piping in sandy gravels. *Géotechnique*, 44, 3, pp. 449–460.

Skempton, A W and Chrimes, M M. (1994). Thames Tunnel: Geology, site investigation and geotechnical problems. *Géotechnique*, 44, 2, pp. 191–216.

Skipp, B O. (1993). Keynote paper: Setting the scene. *Groundwater Problems in Urban Areas* (Wilkinson, W B, ed). Thomas Telford, London, pp. 3–16.

Slocombe, R, Buchanan J and Lamont, D. (2003). *Engineering and Health in Compressed Air Work*. Thomas Telford, London.

Smith, A. (2018). Caissons and shafts. *Temporary Works: Principles of Design and Construction*, 2nd edition (Filip, R and Pallett, P F, eds). ICE Publishing, London, pp. 207–216.

Soler, R, Colace, A, Stärk, A, Zeizig, W and Roberts, T O L. (2016). In-tunnel depressurisation for SCL tunnels at Whitechapel and Liverpool street stations. *Crossrail Project: Infrastructure Design and Construction* (Black, M, ed). ICE Publishing, London, pp. 33–51.

Somerville, S H. (1986). *Control of Groundwater for Temporary Works*. Construction Industry Research and Information Association, CIRIA Report 113, London.

Sopko, J A. (2017). Forensics by freezing. *Grouting 2017: Case Histories, Selected Papers from Sessions of Grouting 2017* (Bruce, D A, El Mohtar, C, Byle, M J, Gazzarrini, P, Johnsen, L F and Richards, T D, eds). Geotechnical Special Publication Number 287, American Society of Civil Engineers and Curran Associates, Inc., Red Hook, New York, pp. 96–105.

Sopko, J A. (2019). Design of ground freezing for cross passages and tunnel adits. *Tunnels and Underground Cities: Engineering and Innovation meet Archaeology, Architecture and Art*. Proceedings of the WTC 2019 ITA-AITES World Tunnel Congress (WTC 2019), Naples, Italy. CRC Press, Boca Raton, Florida, pp. 1549–1558.

Soudain, M. (2002). Terms of internment. *Ground Engineering*, 35, March, pp. 26–27.

Spink, T, Duncan, I, Lawrance, A and Todd, A. (2014). *Transport Infrastructure Drainage: Condition Appraisal and Remedial Treatment*. Construction Industry Research and Information Association, CIRIA Report C714, London.

Starr, M R, Skipp, B O and Clarke, D A. (1969). Three-dimensional analogue used for relief well design in the Mangla Dam project. *Géotechnique*, 19, 1, pp. 87–100.

Steenfelt, J S and Hansen, H K. (1995). *Key Note Address: The Storebaelt Link – A Geotechnical View*. Aalborg: Geotechnical Engineering Group. AAU Geotechnical Engineering Papers: Foundation Engineering Paper, No. 2, Vol. R 9509.

Stephens, D B and Ankeny, M D. (2004). The missing link in the historical development of hydrogeology. *Groundwater*, 42, pp. 304–309.

Stephens, D B and Stephens, D A. (2006). British land drainers: Their place among pre-Darcy forefathers of applied hydrogeology. *Hydrogeology Journal*, 14, pp. 1367–1376.

Sterrett, R. (2008). *Groundwater and Wells*, 3rd edition. Johnson Division, St Paul Minesota.

Stone, K, Murray, A, Cooke, S, Foran, J and Gooderham, L. (2009). *Unexploded Ordnance (UXO) A Guide for the Construction Industry.* Construction Industry Research and Information Association, CIRIA Report C681, London.

Stow, G R S. (1962). Modern water well drilling techniques in use in the United Kingdom. *Proceedings of the Institution of Civil Engineers,* 23, September, pp. 1–14.

Stow, G R S. (1963). Discussion of Modern water well drilling techniques in use in the United Kingdom. *Proceedings of the Institution of Civil Engineers,* 25, May, pp. 219–241.

Streetly, M. (1998). Dewatering and environmental monitoring for the extractive industry. *Quarterly Journal of Engineering Geology,* 31, pp. 125–127.

Sumbler, M G. (1996). *British Regional Geology: London and the Thames Valley,* 4th edition. HMSO, London.

Sutton, J S. (1996). Hydrogeological testing in the Sellafield area. *Quarterly Journal of Engineering Geology,* 29, pp. S29–S38.

Taylor, H F, O'Sullivan, C and Sim, W W. (2016). Geometric and hydraulic void constrictions in granular media. *Journal of Geotechnical and Geoenvironmental Engineering,* 142, 11.

Terzaghi, K. (1939). Soil mechanics – a new chapter in engineering science, the 45th James Forrest Lecture, 1939. *Journal of the Institution of Civil Engineers,* 12, 7, June, pp. 106–142.

Terzaghi, K. (1960). Land forms and subsurface drainage in the Gacka region in Yugoslavia. *From Theory to Practice in Soil Mechanics: Selections from the Writings of Karl Terzaghi.* Wiley, New York.

Terzaghi, K, Peck, R B and Mesri, G. (1996). *Soil Mechanics in Engineering Practice,* 3rd edition. Wiley, New York.

Theis, C V. (1935). The relation between the lowering of the piezometric surface and the rate and duration of discharge of a well using groundwater storage. *Transactions of the American Geophysical Union,* 16, pp. 519–524.

Thomas, A. (2019). *Sprayed Concrete Lined Tunnels,* 2nd edition. CRC Press, Boca Raton, Florida.

Thomas, J B and Fisher, A W. (1974). St. Paul's Cathedral – measurement of settlements and movements of the structure. *Proceedings of the Conference on Settlement of Structures, Cambridge,* British Geotechnical Society, Pentech Press, pp. 211–233.

Thomson, J. (1852). On a jet pump or apparatus for drawing water up by the power of a jet. *Report of the British Association,* pp. 130–131.

Thorn, B. (1996). Dewatering: environmental issues. *Temporary Works: Dewatering and Stability,* seminar notes, Institution of Civil Engineers, London, 2nd July.

Tolman, C F. (1937). *Ground Water.* McGrawHill, New York.

Trenter, N A. (1999). A note on the estimation of permeability of granular soils. *Quarterly Journal of Engineering Geology,* 32, pp. 383–388.

Troughton, V M. (1987). Groundwater control by pressure relief and recharge. *Groundwater Effects in Geotechnical Engineering* (Hanrahan, E T, Orr, T L L and Widdis, T F, eds), Balkema, Rotterdam, pp. 259–264.

Unified Facilities Criteria. (2004). *Dewatering and Groundwater Control.* UFC 3-220-05, 16 January 2004 (Formerly Joint Departments of the Army, the Air Force, and the Navy, USA, Technical Manual TM 5-818-5/AFM 88-5, Chap 6/NAVFAC P-418, Dewatering and Groundwater Control).

US Army Corps of Engineers. (2000). *Operation and Maintenance of Extraction and Injection Wells at HTRW Sites.* Department of the Army, Washington, DC.

USBR. (2001). *Water Measurement Manual,* 3rd edition (revised, reprinted). United States Department of the Interior, Bureau of Reclamation, Washington, DC.

Van Zyl, D and Harr, M E. (1981). Seepage erosion analyses of structures. *Proceedings of the 10th International Conference on Soil Mechanics and Foundation Engineering,* Stockholm, Sweden, pp. 503–509.

Viggiani, G and Casini, F. (2015). Artificial ground freezing: From applications and case studies to fundamental research. *Proceedings of the XVI ECSMFGE, Geotechnical Engineering for Infrastructure and Development* (Winter, M C, Smith, D M, Eldred, P J L and Toll, D G, eds). ICE Publishing, London, pp. 4129–4134.

Vukovic, M and Soro, A. (1992). *Determination of Hydraulic Conductivity of Porous Media from Grain-Size Composition.* Water Resources Publications, Littleton, Colorado.

Walthall, S. (1990). Packer testing in geotechnical engineering. *Field Testing in Engineering Geology* (Bell, F G, Culshaw, M G, Cripps, J C and Coffey, J R, eds). Geological Society Engineering Geology Special Publication No. 6, London, pp. 345–350.

Waltham, T. (2016). Control the drainage: The gospel accorded to sinkholes. *Quarterly Journal of Engineering Geology and Hydrogeology*, 49, pp. 5–20.

Ward, W H. (1948). A coastal landslip. *Proceedings of the 2nd International Conference on Soil Mechanics and Foundation Engineering*, Rotterdam, Vol. 2, pp. 33–38.

Ward, W H. (1957). The use of simple relief wells in reducing water pressure beneath a trench excavation. *Géotechnique*, 7, 3, pp. 134–139.

Warren, C D, Hughes, P, Swift, A and Mielenz, J. (2003). Cross passage ground treatment for the CTRL Thames Tunnel. *Proceedings of Underground Construction 2003*, British Tunnelling Society, London, pp. 253–265.

Warren, C D, Newman, T and Hadlow, N W. (2018). Comparison of earth pressure balance and slurry tunnel boring machines used for tunnelling in chalk. *Engineering in Chalk: Proceedings of the Chalk 2018 Conference* (Lawrence, J A, Preene, M, Lawrence, U L and Buckley, R, eds). ICE Publishing, London, pp. 617–628.

Weber, H. (1928). *Die Reichweite von Grundwasserabsenkungen Mittels Rohrbunnen*. Springer, Berlin.

Werblin, D A. (1960). Installation and operation of dewatering systems. *Journal of the Soil Mechanics and Foundations Division, Proceedings of the American Society of Civil Engineers*, 86, SM1, pp. 47–66.

Wesley, L D. (2014). Unconfined seepage behaviour in coarse and fine grained soils. *Proceedings of the 12th Australia New Zealand Conference on Geomechanics*, Wellington, New Zealand.

Wesley, L D and Preene, M. (2019). A historical perspective on unconfined seepage: Correcting a common fallacy. *Proceedings of the Institution of Civil Engineers, Engineering History and Heritage*, 172, 2, May, pp. 57–69.

Westcott, F J, Lean, C M B and Cunningham, M L. (2001). *Piling and Penetrative Ground Improvement Methods on Land Affected by Contamination: Guidance on Pollution Protection*. Environment Agency, National Groundwater and Contaminated Land Centre Report NC/99/73, Solihull.

Wheelhouse, P J, Belton, J A and Auld, F A. (2001). Ground freezing at Dorney Bridge box jacking. *Proceedings of Underground Construction 2001*, The Hemming Group, London, pp. 521–532.

Whitaker, D. (2002). Deep-well dewatering on CTRL. *Tunnels & Tunnelling International*, 34, March, pp. 50–52.

Whitaker, D. (2004). Groundwater control for the Stratford CTRL station box. *Proceedings of the Institution of Civil Engineers, Geotechnical Engineering*, 157, October, pp. 183–191.

White, M and Capps, C T F. (1997). The design and construction of the casting basin for the River Lee Tunnel. *Immersed Tunnel Techniques 2*, Thomas Telford, London, pp. 320–330.

White, J K and Roberts, T O L. (1993). The significance of groundwater tidal fluctuations. *Groundwater Problems in Urban Areas* (Wilkinson, W B, ed). Thomas Telford, London, pp. 31–42.

Wiese, B, Böhner, J, Enachescu, C, Würdemann, H, Zimmermann, G. (2010). Hydraulic characterisation of the Stuttgart formation at the pilot test site for CO_2 storage, Ketzin, Germany. *International Journal of Greenhouse Gas Control*, 4, pp. 960–971.

Wilkinson, W B. (1986). Design of boreholes and wells. *Groundwater, Occurrence, Development and Protection* (Brandon, T W, ed). Institution of Water Engineers and Scientists, Water Practice Manual No. 5, London, pp. 385–406.

Wilkinson, W B, ed. (1994). *Groundwater Problems in Urban Areas*. Thomas Telford, London.

Wilkinson, W B and Brassington, F C. (1991). Rising groundwater levels – an international problem. *Applied Groundwater Hydrology* (Downing, R A and Wilkinson, W B, eds). Clarendon Press, Oxford, pp. 35–53.

Williams, B. (2008). Cardiff Bay barrage: Management of groundwater issues. *Proceedings of the Institution of Civil Engineers, Water Management*, 161, December, pp. 313–321.

Williams, B P and Waite, D. (1993). *The Design and Construction of Sheet-Piled Cofferdams.* Construction Industry Research and Information Association, CIRIA Special Publication 95, London.

Williams, R E, ed. (2010). *Rudolph Glossop and the Rise of Geotechnology.* Whittles Publishing, Caithness, Scotland.

Williams, R E and Norbury, D R. (2008). Rudolph Glossop and the development of geotechnology. *Quarterly Journal of Engineering Geology and Hydrogeology*, 41, pp. 189–200.

Williamson, S and Vogwill, R I J. (2001). Dewatering in the hot groundwater conditions at Lihir Gold. *Proceedings of the International Mine Water Association Symposium – Mine Water and the Environment*, Belo Horizonte, Brazil.

Woods, M A, Allen, D J, Forster, A, Pharoah, T C and King, C. (2004). *The Geology of London.* British Geological Survey, Keyworth, Nottingham.

Woodward, J. (2005). *An Introduction to Geotechnical Processes.* Spon, Abingdon.

Wrench, B P and Stacey, T R. (1993). Prediction of inflows into tunnels. *TUNCON '93: Aspects of Control in Tunnelling Projects*, Johannesburg, pp. 25–28.

Wroth, C P. (1984). 24th Rankine Lecture: The interpretation of in situ soil tests. *Géotechnique*, 34, 4, pp. 449–488.

Younger, P L. (2004). 'Making water': The hydrogeological adventures of Britain's early mining engineers. *200 Years of British Hydrogeology* (Mather, J D, ed). Geological Society Special Publication 225, London, pp. 121–157.

Younger, P L. (2007). *Groundwater in the Environment: An Introduction.* Blackwell, Oxford.

Yulan, J, Holzbecher, E and Sauter, M. (2016). Dual-screened vertical circulation wells for groundwater lowering in unconfined aquifers. *Groundwater*, 54, pp. 15–22.

Yungwirth, G, Preene, M, Dobr, M and Forero Garcia, F. (2013). Practical application and design considerations for fully grouted vibrating wire piezometers in minewater investigations. *International Mine Water Association Conference – Reliable Mine Water Technology*, Denver, Colorado, August 2013, Vol. I, pp. 229–237.

Index

Printed and bound by CPI Group (UK) Ltd, Croydon, CR0 4YY

24/10/2024

01778286-0019